T0328326

Applications of Geospatial Technology and Modeling for River Basin Management

Modern Cartography Series

Applications of Geospatial Technology and Modeling for River Basin Management

Volume 12

Edited by

Subodh Chandra Pal

Department of Geography, The University of Burdwan, Golapbag, Purba Bardhaman, West Bengal, India

Uday Chatterjee

Department of Geography, Bhatter College, Dantan (Affiliated to Vidyasagar University), Paschim Medinipore, West Bengal, India

Rabin Chakrabortty

Department of Geography, The University of Burdwan, Golapbag, Purba Bardhaman, West Bengal, India

Series Editor

D. R. Fraser Taylor

ELSEVIER

Elsevier
Radarweg 29, PO Box 211, 1000 AE Amsterdam, Netherlands
125 London Wall, London EC2Y 5AS, United Kingdom
50 Hampshire Street, 5th Floor, Cambridge, MA 02139, United States

Notices

Knowledge and best practice in this field are constantly changing. As new research and experience broaden our understanding, changes in research methods, professional practices, or medical treatment may become necessary.

Practitioners and researchers must always rely on their own experience and knowledge in evaluating and using any information, methods, compounds, or experiments described herein. In using such information or methods they should be mindful of their own safety and the safety of others, including parties for whom they have a professional responsibility.

To the fullest extent of the law, neither the Publisher nor the authors, contributors, or editors, assume any liability for any injury and/or damage to persons or property as a matter of products liability, negligence or otherwise, or from any use or operation of any methods, products, instructions, or ideas contained in the material herein.

ISBN: 978-0-443-23890-1

ISSN: 1363-0814

For Information on all Elsevier publications
visit our website at https://www.elsevier.com/books-and-journals

Publisher: Candice Janco
Acquisitions Editor: Peter Llewellyn
Editorial Project Manager: Sara Valentino
Production Project Manager: Paul Prasad Chandramohan
Cover Designer: Greg Harris

Typeset by MPS Limited, Chennai, India

Working together
to grow libraries in
developing countries

www.elsevier.com • www.bookaid.org

Contents

Section I Hydrology

**CHAPTER 8 Quantifying land use/land cover dynamics
and fragmentation analysis along the stretch of
Bhagirathi River in Murshidabad district,
West Bengal** **197**
*Anukul Chandra Mandal, Raja Majumder, Partha Gorai
and Gouri Sankar Bhunia*

Section II Edaphology

CHAPTER 16 Geographic information system-based statistical mapping of socioeconomic vulnerability in the Upper Citarum River, West Java Province, Indonesia 413

Setiawan Hari Harjanto, Tanjung Mahdi Ibrahim, Abdullah Abdullah, Djaenudin Djaenudin and Siswoyo Suhandy

CHAPTER 23 Estimation of soil erosion risk and vulnerable zone using the revised universal soil loss equation and geographic information system approaches............ 597

Rahul Kumar, Shambhu Nath Mishra, Rajiv Pandey and Vijender Pal Panwar

CHAPTER 24 Monitoring land degradation and desertification using the state-of-the-art methods and remote sensing data... 627

Debaaditya Mukhopadhyay and Gaurav Mishra

List of contributors

Abdullah Abdullah
School of Economics and Business, Department of Management, Telkom University, Bandung, Indonesia

Muhammad Azam
Department of Structures and Environmental Engineering, Faculty of Agricultural Engineering & Technology, PMAS Arid Agriculture University Rawalpindi, Rawalpindi, Pakistan

Sara Izrar Aziz
Interdisciplinary Program in Landscape Architecture, Seoul National University, Seoul, South Korea

Arnab Bandyopadhyay
North Eastern Regional Institute of Science and Technology, Agricultural Engineering, Itanagar, Arunachal Pradesh, India

Jatisankar Bandyopadhyay
Remote Sensing and GIS, Vidyasagar University, Midnapore, West Bengal, India

Aditi Bhadra
North Eastern Regional Institute of Science and Technology, Agricultural Engineering, Itanagar, Arunachal Pradesh, India

Bimal Bhattacharya
Space Applications Centre, ISRO, Ahmedabad, Gujarat, India

Gouri Sankar Bhunia
Department of Geography, Seacom Skills University, Kendradangal, Saturia, West Bengal, India

Ana Carolina Borges Monteiro
Renato Archer Information Technology Center (CTI), Campinas, São Paulo, Brazil

Sandip Chand
Indian Institute of Technology Kharagpur, Kharagpur, West Bengal, India

Santasmita Das Bhattacharya
Department of Geography, Amity School of Social Sciences, Amity University, Kolkata, West Bengal, India

Ferrucio de Franco Rosa
Renato Archer Information Technology Center (CTI), Campinas, São Paulo, Brazil

Herica Fernanda de Sousa Carvalho
Renato Archer Information Technology Center (CTI), Campinas, São Paulo, Brazil

Pratik Deb
CLWRM (Centre for Land and Water Resource Management), GBPNIHE,
Garhwal Regional Centre, Pauri Garhwal, Uttarakhand, India

Djaenudin Djaenudin
National Research and Innovation Agency (BRIN), Center for Environmental
Research and Clean Technology, Jakarta, Indonesia

Hamza EL Fadili
Laboratory of Spectroscopy, Molecular Modeling, Materials, Nanomaterials,
Water and Environment, Materials for Environment Team, ENSAM,
Mohammed V University in Rabat, Rabat, Morocco

Tika Ainnunisa Fitria
Architecture Study Program, Universitas 'Aisyiyah Yogyakarta, Yogyakarta, Indonesia

Kaushik Ghosal
Department of Mining Engineering, Indian Institute of Engineering Science and
Technology Shibpur, Howrah, West Bengal, India

Argha Ghosh
Agricultural Meteorology, Odisha University of Agriculture and Technology,
Bhubaneswar, Odisha, India

Partha Gorai
Department of Geography, Seacom Skills University, Kendradangal, Saturia,
West Bengal, India

Jayanta Gour
Department of Geography, Sambhu Nath College, Labpur, Birbhum,
West Bengal, India

Akhilesh Kumar Gupta
Agricultural Statistics, Odisha University of Agriculture and Technology,
Bhubaneswar, Odisha, India

Setiawan Hari Harjanto
National Research and Innovation Agency (BRIN), Research Center for Social
Welfare, Villages and Connectivity, Jakarta, Indonesia

Archita Hazarika
Centre for Studies in Geography, Dibrugarh University, Dibrugarh, Assam, India

Manas Hudait
Amity University, Kolkata, West Bengal, India

Zoheb Islam
Department of Geography, Hari-Har Mahavidyalaya, Bansra, West Bengal, India

Maeng Seung Jin
Department of Agricultural and Rural Engineering, Chungbuk National
University, Cheongju-si, Chungcheongbuk-do, South Korea

Amrit Kamila
Coastal Observatory and Outreach Centre (COOC), Vidyasagar University,
Midnapore, West Bengal, India

Daye Kim
Department of Agricultural and Rural Engineering, Chungbuk National University, Cheongju-si, Chungcheongbuk-do, South Korea

Rahul Kumar
Forest Ecology and Climate Change Division, ICFRE-Institute of Forest Productivity, Ranchi, Jharkhand, India

Naba Kumar Mondal
Environmental Chemistry Laboratory, Department of Environmental Science, The University of Burdwan, Burdwan, West Bengal, India

Seungwook Lee
Chungbuk Research Institute, Choengju-si, Chungcheongbuk-do, South Korea

Tanjung Mahdi Ibrahim
Ministry of Public Works and Housing, Water Resources, Bandung, Indonesia

Raja Majumder
Department of Geography, State Aided College Teacher Dinabandhu Mahavidyalaya, Bangaon, West Bengal, India

Anukul Chandra Mandal
Department of Geography, Manipur International University, Imphal, Manipur, India

Sameer Mandal
North Eastern Regional Institute of Science and Technology, Agricultural Engineering, Itanagar, Arunachal Pradesh, India

Debojyoti Mishra
Environmental Chemistry Laboratory, Department of Environmental Science, The University of Burdwan, Burdwan, West Bengal, India

Gaurav Mishra
Centre of Excellence on Sustainable Land Management, Indian Council of Forestry Research and Education, Dehradun, Uttarakhand, India

Shambhu Nath Mishra
Forest Ecology and Climate Change Division, ICFRE-Institute of Forest Productivity, Ranchi, Jharkhand, India

Tapas Mistri
The University of Burdwan, Burdwan, West Bengal, India

Rohana Mohd Firdaus
Program of Landscape Architecture, Universiti Teknologi Malaysia, Skudai, Malaysia

Momsona Mondal
Agricultural Meteorology and Physics, Bidhan Chandra Krishi Viswavidyalaya, Mohanpur, West Bengal, India

Debaaditya Mukhopadhyay
ICFRE—Rain Forest Research Institute, Jorhat, Assam, India

Manoj Kumar Nanda
Agricultural Meteorology and Physics, Bidhan Chandra Krishi Viswavidyalaya, Mohanpur, West Bengal, India

N.M. Refat Nasher
Department of Geography and Environment, Faculty of Life and Earth Sciences, Jagannath University, Dhaka, Bangladesh

Uttara Nath
University of North Bengal, West Bengal, India

Reinaldo Padilha França
Renato Archer Information Technology Center (CTI), Campinas, São Paulo, Brazil

Dhruv Pandey
CLWRM (Centre for Land and Water Resource Management), GBPNIHE, Garhwal Regional Centre, Pauri Garhwal, Uttarakhand, India

Rajiv Pandey
Indian Council of Forestry Research and Education, Dehradun, Uttarakhand, India

Vijender Pal Panwar
ICFRE-Forest Research Institute, Dehradun, Uttarakhand, India

Hemlata Patel
Department of Geography, SP Pune University, Pune, Maharashtra, India

Sribas Patra
Department of Geography, Ravenshaw University, Cuttack, Odisha, India

Pulakesh Pradhan
Department of Geography, Ravenshaw University, Cuttack, Odisha, India

Md Naimur Rahman
Department of Geography, Hong Kong Baptist University, Kowloon, Hong Kong, P.R. China; David C Lam Institute for East-West Studies, Hong Kong Baptist University, Kowloon, Hong Kong, P.R. China; Department of Development Studies, Daffodil International University, Dhaka, Bangladesh

Kazi Jihadur Rashid
Sensing Earth and Environment (SEE) Lab, Bangladesh; Center for Environmental and Geographic Information Services, Dhaka, Bangladesh

Rodrigo Rodrigo
Renato Archer Information Technology Center (CTI), Campinas, São Paulo, Brazil

Jyoti Saikia
Department of Geography, DHSK College, Dibrugarh, Assam, India

Sailajananda Saikia
Department of Geography, Rajiv Gandhi University, Itanagar, Arunachal Pradesh, India

Md Mushfiqus Saleheen
Department of Geography and Environmental Science, Begum Rokeya University, Rangpur, Bangladesh

Debolina Sarkar
Agricultural Meteorology and Physics, Bidhan Chandra Krishi Viswavidyalaya, Mohanpur, West Bengal, India

Md Nazirul Islam Sarker
Miyan Research Institute, International University of Business Agriculture and Technology, Dhaka, Bangladesh

Mathanraj Seevarethnam
Department of Geography, Eastern University, Batticaloa, Sri Lanka

Kamalesh Sen
Environmental Chemistry Laboratory, Department of Environmental Science, The University of Burdwan, Burdwan, West Bengal, India

Rajesh Kumar Shah
Department of Zoology, D.H.S.K.College, Dibrugarh, Assam, India

Rani Kumari Shah
Department of Geography, Cotton University, Guwahati, Assam, India

Shafkat Sharif
Hydroscience and Engineering, Civil and Environmental Engineering, Technische Universität Dresden, Dresden, Germany

Sweety Singh
Department of Geography, SP Pune University, Pune, Maharashtra, India

Siswoyo Suhandy
Indonesian University of Education (UPI), Architecture Study Program, Bandung, Indonesia

Akter Tahmina
Center for Environmental and Geographic Information Services, Dhaka, Bangladesh

Soukhin Tarafdar
CLWRM (Centre for Land and Water Resource Management), GBPNIHE, Garhwal Regional Centre, Pauri Garhwal, Uttarakhand, India

Manisha Tikader
Govt. Bilasa Girls P.G. College, Atal Bihari Vajpayee Vishwavidyalaya, Bilaspur, Chhattisgarh, India

Rajsree Das Tuli
Sensing Earth and Environment (SEE) Lab, Bangladesh; Food and Agriculture Organization of the United Nations, Dhaka, Bangladesh

Swapnil S. Vyas
Department of Geography (Geoinformatics), SP Pune University, Pune, Maharashtra, India

Jung Yongbae
K-water Seomjingang Dam Office, Deajeon, Republic of Korea

Rubaiya Zumara
Department of Geography and Environment, Jagannath University, Dhaka, Bangladesh

Foreword

I am really delighted to write a foreword for the book entitled *Applications of Geospatial Technology and Modeling for River Basin Management*, edited by Subodh Chandra Pal, Uday Chatterjee, and Rabin Chakrabortty, which is going to be published by Elsevier.

Drainage basin being an integral part of riverine ecosystem represents an essential component of our society and surrounding environment, which has continuously been undergoing deterioration ecologically because of both natural and man-made reasons. In this book, the three ecological aspects pertaining to riverine drainage basin are being investigated through spatial modeling, as well as risk assessment of drainage basins in the context of environment and also on several social, cultural-religious issues. There are many challenges and ambiguities in analyzing the Earth's existing environmental status as a result of least knowledge about the ongoing climate change and anti-environmental man-made activities, which have made it extremely difficult to analyze and draw reliable research outcomes regarding environment resources in respect of ecological changes. Many difficulties have been emerged from the ineffective management of the present and future land, water, and forest resources. It is also very much incomponent to use new technologies and methods to reinforce existing environmental regulations and improve the environmental management strategies. In this regard, proper understanding of the relationships among the three devices, namely remote sensing, geographical information system (GIS), and the R programming interface, is very important as they are in use to measure the land conservation and also to control the deteriorating quality of soil and water. In such context, new legislation is being developed, based on reliable measurements and estimates, on three such technologies having an open-access quantitative forecasting approach, which is supposed to help realize the effects releted to the ongoing climate change and relevant environmental management rules. Nonetheless, this edited book has provided a useful framework for analyzing current advancements in geospatial artificial intelligence technologies and their relevance to the planet's multidimensional environmental and socioeconomic components in a single volume. In an era of globalization coupled with liberalization, the sustainable management of drainage basins has assumed paramount importance, and this book has attempted to explore the intersection of geospatial technology and modeling to address the complex challenges faced by our society as well as other components of Mother Earth. The degradation of drainage basins, being a pressing concern, has necessitated a multidisciplinary approach that integrates spatial modeling, risk assessment, and effective management strategies. This edited book contains 27 chapters organized into three sections namely Section I focusing on Hydrology, Section II dealing with Edaphology, and Section III discussing on Environment and Sustainability. Each section has meticulously highlighted the critical components of the bio-geographical environment of river basins and emphasized on proper strategies for sustainable management.

I extend my appreciation to the editors, contributors, and publication house, the Elsevier for their commitments to advancing the discourse on geospatial technology and modeling. This edited book has enormously contributed and upgraded the current literature on land reclamation and environmental restoration using contemporary remote sensing and GIS methodology. It is my hope that this book will find a wide audience among researchers, educators, environmental planners and practitioners, contributing to the ongoing dialogue on sustainable river basin management.

Professor Susanta Kumar Chakraborty
Vice-Chancellor
Vidyasagar University, Midnapore,
West Bengal, India

Professor Susanta Kumar Chakraborty

Preface

Geospatial information technologies have garnered increasing interest in Earth's environmental and social scientific research groups in recent decades due to their effective capacity to solve and notice global challenges and generate innovative solutions for a healthy Earth and human civilization. This book investigates the combination of water resource management for sustainable consumption with appropriate water consumption allocation for regional economic growth. It also helps with studies on water resource management in the face of climate change and environmental resilience. It will serve as a reference guide for all scholars working on integrated river basin management. Our book offers thorough research methodologies, challenges, and future research concerns, as well as our replies, for anybody concerned in this study area. The socioeconomic transaction of water usage planning, in particular, is especially significant to the everyday lives of individuals and their standard of existence in the context of environmental difficulties. This book's scholarly contributions provide a detailed description of words, interactions, linkages, and the consequences of environmental problems from the perspective of integrated water management. The book will promote synergistic practices among scientists and technicians working in fields such as data mining and machine learning by integrating spatial computational intelligence technologies into social and environmental concerns. It takes a GIS technology approach to data mining techniques, data analysis, modeling, risk assessment, and visualization, as well as management solutions for many aspects of river basin concerns. The book explores cutting-edge methodologies based on open-source software and R statistical programming, Google Earth Engine, and modeling in contemporary artificial intelligence techniques, with a focus on recent advancements in data mining techniques and robust modeling in river basin management. The book covers a wide range of topics, including experimental studies, pilot studies for the development of river basin management plans, people's participation in the planning process, modeling and analysis, and detecting and meeting environmental goals using advanced geospatial techniques with ground validation. Significant progress has been achieved in the monitoring, analysis, and prediction of riverine system behavior, resulting in a better understanding of their mechanisms and the prospect of optimizing these resources, as well as controlling or mitigating the consequences of severe disasters. These advancements have resulted from improved surveying and measurement procedures, as well as the usage of increasingly precise computer codes. They are also the result of improved contact between practitioners and academics as a result of their participation in collaborative research initiatives. The book provides a variety of experiences from many countries throughout the world in implementing the SDGs' environmental goals. It is a wonderful source of knowledge and ideas for the wide drafting of river basin management plans currently taking place across the world.

The contributions in this book have been divided into the following sections: Section I: Hydrology, Section II: Edaphology, and Section III: Environment and Sustainability, which encompasses river and watershed management, flood research, hydrological modeling, river restoration, and environmental effect erosion and sediment movement, water resource management, and environmental modeling. Specific chapters primarily focus on the key principles of science and information, as well as some issues that have received less attention (risk evaluation and management). Each chapter will give an overview of the subject's present interpretation, outlining the nature of the study and exploring where future attempts should be concentrated. However, we believe that this book has significant potential for teaching novices and researchers on a range of essential concepts at the core of extant research on environmental and social hazards, risk assessment, and management. It is anticipated that the book as a whole not only provides a timely review of an expanding and major area of study but will also present new and intriguing notions that will build a unified and productive framework for future practice. We hope that this book will be of great interest to a wide range of academics, biologists, geographers, remote sensing and GIS specialists, activists, meteorologists, and computational experts worldwide who are interested in geospatial artificial intelligence technologies in the environmental, human, and social sciences.

Subodh Chandra Pal, *Burdwan, West Bengal, India*
Uday Chatterjee, *Midnapore, West Bengal, India*
Rabin Chakrabortty, *Burdwan, West Bengal, India*

Acknowledgment(s)

We extend our sincere gratitude to all those who contributed to the creation and publication of the book, "Application of Geospatial Technology and Modeling for River Basin Management," by Elsevier. First and foremost, we would like to thank the contributors, colleagues, researchers, and readers whose valuable contributions, expertise, support, and dedication were instrumental in shaping this comprehensive book. Their invaluable insights and contributions have enriched the content and facilitated a deeper understanding of the challenges associated with river basin management. Special recognition is also due to our families for their salient support, patience, and understanding throughout the entire process of publishing this book. Their encouragement has been a constant source of motivation. We would like to thank Elsevier, the publisher, for the opportunity, and we express our appreciation to Peter Llewellyn, Senior Acquisitions Editor, Space and Planetary Science/Geology; Series Editor, Dr. Fraser Taylor; Senior Editorial Project Manager, Sara Valentino, and her publishing team at Elsevier, for their exceptional guidance, expertise, and commitment to excellence. Their diligent efforts in editing, design, and production have been crucial in ensuring the success of this publication. Thank you to everyone who contributed to making this book a reality. Your collective efforts have made a significant impact in the field of river basin management, and we are truly grateful for your contributions.

Subodh Chandra Pal, *Burdwan, West Bengal, India*
Uday Chatterjee, *Midnapore, West Bengal, India*
Rabin Chakrabortty, *Burdwan, West Bengal, India*

Hydrology

Recent land use land cover changes, demographic transition, and rainfall trends in middle Himalayan watershed

Soukhin Tarafdar, Pratik Deb and Dhruv Pandey

CLWRM (Centre for Land and Water Resource Management), GBPNIHE, Garhwal Regional Centre, Pauri Garhwal, Uttarakhand, India

1.1 Introduction

Systematic mapping and monitoring of land use transition are important for natural resource management and planning strategies to safeguard from any alteration caused by natural and human-induced factors, which may severely impact the functioning of the watershed. Land use changes are known to have impacts on freshwater resources by altering the hydrological cycle (Foley et al., 2005) as well as on groundwater recharge (Scanlon, Reedy, Stonestrom, Prudic, & Dennehy, 2005). Monitoring the land use land cover (LULC) changes at regional and local scales over time is essential to gain insight into the extent of LULC and their dynamics. Remote sensing and geographical information system (GIS) have been widely applied for quantification of change in forest cover (Dasgupta, Kumar, & Tripathi, 2010), as well as in cropland monitoring and change detection (Ramankutty & Foley, 1998). The applicability of satellite data for LULC mapping is demonstrated in many studies from Kumaon Lesser Himalaya (Tiwari & Joshi, 2012; Rao & Pant, 2001) and Garhwal mid-Himalaya (Semwal et al., 2004; Batar, Watanabe, & Kumar, 2017). Several studies on hydrological modeling indicate that land use change can be an important driver in altering the water balance at the local and regional scales (Seibert, McDonnell, & Woodsmith, 2010; Wagner, 2013). The interlinkages between land use, soil erosion, stream flow, and sediment loads were reported by Tiwari (2000) from watershed-based investigation in three distinctly different physiographic settings of Indian Himalaya. The study highlights the subsequent downstream consequences like rise in riverbeds, diminishing stream flow, spring flow, and river flow, and rise in the flood frequencies as a result of land use change in terms of loss of forest, expansion of agriculture, and unplanned development in the Kumaon region. The ongoing

Applications of Geospatial Technology and Modeling for River Basin Management.
DOI: https://doi.org/10.1016/B978-0-443-23890-1.00001-3

abandonment of rural agricultural land from the mountainous region of middle Himalaya may have its driver in the demographic and socioeconomic factors. The assessment of terraced cropland abandonment is witnessed in many regions of the world, especially in European countries, for example, Spain, Italy, France, Switzerland, Germany, and Norway (MacDonald et al., 2000; Quintas-Soriano, Buerkert, & Plieninger, 2022) as well as adjoining countries of Nepal (Subedi, Kristiansen, & Cacho, 2022) and China (Zhang, Li, & Song, 2014). Although a general understanding of the reasons for such a widespread abandonment of traditional agriculture landscape in Uttarakhand has emerged (Mamgain & Reddy, 2017), however, clarity in the social and ecological consequences of rapid land abandonment and subsequent regrowth of secondary forest leading to the rise in the forest cover area (reported from the European regions also) associated with land use change is not fully realized and recognized in the middle mountains of Indian Himalaya.

The estimation of temporal and spatial distribution of precipitation is indispensable for water resource management, soil moisture modeling, hydrological modeling, and climate change studies. The applicability of gridded satellite-based rainfall estimates in the data-scarce region of the Himalaya is demonstrated by several workers (Bookhagen & Burbank, 2006; Khandelwal, Gupta, & Chauhan, 2015). The satellite-based long-term, high-resolution datasets can be analyzed and compared with other data sources to find out the trends in daily precipitation. PERSIANN CDR (Precipitation Estimation from Remotely Sensed Information using Artificial Neural Networks—Climate Data Record) uses longwave infrared imagery for the estimation of half-hourly rainfall at $0.25° \times 0.25°$ resolution (Ashouri et al., 2015). The report on state-level climatic change trends in Uttarakhand based on Indian Meteorological Department (IMD) rainfall data (1951−2010) suggests a negative trend in annual and monsoon rainfall (Rathore, Attri, & Jaswal, 2013). The trend analysis using the Mann−Kendall trend test (Mann, 1945; Kendall, 1975) is widely used as nonparametric test to detect significant trends (linear or nonlinear) in time series data. The Mann−Kendall trend test, being a function of the ranks of the observations rather than their actual values, is not affected by the actual distribution of the data and is less sensitive to outliers. Findings from trend analysis of records over 80 years (1901−80) based on 30 rain gage stations maintained by the IMD in Uttarakhand (Basistha, Arya, & Goel, 2009) indicate a declining trend in monsoon rainfall in the 1965−80 period (Singh & Mal, 2014). The observation from the very limited station-wise results from Uttarakhand using monthly IMD rainfall data does not suggest any trend in the monsoon rainfall. However, Bhutiyani, Kale, and Pawar (2010) found a statistically significant decreasing trend in monsoon as well as annual rainfall for the 1866−2006 period from northwestern Himalaya. The objectives of the present study are to develop LULC maps using high-resolution satellite images for 2008 and 2017 and to quantify LULC change for the Ir-gad watershed, to integrate the nonspatial demographic census data for 2001 and 2011 with the village boundary vector map and mapping of spatiotemporal decadal changes in the two blocks falling in the study area, and to perform nonparametric trend analysis at the seasonal and annual scale, using high accuracy gridded precipitation data and comparing PERSIANN CDR $0.25° \times 0.25°$ gridded rainfall estimates to the IMD $0.5° \times 0.5°$ gridded database falling in the Ir-gad watershed. This study unravels some of the driving forces that could be responsible for bringing the recent conversion in the rural watersheds, rather the associated impacts per se, with little or no direct observations which may have been caused by such a regional-scale transformation.

1.2 Description of the study area

The Ir-gad watershed is a part of the Pashchimi Nayar River Basin (Fig. 1.1) situated in Pauri district, Uttarakhand. The Ir-gad watershed is located between longitudes 78°44′E and 78°50′E and latitudes 30°00′N−30°07′N covering an area of nearly 61 km². The elevation ranges from 600 to 2000 m in the mid-Himalayan region. Forest covers almost 53% of the watershed area and falls under the reserved forest and village forest. The dominant LULC of the watershed is a pine-dominated forest, oak-dominated forest, mixed forest, scrub forest, agricultural land, barren land, rural settlement, and streams. The higher elevation region is dominated by forest, whereas the valley portion is the region occupied by traditional agricultural land. The kharif (monsoon crop) and rabi (winter crop) are the two main cropping seasons in the mid-Himalayan region. Major cereal crops grown in the basin are paddy, millet, wheat, and mustard. The agricultural land is being converted into fallow land due to out-migration from the Pauri district. The long-term average annual rainfall for the district is 1540 mm (IMD), where more than 75% of rain is due to the SW monsoon that prevails from the middle of June to late September. A dendritic drainage pattern is observed in the catchment area and Ghat-gad and Paidul-gad join the central Ir-gad stream, which is a tributary to the Pashchimi Nayar River.

1.3 Materials and methods

The integrated approach of combining the LULC change, the demographic transition, and the recent trend in the rainfall are elaborated in the methodological flow-chart (Fig. 1.2).

1.3.1 Land use land cover classification

High-resolution multispectral satellite images (LISS-IV) for 2008 and 2017 and the Survey of India topographic map (53 J/16) is analyzed for mapping the recent LULC. LISS-IV data, having a spatial resolution of 5.4 m, was procured from the National Remote Sensing Centre, Hyderabad, for May 2008 and 2017 to maintain

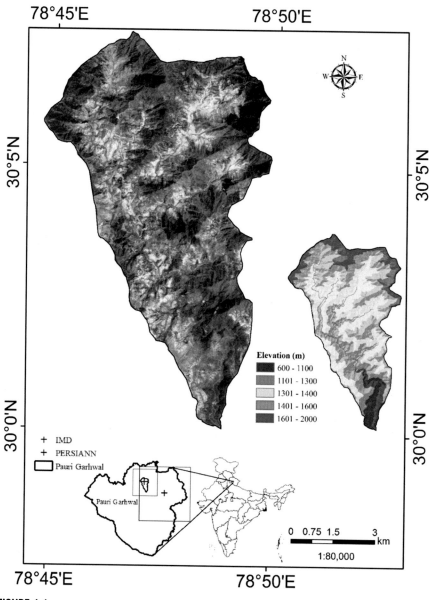

FIGURE 1.1

Study area map of the Ir-gad watershed in the Pauri district with elevation zones. The index map shows the grid size of the Indian Meteorological Department (IMD) and Precipitation Estimation from Remotely Sensed Information using Artificial Neural Networks—Climate Data Record (PERSIANN CDR) gridded data.

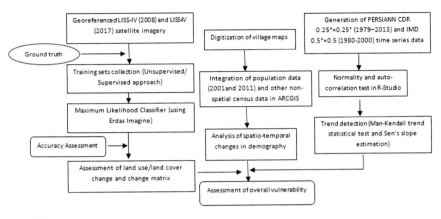

FIGURE 1.2

Flowchart of the general methodological approach for assessment of vulnerability. Schematic diagram to explain the methodological approach followed for assessment of overall vulnerability of the watershed.

the compatibility in the data set. The satellite images were geocorrected with ground control points collected through Trimble GPS and projected to Universal Transverse Mercator projection zone with WGS 1984 datum. A supervised classification method using a maximum likelihood classifier was used to classify satellite images and to develop the LULC maps for 2008 and 2017. The remote sensing data processing and classification were performed using ERDAS IMAGINE software. The training samples were selected for each class, and separability between each class was analyzed through feature space plots. The settlements, river, and roads were digitized using Google Earth using very high-resolution satellite data from both periods. Accuracy assessment was carried out to assess the overall accuracy of the classified map using a random sampling approach.

1.3.2 Determination of change matrix

Change matrices were generated using ERDAS software to analyze the broad changes in the area covered by different LULC classes for the two periods (Munsi, Malaviya, Oinam, & Joshi, 2010).

$$\text{Change in classes} = \text{Matrix (time 1, time 2)} \tag{1.1}$$

The rate of change of different classes were calculated through compound interest formula.

$$r = 2.303x \left[\frac{1}{(t_1 - t_2)} \right] \times \left[\log \left(\frac{A_2}{A_1} \right) \right] \tag{1.2}$$

where r is the rate of LULC change, and A_1 and A_2 are the area of each class.

1.3.3 Analysis of demographic drivers in Pauri and Kaljikhal blocks

One of the drivers for the ongoing land transformation is demographic change. The depopulation from the rural villages has impaired the rural socioeconomy and made the existing rural population more vulnerable. The approach from the study of two blocks, although smaller in extent, reflects a significant part of the large-scale demographic transition that the mountainous districts have been in the process of continual transformation for more than four decades or so. The study focuses on two of the closest blocks to the district headquarters (Pauri town) of Pauri district. The two administrative blocks, namely, Pauri and Kaljikhal partly cover the study area of Ir-Gad watershed located in the headwaters of the Nayar River basin. The basic population numeration data for population statistics of each village of two blocks was acquired from the "District Census Handbook (DCHB)," the official census records, accessible from the official website of the "Census of India" (https://censusindia.gov. in/2011census/dchb/DCHB.html). The data contained the population for both the census years 2001 and 2011. The data was processed and further tabulated to find the variation in the population trend of the blocks over the decade. This change analysis was further carried out spatially with each single nonspatial data linking with the vectorized village map of the blocks. With the help of ArcGIS, the population statistics of the census years 2001 and 2011 are linked to each village and a thematic population change map was generated showing the change in population in the block between 2001 and 2011. The total worker population was further analyzed to gain an insight into the terraced cropland abandonment.

1.3.4 Rainfall time series analysis

To identify the trends in long-term rainfall patterns, gridded daily precipitation data were analyzed from two different sources that is, PERSIANN CDR $0.25° \times 0.25°$ (1979−2013) and IMD $0.5° \times 0.5°$ (1980−2000) database. The seasonal (premonsoon, monsoon, postmonsoon, and winter) and annual rainfall time series in the study area were analyzed for detecting trends by applying the Mann−Kendall trend statistical test and Sen's slope estimation after assessing the normality and autocorrelation of the rainfall time series.

The Mann−Kendall test is a statistical test widely used for the analysis of trends in hydrologic and climatologic time series. The Mann−Kendall test is a nonparametric rank-based test. The details regarding the statistical test can be found in Kendall (1975); Mann (1945); and Hirsch, Slack, and Smith (1982).

1.4 Results

1.4.1 Land use land cover pattern for 2008

The map and detailed area statistics of each LULC class for the year 2008 are shown in Fig. 1.3 and Table 1.1. The LULC classes are agricultural land,

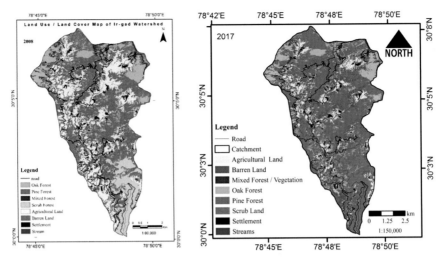

FIGURE 1.3

Spatial distributions of land use land cover (LULC) over the Ir-gad watershed for the year 2008 and 2017. Increase in the forest cover area and reduction in agricultural land can be marked from the LULC maps.

Table 1.1 Details of area under each land use land cover class and net change in the study area for the period 2008–17.

Land use land cover	2008		2017		Change
	Area (km²)	Area (%)	Area (km²)	Area (%)	
Oak forest	4.67	7.63	4.51	7.37	−3.42
Pine forest	21.75	35.52	30.43	49.69	39.89
Mixed forest	1.67	2.72	0.97	1.58	−42.06
Scrub forest	4.48	7.32	4.45	7.26	−0.71
Agricultural land	14.41	23.54	7.55	12.32	−47.64
Barren land	12.52	20.45	11.69	19.09	−6.67
Settlement	1.22	1.99	1.20	1.96	−1.53
Stream	0.50	0.82	0.44	0.72	−12.60

barren land, chir pine (*Pinus roxburghii*) forest, banj oak (*Quercus leucotrichopora*) forest, mixed forest, scrub forest, streams, and settlements. The total area under forest cover is 32.57 km² (about 53.20% of the total catchment area), which is mainly dominated by chir pine, followed by banj oak, mixed forest, and scrub forest. The area under the traditional agriculture land is around 14.41 km², whereas barren land (wasteland) covers 12.52 km² of the area. Barren land constitutes a 20.45% area of the catchment. These areas are sloping terrain and rocky and devoid of vegetation cover during summer but are converted into grassland during the monsoon period. The fallow land could not be separated from barren land using supervised classification due to poor class separability.

Overall accuracy and kappa statistics were achieved as 68.09% and 0.60, respectively. Most of the forest cover area falls into the village forest category (Civil Forest), and the northwestern part of the catchment area is occupied by Panchokhali Reserve Forest, managed by the State Forest Department. These forests serve a very crucial role in the supply of fuel wood and fodder for the rural population in Ir-gad watershed.

1.4.2 Land use land cover pattern 2017

The spatial distribution of classified LULC for 2017 over the Ir-gad watershed is shown in Fig. 1.3. The overall accuracy and the kappa statistic in LULC classification were 66.67% and 0.54%, respectively. Table 1.2 shows the change matrix of spatially distributed LULC for 2008 and 2017 year in the Ir-gad watershed. The diagonal elements of the change matrix represent the area that has remained unchanged in both periods. Other elements except diagonal elements indicate the percentage of area that changed from one class to another class over the period in the study period.

A significant change in the area under forest cover (mainly pine and mixed forest) and agricultural land has been observed over a decade. The area distribution of LULC classes during the period 2008 and 2017 and their net change in percentage over the watershed is represented in Table 1.1 and Fig. 1.4. The total forest cover has marked the highest growth from 53.20% to 65.91% of the total watershed area over the 2008−17 period. The area under the pine forest has increased from 21.75 to 30.43 km². The agricultural land has experienced negative growth from 23.54% to 12.32% in the area over the watershed. A significant change was also observed due to expansion of the road network from 2008 to 2017. The change matrix shown in Table 1.2 reveals that significant portions of agricultural land and barren land are being converted to pine forests.

1.4.3 Demographic changes

The Pauri block encompasses an area of 123.84 sq. km, while Kaljikhal covers an area of 188.81 sq. km, covering 2.3% and 3.5% of the area, respectively, of the total Pauri Garhwal district.

The Pauri and Kaljikhal blocks have recorded a negative growth rate (−1.5%), although the urban centers have managed significant growth (25%) between the last two census years (2001−11). The rural population of the Pauri block was 28,694 in 2011, contributing only 4.1% of the overall population of the district, while Kaljikhal has a population of 29,287, which makes only 4.2% to the total population. Fig. 1.5 highlights the recent decadal change in the studied two blocks, which are experiencing a negative growth rate in the population.

Table 1.2 Change matrix for 2008 and 2017 (area in km^2).

2008	2017							
	Oak forest	Pine forest	Mixed forest	Scrub land	Agricultural land	Barren land	Settlement	Stream
Oak forest	2.656	1.760	0.125	0.000	0.033	0.099	0.001	0.000
Pine forest	1.080	17.400	0.214	0.003	0.810	2.224	0.008	0.007
Mixed forest	0.373	0.765	0.152	0.000	0.123	0.253	0.002	0.001
Scrub land	0.000	0.004	0.000	4.418	0.026	0.012	0.003	0.011
Agricultural land	0.175	3.481	0.339	0.005	5.002	5.382	0.013	0.010
Barren land	0.224	6.932	0.132	0.013	1.521	3.676	0.008	0.012
Settlement	0.002	0.017	0.002	0.000	0.017	0.015	1.169	0.000
Streams	0.002	0.056	0.001	0.008	0.014	0.022	0.000	0.399

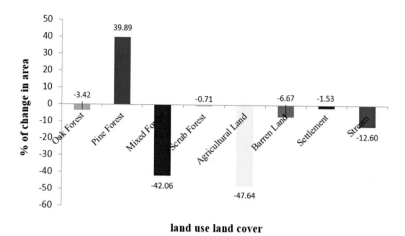

FIGURE 1.4

Percentage of change in land use/land cover classes in the Ir-gad watershed between 2008 and 2017.

The total worker population makes up to 40% of population of district Pauri and registered a minor increase of only 1.6%. The population of agricultural labor increased substantially over a decade and registered a growth of 203% for the Pauri district.

The household industry workers hiked by 63%, while the population of other workers increased by 27.3%. The scenario in the individual blocks was a little different. In the case of Pauri block, the total worker population increased by 8.2%, whereas the Kaljikhal block showed a net decline in the worker population by 9.5%. The Pauri block showed a decline in cultivators and household workers by 5% and 27%, respectively, while agricultural laborers and other worker populations increased by 124% and 16.5%, respectively. In the Kaljiklhal block, a net decline was also observed in cultivators and household workers by 25% and 44%, respectively, while agricultural laborers and other worker populations increased by 138% and 55%, respectively. Fig. 1.6 highlights the overall cultivators' population change from the year 2001 to 2011.

1.4.4 Blockwise population distribution

Table 1.3 elaborates on the typology of the population dynamics over a decade, categorizing the number of villages in a specified population range in both census years. A closer look at Table 1.3 highlights an increase in the frequency of villages with less than 200 populations and an overall decrease in the frequency of villages with more than 200 populations with an exception in the Pauri block

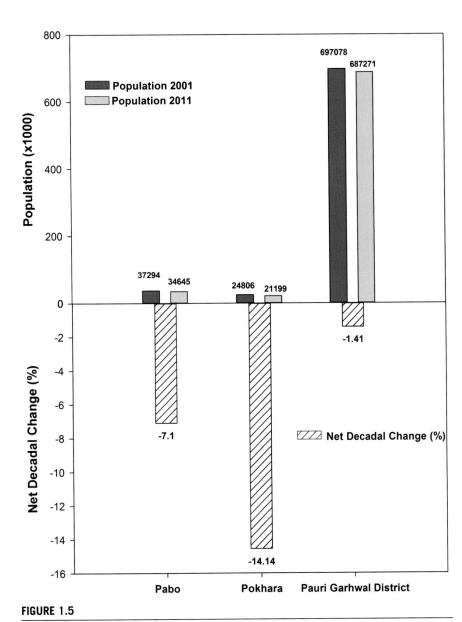

FIGURE 1.5

Population and decadal change graph, highlighting the population and population change for the years 2001 and 2011 of Pauri block, Kaljikhal block, and Pauri district.

(Village Ufalda showed an increase of 51% over a decade). The trends for the district as a whole are similar to the two blocks with an increase in the number of villages above 1000 population.

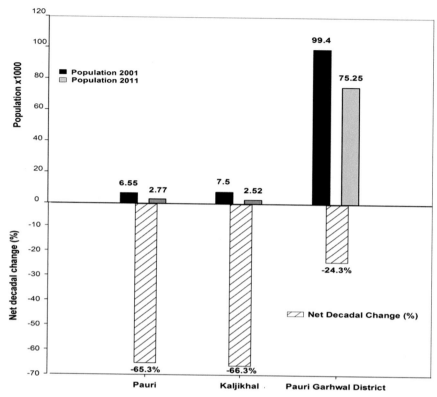

FIGURE 1.6

The graph showing prominent change in the cultivators' population for the years 2001 and 2011, in effect causing conversion of agriculture land to fallow land of Pauri block, Kaljikhal block, and Pauri.

1.4.5 Spatial-temporal mapping through a geographical information system for decadal change analysis

The trends of population change over a decade are represented by shades of red showing decline; green shades display an increase in population, while the forest villages represented in hatch have a very nominal population (Fig. 1.7).

In the Pauri block, 158 villages including 40 uninhabited villages out of a total of 214 villages showed a decline in population, while 54 villages showed a net increase in the population. However, only 14 villages showed a net decline of more than 50% and 12 villages registered a population growth of more than 50%. As evident from the map (Fig. 1.7) area-wise statistics indicate a preponderance of villages with a negative trend in both the blocks. Around 97 sq. km, that is, about 79% of the Pauri block had shown an overall negative trend, whereas the remaining 26 sq. km, that is, about 21% of the total area had shown a net positive

Table 1.3 Number and percentage of villages 2001 and 2011.

Block/District name	Number of villages		Villages inhabited		Population less than 200		Population 200–499		Population 500–999		Population 1000–1999		Population 2000–4999	
	2001	2011	2001	2011	2001	2011	2001	2011	2001	2011	2001	2011	2001	2011
Pauri block	214	214	180 (84%)	177 (82%)	132 (73.3%)	133 (75%)	35 (19.4%)	34 (19%)	12 (6.7%)	9 (5%)	1 (0.6%)	Nil	Nil	1 (1%)
Kaljikhal block	274	274	242 (88%)	242 (88%)	187 (77.3%)	195 (81%)	50 (20.7%)	42 (17%)	5 (2%)	4 (2%)	Nil	Nil	Nil	Nil
Pauri district	473	3483	3151 (90%)	3142 (90%)	2190 (69.5%)	2303 (73%)	790 (25.1%)	683 (22%)	136 (4.3%)	109 (3%)	20 (0.6%)	28 (1%)	13 (0.4%)	17 (1%)

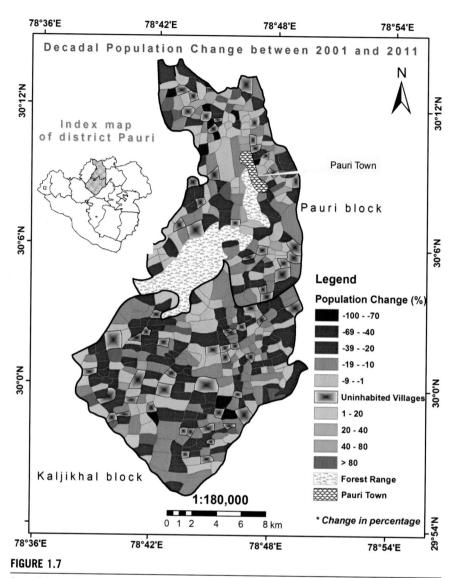

FIGURE 1.7

Decadal population change for 2001–11 in Pauri and Kaljikhal blocks.

trend. The Kaljikhal block has 274 villages, out of which 228 villages showed a population decline (including 32 uninhabited villages "Ghost Village"), while 46 villages showed an increment. Out of these 228 villages, only 13 villages showed a positive trend of more than 50%, while 16 villages registered a negative growth of more than 50% over a decade. Around 156 sq. km (83%) of the area shows a negative trend in the overall population and only 32 sq. km (17%) of the

Kaljikhal block has shown a positive trend. Overall, the majority of villages have shown an overall decline in the population, which is evident from the map with the dominance of red shades.

1.4.6 Trend analysis of the rainfall time series datasets

The seasonal and annual time series data of gridded rainfall time series is analyzed using the nonparametric Mann–Kendal test for the Ir-gad catchment. In the Mann–Kendall test, parameters like S statistic, Z_{mk}, Q, Q_{min95}, and Q_{max95} were considered to identify the increasing or decreasing trend in the time series of daily rainfall from two gridded data sources. The results of the Mann–Kendall trend analysis of seasonal and annual precipitation are summarized in Fig. 1.8 and Table 1.4, and it is observed that the calculated test statistics values of the Mann–Kendall test indicate the presence of a trend in the annual and seasonal rainfall time series data over the studied catchment. Based on the PERSIANN CDR $0.25° \times 0.25°$ gridded (1983–2017) rainfall database, a statistically significant trend and negative value of Sen's slope are observed for annual ($Q = -7.62$ mm/year), monsoon ($Q = -7.72$ mm/year), and winter ($Q = -1.92$ mm/year) seasons,

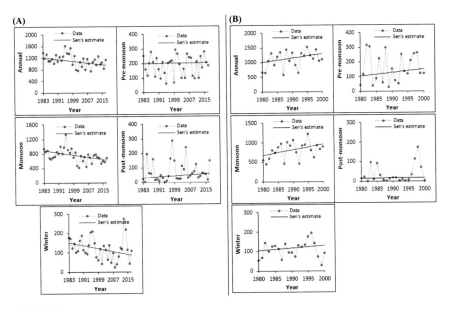

FIGURE 1.8

The seasonal and annual rainfall variation time over the upper Ir-gad catchment, based on (A) Precipitation Estimation from Remotely Sensed Information using Artificial Neural Networks—Climate Data Record $0.25° \times 0.25°$ gridded (1983–2017) database and (B) the Indian Meteorological Department $0.5° \times 0.5°$ gridded (1980–2000) database.

Table 1.4 The seasonal and annual rainfall variation time over the upper Ir-gad catchment, based on Precipitation Estimation from Remotely Sensed Information using Artificial Neural Networks—Climate Data Record (PERSIANN CDR) 0.25° × 0.25° gridded (1983–2017) database and the Indian Meteorological Department (IMD) 0.5° × 0.5° gridded (1980–2000) database.

| Time series | PERSIANN CDR 0.25° × 0.25° database (1983–2017) | | | | | IMD 0.5° × 0.5° gridded database (1980–2000) | | | | |
| | MK trend | | Sen's slope estimate | | | MK trend | | Sen's slope estimate | | |
	Z_{mk}	Sigf.	Q	Q_{min95}	Q_{max95}	Z_{mk}	Sigf.	Q	Q_{min95}	Q_{max95}
Annual	−2.50	*	−7.62	−14.976	−1.663	1.24		14.65	−9.09	39.47
Premonsoon	0.20		0.18	−2.068	2.596	0.57		2.37	−5.22	9.46
Monsoon	−2.87	**	−7.72	−12.969	−3.101	1.72	+	14.43	−3.82	31.18
Postmonsoon	1.08		0.85	−0.823	2.272	0.57		0.17	−1.17	3.26
Winter	−1.87	+	−1.92	−3.455	0.048	0.63		1.29	−2.31	4.88

Signf.: Smallest significance level α with which the test shows that the null hypothesis of no trend should be rejected. *** If the trend is at the α = 0.001 level of significance, ** if the trend is at the α = 0.01 level of significance, * if the trend is at the α = 0.05 level of significance, + if the trend is at the α = 0.1 level of significance. Sen's slopes estimate Q: Sen's estimator for the true slope of the linear trend, that is, changes per unit time (in this case a year). Q_{min99}: Lower limit of the 99% confidence interval of Q (α = 0.1). Q_{max99}: Upper limit of the 99% confidence interval of Q (α = 0.1). Q_{min95}: Lower limit of the 95% confidence interval of Q (α = 0.05).

indicating a downward trend in the rainfall time series, whereas the positive value of Sen's slope for premonsoon ($Q = +0.18$ mm/year) and postmonsoon ($Q = +0.85$ mm/year) seasons shows an upward trend in the rainfall time series over the studied catchment. However, the Mann−Kendall trend analysis of the IMD $0.5° \times 0.5°$ gridded (1980−2000) precipitation suggested the positive value of Sen's slope for annual ($Q = +14.65$ mm/year), premonsoon ($Q = +2.37$ mm/year), monsoon ($Q = +14.43$ mm/year), postmonsoon ($Q = +0.17$ mm/year), and winter ($Q = +1.29$ mm/year) seasons suggests an upward trend in the rainfall time series. The trends in the two gridded datasets do not agree, which could be due to the inherent complexity of rainfall distribution in rugged mountainous topography and the difference in the resolution of the gridded datasets.

1.5 Discussion

Slow and gradual demographic transition mainly caused by migration of the workforce from rural-to-urban centers in the state of Uttarakhand over the past decades has led to rapid local and regional-scale changes in the LULC in the watersheds of middle Himalayan basins. The widespread rural abandonment of cultivable land is a consequence caused by depopulation from the rural villages. Hence, an accurate assessment of dominant patterns of mid-Himalayan LULC change becomes very important as long-term land and water resource sustainability depends on understanding the recent dynamics and its future consequences. Expansion of the forest cover area is the most dominant change between 2007 and 2017 in the LULC of Ir-gad watershed in the middle Himalayan region. Over 11 years, the forest cover area has increased by 13%, which can be primarily attributed to the increase in chir pine forest, caused by colonization of fast-growing *P. roxburghii* in the terraced agricultural land and barren land. This significant increase in the *P. roxburghii* forest cover area from nearly 36% area (as in 2007) to nearly 50% area in 2017 of the 61 km^2 watershed is a result of the dominance of the conifer in the forested landscapes, its fire-resistant characteristics, and plantation of Pine (*Pinus* spp.) by Department of Forest, Uttarakhand, till 2005 and major dependency of people on the forest resources, especially pine for fuel wood in this region of middle Himalaya. The progressive decline in the traditional agricultural practices has resulted in the reduction in the agricultural area by almost half (6.86 km^2 of area lost over 11 years) of the existing area under cultivated land in the year 2007. The satellite-based mapping shows that the irrigated system in the valleys as well as terraced rain-fed agricultural land are abandoned at an alarming rate. The people opt to shift to cities for better livelihood in secondary and tertiary sectors than in the age-old primary sector of subsistence traditional agriculture, causing abandonment of fertile agricultural land and regrowth of secondary forest. A household survey-based study from eight villages in Garhwal Himalaya also points toward (−) 50% decline in the area under traditional agricultural land over past two decades (Sati & Kumar, 2023). Case studies of decline in traditional farming practices are reported from the mountain areas of Europe, China, and Nepal (Quintas-Soriano et al., 2022; Subedi et al., 2022; Zhang et al., 2014). The consequences of such a LULC transformation are many, not only in the immediate vicinity but also for the downstream population; for instance, a very recent study highlights that an expansion in tree cover could have spatially divergent and complex outcome, leading to reduced water availability (Hoek van Dijke et al., 2022). The social and environmental implications of arable land abandonment and possible corrective measures through land consolidation of fragmented agricultural land are being suggested (Joshi, 2018; Shukla, Chakraborty, Sachdeva, & Joshi, 2018). The present study shows that out of 102 villages of the Pauri and Kaljikhal block, which are falling in the studied Ir-gad watershed, around 67% of the villages are experiencing a decline in the population. Nonetheless, the apparent demographic transition through comparative analysis over the past decade till 2011 (last census year) indicates a rising vulnerability caused by not only the dominance of villages registering a significant negative trend in population but also the shrinking livelihood opportunities for the residents of rural communities. Additionally, a series of past studies on rainfall trend analysis from the mountainous region of Uttarakhand (Basistha et al., 2009; Dash, Nair, Kulkarni, & Mohanty, 2011; Tarafdar & Dutta, 2023) as well as the present study point toward compounded vulnerability for the land and water ecosystems in the rural landscapes of middle Himalaya.

1.6 Conclusions

Rapid land use change as well as modifications in the rainfall distribution and its long-term trend can have a significant impact on the hydrology of the watershed. The classified output from high-resolution satellite data over 10 years highlights that the area under the pine forest has increased by (+) 40%, whereas the area under the agricultural land has reduced by (−) 48% from the study area falling in the middle Himalayan region. The change matrix indicates that a significant area of barren land and agricultural land have been converted into pine forests. Such a rapid transformation in land use may bring slow changes in the surface hydraulic conductivity of soil and make the watershed even drier and water-limited. The contextual clues for traditional terraced cropland abandonment and fast regrowth of *Pinus* species leading to forest area expansion is an outcome of a negative growth rate in total population and more than 65% decline in cultivators' population from the two blocks. Permanent fallow land could not be separated and is classified as barren land due to poor separability between the two classes. Future work at a regional scale using an object-based classification approach of very high-resolution satellite data (<1 m) will throw more light on the trends and spatial distribution of LULC change. The analysis of long-term precipitation of the

PERSIANN CDR $0.25° \times 0.25°$ gridded (1983–2017) database over 35 years indicates a statistically significant downward trend in the annual, monsoonal, and winter rainfall, which is in agreement with the findings from the western Himalaya. A rising trend of precipitation in seasonal as well as annual rainfall was observed in IMD gridded data during the 1980–2000 periods. Disagreement between the two rainfall estimates could be due to the difference in spatial resolution of gridded rainfall data and the complexity of rugged Himalayan topography. Research findings could provide useful information for adaptive land use policy for wise natural resource management at the smallest administrative units.

Acknowledgments

The authors express gratitude to National Mission on Himalayan Studies (NMHS), implemented by the Ministry of Environment, Forest & Climate Change (MoEF&CC), New Delhi Nodal and Serving hub with G.B. Pant National Institute of Himalayan Environment & Sustainable Development (GBPNIHESD), Almora, Uttarakhand, for providing the funding and research facilities during the research work.

References

Ashouri, H., Hsu, K. L., Sorooshian, S., Braithwaite, D. K., Knapp, K. R., Cecil, L. D., . . . Prat, O. P. (2015). PERSIANN-CDR: Daily precipitation climate data record from multisatellite observations for hydrological and climate studies. *Bulletin of the American Meteorological Society*, *96*(1), 69–83. Available from https://doi.org/10.1175/BAMS-D-13-00068.1, http://journals.ametsoc.org/doi/pdf/10.1175/BAMS-D-13-00068.1.

Basistha, A., Arya, D. S., & Goel, N. K. (2009). Analysis of historical changes in rainfall in the Indian Himalayas. *International Journal of Climatology*, *29*(4), 555–572. Available from https://doi.org/10.1002/joc.1706.

Batar, A., Watanabe, T., & Kumar, Ajay (2017). Assessment of land-use/land-cover change and forest fragmentation in the Garhwal Himalayan Region of India. *Environments*, *4*(2). Available from https://doi.org/10.3390/environments4020034.

Bhutiyani, M. R., Kale, V. S., & Pawar, N. J. (2010). Climate change and the precipitation variations in the northwestern Himalaya: 1866–2006. *International Journal of Climatology*, *30*(4), 535–548. Available from https://doi.org/10.1002/joc.1920.

Bookhagen, B., & Burbank, D. W. (2006). Topography, relief, and TRMM-derived rainfall variations along the Himalaya. *Geophysical Research Letters*, *33*(8). Available from https://doi.org/10.1029/2006GL026037.

Dasgupta, S., Kumar, R., & Tripathi, S. (2010). Temporal change detection in two watershed areas of Kumaon region and its impact on livelihood of forest fringe communities. *Current Science*, *98*(10), 1349–1353. Available from http://www.ias.ac.in/currsci/25may2010/1349.pdf.India.

Dash, S. K., Nair, A. A., Kulkarni, M. A., & Mohanty, U. C. (2011). Characteristic changes in the long and short spells of different rain intensities in India. *Theoretical and*

Applied Climatology, 105(3), 563–570. Available from https://doi.org/10.1007/s00704-011-0416-x, http://link.springer.de/link/service/journals/00704/.

Foley, J. A., DeFries, R., Asner, G. P., Barford, C., Bonan, G., Carpenter, S. R., ... Snyder, P. K. (2005). Global consequences of land use. *Science (New York, N.Y.), 309* (5734), 570–574. Available from https://doi.org/10.1126/science.1111772.

Hirsch, R. M., Slack, J. R., & Smith, R. A. (1982). Techniques of trend analysis for monthly water quality data. *Water Resources Research, 18*(1), 107–121. Available from https://doi.org/10.1029/WR018i001p00107.

Hoek van Dijke, A. J., Herold, M., Mallick, K., Benedict, I., Machwitz, M., Schlerf, M., ... Teuling, A. J. (2022). Shifts in regional water availability due to global tree restoration. *Nature Research, Luxembourg Nature Geoscience, 15*(5), 363–368. Available from https://doi.org/10.1038/s41561-022-00935-0, http://www.nature.com/ngeo/index.html.

Joshi, B. (2018). Recent trends of rural out-migration and its socio-economic and environmental impacts in Uttarakhand Himalaya. *Journal of Urban and Regional Studies on Contemporary India, 4*(2), 1–14.

Kendall, M. G. (1975). *Rank correlation methods.* Google Scholar.

Khandelwal, D. D., Gupta, A. K., & Chauhan, V. (2015). Observations of rainfall in Garhwal Himalaya, India during 2008–2013 and its correlation with TRMM data. *Indian Academy of Sciences, India Current Science, 108*(6), 1146–1150. Available from http://www.currentscience.ac.in/Volumes/108/06/1146.pdf.

MacDonald, D., Crabtree, J. R., Wiesinger, G., Dax, T., Stamou, N., Fleury, P., ... Gibon, A. (2000). Agricultural abandonment in mountain areas of Europe: Environmental consequences and policy response. *Journal of Environmental Management, 59*(1), 47–69. Available from https://doi.org/10.1006/jema.1999.0335.

Mamgain, R. P., & Reddy, D. N. (2017). Out-migration from the Hill Region of Uttarakhand: Magnitude, challenges, and policy options. In D. N. Reddy, & K. Sarap (Eds.), *Rural labour mobility in times of structural transformation: Dynamics and perspectives from Asian economies* (pp. 209–235). India: Springer Singapore. Available from http://www.springer.com/in/book/9789811056277.

Mann, Henry B. (1945). Nonparametric tests against trend. *Econometrica: Journal of the Econometric Society, 13*(3). Available from https://doi.org/10.2307/1907187.

Munsi, M., Malaviya, S., Oinam, G., & Joshi, P. K. (2010). A landscape approach for quantifying land-use and land-cover change (1976–2006) in middle Himalaya. *Regional Environmental Change, 10*(2), 145–155. Available from https://doi.org/10.1007/s10113-009-0101-0.

Quintas-Soriano, C., Buerkert, A., & Plieninger, Tobias. (2022). Effects of land abandonment on nature contributions to people and good quality of life components in the Mediterranean region: A review. *Land use Policy, 116.* Available from https://doi.org/10.1016/j.landusepol.2022.106053.

Ramankutty, N., & Foley, J. A. (1998). Characterizing patterns of global land use: An analysis of global croplands data. *Global Biogeochemical Cycles, 12*(4), 667–685. Available from https://doi.org/10.1029/98GB02512, http://onlinelibrary.wiley.com/journal/10.1002/(ISSN)1944-9224.

Rao, K. S., & Pant, Rekha (2001). Land use dynamics and landscape change pattern in a typical micro watershed in the mid elevation zone of central Himalaya, India. *Agriculture, Ecosystems & Environment, 86*(2), 113–124. Available from https://doi.org/10.1016/s0167-8809(00)00274-7.

Rathore, L. S., Attri, S. D., & Jaswal, A. K. (2013). State levelclimate change trends in India. Chapter 3. Meteorological Monograph No. ESSO/IMD/EMRC/02/2013, p.11.

Sati, V. P., & Kumar, S. (2023). Declining agriculture in Garhwal Himalaya: Major drivers and implications. *Cogent OA, India Cogent Social Sciences*, *9*(1). Available from https://doi.org/10.1080/23311886.2023.2167571, http://www.cogentoa.com/journal/social-sciences.

Scanlon, B. R., Reedy, R. C., Stonestrom, D. A., Prudic, D. E., & Dennehy, K. F. (2005). Impact of land use and land cover change on groundwater recharge and quality in the southwestern US. *Global Change Biology*, *11*(10), 1577−1593. Available from https://doi.org/10.1111/j.1365-2486.2005.01026.x.

Seibert, J., McDonnell, J. J., & Woodsmith, R. D. (2010). Effects of wildfire on catchment runoff response: A modelling approach to detect changes in snow-dominated forested catchments. *Hydrology Research*, *41*(5), 378−390. Available from https://doi.org/10.2166/nh.2010.036.

Semwal, R. L., Nautiyal, S., Sen, K. K., Rana, U., Maikhuri, R. K., Rao, K. S., & Saxena, K. G. (2004). Patterns and ecological implications of agricultural land-use changes: A case study from central Himalaya, India. *Agriculture, Ecosystems & Environment*, *102* (1), 81−92. Available from https://doi.org/10.1016/s0167-8809(03)00228-7.

Shukla, R., Chakraborty, A., Sachdeva, K., & Joshi, P. K. (2018). Agriculture in the western Himalayas—An asset turning into a liability. *Development in Practice*, *28*(2), 318−324. Available from https://doi.org/10.1080/09614524.2018.1420140, http://www.tandf.co.uk/journals/titles/09614524.asp.

Singh, R. B., & Mal, Suraj (2014). Trends and variability of monsoon and other rainfall seasons in Western Himalaya, India. *Atmospheric Science Letters*, *15*(3), 218−226. Available from https://doi.org/10.1002/asl2.494.

Subedi, Y. R., Kristiansen, P., & Cacho, O. (2022). Reutilising abandoned cropland in the Hill agroecological region of Nepal: Options and farmers' preferences. *Land use Policy*, *117*. Available from https://doi.org/10.1016/j.landusepol.2022.106082.

Tarafdar, S., & Dutta, S. (2023). Long-term decline in rainfall causing depletion in groundwater aquifer storage sustaining the springflow in the middle-Himalayan headwaters. *Journal of Earth System Science*, *132*(3). Available from https://doi.org/10.1007/s12040-023-02136-8, https://www.springer.com/journal/12040.

Tiwari, P. C. (2000). Land-use changes in Himalaya and their impact on the plains ecosystem: Need for sustainable land use. *Land Use Policy*, *17*(2), 101−111. Available from https://doi.org/10.1016/s0264-8377(00)00002-8.

Tiwari, P. C., & Joshi, B. (2012). Environmental changes and sustainable development of water resources in the Himalayan headwaters of India. *Water Resources Management*, *26*(4), 883−907. Available from https://doi.org/10.1007/s11269-011-9825-y.

Wagner, P. D. (2013). An assessment of land use change impacts on the water resources of the Mula and Mutha Rivers catchment upstream of Pune, India. *HESS*. Available from https://doi.org/10.5194/hess-17-2233-2013.

Zhang, Y., Li, X., & Song, W. (2014). Determinants of cropland abandonment at the parcel, household and village levels in mountain areas of China: A multi-level analysis. *Land Use Policy*, *41*, 186−192. Available from https://doi.org/10.1016/j.landusepol.2014.05.011, http://www.elsevier.com/inca/publications/store/3/0/4/5/1/.

Hypsometric analysis for determining erosion susceptibility of Karnaphuli Watershed, Bangladesh, using remote sensing and GIScience

Rubaiya Zumara[1] and N.M. Refat Nasher[2]

[1]*Department of Geography and Environment, Jagannath University, Dhaka, Bangladesh*
[2]*Department of Geography and Environment, Faculty of Life and Earth Sciences, Jagannath University, Dhaka, Bangladesh*

2.1 Introduction

A watershed refers to a geographically defined area of land that encompasses the collection and subsequent downward movement of all water within its boundaries toward distinct water bodies like oceans, lakes, and rivers (Mahalingam, 2012; Shekar et al., 2023). The study of hypsometry involves the analysis and evaluation of the correlation between elevation and the size of a basin, with the aim of determining the degree of erosion and the developmental stage of a landform. This study assesses the long-term influence of tectonic and geological factors. Hypsometric analysis is a method used to establish the correlation between the horizontal and vertical elevation ratios inside a watershed. This enables the contrast of subwatersheds, regardless of their scale components (Al-Hantouli, Awawdeh, & Obeidat, 2023; Dowling, Walker, Richardson, Sullivan, & Summerell, 1998; Enea, Iosub, & Stoleriu, 2023; Rakesh, Govindaraju, Lokanath, & Kumar, 2023; Shekar et al., 2023).

The application of hypsometric analysis was employed in order to ascertain the mechanisms of geomorphic processes and the formation of landforms. The configuration of the catchment and drainage network exerts a notable influence on the hypsometric integrals (HIs). In terms of computation, it relates to determining how the elevation is distributed in relation to the catchment area. Strahler (1952) proposed a comprehensive and more accurate hypsometric curve (HC) based on Langbein's (1947) concept. A contemporary HC was developed (Ritter, Kochel, & Miller, 2002), as shown in Fig. 2.1.

Applications of Geospatial Technology and Modeling for River Basin Management.
DOI: https://doi.org/10.1016/B978-0-443-23890-1.00002-5

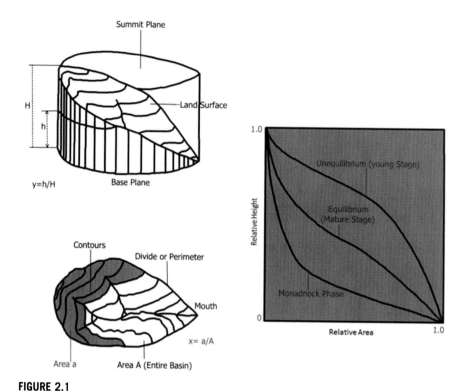

FIGURE 2.1

Schematic diagram of hypsometric curve after Ritter et al. (2002).

From Ritter, D. F., Kochel, R. C., & Miller, J. R. (2002). Process geomorphology. *McGraw Hill. (Original work published 2002).*

The HC was assessed and explained by applying dimensionless parameters, without consideration of the true scale. In the portion of the watershed above a particular altitude, HCs are unit less (Strahler, 1952). Correlations between HCs and the landscape and geological development of the catchment area can be found (Hurtrez, Sol, & Lucazeau, 1999; Schumm, 1956). Hypsometric curves have been employed as a means to ascertain the age of landforms, as demonstrated by Strahler in 1964. The form of the HC and the value of the HI are crucial indications for determining the influence of the topography of the watershed (Marani, Eltahir, & Rinaldo, 2001). The HC exhibited analogous hydrogeologic circumstances and offered valuable insights into the dynamics of soil movement, specifically erosion and accretion within the basin area (Stepinski & Stepinski, 2005). Hypsometric analysis, as demonstrated by Mahmood and Gloaguen (2011), can provide valuable insights into the volume of soil and the extent of erosion that has taken place in the basin area, compared to the remaining mass. The evaluation of the soil's temporal displacement within the basin, assuming consistent hydrogeological circumstances, can be accomplished through a comparative analysis of the HC of several watersheds (Gajbhiye, Mishra, & Pandey, 2014). Therefore, the disparities in the mean gradient of the initial catchment can be elucidated by the arrangement of the HCs (Van Liem et al., 2016). The HC serves as a quantitative representation of the probable rock weathering within a basin (Bishop, Shroder, Bonk, & Olsenholler, 2002).

The HIs and curves were considered significant parameters for assessing drainage basin conditions (Ritter et al., 2002). The HC and HI represent the degree of erosion and geological forces (Weissel, Pratson, & Malinverno, 1994). A comprehensive comprehension of the soil displacement in past eras of the watershed can be achieved by examining and contrasting the HCs' forms for various drainage basins within similar hydrological conditions (Singh & Singh, 2018). Therefore, the HC suggested the chronological differences of the slope for early basin characteristics. Strahler (1952) categorized the drainage basins into three groups early (highly active), matured (less active), peneplain or old analyzing the HCs. These HCs represent the phases of landscape development and provide information about the grade of loss of the drainage basins (Singh & Singh, 2018). The HCs undergo significant changes throughout the early geomorphic stages of growth, but experience little modifications after the basin area has reached the mature stage. The HC with a convex shape indicated the stable state of the watersheds, whereas concave indicated more vulnerability to erosion processes of the watershed (Hurtrez et al., 1999).

A number of experts have used the hypsometric method to study erosional topography (Singh and Singh, 2018). Based on these study results, we can say that the HI was calculated from the HCs that were made using the usual graph-based area estimation methods. The HI value of the catchment is inversely proportional to how much denudation is happening on the slopes. This shows that the plano-altimetric formations of the watershed rely a lot on how well the denudation processes work (Strahler, 1952). They did not talk about how the HI affects the erosion state of the basin.

The purpose of this chapter was to develop and assess hypsometric and statistical analysis using a geographic information system (GIS) to assess erosion in the Karnaphuli river basin through the study of morphometric features. This research has the potential to inform soil and water conservation efforts by shedding light on morphological shifts and denudational processes. The primary goals are (1) GIScience-based hypsometric analysis and (2) HC-based erosion risk assessment.

2.2 Limitations of the study

Hypsometric analysis-based research for big river basins like the Karnaphuli River had constraints due to the time-consuming procedure of data gathering and the evaluation needed for approximation in order to investigate watershed conditions. However, there are caveats to this study as well. Very little time was allotted to complete this investigation. Due to time constraints, we will be unable to complete the necessary sample collection and field visit. Also, because this study relies on secondary data sources, there may be gaps in coverage or inability to clearly denote changes in slope because of this. Using sedimentation data and various methods, this can be done with more precision. Due to the lack of high-resolution digital elevation model (DEM) data in freely available data sources, the resolution of images is also restricted. A scarcity of hypsometric analysis-based research evaluating the watershed health in this basin was also noted. This could be because estimating the HI requires a lengthy method of data collection and analysis. The investigation cannot evaluate the sedimentation rate or other morphological factors due to a lack of funding.

2.3 Methods and materials

2.3.1 Geological and climatic conditions of the study area

The Karnaphuli River holds significant importance as one of the longest rivers within the Chattogram Hill Tracts region. The river has a catchment area of around 11,000 sq. km, and its source may be traced back to the Lushai Hills in Mizoram State, India (Ahmed et al., 2013). The river meanders through approximately 180 km of the hilly region of Rangamati in Bangladesh, followed by a further 170 km, as it courses through the port city of Chattogram, ultimately emptying into the Bay of Bengal. From a geological perspective, the river basin is primarily comprised of tertiary bedrock, which is interspersed with alluvial deposits. Additionally, the uppermost levels of sediment consist of many strata of mud and sand. The entirety of the basin was composed of sedimentary deposits from the Tertiary epoch (Rizvi, 1975). The Karnaphuli Estuary is one of the notable estuaries in Bangladesh situating near Patenga seaport in Chattogram City between latitude 22°53′ and longitude 91°47′E. The estuary exhibits semidiurnal tidal patterns, with tidal fluctuations ranging from 2 to 4 m. The average depth of the channel in the outer zone is approximately 8−10 m, as reported by Lara et al. (2009). The significant impact of the Indian monsoon results in periodic alterations in the environmental factors of the Karnaphuli estuary (Alam & Zafar, 2012).

The geographical location of the Karnafuli reservoir watershed is situated inside the tropical region. The primary hydrological divide of the river exhibits an extended orientation in the north−south axis, situated inside constricted valleys flanked by parallel hill ridges. These valleys serve as the conduit through which the river has eroded its course. The morphology of the Karnafuli river basin has a relatively flat and low terrain in its downstream region, while displaying a more rocky and uneven landscape in its upstream region. The river traverses a considerable distance in order to reach its final destination, the sea. Throughout the duration of the route, the meandering river in question traverses notable landmarks such as the Kaptai hydroelectric power plant, the Halda—Karnafuli confluence, and many bridges (Ahmed et al., 2013). The construction of a dam on the Karnafuli River in Kaptai took place in 1962. The primary objective of this system is the production of electrical energy. The formation of the Kaptai Lake reservoir has occurred as a result of the construction of a dam. The distance between the Kaptai Dam and the Halda—Karnafui confluence measures approximately 45 km, while the distance from the Halda—Karnafuli confluence to BN Academy spans approximately 30 km. Fig. 2.2 depicts the geographical positioning of the Karnafuli river basin with subwatersheds (Ahmed et al., 2013). At the upstream of the Kaptai dam, major tributaries of the Karnafuli River are Kasalong, Maini, Chengi, Cholok, and Rainkhiang rivers. At the downstream of the Kaptai dam, the main tributaries at its left bank are Kaptai, Shilok, Chondaria, Raikhali, Boalkhali, and Shikalbaha and its right bank are the Ichamati and Halda River along with the Chaktai Khal and Mohesh Khal.

The study area can be split into three main geomorphological groups: (1) the hilly area, (2) the fluvio-tidal plain, and (3) the tidal plain. The flow of this river changes a lot because of the big changes in rainfall between the rainy and dry seasons. It rains a lot in Chittagong and the surrounding hills during the tropical monsoon season. From November to February, it is dry and cool. From March to May, it is very hot because it is before the rainy season. From June to

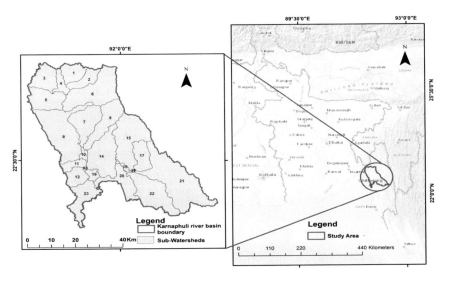

FIGURE 2.2

Study area and its location on Bangladesh map with subwatersheds.

October, it rains a lot and is warm during the rainy season. Chattogram city has an average temperature of 25.3°C (77.6°F), and it rains about 2777 mm (109.3 in.) a year.

2.3.2 Data collection and processing

The primary DEM of the research area was generated using photographs from the Shuttle Radar Topographic Mission (SRTM). The image possessed a resolution of 30 m. The photos were acquired via the USGS Earth Explorer data portal (https:// earthexplorer.usgs.gov/) and georeferenced using the Universal Transverse Mercator projection (WGS 1984, Zone 46N). This process was conducted on February 15, 2023. Additionally, the basin border was extracted using ArcSWAT and Hydrology tools. Using the ArcGIS software packages, topographic elevation maps were constructed based on the available data. Subsequently, the drainage networks for the Karnaphuli subbasins were generated. The initial delineation of the fill, flow direction, and flow accumulation grid was performed using the hydrology tool within the ArcGIS software. The computation of stream order, stream length, and basin size was performed using ArcGIS software, utilizing the flow direction and flow accumulation grid. The fundamental parameters, namely, stream length, area, perimeter, number of streams, and basin length, are determined from the drainage layer. Each stream was assigned a stream order using the stream ordering system devised by Strahler. The Karnaphuli river basin was determined to consist of six orders in its whole. The primary watershed region, boundaries, and length of the watershed were assessed utilizing GIS software. The HC, maximum elevation, and minimum elevation were computed at the stream order level and along the entire length of the main stream. The attribute feature classes that were employed were used to generate HCs for the watersheds under investigation. These curves were then used to estimate the HI values for the subbasin of the Karnaphuli river. In this study, the calculation of HI values was conducted using the elevation-relief ratio method, as proposed by Pike and Wilson (1971). Various approaches have been developed to estimate the HI, but for the purposes of this investigation, the method put forth by Pike and Wilson was employed. The utilization of the elevation-relief ratio approach has been determined to be both more precise and straightforward to compute within the context of a GIS.

The utilization of the GIS and remote sensing techniques has demonstrated their effectiveness and efficiency as tools for accurately delineating drainage systems and updating them (Waikar & Nilawar, 2014). The erosion status of the Karnaphuli watershed was indirectly assessed using the HI value. This assessment was conducted by applying the Soil and Water Assessment Tool (SWAT) in the ArcGIS environment. The final calculation of the HIs and HCs was performed in Excel. The overall methodological flowchart has been given in Fig. 2.3.

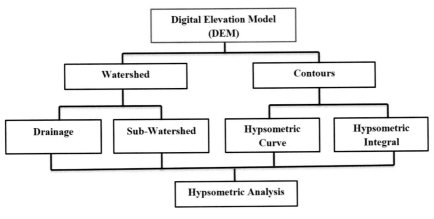

FIGURE 2.3

Methodological flowchart adopted for this study.

2.4 Results and discussion

The geological stage of the watershed was determined by the shape of the HCs (Fig. 2.4). The findings of this study are analyzed and presented under the following subcategories: Estimation of the HC, estimation of the HI, and the correlation between the watershed area and HI.

2.4.1 Estimation of the hypsometric curve

The HC is a brief, yet significant and effective tool that captures the essential elements required to depict the shape and growth of a drainage basin (Singh & Singh, 2018). The characteristics of HCs are not resolution-dependent; therefore, it is not necessary to use high-resolution images in such a study (Keller et al., 2019). The research region was cut off with a reference to the defined boundary and processed using ArcGIS 10.5 version. The hydrology and spatial analyst tools were used to create contour lines at intervals of 100 m, extracted natural drainage patterns, and defined subwatershed boundaries (Fig. 2.5).

The HC for the Karnaphuli river basin was generated by utilizing 30-m SRTM DEM data. The HC is generated by graphing the ratio of height to the maximum elevation (h/H) on the y-axis and the ratio of the area to the total area (a/A) on the x-axis. Digital contour maps have been utilized to generate the necessary data for conducting investigations on relative area and elevation. The relative area is obtained by dividing the area above a specific contour (a) by the total area of the subwatershed above the outlet (A). The relative elevation is determined by utilizing the maximum basin elevation (H), which extends to the farthest point of the subwatershed from the outflow, as well as the height (h) of a specific contour

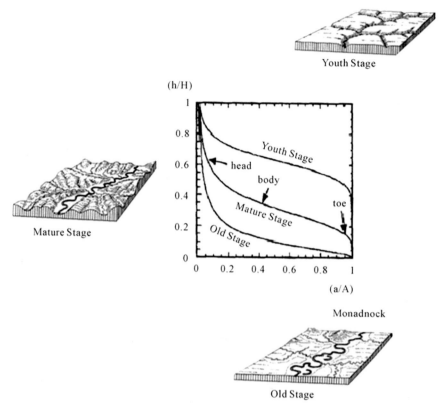

FIGURE 2.4

Different hypsometric curves represent different ages of geomorphic development, as proposed by Strahler (1952).

From Strahler, A. N. (1952). Hypsometric (area-altitude) analysis of erosional topography. Bulletin of the Geological Society of America, 63(11), 1117–1142. https://doi.org/10.1130/0016-7606(1952)63[1117: HAAOET]2.0.CO;2.

relative to the base plane. The HC's slope exhibits variation depending on the stage of watershed development, which significantly influences the erosion features of the watershed and serves as an indicator of the erosion cycle (Garg, 1991; Strahler, 1952). A HC effectively depicts the conflict between eroding and lifting because tectonic waves bring rocks above the mean sea level; however, erosion often pushes them down to the seafloor; which side dominates the competition is shown by the numerical values of HI numerical, which represents the region between 0 and 1 down in the HC (Table 2.1).

The HCs of the subwatersheds indicate that the research area is transitioning toward the mature stage. The comparison of these curves unveiled a little discrepancy in the quantity of mass that was extracted from the subwatersheds within the

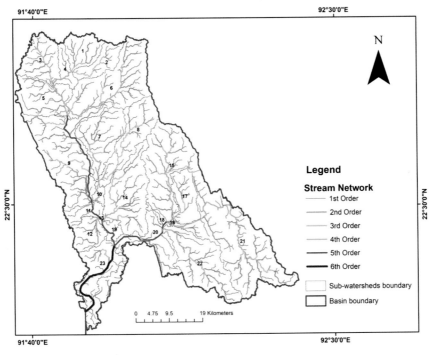

FIGURE 2.5

Drainage network and 23 subwatersheds of study area.

Table 2.1 Hypsometric values with the associated geological stage, curve types, and erosion potential according (Strahler, 1952).

Sr. no.	Geological stages	Hypsometric curve shapes	Hypsometric integrals (HI)	Erosion potential
1	Monadnock (old)	Convex upward	<0.3	Low (fully stabilized)
2	Equilibrium (mature)	S-shaped	0.3–0.6	Medium (moderately stabilized)
3	In-equilibrium (young)	Concave upward	>0.6	High (fully unstabilized)

From Strahler, A. N. (1952). Hypsometric (area-altitude) analysis of erosional topography. Bulletin of the Geological Society of America, 63(11), 1117–1142. https://doi.org/10.1130/0016-7606(1952)63 [1117:HAAOET]2.0.CO;2.

research area. Furthermore, it was observed that the HCs pertaining to the watershed being studied and its subwatersheds had an S-shaped pattern. The alterations in landforms were likewise manifested in various subwatersheds within the designated study area. The observed phenomenon may be attributed to soil erosion occurring within these subwatersheds, which is caused by the incision of channel beds, downward displacement of topsoil and bedrock materials, washout of the soil mass, and the cutting of stream banks. The susceptibility of the Karnaphuli River basin to erosion is attributed to its undulating landscape.

2.4.2 Estimation of the hypsometric integral

The HI is considered as a prominent geomorphological parameter, which corresponds to the geologic phases of basin formation. It is essential for estimating the erosion of a basin and, as a result, plays a role in the prioritizing of watersheds for implementing the protection of water and soil measures. Furthermore, the HI value can serve as an indirect estimate for quantifying erosion within watershed systems. Additionally, it offers valuable insights into the spatial arrangement of land surfaces across different altitudes. According to Fig. 2.6 and Fig. 2.7, the DEM indicates that the maximum elevation is recorded at 546 m, while the minimum elevation is seen at −37 m. The HI is quantified in terms of percentage units. When the HI is modest (about *0.5), lifting and eroding are almost equal in

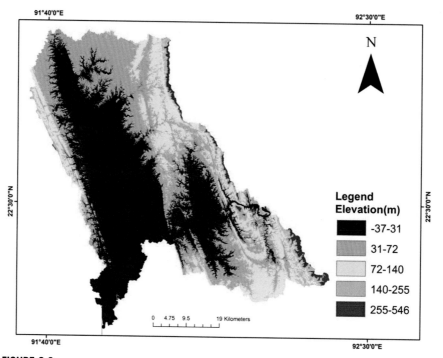

FIGURE 2.6

Digital elevation map of the study area.

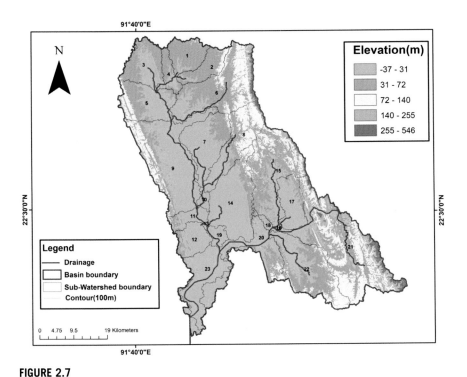

FIGURE 2.7

Subwatershed based elevation map of the study area.

strength. A relatively high value suggests that tectonic activity is dominant. A low value indicates that erosion processes are prominent (Davis, 1899).

In a prior study conducted by Strahler (1952), it was established that the HI exhibited an inverse relationship with many factors, namely, drainage basin height, slope steepness, stream channel gradients, and drainage density. There is a lack of correlation between HIs and both bifurcation ratios and length ratios. On the other hand, a notable correlation exists between the average length of stream segments of a specific order within each region and the corresponding mean HIs. This correlation is characterized by a consistent drop in stream length as the integrals decrease. Singh (2008) identified notable correlations between the HI and the spatial arrangement of watersheds within an active tectonic setting characterized by comparable lithology, climate, and geologic structure. Farhan, Elgaziri, Elmaji, and Ali (2016) found a notable association between the total area encompassed by the respective hypsometric and area categories, and the quantity of basins within the corresponding groups, for both the HI and area classifications.

Furthermore, the HI indicates the "cycle of erosion" (Garg, 1991). The "cycle of erosion" indicates the entire period needed to erode the land's surface to the basal level, which is the lowest point to which streams may decrease the area if every other parameter remained stable except time (Fig. 2.3). Three distinct phases can be identified within this entire "cycle of erosion", (1) monadnock (old) or entirely stabilized watershed ($H_{si} < 0.3$), (2) mature stage or equilibrium (H_{si} $0.3 < H_{si} < 0.6$), (3) young stage or inequilibrium ($H_{si} > 0.6$), which is highly susceptible to erosion (Strahler, 1952). The total area of the volume that holds for the entire basin is connected to the non-dimensional HI (Bishop et al., 2002). In a recent study, Markose and Jayappa (2011) proposed that the HI has a significant role in determining the shape of a HC, thus serving as an indicator for the geomorphic history of drainage basins. The Hydrologic Index (HI) values offer significant insights for determining the geological phase of watershed development. The Hydrologic Index (HI) of the entire river basin is 0.50, indicating a state of maturity or equilibrium. However, the HI values for subbasins 1−23 exhibit variability, as shown in Table 2.2. During the stage of in-equilibrium, the watershed is in the

Table 2.2 The Area, HI, and geological stage of the subwatersheds.

Subwatersheds no.	Area (km²)	Elevation-relief ratio (HI)	Geological stage
1	80.90	0.49	Mature
2	67.99	0.49	Mature
3	111.85	0.48	Mature
4	40.81	0.49	Mature
5	137.49	0.48	Mature
6	269.38	0.47	Mature
7	157.24	0.49	Mature
8	155.35	0.48	Mature
9	324.67	0.48	Mature
10	15.19	0.44	Mature
11	42.33	0.50	Mature
12	68.89	0.50	Mature
13	0.82	0.49	Mature
14	221.25	0.49	Mature
15	258.92	0.48	Mature
16	0.87	0.47	Mature
17	91.56	0.47	Mature
18	12.12	0.48	Mature
19	25.78	0.49	Mature
20	112.39	0.48	Mature
21	428.94	0.48	Mature
22	220.12	0.49	Mature
23	135.95	0.49	Mature

process of development, whereas the equilibrium stage represents the mature phase of watershed development. In the equilibrium stage, the development has reached a steady-state condition. On the other hand, the monadnock phase occurs when isolated bodies of resistant rock from prominent hills are present above the subdued surface. This is indicated by the presence of a distorted HC. Hypsometric curves exhibit variations in both the curvature of their shape and the relative magnitude of the region beneath the curve. The presence of a homogenous rock in various places results in the development of diverse hypsometric shapes during the stages of youth, maturity, and old age. However, it should be noted that the curves of mature and old stages exhibit similarity, unless there are monadnock masses present (Strahler, 1952).

The elevation−relief ratio (E) relationship: The elevation-relief ratio method was used (Pike & Wilson, 1971). The relationship is as follows:

$$E \sim H_{is} = (\text{Elev}_{mean} - \text{Elev}_{min})/(\text{Elev}_{max} - \text{Elev}_{min})$$

where E is the HI H_{si}; Elev_{mean} is the weighted mean elevation of the basin; and Elev_{min} and Elev_{max} are the minimum and maximum elevations within the basin.

The elevation-relief approach has been determined to be the most suitable for calculating HIs, and the values of those integrals for the research area's 23 subwatersheds are reported in Table 2.2.

2.4.3 Relationship between watershed area and hypsometric integral

Surface runoff and sediment losses in basin networks are the two basic hydrologic reactions to precipitation events (Pilgrim, Chapman, & Doran, 1988). As a result, the values of HI were used as a secondary measure of surface erosion of subwatersheds. The HI values are ranging from 0 to 1, with a high number representing the young stage and a low value representing the monadnock stage (Ritter, Kochel, Miller, & Miller, 1995). The young stage of the watershed is more prone to surface erosion than the old stage (Yousaf et al., 2018). HI identifies the erosion potential of any landscape by means of the age that has occurred in the basin as a result of hydrologic events and soil deterioration variables (Gajbhiye et al., 2014).

From the analysis, the HI value range (0.44−0.50) of the subwatersheds showed that the basin is in a mature stage, and the HC is an S-shaped curve, which indicates that the study area is characterized by moderately eroded areas (Fig. 2.8). According to the analysis, the watershed's terrain shifts to an equilibrium stage, meaning that erosion will be slow unless there are extremely powerful precipitation events, including strong winds and considerable rains that cause torrential flow peaks (Ritter et al., 1995). This finding indicates that the soil erosion in these subwatersheds is predominantly caused by the incision of channel beds, the downward movement of topsoil and bedrock material, the washout of the soil

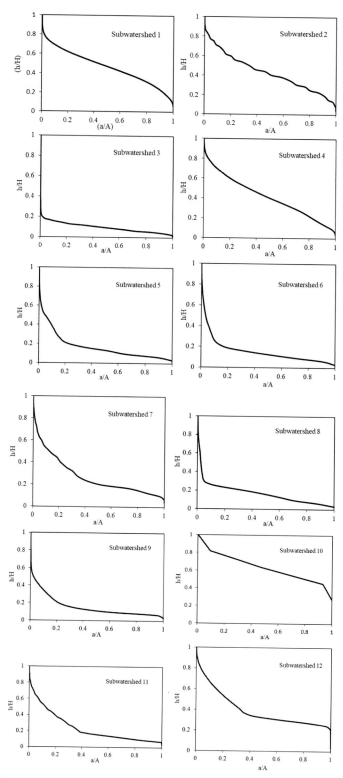

FIGURE 2.8

The hypsometric curves pertaining to the subwatersheds of the Karnaphuli River.

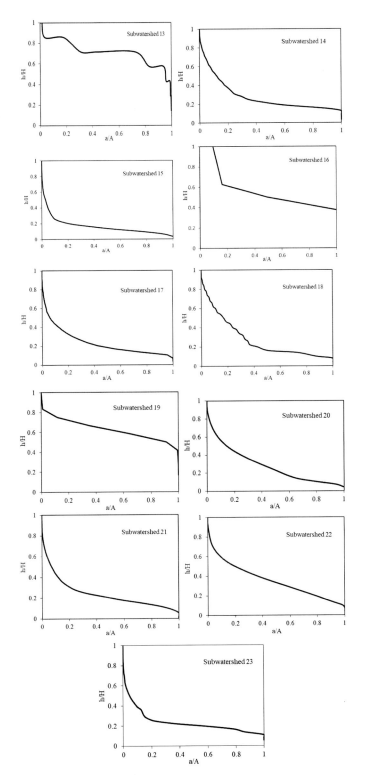

FIGURE 2.8

(Continued).

mass, and the cutting of stream banks. The average HI value for the Karnaphuli River Basin is determined to be 0.50. The implementation of basic mechanical and vegetative techniques is necessary to mitigate sediment loss. However, further water harvesting structures may be needed to effectively preserve water in specific areas of the watershed, facilitating its conjunctive usage.

More or less, the entire basin is affected by geomorphically hazardous conditions such as slope instability, floods, high sediment yield, and erosion, and the other part of the basin is affected by anthropogenic effects such as deforestation, overgrazing, land use/land cover changes, and insufficient conservation measures, all of which maximize soil erosion. Jhum cultivation, a traditional agricultural practice that involves clearing the land of trees and other vegetation, burning it, and then cultivating it for a predetermined amount of time in hilly areas, is another key factor that contributes to the worsening of erosion. Jhum cultivation is one of the primary reasons why hilly areas are losing so much of their topsoil. The large differences in precipitation that occur during the monsoon and the dry season both have a significant impact on the river's flow. It is expected that this variance will become more pronounced as a result of climate change.

2.5 Challenges and solutions

It is imperative to conduct hypsometric analysis-based studies for small dams in order to identify the subwatersheds that are particularly susceptible to the issue of siltation in the dam. Nevertheless, researchers tend to avoid doing studies of this sort due to the repetitive process of collecting field data and the intricate analysis required for calculations (Gajbhiye et al., 2014). However, it is imperative to conduct a thorough assessment of soil erosion risk in order to ensure the sustainability of watershed management practices. Soil erosion has the detrimental consequence of diminishing the quality of land appropriate for agricultural endeavors and urban development. Additionally, the ultimate outcome of soil erosion is siltation, which leads to a reduction in the capacity of water bodies, reservoirs, and dams. Sedimentation is primarily influenced by several key elements, namely, precipitation, soil composition, vegetation cover, as well as the topographic and morphological attributes of the basin (Kothyari & Jain, 1997). In cases where data on precipitation and sediment yield are not accessible, an estimation of the relative vulnerability of watersheds can be made by employing time-independent characteristics such as soil type, topography, and morphology. The topographic features, including elevation and contour maps, were generated using the ArcGIS 10.5 software platform. These maps were produced by utilizing a DEM with a spatial resolution of 30 m. The GIS possesses the capability to effectively examine the aforementioned factors. The morphometric analysis technique places emphasis on subwatersheds within a drainage basin and aims to assess the erosional risk associated with them (Bhat, Danelljan, Van Gool, & Timofte, 2019;

Ghasemlounia & Utlu, 2021). Moreover, morphometric analysis offers a quantitative depiction of the basin's geometry, enabling the comprehension of beginning slopes or disparities in rock hardness, structural influences, recent tectonic activity, as well as the geological and geomorphic evolution of the drainage basin (Strahler, 1964).

An undetected and incorrectly classified image influences the projected water surface elevation in many different situations by modifying the amount of water observed inside the selected zone. Any incorrect or missed water influences not just the predicted surface water area but is also probably to be considered in the proportionate hypsometric relationship, which affects the predicted water surface elevation and leads to greater error and uncertainty (Weekley & Li, 2021).

However, using remotely sensed approaches, synoptic, timely, and recurring data sets throughout the broad spectral range, and other GIS tools can make the estimation process simpler and more efficient than arduous conventional procedures (Gajbhiye et al., 2014). This can be achieved by reducing the amount of time spent on the estimation process. The findings of the research might also be applied in the future to other kinds of research and development. It is envisaged that the findings of this study would be useful in suggesting potential future development activities that may be carried out in the river basin.

2.6 Recommendations

The major problems of the river basin are deforestation, river sedimentation, and navigability during the dry season in the rivers situated upstream of the Kaptai dam. Moreover, river bank erosion, flooding, and the impact of natural and man-made effects are major reasons for concern in the basin. The main interventions recommended for development and management of the Karnaphuli river basin are dredging/excavation of rivers/khals, river bank erosion protection, dismantling of regulators and weir, replacing them with hydraulic elevators dams, and construction of the embankment/flood wall at the upstream of the tributaries of Karnaphuli river. In both the Environmental Impact Assessment and the Environmental Management Plan, it is advised to conduct social consultations in addition to economic and financial studies, which is technically feasible, socially acceptable, environment-friendly, and economically viable. The effectiveness of different options is recommended considering the issues of improvement of fisheries resources and enhancement of irrigation facilities, drainage, and navigability. Furthermore, developing a long-term investment plan is also suggested for environmentally sustainable watershed management in response to human and climate interventions. In addition, the Karnaphuli watershed provides a significant amount of water for local households and crops. Therefore, additional analysis is advised for the next hydrological research and water resource activities, including groundwater reservoirs, land resource development, rainwater harvesting structures, and watershed planning and conservation. These may enable the understanding of the many geomorphological land form types, particularly those related to drainage patterns, morphometric parameter situations, and the geological and topographic settings of the Karnaphuli river basin.

2.7 Conclusion

The HC is a straightforward yet powerful tool that encompasses the fundamental components necessary for depicting the configuration of a drainage basin and its evolutionary trajectory. Additionally, the HI, serving as a functional parameter, offers a quantitative assessment of erosion and the pace at which morphological alterations occur. The current work showcases the efficacy of employing remote sensing and GIS techniques in basin development, specifically through the utilization of hypsometric analysis. This analysis allows for the assessment of denudational processes and the quantification of morphological changes, hence highlighting the challenges associated with these processes and their respective rates. Hence, it is vital to comprehend the erosion condition of watersheds and establish a hierarchy for implementing effective actions to safeguard soil and water resources. The hypsometric analysis suggested 23 subwatersheds for the Karnaphuli watershed. HI values were considered to determine the geologic stage, age, and the erosive processes that were taking place in the watershed under study. Soil erosion is exacerbated by various factors including deforestation, alterations in land use and land cover, excessive grazing practices, and inadequate conservation measures. The primary causes of soil erosion across the entire watershed and its subbasins can be attributed to many factors, including the incision of channel beds, movement of regolith and bedrock materials due to landslides, washout of topsoil, and erosion of stream banks through undercutting. The primary objective of this chapter was to determine the growth stage of the Karnaphuli river basin and its subbasins. The findings of the study indicate that the Karnaphuli River basin exhibits a relatively advanced stage of development, with stabilized erosion stages and surface runoff emerging as the primary process across the basin. The analysis of HIs has revealed that subwatersheds 11 and 12 within the Karnaphuli River basin exhibit a higher vulnerability to erosion in comparison to other subwatersheds. Consequently, it is imperative to implement soil and water conservation strategies at strategic locations within these subwatersheds. This intervention aims to mitigate sediment outflow and ensure the sustainable management of water resources.

Nevertheless, it is imperative to use caution when examining and contrasting HCs due to the intricate nature of the calculations involved. Hence, the successful execution of soil and water conservation measures within the framework of integrated watershed management necessitates a thorough understanding of the erosion conditions prevalent in watersheds, as well as their relative importance. This study aims to identify appropriate locations for promoting groundwater recharge practices and mitigating soil erosion in the region. This will be achieved through the identification of different landform phases, their geomorphic evolution, and the hydrologic response of the entire drainage system. Additionally, implementing measures to effectively manage the subwatersheds can contribute to reducing the susceptibility of sedimentation accumulation within the dam's basin.

Acknowledgments

We would like to thank the USGS for supplying the SRTM data at no cost to us, the anonymous reviewers for their thoughtful and helpful recommendations and feedback, and the Editor-in-Chief of the journal for providing us with a platform to review the article.

References

Ahmed, S., Galagan, S., Scobie, H., Khyang, J., Prue, C. S., Khan, W. A., . . . Sack, D. A. (2013). Malaria hotspots drive hypoendemic transmission in the Chittagong Hill districts of Bangladesh. *PLoS One*, *8*(8). Available from https://doi.org/10.1371/journal. pone.0069713Bangladesh, http://www.plosone.org/article/fetchObjectAttachment.action; jsessionid = 218899C714F05008FE99E01AFE3DB2AC?uri = info%3Adoi%2F10.1371% 2Fjournal.pone.0069713&representation = PDF.

Alam, M. W., & Zafar, M. (2012). Occurrences of Salmonella spp. in water and soil sample of the Karnafuli river estuary. *Microbes and Health*, *1*, 41–45.

Al-Hantouli, F., Awawdeh, M., & Obeidat, M. (2023). Soil erosion prioritization of Yarmouk river basin, Jordan using multiple approaches in a GIS environment. In *Water Resources Management and Sustainability: Solutions for Arid Regions*, (pp. 291–318). Springer.

Bhat, G., Danelljan, M., Van Gool, L., & Timofte, R. (2019). Switzerland Learning discriminative model prediction for tracking. In *Proceedings of the IEEE international conference on computer vision* (pp. 6181–6190). 2019/10/01, Institute of Electrical and Electronics Engineers Inc. https://doi.org/10.1109/ICCV.2019.00628 9781728148038. http://ieeexplore.ieee.org/xpl/conhome.jsp?punumber = 1000149 2019-.

Bishop, M. P., Shroder, J. F., Bonk, R., & Olsenholler, J. (2002). Geomorphic change in high mountains: A western Himalayan perspective. *Global and Planetary Change*, *32*(4), 311–329. Available from https://doi.org/10.1016/S0921-8181(02)00073-5.

Davis, W. M. (1899). The geographical cycle. *The Geographical Journal*, *14*(5). Available from https://doi.org/10.2307/1774538.

Dowling, T., Walker, J., Richardson, D. P., O'Sullivan, A., & Summerell, G. K. (1998). *Application of the hypsometric integral and other terrain based metrics as indicators of catchment health: a preliminary analysis*. Canberra, ACT, Australia: CSIRO, Land and Water.

Enea, A., Iosub, M., & Stoleriu, C. C. (2023). A low-cost, UAV-based, methodological approach for morphometric analysis of Belci Lake Dam Breach, Romania. *Water*, *15*, 1655.

Farhan, Y., Elgaziri, A., Elmaji, I., & Ali, I. (2016). Hypsometric analysis of Wadi Mujib-Wala Watershed (Southern Jordan) using remote sensing and GIS techniques.

International Journal of Geosciences, 07(02), 158−176. Available from https://doi.org/10.4236/ijg.2016.72013.

Gajbhiye, S., Mishra, S. K., & Pandey, A. (2014). Prioritizing erosion-prone area through morphometric analysis: an RS and GIS perspective. *Applied Water Science, 4*, 51−61.

Garg, S. K. (1991). *Geology: The science of earth*. Khanna.

Ghasemlounia, R., & Utlu, M. (2021). Flood prioritization of basins based on geomorphometric properties using principal component analysis, morphometric analysis and Redvan's priority methods: A case study of Harşit River basin. *Journal of Hydrology, 603*. Available from https://doi.org/10.1016/j.jhydrol.2021.127061.

Hurtrez, J.-E., Sol, C., & Lucazeau, F. (1999). Effect of drainage area on hypsometry from an analysis of small-scale drainage basins in the Siwalik Hills (Central Nepal). *Earth Surface Processes and Landforms: The Journal of the British Geomorphological Research Group, 24*, 799−808.

Keller, C. B., Husson, J. M., Mitchell, R. N., Bottke, W. F., Gernon, T. M., Boehnke, P., ... Peters, S. E. (2019). Neoproterozoic glacial origin of the Great Unconformity. *Proceedings of the National Academy of Sciences, 116*, 1136−1145.

Kothyari, U. C., & Jain, S. K. (1997). "Sediment yield estimation using GIS.". *Journal of Hydrological Scienc, 75*.

Lara, R. J., Neogi, S. B., Islam, M. S., Mahmud, Z. H., Yamasaki, S., & Nair, G. B. (2009). Influence of catastrophic climatic events and human waste on Vibrio distribution in the Karnaphuli estuary, Bangladesh. *EcoHealth, 6*(2), 279−286. Available from https://doi.org/10.1007/s10393-009-0257-6.

Mahalingam, B. (2012). Hypsometric properties of drainage basins in Karnataka using geographical information system. *New York Science Journal, 5*, 156−158.

Mahmood, S. A., & Gloaguen, R. (2011). Analyzing spatial autocorrelation for the hypsometric integral to discriminate neotectonics and lithologies using DEMs and GIS. *GIScience & Remote Sensing, 48*, 541−565. Available from https://doi.org/10.2747/1548-1603.48.4.541.

Marani, M., Eltahir, E., & Rinaldo, A. (2001). Geomorphic controls on regional base flow. *Water Resources Research, 37*, 2619−2630. Available from https://doi.org/10.1029/2000WR000119.

Markose, V. J., & Jayappa, K. S. (2011). Hypsometric analysis of Kali River Basin, Karnataka, India, using geographic information system. *Geocarto International, 26*(7), 553−568. Available from https://doi.org/10.1080/10106049.2011.608438.

Pike, R. J., & Wilson, S. E. (1971). Elevation-relief ratio, hypsometric integral, and geomorphic area-altitude analysis. *Bulletin of the Geological Society of America, 82*(4), 1079−1084. Available from https://doi.org/10.1130/0016-7606(1971)82[1079:ERHIAG]2.0.CO;2, http://gsabulletin.gsapubs.org/content/by/year.

Pilgrim, D. H., Chapman, T. G., & Doran, D. G. (1988). Problems of rainfall-runoff modelling in arid and semiarid regions. *Hydrological Sciences Journal, 33*, 379−400.

Rakesh, C. J., Govindaraju, S., Lokanath, A. K., & Kumar. (2023). Prioritization of soil erosion prone sub-watersheds through morphometric analysis using geospatial and weighted sum approach: A case study of Boranakanive reservoir catchment in Tumkur district, Karnataka, India. *Environmental Earth Sciences, 82*(12). Available from https://doi.org/10.1007/s12665-023-10996-y, https://www.springer.com/journal/12665.

Ritter, D. F., Kochel, R. C., & Miller, J. R. (2002). *Process geomorphology*. Boston: McGraw Hill.

Ritter, D. F., Kochel, R. C., Miller, J. R., & Miller, J. R. (1995). *Process geomorphology*. Wm. C. Brown.

Rizvi, S. N. H. (1975). *Bangladesh district Gazetteers: Chittagong*. East Pakistan Government Press.

Schumm, S. A. (1956). Evolution of drainage systems and slopes in badlands at Perth Amboy, New Jersey. *Bulletin of the Geological Society of America, 67*(5), 597–646. Available from https://doi.org/10.1130/0016-7606(1956)67[597:EODSAS]2.0.CO;2, http://gsabulletin.gsapubs.org/content/by/year.

Shekar, P. R., Mathew, A., Abdo, H. G., Almohamad, H., Abdullah Al Dughairi, A., & Al-Mutiry, M. (2023). Prioritizing sub-watersheds for soil erosion using geospatial techniques based on morphometric and hypsometric analysis: A case study of the Indian Wyra River basin. *Applied Water Science, 13*(7). Available from https://doi.org/10.1007/s13201-023-01963-w, https://www.springer.com/journal/13201.

Singh, V., & Singh, S. K. (2018). Hypsometric Analysis Using Microwave Satellite Data and GIS of Naina–Gorma River Basin (Rewa district, Madhya Pradesh, India). *Water Conservation Science and Engineering, 3*(4), 221–234. Available from https://doi.org/10.1007/s41101-018-0053-7, https://www.springer.com/journal/41101.

Stepinski, T. F., & Stepinski, A. P. (2005). Morphology of drainage basins as an indicator of climate on early Mars. *Journal of Geophysical Research: Planets, 110*(E12). Available from https://doi.org/10.1029/2005je002448.

Strahler, A. N. (1952). Hypsometric (area-altitude) analysis of erosional topography. *Bulletin of the Geological Society of America, 63*(11), 1117–1142. Available from https://doi.org/10.1130/0016-7606(1952)63[1117:HAAOET]2.0.CO;2, http://gsabulletin.gsapubs.org/content/by/year.

Strahler, A. N. (1964). *Quantitative geomorphology of drainage basin and channel networks. Handbook of applied hydrology*. McGraw Hill.

Singh, Tejpal (2008). Hypsometric analysis of watersheds developed on actively deforming Mohand anticlinal ridge, NW Himalaya. *Geocarto International, 23*(6), 417–427. Available from https://doi.org/10.1080/10106040801965821.

Van Liem, N., Dat, N. P., Dieu, B. T., Phai, V. V., Trinh, P. T., Vinh, H. Q., & Phong, T. V. (2016). Assessment of geomorphic processes and active tectonics in con VOI mountain range area (Northern Vietnam) using the hypsometric curve analysis method. *Vietnam Journal of Earth Sciences, 38*(2). Available from https://doi.org/10.15625/0866-7187/38/2/8602.

Waikar, M. L., & Nilawar, A. P. (2014). Morphometric analysis of a drainage basin using geographical information system: A case study. *International Journal of Multidisciplinary and Current Research, 2*, 179–184.

Weekley, D., & Li, X. (2021). Tracking lake surface elevations with proportional hypsometric relationships, Landsat imagery, and multiple DEMs. *Water Resources Research, 57*(1). Available from https://doi.org/10.1029/2020wr027666.

Weissel, J. K., Pratson, L. F., & Malinverno, A. (1994). The length-scaling properties of topography. *Journal of Geophysical Research: Solid Earth, 99*, 13997–14012.

Yousaf, W., Mohayud-Din-Hashmi, S. G., Akram, U., Saeed, U., Ahmad, S. R., Umar, M., & Mubashir, A. (2018). Erosion potential assessment of watersheds through GIS-based hypsometric analysis: A case study of Kurram Tangi Dam. *Journal of Geosciences, 11* (22). Available from https://doi.org/10.1007/s12517-018-4059-4, http://www.springer.com/geosciences/journal/12517?cm_mmc = AD-_-enews-_-PSE1892-_-0.

Impact analysis of the 2022 flood event in Sylhet and Sunamganj using Google Earth Engine

Rajsree Das Tuli[1,2], Kazi Jihadur Rashid[1,3] and Akter Tahmina[3]

[1]*Sensing Earth and Environment (SEE) Lab, Bangladesh*
[2]*Food and Agriculture Organization of the United Nations, Dhaka, Bangladesh*
[3]*Center for Environmental and Geographic Information Services, Dhaka, Bangladesh*

3.1 Introduction

Floods are widely recognized as a highly destructive hydrological hazard on a global scale, resulting in significant harm to various aspects of society, including human life, economic activities, environmental integrity, and inanimate entities such as displacement, property damage, loss of livelihood, crop failure, disruptions in transportation, and infrastructure impairment (Ahmed, Rahaman, Kok, & Hassan, 2017). A flash flood can be described as the sudden and intense influx of a large volume of water into an area that is typically arid or the quick elevation of a stream beyond a predetermined flood threshold, occurring within minutes or over an extended interval following a triggering event, such as heavy precipitation or the failure of a dam (Abshire & Mullusky, 2019). Thus flash flooding can be induced by intense downpours that flow upstream of the watershed, as well as in regions characterized by a narrow catchment area and a precipitous gradient (Azmeri & Isa, 2018). Floods have become more frequent worldwide and problematic due to a combination of population growth, anthropogenic activities, and climate change (Malik et al., 2020). Floods are primarily influenced by rainfall, which encompasses a range of climatic factors such as intensity, quantity, and duration. The floods' magnitude and intensity are influenced by fluctuations in precipitation intensity or frequency, as well as rising temperatures (Immerzeel, 2008). Hence, in order to alleviate the detrimental effects of flooding and establish preventative infrastructure and disaster readiness, it is critical to find and evaluate flood-prone areas and the time frame of water extension (Kwak, 2017). The researchers consider characteristics such as land use and land cover (LULC), total population, and population density when conducting a vulnerability assessment (Ghosh et al., 2023). Furthermore, this provides policymakers and decision-makers

Applications of Geospatial Technology and Modeling for River Basin Management.
DOI: https://doi.org/10.1016/B978-0-443-23890-1.00003-7

with valuable information regarding future preparedness and sustainable management (Armenakis, Du, Natesan, Persad, & Zhang, 2017).

Being a ubiquitous catastrophe, around 40% of the disastrous effects of natural disasters worldwide can be attributed to floods (Bich, Quang, Ha, Hanh, & Guha-Sapir, 2011). It alone affected about 2.3 billion global population between 1998 and 2015 with a major share (95%) living in Asia (Wallemacq, Guha-Sapir, & McClean, 2015). River floods affect approximately 21 million people globally each year, but because of climate change and socioeconomic growth, it is anticipated to rise to 54 million within the next decade (Luo, Maddocks, Iceland, Ward, & Winsemius, 2015). The detrimental effects of this phenomenon have been documented by scientists to include disruptions to human settlements, contamination of both natural and built environments, and loss of life (Messner & Meyer, 2006). Around 1 billion individuals reside in close proximity to river basins that are susceptible to annual flooding (United Nations University, 2014). Bangladesh is located in a low-lying, flat terrain that experiences frequent and severe inundation in regions prone to flooding on an annual basis (Baky, Islam, Paul, & Hazard, 2020). Additionally, it is among the most populous and disaster-prone nations globally (Ahmed et al., 2017). The occurrence of substantial inundation is a customary aspect of the monsoon season, which generally endures from June to September, owing to the distinctive topography of this area (Islam, Haque, & Bala, 2010). On the contrary, in exceptionally severe years, flooding may persist until October (Monirul Qader Mirza, 2002). Furthermore, due to the steep slope of the rivers draining the northeastern part (seasonal wetlands) region and the hilly topography, flash flooding in that area could potentially occur as early as April (Abedin & Khatun, 2020). In extreme circumstances, flooding along riverbanks or precipitation can inundate up to 70% of the nation annually, resulting in substantial economic devastation (Monirul Qader Mirza, 2002).

The first and second waves of flooding in Bangladesh occurred in May and June 2022, respectively, as a result of the excessive rainfall. A cumulative sum of around 7.2 million people have been adversely affected by the disastrous flash floods that have struck nine districts in the northeastern area of Bangladesh. These districts are Mymensingh, Sherpur, Sunamganj, Moulvibazar, Habiganj, Kishoreganj, Netrakona, Brahmanbaria, and Mymensingh. The profound inundation of the Sylhet division by heavy monsoon rainfall and water flowing downstream from India's northeast has resulted in the entrapment of millions of individuals and the outbreak of a humanitarian crisis. Despite the continuous and consistent decline in water levels, the humanitarian situation in the flood-affected regions continues to be extremely distressing, especially throughout the Sunamganj and Sylhet districts (IRFC, 2022; UNICEF, 2022). Flood-related isolation has confined numerous households, whereas others have sought refuge in open areas, thereby significantly jeopardizing the security and safety of women and adolescent girls residing in such households. The incident resulted in the destruction of an estimated 106,727 water points, 283,355 latrines, 663,534 cattle, and 254,251 ha of croplands (UN RC Bangladesh, & UNCT Bangladesh, 2022).

In the regions affected by the floods, clean water is in limited supply. Severe floods have compromised dams in various regions, and the power has been disconnected in both districts. The socially vulnerable and marginalized population groups, including children, adolescent females, women, older persons, and disabled individuals, were disproportionately affected by this flood.

The utilization of geospatial technology has become prevalent in contemporary times for the purpose of ascertaining areas affected by flooding and conducting impact analyses to map, monitor, and analyze geographic data and generate spatiotemporal scenarios (Moniruzzaman et al., 2021). The Google Earth Engine (GEE) has gained significant popularity due to its extensive utilization in enhancing data processing and analysis efficiency. By effectively implementing parallel processing techniques, the infrastructure of the platform enables efficient resource allocation, data distribution, and streamlined data management and distribution (Pandey, Kaushik, & Parida, 2022). The inability of optical satellite remote sensing to acquire images of the monsoon in Asia is typically attributed to high cloud cover presence and overcast atmospheric conditions (Qadir & Mondal, 2020). On the other hand, Sentinel-1 synthetic aperture radar (SAR) data can dependably provide global observations of the earth's surface throughout the day and night, regardless of the weather (ESA, 2023). As a result, SAR sensors may accurately discern the ground surface, including surface water, even when passing through clouds. These sensors providing less error data are typically more complicated for mapping flood scenarios than those utilizing optical sensors (Shen, Wang, Mao, Anagnostou, & Hong, 2019).

Numerous studies examined recent flood scenarios in Bangladesh, particularly those pertaining to flood hazards in the northeastern regions (Akter, Islam, Karim, Miah, & Rahman, 2023; Aldhshan, Mohammed, & Shafri, 2019; Choudhury & Haque, 2016; Dey, Parvez, & Islam, 2021; Hoque, Nakayama, Matsuyama, & Matsumoto, 2011; Kamruzzaman & Shaw, 2018; Quader, Dey, Malak, & Rahman, 2023; Uddin, Matin, & Meyer, 2019). In their study, Uddin et al. (2019) utilized Sentinel-1 SAR data to map the complete inundated region that occurred in Bangladesh during the 2017 flash floods and the compared the resulting maps with Landsat-8 OLI data to validate their findings. Quader et al. (2023) examined the temporal expansion of the water surface during the flood season using geospatial techniques northeastern depressed basin, Bangladesh, during 2017 and 2019. Hoque et al. (2011) conducted a study to generate inundation maps from 2000 to 2004 using RADARSAT images for the northeastern region of Bangladesh. Aldhshan et al. (2019), employing Sentinel-1 SAR and Landsat-8 OLI/TIRS images, identified damaged croplands in eight upazilas of the Sunamganj district while mapping flash flood zones. In order to evaluate the spatiotemporal variability of rainfall and the occurrence of flash flood events in the Sylhet haor region of Bangladesh, Akter et al. (2023) analyzed April precipitation data spanning the years 1995–2022. While notable studies were conducted for flood mapping in the region, no prior research addressed the most significant flood occurred in 2022, providing a holistic assessment of settlement and cropland losses, coupled with an estimation of affected population. Given the identified research gap pertaining to flood occurrences in the year 2022, the primary aim of this chapter is to convey an integrated GEE-based approach to measure the areas of flooded settlement and cropland, as well as understanding demographic consequences. The first objective of this chapter is to extract the extent of flood inundation in the regions of Sylhet and Sunamganj by utilizing GEE. Afterward, it intends to measure the area coverage for flooded land classes emphasizing settlement and cropland. Finally, it provides an estimation of affected population along with administrative union/ward based on the four levels of inundation. This approach can be helpful for NGO officials, humanitarian workers, and local governments in understanding real-time flood scenarios and make early interventions.

3.2 2 Materials and methods

3.2.1 Study area

The study area location is situated between 24°30′N−25°15′N latitude and 90°45′E−92°00′E longitude with a total area of 7104.9 sq. km, consisting of two administrative districts of Bangladesh, namely, Sylhet and Sunamganj. It is surrounded by hills ranging from Assam and Meghalaya in India from north to southeast and floodplains of Bangladesh in the west and south. This area is mostly known for its vast geological depression called haors, characterized by seasonal submergence (Akter et al., 2023). During the dry season, these lands are covered by crops. The area has an average annual rainfall of ranges between 2650 and 4050 mm (Khatun, Rashid, & Hygen, 2016), and the pattern shifts from low in the southwest to high in the northeast. The upper stream hills in Meghalaya and Assam experience heavy rainfall as well as rainfall-induced disasters (Akter et al., 2023). Due to the geographical location, cumulated rainfall-runoff water transported through the rivers often results in flash floods in the area. With a total population of 6,539,788 and an average density of 921 persons per sq. km, this region is highly susceptible to inundation due to flooding (Kamal et al., 2018; Nowreen, Murshed, Islam, Bhaskaran, & Hasan, 2015; Quader et al., 2023). Fig. 3.1 shows the geographical position of the area of this study.

3.2.2 Data description

This study utilized Sentinel-1 SAR data to prepare flood extent mapping in the regions of Sylhet and Sunamganj in the year 2022. The Sentinel-1 mission offers satellite data obtained from a dual-polarization C-band SAR sensor operating at a frequency of 5.405 GHz. This study used the Sentinel-1 ground range detected (GRD) scenes, which have undergone elimination of thermal noise, calibration of radiometric measurements, and correction of terrain distortions processing through the Sentinel-1 Toolbox, resulting in calibrated and *ortho*-corrected

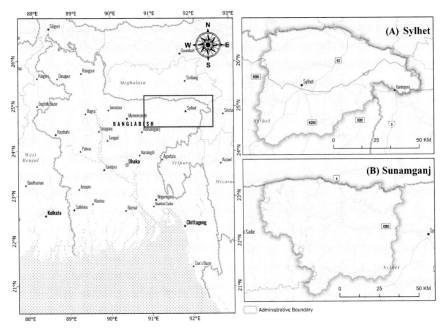

FIGURE 3.1 Study area map.

The geographical position of the area studied. Map lines delineate study areas and do not necessarily depict accepted national boundaries.

products. This extensive collection comprises all the GRD scenes, each possessing one of the available resolutions of 10, 25, or 40 m, and various band combinations are available for different scene polarizations, operating inside one of three instrument modes. For this study, Sentinel-1 images were processed in GEE from April to June 2022 and used to identify submerged pixels (Attema et al., 2010; De Zan & Monti Guarnieri, 2006). The process of extracting flood extent from SAR images presents challenges stemming from the inherent difficulty in distinguishing between flood areas and shaded regions due to similar gray values. This work used the WWF HydroSHEDS Void-Filled DEM for the purpose of masking the misleading flood extent due to shadow influence from satellite radar images. The dataset offers hydrographic data for regional and global scale applications such as river systems, watershed perimeters, drainage directions, and stream accumulations. The data is obtained from elevation measurements collected during NASA's Shuttle Radar Topography Mission (SRTM). The presence of spikes and wells in the SRTM data was identified and subsequently removed. The interpolation of neighboring elevations is utilized to fill small holes. The dataset is provided at a resolution of 3 arc seconds (Lehner, Verdin, & Jarvis, 2008). Finally, the JRC Global Surface Water Mapping Layers dataset was used to eliminate permanent surface water bodies from the water extent to extract the actual inundated area. This dataset was created leveraging Landsat 5, 7, and 8 satellites, acquired between 1984 and 2021, to provide spatiotemporal distribution of surface water at 30-m resolution. Each individual pixel underwent classification as either water or nonwater (Pekel, Cottam, Gorelick, & Belward, 2016).

To assess the extent of flood damage, the study integrated LULC data with the population dataset. The population data was used to locate regions with high population density and assess the impacted population. Population information was extracted from the 2021 census dataset at the union level and was employed to identify flood-impacted unions and the magnitude of impact. The ESRI Annual Land Use Land Cover dataset was used to extract the different land classes in the regions for damage assessment. The LULC maps are generated using the European Space Agency's Sentinel-2 satellite imagery at a resolution of 10 m. The composite maps are created by incorporating LULC estimates for nine distinct classes over the course of a year to produce an accurate LULC layer of each annual period. The Impact Observatory employed a vast number of human-labeled pixels, which were done by the National Geographic Society. These labeled pixels were then utilized to train a deep learning model designed explicitly for land classification through the application of the DL model to the annual scene collections of Sentinel-2. Each of the maps exhibits an evaluated mean accuracy surpassing 75% (Karra et al., 2021).

3.2.3 Flood extent estimation

The flood extent map was generated by analyzing SAR data using a change detection approach by a comparative assessment of a preflood composite image and a during-flood composite image. Prior to flood extent extraction, images were preprocessed within the GEE platform, including conversion between decibels and natural units and the application of a speckle filter (Refined Lee). This filter is used to reduce speckle noise in the Sentinel 1 imagery. This study used vertical-horizontal (VH) polarization favored over vertical-vertical (VV) polarization for flood mapping due to VH's reduced tendency for overestimation. The VV mode transmits and receives vertical waves to generate the SAR image, while the VH mode transmits vertical waves and receives horizontal waves for SAR imaging. Various polarization combinations exhibit differing sensitivities to diverse land cover features and phenomena. The VV polarization is sensitive to vertical structures, whereas the VH polarization enhances the ability to detect and measure variations on the Earth's surface. The VH polarization was found to be more suitable for identifying flood-affected regions, characterized by darker and black tones (Conde & De Mata Muñoz, 2019; Martinis & Rieke, 2015). In order to measure the extent of alterations caused by the flood, the mosaics captured during the flood were divided by the mosaics captured before the event. A threshold value was employed to create a binary raster layer to separate flooded and non-flooded areas; the pixels greater than the threshold values were the flooded pixels. Pixels were assigned a value of 1, indicating flood-affected areas, and a value of 0 representing nonflood areas. After determining the initial flood inundation extent with the pixels assigned 1, permanent water bodies were eliminated using these pixels to detect the dry areas, which are actually flood-affected. Next, the pixels with slopes over 5% were identified to remove incorrect pixels as these steep areas will not be flooded. Finally, isolated pixels were masked using a threshold to remove noisy pixels. In this study, any pixels identified that were not connected with 8 pixels were excluded. Then, the flooded area for both regions was calculated in hectares, and the percentage of the inundated area was determined.

3.2.4 Flood impact analysis

To assess the impact on various land classes, the pixels corresponding to specific land classes coinciding with the area affected by flooding were extracted to estimate the percentage of the land area affected by the total inundation. This spatial analysis helped identify patterns and trends in the affected land uses and have deeper insight into the spatial distribution of inundated areas and the specific type of land use that has been damaged. This information will help relevant bodies comprehensively understand the land use changes induced by flash flood inundation, providing valuable information for flood management and informed land-use planning. Furthermore, population density calculations were carried out for the entire region, utilizing Census 2021 data. The census record provided information at various levels, such as population counts for each administrative unit; in this present study, records at the union level have been used. Later, the population density was calculated for each union within the two districts for a more comprehensive understanding of population distribution. Then, the overlay technique was used on this population data combined with the inundation extent to ascertain the population threatened by the flood. This result will help assess such event's adverse impact on the population and their vulnerability, providing crucial information for emergency response planning, risk assessment, and the development of resilient communities.

3.3 3 Results

3.3.1 Flood inundation extent

In this present study, an analysis of flood inundation extent is employed using Sentinel-1A satellite data between April and June 2022, revealing that the flood inundated a large extent of the area with a substantial impact in both districts, which is evident from Fig. 3.2 and Fig. 3.3. Fig. 3.2 presents a visual representation of the study area using Red-Green-Blue (RGB) composite SAR images of April (before Flood) and May–June (During Flood), illustrating the landscape before and after the hydroclimatic phenomenon. These RGB composite images are created with three bands, "VV," "VH," and "VV/VH." The figure's top part depicts the region's condition before the event, serving as a reference point for

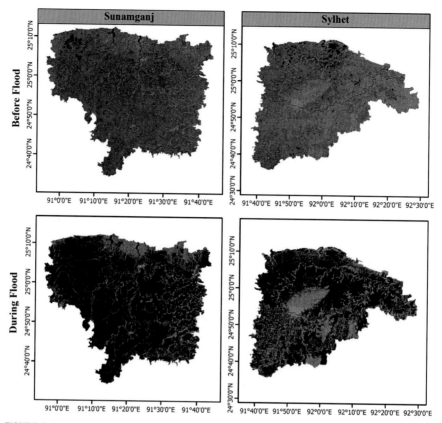

FIGURE 3.2 RGB composite map.

The RGB composite synthetic aperture radar images before the flood in April and during a flood in May and June. Map lines delineate study areas and do not necessarily depict accepted national boundaries.

comprehending the initial landscape. On the bottom part, we can see the identical area following the occurrence of the flooding event and the appearance of darker areas representing affected areas. These before-after RGB imageries function as a visual aid to the quantitative data in Fig. 3.3 and Table 3.1, augmenting our comprehension of the spatial extent and progression of the flooding incident.

The areas affected in each region are presented in Table 3.1, and the flood extent is shown in Fig. 3.3. The composite flood inundation analysis indicated that around 152,252.81 and 210,670.35 ha of Sylhet and Sunamganj were flooded, respectively. Sunamganj had a comparatively worse impact, as 57% of the district got flooded, whereas for Sylhet, it was 44%. From Fig. 3.3, few high elevated patches were not submerged; however, they became disconnected islands and the livelihood of the people living there was severely affected.

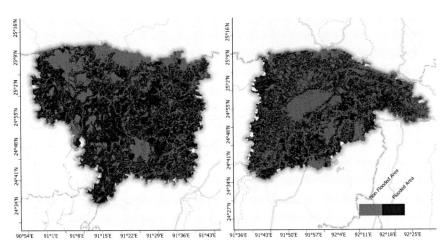

FIGURE 3.3 Flood extent map.

The extent of the inundated area in the studied region. Map lines delineate study areas and do not necessarily depict accepted national boundaries.

Table 3.1 Provides the calculation of areas and percentages of flooded sites in the region.

Region	Total area (ha)	Flooded area (ha)	Percentage of flooded area
Sylhet	342,945.0469	152,252.81	44
Sunamganj	367,544.6973	210,670.35	57

3.3.2 Flood impact on cropland and settlement

The spatial distribution of the landscape classification in the depression setting of Sylhet and Sunamganj is presented in Fig. 3.4. Inundated land classes during the event were identified by overlaying with the flood extent layer in Fig. 3.5 and Fig. 3.6. This provides a detailed breakdown of the impact of flooding within each category.

The study showed that the croplands are the worst affected by the floods in both the districts; 66% and 76% went under water of total cropland in Sylhet and Sunamganj, respectively, followed by rangeland and trees. In Sylhet, 6% of built-up areas and 25% of rangeland were flooded, while in Sunamganj, these fractions were 7% for built-up areas and 45% for rangeland. Additionally, in Sylhet, 10% of bare ground and 8% of trees were affected by flooding, whereas in Sunamganj, these percentages were 2% for bare ground and 19% for trees. The percentages represent the proportion of each class that is submerged, offering insights into the particular susceptibility of different areas to the flooding event.

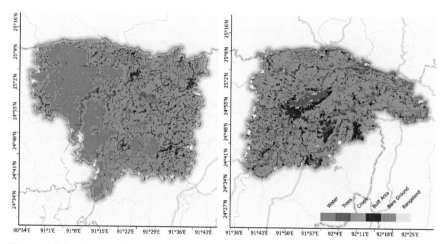

FIGURE 3.4 Land use and land cover (LULC) map.

The land use land cover for the two areas prepared using the ESRI's Sentinel-2 10-m LULC with a 10 m spatial resolution. Map lines delineate study areas and do not necessarily depict accepted national boundaries.

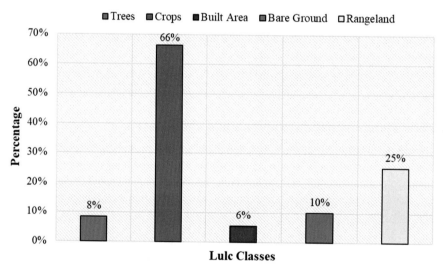

FIGURE 3.5 Land use and land cover (LULC)-wise percentage of areas flooded in Sylhet

Represents the LULC-wise percentage of classes inundated during the event of 2022 in Sylhet.

3.3.3 Flood inundation impact on population

Fig. 3.7 shows the flooded inundation level at the union/ward level. It is evident that the city corporation and municipality areas were comparatively less affected

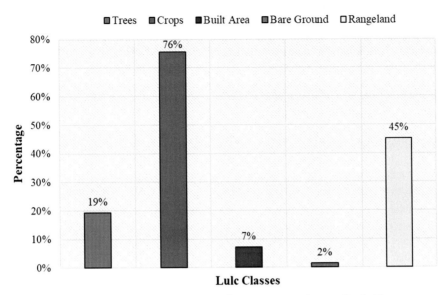

FIGURE 3.6 Land use and land cover (LULC)-wise percentage of areas flooded in Sunamganj.

Represents the LULC-wise percentage of classes inundated during the event of 2022 in Sunamganj.

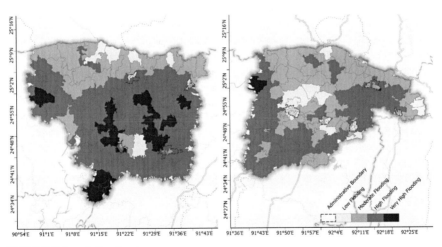

FIGURE 3.7 Inundation level map.

Union/ward-wise flooding where low, moderate, high, and very high flooding refer to percentage of areas submerged of a union below 25%, 25%−50%, 50%−75%, and above 75%, respectively. Map lines delineate study areas and do not necessarily depict accepted national boundaries.

than the rest of the districts as most wards suffered from low to moderate flooding; except for Derai municipality of Sunamganj which experienced very high flooding. This could be improved drainage system inside the city corporation and municipality area along with these locations having higher elevation.

Moreover, Sylhet has a total of 181 unions/wards, whereas Sunamganj has 124; all of these areas were inundated to an extent during the event. Table 3.2 shows that among the unions/wards of Sylhet, 2 were marked as very high flooded zone and 61 as high flooded zone. In the case of Sunamganj, 13 were marked as very high flooded zone and 66 as high flooded zone. Moderate and low-flooded union/ward areas numbered 64 and 54, respectively, in Sylhet, and in Sunamganj, these figures were 28 for moderate flooding and 17 for low flooding. In Sylhet and Sunamganj, around 3,849,638 and 2,690,150 people were affected by 2022 flood, respectively. The population density map for the district is shown in Fig. 3.8.

From Fig. 3.8, it is evident that the municipality and Sylhet City Corporation area have comparatively higher population densities. In Sylhet, the lowest population density is 393 per sq. km in Dakkhin Ranikhai Union, while the highest density is 37,066 per sq. km in Ward 6 of the City Corporation. On the other hand, in Sunamganj, the lowest population density is 190 per sq. km in Rajapur, and the highest population density of 25,559 per sq. km is found in Ward 6 of Chhatak Municipality. Table 3.2 presents the inundation levels in the Sylhet and Sunamganj regions along with the corresponding number of Union/Ward areas affected and the population impacted at each flooding level. It shows that the affected populations in Sylhet were estimated to be 21,053 approx. as very high flooded zone and 1,409,187 approx. as high flooded zone. In the case of Sunamganj, 291,289 approx. were marked as a very high flooded zone and 1,532,948 approx. as a high flooded zone. Moderate and low-flood-affected populations were 1,457,116 and 962,282, respectively, in Sylhet and Sunamganj; these figures were 609,762 for moderate flooding and 256,151 for low flooding.

To conclude, this research work aimed to analyze the spatial distribution of land cover, inundation levels, and their implications for flood vulnerability in the

Table 3.2 Inundation levels and population impact in Sylhet and Sunamganj regions.

Regions	Inundation level	Number of union/ward	Population affected
Sylhet	Low flooding	54	962,282
	Moderate flooding	64	1,457,116
	High flooding	61	1,409,187
	Very high flooding	2	21053
Sunamganj	Low flooding	17	256,151
	Moderate flooding	28	609,762
	High flooding	66	1,532,948
	Very high flooding	13	291,289

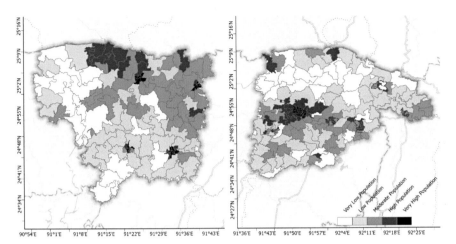

FIGURE 3.8 Population density map.

The population density at five density levels such as very low, low, moderate, high, and very high. Map lines delineate study areas and do not necessarily depict accepted national boundaries.

districts of Sylhet and Sunamganj, Bangladesh. In conclusion, the disparity in flood vulnerability emphasizes the need for region-specific flood risk management strategies that account for the unique characteristics of each district. Furthermore, the examination of inundation levels revealed varying degrees of flood severity across different regions within Sylhet and Sunamganj. Moderate and high flooding were prevalent in both districts, with differences observed in the frequency and distribution of these inundation levels. The spatial heterogeneity of flood risk calls for the importance of targeted interventions to mitigate the impacts of flooding on communities and infrastructure. In addition, this study highlights the critical role of population demographics in shaping flood vulnerability. Despite differences in flood susceptibility, both Sylhet and Sunamganj sustain substantial populations, indicating the potential socioeconomic repercussions of flooding events. This implied the need for integrated approaches to flood risk management that consider not only the spatial distribution of flood vulnerability but also the demographic characteristics of affected populations.

3.4 Discussion

Bangladesh frequently experiences hydrometeorological risks, specifically floods, and coastal flooding caused by extreme rainfall and storm surges, accordingly. These hazards put human lives at risk and have far-reaching advance consequences on livelihoods as such phenomena result in extensive devastation and harm to agricultural land and infrastructure (Kamal et al., 2018). Floods are often recognized as the most destructive natural disasters globally due to their significant impact on human lives surpassing the catastrophic effects of other natural occurrences (Jeyaseelan, 2004). Each year low-lying parts of Bangladesh, which encompass around 20% of the country, experience flash floods specially the northeastern region, which can be attributed to the country's geographical positioning as a downstream region of the Ganges, Brahmaputra, and Meghna rivers and its subtropical monsoon climate. The northeastern haor region has around 400 wetlands and is characterized by the presence of 23 transboundary rivers, which originate from the bordering country of India and flow into the region (Suman & Bhattacharya, 2015). Haors are geographical features characterized by their low-lying nature and physiographic depressions, often referred to as floodplain wetlands, are frequently observed in the northeastern region of Bangladesh (Kamal et al., 2018). These regions exhibit a higher magnitude of annual precipitation compared to other parts of the nation as Cherrapunji, is situated in close proximity to the northeastern border of Bangladesh, which is renowned for experiencing the most significant amount of precipitation on the planet (Das, 1951; Ohar & Nandargi, 1996). The considerable yearly rainfall, in conjunction with the rugged mountainous landscape, results in the rivers originating from the northeastern border exhibiting a substantial flow velocity due to the steep terrain. Hence due to this excessive rainfall prior to the monsoon season, the northeastern rivers through the downstream river network receive abnormally high volumes of water flow in such as Someswari, Kushiyara, Manu, Khowai, and Surma (Adnan, Dewan, Zannat, & Abdullah, 2019). Upon entering the territory of Bangladesh, the water expands across a significant area, resulting in recurrent flood occurrences of varying magnitudes inside the country. Furthermore, the exacerbation of flood conditions might be attributed to the occurrence of locally concentrated and protracted heavy rainfall (Uddin et al., 2019). Generally, this kind of sudden flash flood occurs due to premonsoon storms in April and May (Haque, Islam, Islam, Salehin, & Khan, 2017). Subsequently, with the advent of the monsoon season in June, riverine flooding is commonly observed. Hence, the inhabitants of haor are susceptible to both riverine and flash floods due to their distinct physical circumstances.

As per the Bangladesh Water Development Board, an increased frequency of floods has been seen in the haor area. Consequently, this has resulted in the entrapment of a considerable number of individuals in these affected regions (Akter et al., 2023). According to Masood, Yeh, Hanasaki, and Takeuchi (2015), it was anticipated that there may be an increase in rainfall during the premonsoon seasons in the haor region in the near future (2015–99). Furthermore, other research investigations have projected an escalation in the depth of flash floods inside the haor region as a consequence of irregular precipitation patterns (Choi, Campbell, Aldridge, & Eltahir, 2021). Furthermore, it is conceivable that the timing of flash floods may undergo alterations in subsequent periods (Masood & Takeuchi, 2016).

3.4.1 Impact of flash flood

Flash floods, due to their very limited warning periods, exhibit the greatest average mortality rates per occurrence and are accountable for most flood-related fatalities globally (Jonkman & Kelman, 2005; Jonkman, 2005). The uncontrolled population expansion and alterations in land use patterns as well as changing climatic patterns due to anthropogenic impacts have resulted in an elevated susceptibility of human populations to flood events (Doocy, Daniels, Murray, & Kirsch, 2013; Malik et al., 2020). The detrimental consequences such as immediate loss of life and health complications, as well as subsequent relocation and extensive harm to agricultural produce, infrastructure, and assets. The primary factors contributing to mortality in flood events are drowning and physical trauma or injury (Beinin, 1985; Jonkman & Kelman, 2005). Over a prolonged duration, there is a potential for heightened mortality rates as a result of infectious diseases (Alajo, Nakavuma, & Erume, 2006; French, Ing, Von Allmen, & Wood, 1983; Li, Zhu, Zhao, Zhang, & Chen, 2010). Hence, in this study, we aimed to analyze the spatial distribution of land cover, inundation levels, and their implications for flood vulnerability in the districts of Sylhet and Sunamganj, Bangladesh. Our findings reveal important insights into the complex land cover dynamics, flood susceptibility, and population demographics in these regions.

3.4.2 Impact of flash flood 2022 on population and health

This study found that almost half of both districts were flooded in the flash flood of 2022, posing a threat to the lives and livelihoods of over 3,849,638 and 2,690,150 people in Sylhet and Sunamganj, respectively. In both the hoar regions, croplands were the most affected land class. The present study reveals that, in the Sylhet region, a total of 3996.81 ha of trees, 146,499.07 ha of crops, 2621.53 ha of built area, 39.74 ha of bare ground, and 821.49 ha of rangeland area were subjected to inundation. In the case of Sunamganj, the recorded damage levels were as follows: 2301.40 ha for trees, 152,210.35 ha for crops, 2044.31 ha for built areas, 3.36 ha for bare ground, and 405.14 ha for rangeland. During these 2 months, the northeastern region of Bangladesh experienced significant precipitation in the form of premonsoon and monsoon rains, resulting in flash flooding. While such excessive rainfall is a recurrent phenomenon in this region, however, the current monsoon season has been characterized as the most severe in the past 122 years, resulting in extensive flooding across this part of the country. Consequently, the impact of this event has led to the displacement of approximately 500,000 individuals from their residences, affecting a staggering population of over 7 million residents (IRFC, 2022; Raulerson, 2022; UNICEF, 2022) (Matt), out of which 3.5 million were children (UNICEF, 2022). According to the Ministry of Disaster Management and Relief (MoDMR), a total of 472,856 individuals were transported to approximately 1605 shelter centers through a collaborative endeavor involving the Army, Navy, Fire Service, and local authorities.

Based on the findings of the MoDMR, the housing sector is projected to incur a substantial financial loss amounting to USD 176.0 million. According to the estimations made by the Department of Public Health Engineering, a significant number of 106,727 water points and 283,355 sanitation facilities have suffered extensive damage, requiring urgent restoration and reconstruction while the potential for outbreaks of Acute Watery Diarrhea remains significant, alongside the persistent prevalence of skin and respiratory ailments (IRFC, 2023; OCHA, 2022). A total of 12,503 cases of waterborne diseases and other related ailments have been documented within the Sylhet division (UNICEF, 2022).

3.4.3 Impact of flash flood 2022 on crop cultivation and livelihood

The analysis of land cover distribution highlighted significant variations between Sylhet and Sunamganj districts. In both districts, crops emerged as the dominant land cover class, underscoring the importance of agricultural activities in the region. However, Sunamganj exhibited a higher susceptibility to flooding compared to Sylhet, as evidenced by the larger proportion of flooded areas relative to its total land area. Ministry of Agriculture reported that the flash flood event resulted in significant damage to a total of 254,251 ha of croplands. Additionally, the Ministry of Livestock and Fisheries reported that it had a profound impact on the local livestock population, affecting a staggering 663,534 cattle. Furthermore, the floodwaters inundated a substantial number of ponds, reaching a total of 106,000, and resulted in the unfortunate loss of 17,000 MT of fish (IRFC, 2023; OCHA, 2022; UN RC Bangladesh, & UNCT Bangladesh, 2022UN RC Bangladesh, & UNCT Bangladesh, 2022). According to the comprehensive household assessment carried out by the Bangladesh Red Crescent Society (BDRCS) in December 2022, a mere 2.3% of the households surveyed in the districts of Sylhet and Sunamganj possess the capacity to fully restore their means of living. Furthermore, a mere 7% of the families assessed have the ability to reconstruct their dwellings, which suffered damage, while 9% of the evaluated households are capable of rebuilding their impaired latrines (IRFC, 2023; OCHA, 2022). Agriculture, livestock husbandry, and fisheries serve as the primary sources of income for the population residing in the northeastern region (IRFC, 2023). The northeastern haor region has distinctive physiographic characteristics and hydroclimatological conditions influenced by subtropical monsoon climate, which create a diverse range of earning opportunities for the inhabitants and suitable conditions for crop cultivation (Kamruzzaman & Shaw, 2018; Nowreen et al., 2015). Approximately 85% of the haor areas are dedicated to Boro rice cultivation, while only 15% of the site is allocated for Aman rice and rabi crops (Khan, Hossain, & Rahman, 2020). In recent times, the cultivation of high-yielding Boro rice has become prevalent in a significant portion of the floodplains of the haor region, facilitated by the utilization of additional irrigation sourced from shallow alluvial aquifers. It is practiced in these wetlands during both the wet and dry seasons. However, the cultivation of crops in these areas during the dry season is primarily limited to Boro rice, with minimal opportunities for the growth of other crops as the water bodies predominantly experience drying up during the postmonsoon period. Consequently, the predominant occupations among the haor inhabitants include agricultural cultivation, fishing, poultry and cattle farming, and engagement in daily wage employment (Kamal et al., 2018). The production of Boro rice is heavily reliant on natural factors, rendering this crop susceptible to varying degrees of harm, ranging from partial to complete destruction, due to premonsoon flash floods that occur shortly before the harvesting period (Hellin et al., 2020; Shahe Alam, Quayum, & Islam, 2011). Regrettably, this region is being subjected to significant climatic hazards, particularly early occurrences of flash floods with the escalating climate change, and it is anticipated that these occurrences will intensify in the coming years, and consequently, the viability of production and farming systems in this region will become increasingly precarious, posing significant challenges to their sustainability in the foreseeable future (Campbell et al., 2018; Dey et al., 2021; Hansen et al., 2019). Consequently, we expect to observe substantial reductions in Boro rice yields as a result of catastrophic flash flood events (Haque et al., 2017; Kamal et al., 2018), as such events destroy thousands of hectares of staple crops (Quader et al., 2023). Consequently, during the flood period, agriculture wages were reduced in those affected areas (Banerjee, 2010). Hence, floods pose a significant risk not only to human lives but also to the provision of livelihoods and nutritional well-being. Despite this inherent susceptibility, individuals residing in the haor region have endured diverse hydrometeorological perils such as floods, tropical storms, and waterlogging for generations, and their capacity to withstand and recover from these natural hazards is evident.

Thus this study provides valuable insights into the spatial distribution of land cover, inundation levels, and their implications for flood vulnerability in the districts of Sylhet and Sunamganj. Hence, comprehensive flood risk assessment and management strategies for these regions should incorporate land cover dynamics, inundation levels, and population demographics. Moreover, this study's findings and approach can contribute to the development of evidence-based flood risk management strategies aimed at enhancing the resilience of communities in flood-prone regions.

3.5 Conclusion

This chapter provides a comprehensive discussion on the impact of the flash flood in 2022 in the northeastern hoar region of Bangladesh, particularly the Sylhet and Sunamganj districts. The devastating effect of this flood not only took lives but also severely impacted the livelihoods of people living in this region. Sylhet, encompassing a total area of 342,945.0469 ha, experiences a significant portion affected by flooding, with 152,252.81 ha inundated, accounting for approximately

44% of its total area. Correspondingly, the region accommodates a substantial population of 3,849,638 individuals. In contrast, Sunamganj exhibits a larger total area of 367,544.6973 ha, with a greater extent of land subject to flooding, totaling 210,670.35 ha or approximately 57% of its total area, while this area is home to a population of 2,690,150 inhabitants. A significant share of croplands was inundated, affecting the food security of this region. 146,499.07 ha of cropland in Sylhet and 152,210.3547 ha in Sunamganj were inundated during this time affecting 66% and 76% of the total agriculture land of the respective districts. Conversely, in Sylhet the smallest land cover class, Bare Ground, exhibits a relatively higher vulnerability to flooding, with 10% of its total area of 392.928669 ha experiencing inundation, equivalent to 39.740323 ha. Notably, trees, built area, and rangeland classes display varying degrees of susceptibility to flooding, with percentages ranging from 6% to 25%. In the Sunamganj district, the land cover class bare ground exhibits a notably lower susceptibility to flooding, with only 2% of its total area of 220.341222 ha affected. Noteworthy variations in flood susceptibility are observed across different land cover classes of trees, built area, and rangeland, with percentages ranging from 2% to 45%, indicating heterogeneous vulnerability patterns within the district. Furthermore, in this study, the inundation levels are classified into four categories, low flooding, moderate flooding, high flooding, and very high flooding, and the corresponding counts of union/wards affected by flooding in the Sylhet and Sunamganj districts of Bangladesh were extracted. In the district of Sylhet, among these categories, the region experiences the highest frequency of moderate flooding, with 64 union/ward areas affected. Following closely are high flooding areas, comprising 61 union/ward regions. Low flooding and very high flooding areas exhibit relatively lower occurrences, with 54 and 2 union/ward regions affected, respectively. Conversely, in the Sunamganj district, similar inundation levels are observed. Notably, the district experiences the highest incidence of high flooding, affecting 66 union/ward areas. Moderate flooding follows, affecting 28 union/ward regions. Low flooding and very high flooding areas exhibit comparatively lower occurrences, with 17 and 13 union/ward regions affected, respectively. It is a stark example of the helplessness of humans in the face of natural disturbances. Though there are limitations in moderate resolution land use and population data, this study served as an arbitrary portrayal of the flood scenarios in a fast and feasible manner. This study can be helpful and provide valuable information for swift action and early precaution in times of crisis. Additionally, future research can further explore the underlying drivers of flood vulnerability in these regions, including factors such as land use change, hydrological dynamics, and socioeconomic factors.

References

Abedin, J., & Khatun, H. (2020). Impacts of flash flood on livelihood and adaptation strategies of the haor inhabitants: A study in Tanguar Haor of Sunamganj, Bangladesh. *Journal of Earth and Environmental Sciences*, 8(1), 41−51. Available from https://doi.org/10.3329/dujees.v8i1.50757.

Abshire, K., & Mullusky, M. (2019). *National Weather Service Manual (p. 10-950)*. Retrieved from https://www.nws.noaa.gov/directives/sym/pd01009050curr.pdf.

Adnan, M. S. G., Dewan, A., Zannat, K. E., & Abdullah, A. Y. M. (2019). The use of watershed geomorphic data in flash flood susceptibility zoning: A case study of the Karnaphuli and Sangu river basins of Bangladesh. *Natural Hazards*, *99*(1), 425−448. Available from https://doi.org/10.1007/s11069-019-03749-3, http://www.wkap.nl/journalhome.htm/0921-030X.

Ahmed, M., Rahaman, K., Kok, A., & Hassan, Q. (2017). Remote sensing-based quantification of the impact of flash flooding on the rice production: A case study over northeastern Bangladesh. *Sensors*, *17*(10). Available from https://doi.org/10.3390/s17102347.

Akter, N., Islam, M. R., Karim, M. A., Miah, M. G., & Rahman, M. M. (2023). Spatiotemporal rainfall variability and its relationship to flash flood risk in northeastern Sylhet Haor of Bangladesh. *Journal of Water and Climate Change*, 2040−2244. Available from https://doi.org/10.2166/wcc.2023.165.

Alajo, S. O., Nakavuma, J., & Erume, J. (2006). Cholera in endemic districts in Uganda during El Niño rains: 2002−2003. *African Health Sciences*, *6*(2), 93−97. Available from http://www.atypon-link.com/MMS/doi/pdf/10.5555/afhs.2006.6.2.93.

Aldhshan, S. R. S., Mohammed, O. Z., & Shafri, H. Z. M. (2019). Flash flood area mapping using sentinel-1 SAR data: a case study of eight upazilas in Sunamganj district, Bangladesh. *IOP Conference Series: Earth and Environmental Science*, *357*(1). Available from https://doi.org/10.1088/1755-1315/357/1/012034.

Armenakis, C., Du, E., Natesan, S., Persad, R., & Zhang, Y. (2017). Flood risk assessment in urban areas based on spatial analytics and social factors. *Geosciences*, *7*(4). Available from https://doi.org/10.3390/geosciences7040123.

Attema, E., Cafforio, C., Gottwald, M., Guccione, P., Guarnieri, A. M., Rocca, F., & Snoeij, P. (2010). Flexible dynamic block adaptive quantization for Sentinel-1 SAR missions. *IEEE Geoscience and Remote Sensing Letters*, *7*(4), 766−770. Available from https://doi.org/10.1109/lgrs.2010.2047242.

Azmeri, A., & Isa, A. H. (2018). An analysis of physical vulnerability to flash floods in the small mountainous watershed of Aceh Besar Regency, Aceh province, Indonesia. *Jàmbá: Journal of Disaster Risk Studies*, *10*(1). Available from https://doi.org/10.4102/jamba.v10i1.550.

Baky, M. A. A., Islam, M., Paul, S., & Hazard, F. (2020). Vulnerability and risk assessment for different land use classes using a flow model. *Earth Systems and Environment*, *4*(1), 225−244. Available from https://doi.org/10.1007/s41748-019-00141-w, https://link.springer.com/journal/41748.

Banerjee, L. (2010). Effects of flood on agricultural productivity in Bangladesh. *Oxford Development Studies*, *38*(3), 339−356. Available from https://doi.org/10.1080/13600818.2010.505681.

Beinin, L. (1985). *Medical consequences of natural disasters*. Springer. Available from 10.1007/978-3-642-70532-8.

Bich, T. H., Quang, L. N., Ha, L. T. T., Hanh, T. T. D., & Guha-Sapir, D. (2011). Impacts of flood on health: Epidemiologic evidence from Hanoi, Vietnam. *Global Health Action*, *4*. Available from https://doi.org/10.3402/gha.v4i0.6356.

Campbell, B. M., Hansen, J., Rioux, J., Stirling, C. M., Twomlow, S., & Wollenberg, Eva (Lini) (2018). Urgent action to combat climate change and its impacts (SDG 13): Transforming agriculture and food systems. *Current Opinion in Environmental Sustainability*, *34*, 13−20. Available from https://doi.org/10.1016/j.cosust.2018.06.005.

Choi, Y. W., Campbell, D. J., Aldridge, J. C., & Eltahir, E. A. B. (2021). Near-term regional climate change over Bangladesh. *Climate Dynamics, 57*(11−12), 3055−3073. Available from https://doi.org/10.1007/s00382-021-05856-z, http://link.springer.de/link/service/journals/00382/index.htm.

Choudhury, M. U. I., & Haque, C. E. (2016). "We are more scared of the power elites than the floods": Adaptive capacity and resilience of wetland community to flash flood disasters in Bangladesh. *International Journal of Disaster Risk Reduction, 19*, 145−158. Available from https://doi.org/10.1016/j.ijdrr.2016.08.004, http://www.journals.elsevier.com/international-journal-of-disaster-risk-reduction/.

Conde, F. C., & De Mata Muñoz, M. (2019). Flood monitoring based on the study of Sentinel-1 SAR images: The Ebro River case study. *Water, 11*(12). Available from https://doi.org/10.3390/w11122454.

Das, J. C. (1951). On certain aspects of rainfall at Cherrapunji. *Mausam, 2*(3), 197−202. Available from https://doi.org/10.54302/mausam.v2i3.4681.

De Zan, F., & Monti Guarnieri, A. (2006). TOPSAR: Terrain observation by progressive scans. *IEEE Transactions on Geoscience and Remote Sensing, 44*(9), 2352−2360. Available from https://doi.org/10.1109/TGRS.2006.873853.

Dey, N. C., Parvez, M., & Islam, M. R. (2021). A study on the impact of the 2017 early monsoon flash flood: Potential measures to safeguard livelihoods from extreme climate events in the haor area of Bangladesh. *International Journal of Disaster Risk Reduction, 59*. Available from https://doi.org/10.1016/j.ijdrr.2021.102247, http://www.journals.elsevier.com/international-journal-of-disaster-risk-reduction/.

Doocy, S., Daniels, A., Murray, S., & Kirsch, T. D. (2013). The human impact of floods: A historical review of events 1980−2009 and systematic literature review. *PLoS Currents*. Available from https://doi.org/10.1371/currents.dis.f4deb457904936b07c09-daa98ee8171a, http://currents.plos.org/disasters/article/the-human-impact-of-floods-a-historical-review-of-events-1980-2009-and-systematic-literature-review/pdf/.

ESA. (2023). *User guides—Sentinel-1 SAR*. Sentinel Online.

French, J., Ing, R., Von Allmen, S., & Wood, R. (1983). Mortality from flash floods: A review of National Weather Service reports, 1969-81. *Public Health Reports, 98*(6), 584−588.

Ghosh, A., Chatterjee, U., Pal, S. C., Towfiqul Islam, A. R. M., Alam, E., & Islam, M. K. (2023). Flood hazard mapping using GIS-based statistical model in vulnerable riparian regions of sub-tropical environment. *Geocarto International, 38*(1). Available from https://doi.org/10.1080/10106049.2023.2285355, http://www.tandfonline.com/toc/tgei20/current.

Hansen, J., Hellin, J., Rosenstock, T., Fisher, E., Cairns, J., Stirling, C., ... Campbell, B. (2019). Climate risk management and rural poverty reduction. *Agricultural Systems, 172*, 28−46. Available from https://doi.org/10.1016/j.agsy.2018.01.019.

Haque, S., Islam, A. K. M. S., Islam, G. T., Salehin, M., & Khan, M. J. U. (2017). Event based flash flood simulation at Sunamganj using HEC-HMS. In 6th international conference on water & flood management (ICWFM-2017).

Hellin, J., Balié, J., Fisher, E., Kohli, A., Connor, M., Yadav, S., ... Gummert, M. (2020). Trans-disciplinary responses to climate change: Lessons from rice-based systems in Asia. *MDPI AG, Philippines Climate, 8*(2). Available from https://doi.org/10.3390/cli8020035, https://res.mdpi.com/d_attachment/climate/climate-08-00035/article_deploy/climate-08-00035.pdf.

Hoque, R., Nakayama, D., Matsuyama, H., & Matsumoto, J. (2011). Flood monitoring, mapping and assessing capabilities using RADARSAT remote sensing, GIS and ground data for Bangladesh. *Natural Hazards, 57*(2), 525−548. Available from https://doi.org/10.1007/s11069-010-9638-y.

Immerzeel, W. (2008). Historical trends and future predictions of climate variability in the Brahmaputra basin. *International Journal of Climatology, 28*(2), 243−254. Available from https://doi.org/10.1002/joc.1528.

IRFC. (2022). *Millions in Bangladesh impacted by one of the worst floodings ever seen.* IFRC.

IRFC. (2023). Bangladesh: Floods − Operation update #2 (6-month update), In *Emergency appeal no. MDRBD028.* ReliefWeb.

Islam, A. S., Haque, A., & Bala, S. K. (2010). Hydrologic characteristics of floods in Ganges-Brahmaputra-Meghna (GBM) delta. *Natural Hazards, 54*(3), 797−811. Available from https://doi.org/10.1007/s11069-010-9504-y.

Jeyaseelan, A. (2004). *Droughts & floods assessment and monitoring using remote sensing and GIS satellite remote sensing and GIS applications in agricultural meteorology.*

Jonkman, S. N. (2005). Global perspectives on loss of human life caused by floods. *Natural Hazards, 34*(2), 151−175. Available from https://doi.org/10.1007/s11069-004-8891-3.

Jonkman, S. N., & Kelman, I. (2005). An analysis of the causes and circumstances of flood disaster deaths. *Disasters, 29*(1), 75−97. Available from https://doi.org/10.1111/j.0361-3666.2005.00275.x, http://onlinelibrary.wiley.com/journal/10.1111/(ISSN)1467-7717.

Kamal, A. S. M. M., Shamsudduha, M., Ahmed, B., Hassan, S. M. K., Islam, M. S., Kelman, I., & Fordham, M. (2018). Resilience to flash floods in wetland communities of northeastern Bangladesh. *International Journal of Disaster Risk Reduction, 31,* 478−488. Available from https://doi.org/10.1016/j.ijdrr.2018.06.011, http://www.journals.elsevier.com/international-journal-of-disaster-risk-reduction/.

Kamruzzaman, M., & Shaw, R. (2018). Flood and sustainable agriculture in the haor basin of Bangladesh: A review paper. *Universal Journal of Agricultural Research, 6*(1), 40−49. Available from https://doi.org/10.13189/ujar.2018.060106.

Karra, K., Kontgis, C., Statman-Weil, Z., Mazzariello, J.C., Mathis, M., & Brumby, S.P. (2021). Global land use/land cover with sentinel 2 and deep learning 2021. In *International Geoscience and Remote Sensing Symposium (IGARSS)* (pp. 4704−4707). Institute of Electrical and Electronics Engineers Inc. https://doi.org/10.1109/IGARSS47720.2021.9553499 9781665403696.

Khan, A., Hossain, M., & Rahman, M. (2020). *JICA Haor Flood Management and Livelihood Improvement Project (HFMLIP) Quarterly progress report.*

Khatun, M.A., Rashid, M.B., & Hygen, H.O. (2016). *Climate of Bangladesh (08/2016).* Bangladesh Meteorological Department.

Kwak, Y.-J. (2017). Nationwide flood monitoring for disaster risk reduction using multiple satellite data. *ISPRS International Journal of Geo-Information, 6*(7). Available from https://doi.org/10.3390/ijgi6070203.

Lehner, B., Verdin, K., & Jarvis, A. (2008). New global hydrography derived from space-borne elevation data. *Eos, Transactions American Geophysical Union, 89*(10), 93−94. Available from https://doi.org/10.1029/2008eo100001.

Li, H. J., Zhu, T., Zhao, D. F., Zhang, Z. F., & Chen, Z. M. (2010). Kinetics and mechanisms of heterogeneous reaction of NO_2 on $CaCO_3$ surfaces under dry and wet conditions. *Atmospheric Chemistry and Physics, 10*(2), 463−474. Available from https://doi.org/10.5194/acp-10-463-2010, http://www.atmos-chem-phys.net/volumes_and_issues.html.

Luo, T., Maddocks, A., Iceland, C., Ward, P., & Winsemius, H. (2015). *World's 15 countries with the most people exposed to river floods*. Retrieved from https://reliefweb.int/report/world/world-s-15-countries-most-people-exposed-river-floods-0?gad_source = 1&gclid = CjwKCAjwi_exBhA8EiwA_kU1MlD8wPnj0C_iiee67vV8XQkhgWmDn5Zod-FNR1WNfVpXesDQ36MF-RoCqAwQAvD_BwE.

Malik, S., Pal, S. C., Chowdhuri, I., Chakrabortty, R., Roy, P., & Das, B. (2020). Prediction of highly flood prone areas by GIS based heuristic and statistical model in a monsoon dominated region of Bengal Basin. *Remote Sensing Applications: Society and Environment*, *19*. Available from https://doi.org/10.1016/j.rsase.2020.100343.

Martinis, S., & Rieke, C. (2015). Backscatter analysis using multi-temporal and multi-frequency SAR data in the context of flood mapping at River Saale, Germany. *Remote Sensing*, *7*(6), 7732−7752. Available from https://doi.org/10.3390/rs70607732.

Masood, M., & Takeuchi, K. (2016). Climate change impacts and its implications on future water resource management in the Meghna Basin. *Futures*, *78−79*, 1−18. Available from https://doi.org/10.1016/j.futures.2016.03.001, http://www.elsevier.com/inca/publications/store/3/0/4/2/2/.

Masood, M., Yeh, P. J.-F., Hanasaki, N., & Takeuchi, K. (2015). Model study of the impacts of future climate change on the hydrology of Ganges−Brahmaputra−Meghna basin. *Hydrology and Earth System Sciences*, *19*(2), 747−770. Available from https://doi.org/10.5194/hess-19-747-2015.

Messner, F., & Meyer, V. (2006). Flood damage, vulnerability and risk perception−challenges for flood damage research. In *Flood risk management: hazards, vulnerability and mitigation measures*, (pp. 149−167). Dordrecht: Springer Netherlands. Available from https://doi.org/10.1007/978-1-4020-4598-1_13.

Monirul Qader Mirza, M. (2002). Global warming and changes in the probability of occurrence of floods in Bangladesh and implications. *Global Environmental Change*, *12*(2), 127−138. Available from https://doi.org/10.1016/S0959-3780(02)00002-X, http://www.elsevier.com/inca/publications/store/3/0/4/2/5.

Moniruzzaman, M., Thakur, P. K., Kumar, P., Ashraful Alam, M., Garg, V., Rousta, I., & Olafsson, H. (2021). Decadal urban land use/land cover changes and its impact on surface runoff potential for the Dhaka City and surroundings using remote sensing. *Remote Sensing*, *13*(1), 83. Available from https://doi.org/10.3390/rs13010083.

Nowreen, S., Murshed, S. B., Islam, A. K. M. S., Bhaskaran, B., & Hasan, M. A. (2015). Changes of rainfall extremes around the haor basin areas of Bangladesh using multi-member ensemble RCM. *Theoretical and Applied Climatology*, *119*(1−2), 363−377. Available from https://doi.org/10.1007/s00704-014-1101-7, https://rd.springer.com/journal/volumesAndIssues/704.

OCHA. (2022). *Bangladesh: Floods and landslides − May 2022*. ReliefWeb.

Ohar, O. N., & Nandargi, S. (1996). Which is the rainiest station in India—Cherrapunji or Mawsynram? *Weather*, *51*(9), 314−316. Available from https://doi.org/10.1002/j.1477-8696.1996.tb06233.x.

Pandey, A. C., Kaushik, K., & Parida, B. R. (2022). Google Earth Engine for large-scale flood mapping using SAR data and impact assessment on agriculture and population of Ganga-Brahmaputra Basin. *Sustainability*, *14*(7). Available from https://doi.org/10.3390/su14074210.

Pekel, J. F., Cottam, A., Gorelick, N., & Belward, A. S. (2016). High-resolution mapping of global surface water and its long-term changes. *Nature*, *540*(7633), 418−422. Available from https://doi.org/10.1038/nature20584, http://www.nature.com/nature/index.html.

Qadir, A., & Mondal, P. (2020). Synergistic use of radar and optical satellite data for improved monsoon cropland mapping in India. *Remote Sensing, 12*(3). Available from https://doi.org/10.3390/rs12030522.

Quader, M. A., Dey, H., Malak, A., & Rahman, Z. (2023). A geospatial assessment of flood hazard in north-eastern depressed basin, Bangladesh. *Journal of Tropical Geography, 44*(2), 277–299. Available from https://doi.org/10.1111/sjtg.12476, http://onlinelibrary.wiley.com/journal/10.1111/(ISSN)1467-9493.

Raulerson, M. (2022). *Bangladesh's flood displacement: Yet another case for loss & damage*. Climate Refugees.

Shahe Alam, M., Quayum, M. A., & Islam, M. A. (2011). Crop production in the haor areas of Bangladesh: Insights from farm level survey. *The Agriculturists, 8*(2), 88–97. Available from https://doi.org/10.3329/agric.v8i2.7582.

Shen, X., Wang, D., Mao, K., Anagnostou, E., & Hong, Y. (2019). Inundation extent mapping by synthetic aperture radar: A review. *Remote Sensing, 11*(7). Available from https://doi.org/10.3390/rs11070879.

Suman, A., & Bhattacharya, B. (2015). Flood characterisation of the haor region of Bangladesh using flood index. *Hydrology Research, 46*(5), 824–835. Available from https://doi.org/10.2166/nh.2014.065, http://www.iwaponline.com/nh/046/0824/0460824.pdf.

Uddin, K., Matin, M. A., & Meyer, F. J. (2019). Operational flood mapping using multi-temporal Sentinel-1 SAR images: A case study from Bangladesh. *Remote Sensing, 11*(13). Available from https://doi.org/10.3390/rs11131581, https://res.mdpi.com/remotesensing/remotesensing-11-01581/article_deploy/remotesensing-11-01581-v2.pdf?filename = &attachment = 1.

UN RC Bangladesh, & UNCT Bangladesh. (2022). Flash flood humanitarian response plan 2022 humanitarian coordinator task team (HCTT)—Monitoring dashboard as of 20 october 2022—Bangladesh (p. 17) [Situation Report]. United Nations Bangladesh.

UNICEF. (2022). Bangladesh Humanitarian Situation Report, January to June 2022 (Situation Report 60; p. 11). UNICEF.

United Nations University. (June 14, 2014). Two billion vulnerable to floods by 2050; number expected to double or more in two generations. *ScienceDaily*. http://www.sciencedaily.com/releases/2004/06/040614081820.htm.

Wallemacq, P., Guha-Sapir, D., & McClean, D. (2015). *CRED, UNISDR, the human cost of weather related disasters—1995–2015*. https://doi.org/10.13140/RG.2.2.17677.33769.

Geographical appraisal of the basin hydrological phenomenon using Google Earth Engine

Sameer Mandal, Aditi Bhadra and Arnab Bandyopadhyay

North Eastern Regional Institute of Science and Technology, Agricultural Engineering, Itanagar, Arunachal Pradesh, India

4.1 Introduction

The understanding of intricate movement of water between phases and systems is referred as basin hydrology (Smith, 1984; Stewardson, Walker, & Coleman, 2021). The atmospheric, surface, and subsurface subsystems serve as conduits for water circulation, sustaining biodiversity, and meeting diverse needs throughout the year (Yang, Yang, & Xia, 2021). Hydrological processes, including precipitation, infiltration, runoff, stream flow, and evapotranspiration (ET), are essential components influencing water dynamics, influenced by basin topography, soil composition, vegetation, and other factors (Chaponnière, Boulet, Chehbouni, & Aresmouk, 2008; Gao et al., 2018). Among these factors, the physical attributes of the basin hold particular significance, which can be interpreted through morphometric analysis (Mahala, 2020). Basin morphometry, encompassing characteristics like shape, size, drainage density, and overland flow patterns, provides essential insights about the physical features of the basin. These morphometric properties play a crucial role in environmental studies, informing land use planning, soil erosion assessment, groundwater potential zone identification, and river morphology analysis. Thus, morphometric evaluation is crucial for understanding and estimating the complex behavior of hydrological systems within any given basin (Kottagoda & Abeysingha, 2017; Sreedevi, Subrahmanyam, & Ahmed, 2005).

Additionally, hydroclimatic characteristics play an important role in shaping the behavior and response of the basin. The historical precipitation patterns and resulting runoff from the basin are essential factors that influence the future scenario of the basin (Chen et al., 2023; Rashid et al., 2022). The intensity and distribution of rainfall signify both the amount and duration of the runoff, which is considered one of the most critical components of the hydrological cycle

Applications of Geospatial Technology and Modeling for River Basin Management.
DOI: https://doi.org/10.1016/B978-0-443-23890-1.00004-9

(Wu, Peng, Qiao, & Ma, 2018; Yan et al., 2018). More intense rainfall for a shorter duration leads to a larger runoff, and conversely, less intense rainfall typically generates less runoff. Larger runoff for a shorter duration is indeed challenging to manage and can lead to disastrous incidents. The rapid flow of surface runoff provides less opportunity time for the water to infiltrate into the soil, causing a rapid depletion of soil moisture and groundwater resources (Owuor et al., 2016; Qi et al., 2020). Another crucial parameter is ET, which encompasses both evaporation and transpiration from plants, soil, and water bodies. Potential ET is primarily affected by weather elements such as temperature, humidity, sunshine hours, humidity, and so forth, whereas actual ET depends on a combination of weather conditions, soil properties, and vegetation stages and types (Feng et al., 2020; Luo, Gao, & Mu, 2021; Yin, Wu, & Dai, 2010). Furthermore, the ratio of potential to actual ET provides insights into the stress on soil moisture and the occurrence of drought phenomena (Rehana & Monish, 2021; Stoyanova, Georgiev, & Neytchev, 2023). This interdependence of various hydrological phenomena highlights how one component can impact others and potentially lead to catastrophic events. It underscores the need for a comprehensive geographical appraisal of the hydrological phenomena within a basin to better understand and manage these complex interactions.

Understanding the complexities of land use and land cover (LULC) provides valuable insights into the available resources within a watershed. LULC represents the delicate balance between the natural features and utilization of land on the surface of the earth by human beings for different purposes. Achieving the right equilibrium is crucial to minimizing the exploitation of natural resources and maintaining healthier ecosystems for all living beings. However, certain features like forests, water bodies, and agriculture not only serve essential human needs but also act as a crucial tool for mitigating natural disasters, such as flood and drought control, and ensuring the supply of irrigation and drinking water and so forth. Negative changes in LULC, such as encroachment of forest and water bodies, or unplanned urban expansion, can contribute to climate change, leading to increased temperatures, reduced rainfall, and declining water levels in both surface and groundwater (Samal & Gedam, 2021; Thapa, 2022). This is particularly significant in developing countries such as Indian Subcontinent and Africa, where climate change poses a significant challenge. Therefore, it is a need of the time to consider all the factors that can trigger the climate change phenomenon when formulating action plans. This study focuses on a comprehensive geographical appraisal of hydrological phenomena, incorporating morphometric analysis, hydrometeorological characteristics, LULC information, and future climate projections applying the Google Earth Engine (GEE) platform. The study offers an overview of the watershed and lays the groundwork for conducting more in-depth analysis. It also underscores the importance of considering these critical factors when formulating policies and plans.

4.2 Study area and dataset used

4.2.1 Study area description

The Dikrong River basin is situated along the border of Arunachal and Assam states in North Eastern India. This region, being a part of the Eastern Himalayan, is abundant in natural resources and diverse habitats but faces challenges related to accessibility. The outlet location of the basin is at Dikrong River, with a latitude of 26°57′43.508″N and a longitude of 93°59′31.111″E. The Dikrong River eventually flows into the Subansiri River, and traveling after just about 8 km from that confluence, it joins the river Brahmaputra. The extent of the basin varies from a latitude of 26°57′22.855″N to 27°22′11.03″N and a longitude from 93°13′15.02″E to 93°59′ 58.56″E. The altitude of the basin ranges from 76 to 2898 m above the mean sea level (MSL). The geographical area of the basin is about 1409.50 sq. km, out of which approx. 89% of the area falls within Arunachal Pradesh and the remaining portion is in Assam. Being a part of the Eastern Himalayan region, the area experiences high humidity and has unique hydrological characteristics. The major portion of the area is occupied by dense evergreen forests. The upland area of the basin exhibits significant variations in the topography and is often inaccessible, while the lowland areas are relatively flat and feature major settlements and agricultural lands. Most of the settlements are located in the lowland areas including the capital of Arunachal Pradesh; therefore, the drastic LULC change has been observed in this area as well. The average temperature was recorded as 15.15°C in the month of January (moderate cold) and 26.96°C in the month of July (Summer). The peak of the monsoon was seen in the month of July with a rainfall of 602−986 mm, sometimes April also receives a good amount of rainfall (Huda & Singh, 2012; LUP NBSS, 2004). The study area map with its location is presented in Fig. 4.1.

4.2.2 Dataset used

The study primarily focused on the application of GEE to hydrological analysis. This work aims to showcase the capabilities of GEE and the datasets accessible on this platform. The study began with the selection of the basin and subbasins. Once the basin was chosen, the basin and watershed boundaries were derived from the WWF HydroSHEDS and HydroRIVERS datasets. These datasets are available in the vector format and provide comprehensive information about the watershed and its associated streams, including details such as watershed ID, area, perimeter, stream length, stream orders, and so forth. HydroRIVERS v1.0 consists of all the rivers with a watershed area of at least 10 sq. km or an average flow of the river as 0.1 cumecs, or both (WWF, 2021). In GEE, HydroSHEDS data are categorized into 12 different levels based on the mapping scale, where

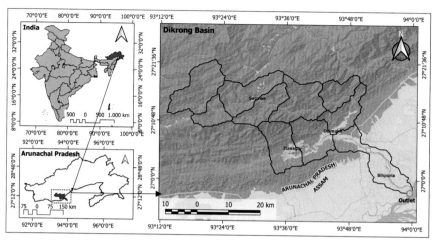

FIGURE 4.1

Location map of the study area. Map lines delineate study areas and do not necessarily depict accepted national boundaries.

level-1 represents coarser resolution and level-12 represents a finer solution with all subwatersheds (SWs). This study utilizes level-12 HydroSHEDS datasets.

The administrative boundary was used as a reference to mark the boundary of Arunachal Pradesh and Assam, which was acquired from the Survey of India. The hydrometeorological data were acquired from the European Centre for Medium-Range Weather Forecasts (ECMWF) platform. Specifically, the study used the ERA5 Land dataset, which is derived from the fifth generation of ECMWF reanalysis (ERA5), along with Terra MODIS (MOD16A2 ver. 6.1) for determining watershed hydrometeorological characteristics. Data on total precipitation, runoff, and temperature were extracted from the ERA5 Land dataset, while actual ET and reference ET were acquired from the MODIS dataset. Reanalysis data are estimated by combining model and observational data from around the world by applying the laws of governing physics. The dataset used in this study is a monthly aggregate and is available on the GEE platform, derived from the ERA5 Land hourly data.

LULC data were acquired from the Copernicus Global Land Cover Services. Presently, the dataset was available from 2015 to 2019 with a pixel size of 100 m (spatial resolution). Climate data for future scenarios was also integrated into the basin assessment, and this data was acquired from the Coupled Model Intercomparison Project Phase 6 (CMIP6) dataset. These datasets were obtained from NASA-GDDP, which downscaled global circulation model data under various climate scenarios. The CMIP6 dataset available on GEE is provided on a daily scale with a spatial resolution of 27,830 m. It comprises two different scenarios "SSP245" and "SSP585" along with historical data for 34 different models. The description of the data used is illustrated in Table 4.1.

Table 4.1 Details of data used.

S. no.	Dataset category	Details of the dataset	Spatial resolution
1	Basin and watershed	WWF HydroSHEDS Basins Level 12 (Lehner, Verdin, & Jarvis, 2008)	15 arc-seconds (approx. 500 m)
2	Rivers and streams	WWF HydroRIVERS (Lehner et al., 2008)	15 arc-seconds (approx. 500 m)
3	Administrative boundary	Survey of India	
4	Digital elevation model	NASA DEM (NASA JPL, 2020)	1 arc-seconds (approx. 30 m)
5	Hydrometeorological data	ERA5 Monthly Aggregates (Copernicus Climate Change Service C3S, 2017) Terra MODIS (MOD16A2) ver. 6.1	27,830 m500 m
6	Land use/land cover	Copernicus Global Land Cover Layers (Buchhorn et al., 2020)	100 m
7	Future projected climate data	NEX-GDDP-CMIP6 (Thrasher et al., 2012)	27,830 m

4.3 **Methodology**

The basic aim of this study is to conduct a geographical appraisal of basin hydrological phenomena within the GEE environment. All the datasets used in this study are readily available on the GEE platform at no cost. GEE is a web-based platform that enables the analysis of complex environmental, remote sensing, and geographic information system datasets. The developed program serves as a valuable tool for users to achieve a fundamental understanding of the basin and its SWs. Input datasets, such as topographical data, hydrometeorological data, land cover data, and future climate data, are further analyzed, and real insights are estimated using graphs, charts, and tables. These inputs provide critical information, allowing for the characteristics of morphometric features, hypsometric analysis, the establishment of rainfall-runoff relationships, minimum and maximum temperature profile, LULC distribution, and future climatic conditions at a SW scale. The detailed methodology is given below under different sections as per the output of the data, and the methodology description is shown in Fig. 4.2.

4.3.1 **User input**

The developed program required few inputs from users to perform the analysis, which are designated as variables, viz, N, H, T, Y, and M in the script. Here, N is a 10-digit integer, representing the watershed ID for which the user wishes to conduct the analysis; H is an integer, which denotes the number of classes to be used for calculating the relative area and relative height in order to perform the

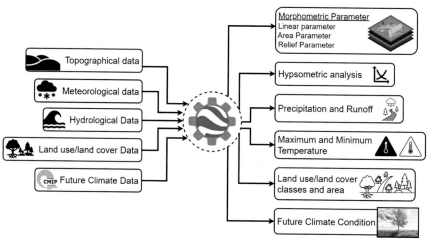

FIGURE 4.2

Process flow diagram of methodology.

hypsometric analysis; T is date time input (*yyyy:mm*), indicating the time duration for which hydrometeorological data should be visualized on a map to observe its spatial distribution; Y is the year for which LULC assessment and visualization is to be carried out. Since the LULC data is available from 2015 to 2019, the user should select a year within this timeframe, and the last one is M, which represents the model name to be used as input for future climate data analysis. It is a string input, and a list of available models can be found in the CMIP6 data portal or the GEE data catalog.

4.3.2 Morphometric analysis

The morphometric characteristics of a basin have a major impact on its hydrological behavior. These parameters, when combined with hydrological variables, provide a basis for quantifying and simulating the hydrological characteristics of the basin. In this study, morphometric variables have been classified into three groups: geometric characteristics, relief characteristics, and drainage network characteristics.

4.3.2.1 Geometric characteristics

Geometric characteristics pertain to the linear attributes of the watershed. These characteristics are derived from the watershed area, length, and perimeter. They include factors such as shape factor, form factor, compactness coefficient, and elongation ratio. These parameters provide insight into the overall geometry of the basin and how it might respond to different hydrological events. Details of the geometric parameters of the basin and formulae used are depicted in Table 4.2.

Table 4.2 Geometric characteristics.

Geometric parameters	Details and formula	Unit	References
Watershed area (a)	A = Watershed area	sq. km	Horton (1945)
Watershed perimeter (w_p)	P = Watershed perimeter	km	Horton (1945)
Basin length (l_b)	$L_b = 1.312\,a^{0.568}$	km	Nooka Ratnam, Srivastava, Venkateswara Rao, Amminedu, and Murthy (2005)
Form factor (f_f)	$F_f = a/w_p^2$	–	Horton (1932, 1945)
Shape factor (B_s)	$B_s = l_b^2/a$	–	Horton (1932)
Elongation ratio (R_e)	$R_e = 1.128\sqrt{a}/l_b$	–	Schumm (1956)
Circulatory ratio (C_r)	$C_r = 4\pi a/w_p^2$	–	Miller (1953) and Strahler (1964)
Compactness co-efficient (C_c)	$C_c = 0.2821\,w_p/a^{0.5}$	–	Gravelius (1914)

Table 4.3 Watershed relief characteristics.

Relief parameters	Details and formula	Unit	Reference
Mean watershed elevation	h_{mean}	m	Wilson (2018)
Maximum watershed elevation	h_{max}	m	Wilson (2018)
Minimum watershed elevation	h_{min}	m	Wilson (2018)
Average slope (s_{avg})	$s_{avg} = (M \times N \times 100)/A$	%	Strahler (1956)
Basin relief (r)	$r = h_{max} - h_{min}$	m	Hadley and Schumm (1961)
Relief ratio (r_r)	$r_r = r/l_b$	–	Schumm (1956)
Relative relief (r_{hp})	$r_{hp} = h \times 100/w_p$	–	Melton (1957)
Ruggedness number (r_n)	$r_n = r \times d_d$	–	Schumm (1956)
Dissection index (d_{in})	$d_{in} = r/h_{max}$	–	Gravelius (1914)
Geometric number (g_m)	$g_m = r_n/s_{avg}$	–	

M is the total length of the contour line (km); N is the contour interval (m); a is an area of the basin (sq. km); l_b is basin length (km); h is watershed relief (m); w_p is watershed perimeter (km); d_d is drainage density; f_s is stream frequency.

4.3.2.2 Relief characteristics

Relief characteristics of the basin largely depend on features such as underlying geology, geomorphology, and drainage patterns of the region. These characteristics help to identify the erosion stages of the river and land surface. Topographical information, obtained from a digital elevation model, serves as a crucial input in the identification of relief characteristics. The details of different parameters of relief characteristics and their corresponding formulae are given in Table 4.3.

4.3.2.3 Drainage network characteristics

Drainage network characteristics provide information about the structure of a stream network and how it responds to different governing factors when subjected to varying magnitudes of rainfall. These parameters offer systematic insights about the stream all the way from its origin to the outlet of the basin. For instance, the origin of a stream is referred to as a first order stream, while a stream that flows to the outlet of the basin is termed a trunk order stream. This information allows for the identification of the capacity of the first order stream, which can carry less amount of runoff compared to trunk order stream. Details of other drainage network characteristics and their corresponding formulae are given in Table 4.4.

4.3.3 Hypsometric analysis of the basin

Hypsometric characterization involves the identification of the association between the relative horizontal area and the relative height within a basin. This analysis is conducted through the construction of a curve, known as the hypsometric curve, driven by both the relative area and relative height of the basin. The curve is constructed with the abscissa, indicating the relative area and the ordinate indicating relative height, both of which are dimensionless quantities. The shape of the curve serves as an indicator of the developmental stage, with three typical shapes used for reference: (1) Inequilibrium stage, denoting a basin in the process of development

Table 4.4 Drainage network characteristics.

Drainage network parameters	Details and formula	Unit	References
Stream order (u)	Hierarchical rank	–	Strahler (1964)
Stream length (l_u)	Length of the stream of order "u"	km	Horton (1945)
Stream number (n_u)	$n_u = \sum n_1 + n_2 + \ldots n_n$	–	Horton (1945)
Mean stream length (l_{sm})	$l_{sm} = l_u/n_u$	km	Strahler (1964)
Drainage density (d_d)	$d_d = l_u/a$	–	Horton (1932, 1945)
Stream frequency (f_s)	$f_s = \sum n_u/a$	–	Horton (1932, 1945)
Drainage texture (d_t)	$d_t = \sum n_u/w_p$	–	Horton (1945)
Length of overland flow (l_f)	$l_f = 1/2d_d$	km	Horton (1945)
Bifurcation ratio (r_b)	$r_b = n_u/n_{u+1}$	–	Schumm (1956)
Mean bifurcation ratio (r_{bm})	r_{bm} = Mean bifurcation ratio	–	Strahler (1956)
Drainage intensity (d_i)	$d_i = f_s/d_d$	–	Horton (1945)
Infiltration number (l_f)	$l_f = d_d f_s$	–	Faniran (1969)

$n_1, n_2, \ldots, n_n$ is number of streams in 1st, 2nd, and nth order streams; segment of order "u"; a is area of the basin (sq. km); $\sum n_u$ is total number of streams of all orders; d_d is drainage density; f_s is stream frequency; n_{u+1} is number of segments of next higher order.

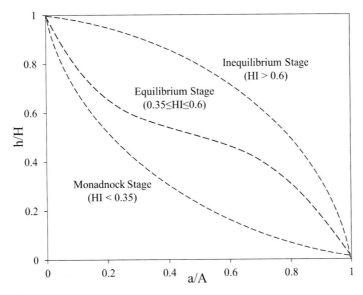

FIGURE 4.3

Typical hypsometric curve with a range of hypsometric integral (HI) values (Singh *et al.* 2008).

and referred to as the young stage; (2) equilibrium stage, representing the mature stage where development has reached a steady condition; and (3) monadnock stage, also known as the old stage, formed when isolated bodies of resistant rock from prominent hills are found above the subdued surface (Singh, Sarangi, & Sharma, 2008). Typical curves for all three stages are presented in Fig. 4.3. The phase of development of the watershed can be determined by evaluating the hypsometric integral (HI), which is the area under the curve. A typical curve with a range of values for different watershed development stages is depicted in Fig. 4.3. Specifically, a value larger than 0.6 signifies a young stage, a value between 0.35 and 0.6 suggests a mature stage, and a value less than 0.35 suggests an old stage of watershed development.

4.3.4 **Hydrometeorological condition**

Hydrometeorological data encompasses both flow and climate data. In the present study, ERA-5 Land reanalysis data is utilized for analyzing the characteristics of the basin. The study makes use of monthly aggregate data for total precipitation and runoff, which are available on the GEE platform. These monthly datasets were extracted from the year 2000 to 2020 and plotted on graphs to observe the trend of change in the data over this time period. A similar approach was adopted for assessing the minimum and maximum temperatures within the watershed. For the evaluation of actual evapotranspiration (AET) and potential evapotranspiration (PET), monthly data was obtained from Terra MODIS (MOD16A2 ver.

6.1). The spatial resolution of MODIS data is finer than the ERA-5 data; however, it is available only after the year 2021. This comprehensive data analysis approach provides valuable insights about the hydrometeorological characteristics of the basin and allows for the observation of trends and patterns over the specified timeframe.

4.3.5 Land use and land cover analysis

The Copernicus land cover data consists of a discrete classification system to depict the land cover characteristics (Buchhorn et al., 2020). This dataset comprises a total of 25 land cover classes, and each land cover class is assigned distinct pixel values. Notably, two specific values, 0 and 200, are assigned to represent unknown areas and ocean or sea, respectively. The dataset includes various color palettes to visualize different classes of the LULC. These palettes are used in the present study to visualize the land cover within the basin. The analysis involves the categorization of the basin into different LULC classes and estimating the area covered by each class in sq. km. Users have the option to visualize these land classes on the map, as well as view the area and percentage distribution of these classes through a pie chart displayed in the console. This approach provides valuable insights into the land cover properties and their spatial distribution within the basin.

4.3.6 Future climatic scenario

Interpretation of the potential effects of climate change in the future is essential, and based on this comprehension, it becomes imperative to undertake necessary measures for the watershed. The current study is centered on two climate scenarios: SSP245 and SSP585 (Thrasher, Maurer, McKellar, & Duffy, 2012). SSP245 is an advancement of RCP4.5, encompassing surplus radiative forcing of 4.5 W/m^2 by the year 2100, and represents a medium pathway for future greenhouse gas emissions. In contrast, SSP585, an advancement of RCP 8.5, integrates surplus radiative forcing of 8.5 W/m^2 by the year 2100, depicting a worst-case scenario by avoiding any precautionary measures. It is viewed as an enhancement of the CMIP5 scenarios, with CMIP6 incorporating socioeconomic factors. The present study gathered data up to the year 2050, transforming it into a monthly dataset to identify and evaluate potential future climate scenarios. This analysis will help in strategic planning and preparedness for the potential effects of climate change on the watershed.

4.4 Results

4.4.1 Morphometric analysis of basin

Watershed morphometry comprises geometric characteristics, relief characteristics, and drainage network characteristics. All these variables were evaluated using the GEE for all the SWs of the Dikrong basin. A total of 10 SWs are demarcated within the Dikrong basin and named as WS0, WS1, ..., WS9. Their morphometric properties are shown in Table 4.5.

4.4.1.1 Geometric characteristics

Geometric characteristics include various parameters such as area, perimeter, basin length, form factor, shape factor, circulatory ratio, elongation ratio, and compactness coefficient. The approximate basin area was found to be 1409.50 sq. km. The maximum and minimum area were observed in SW9 and SW1 as 192.70 and 75.10 sq. km with a perimeter of 86.98 and 41.63 km, respectively. Larger SWs are assumed to have a greater potential for runoff generation compared to smaller ones, assuming similar shapes. Basin length, representing the distance from the farthest point to the outlet, provides preliminary information about the shape of the basin. In the present study, the largest SW (SW9) has a longer basin length (26.05 km), while the smallest SW (SW1) has a shorter basin length (15.23 km). Basin length serves as an indicator of watershed shape, and a more comprehensive understanding of the shape is obtained through parameters such as form factor, shape factor, elongation ratio, circulatory ratio, and compactness coefficient. The shape of the watershed characterized as an important factor gives an understanding of the rate and duration of peak runoff resulting from rainfall. The range of form factor, shape factor, and elongation ratio in the present basin varies from 0.28 to 0.32, 3.10 to 3.52, and 0.60 to 0.64, respectively, with average values of 0.30, 3.36, and 0.62. A lower value of elongation ratio and form factor indicates an elongated shape and vice versa. SW0 and SW9 show lower values for form factor and elongation ratio, indicating their comparatively elongated shapes. Though the direction of elongation for both SW0 and SW9 is different, SW0 is elongated along the y-axis, while SW9 is elongated along the x-axis.

A higher value of form factor and elongation ratio was observed in SW1. Additionally, the maximum and minimum value of the shape factor was observed in SW9 and SW1, respectively, elaborating that the SW9 is elongated compared to other SWs. The circulatory ratio and compactness coefficient is also a measure of basin shape by comparing the basin's shape with a circle. The circulatory ratio and compactness coefficient found in this study range from 0.32 to 0.62 and 1.82 to 2.50 with a mean of 0.50 and 2.03, respectively. The value of the circulatory ratio close to 1 shows the circular shape of the watershed, and values close to 0 represent an elongated shape. On the other hand, the higher value of the compactness coefficient represents the circle shape of the watershed and vice versa. The present study shows that SW6 is close to the circular shape with higher values of both compactness coefficient and circulatory ratio, whereas SW9 has an elongated shape with the least values of both of those parameters across all the SWs. In general, the elongated shape of the watershed would produce lower and extended peak flow, whereas circular shape would produce higher and sharp peak flow.

Table 4.5 Morphometric characteristics of watershed.

Watershed characteristics	Unit	SW0	SW1	SW2	SW3	SW4	SW5	SW6	SW7	SW8	SW9
Geometric characteristics											
Watershed area (a)	sq. km	189.20	75.10	158.70	114.20	173.20	154.50	151.40	93.80	106.70	192.70
Perimeter (w_p)	km	66.95	41.63	63.48	55.11	69.00	56.74	55.60	45.91	59.68	86.98
Basin length (l_b)	m	25.78	15.25	23.33	19.35	24.51	22.97	22.71	17.30	18.62	26.05
Form factor (f_f)	–	0.28	0.32	0.29	0.31	0.29	0.29	0.29	0.31	0.31	0.28
Shape factor (B_s)	–	3.51	3.10	3.43	3.28	3.47	3.42	3.41	3.19	3.25	3.52
Elongation ratio (R_e)	–	0.60	0.64	0.61	0.62	0.61	0.61	0.61	0.63	0.63	0.60
Circulatory ratio (C_r)	–	0.53	0.54	0.49	0.47	0.46	0.60	0.62	0.56	0.38	0.32
Compactness coefficient (C_c)	–	1.94	1.92	2.01	2.06	2.09	1.82	1.80	1.89	2.31	2.50
Watershed relief characteristics											
Mean elevation (h_{mean})	m	1862.63	1472.06	1472.96	1357.05	1099.74	697.61	436.38	900.43	424.02	118.14
Maximum elevation (h_{max})	m	2893.00	2265.00	2246.00	2360.00	2107.00	2351.00	1435.00	2178.00	1329.00	399.00
Minimum elevation (h_{min})	m	1065.00	994.00	845.00	673.00	376.00	160.00	114.00	235.00	117.00	72.00
Mean slope (S_{avg})	%	21.93	22.09	20.69	23.08	27.29	22.11	19.35	28.01	20.86	4.68
Basin relief (r)	m	1828.00	1271.00	1401.00	1687.00	1731.00	2191.00	1321.00	1943.00	1212.00	327.00
Relief ratio (r_r)	–	70.91	83.34	60.05	87.18	70.62	95.39	58.17	112.31	65.09	12.55
Relative relief (r_{hp})	–	2.73	3.05	2.21	3.06	2.51	3.86	2.38	4.23	2.03	0.38
Ruggedness number (r_n)	–	859.16	686.34	784.56	995.33	917.43	1314.60	779.39	1107.51	690.84	209.28
Dissection index (d_{in})	–	0.63	0.56	0.62	0.71	0.82	0.93	0.92	0.89	0.91	0.82
Geometric number (g_m)	–	39.18	31.07	37.92	43.13	33.62	59.46	40.28	39.54	33.12	44.72

Drainage network characteristics

Stream length (l_u)	km	89.52	40.58	88.84	67.77	91.63	92.42	90.04	53.02	61.08	122.83
Stream number (n_u)	–	53.00	31.00	65.00	47.00	62.00	57.00	60.00	30.00	42.00	65.00
Mean stream length (l_{sm})	km	1.69	1.31	1.37	1.44	1.48	1.62	1.50	1.77	1.45	1.89
Drainage density (d_d)	–	0.47	0.54	0.56	0.59	0.53	0.60	0.59	0.57	0.57	0.64
Stream frequency (f_s)	–	0.28	0.41	0.41	0.41	0.36	0.37	0.40	0.32	0.39	0.34
Drainage texture (d_t)	–	0.79	0.74	1.02	0.85	0.90	1.00	1.08	0.65	0.70	0.75
Length of overland flow (l_f)	km	1.06	0.93	0.89	0.85	0.94	0.83	0.85	0.88	0.88	0.78
Mean bifurcation ratio (r_{bm})	–	0.00	0.00	0.00	0.00	0.00	0.00	0.00	0.00	0.00	0.00
Drainage intensity (d_i)	–	0.60	0.76	0.73	0.69	0.68	0.62	0.68	0.56	0.68	0.53
Infiltration number (I_f)	–	0.13	0.22	0.23	0.24	0.19	0.22	0.24	0.18	0.22	0.22
Trunk order (T_o)	–	IV	IV	IV	IV	IV	IV	IV	IV	IV	V

4.4.1.2 Relief characteristics

Relief characteristics describe the topographical feature and its variations within the watershed. It includes elevation, slope, relief, and its components. This gives preliminary information about the flow of watershed and sediment movement. The higher mean elevation values were found in the upper SWs and vice versa with a range of 118.14−1862.63 m above the MSL. The maximum elevation was found in SW0 and the lowest in SW9. The mean percent slope was also calculated for all the SWs, which varies from 4.68% to 28.01% with a mean percent slope for the entire Dikrong basin as 21.01%. Similar to lower elevation, the lowest percent slope was also found in the SW9. However, the maximum mean slope (28.01%) was found in SW7. Basin relief is the parameter that informs about the elevation variation between the outlet and watershed divide. On the other hand, the relief ratio and relative relief can be derived with the help of basin relief, basin length, and basin perimeter. Higher values of relative relief represent higher steepness and vice versa. Both infiltration capability and surface runoff potential of a basin are a function of the steepness of the basin, and higher steepness tends to create more runoff and low infiltration and vice versa. The ranges of basin relief, relief ratio, and relative relief for all the SWs of Dikrong River were found as 327−2191, 12.55−112.31, and 0.38−4.23 m, respectively. The minimum value of all three relief parameters was found in SW9, whereas the maximum value of basin relief was found in SW5, and the maximum relief ratio and relative relief were found in SW7.

The ruggedness number is calculated as the product of basin relief and drainage density, providing information about both the steepness and the length of steep terrain within the basin. The higher ruggedness number depicts the higher and longer steepness. The range of ruggedness number was found as 209.28−1314.60 in SW9 and SW5, respectively, in the present basin. The dissection index is calculated by dividing the basin relief by the absolute relief, which describes the potential of vertical erosion in the region and ranges from 0 to 1, where a value near 0 indicates no vertical erosion or flat region. The study found that the dissection index varies from 0.56 to 0.93 in SW1 and SW5, respectively. Similar observations were found with the geometric number of the watersheds. The geometric number is a function of ruggedness number and ground slope. The value of geometric numbers varies from 31.07 to 59.46 in SW1 and SW5, respectively.

4.4.1.3 Drainage network characteristics

The drainage network within the basin serves as the conduit for runoff flow, carrying it out of the basin. This characteristic reflects the stage of basin development and its capacity to generate runoff. Strahler stream orders and their corresponding lengths were obtained from the HydroRIVERS dataset, while other parameters such as stream number, mean stream length, drainage density, stream frequency, drainage texture, length of overland flow, mean bifurcation ratio, drainage intensity, and infiltration number were estimated. Except for SW9, all

SWs have a trunk order stream of IV. SW9 has a trunk order stream of V since the basin's outlet is located in this SW. The mean stream length for each SW was computed using the combined length of all streams within that subcatchment and the total count of streams it contains. The total stream length across all the SWs varied from 40.55 to 122.76 km, with an average of 79.72 km. Meanwhile, the mean stream length ranged from 1.31 to 1.89 km, with an average value of 1.55 km. The maximum and minimum value of total stream length as well as mean stream length was found in SW9 and SW1, respectively. The SW9 also has the highest stream counts across all the SWs, indicating potential for increased runoff generation.

The drainage density is computed by dividing the total length of streams by the basin area, providing an understanding about the permeability and porosity of the watershed. An area characterized by high permeability and flow resistance exhibits low drainage density, often associated with dense vegetation cover and low relief, implying higher resistance. Conversely, areas with low permeability, sparse vegetation, and higher relief exhibit higher drainage density. In the SWs of the Dikrong basin, the range of drainage density varied from 0.47 to 0.64, with the minimum and maximum values observed in SW0 and SW9, respectively. Stream frequency, representing the number of streams per unit area, indicates the runoff capacity of the watershed. Stream frequency ranged from 0.28 to 0.41, with the minimum observed in SW0 and the maximum in SW1, SW2, and SW3. Drainage texture, a product of stream frequency and drainage density, reflects the lithology, infiltration capacity, and relief aspects of the watershed. The range of drainage texture was 0.65−1.08, with SW7 and SW6 indicating the maximum and minimum values, respectively, suggesting that SW6 has higher permeability and better groundwater recharge potential compared to SW7. The length of overland flow, influencing the erosion status of the watershed, varied from 0.78 to 1.06. SW9 and SW0 showed the minimum and maximum values, respectively, with an average value of 0.89. The higher overland flow value for SW0 indicates a higher potential for sheet erosion. The bifurcation ratio, a metric indicating watershed relief and dissection, is defined as the ratio between the stream number of a particular order to the stream number of the next higher order. Calculations revealed a minimum bifurcation ratio of 1.64 for SW1 and a maximum bifurcation ratio of 5.18 for SW0, with a mean of 2.43.

Drainage intensity is calculated by dividing stream frequency by drainage density, giving a value range from 0.53 to 0.76 in the present study. SW9 showed the minimum drainage intensity value, suggesting higher infiltration and lower surface runoff. In contrast, SW1 showed the highest value, suggesting less infiltration and more surface runoff. The infiltration number, calculated as the product of drainage density and stream frequency, gives insight into the watershed's infiltration capacity. The range of infiltration number in the present study varies from 0.13 to 0.24, with higher values for both SW3 and SW6 and a lower value for SW0. This suggests that SW0 has a higher potential for infiltration and a lower capacity for runoff generation. SW9, with an infiltration number of 0.22 close to

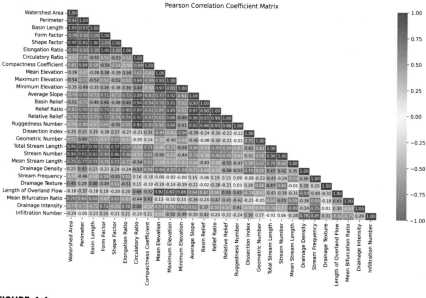

FIGURE 4.4

Pearson correlation coefficient matrix for all the morphometric variables.

the maximum value and given that the outlet of the basin is situated in this SW, indicates a significant capacity for runoff generation. Pearson correlation analysis was conducted for all morphometric parameters, and the results are presented in Fig. 4.4. The high positive correlations are denoted in green, while high negative correlations are denoted in red.

4.4.2 Hypsometric analysis

Hypsometry was assessed on the watershed scale for all the SWs as well as on the basin scale for the Dikrong River basin. The plot of all hypsometric curves is visible on the console of GEE along with the value of HI. The hypsometric curve and its HI value are shown in Fig. 4.5 and Table 4.6. The hypsometric curve and HI value of all the SWs can be compared with the hypsometric curve and HI value of the Dikrong River basin in Fig. 4.4 and Table 4.6. The range of HIs varies from 0.23 to 0.39. The maximum value of HI was observed in SW2, and the minimum value was observed in SW8. The lower HI value of SW8 shows the equilibrium stage of the SW, whereas higher HI was found in SW2. All the SWs fall under the equilibrium and monadnock stage of watershed development. A total of four SWs, namely, SW0, SW2, SW3, and SW4 are under the equilibrium stage, and SWs other than these are under the monadnock stage. The equilibrium SWs are located on upland areas with undulating topography and steep slopes. On the other hand,

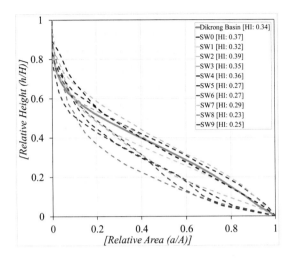

FIGURE 4.5

Hypsometric curve with the hypsometric integral.

Table 4.6 Development stage of all the subwatersheds.

Subwatershed	HI value	Stage of development
SW0	0.37	Equilibrium stage
SW1	0.32	Monadnock/old stage
SW2	0.39	Equilibrium stage
SW3	0.35	Equilibrium stage
SW4	0.36	Equilibrium stage
SW5	0.27	Monadnock/old stage
SW6	0.27	Monadnock/old stage
SW7	0.29	Monadnock/old stage
SW8	0.23	Monadnock/old stage
SW9	0.25	Monadnock/old stage
Dikrong basin	0.34	Monadnock/old stage

monadnock SWs are located near outlets or lowland areas, where terrains are relatively flattened. The HI value of the Dikrong River basin was found to be 0.34, which shows that the basin is under monadnock or old stage.

4.4.3 Hydrometeorological analysis

4.4.3.1 Total precipitation and runoff

Total rainfall and runoff data were acquired from the ERA-5 Land dataset. The long-term monthly average of total rainfall and surface runoff from 2000 to 2020 was also calculated for the entire Dikrong basin and found that the total rainfall in a year was 3663.47 mm and the resulting runoff was 899.12 mm (around 25%). The rainfall and resulting runoff usually start from March to April month and end around September to October. The long-term averages of total rainfall and surface runoff are shown in Fig. 4.6. Subbasin data of total rainfall and surface runoff reveal some interesting insights. The SW-wise patterns of total rainfall and surface runoff are shown in Fig. 4.7, representing that the SW0 receives more rainfall and generates higher runoff. The long-term mean of peak rainfall observed in SW0 is 825.41 mm for the month of June, and higher runoff was observed as 444.61 mm in the same month. The graph also reveals that the amounts of total

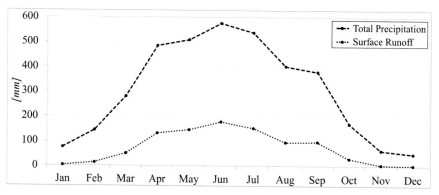

FIGURE 4.6

Long-term average of total rainfall and surface runoff from Dikrong basin.

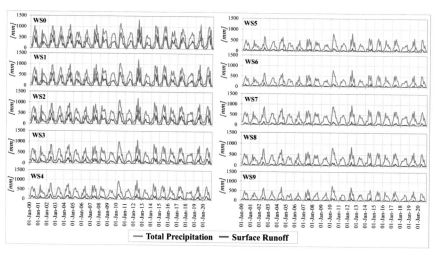

FIGURE 4.7

Total precipitation and resulting runoff in all the subwatersheds from 2000 to 2020.

rainfall and surface runoff are decreasing toward the lowland areas and found the lowest values in the SW9 among all the SWs. The range of long-term averages of total rainfall and surface runoff for SW9 was found to be 9.79–394.94 mm and 0.07–57.13 mm, respectively. However, the maximum rainfall in SW9 was observed in the month of July, whereas maximum runoff was observed in the month of June.

4.4.3.2 Minimum and maximum temperatures

The Dikrong basin experiences a range of temperature values, the minimum temperatures of long-term the basin varying from 5.44°C to 20.36°C and maximum temperatures ranging from 19.66°C to 29.41°C. The lowest minimum and maximum temperatures were recorded in January, while the highest values were observed in August and December, respectively. Fig. 4.8 illustrates the long-term average of minimum and maximum temperatures based on ERA-5 Land data. Analysis of long-term data from 2000 to 2020 reveals that the upland area (SW0) exhibits lower values for both minimum and maximum temperatures. These values tend to increase with a decrease in elevation, with SW9 demonstrating higher values for both minimum and maximum temperatures. Fig. 4.9 presents the long-term minimum and maximum temperatures for all SWs, representing a comprehensive view of the temperature variations across the basin.

4.4.3.3 Evapotranspiration

ET is a function of soil type, vegetation characteristics, and climatic conditions of the region and affects the watershed characteristics significantly. ET data was acquired from the MODIS dataset from January 2021 to September 2023. The plot of potential and actual ET is shown in Fig. 4.10. The range of potential ET varies from 10 to 60 mm, whereas the actual ET ranges from 7 to 37 mm in SW0, whose values increase with a decrease in elevation and are found to be highest in SW9 with

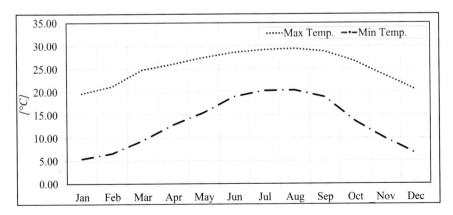

FIGURE 4.8

Long-term average of minimum and maximum temperature of Dikrong basin.

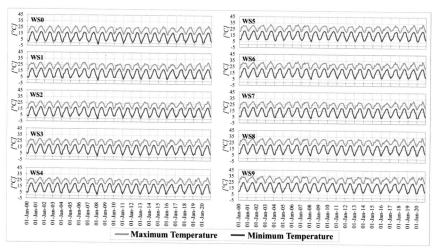

FIGURE 4.9

Maximum and minimum temperature profiles of all the subwatersheds from 2000 to 2020.

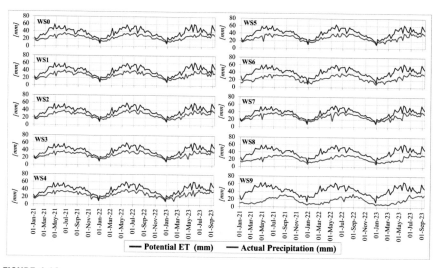

FIGURE 4.10

Potential and actual evapotranspiration of all the subwatersheds from January 2021 to September 2023.

a range of PET from 13 to 66 mm and range of AET from 5 to 35 mm. The maximum and minimum values of PET and AET are presented in Fig. 4.11A and B. In Fig. 4.10, it is evident that SW9 shows the highest maximum and minimum PET values, while simultaneously having the lowest values for both maximum and

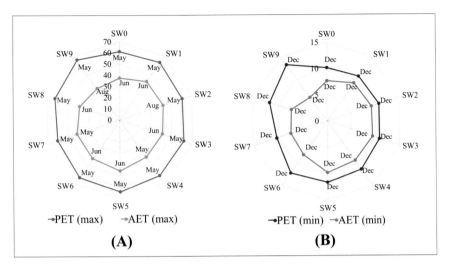

FIGURE 4.11

Evapotranspiration pattern from January 2021 to September 2023 of all the subwatersheds. (A) Maximum PET and AET and (B) minimum AET and PET.

minimum AET among all the SWs. This observation indicates that the variation between PET and AET, for both maximum and minimum values, is most pronounced in SW9. Furthermore, the analysis reveals a consistent pattern across all SWs, wherein the minimum PET and AET occur in the month of December. Conversely, the maximum PET is consistently observed in the month of May for all SWs. On the other hand, the maximum value of AET is observed in the month of May for SW4 and SW7, June for SW0, SW1, SW3, SW5, SW6, and SW8, and August for SW2 and SW9.

4.4.4 Land use and land cover pattern

LULC was acquired from the Copernicus data portal with a spatial resolution of 100 m. There are a total of 23 land cover classes in the Copernicus dataset, out of which 14 different land cover classes were found in different SWs of the Dikrong basin. A detailed description of LULC classes and area is given in Table 4.7. To understand the distribution of major land cover classes, the LULC classes were again reclassified into seven different major LULC classes. Four different close forest types were merged into a single close forest class, three different open forests were merged into one open forest class, herbaceous vegetation and bare/sparse vegetation were merged into vegetation/bareland class, and permanent water bodies were combined with herbaceous wetland as wetland/water bodies class. The distribution of reclassified land cover and their area is given in Fig. 4.12. The analysis represented that all the SWs are dominated by close forest except SW9

Table 4.7 Land use and land cover (LULC) statistics of all the subwatersheds.

S. no.	LULC classes	Area (sq. km)									
		SW0	SW1	SW2	SW3	SW4	SW5	SW6	SW7	SW8	SW9
		21.7	9.59	7.13	10.64	12.12	4.24	2.39	4.36	2.75	0.12
1	Closed forest, evergreen needle leaf	164.79	61.24	146.04	100.31	154.01	123.75	114.27	82.48	84.98	36.86
2	Closed forest, evergreen broad leaf	N/P	N/P	0.02	N/P	N/P	0.02	N/P	N/P	N/P	N/P
3	Closed forest, deciduous broad leaf	0.12	0.36	0.24	0.42	1.05	1.36	2.84	2.06	1.55	1.47
4	Closed forest, not matching any of the other definitions	0.07	0.09	0.03	0.05	0.04	N/P	0.02	N/P	N/P	N/P
5	Open forest, evergreen needle leaf	0.61	0.34	0.96	0.55	1.08	1.72	1.57	1.38	1.14	0.99
6	Open forest, evergreen broad leaf	1.08	1.69	2.71	1.79	3.58	9.91	8.62	2.87	7.84	15.27
7	Open forest, not matching any of the other definitions	N/P	0.02	0.01	N/P	N/P	0.18	0.13	0.09	0.14	0.55
8	Shrubs	N/P	0.09	0.01	N/P	N/P	0.09	0.91	0.02	0.18	7.21
9	Herbaceous vegetation	0.41	1.5	0.74	0.25	1	2.77	7.94	0.19	5.23	107.28
10	Cultivated and managed vegetation/agriculture	0.11	0.06	0.43	0.04	0.05	10.17	11.86	0.14	2.01	12.83
11	Urban/built-up	N/P	N/P	N/P	N/P	N/P	N/P	N/P	N/P	N/P	0.97
12	Bare/sparse vegetation	N/P	N/P	N/P	N/P	N/P	0.01	N/P	N/P	N/P	2.15
13	Permanent water bodies	N/P	N/P	0.09	N/P	0.02	0.01	0.64	N/P	0.73	6.63
14	Herbaceous wetland										

N/P, Not present.

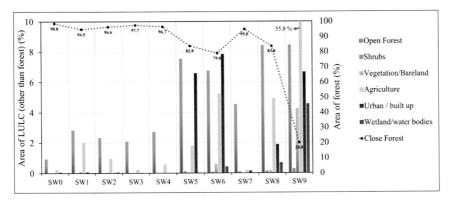

FIGURE 4.12

Land use and land cover (LULC) pattern of all the subwatersheds for the year 2019.

where agriculture is more dominated. The highest close forest is observed in SW0 (186.61 sq. km), whereas the least is in SW9 (38.45 sq. km). The area under close forest shows a decreasing trend as the elevation decreases and found very less forest cover area in SW9 which is in a lowland area where the outlet of the basin is located. The basin upland area does not have much water resources such as lakes, ponds, reservoirs, etc. However, few evidences of spring were found in that region. The lowland area has much more water resources available and is highest in the SW9. The highest urban/built-up was found in the SW9 followed by SW6 and SW5. The total built-up area in the basin is 37.3 sq. km, and 93% of it was found in the SW9, SW6, and SW5.

4.4.5 Future climate scenario of the basin

To measure the future climatic situation of the basin, projected climate data needs to be assessed. The future climatic condition gives an idea of what actions should be taken to combat the effect of climate change in the basin. The future projected precipitation and maximum and minimum temperatures data of the basin were acquired and converted to monthly data for each SW. The plots of future projected precipitation and temperature are shown in Figs. 4.13–4.15. Linear trend lines were drawn for both precipitation and temperature data, and equations were derived. The value of the slope of the trend line gives a preliminary understanding about the trend of precipitation and temperature in the future. The slope values of precipitation found in the study are 0.0006 and 0.0007 for almost all the SWs except WS2 (0.002) for the SSP245 scenario, whereas these values are slightly higher in SSP585 (varies from 0.0006 to 0.0021). Similar observations were made for both maximum and minimum temperatures. The minimum temperature shows a rising trend with the decrease in elevation and larger values of minimum temperature are found in SW9. All the SWs show a constant slope value of 0.0001 for

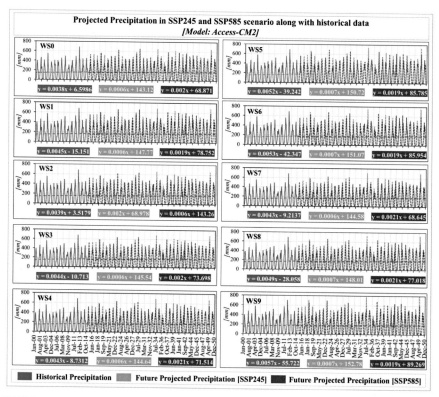

FIGURE 4.13

Projected monthly precipitation in SSP245 and SSP585 scenarios from 2015 to 2050 along with historical data from 2000 to 2014.

both maximum and minimum temperatures. With this, it can be said that the temperature is rising at a constant rate from 2015 to 2050 under both SSP425 and SSP585 scenarios. The plot of maximum temperature shows that the SW5, SW6, and SW8 have increasing maximum temperature values compared to all other SWs. The reason behind the increase in the maximum temperature in these SWs could be an increase in urban and built-up areas along with global climate change.

4.5 Discussion

The study aims to analyze the geographical appraisal of basin hydrology using GEE (Google Earth Engine, n.d.). The GEE environment is a seamless cloud-based platform that can calculate the different characteristics of the basin in few minutes (Şener & Arslanoğlu, 2023). All these characteristics are important in terms of

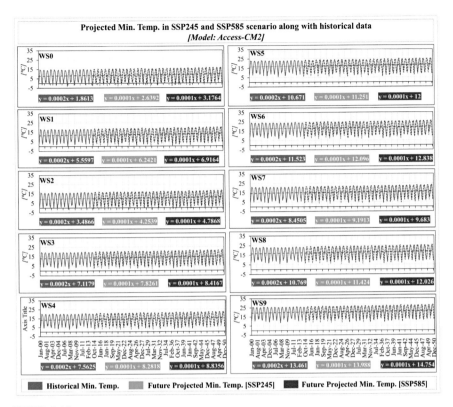

FIGURE 4.14

Projected monthly min temp. in SSP245 and SSP585 scenarios from 2015 to 2050 along with historical data from 2000 to 2014.

planning watershed management. The current study was conducted on the Dikrong River basin, which drains in the Subansiri River in the downstream and meets Brahmaputra River at the end. The basin is prone to flood inundation in the downstream (Bhadra, Choudhury, & Kar, 2011). The basin has 10 SWs and covers an area of around 1409.50 sq. km, which extends from the Papum Pare district of Arunachal Pradesh to the North Lakhimpur district of Assam. The region falls under the Eastern Himalayan Region and covers various land cover classes from dense forest in the upstream to agricultural lands in the downstream. The study area also has major settlements including the capital of Arunachal Pradesh Itanagar, Naharlagun, Doimukh, Nirjuli, and Lakhimpur City of Assam.

The present study of geographical appraisal of basin hydrology comprises a morphometric assessment of the basin, hydroclimatic characteristics, LULC variations, and future climatic conditions. The morphometric assessment of the basin was categorized into three classes, viz., geometric characteristics, relief characteristics, and drainage network characteristics. The geometric parameter of the basin

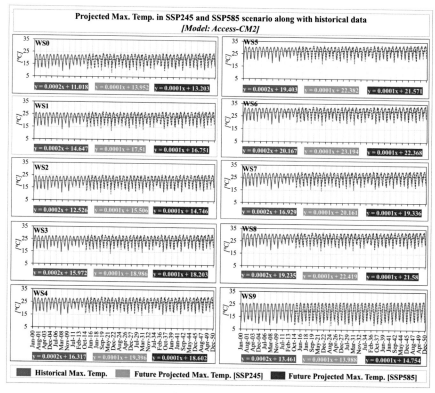

FIGURE 4.15

Projected monthly max. temp. in SSP245 and SSP585 scenarios from 2015 to 2050 along with historical data from 2000 to 2014.

estimates the geometry of the watershed and gives an idea about the shape and size of the watershed (Sassolas-Serrayet, Cattin, & Ferry, 2018). The size and shape of the watershed are directly related to the potential water yield of the watershed and the comparative duration of generating peaks from that watershed (Harlin, 1984). The shape of the watershed can be broadly described as elongated and circular; elongated watershed generates a flatter hydrograph than a circular shape (Abrishamchi, Dashti, Alamdari, & Salavitabar, 2011). Similarly, a larger watershed size has the potential to generate more discharge than the smaller watershed. In the present study, SW6 shows a circular shape compared to all other SWs with an area of 154.60 sq. km. However, the largest area of the SW was estimated as 192.70 sq. km for SW9. Watershed relief characteristics consist of topographical variation within the watershed and indicate the flow direction of the water. The difference in mean elevation of the basin was significant with a value of about 1745 m among all the SWs. The highest elevation was observed in SW0, and the lowest was in SW9. This variation in the elevation gives a high value of basin relief in almost all the SWs. The maximum value of basin relief was observed in SW6 as 2191 m. Higher basin relief leads to a high mean slope. The average mean slope for all the SWs was observed as 21% with a maximum of 28% in SW7 and a minimum of 4.7% in SW0. The impact of these variations in the topography can be seen in the LULC pattern of the watershed. Relatively falter land was utilized more for agriculture and settlement purposes, whereas most of the SWs were occupied with a dense forest area. The LULC change study also revealed that the urban expansion and increase in agriculture activities in this basin (Mandal, Bandyopadhyay, & Bhadra, 2023).

Basin drainage network characteristics are an essential factor in determining the potential to generate discharge of the basin (Altaf, Meraj, & Romshoo, 2013; Singh, Kanhaiya, Singh, & Chaubey, 2018). In this study, SW2 and SW9 have the highest number of streams; however, SW9 gives a higher value of total stream length, which leads to higher drainage density. This signifies that SW9 has more water resources, and evidence of findings is visible in the land cover statistics as well with the highest percentage area of wetland/water bodies class among all the SWs. The drainage network analysis also reveals that the length of overland flow is high for SW0 where the land topographical variation is high with a high mean bifurcation ratio and very low infiltration number. This illustrates that the SW0 is very poor in infiltration and rapid overland flow is occurring, which may lead to high soil erosion. Person correlation analysis was also conducted between the morphometric variables and found that most of the geometric parameters are highly correlated to each other (either positive or negative) except circulatory ratio and compactness coefficient. These two variables have a comparatively low correlation to other geometric variables; however, they are highly correlated to each other. Mean slope, basin relief, relief ratio, and relative relief are highly correlated to each other showing more than 0.8 correlation coefficient values. Correlation among the drainage network characteristics is poor except for stream number, length of overland flow, drainage intensity, and infiltration number; these four variables have significant correlation with total stream length, drainage density, mean stream length, and stream frequency, respectively. Relief ratio and relative relief of relief characteristics show high correlation coefficient values with watershed perimeter and compactness coefficient of geometric characteristics. The characteristics of the drainage network, including total stream length and stream number, exhibit strong correlations with geometric parameters such as watershed area, perimeter, length, form factor, shape factor, and elongation ratio. The correlation between relief characteristics and drainage network characteristics is not significant. These parameters along with hypsometric analysis give an idea of the watershed development stage. Hypsometric analysis is the primary measure to investigate the erosion status of the watershed (Sharma, Gajbhiye, Tignath, & Patil, 2018; Shekar & Mathew, 2022). SWs of the present study are under equilibrium and monadnock stages, which is also referred to as the mature and old stage.

Hydroclimatic characteristics and LULC pattern are important information about the basin. Hydroclimatic characteristics such as precipitation, runoff, temperature, and ET are considered in this study. These variables are affected by the LULC pattern of the region (Berihun et al., 2019; Kumar, Denis, Kundu, Joshi, & Suryavanshi, 2022). The precipitation amount and resulting runoff are dependent on forest or vegetation coverage of the area; similarly, temperature and ET are also affected by the vegetation, settlements, agriculture, and waterbody areas (Ding et al., 2022; Li, Li, Di, Niu, & Zhang, 2022; Qiu et al., 2013; Yu et al., 2015). The health status of crops and other vegetation, which is driven by available water resources in the region, leads to the potential of actual ET (Zhang et al., 2020). The present study shows that the WS0 receives more precipitation, resulting of that generates more runoff compared to all other SWs. On the other hand, SW9 receives less precipitation and generates very runoff. The land cover conditions of WS0 and WS9 are also very different from each other, WS0 has more than 98% of close forest cover, and WS9 is dominated by agricultural land with more than 55% of agricultural land. The impact of these characteristics can be seen in both the minimum and maximum temperature profiles of the SWs. Both minimum and maximum temperatures are low in SW0 compared to SW9. Temperature is increasing with the decrease in mean elevation of the SWs. These trends are continuing to future climate scenarios as well. Both minimum and maximum temperature are increasing with the elevation and the rate of increase in temperature on temporal scale is also constant. The future climatic conditions of the basin can be helpful in the preparation of a watershed management plan (Worku, Teferi, Bantider, & Dile, 2020).

4.6 **Limitations of the study**

The present study involves a geographical appraisal of basin hydrology using the GEE environment. The developed program is designed to estimate various parameters, including morphometric analysis, hydroclimatic characteristics, LULC pattern, and future climate scenarios, encompassing precipitation and temperature. It is essential to observe that the present study focuses on estimating these parameters for one SW at a time. To estimate all these parameters for multiple SWs, users will need to manually iterate the process by providing the serial number of each SW ID. Additionally, few more user inputs also need to be given. The details of the users' inputs are given in the user input section under methodology. This input from the user is essential for conducting a comprehensive assessment of basin hydrology and its associated components in the study region.

4.7 **Conclusion**

The comprehensive assessment of basin geographical appraisal including morphometric analysis, hydroclimatic condition, LULC pattern, and future climate scenario are crucial to know before going to implement any watershed management activity. Watershed management activities involves optimum utilization of natural resources through a combination of engineering and social measures. Parameters derived from basin appraisal offer valuable insights into the overall condition of the watershed. Morphometric analysis and LULC patterns explain the physical characteristics, while hydroclimatic conditions provide essential information about the climatic influences on the hydrology of the watershed. Additionally, understanding the future climatic scenario allows for a forward-looking perspective on potential changes. Traditionally, evaluating these characteristics has been a complex and resource-intensive process, often requiring expertise and the use of multiple software platforms. The advent of GEE represents a significant advancement in spatial data analysis and remote sensing applications. Leveraging cloud-based resources, GEE enables the seamless execution of complex calculations and analyses by directly accessing cloud-based datasets. The proposed approach streamlines the assessment of morphometric parameters, hydroclimatic conditions, LULC patterns, and future climate scenarios simultaneously for any given basin, offering a more efficient and integrated approach. This analysis, facilitated by GEE, holds the potential to aid in prioritizing watersheds based on vulnerability and developing watershed management plans that incorporate considerations for future climate effects. The future prospect of the current study is to build a web-based application for basin appraisal, which can enable the user to get the information about the basin by a single mouse click. The web-based application can be prepared in the GEE platform itself by providing the provision to add user defined datasets additionally.

References

Abrishamchi, A., Dashti, M., Alamdari, N., & Salavitabar, A. (2011). A GIS-Google earth based approach to estimating the flood damage function in large river basins. In *Proceedings of the 2011 world environmental and water resources Congress 2011: Bearing knowledge for sustainability* (pp. 3811–3821). 10.1061/41173(414)399.

Altaf, F., Meraj, G., & Romshoo, S. A. (2013). Morphometric analysis to infer hydrological behaviour of Lidder watershed, Western Himalaya, India. *Geography Journal, 2013*, 1–14. Available from https://doi.org/10.1155/2013/178021.

Berihun, M. L., Tsunekawa, A., Haregeweyn, N., Meshesha, D. T., Adgo, E., Tsubo, M., . . . Ebabu, K. (2019). Hydrological responses to land use/land cover change and climate variability in contrasting agro-ecological environments of the Upper Blue Nile basin, Ethiopia. *Science of the Total Environment, 689*, 347–365. Available from https://doi.org/10.1016/j.scitotenv.2019.060.338, http://www.elsevier.com/locate/scitotenv.

Bhadra, A., Choudhury., & Kar, D. (2011). Flood hazard mapping in Dikrong Basin of Arunachal Pradesh (India). *International Journal of Environmental, Chemical, Ecological, Geological and Geophysical Engineering, 5*(12), 861–866.

Buchhorn, M., Lesiv, M., Tsendbazar, N.-E., Herold, M., Bertels, L., & Smets, B. (2020). Copernicus global land cover layers—Collection 2. *Remote Sensing, 12*(6). Available from https://doi.org/10.3390/rs12061044.

Chaponnière, A., Boulet, G., Chehbouni, A., & Aresmouk, M. (2008). Understanding hydrological processes with scarce data in a mountain environment. *Hydrological Processes, 22*(12), 1908−1921. Available from https://doi.org/10.1002/hyp.6775.

Chen, T., Ye, Y., Yang, K., Zhang, X., Ao, T., & Rathnayake, U. (2023). Study on the impact of future climate change on extreme meteorological and hydrological elements in the upper reaches of the Minjiang River. *Advances in Meteorology, 2023*, 1−18. Available from https://doi.org/10.1155/2023/9458678.

Copernicus Climate Change Service (C3S). (2017). ERA5: Fifth generation of ECMWF atmospheric reanalyses of the global climate. In *Copernicus climate change service climate data store (CDS)*. Copernicus Climate Change Service.

Ding, B., Zhang, Y., Yu, X., Jia, G., Wang, Y., Wang, Y., . . . Li, Z. (2022). Effects of forest cover type and ratio changes on runoff and its components. *International Soil and Water Conservation Research, 10*(3), 445−456. Available from https://doi.org/10.1016/j.iswcr.2022.010.006, http://www.keaipublishing.com/en/journals/international-soil-and-water-conservation-research/.

Faniran, A. (1969). The Index of drainage intensity: A provisional new drainage factor. *Australian Journal of Science, 31*, 328−330.

Feng, S., Liu, J., Zhang, Q., Zhang, Y., Singh, V.P., Gu, X., Sun, P. (2020). A global quantitation of factors affecting evapotranspiration variability. Journal of Hydrology. **58**4. Available from https://doi.org/10.1016/j.jhydrol.2020.124688.

Gao, H., Sabo, J. L., Chen, X., Liu, Z., Yang, Z., Ren, Z., & Liu, M. (2018). Landscape heterogeneity and hydrological processes: A review of landscape-based hydrological models. *Landscape Ecology, 33*(9), 1461−1480. Available from https://doi.org/10.1007/s10980-018-0690-4, http://www.springerlink.com/content/103025/.

Google Earth Engine. (n.d.). *Link of the developed script for geographical appraisal of basin hydrology using GEE*. https://code.earthengine.google.com/b32e30834bee1-b24c7d8ab9ff9c33305.

Gravelius, H. (1914). Grundrifi der gesamten Gewcisserkunde. In *Band I: Flufikunde (compendium of hydrology)*.

Hadley, R. F., & Schumm, S. A. (1961). *Sediment sources and drainage basin characteristics in upper Cheyenne river basin water-supply paper*. Washington, DC: US Geological Survey. Available from 10.3133/wsp1531.

Harlin, J. M. (1984). Watershed morphometry and time to hydrograph peak. *Journal of Hydrology, 67*(1−4), 141−154. Available from https://doi.org/10.1016/0022-1694(84)90238-5.

Horton, R. (1932). Drainage-basin characteristics. *Eos, Transactions American Geophysical Union, 13*(1), 350−361. Available from https://doi.org/10.1029/tr013i001p00350.

Horton, R. (1945). Erosional development of streams and their drainage basins; hydrophysical approach to quantitative morphology. *Geological Society of America Bulletin, 56*(3), 10.1130/0016−7606(1945)56[275:EDOSAT]2.0.CO;2.

Huda, M. E. A., & Singh, S. (2012). Assessment of runoff in the high humid foot-hill areas of Arunachal Himalayas using Thornthwaite equation. *International Journal of GEOMATE, 3*(2), 397−401. Available from https://geomatejournal.com/geomate/article/view/1649.

NASA JPL. (2020). *NASADEM Merged DEM Global 1 arc second V001 [Data set NASA EOSDIS land processes DAAC]*. Available from https://doi.org/10.5067/MEaSUREs/NASADEM/NASADEM_HGT.001.

Kottagoda, S. D., & Abeysingha, N. S. (2017). Morphometric analysis of watersheds in Kelani river basin for soil and water conservation. *Journal of the National Science Foundation of Sri Lanka*, *45*(3). Available from https://doi.org/10.4038/jnsfsr.v45i3.8192.

Kumar, M., Denis, D. M., Kundu, A., Joshi, N., & Suryavanshi, S. (2022). Understanding land use/land cover and climate change impacts on hydrological components of Usri watershed, India. *Applied Water Science*, *12*(3). Available from https://doi.org/10.1007/s13201-021-01547-6, http://www.springer.com/earth + sciences + and + geography/hydrogeology/journal/13201.

Lehner, B., Verdin, K., & Jarvis, A. (2008). New global hydrography derived from spaceborne elevation data. *Eos, Transactions American Geophysical Union*, *89*(10), 93−94. Available from https://doi.org/10.1029/2008eo100001.

Li, X., Li, Y., Di, S., Niu, Y., & Zhang, C. (2022). Evapotranspiration and land surface temperature of typical urban green spaces in a semi-humid region: Implications for green management. *Frontiers in Environmental Science*, *10*. Available from https://doi.org/10.3389/fenvs.2022.977084, journal.frontiersin.org/journal/environmental-science.

Luo, Y., Gao, P., & Mu, X. (2021). Influence of meteorological factors on the potential evapotranspiration in Yanhe River Basin, China. *Water*, *13*(9). Available from https://doi.org/10.3390/w13091222.

LUP NBSS. (2004). *The soils of Arunachal Pradesh at R.F. 1: 250,000, prepared for Government of Arunachal Pradesh, Itanagar by the National Bureau of Soil Survey and Land Use Mapping*. Regional Centre.

Mahala, A. (2020). The significance of morphometric analysis to understand the hydrological and morphological characteristics in two different morpho-climatic settings. *Applied Water Science*, *10*(1). Available from https://doi.org/10.1007/s13201-019-1118-2, https://www.springer.com/journal/13201.

Mandal, S., Bandyopadhyay, A., & Bhadra, A. (2023). Dynamics and future prediction of LULC on Pare River basin of Arunachal Pradesh using machine learning techniques. *Environmental Monitoring and Assessment*, *195*(6). Available from https://doi.org/10.1007/s10661-023-11280-z, https://www.springer.com/journal/10661.

Melton, M. A. (1957). An analysis of the relations among elements of climate. *Surface Properties and Geomorphology*, *11*. Available from https://doi.org/10.21236/ad0148373.

Miller, V. C. (1953). A quantitative geomorphic study of drainage basin characteristics in the Clinch Mountain area, Virginia and Tennessee. *Project NR*, 389−402.

Nooka Ratnam, K., Srivastava, Y. K., Venkateswara Rao, V., Amminedu, E., & Murthy, K. S. R. (2005). Check dam positioning by prioritization of micro-watersheds using SYI model and morphometric analysis—Remote sensing and GIS perspective. *Journal of the Indian Society of Remote Sensing*, *33*(1), 25−38. Available from https://doi.org/10.1007/bf02989988.

Owuor, S. O., Butterbach-Bahl, K., Guzha, A. C., Rufino, M. C., Pelster, D. E., Díaz-Pinés, E., & Breuer, L. (2016). Groundwater recharge rates and surface runoff response to land use and land cover changes in semi-arid environments. *Ecological Processes*, *5*(1). Available from https://doi.org/10.1186/s13717-016-0060-6.

Qi, J., Lee, S., Zhang, X., Yang, Q., McCarty, G. W., & Moglen, G. E. (2020). Effects of surface runoff and infiltration partition methods on hydrological modeling: A comparison of four schemes in two watersheds in the Northeastern US. *Journal of Hydrology*, *581*. Available from https://doi.org/10.1016/j.jhydrol.2019.124415, http://www.elsevier.com/inca/publications/store/5/0/3/3/4/3.

Qiu, G.-yu, Li, H.-yong, Zhang, Q.-tao, Chen, W., Liang, X.-jian, & Li, X.-ze (2013). Effects of evapotranspiration on mitigation of urban temperature by vegetation and urban agriculture. *Journal of Integrative Agriculture*, *12*(8), 1307−1315. Available from https://doi.org/10.1016/s2095-3119(13)60543-2.

Rashid, H., Yang, K., Zeng, A., Ju, S., Rashid, A., Guo, F., & Lan, S. (2022). Predicting the hydrological impacts of future climate change in a humid-subtropical watershed. *Atmosphere*, *13*(1). Available from https://doi.org/10.3390/atmos13010012.

Rehana, S., & Monish, N. T. (2021). Impact of potential and actual evapotranspiration on drought phenomena over water and energy-limited regions. *Theoretical and Applied Climatology*, *144*(1−2), 215−238. Available from https://doi.org/10.1007/s00704-021-03521-3.

Samal, D. R., & Gedam, S. (2021). Assessing the impacts of land use and land cover change on water resources in the Upper Bhima river basin, India. *Environmental Challenges*, *5*. Available from https://www.journals.elsevier.com/environmental-challenges, 10.1016/j.envc.2021.100251.

Sassolas-Serrayet, T., Cattin, R., & Ferry, M. (2018). The shape of watersheds. *Nature Communications*, *9*(1). Available from https://doi.org/10.1038/s41467-018-06210-4, http://www.nature.com/ncomms/index.html.

Schumm, S. A. (1956). Evolution of drainage systems and slopes in badlands at Perth Amboy, New Jersey. *Bulletin of the Geological Society of America*, *67*(5), 597−646. Available from https://doi.org/10.1130/0016-7606(1956)67[597:EODSAS]2.0.CO;2, http://gsabulletin.gsapubs.org/content/by/year.

Şener, M., & Arslanoğlu, M. C. (2023). Morphometric analysis in Google Earth Engine: An online interactive web-based application for global-scale analysis. *Environmental Modelling and Software*, *162*. Available from https://doi.org/10.1016/j.envsoft.2023.105640, http://www.elsevier.com/inca/publications/store/4/2/2/9/2/1.

Sharma, S. K., Gajbhiye, S., Tignath, S., & Patil, R. J. (2018). *Hypsometric analysis for assessing erosion status of watershed using geographical information system* (pp. 263−276). Springer Science and Business Media LLC. Available from 10.1007/978-981-10-5801-1_19.

Shekar, P. R., & Mathew, A. (2022). Evaluation of morphometric and hypsometric analysis of the Bagh River Basin using remote sensing and geographic information system techniques. *Energy Nexus*, *7*. Available from https://doi.org/10.1016/j.nexus.2022.100104.

Singh, O., Sarangi, A., & Sharma, M. C. (2008). Hypsometric integral estimation methods and its relevance on erosion status of North-Western Lesser Himalayan watersheds. *Water Resources Management*, *22*(11), 1545−1560. Available from https://doi.org/10.1007/s11269-008-9242-z.

Singh, S., Kanhaiya, S., Singh, A., & Chaubey, K. (2018). Drainage network characteristics of the Ghaghghar River Basin (GRB), Son Valley, India. *Geology, Ecology, and Landscapes*, *3*(3), 159−167. Available from https://doi.org/10.1080/24749508.2018.1525670.

Smith, K. (1984). *Hydrology* (pp. 296−303). Springer Science and Business Media LLC. Available from 10.1007/0-387-30842-3_35.

Sreedevi, P. D., Subrahmanyam, K., & Ahmed, S. (2005). The significance of morphometric analysis for obtaining groundwater potential zones in a structurally controlled terrain. *Environmental Geology*, *47*(3), 412−420. Available from https://doi.org/10.1007/s00254-004-1166-1.

Stewardson, M. J., Walker, G., & Coleman, M. (2021). *Hydrology of the Murray−Darling basin* (pp. 47−73). Elsevier BV. Available from 10.1016/b978-0-12-818152-2.00003-6.

Stoyanova, J. S., Georgiev, C. G., & Neytchev, P. N. (2023). Drought monitoring in terms of evapotranspiration based on satellite data from Meteosat in areas of strong land−Atmosphere coupling. *MDPI, Bulgaria Land*, *12*(1). Available from https://doi.org/10.3390/land12010240, http://www.mdpi.com/journal/land/.

Strahler, A. N. (1956). Quantitative slope analysis. *Geological Society of America Bulletin*, *67*(5), 10.1130/0016-7606(1956)67[571:QSA]2.0.CO;2.

Strahler, A. N. (1964). *Quantitative geomorphology of drainage basins and channel networks* (pp. 39−76). McGraw Hill.

Thapa, P. (2022). *The relationship between land use and climate change: A case study of Nepal*. IntechOpen. Available from https://doi.org/10.5772/intechopen.98282.

Thrasher, B., Maurer, E. P., McKellar, C., & Duffy, P. B. (2012). Technical Note: Bias correcting climate model simulated daily temperature extremes with quantile mapping. *Hydrology and Earth System Sciences*, *16*(9), 3309−3314. Available from https://doi.org/10.5194/hess-16-3309-2012.

WWF. (2021). *HydroSHEDS*. Available from: https://www.hydrosheds.org/page/hydrorivers (accessed 05.10.23).

Wilson, J. P. (2018). *Environmental applications of digital terrain modeling*. Wiley. Available from https://doi.org/10.1002/9781118938188.

Worku, G., Teferi, E., Bantider, A., & Dile, Y. T. (2020). Prioritization of watershed management scenarios under climate change in the Jemma sub-basin of the Upper Blue Nile Basin, Ethiopia. *Ethiopia Journal of Hydrology: Regional Studies*, *31*. Available from https://doi.org/10.1016/j.ejrh.2020.100714, https://www.journals.elsevier.com/journal-of-hydrology-regional-studies.

Wu, L., Peng, M., Qiao, S., & Ma, Xy (2018). Effects of rainfall intensity and slope gradient on runoff and sediment yield characteristics of bare loess soil. *Environmental Science and Pollution Research*, *25*(4), 3480−3487. Available from https://doi.org/10.1007/s11356-017-0713-8, http://www.springerlink.com/content/0944-1344.

Yan, Y., Dai, Q., Yuan, Y., Peng, X., Zhao, L., & Yang, J. (2018). Effects of rainfall intensity on runoff and sediment yields on bare slopes in a karst area, SW China. *Geoderma*, *330*, 30−40. Available from https://doi.org/10.1016/j.geoderma.2018.050.026, http://www.elsevier.com/inca/publications/store/5/0/3/3/3/2.

Yang, D., Yang, Y., & Xia, J. (2021). Hydrological cycle and water resources in a changing world: A review. *Geography and Sustainability*, *2*(2), 115−122. Available from https://doi.org/10.1016/j.geosus.2021.050.003, http://www.journals.elsevier.com/geography-and-sustainability.

Yin, Y. H., Wu, S. H., & Dai, E. F. (2010). Determining factors in potential evapotranspiration changes over China in the period 1971−2008. *Chinese Science Bulletin*, *55*(29), 3329−3337. Available from https://doi.org/10.1007/s11434-010-3289-y.

Yu, P., Wang, Y., Coles, N., Xiong, W., Xu, L., & Piao, S. (2015). Simulation of runoff changes caused by cropland to forest conversion in the Upper Yangtze River Region, SW China. *PLoS One*, *10*(7). Available from https://doi.org/10.1371/journal.pone.0132395.

Zhang, F., Liu, Z., Zhangzhong, L., Yu, J., Shi, K., & Yao, L. (2020). Spatiotemporal distribution characteristics of reference evapotranspiration in Shandong Province from 1980 to 2019. *Water*, *12*(12). Available from https://doi.org/10.3390/w12123495.

Arsenic enrichment in the groundwater mechanism through hydrogeochemical control, mobilization, and sorption in the Lower Gangetic Basin in West Bengal

Kamalesh Sen, Debojyoti Mishra and Naba Kumar Mondal

Environmental Chemistry Laboratory, Department of Environmental Science, The University of Burdwan, Burdwan, West Bengal, India

5.1 Introduction

Arsenic (As) contamination of groundwater has emerged as a global environmental crisis, affecting numerous regions worldwide. However, it is most notably pronounced in the Lower Gangetic Basin in West Bengal, where millions of people are at risk of exposure to elevated As levels through their drinking water source (Farooq et al., 2011; Mishra, Chakrabortty, Sen, Pal, & Mondal, 2023; Mishra, Sen, Mondal, Kundu, & Mondal, 2022; Mukherjee et al., 2018; Rahman et al., 2014). This review aims to elucidate the multifaceted processes underlying As enrichment in groundwater in this region, shedding light on the hydrogeochemical controls, mobilization mechanisms, and sorption processes involved (Chakraborty, Mukherjee, & Ahmed, 2015; Farooq et al., 2011; Khan & Rai, 2023).

As contamination in the Lower Gangetic Basin is a matter of immense concern, primarily due to its severe health implications for the population relying on groundwater for drinking (Khan, Haque, et al., 2023; Khan, Musahib, Vishwakarma, Rai, & Jahan, 2023; Panda, Tripathi, D.j, & Sharma, 2022). Millions of people in this region face the daily risk of exposure to As concentrations that far exceed permissible levels (Banerjee et al., 2023; Chakraborty, Mishra, & Mukherjee, 2022; Chakraborty, Mukherjee, & Ahmed, 2022). Long-term exposure to As has been associated with a spectrum of health concerns, encompassing skin lesions, cardiovascular conditions, and multiple types of cancer, presenting a considerable public health emergency (Hong, Song, & Chung,

Applications of Geospatial Technology and Modeling for River Basin Management.
DOI: https://doi.org/10.1016/B978-0-443-23890-1.00005-0

2014; Martinez, Vucic, Becker-Santos, Gil, & Lam, 2011). In this review, the role of hydrogeochemical controls in shaping As contamination in the Lower Gangetic Basin has been explored. The presence of reducing conditions, high organic matter content, and specific geological characteristics have been identified as critical factors. These controls influence the redox state of As, promoting the more toxic As(III) form under reducing conditions, which is more mobile in groundwater (Mishra et al., 2023; Panda et al., 2022; Rahman et al., 2014). Understanding these hydrogeochemical controls is essential for pinpointing areas at higher risk of contamination (Shaji et al., 2021).

As contamination in the groundwater of the Lower Gangetic Basin in West Bengal is often associated with specific rocks and minerals that influence the release and mobilization of As (Mazumder et al., 2010; Mukherjee, Verma, Gupta, Henke, & Bhattacharya, 2014). The primary geological factors contributing to pollution in the lower Gangetic basin region stem from the predominant geological formation, characterized by alluvial sediments comprising a blend of sand, silt, clay, and organic material (Banerji, Kalawapudi, Salana, & Vijay, 2019; Ghosh & Donselaar, 2023). These sediments act as the primary aquifer materials in the region. As contamination often occurs when As-rich minerals are present within these sediments. On the other hand, iron oxides and hydroxides, such as goethite and hematite, are commonly found in the aquifer sediments of the region (Mukherjee et al., 2014). These minerals have a strong affinity for As and can adsorb it effectively, reducing its mobility. However, under reducing conditions (anaerobic), these iron oxides can be reduced, potentially releasing As into the groundwater. Manganese oxides, like pyrolusite, are also present in the aquifer sediments (Anawar et al., 2011). Similar to iron oxides, manganese oxides can adsorb As. Under reducing conditions, manganese oxides may undergo reduction, impacting their capacity to retain As. In some areas, the presence of sulfide minerals, such as pyrite (iron sulfide), can contribute to As contamination (Feng et al., 2023; Stolze, Battistel, & Rolle, 2022). Sulfide minerals can generate acidity when they react with water and oxygen, which can lead to the dissolution of As-bearing minerals and the release of As into the groundwater (Stolze et al., 2022). Clay minerals, like kaolinite and montmorillonite, are often present in the aquifer sediments and can influence As mobility. These minerals can adsorb As, but their effectiveness varies depending on local geochemical conditions (Herath, Vithanage, Bundschuh, Maity, & Bhattacharya, 2016; Mukhopadhyay, Manjaiah, Datta, Yadav, & Sarkar, 2017). High organic matter content in sediments can create reducing conditions and influence As mobilization. Organic matter can serve as an electron donor for microbial activity, promoting the reduction of As(V) to the more mobile As(III) form (Shaji et al., 2021). The complex interplay between these geological factors and hydrogeochemical conditions, such as redox potential and pH, determines the extent of As contamination in groundwater in the Lower Gangetic Basin (Shankar, Shanker, & Shikha, 2014). It is important to note that the specific mineralogical composition of aquifer sediments can vary from one location to another within the basin, leading to spatial variability in As contamination levels (Mazumder et al., 2010; Rahman et al., 2014). Understanding these geological factors is critical for assessing and managing As contamination in the region.

Groundwater contamination of As in the Lower Gangetic Basin in West Bengal is particularly prevalent in specific regions within the basin (Chakraborty, Mishra, et al., 2022; Mishra et al., 2023). While As contamination can occur in various areas of the basin, some regions have been more severely affected. Here are some of the main regions where As groundwater contamination is prominent in the Lower Gangetic Basin; Nadia is one of the worst-affected districts in West Bengal (Chakraborty, Mukherjee, & Ahmed, 2022; Chatterjee, 2022; Rahman et al., 2014). Several blocks within Nadia, including Tehatta, Chapra, and Krishnanagar, have reported high levels of As contamination in groundwater (Rahman et al., 2014). Murshidabad is another heavily affected district in the Lower Gangetic Basin. Various blocks in this district, such as Baharampur and Lalgola, have recorded elevated As concentrations in groundwater (Mishra et al., 2022). Some areas in North 24 Parganas and South 24 Parganas districts have also reported As contamination (Ahmad & Khan, 2015; Chowdhury et al., 1997; Sarkar, Paul, & Darbha, 2022; Shrivastava, Barla, Yadav, & Bose, 2014). These districts are in the southern part of the Lower Gangetic Basin. Parts of Bardhaman district have documented As contamination, particularly in areas around Katwa, Purbasthali, and so forth (Chatterjee, 2022). Hooghly district, including regions near Howrah, has experienced As contamination issues. As contamination is not limited to these specific districts but can also be found in other areas across the Lower Gangetic Basin, depending on local geological and hydrogeological conditions (Chatterjee, 2022). It is important to note that within these districts, the severity of As contamination can vary from one block or village to another (Chatterjee, 2022; Mazumder et al., 2010; Mishra et al., 2022, 2023). The presence of certain geological factors, such as the composition of aquifer sediments and prevailing redox conditions, plays a crucial role in determining the extent of contamination in specific regions (Moore et al., 2023). Extensive efforts have been made to assess, monitor, and address As contamination in these areas, including the installation of As-safe drinking water wells and community-based initiatives to mitigate the health risks associated with As exposure (Chatterjee, 2022; Sarkar et al., 2022). Table 5.1 presents a compilation of areas affected by As in the Gangetic Basin, West Bengal.

The WHO standard permissible level is 0.01 mg/L.

The mobilization of As from solid phases into groundwater is a complex process driven by various mechanisms (Rios-Valenciana et al., 2020). Microbial activity, reductive dissolution of iron oxides, and competitive ion exchange all contribute to the release of As into the aquifers (Rios-Valenciana et al., 2020; Zheng et al., 2020). These processes have been particularly pronounced in this

Table 5.1 List of districts in the Bengal basin affected by As contamination in groundwater.

List of districts	As ranges (mg/L)	No. of affected blocks	Similar references
1. North 24 Parganas	0.06–1.28	20–23	Mandal (2017) and Mishra et al. (2022)
2. South 24 Parganas	0.06–3.20	25–29	
3. Malda	0.05–1.454	7–9	
4. Murshidabad	0.05–0.96	19–22	
5. Nadia	0.05–1.00	17	
6. Hooghly	0.11–0.50	8–9	
7. Howrah	0.09–0.11	2	
8. Birbhum	0.03		
9. Bardhaman (Purba)	0.01–0.5	5	

From Mandal, V. (2017). Status of arsenic toxicity in ground water in West Bengal, India: A review. MOJ Toxicology, 3(5). https://doi.org/10.15406/mojt.2017.03.00063.

region due to the prevailing reducing conditions. Recognizing these mobilization mechanisms is crucial for developing effective strategies to prevent further contamination.

These objectives underscore the crucial role of sorption processes in mitigating or exacerbating As contamination in groundwater. They recognize the complexity of As contamination in the Lower Gangetic Basin and its significant impact on public health and the environment. The review investigates the interaction between hydrogeochemical controls, mobilization mechanisms, and sorption processes that influence As accumulation in groundwater. It aims to contribute to broader efforts to mitigate As contamination and protect the health of millions in the affected region. The review has also highlighted the role of sorption processes in mitigating or exacerbating As contamination. Minerals like iron and manganese oxides, clay minerals, and natural organic matter (NOM) have been identified as sorbents that can either retain or release As in groundwater. Understanding how these processes work can guide efforts to design remediation strategies that leverage the sorptive capacity of aquifer materials (Mazumder & Bhunia, 2022; Shrivastava et al., 2014; Singh et al., 2022). As contamination in the Lower Gangetic Basin represents a complex and multifaceted challenge with far-reaching implications for public health and the environment. The processes governing its enrichment in groundwater involve a delicate interplay of hydrogeochemical controls, mobilization mechanisms, and sorption processes. Addressing this crisis requires a multidisciplinary approach that combines hydrogeology, geochemistry, and public health initiatives. By shedding light on these processes, this review contributes to the broader effort to mitigate As contamination and safeguard the well-being of millions of people in this region.

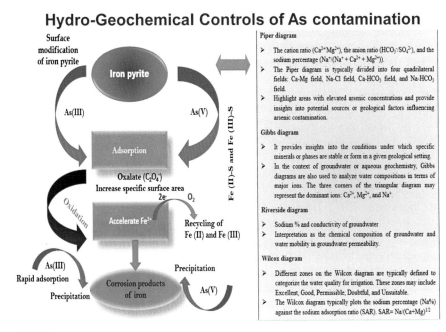

Hydro-Geochemical Controls of As contamination

Piper diagram

➢ The cation ratio (Ca^{2+}/Mg^{2+}), the anion ratio (HCO_3^-/SO_4^{2-}), and the sodium percentage ($Na^+/(Na^+ + Ca^{2+} + Mg^{2+})$).

➢ The Piper diagram is typically divided into four quadrilateral fields: Ca-Mg field, Na-Cl field, Ca-HCO₃ field, and Na-HCO₃ field.

➢ Highlight areas with elevated arsenic concentrations and provide insights into potential sources or geological factors influencing arsenic contamination.

Gibbs diagram

➢ It provides insights into the conditions under which specific minerals or phases are stable or form in a given geological setting.

➢ In the context of groundwater or aqueous geochemistry, Gibbs diagrams are also used to analyze water compositions in terms of major ions. The three corners of the triangular diagram may represent the dominant ions: Ca^{2+}, Mg^{2+}, and Na^+.

Riverside diagram

➢ Sodium % and conductivity of groundwater

➢ Interpretation as the chemical composition of groundwater and water mobility in groundwater permeability.

Wilcox diagram

➢ Different zones on the Wilcox diagram are typically defined to categorize the water quality for irrigation. These zones may include Excellent, Good, Permissible, Doubtful, and Unsuitable.

➢ The Wilcox diagram typically plots the sodium percentage (Na%) against the sodium adsorption ratio (SAR). $SAR = Na/(Ca+Mg)^{1/2}$

FIGURE 5.1

Mechanism of hydrogeochemical controls of As contamination.

5.2 Hydrogeochemical controls

As enrichment in groundwater is profoundly influenced by the hydrogeochemical characteristics of the aquifers. The presence of reducing conditions, high organic matter content, and the geologic composition of the aquifers play pivotal roles in controlling As concentrations. This section delves into the geochemical factors that render aquifers in the Lower Gangetic Basin susceptible to As contamination (Mazumder & Bhunia, 2022; Mishra et al., 2022; Sarkar et al., 2022). Hydrogeochemical controls play a crucial role in As enrichment in groundwater. Understanding these controls is essential for assessing the vulnerability of aquifers to As contamination. Here, we delve deeper into the hydrogeochemical factors that influence As levels in groundwater in Fig. 5.1.

5.3 Redox conditions influence the groundwater hydrogeochemical control

As exists in different chemical forms, or oxidation states, with arsenite (As(III)) and arsenate (As(V)) being the most prevalent in groundwater. Redox conditions in the aquifer influence which form of As predominates (Mapoma, Xie, & Zhang, 2014). Under reducing conditions, such as those found in many groundwater systems, As tends to exist in the more mobile and toxic As(III) form. Oxidizing conditions, on the other hand, promote the conversion of As(III) to the less mobile As(V) form (Bhatt, Kumar, & Singh, 2022). The presence of organic matter and certain minerals can create reducing conditions in the aquifer, facilitating the mobilization of As. Indeed, redox conditions are a pivotal component of groundwater hydrogeochemical controls, and they exert a significant influence on the behavior and mobility of various elements, including As. Redox conditions refer to the prevailing state of oxidation and reduction reactions occurring within the groundwater and aquifer sediments (Patel, Das, & Kumar, 2019; Shrivastava et al., 2014; Singh et al., 2022). In the framework of As contamination, redox conditions are crucial for several reasons: (1) As exists in different chemical forms or species, primarily arsenite (As(III)) and arsenate (As(V)). These two forms have distinct solubilities and behaviors. Under reducing conditions, such as in anaerobic environments, As tends to be present predominantly as the more mobile and toxic As(III) species (Shrivastava, Ghosh, Dash, & Bose, 2015; Zhang et al., 2015). In contrast, under oxidizing conditions, As(V) is favored. The speciation of As has a direct impact on its mobility and bioavailability in groundwater. (2) As in its As(III) form is more soluble and mobile than As(V). Therefore in reducing environments characterized by low oxygen levels (anaerobic conditions), As(III) is less likely to be adsorbed onto solid surfaces and more likely to be transported with groundwater. This mobility can result in elevated As concentrations in groundwater (Patel et al., 2019; Thi Hoa Mai et al., 2014; Viacava et al., 2022; Xue, Ran, Tan, Peacock, & Du, 2019). (3) Iron and manganese oxides are common constituents of aquifer sediments and are known to adsorb As effectively. Under oxidizing conditions, these oxides are more stable and capable of retaining As through adsorption. However, in reducing conditions, these oxides can be reduced themselves (e.g., ferrous iron formation), potentially releasing adsorbed As back into the groundwater. (4) Reducing conditions often promote the growth of certain bacteria, which can contribute to the reduction of As(V) to As(III). This microbial activity can further enhance the mobility of As in groundwater (Viacava et al., 2022). (5) Redox conditions also influence the redox potential (Eh) of the groundwater system. Changes in Eh can lead to the transformation of As species and their subsequent release or immobilization. For example, the reduction of As(V) to As(III) is a common redox-driven transformation (Raju, 2022). In regions where reducing conditions prevail, such as in the presence of organic-rich sediments or in groundwater systems influenced by geological factors, As enrichment and mobilization are more likely (Shankar et al., 2014). Understanding the redox conditions within an aquifer is essential for assessing the vulnerability of groundwater to As contamination and designing appropriate mitigation strategies (Chakrabarti, Singh, Rashid, & Rahman, 2019; Chattopadhyay et al., 2020; Mazumder et al., 2010; Mishra et al., 2022). Efforts to manage and remediate As-contaminated groundwater often involve measures to control redox conditions, such as the introduction of oxygen to oxidize As(III) to the less mobile As(V) form or the promotion of adsorption through the addition of materials like iron-based sorbents. Therefore a comprehensive understanding of redox conditions is critical for addressing As contamination challenges effectively (Chakraborty, Mukherjee, & Ahmed, 2022; Chowdhury et al., 1997; Duan et al., 2021; Mishra et al., 2022, 2023).

Indeed, redox conditions play a crucial role in influencing the groundwater hydrogeochemical control in the Indo-Gangetic Basin, particularly concerning As contamination. The Indo-Gangetic Basin, which spans across India, Bangladesh, Nepal, and Pakistan, has been extensively studied for its high levels of naturally occurring As in groundwater. The Indo-Gangetic Basin is characterized by extensive alluvial aquifers, where reducing (anaerobic) conditions are prevalent (Rahman, Mondal, & Fauzia, 2021). These reducing conditions are conducive to the release and mobilization of As, primarily in the more soluble and toxic As (III) form. Under reducing conditions, As is less likely to be adsorbed onto solid surfaces, making it more mobile in groundwater (Chattopadhyay et al., 2020). Iron and manganese oxides are common constituents of the aquifer sediments in the Indo-Gangetic Basin (Islam & Mostafa, 2023). These minerals have a strong affinity for As and can adsorb it effectively, reducing its mobility. However, under reducing conditions, these oxides can undergo dissolution or reduction, potentially releasing adsorbed As back into the groundwater. This dissolution and re-adsorption cycle is influenced by the fluctuating redox conditions in the aquifers. Anaerobic conditions often promote the growth of specific microbial communities that are capable of reducing As(V) to As(III) (Viacava et al., 2022). This microbial activity can further enhance the mobility of As in groundwater by converting it into the more soluble As(III) form. The redox potential (Eh) of groundwater in the Indo-Gangetic Basin can fluctuate due to seasonal variations in water levels, agricultural practices, and other factors. These fluctuations can drive redox-driven transformations of As species, such as the reduction of As(V) to As (III) and vice versa (Richards, Kumari, et al., 2022). These transformations influence the bioavailability and mobility of As. Sulfur compounds, including sulfides, are also common in the aquifer sediments of the Indo-Gangetic Basin. Sulfur cycling can affect redox conditions and interact with As in various ways, potentially leading to the release of As into groundwater.

Given the prevalence of reducing conditions in the Indo-Gangetic Basin, As contamination in groundwater is a significant concern in this region. The release and mobilization of As in groundwater under these conditions pose substantial health risks to the population that relies on these contaminated water sources for drinking (Barringer & Reilly, 2013; Mazumder & Bhunia, 2022). Efforts to manage and mitigate As contamination in the Indo-Gangetic Basin often involve interventions aimed at controlling or altering redox conditions. These interventions may include the installation of oxidation-based treatment systems, the promotion of aerobic conditions in wells, or the development of strategies to enhance As adsorption onto iron oxides (Hao, Liu, Wang, & Li, 2018). Understanding the interplay between redox conditions and As mobility is essential for effective As contamination management in this region.

5.4 pH levels influencing arsenic mobility

The pH of groundwater is a critical factor influencing As mobility. As solubility increases under low-pH (acidic) conditions, favoring its release into groundwater. In areas where the pH is naturally low or becomes acidic due to anthropogenic activities, such as acid mine drainage, As is more likely to be mobilized. The pH of groundwater is indeed a critical factor influencing the mobility of As. As is a naturally occurring element that can be found in rocks and minerals (Islam & Mostafa, 2023). When groundwater comes into contact with these As-containing geological materials, it can dissolve some of the As, leading to its presence in the groundwater.

The pH of groundwater, which measures its acidity or alkalinity, plays a significant role in determining the solubility and mobility of As. Under acidic conditions (pH below 7), As tends to be more soluble and mobile in groundwater (Mandal, 2017; Raju, 2022; Rios-Valenciana et al., 2020). This means that As is more likely to leach from geological materials and enter the groundwater when the pH is low. As pH becomes more neutral (around pH 7) and especially in alkaline conditions (pH > 7), As tends to be less soluble and less mobile. This reduces the risk of As contamination in groundwater. Therefore maintaining a neutral to slightly alkaline pH in groundwater can help mitigate the mobility of As and reduce the risk of As contamination in drinking water sources (Islam & Mostafa, 2023). Various treatment methods, such as adding alkaline substances or using ion-exchange resins, can be employed to raise the pH of groundwater and reduce As solubility. The Indo-Gangetic Basin, which covers parts of India, Pakistan, Nepal, and Bangladesh, is known for its significant As contamination issues in groundwater (Thakur, Gupta, Bhattacharya, Jakariya, & Tahmidul Islam, 2021). The relationship between pH levels and As mobility in this region can vary depending on the specific mineralogy of the aquifer systems (Chattopadhyay et al., 2020). In the Indo-Gangetic Basin, several minerals can influence As mobility and solubility. Iron oxides and hydroxides are common minerals in the sediments of the Indo-Gangetic Basin. As often binds to these iron minerals. In low-pH (acidic) conditions, such as those found in some parts of the basin, these iron minerals can dissolve, releasing As into the groundwater. So, low pH levels in such areas can lead to increased As mobility (Banerji et al., 2019; Chakraborty, Mishra, et al., 2022; Stolze, Zhang, Guo, & Rolle, 2019). In some areas with sulfide-rich geological formations, As can be associated with sulfide minerals. Under reducing (low oxygen) conditions, such as those found in parts of the Indo-Gangetic Basin, sulfide minerals can release As when they react with groundwater, regardless of the pH level. In regions where there are carbonate-rich formations, the pH of groundwater may naturally be more alkaline due to the buffering capacity of carbonate minerals (Sikdar & Chakraborty, 2008). In such cases, As tends to be less soluble and mobile because the alkaline pH reduces its solubility. Clay minerals can also play a role in adsorbing As, reducing its mobility in groundwater (Islam & Mostafa, 2023; Barringer & Reilly, 2013; Sikdar & Chakraborty, 2008). The behavior of clay minerals in relation to As can be influenced by pH, as well as other factors like organic matter content. In the Indo-Gangetic Basin, it is essential to recognize that there is considerable geological heterogeneity (Farooq et al., 2011; Mishra et al., 2022, 2023; Sikdar & Chakraborty, 2008). Different regions within the basin may have varying mineral compositions and groundwater chemistry, which can lead to different As mobility patterns. Therefore while pH is an important factor influencing As mobility, the specific mineralogy and geochemical conditions of each area within the basin must be considered (Islam & Mostafa, 2023). Monitoring and managing pH levels in conjunction with other geochemical factors is crucial for mitigating As contamination in the Indo-Gangetic Basin and ensuring safe drinking water supplies for the affected populations.

5.5 Geologic composition and mineral dissolution

The type of geological materials that make up the aquifer can significantly impact As enrichment. As is often associated with certain minerals, such as iron and manganese oxides, sulfides, and clay minerals (Chakraborty, Mishra, et al., 2022). Iron oxides, in particular, can adsorb As, reducing its mobility. The presence of these minerals in the aquifer matrix can act as a source or sink for As, depending on local conditions (Biswas et al., 2023). In the Indo-Gangetic Basin, particularly in West Bengal, As contamination in groundwater is a significant concern, and it is closely related to the dissolution of specific minerals in the aquifers (Banerji et al., 2019; Chowdhury et al., 1997). The primary mineral associated with As contamination in this region is iron sulfides (commonly pyrite, FeS_2). Pyrite is a common mineral found in the sedimentary rocks of the Indo-Gangetic Basin (Neil, Yang, Schupp, & Jun, 2014; Rahman et al., 2019). When pyrite comes into contact with water, it can undergo oxidation, leading to the release of iron (Fe) and sulfate ions (SO_4^{2-}) into the groundwater (Banerji et al., 2019; Rahman et al., 2019). This process lowers the pH of the groundwater and creates reducing conditions, as oxygen is consumed during pyrite oxidation. In reducing conditions, As that is naturally adsorbed onto iron hydroxide minerals or coprecipitated with iron can be desorbed or released into the groundwater. This is because the presence of reduced iron (Fe^{2+}) can compete with As for adsorption sites on iron minerals, making As more mobile and soluble in the groundwater (Hao et al., 2018; Xia, Teng, & Zhai, 2022). In some parts of West Bengal, there are also alkaline earth carbonates like calcite ($CaCO_3$) and dolomite ($CaMg(CO_3)_2$)

present in the aquifers (Mishra et al., 2023). These carbonates can buffer the pH of the groundwater, making it more alkaline. In alkaline conditions, As tends to be less soluble and less mobile. So, in West Bengal, the dissolution of iron sulfides like pyrite and the resulting changes in pH and redox conditions are the primary factors influencing As mobility in groundwater (Alarcón-Herrera & Gutiérrez, 2022; Kumar, Patel, & Singh, 2022; Raju, 2022). When pyrite oxidizes and releases iron, it can create reducing conditions that mobilize As, increasing its concentration in the water. To address As contamination in West Bengal's groundwater, it is essential to understand the local geological and hydrogeochemical conditions, including the presence of pyrite and other minerals, as well as pH levels (Khan, Musahib, et al., 2023; Raju, 2022). Various strategies, such as the installation of safe drinking water wells, promoting aerobic conditions to prevent pyrite oxidation, and treating groundwater to raise pH levels, are used to mitigate As contamination and provide safe drinking water to affected communities.

5.6 Ion exchange and competition

The presence of other ions, such as phosphate, silicate, and bicarbonate, can compete with As for sorption sites on mineral surfaces. High concentrations of these ions can reduce the capacity of the aquifer to retain As, leading to increased As mobility (Khan & Rai, 2023). Ion exchange is a water treatment process that can be employed to remove As from groundwater with contamination. It involves the exchange of ions in the groundwater with ions attached to a solid phase, typically a resin or other adsorbent material. Ion exchange and competition for As mobility in the Indo-Bengal aquifer system are important factors to consider when addressing As contamination in groundwater in the region (Chattopadhyay et al., 2020; Mukhopadhyay et al., 2017). The Indo-Bengal aquifer system, which includes parts of West Bengal in India and Bangladesh, is known for having high levels of naturally occurring As in groundwater. As in the Indo-Bengal aquifer system is often present as both arsenate (As(V)) and arsenite (As(III)). Arsenate is negatively charged and has a similar charge to other common ions in groundwater, such as sulfate (SO_4^{2-}) and bicarbonate (HCO_3^-) (Chakraborty et al., 2015). Arsenite, on the other hand, is neutral but can undergo oxidation to become arsenate under certain conditions. Ion exchange materials present in the aquifer, such as iron hydroxides and clay minerals, can adsorb As ions (both arsenate and arsenite) from the groundwater (Alarcón-Herrera & Gutiérrez, 2022; Singh et al., 2022). These materials have charged surfaces that attract and bind As ions through electrostatic interactions. In groundwater, there are multiple ions and solutes present, including sulfate, bicarbonate, and other anions, which can compete with As ions for adsorption sites on the mineral surfaces (Chakraborty et al., 2015; Singh et al., 2022). The competition for adsorption sites is influenced by factors such as pH, redox conditions, and the concentrations of these competing ions. The pH of groundwater in the Indo-Bengal aquifer system can vary, and it plays a critical role in As mobility. Under reducing (low oxygen) conditions and lower pH levels, arsenite (As(III)) can dominate, and its mobility can increase as it competes with other anions for adsorption sites (Alarcón-Herrera & Gutiérrez, 2022; Singh et al., 2022). Under more oxidizing conditions and higher pH levels, arsenate (As(V)) may be the predominant form, which can also compete for adsorption sites. Some minerals have a higher affinity for As ions compared to other competing ions (Deng et al., 2018). For example, iron hydroxides are known to selectively adsorb As ions (Datta et al., 2011; Kumar et al., 2022). This selective adsorption can affect the relative mobility of As and other ions in the groundwater. Over time, as As-contaminated groundwater is pumped from wells, the competition for adsorption sites can result in changes in the composition of the aquifer, potentially leading to shifts in As mobility patterns and the migration of As within the aquifer (Alarcón-Herrera & Gutiérrez, 2022; Ghosh & Donselaar, 2023).

Understanding the ion exchange processes and competition for adsorption sites is crucial when designing As remediation strategies in the Indo-Bengal aquifer system (Hao et al., 2018; Neil et al., 2014; Singh et al., 2022). Various treatment methods, such as installing safe drinking water wells that tap into deeper, lower-As aquifers, enhancing oxidation to convert arsenite to arsenate for better removal, and employing selective adsorbents like iron-based materials, can be used to mitigate As contamination and ensure the provision of safe drinking water in the region. Monitoring the groundwater chemistry and As levels is essential to assess the effectiveness of these strategies and protect public health.

5.7 Microbial activity for arsenic mobilization

Microbial processes can influence As mobilization. Certain bacteria can facilitate the reduction of As(V) to As(III), increasing its mobility in groundwater. Microbial activity is often linked to the presence of organic matter and specific redox conditions (Islam & Mostafa, 2023; Mishra et al., 2022). Microbial processes can indeed play a significant role in As mobilization in groundwater. Certain bacteria can transform As species and influence their mobility, primarily by facilitating the reduction of arsenate (As(V)) to arsenite (As(III)) (Chen et al., 2017; Zecchin et al., 2021; Zheng et al., 2020). Some bacteria, known as arsenate-reducing bacteria (ARB), are capable of using arsenate as an electron acceptor in their metabolic processes. In doing so, these bacteria can reduce As(V) to As(III) (Cai et al., 2022; Zecchin et al., 2021). This microbial reduction of As changes its oxidation state from $+5$ to $+3$, which alters its chemical behavior and solubility (Cai et al., 2022; Sathe, Goswami, & Mahanta, 2021; Zecchin et al., 2021). The presence of organic matter in groundwater can support the growth and activity of ARB. Organic matter serves as a carbon source for these bacteria, providing them with the energy needed for arsenate reduction (Chen et al., 2017; Deng et al., 2018). Organic matter can originate from sources such as decaying plant material, sewage, or agricultural runoff. Microbial activity and the reduction of arsenate are closely tied to specific redox conditions in the groundwater (Cai et al., 2022; Raturi et al., 2023; Yang et al., 2022). These conditions are often described as reducing or anoxic, where oxygen is limited or absent. In such environments, microorganisms can utilize alternative electron acceptors like arsenate, facilitating its reduction (Guo et al., 2015). Arsenite (As(III)) is typically more mobile and soluble in groundwater compared to arsenate (As(V)) (Jia, Guo, Jiang, Wu, & Zhou, 2014). This is because As(III) is less prone to adsorption onto mineral surfaces and can more easily migrate through porous geological formations (Manning & Goldberg, 1997; Yang et al., 2022). As a result, when ARB reduce As(V) to As(III), it can lead to an increase in the mobility of As in the groundwater. The behavior of As in groundwater is complex and can be influenced by various factors, including pH, the presence of other chemical species, and the specific types of bacteria present. Microbial processes are just one aspect of this complexity (Cai et al., 2022; Fang et al., 2023; Zanzo et al., 2017; Zhang et al., 2015; Zheng et al., 2020).

Understanding the microbial processes involved in As mobilization is crucial for managing and mitigating As contamination in groundwater. It is also important for designing effective water treatment strategies (Deng et al., 2018; Sathe et al., 2021; Zheng et al., 2020). For example, in areas where microbial As reduction is a concern, water treatment may include oxidation of As(III) back to As(V) before removal through methods such as adsorption, coagulation, or precipitation (Ghosh & Donselaar, 2023; Liu & Qu, 2021; Rahidul Hassan, 2023; Rahman et al., 2019). Additionally, maintaining aerobic (oxygen-rich) conditions in groundwater can help inhibit microbial As reduction and decrease As mobility.

5.8 Sulfide minerals effective on mobility of arsenic for groundwater

In some aquifers, the presence of sulfide minerals can lead to the release of As. Sulfide minerals can react with groundwater, producing sulfuric acid and releasing As from mineral surfaces. In certain aquifers, the presence of sulfide minerals can indeed lead to the release of As into groundwater through a series of chemical reactions (Rios-Valenciana et al., 2020; Zheng et al., 2020). This process is often referred to as "sulfide-driven As mobilization" and is linked to the oxidation of sulfide-related minerals. Sulfide minerals, such as pyrite (iron sulfide, FeS_2), are commonly found in geological formations (Duan et al., 2021; Islam & Mostafa, 2023). These minerals contain sulfur (S) in combination with other elements. When sulfide minerals come into contact with oxygen (O_2) in groundwater or in the presence of air, they can undergo oxidation (Stolze et al., 2022). This oxidation reaction produces sulfuric acid (H_2SO_4) and releases sulfate ions ($SO_4{}^{2-}$) into the water:

$$FeS_2 + 7/2O_2 + H_2O \rightarrow Fe^{2+} + 2\ SO_4^{2-} + 2H^+$$

This reaction generates acidity in the groundwater, lowering its pH. The release of sulfate ions also contributes to an increase in sulfate concentration. As is often naturally associated with sulfide minerals, either adsorbed onto their surfaces or as a segment of the mineral structure. When sulfide minerals oxidize and release sulfuric acid, the acidic conditions can promote the desorption or dissolution of As from the mineral surfaces (Battistel, Stolze, Muniruzzaman, & Rolle, 2021; Stolze et al., 2019). This process leads to the mobilization of As into the groundwater (Fig. 5.2). The mobilized As can lead to increased As concentrations in the groundwater, posing a risk to drinking water sources and potentially causing As contamination issues (Stolze et al., 2019).

Several factors, such as the presence of other minerals, groundwater pH, and dissolved oxygen concentration, can impact the extent of As mobilization driven by sulfide processes (Bondu, Cloutier, Benzaazoua, Rosa, & Bouzahzah, 2017). Lower pH values and higher oxygen levels tend to promote the oxidation of sulfide minerals and the release of As. Sulfide-driven As mobilization is particularly common in areas with sulfide-rich geological formations and can be a significant concern for

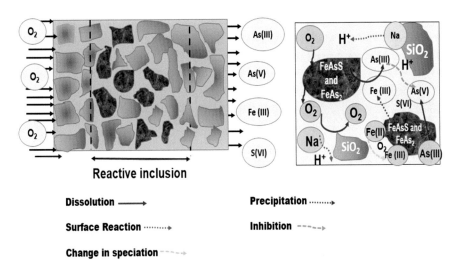

FIGURE 5.2

Dissolution of arsenic-containing sulfide minerals in groundwater: influence of hydrochemical and hydrodynamic conditions on arsenic release and surface changes.

From Stolze, L., Battistel, M., & Rolle, M. (2022). Oxidative dissolution of arsenic-bearing sulfide minerals in groundwater: Impact of hydrochemical and hydrodynamic conditions on arsenic release and surface evolution. Environmental Science and Technology, 56(8), 5049–5061. https://doi.org/10.1021/acs.est.2c00309.

As contamination in groundwater (Bondu et al., 2017). Managing and mitigating As in such areas may involve strategies like maintaining anaerobic (low oxygen) conditions in wells, which can help prevent the oxidation of sulfide minerals and the subsequent release of As (Battistel et al., 2021; Bondu et al., 2017; Duan et al., 2021). Additionally, water treatment methods that address the elevated As levels may be necessary to provide safe drinking water to affected communities.

5.9 Hydraulic conductivity depends on Bengal aquifer and its flux for arsenic

The hydraulic properties of the aquifer, such as its permeability and porosity, can affect the movement of groundwater and, subsequently, the transport of As. Aquifers with high hydraulic conductivity may allow As-contaminated water to spread more rapidly (Khan, Haque, et al., 2023). The hydraulic conductivity of aquifers is a vital characteristic that significantly influences the flow of groundwater, playing a critical role in the transport of contaminants such as As. In the perspective of the Gangetic Bengal aquifer and its flux for As, hydraulic conductivity is a key factor to consider (Lu et al., 2022). Hydraulic conductivity (K) is a measure of the ability of an aquifer material to transmit water (Dewan et al., 2017). It quantifies the ease with which water can flow through the porous spaces and fractures within the aquifer. It is typically measured in units of velocity (e.g., meters per day) and depends on factors like the size and connectivity of pore spaces and the properties of the aquifer material (Águila et al., 2023). The hydraulic conductivity of the Gangetic Bengal aquifer determines how easily groundwater can move within the aquifer system. High hydraulic conductivity indicates that groundwater can flow more readily, whereas low hydraulic conductivity means that groundwater flow is slower (Bleam, 2012; Harvey & Wagner, 2000). In regions with high hydraulic conductivity, groundwater flow rates can be relatively high, allowing contaminants to be transported more quickly. As (III and V) contamination in the Gangetic Bengal aquifer is frequently linked to the existence of naturally occurring As in geological formations (Barrera Olivarez et al., 2022). The hydraulic conductivity of the aquifer materials can influence how As-contaminated water migrates within the aquifer. Higher hydraulic conductivity can result in the faster movement of As-contaminated water through the aquifer, potentially affecting a larger area (Chakraborty et al., 2015). The flux of As refers to the rate at which As is transported through the aquifer. It depends on hydraulic conductivity, groundwater flow rates, and the concentration of As in the groundwater. In areas with high hydraulic conductivity, the flux of As can be relatively high, potentially leading to the spread of As contamination over a larger region. Understanding the hydraulic conductivity of the Gangetic Bengal aquifer is crucial for managing and mitigating As contamination (Chakraborty et al., 2015; Chakraborty, Mukherjee, & Ahmed, 2022; Farooq et al., 2011). It can inform decisions about well placement, pumping rates, and the design of remediation systems. Lowering groundwater flow rates in areas with As contamination can be an effective strategy to limit the flux of As and reduce the extent of contamination. In summary, hydraulic conductivity is a fundamental property of aquifers that affects the movement of groundwater and, consequently, the transport of contaminants like As. In the framework of the Gangetic Bengal aquifer, it is important for assessing the potential spread of As contamination and implementing strategies to address this environmental and public health concern (Chakraborti et al., 2018; Chattopadhyay et al., 2020; Hong et al., 2014).

In summary, hydrogeochemical controls for As in groundwater are complex and interrelated. The presence of reducing conditions, specific minerals, pH levels, and competing ions all influence the mobility and distribution of As in aquifers. Understanding these controls is essential for assessing As contamination risks and developing effective mitigation strategies to ensure safe drinking water sources in affected regions.

5.10 The role of aquifer types, sediment characteristics, and hydrostratigraphy

The role of aquifer types, sediment characteristics, and hydrostratigraphy plays a crucial role in understanding and addressing As enrichment in groundwater in regions like the Lower Gangetic Basin in West Bengal (Battistel et al., 2021; Mukherjee, Fryar, & Howell, 2007). These geological and hydrogeological factors have a direct impact on the distribution and mobilization of As in groundwater. Aquifers are underground geological formations that store and transmit groundwater. In the context of As contamination, the type of aquifer present in an area can significantly influence the occurrence of As. In the Lower Gangetic Basin, two primary types of aquifers are encountered. These are composed of unconsolidated sediments such as sand, silt, and clay, which are deposited by rivers over time (Das & Kumar, 2015; Kumar, Kumar, Ramanathan, & Tsujimura, 2010). Alluvial aquifers are highly permeable and have a significant role in the As contamination issue in the region (Mukherjee et al., 2007; Rahman et al., 2021). As-rich sediments can be leached into the groundwater due to various hydrogeochemical processes in these aquifers. These aquifers are often associated with the delta regions of rivers, such as the Ganges and Brahmaputra (Islam & Mostafa, 2023). They are typically a mix of alluvial and deltaic sediments. Fluvio-deltaic aquifers are known to be prone to As contamination due to their complex hydrostratigraphy and the presence of fine-grained sediments that facilitate As mobility (Chakraborty, Mukherjee, Ahmed, Fryar, et al., 2022; Das & Kumar, 2015). Understanding the type of aquifer present is crucial because it dictates groundwater flow patterns, hydraulic conductivity, and the potential for As release into the groundwater.

The characteristics of sediments in the subsurface play a significant role in As enrichment. In the Lower Gangetic Basin, As is often associated with iron-rich sediments (Banerji et al., 2019). The following sediment characteristics are relevant where as high iron content in sediments is a common feature in As-affected areas. Iron oxides and hydroxides can adsorb and release As depending on the prevailing redox conditions (Banerji et al., 2019; Mukherjee et al., 2007). Under reducing conditions, As can be released into the groundwater when it competes with other ions for sorption sites on iron minerals. Fine-grained sediments, such as clay and silt, have a higher surface area, making them more effective in adsorbing and retaining As (Chakraborty et al., 2015). When groundwater interacts with these fine sediments, it can lead to the release of adsorbed As, particularly in reducing environments (Das & Kumar, 2015). Organic matter in sediments can influence As mobility. It can facilitate the reduction of arsenate (As(V)) to more mobile arsenite (As(III)), which is often more toxic and mobile in groundwater (Islam & Mostafa, 2023).

Hydrostratigraphy refers to the arrangement and distribution of different aquifer layers in the subsurface. In the Lower Gangetic Basin, the hydrostratigraphy can be quite complex due to the deltaic environment. The presence of multiple aquifer layers, both confined and unconfined, can lead to the interconnection of groundwater bodies (Biswas et al., 2014; Mukherjee et al., 2007). This interconnection can facilitate the movement of As-contaminated water between different aquifers. Impermeable layers or confining units can act as barriers to the vertical movement of groundwater. Understanding the location and properties of these layers is essential for assessing the potential for As migration. The depth of the water table or groundwater level can affect the extent of As contamination (Das & Kumar, 2015). Shallow groundwater tables may be more susceptible to contamination, as they are more likely to come into contact with As-rich sediments. Thus aquifer types, sediment characteristics, and hydrostratigraphy are integral components in the complex puzzle of As contamination in the Lower Gangetic Basin in West Bengal (Biswas et al., 2014; Manning & Goldberg, 1997; Mukherjee et al., 2007). A comprehensive understanding of these factors is crucial for devising effective mitigation strategies and ensuring the provision of safe drinking water to the affected communities in the region.

5.11 Influencing of climatic factors and their impact on arsenic mobility

Climatic factors have a significant influence on the mobility of As in groundwater in regions like the Lower Gangetic Basin in West Bengal. While geological and hydrogeological factors play a fundamental role, climatic conditions can exacerbate or alleviate the As contamination problem (Chakraborti et al., 2018; Das & Kumar, 2015). Here is a closer look at how climatic factors affect As mobility.

The Lower Gangetic Basin experiences a distinct monsoon season with heavy rainfall. During this period, there is a substantial recharge of groundwater (Khan & Rai, 2023). Fig. 5.3 presents a comprehensive conceptual framework linking climate change drivers to As contamination in the Indian groundwater system. It elucidates the intricate interactions at play, where variations in weather patterns can release As from geological formations into groundwater (Taylor et al., 2013; Van Geen et al., 2008). Altered precipitation patterns impact river flow and surface runoff, influencing the transport of As to groundwater (Swain, Taloor, Dhal, Sahoo, & Al-Ansari, 2022). The framework emphasizes how changes in surface water dynamics affect groundwater recharge, with shifts in recharge rates contributing to the infiltration of As-laden water into aquifers, exacerbating groundwater contamination. The influx of rainwater can alter the subsurface hydrology and influence the movement of As-rich groundwater. Increased water levels can dilute the concentration of As, potentially reducing its impact on drinking water sources. Conversely, during the dry season, groundwater levels may decrease, causing

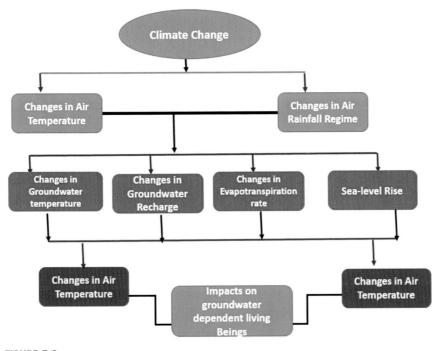

FIGURE 5.3

Diagram illustrating the conceptual framework of natural impacts of climate change on the groundwater system.

Swain, S., Taloor, A. K., Dhal, L., Sahoo, S., Al-Ansari, N. (2022). Impact of climate change on groundwater hydrology: A comprehensive review and current status of the Indian hydrogeology. Applied Water Science, 12, *120. https://doi.org/10.1007/s13201-022-01652-0.*

increased concentrations of As in the remaining groundwater. Lower water tables can lead to changes in redox conditions within the aquifer, favoring the release of As from sediments (Swain et al., 2022). Temperature fluctuations between seasons can impact microbial activity in the subsurface. Microbes play a role in the biogeochemical processes that affect As mobility. Warmer temperatures can stimulate microbial activity, potentially influencing the transformation and mobility of As species in groundwater. Climate-related changes in groundwater levels can alter the availability of oxygen in the subsurface (Swain et al., 2022; Yuan, Wei, Xu, & Cao, 2023). Oxygen is a key factor in determining whether As exists in its less mobile oxidized form (arsenate, As(V)) or its more mobile reduced form (arsenite, As(III)) (Duan et al., 2017). Higher oxygen levels promote the formation of arsenate, which tends to be less mobile and less toxic. Heavy monsoon rains can lead to soil erosion and sediment transport, potentially redistributing As-rich sediments (Kumar et al., 2018). Sedimentation and deposition can impact the As content of various geologic formations, affecting groundwater quality. Climatic factors can alter groundwater flow patterns and velocities. Faster groundwater flow can influence the contact time between groundwater and As-bearing sediments, affecting the degree of As mobilization (Islam & Mostafa, 2023).

In coastal areas like the Lower Gangetic Basin, rising sea levels and changes in salinity can affect the geochemistry of groundwater. High salinity levels can influence the adsorption and mobility of As, potentially exacerbating contamination issues (Mukherjee et al., 2014; Mukherjee et al., 2018; Rahman et al., 2014). The effects of long-term climate change, such as altered rainfall patterns, increased temperatures, and sea-level rise, can have profound implications for As contamination (Swain et al., 2022). These changes can impact the hydrological cycle, groundwater recharge rates, and the availability of fresh groundwater sources.

In summary, climatic factors, especially seasonal variations in precipitation and temperature, can influence the mobility of As in groundwater in the Lower Gangetic Basin and similar regions (Mahmud, Sultana, Hasan, & Ahmed, 2017). The interplay between climatic conditions and geological/hydrogeological factors is complex, and a holistic understanding of these dynamics is essential for effectively managing and mitigating As contamination in affected areas (Kumar et al., 2018). Strategies for combating As contamination should consider the dynamic nature of both natural and anthropogenic factors.

5.12 Mobilization mechanisms of arsenic controlling in the Bengal basin

Understanding how As is mobilized from solid phases into groundwater is crucial for identifying contamination sources and predicting its behavior. Microbial processes, reductive dissolution of iron oxides, and competitive ion exchange are some of the key mechanisms responsible for As mobilization in the region (Biswas et al., 2014; Mukherjee et al., 2007; Rahman et al., 2014). This section explores these processes in detail and highlights their significance. In the framework of West Bengal's groundwater (Islam & Mostafa, 2023; Khan & Rai, 2023). A comprehensive analysis of water chemistry, including major ions, trace elements (particularly As), pH, redox potential (Eh), and other relevant parameters, is conducted. This analysis will provide the data needed for plotting on the diagrams. Controlling the mobilization of As in the Bengal Basin is a critical issue due to the widespread occurrence of naturally elevated As levels in groundwater, which poses a significant health risk to the population (Islam & Mostafa, 2023; Thakur et al., 2021). Various hydrogeochemical diagrams, including Piper and Gibbs diagrams, can help in understanding and managing As mobilization in groundwater. The Piper diagram is a graphical representation of water chemistry that helps classify groundwater into different hydrochemical facies based on major ion concentrations (Nayak, Matta, & Uniyal, 2023). To control As mobilization, the Piper diagram can be used in the following ways. Piper diagrams categorize groundwater samples into different hydrochemical facies based on the concentrations of major ions, such as bicarbonate (HCO_3^-), sulfate (SO_4^{2-}), chloride (Cl^-), sodium (Na^+), potassium (K^+), calcium (Ca^{2+}), and magnesium (Mg^{2+}) (Islam & Mostafa, 2023; Mishra et al., 2022; Zhang, Xiao, Adeyeye, Yang, & Liang, 2020). By plotting water chemistry data on a Piper diagram, hydrogeochemists can identify the dominant facies in a region. This is crucial for understanding the baseline water quality and recognizing any deviations caused by contamination or natural processes. It is clear provided that the sodium-chloride and calcium-chloride-rich conditions in the previous study area can be attributed to the composition of the parent rocks, specifically Fissile Hornblende Biotite gneiss and Charnockite, which are known to contain sodium and calcium-rich minerals (Biswas et al., 2014; Singh & Taylor, 2019; Singh et al., 2022). The geological composition plays a crucial role in influencing the chemistry of groundwater and surface water in a given region. Regarding the use of series and time series plots in visualizing water analysis results, these tools are indeed effective for tracking changes in water quality over time (Mazumder et al., 2010). The hydrochemical facies of groundwater samples in the Bengal Basin are determined to understand the dominant ions and water types. A Gibbs diagram is created using the same water chemistry data. The Gibbs diagram will help assess the dominant geochemical processes controlling groundwater chemistry (Mishra et al., 2022). The position of the data points on the diagram should be paid attention, as it can provide insights into processes like ion exchange, mineral dissolution, and mixing. As mobilization often correlates with specific ion compositions. The Piper diagram can provide insights into the redox conditions of groundwater (Singh et al., 2022). As mobility is often linked to reducing environments. If the water samples cluster in a specific part of the Piper diagram indicative of reducing conditions, it can be a clue that As mobilization may be a concern (Mishra et al., 2022). Regular monitoring and plotting of groundwater chemistry on the Piper diagram can help track changes in hydrochemical facies over time. Sudden shifts in facies may indicate changes in geochemical processes influencing As mobilization. In the context of As mobilization, the Gibbs diagram proves useful in several applications. By graphically representing the chemical makeup of groundwater samples on the Gibbs diagram, it becomes possible to discern the predominant processes dictating water chemistry. As mobilization is frequently linked to specific mechanisms, such as ion exchange and mineral dissolution (Wan, Wang, Shi, Qu, & Zhang, 2023). The application of Gibbs diagrams extends to the assessment of saturation indices for minerals such as iron and As-bearing minerals (Mapoma et al., 2014; Marandi & Shand, 2018; Mishra et al., 2022, 2023; Nayak et al., 2023). If groundwater is found to be undersaturated concerning these minerals, it indicates the potential for mineral dissolution, potentially leading to the release of As. This underscores the Gibbs diagram's significance in understanding and managing As mobilization in groundwater (Singh et al., 2022). If groundwater samples plot along the mixing line on the Gibbs diagram, it indicates the mixing of different water sources. Understanding mixing can help in identifying potential sources of As contamination (Chakraborty et al., 2015; Das & Kumar, 2015; Dewan et al., 2017; Kumar et al., 2018). To effectively control As mobilization in the Bengal Basin, it's crucial to combine the information from Eh−pH diagram with detailed hydrogeological and geochemical contamination (Fig. 5.4). Factors such as groundwater flow patterns, redox conditions, and the presence of specific minerals play a significant role in As

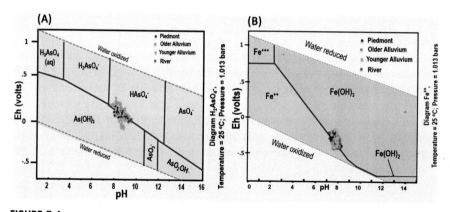

FIGURE 5.4

Eh−pH diagram for groundwater, showcasing the (A) arsenic (As) and (B) iron (Fe) species present in the previous study.

From Kumar, M., Ramanathan, A. L., Mukherjee, A., Verma, S., Rahman, M. M., & Naidu, R. (2018). Hydrogeo-morphological influences for arsenic release and fate in the central Gangetic Basin, India. Environmental Technology and Innovation, 12, 243−260. https://doi.org/10.1016/j.eti.2018.09.004.

mobilization and need to be considered in any remediation or management strategy (Kumar et al., 2018; Moore et al., 2023). Additionally, implementing appropriate treatment technologies and ensuring access to safe drinking water sources is essential to mitigate the health risks associated with As contamination in groundwater. The Piper and Gibbs diagrams are analyzed to identify potential As mobilization mechanisms. Correlations are identified between the hydrochemical facies and dominant geochemical processes (Biswas et al., 2014; Kumar et al., 2018). For example, if reducing conditions are prevalent in the region and As is present in groundwater, this suggests that reductive dissolution of As-bearing minerals may be a key mechanism (Biswas et al., 2014; Chakraborti et al., 2018; Khan & Rai, 2023; Mishra et al., 2023). Geological mapping in West Bengal unveils a variety of rock formations and soil compositions throughout the region. Specific geological formations, especially those related to deltaic sedimentation in the Gangetic Basin, frequently correlate with As contamination in groundwater. Regions dominated by Holocene alluvial sediments exhibit an increased vulnerability to As mobilization due to distinct geochemical conditions, as illustrated in Fig. 5.5. The significance of iron and manganese oxides in the geological composition is pivotal for the sorption of As.

5.13 Sorption processes and arsenic mobility

Sorption onto mineral surfaces and organic matter is a fundamental process that can either attenuate or exacerbate As contamination in groundwater (Thi Hoa Mai et al., 2014). The review examines the role of iron and manganese oxides, clay minerals, and NOM in sorbing As species and discusses how sorption processes can be harnessed for remediation efforts (Águila et al., 2023; Subdiaga, Orsetti, & Haderlein, 2019; Xue et al., 2019; Yang et al., 2022; Zhang et al., 2015). Sorption onto mineral surfaces and organic matter is indeed a fundamental process that plays a critical role in determining the fate and transport of As in groundwater (Islam & Mostafa, 2023). This process can either mitigate or worsen As contamination, depending on various factors. Therefore it delves into the role of different sorbents, such as iron and manganese oxides, clay minerals, and NOM, in sorbing As species and how these processes can be harnessed for remediation efforts (He & Hering, 2009; Marghade, Mehta, Shelare, Jadhav, & Nikam, 2023; Mohammadian et al., 2022; Shankar et al., 2014): (1) Iron and manganese oxides, commonly found in subsurface environments, are known to strongly adsorb As species. The adsorption of As onto these oxides is often favored under oxidizing conditions (Hao et al., 2018; Hare, Chowdhary, & Singh, 2020; Zanzo et al., 2017). (2) As sorption onto iron and manganese oxides occurs through surface complexation reactions, where As forms inner-sphere complexes with metal ions on the oxide surfaces (Fig. 5.6). This sorption process can be harnessed for remediation by introducing iron or manganese oxides into the aquifer, either

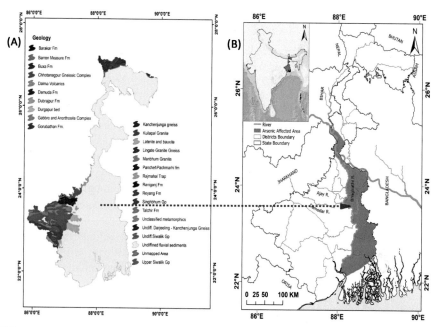

FIGURE 5.5

Geological mapping of West Bengal and comparison with arsenic contamination areas.
(A) Mapping the geology of West Bengal, (B) Comparing the geological position within the
lower Ganga basin to exact locations.

*Acharyya, S. K., & Shah, B. A. (2007). Arsenic-contaminated groundwater from parts of Damodar fan-delta
and west of Bhagirathi River, West Bengal, India: Influence of fluvial geomorphology and quaternary
morphostratigraphy.* Environmental Geology, 52, *489–501. https://doi.org/10.1007/s00254-006-0482-z.*

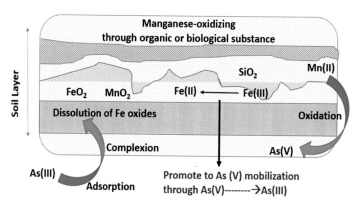

FIGURE 5.6

The process of arsenic sorption onto iron and manganese oxides that occurs via surface
complexation reactions.

naturally or through engineered approaches (Hao et al., 2018; Zhang et al., 2020). These oxides can act as adsorbents, reducing the concentration of As in groundwater (Singh et al., 2022; Zanzo et al., 2017). (3) Clay minerals, such as kaolinite, montmorillonite, and illite, can also influence As mobility in groundwater. Their surfaces possess negatively charged sites that can interact with positively charged As species. As adsorption onto clay minerals is influenced by factors like pH, ionic strength, and the presence of competing ions (Islam & Mostafa, 2023; Mukherjee et al., 2018). Under certain conditions, clay minerals can attenuate As migration by adsorption. In remediation efforts, understanding the interactions between clay minerals and As is crucial. Techniques like capping contaminated aquifers with clay-rich materials or adding clay minerals to contaminated sites can help immobilize As (Subdiaga et al., 2019; Zhang et al., 2020). (4) NOM, consisting of humic and fulvic acids, can also influence As sorption. NOM can either compete with As for binding sites or enhance sorption by forming ternary complexes with metal oxides (Li et al., 2017; Rahim, Allevato, Vaccari, & Stazi, 2023; Subdiaga et al., 2019; Swain et al., 2022). The role of NOM in As sorption is complex and can vary depending on the NOM content, composition, and pH of the groundwater. In some remediation strategies, adding organic-rich materials can be beneficial. For example, using organic-rich amendments in permeable reactive barriers (PRBs) can promote As immobilization through sorption onto NOM (Singh et al., 2022; Subdiaga et al., 2019; Thakur et al., 2021). (5) Several remediation technologies leverage sorption processes to mitigate As contamination. These include PRBs, where sorbents like iron oxides or organic-rich materials are placed in the flow path of contaminated groundwater (Singh et al., 2022). Another approach is the use of adsorbent materials, such as iron-based adsorbents, which can be introduced into wells or treatment systems to capture As from groundwater. Careful consideration of site-specific conditions, including groundwater chemistry and hydrogeology, is essential when designing and implementing these remediation strategies (Subdiaga et al., 2019).

In summary, the sorption of As onto mineral surfaces and organic matter is a complex process that can significantly impact the mobility and bioavailability of As in groundwater (Tahri, Bahafid, Sayel, & El Ghachtouli, 2013; Zanzo et al., 2017). Understanding these processes is crucial for both assessing the risk of contamination and designing effective remediation strategies to protect human health and the environment. Successful remediation efforts often involve a combination of engineering and geochemical approaches tailored to the specific site conditions.

5.14 Impacts on health and the crisis in water resource utilization

The consequences of As contamination on human health are severe, ranging from skin lesions to various forms of cancer. This section briefly outlines the health risks associated with long-term exposure to elevated As levels in drinking water and emphasizes the urgency of mitigating this crisis (Chakraborti et al., 2018; Singh & Taylor, 2019). The consequences of As contamination in drinking water on human health are indeed severe and can have devastating long-term effects. Long-term exposure to As-contaminated water is often associated with the development of skin lesions. These can include discolored patches, warts, and hyperkeratosis (thickening of the skin) (Ahmad & Khan, 2015; Chakrabarti et al., 2019; Chakraborti et al., 2018). Skin lesions are often one of the earliest visible signs of chronic As exposure. Perhaps the most alarming health risk associated with As contamination is the increased risk of various forms of cancer. As is classified as a Group 1 carcinogen by the International Agency for Research on Cancer (Ahmad & Khan, 2015; Martinez et al., 2011; Mishra et al., 2022, 2023; Rahman et al., 2014; Stolze et al., 2022). Prolonged exposure to As has been linked to particularly basal cell carcinoma and squamous cell carcinoma, an increased risk of lung cancer, especially in regions with high As levels in drinking water and elevated As levels in drinking water have been subjected with an rising risk of cancer (like bladder) (Wu et al., 2023; Yamauchi, Yoshida, & Takata, 2023). Chronic As exposure has been linked to cardiovascular diseases, including hypertension, atherosclerosis, and an increased risk of heart attacks. It can also lead to respiratory issues, such as chronic obstructive pulmonary disease (Chowdhury et al., 2023; Moon, Guallar, & Navas-Acien, 2012). Some studies suggest that As exposure may have adverse effects on cognitive function and child development, particularly in regions with high As levels (Khan & Rai, 2023). There is evidence to suggest that prolonged exposure to As may increase the risk of developing diabetes mellitus. As can weaken the immune system, making individuals more susceptible to infections and diseases. Long-term exposure to As in drinking water has been associated with higher mortality rates due to the aforementioned health effects (Moon et al., 2012). The urgency of mitigating As contamination cannot be overstated. Millions of people worldwide are at risk of As exposure due to contaminated groundwater sources (Chakrabarti et al., 2019; Chowdhury et al., 2023). It is a silent, slow-acting crisis that can take years to manifest its health effects. As such, proactive measures to provide safe drinking water are essential to prevent further harm (Khatisashvili et al., 2022). Efforts to mitigate As contamination include the development of safe drinking water sources, the implementation of water treatment technologies, and public health awareness campaigns (Wu et al., 2023). Local governments, international organizations, and communities need to work together to address this pressing issue and ensure access to As-free drinking water to protect the health and well-being of affected populations.

5.15 Mitigation and remediation strategies

Efforts to combat As contamination in the Lower Gangetic Basin encompass a range of mitigation and remediation approaches. These include the installation of As-safe drinking water wells, chemical treatments, and bioremediation techniques

(Khatisashvili et al., 2022). The review provides insights into the effectiveness of these strategies and highlights the need for ongoing research and community engagement (Fang et al., 2023; Islam & Mostafa, 2023). Efforts to combat As contamination in the Lower Gangetic Basin, a region known for its significant As groundwater contamination problems, indeed involve a range of mitigation and remediation approaches (Bhardwaj, Misra, & Rajput, 2019; Chakraborti et al., 2018). These strategies aim to provide safe drinking water to affected communities and reduce the health risks associated with As exposure (Chakraborty et al., 2015; Richards, Parashar, et al., 2022). An overview of some key approaches and considerations as one of the primary strategies is the installation of As-safe drinking water wells. These wells tap into deeper aquifers with lower As concentrations, providing a source of safe drinking water. The effectiveness of this approach depends on proper siting and regular monitoring to ensure that the newly installed wells remain As-free over time. Chemical treatments involve adding substances that can precipitate or adsorb As from contaminated water (Bhardwaj et al., 2019; Richards, Parashar, et al., 2022). Common chemicals used include iron salts (e.g., ferric chloride) and aluminum salts (e.g., alum). Coagulation-flocculation processes can be used to form As-containing flocs that can be separated from water through sedimentation or filtration. The effectiveness of chemical treatments depends on water chemistry and the availability of suitable chemicals. It requires careful monitoring and adjustment to optimize treatment performance (Bhardwaj et al., 2019; Khatisashvili et al., 2022).

Bioremediation approaches harness the activity of microorganisms to transform or immobilize As in the environment. In situ bioremediation techniques can involve stimulating naturally occurring microorganisms to reduce As mobility through processes like microbial sulfate reduction (Hare et al., 2020; Raturi et al., 2023). Ex situ bioremediation techniques may include using microbial cultures or biofilms to treat As-contaminated water in controlled settings (Fang et al., 2023; He & Hering, 2009). Adsorbent materials, such as iron-based materials, activated carbon, and clay minerals, can be used to remove As from water(Duan et al., 2017; Hao et al., 2018). These materials are added to water treatment systems or can be placed in PRBs to capture As as water passes through (Da'ana et al., 2021). Potential mitigation measures have been proposed, as illustrated in Fig. 5.7.

5.16 Ongoing research and monitoring

Continuous research is crucial to developing and improving mitigation and remediation strategies. This includes studies on the effectiveness of various treatment methods, their long-term performance, and their environmental impacts (Chen et al., 2017; Feng et al., 2023; Raju, 2022). Regular monitoring of water quality and well conditions is essential to ensure that mitigation efforts remain effective over time. Community involvement is vital to the success of As mitigation efforts.

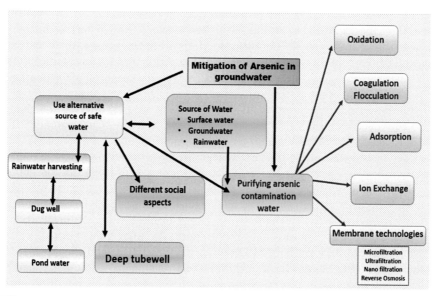

FIGURE 5.7

Diagram outlining arsenic mitigation strategies.

Raising awareness about the risks of As contamination and promoting safe water practices is essential. Engaging with local communities to understand their needs, concerns, and preferences helps in designing effective and sustainable solutions. Governments and regulatory bodies play a crucial role in addressing As contamination. Establishing and enforcing water quality standards is essential to safeguard public health (Ahmad & Khan, 2015; Alarcón-Herrera & Gutiérrez, 2022; Chakrabarti et al., 2019; Hare et al., 2020; Kumar et al., 2010). Policies that promote the use of safe water sources and the regular testing of well water for As are important components of a comprehensive approach. Therefore addressing As contamination in the lower Gangetic Basin and similar regions requires a multi-faceted approach that encompasses safe water supply, treatment technologies, bioremediation, ongoing research, and community engagement. Sustainable solutions must be tailored to the specific local conditions and involve a collaborative effort among governments, researchers, NGOs, and affected communities to ensure the provision of safe drinking water and the reduction of health risks associated with As exposure.

5.17 Conclusions

As contamination in the groundwater of West Bengal's Lower Gangetic Basin poses a multifaceted challenge, influenced by a combination of hydrogeochemical factors, mobilization mechanisms, and sorption processes. The hydrogeological attributes of the area, encompassing aquifer properties like lithology, mineralogy, and hydraulic conductivity, dictate the presence and movement of As in groundwater. Geological formations housing As-containing minerals or sediments act as potential sources of contamination, with factors such as groundwater flow patterns and hydraulic gradients influencing distribution and transport. As mobilization occurs as naturally occurring As is released from geological formations into groundwater, often facilitated by redox reactions in anaerobic conditions. Biological processes, including microbial activity, may further contribute to mobilization by promoting the reduction of As-bearing minerals or organic matter. Sorption, involving the interaction of As with aquifer materials, depends on factors like mineral composition, surface area, pH, and the presence of competing ions. For instance, iron oxides and oxyhydroxides exhibit a high affinity for As sorption. Environmental changes, such as variations in pH or redox potential, can alter the sorption-desorption equilibrium, impacting As mobility in groundwater. Addressing As enrichment necessitates a comprehensive understanding of the intricate interplay between hydrogeochemical controls, mobilization mechanisms, and sorption processes. Mitigation strategies encompass groundwater management practices like aquifer recharge and abstraction management, along with engineered solutions such as As removal technologies (e.g., bioremediation, adsorption, precipitation, and membrane filtration) to ensure a safe drinking water supply in affected regions. Regular monitoring and testing of groundwater quality are essential for early detection and effective management of As contamination in the Lower Gangetic Basin and similar areas.

References

Águila, J. F., McDonnell, M. C., Flynn, R., Butler, A. P., Hamill, G. A., Etsias, G., ... Donohue, S. (2023). Comparison of saturated hydraulic conductivity estimated by empirical, hydraulic and numerical modeling methods at different scales in a coastal sand aquifer in Northern Ireland. *Environmental Earth Sciences*, *82*(13). Available from https://doi.org/10.1007/s12665-023-11019-6, https://www.springer.com/journal/12665.

Ahmad, S. A., & Khan, M. H. (2015). *Ground water arsenic contamination and its health effects in Bangladesh. Handbook of arsenic toxicology* (pp. 51−72). Elsevier Inc. Available from http://www.sciencedirect.com/science/book/9780124186880, 10.1016/B978-0-12-418688-0.00002-2.

Alarcón-Herrera, M. T., & Gutiérrez, M. (2022). Geogenic arsenic in groundwater: Challenges, gaps, and future directions. *Current Opinion in Environmental Science and Health*, *27*. Available from https://doi.org/10.1016/j.coesh.2022.100349, http://www.journals.elsevier.com/current-opinion-in-environmental-science-and-health.

Anawar, H. M., Akai, J., Mihaljevič, M., Sikder, A. M., Ahmed, G., Tareq, S. M., & Rahman, M. M. (2011). Arsenic contamination in groundwater of Bangladesh: Perspectives on geochemical, microbial and anthropogenic issues. *Water (Switzerland)*,

3(4), 1050−1076. Available from https://doi.org/10.3390/w3041050, https://res.mdpi.com/d_attachment/water/water-03-01050/article_deploy/water-03-01050.pdf.

Banerjee, S., Dhar, S., Sudarshan, M., Chakraborty, A., Bhattacharjee, S., & Bhattacharjee, P. (2023). Investigating the synergistic role of heavy metals in Arsenic-induced skin lesions in West Bengal, India. *Journal of Trace Elements in Medicine and Biology*, *75*, 127103. Available from https://doi.org/10.1016/j.jtemb.2022.127103.

Banerji, T., Kalawapudi, K., Salana, S., & Vijay, R. (2019). Review of processes controlling arsenic retention and release in soils and sediments of Bengal basin and suitable iron based technologies for its removal. *Groundwater for Sustainable Development*, *8*, 358−367. Available from https://doi.org/10.1016/j.gsd.2018.11.012, http://www.journals.elsevier.com/groundwater-for-sustainable-development/.

Barrera Olivarez, M., Alfonso Murillo Tovar, M., Vergara Sánchez, J., Luisa García Betancourt, M., Martín Romero, F., María Ramírez Arteaga, A., . . . Albeiro Saldarriaga Noreña, H. (2022). *Mobility of heavy metals in aquatic environments impacted by ancient mining-waste*. IntechOpen. Available from 10.5772/intechopen.98693.

Battistel, M., Stolze, L., Muniruzzaman, M., & Rolle, M. (2021). Arsenic release and transport during oxidative dissolution of spatially-distributed sulfide minerals. *Journal of Hazardous Materials*, *409*, 124651. Available from https://doi.org/10.1016/j.jhazmat.2020.124651.

Bhardwaj, A., Misra, K., & Rajput, R. (2019). *Status of arsenic remediation in India advances in water purification techniques: Meeting the needs of developed and developing countries* (pp. 219−258). Elsevier. Available from: https://www.sciencedirect.com/book/9780128147900. doi: 10.1016/B978-0-12-814790-0.00009-0.

Bhatt, A. G., Kumar, A., & Singh, S. K. (2022). Hydro-geochemical evolution of groundwater and associated human health risk in River Sone subbasin of Middle-Gangetic floodplain, Bihar, India. *Arabian Journal of Geosciences*, *15*(5). Available from https://doi.org/10.1007/s12517-021-09269-4.

Biswas, A., Bhattacharya, P., Mukherjee, A., Nath, B., Alexanderson, H., Kundu, A. K., . . . Jacks, G. (2014). Shallow hydrostratigraphy in an arsenic affected region of Bengal Basin: Implication for targeting safe aquifers for drinking water supply. *Science of the Total Environment*, *485−486*(1), 12−22. Available from https://doi.org/10.1016/j.scitotenv.2014.03.045, http://www.elsevier.com/locate/scitotenv.

Biswas, T., Pal, S. C., Saha, A., Ruidas, D., Islam, A. R. M. T., & Shit, M. (2023). Hydrochemical assessment of groundwater pollutant and corresponding health risk in the Ganges delta, Indo-Bangladesh region. *Journal of Cleaner Production*, *382*. Available from https://doi.org/10.1016/j.jclepro.2022.135229, https://www.journals.elsevier.com/journal-of-cleaner-production.

Bleam, W. F. (2012). *Soil moisture and hydrology* (pp. 41−84). Elsevier BV. Available from 10.1016/b978-0-12-415797-2.00002-9.

Bondu, R., Cloutier, V., Benzaazoua, M., Rosa, E., & Bouzahzah, H. (2017). The role of sulfide minerals in the genesis of groundwater with elevated geogenic arsenic in bedrock aquifers from western Quebec, Canada. *Chemical Geology*, *474*, 33−44. Available from https://doi.org/10.1016/j.chemgeo.2017.10.021, http://www.sciencedirect.com/science/journal/00092541.

Cai, X., Zhang, Z., Yin, N., Lu, W., Du, H., Yang, M., . . . Cui, Y. (2022). Controlling microbial arsenite oxidation and mobilization in arsenite-adsorbed iron minerals: The Influence of pH conditions and mineralogical composition. *Journal of Hazardous Materials, 433,* 128778. Available from https://doi.org/10.1016/j.jhazmat.2022.128778.

Chakrabarti, D., Singh, S. K., Rashid, M. H., & Rahman, M. M. (2019). *Arsenic: Occurrence in groundwater. Encyclopedia of environmental health* (pp. 153−168). Elsevier. Available from http://doi.org/10.1016/B978-0-12-409548-9.10634-7, 10.1016/B978-0-12-409548-9.10634-7.

Chakraborti, D., Singh, S. K., Rahman, M. M., Dutta, R. N., Mukherjee, S. C., Pati, S., & Kar, P. B. (2018). Groundwater arsenic contamination in the ganga river basin: A future health danger. *International Journal of Environmental Research and Public Health, 15*(2). Available from https://doi.org/10.3390/ijerph15020180, http://www.mdpi.com/1660-4601/15/2/180/pdf.

Chakraborty, M., Mishra, A. K., & Mukherjee, A. (2022). Influence of hydrogeochemical reactions along flow paths on contrasting groundwater arsenic and manganese distribution and dynamics across the Ganges River. *Chemosphere, 287.* Available from https://doi.org/10.1016/j.chemosphere.2021.132144, http://www.elsevier.com/locate/chemosphere.

Chakraborty, M., Mukherjee, A., Ahmed, K. M., Fryar, A. E., Bhattacharya, A., ... Chattopadhyay, S. (2022). Influence of hydrostratigraphy on the distribution of groundwater arsenic in the transboundary Ganges River delta aquifer system, India and Bangladesh. *GSA Bulletin, 134*(9−10), 2680−2692. Available from https://doi.org/10.1130/b36068.1.

Chakraborty, M., Mukherjee, A., & Ahmed, K. M. (2015). A review of groundwater arsenic in the Bengal Basin, Bangladesh and India: From source to sink. *Current Pollution Reports, 1*(4), 220−247. Available from https://doi.org/10.1007/s40726-015-0022-0, http://springer.com/environment/pollution + and + remediation/journal/40726.

Chakraborty, M., Mukherjee, A., & Ahmed, K. M. (2022). Regional-scale hydrogeochemical evolution across the arsenic-enriched transboundary aquifers of the Ganges River Delta system, India and Bangladesh. *Science of the Total Environment, 823,* 153490. Available from https://doi.org/10.1016/j.scitotenv.2022.153490.

Chatterjee, S. (2022). *Spatial pattern of arsenic contamination in floodplain aquifers, western bank of Bhagirathi River, lower Ganges delta, West Bengal, India* (pp. 245−272). Springer Science and Business Media LLC. Available from 10.1007/978-981-16-6966-8_13.

Chattopadhyay, A., Singh, A. P., Singh, S. K., Barman, Λ., Patra, A., Mondal, B. P., & Banerjee, K. (2020). Spatial variability of arsenic in Indo-Gangetic basin of Varanasi and its cancer risk assessment. *Chemosphere, 238.* Available from https://doi.org/10.1016/j.chemosphere.2019.124623, http://www.elsevier.com/locate/chemosphere.

Chen, X., Zeng, X. C., Wang, J., Deng, Y., Ma, T., Guoji, E., ... Wang, Y. (2017). Microbial communities involved in arsenic mobilization and release from the deep sediments into groundwater in Jianghan plain, Central China. *Science of the Total Environment, 579,* 989−999. Available from https://doi.org/10.1016/j.scitotenv.2016.11.024, http://www.elsevier.com/locate/scitotenv.

Chowdhury, N. R., Joardar, M., Das, A., Mridha, D., Majumder, S., Mondal, M., ... Roychowdhury, T. (2023). Appraisal of acute and chronic arsenic exposure in differently exposed school children with special reference to micronuclei formation in urine epithelial cells: A comparative study in West Bengal, India. *Groundwater for Sustainable Development, 21,* 100917. Available from https://doi.org/10.1016/j.gsd.2023.100917.

Chowdhury, T. R., Mandal, B. K., Samanta, G., Basu, G. K., Chowdhury, P. P., Chanda, C. R., ... Chakraborti, D. (1997). *Arsenic in groundwater in six districts of West Bengal, India: The biggest arsenic calamity in the world: The status report up to August, 1995* (pp. 93–111). Springer Science and Business Media LLC. Available from 10.1007/978-94-011-5864-0_9.

Da'ana, D. A., Zouari, N., Ashfaq, M. Y., Abu-Dieyeh, M., Khraisheh, M., Hijji, Y. M., & Al-Ghouti, M. A. (2021). Removal of toxic elements and microbial contaminants from groundwater using low-cost treatment options. *Current Pollution Reports*, *7*(3), 300–324. Available from https://doi.org/10.1007/s40726-021-00187-3, http://springer.com/environment/pollution + and + remediation/journal/40726.

Das, A., & Kumar, M. (2015). Arsenic enrichment in the groundwater of Diphu, Northeast India: Coupled application of major ion chemistry, speciation modeling, and multivariate statistical techniques. *Clean − Soil, Air, Water*, *43*(11), 1501–1513. Available from https://doi.org/10.1002/clen.201400632, http://onlinelibrary.wiley.com/journal/10.1002/%28ISSN%291863-0669.

Datta, S., Neal, A. W., Mohajerin, T. J., Ocheltree, T., Rosenheim, B. E., White, C. D., & Johannesson, K. H. (2011). Perennial ponds are not an important source of water or dissolved organic matter to groundwaters with high arsenic concentrations in West Bengal, India. *Geophysical Research Letters*, *38*(20). Available from https://doi.org/10.1029/2011GL049301, http://onlinelibrary.wiley.com/journal/10.1002/(ISSN)1944-8007/issues?year = 2012.

Deng, Y., Zheng, T., Wang, Y., Liu, L., Jiang, H., & Ma, T. (2018). Effect of microbially mediated iron mineral transformation on temporal variation of arsenic in the Pleistocene aquifers of the central Yangtze River basin. *Science of the Total Environment*, *619–620*, 1247–1258. Available from https://doi.org/10.1016/j.scitotenv.2017.11.166, http://www.elsevier.com/locate/scitotenv.

Dewan, A., Corner, R., Saleem, A., Rahman, M. M., Haider, M. R., Rahman, M. M., & Sarker, M. H. (2017). Assessing channel changes of the Ganges-Padma River system in Bangladesh using Landsat and hydrological data. *Geomorphology*, *276*, 257–279. Available from https://doi.org/10.1016/j.geomorph.2016.10.017, http://www.elsevier.com/inca/publications/store/5/0/3/3/3/4/.

Duan, L., Song, J., Yin, M., Yuan, H., Li, X., Zhang, Y., & Yin, X. (2021). Dynamics of arsenic and its interaction with Fe and S at the sediment-water interface of the seasonal hypoxic Changjiang Estuary. *Science of the Total Environment*, *769*, 145269. Available from https://doi.org/10.1016/j.scitotenv.2021.145269.

Duan, Y., Gan, Y., Wang, Y., Liu, C., Yu, K., Deng, Y., ... Dong, C. (2017). Arsenic speciation in aquifer sediment under varying groundwater regime and redox conditions at Jianghan Plain of Central China. *Science of the Total Environment*, *607–608*, 992–1000. Available from https://doi.org/10.1016/j.scitotenv.2017.07.011, http://www.elsevier.com/locate/scitotenv.

Fang, Y., Chen, M., Liu, C., Dong, L., Zhou, J., Yi, X., ... Tong, H. (2023). Arsenic release from microbial reduction of scorodite in the presence of electron shuttle in flooded soil. *Journal of Environmental Sciences*, *126*, 113–122. Available from https://doi.org/10.1016/j.jes.2022.05.018.

Farooq, S. H., Chandrasekharam, D., Norra, S., Berner, Z., Eiche, E., Thambidurai, P., & Stüben, D. (2011). Temporal variations in arsenic concentration in the groundwater of Murshidabad District, West Bengal, India. *Environmental Earth Sciences*, *62*(2),

223−232. Available from https://doi.org/10.1007/s12665-010-0516-4, https://link. springer.com/journal/12665.

Feng, Y., Dong, S., Ma, M., Hou, Q., Zhao, Z., & Zhang, W. (2023). The influence mechanism of hydrogeochemical environment and sulfur and nitrogen cycle on arsenic enrichment in groundwater: A case study of Hasuhai basin, China. *Science of the Total Environment*, *858*, 160013. Available from https://doi.org/10.1016/j. scitotenv.2022.160013.

Ghosh, D., & Donselaar, M. E. (2023). Predictive geospatial model for arsenic accumulation in Holocene aquifers based on interactions of oxbow-lake biogeochemistry and alluvial geomorphology. *Science of the Total Environment*, *856*, 158952. Available from https://doi.org/10.1016/j.scitotenv.2022.158952.

Guo, H., Liu, Z., Ding, S., Hao, C., Xiu, W., & Hou, W. (2015). Arsenate reduction and mobilization in the presence of indigenous aerobic bacteria obtained from high arsenic aquifers of the Hetao basin, Inner Mongolia. *Environmental Pollution*, *203*, 50−59. Available from https://doi.org/10.1016/j.envpol.2015.03.034, https://www.journals.elsevier.com/environmental-pollution.

Hao, L., Liu, M., Wang, N., & Li, G. (2018). A critical review on arsenic removal from water using iron-based adsorbents. *RSC Advances*, *8*(69), 39545−39560. Available from https://doi.org/10.1039/c8ra08512a, http://pubs.rsc.org/en/journals/journal/ra.

Hare, V., Chowdhary, P., & Singh, A. K. (2020). *Arsenic toxicity: Adverse effect and recent advance in microbes mediated bioremediation. Microorganisms for Sustainable Environment and Health* (pp. 53−80). India: Elsevier. Available from https://www. sciencedirect.com/book/9780128190012/microorganisms-for-sustainable-environment-and-health, 10.1016/B978-0-12-819001-2.00004-8.

Harvey, J. W., & Wagner, B. J. (2000). *Quantifying hydrologic interactions between streams and their subsurface hyporheic zones* (pp. 3−44). Elsevier BV. Available from 10.1016/b978-012389845-6/50002-8.

He, Y. T., & Hering, J. G. (2009). Enhancement of arsenic(III) sequestration by manganese oxides in the presence of iron(II). *Water, Air, and Soil Pollution*, *203*(1−4), 359−368. Available from https://doi.org/10.1007/s11270-009-0018-8, http://www.kluweronline. com/issn/0049-6979/.

Herath, I., Vithanage, M., Bundschuh, J., Maity, J. P., & Bhattacharya, P. (2016). Natural arsenic in global groundwaters: Distribution and geochemical triggers for mobilization. *Current Pollution Reports*, *2*(1), 68−89. Available from https://doi.org/10.1007/ s40726-016-0028-2, http://springer.com/environment/pollution + and + remediation/ journal/40726.

Hong, Y. S., Song, K. H., & Chung, J. Y. (2014). Health effects of chronic arsenic exposure. *Journal of Preventive Medicine and Public Health*, *47*(5), 245−252. Available from https://doi.org/10.3961/jpmph.14.035, http://www.ncbi.nlm.nih.gov/pmc/articles/ PMC4186552/pdf/jpmph-47-5-245.pdf.

Islam, M. S., & Mostafa, M. G. (2023). Occurrence, source, and mobilization of iron, manganese, and arsenic pollution in shallow aquifer. *Geofluids*, *2023*. Available from https://doi.org/10.1155/2023/6628095, https://www.hindawi.com/journals/geofluids/.

Jia, Y., Guo, H., Jiang, Y., Wu, Y., & Zhou, Y. (2014). Hydrogeochemical zonation and its implication for arsenic mobilization in deep groundwaters near alluvial fans in the Hetao Basin, Inner Mongolia. *Journal of Hydrology*, *518*, 410−420. Available from https://doi.org/10.1016/j.jhydrol.2014.02.004, http://www.elsevier.com/inca/publications/store/5/0/3/3/4/3.

Khan, M. S. H., Haque, M. E., Ahmed, M., Mallick, J., Islam, A. R. M. T., & Fattah, M. A. (2023). Quantitative analysis and modeling of groundwater flow using visual MODFLOW: A case from subtropical coal mine, northwest Bangladesh. *Environment, Development and Sustainability.* Available from https://doi.org/10.1007/s10668-023-04052-9, https://www.springer.com/journal/10668.

Khan, M. U., Musahib, M., Vishwakarma, R., Rai, N., & Jahan, A. (2023). Hydrochemical characterization, mechanism of mobilization, and natural background level evaluation of arsenic in the aquifers of upper Gangetic plain, India. *Geochemistry, 83*(2), 125952. Available from https://doi.org/10.1016/j.chemer.2023.125952.

Khan, M. U., & Rai, N. (2023). Distribution, geochemical behavior, and risk assessment of arsenic in different floodplain aquifers of middle Gangetic basin, India. *Environmental Geochemistry and Health, 45*(5), 2099−2115. Available from https://doi.org/10.1007/s10653-022-01321-w, https://www.springer.com/journal/10653.

Khatisashvili, G., Varazi, T., Kurashvili, M., Pruidze, M., Bunin, E., Didebulidze, K., . . . Sapojnikova, N. (2022). *Remedial approaches against arsenic pollution.* IntechOpen. Available from 10.5772/intechopen.98779.

Kumar, M., Patel, A. K., & Singh, A. (2022). Anthropogenic dominance on geogenic arsenic problem of the groundwater in the Ganga-Brahmaputra floodplain: A paradox of origin and mobilization. *Science of the Total Environment, 807.* Available from https://doi.org/10.1016/j.scitotenv.2021.151461, http://www.elsevier.com/locate/scitotenv.

Kumar, M., Ramanathan, A. L., Mukherjee, A., Verma, S., Rahman, M. M., & Naidu, R. (2018). Hydrogeo-morphological influences for arsenic release and fate in the central Gangetic Basin, India. *Environmental Technology and Innovation, 12,* 243−260. Available from https://doi.org/10.1016/j.eti.2018.09.004, http://www.journals.elsevier.com/environmental-technology-and-innovation/.

Kumar, P., Kumar, M., Ramanathan, A. L., & Tsujimura, M. (2010). Tracing the factors responsible for arsenic enrichment in groundwater of the middle Gangetic Plain, India: A source identification perspective. *Environmental Geochemistry and Health, 32*(2), 129−146. Available from https://doi.org/10.1007/s10653-009-9270-5.

Barringer, J. L., & Reilly, P. A. (2013). *Arsenic in groundwater: A summary of sources and the biogeochemical and hydrogeologic factors affecting arsenic occurrence and mobility.* InTech. Available from 10.5772/55354.

Li, F., Guo, H., Zhou, X., Zhao, K., Shen, J., Liu, F., & Wei, C. (2017). Impact of natural organic matter on arsenic removal by modified granular natural siderite: Evidence of ternary complex formation by HPSEC-UV-ICP-MS. *Chemosphere, 168,* 777−785. Available from https://doi.org/10.1016/j.chemosphere.2016.10.135, http://www.elsevier.com/locate/chemosphere.

Liu, R., & Qu, J. (2021). Review on heterogeneous oxidation and adsorption for arsenic removal from drinking water. *Journal of Environmental Sciences, 110,* 178−188. Available from https://doi.org/10.1016/j.jes.2021.04.008.

Lu, C., Richards, L. A., Wilson, G. J. L., Krause, S., Lapworth, D. J., Gooddy, D. C., . . . Niasar, V. J. (2022). Quantifying the impacts of groundwater abstraction on Ganges river water infiltration into shallow aquifers under the rapidly developing city of Patna, India. *Journal of Hydrology: Regional Studies, 42.* Available from https://doi.org/10.1016/j.ejrh.2022.101133, https://www.journals.elsevier.com/journal-of-hydrology-regional-studies.

Mahmud, M. I., Sultana, S., Hasan, M. A., & Ahmed, K. M. (2017). Variations in hydrostratigraphy and groundwater quality between major geomorphic units of the Western

Ganges Delta plain, SW Bangladesh. *Applied Water Science*, 7(6), 2919–2932. Available from https://doi.org/10.1007/s13201-017-0581-x, https://www.springer.com/journal/13201.

Mandal, V. (2017). Status of arsenic toxicity in ground water in West Bengal, India: A review. *MOJ Toxicology*, 3(5). Available from https://doi.org/10.15406/mojt.2017.03.00063.

Manning, B. A., & Goldberg, S. (1997). Adsorption and stability of arsenic(III) at the clay mineral-water interface. *Environmental Science and Technology.*, 31(7), 2005–2011. Available from https://doi.org/10.1021/es9608104.

Mapoma, H. W. T., Xie, X., & Zhang, L. (2014). Redox control on trace element geochemistry and provenance of groundwater in fractured basement of Blantyre, Malawi. *Journal of African Earth Sciences*, 100, 335–345. Available from https://doi.org/10.1016/j.jafrearsci.2014.07.010, http://www.sciencedirect.com/science/journal/1464343X.

Marandi, A., & Shand, P. (2018). Groundwater chemistry and the Gibbs diagram. *Applied Geochemistry*, 97, 209–212. Available from https://doi.org/10.1016/j.apgeochem.2018.07.009, http://www.journals.elsevier.com/applied-geochemistry.

Marghade, D., Mehta, G., Shelare, S., Jadhav, G., & Nikam, K. C. (2023). Arsenic contamination in Indian groundwater: From origin to mitigation approaches for a sustainable future. *Water*, 15(23), 4125. Available from https://doi.org/10.3390/w15234125.

Martinez, V. D., Vucic, E. A., Becker-Santos, D. D., Gil, L., & Lam, W. L. (2011). Arsenic exposure and the induction of human cancers. *Journal of Toxicology*, 2011. Available from https://doi.org/10.1155/2011/431287.

Mazumder, A., & Bhunia, G. S. (2022). Arsenic contamination in shallow groundwater in Karimpur block of Nadia district (West Bengal, India)—A spatial and geostatistical approach. *KN – Journal of Cartography and Geographic Information*, 72(2), 173–182. Available from https://doi.org/10.1007/s42489-022-00103-9, https://link.springer.com/journal/42489.

Mazumder, D. N., Ghosh, A., Majumdar, K., Ghosh, N., Saha, C., & Mazumder, R. N. (2010). Arsenic contamination of ground water and its health impact on population of district of Nadia, West Bengal, India. *Indian Journal of Community Medicine*, 35(2), 331–338. Available from https://doi.org/10.4103/0970-0218.66897.

Mishra, D., Chakrabortty, R., Sen, K., Pal, S. C., & Mondal, N. K. (2023). Groundwater vulnerability assessment of elevated arsenic in Gangetic plain of West Bengal, India; Using primary information, lithological transport, state-of-the-art approaches. *Journal of Contaminant Hydrology*, 256. Available from https://doi.org/10.1016/j.jconhyd.2023.104195, http://www.elsevier.com/locate/jconhyd.

Mishra, D., Sen, K., Mondal, A., Kundu, S., & Mondal, N. K. (2022). Geochemical appraisal of groundwater arsenic contamination and human health risk assessment in the Gangetic Basin in Murshidabad District of West Bengal, India. *Environmental Earth Sciences*, 81(5). Available from https://doi.org/10.1007/s12665-022-10273-4, https://link.springer.com/journal/12665.

Mohammadian, S., Tabani, H., Boosalik, Z., Asadi Rad, A., Krok, B., Fritzsche, A., ... Meckenstock, R. U. (2022). In situ remediation of arsenic-contaminated groundwater by injecting an iron oxide nanoparticle-based adsorption barrier. *Water*, 14(13), 1998. Available from https://doi.org/10.3390/w14131998.

Moon, K., Guallar, E., & Navas-Acien, A. (2012). Arsenic exposure and cardiovascular disease: An updated systematic review. *Current Atherosclerosis Reports*, 14(6), 542–555. Available from https://doi.org/10.1007/s11883-012-0280-x.

Moore, O. C., Xiu, W., Guo, H., Polya, D. A., van Dongen, B. E., & Lloyd, J. R. (2023). The role of electron donors in arsenic-release by redox-transformation of iron oxide minerals — A review. *Chemical Geology, 619*. Available from https://doi.org/10.1016/j.chemgeo.2023.121322, http://www.sciencedirect.com/science/journal/00092541.

Mukherjee, A., Fryar, A. E., Eastridge, E. M., Nally, R. S., Chakraborty, M., & Scanlon, B. R. (2018). Controls on high and low groundwater arsenic on the opposite banks of the lower reaches of River Ganges, Bengal basin, India. *Science of the Total Environment, 645*, 1371−1387. Available from https://doi.org/10.1016/j.scitotenv.2018.06.376, http://www.elsevier.com/locate/scitotenv.

Mukherjee, A., Fryar, A. E., & Howell, P. D. (2007). Regional hydrostratigraphy and groundwater flow modeling in the arsenic-affected areas of the western Bengal basin, West Bengal, India. *Hydrogeology Journal, 15*(7), 1397−1418. Available from https://doi.org/10.1007/s10040-007-0208-7.

Mukherjee, A., Verma, S., Gupta, S., Henke, K. R., & Bhattacharya, P. (2014). Influence of tectonics, sedimentation and aqueous flow cycles on the origin of global groundwater arsenic: Paradigms from three continents. *Journal of Hydrology, 518*, 284−299. Available from https://doi.org/10.1016/j.jhydrol.2013.10.044, http://www.elsevier.com/inca/publications/store/5/0/3/3/4/3.

Mukhopadhyay, R., Manjaiah, K. M., Datta, S. C., Yadav, R. K., & Sarkar, B. (2017). Inorganically modified clay minerals: Preparation, characterization, and arsenic adsorption in contaminated water and soil. *Applied Clay Science, 147*, 1−10. Available from https://doi.org/10.1016/j.clay.2017.07.017, http://www.elsevier.com/inca/publications/store/5/0/3/3/2/2/index.htt.

Nayak, A., Matta, G., & Uniyal, D. P. (2023). Hydrochemical characterization of groundwater quality using chemometric analysis and water quality indices in the foothills of Himalayas. *Environment, Development and Sustainability, 25*(12), 14229−14260. Available from https://doi.org/10.1007/s10668-022-02661-4, https://www.springer.com/journal/10668.

Neil, C. W., Yang, Y. J., Schupp, D., & Jun, Y. S. (2014). Water chemistry impacts on arsenic mobilization from arsenopyrite dissolution and secondary mineral precipitation: Implications for managed aquifer recharge. *Environmental Science and Technology, 48*(8), 4395−4405. Available from https://doi.org/10.1021/es405119q, http://pubs.acs.org/journal/esthag.

Panda, S., Tripathi, V. K., D.j, S., & Sharma, R. (2022). *Arsenic contamination of groundwater in Indo-Gangetic Plain. Hydrogeochemistry of Aquatic Ecosystems* (pp. 85−95). Wiley. Available from https://onlinelibrary.wiley.com/doi/book/10.1002/9781119870562, 10.1002/9781119870562.ch4.

Patel, A. K., Das, N., & Kumar, M. (2019). Multilayer arsenic mobilization and multimetal co-enrichment in the alluvium (Brahmaputra) plains of India: A tale of redox domination along the depth. *Chemosphere, 224*, 140−150. Available from https://doi.org/10.1016/j.chemosphere.2019.02.097, http://www.elsevier.com/locate/chemosphere.

Rahidul Hassan, H. (2023). A review on different arsenic removal techniques used for decontamination of drinking water. *Environmental Pollutants and Bioavailability, 35*(1). Available from https://doi.org/10.1080/26395940.2023.2165964, https://www.tandfonline.com/loi/tcsb21.

Rahim, H. U., Allevato, E., Vaccari, F. P., & Stazi, S. R. (2023). Biochar aged or combined with humic substances: Fabrication and implications for sustainable agriculture and environment-a review. *Journal of Soils and Sediments*. Available from https://doi.org/10.1007/s11368-023-03644-2, https://www.springer.com/journal/11368.

Rahman, A., Mondal, N. C., & Fauzia, F. (2021). Arsenic enrichment and its natural background in groundwater at the proximity of active floodplains of Ganga River, northern India. *Chemosphere*, *265*. Available from https://doi.org/10.1016/j.chemosphere.2020.129096, http://www.elsevier.com/locate/chemosphere.

Rahman, M. M., Mondal, D., Das, B., Sengupta, M. K., Ahamed, S., Hossain, M. A., … Chakraborti, D. (2014). Status of groundwater arsenic contamination in all 17 blocks of Nadia district in the state of West Bengal, India: A 23-year study report. *Journal of Hydrology*, *518*, 363–372. Available from https://doi.org/10.1016/j.jhydrol.2013.10.037, http://www.elsevier.com/inca/publications/store/5/0/3/3/4/3.

Rahman, M. S., Clark, M. W., Yee, L. H., Comarmond, M. J., Payne, T. E., & Burton, E. D. (2019). Effects of pH, competing ions and aging on arsenic(V) sorption and isotopic exchange in contaminated soils. *Applied Geochemistry*, *105*, 114–124. Available from https://doi.org/10.1016/j.apgeochem.2019.04.016, http://www.journals.elsevier.com/applied-geochemistry.

Raju, N. J. (2022). Arsenic in the geo-environment: A review of sources, geochemical processes, toxicity and removal technologies. *Environmental Research*, *203*, 111782. Available from https://doi.org/10.1016/j.envres.2021.111782.

Raturi, G., Chaudhary, A., Rana, V., Mandlik, R., Sharma, Y., Barvkar, V., … Dhar, H. (2023). Microbial remediation and plant-microbe interaction under arsenic pollution. *Science of the Total Environment*, *864*, 160972. Available from https://doi.org/10.1016/j.scitotenv.2022.160972.

Richards, L. A., Kumari, R., Parashar, N., Kumar, A., Lu, C., Wilson, G., … Gooddy, D. C. (2022). Environmental tracers and groundwater residence time indicators reveal controls of arsenic accumulation rates beneath a rapidly developing urban area in Patna, India. *Journal of Contaminant Hydrology*, *249*. Available from https://doi.org/10.1016/j.jconhyd.2022.104043, http://www.elsevier.com/locate/jconhyd.

Richards, L. A., Parashar, N., Kumari, R., Kumar, A., Mondal, D., Ghosh, A., & Polya, D. A. (2022). Household and community systems for groundwater remediation in Bihar, India: Arsenic and inorganic contaminant removal, controls and implications for remediation selection. *Science of the Total Environment*, *830*. Available from https://doi.org/10.1016/j.scitotenv.2022.154580, http://www.elsevier.com/locate/scitotenv.

Rios-Valenciana, E. E., Briones-Gallardo, R., Chazaro-Ruiz, L. F., Lopez-Lozano, N. E., Sierra-Alvarez, R., & Celis, L. B. (2020). Dissolution and final fate of arsenic associated with gypsum, calcite, and ferrihydrite: Influence of microbial reduction of As(V), sulfate, and Fe(III). *Chemosphere*, *239*. Available from https://doi.org/10.1016/j.chemosphere.2019.124823, http://www.elsevier.com/locate/chemosphere.

Sarkar, A., Paul, B., & Darbha, G. K. (2022). The groundwater arsenic contamination in the Bengal Basin—A review in brief. *Chemosphere*, *299*, 134369. Available from https://doi.org/10.1016/j.chemosphere.2022.134369.

Sathe, S. S., Goswami, L., & Mahanta, C. (2021). Arsenic reduction and mobilization cycle via microbial activities prevailing in the Holocene aquifers of Brahmaputra flood plain. *Groundwater for Sustainable Development*, *13*, 100578. Available from https://doi.org/10.1016/j.gsd.2021.100578.

Shaji, E., Santosh, M., Sarath, K. V., Prakash, P., Deepchand, V., & Divya, B. V. (2021). Arsenic contamination of groundwater: A global synopsis with focus on the Indian

Peninsula. *Geoscience Frontiers*, *12*(3), 101079. Available from https://doi.org/10.1016/j.gsf.2020.08.015.

Shankar, S., Shanker, U., & Shikha. (2014). Arsenic contamination of groundwater: A review of sources, prevalence, health risks, and strategies for mitigation. *The Scientific World Journal*, *2014*, 1−18. Available from https://doi.org/10.1155/2014/304524.

Shrivastava, A., Barla, A., Yadav, H., & Bose, S. (2014). Arsenic contamination in shallow groundwater and agricultural soil of Chakdaha block, West Bengal, India. *Frontiers in Environmental Science*, *2*. Available from https://doi.org/10.3389/fenvs.2014.00050, http://journal.frontiersin.org/article/10.3389/fenvs.2014.00050/pdf.

Shrivastava, A., Ghosh, D., Dash, A., & Bose, S. (2015). Arsenic contamination in soil and sediment in India: Sources, effects, and remediation. *Current Pollution Reports*, *1*(1), 35−46. Available from https://doi.org/10.1007/s40726-015-0004-2, http://springer.com/environment/pollution + and + remediation/journal/40726.

Sikdar, P. K., & Chakraborty, S. (2008). Genesis of arsenic in groundwater of North Bengal Plain using PCA: A case study of English Bazar Block, Malda District, West Bengal, India. *Hydrological Processes*, *22*(12), 1796−1809. Available from https://doi.org/10.1002/hyp.6742.

Singh, S. K., & Taylor, R. W. (2019). Assessing the role of risk perception in ensuring sustainable arsenic mitigation. *Groundwater for Sustainable Development*, *9*. Available from https://doi.org/10.1016/j.gsd.2019.100241, http://www.journals.elsevier.com/groundwater-for-sustainable-development/.

Singh, S., Sharma, P., Mudhulkar, R., Chakravorty, B., Singh, A., & Sharma, S. D. (2022). Assessment of hydrogeochemistry and arsenic contamination in groundwater of Bahraich District, Uttar Pradesh, India. *Arabian Journal of Geosciences*, *15*(1). Available from https://doi.org/10.1007/s12517-021-09222-5.

Stolze, L., Battistel, M., & Rolle, M. (2022). Oxidative dissolution of arsenic-bearing sulfide minerals in groundwater: Impact of hydrochemical and hydrodynamic conditions on arsenic release and surface evolution. *Environmental Science and Technology*, *56*(8), 5049−5061. Available from https://doi.org/10.1021/acs.est.2c00309, http://pubs.acs.org/journal/esthag.

Stolze, L., Zhang, D., Guo, H., & Rolle, M. (2019). Model-based interpretation of groundwater arsenic mobility during in situ reductive transformation of ferrihydrite. *Environmental Science and Technology*, *53*(12), 6845−6854. Available from https://doi.org/10.1021/acs.est.9b00527, http://pubs.acs.org/journal/esthag.

Subdiaga, E., Orsetti, S., & Haderlein, S. B. (2019). Effects of sorption on redox properties of natural organic matter. *Environmental Science and Technology*, *53*(24), 14319−14328. Available from https://doi.org/10.1021/acs.est.9b04684, http://pubs.acs.org/journal/esthag.

Swain, S., Taloor, A. K., Dhal, L., Sahoo, S., & Al-Ansari, N. (2022). Impact of climate change on groundwater hydrology: A comprehensive review and current status of the Indian hydrogeology. *Applied Water Science*, *12*(6). Available from https://doi.org/10.1007/s13201-022-01652-0, http://www.springer.com/earth + sciences + and + geography/hydrogeology/journal/13201.

Tahri, N., Bahafid, W., Sayel, H., & El Ghachtouli, N. (2013). *Biodegradation: Involved microorganisms and genetically engineered microorganisms*. InTech. Available from 10.5772/56194.

Taylor, R. G., Scanlon, B., Döll, P., Rodell, M., Van Beek, R., Wada, Y., ... Treidel, H. (2013). Ground water and climate change. *Nature Climate Change*, *3*(4), 322−329. Available from https://doi.org/10.1038/nclimate1744.

Thakur, B. K., Gupta, V., Bhattacharya, P., Jakariya, M., & Tahmidul Islam, M. (2021). Arsenic in drinking water sources in the Middle Gangetic Plains in Bihar: An assessment of the depth of wells to ensure safe water supply. *Groundwater for Sustainable Development*, *12*, 100504. Available from https://doi.org/10.1016/j.gsd.2020.100504.

Thi Hoa Mai, N., Postma, D., Thi Kim Trang, P., Jessen, S., Hung Viet, P., & Larsen, F. (2014). Adsorption and desorption of arsenic to aquifer sediment on the Red River floodplain at Nam Du, Vietnam. *Geochimica et Cosmochimica Acta*, *142*, 587−600. Available from https://doi.org/10.1016/j.gca.2014.07.014, http://www.journals.elsevier.com/geochimica-et-cosmochimica-acta/.

Van Geen, A., Zheng, Y., Goodbred, S., Horneman, A., Aziz, Z., Cheng, Z., ... Ahmed, K. M. (2008). Flushing history as a hydrogeological control on the regional distribution of arsenic in shallow groundwater of the Bengal Basin. *Environmental Science and Technology*, *42*(7), 2283−2288. Available from https://doi.org/10.1021/es702316k.

Viacava, K., Qiao, J., Janowczyk, A., Poudel, S., Jacquemin, N., Meibom, K. L., ... Bernier-Latmani, R. (2022). Meta-omics-aided isolation of an elusive anaerobic arsenic-methylating soil bacterium. *bioRxiv*. Available from https://doi.org/10.1101/2022.01.25.477449, https://www.biorxiv.org.

Wan, D., Wang, J., Shi, Y., Qu, D., & Zhang, J. (2023). Construction of continuous-flow electrodialysis ion-exchange membrane bioreactor for effective removal of nitrate and perchlorate: Modelling and impact analysis of environmental variables. *Chemical Engineering Journal*, *462*, 142144. Available from https://doi.org/10.1016/j.cej.2023.142144.

Wu, H., Kalia, V., Niedzwiecki, M. M., Kioumourtzoglou, M. A., Pierce, B., Ilievski, V., ... Gamble, M. V. (2023). Metabolomic changes associated with chronic arsenic exposure in a Bangladeshi population. *Chemosphere*, *320*. Available from https://doi.org/10.1016/j.chemosphere.2023.137998, http://www.elsevier.com/locate/chemosphere.

Xia, X., Teng, Y., & Zhai, Y. (2022). Biogeochemistry of iron enrichment in groundwater: An indicator of environmental pollution and its management. *Sustainability*, *14*(12), 7059. Available from https://doi.org/10.3390/su14127059.

Xue, Q., Ran, Y., Tan, Y., Peacock, C. L., & Du, H. (2019). Arsenite and arsenate binding to ferrihydrite organo-mineral coprecipitate: Implications for arsenic mobility and fate in natural environments. *Chemosphere*, *224*, 103−110. Available from https://doi.org/10.1016/j.chemosphere.2019.02.118, http://www.elsevier.com/locate/chemosphere.

Yamauchi, H., Yoshida, T., & Takata, A. (2023). *Arsenic exposure and health effects: Differences by chemical structure, chemical form and arsenic methylation capacity*. IntechOpen. Available from 10.5772/intechopen.1001454.

Yang, Z., Zhang, N., Sun, B., Su, S., Wang, Y., Zhang, Y., ... Zeng, X. (2022). Contradictory tendency of As(V) releasing from Fe−As complexes: Influence of organic and inorganic anions. *Chemosphere*, *286*, 131469. Available from https://doi.org/10.1016/j.chemosphere.2021.131469.

Yuan, C., Wei, Y., Xu, X., & Cao, X. (2023). Transport and transformation of arsenic in coastal aquifer at the scenario of seawater intrusion followed by managed aquifer recharge. *Water Research*, *229*, 119440. Available from https://doi.org/10.1016/j.watres.2022.119440.

Zanzo, E., Balint, R., Prati, M., Celi, L., Barberis, E., Violante, A., & Martin, M. (2017). Aging and arsenite loading control arsenic mobility from ferrihydrite-arsenite coprecipitates. *Geoderma*, *299*, 91–100. Available from https://doi.org/10.1016/j.geoderma.2017.03.004, http://www.elsevier.com/inca/publications/store/5/0/3/3/3/2.

Zecchin, S., Crognale, S., Zaccheo, P., Fazi, S., Amalfitano, S., Casentini, B., ... Cavalca, L. (2021). Adaptation of microbial communities to environmental arsenic and selection of arsenite-oxidizing bacteria from contaminated groundwaters. *Frontiers in Microbiology*, *12*. Available from https://doi.org/10.3389/fmicb.2021.634025, https://www.frontiersin.org/journals/microbiology#.

Zhang, J., Zhou, W., Liu, B., He, J., Shen, Q., & Zhao, F. J. (2015). Anaerobic arsenite oxidation by an autotrophic arsenite-oxidizing bacterium from an arsenic-contaminated paddy soil. *Environmental Science and Technology*, *49*(10), 5956–5964. Available from https://doi.org/10.1021/es506097c, http://pubs.acs.org/journal/esthag.

Zhang, Z., Xiao, C., Adeyeye, O., Yang, W., & Liang, X. (2020). Source and mobilization mechanism of iron, manganese and arsenic in groundwater of Shuangliao City, Northeast China. *Water*, *12*(2), 534. Available from https://doi.org/10.3390/w12020534.

Zheng, T., Deng, Y., Wang, Y., Jiang, H., Xie, X., & Gan, Y. (2020). Microbial sulfate reduction facilitates seasonal variation of arsenic concentration in groundwater of Jianghan Plain, Central China. *Science of the Total Environment*, *735*, 139327. Available from https://doi.org/10.1016/j.scitotenv.2020.139327.

Geospatial technique based flood hazard assessment and mapping: a case study of Orang National Park, Assam, India

Rani Kumari Shah[1] and Rajesh Kumar Shah[2]

[1]*Department of Geography, Cotton University, Guwahati, Assam, India*
[2]*Department of Zoology, D.H.S.K.College, Dibrugarh, Assam, India*

6.1 Introduction

Natural disasters are the result of various geological, hydrological, as well as meteorological events that lead to significant loss of life, property, and damage to the natural landscapes. As estimated by the United Nations office for Disaster Risk Reduction, these natural hazards claim nearly 42 million lives annually on a global scale and result in an average economic loss of 293 billion USD (UNDRR, 2017). Flood is the most frequently occurring hydrometeorological hazards with a profound impact (Gupta & Dixit, 2022). A severe hike in the frequency and severity of flood events has been witnessed in recent years due to various factors such as irregular rainfall (RF) patterns, river overflows, rapid melting of snow, deforestation, unregulated urbanization and unorganized human settlements along the riverbanks and coastal regions, and so forth (Armenakis, Du, Natesan, Persad, & Zhang, 2017). Flood is among the world's most devastating disasters that cause more casualties and damage to property than any other natural phenomenon (Duan et al., 2014; Forkuo, 2011; Hapuarachchi, Wang, & Pagano, 2011; Tsakiris, 2014; Wang, Li, Tang, & Zeng, 2011). Apart from its severe impact on humans and the environment, it is also one of the most complex phenomenon to model and prepare for. Extreme RF, failures of dams, tsunamis, and storm surges are some of the potential drivers that increase its complexities (Glas et al., 2019; Zwenzner & Voigt, 2009; UNDRR, 2019). Thus it is very important to recognize flood as a potential threat due to their potential for extensive destruction (Stefanidis & Stathis, 2013).

India is the second most flood-prone country in the world, next to China. It encounters around 17 flood events annually on average, affecting nearly 345 million individuals (CRED, 2020). The extensive river network found in India and the monsoon system exacerbate its vulnerability, resulting in the flooding of around 5.74 million hectares of its total land surface (Dhar & Nandargi, 2004; Subrahmanyam, 1988). Assam is a north eastern state of India that faces significant flood risks due to the complex and braided nature of the river Brahmaputra. The periodic floods in this region occur every year and cause havoc and are a major concern in Assam. These floods are caused by heavy RF due to monsoon and snow melt from the Himalayas. Other factors like erosion and human settlements along the flood-prone regions contribute to the severity of flooding. Orang National Park (ONP) is situated within the floodplain of the river Brahmaputra. Brahmaputra flows along the southern boundary of the park and forms a complex network of channels, especially during the monsoon season. Small tributaries like Pachnoi and Dhansiri flow along the park's boundaries, eventually joining the Brahmaputra River. The dynamic channels form several *beels* all over the area. Close proximity of the park to the river as well as its low terrain render it highly susceptible to flooding, mostly during the monsoon season (Hazarika & Saikia, 2010). The park is a home to diverse plants and animals that include many rare and threatened species. Some of the important fauna found here are One-horned *Rhinoceros*, Asiatic elephant, Royal Bengal tiger, Indian hog deer, wild buffalo, pygmy hog, Gangetic dolphin, Indian pangolin, hog deer, *Rhesus macaque*, porcupine, Indian fox, small Indian civet, otter, leopard cat, jungle cat, tortoise, turtle, pythons, and cobras. More than 280 species of birds have been recorded in the park by several workers (Chakdar, Singha, & Choudhury, 2019; Choudhury, 2000; Rahmani, Narayan, Rosalind, & Sankaran, 1990). Some important bird species include Bengal florican, Baer's pochard, Greater adjutant, vultures, yellow-throated sparrow, brown-headed barbet, bristled grass-warbler, black-necked stork, greater adjutant stork, spot-billed pelican, great white pelican, Bengal florican, and lesser adjutant. Some important flora of the park include *Saccharum spontaneum*, *Imperata cylindrica*, *Apluda mutica*, *Dysoxylum binectariferum*, *Melia azadirachta*, *Sterculia villosa*, and *Toona ciliate* (Chakdar et al., 2019). The frequent flooding event in Assam causes profound impacts on both humans and wildlife (Debbarma & Deen, 2020). A significant portion of the ONP was inundated during the flood event that took place in 2022, causing substantial damage, and several antipoaching camps were submerged under water.

Due to severe flood events occurring all over the world and the severe damages they cause, flood hazard mapping has become very important. The major objective of flood hazard mapping is to effectively manage floods that result from factors like heavy RF, dam overflows, and so forth, which aims to minimize the adverse impacts on lives and property. The development of a flood hazard map is very useful in assessing and mitigating flood-related risks in vulnerable areas (Zhang, Zhou, Xu, & Watanabe, 2002). Flood hazard mapping is one of the most significant approaches that is used to evaluate the risk of flooding in a particular area. Such an approach utilizes a combination of qualitative and semiquantitative methods considering a wide range of environmental factors such as the shape of the land, soil type, the amount of precipitation, land use in the area, and the hydrological properties of the surrounding watersheds (Hallegatte, Green, Nicholls, & Corfee-Morlot, 2013). Such an assessment is essential because it helps us to predict flood hazard that is important for developing effective strategies to manage floods and to promote sustainable environmental practices. One of the key aspects of assessment of flood hazard is the generation of flood hazard maps that can play an important role in land use planning in flood-prone regions and also help in identifying areas at high risk of flooding (Khaing et al., 2019). Such maps are user-friendly and easily used by the planners to pinpoint areas that need immediate attention and prioritize flood mitigation efforts (Ajin, 2013; Bapalu & Sinha, 2005; Danumah et al., 2016; Forkuo, 2011; Wang et al., 2011; Argaz, 2019). Flood hazard mapping and analysis is also an important part of early warning systems for preventing and mitigating future food situations. Some of the commonly used models for flood hazard mapping include the analytical hierarchy process (AHP), frequency ratio model, artificial neural networks, hydrological simulation program-FORTRAN, hydrological forecasting systems, HEC-RAS, storm water management model, logistic regression, generalized linear models, entropy, k-nearest neighbors, random forest, weight of evidence, Shannon's entropy model, gray decision-making trial and evaluation laboratory, support vector machine, erosion models, and kinematic runoff (Bhatt, Sinha, Deka, & Kumar, 2014; Shah & Shah, 2023a,b; Wiles & Levine, 2002; FLaouacheria, Kechida, & Chabi, 2019; Fonseca, Santos, & Santos, 2018; Rangari, Sridhar, Umamahesh, & Patel, 2019; Youssef, Pradhan, & Sefry, 2016; Rahmati et al., 2019). Use of remote sensing (RS) and geographic information system (GIS) software tools has gained popularity in recent years due to their ability to provide essential justification and fresh perspective for vulnerability assessments. These tools provide significant insights into specific areas, specifically when satellite imagery is analyzed. The integration of RS and GIS for flood hazard assessment is currently prevalent and highly effective and provides highly accurate maps (Ali, Khatun, Ahmad, & Ahmad, 2019). They serve as an important tool for devising flood hazard risk assessment strategies through the multicriteria analysis (MCA) approach (Arya & Singh, 2021; El-Haddad et al., 2021). The MCA approach is found to be multifunctional and valuable by several workers (Antoine, Fischer, & Makowski, 1997; Rikalovic, Cosic, & Lazarevic, 2014; Gil & Kellerman, 1993; Pohekar & Ramachandran, 2004; Saki, Dehghani, Jodeiri Shokri, & Bogdanovic, 2020; Saleh, Aliani, & Amoushahi, 2020; Wang, Jing, Zhang, & Zhao, 2009). The advanced software tools can simplify the process of preparing flood hazard maps that can give valuable results for mapping flood susceptibility and vulnerability. Such maps are widely acceptable and considered to be more credible.

Very limited reports on the flooding events and their impacts are available as far as ONP is concerned. Given these circumstances, there is an urgent need to undertake a comprehensive study on flood risk assessment in order to identify vulnerable areas and contributing factors. Thus the flood hazard mapping of ONP becomes vital in order to minimize and prevent the potential damage to the park and wildlife in future. As ONP is a homeland for various plants and animals including several globally threatened fauna, any damage to their habitat may also lead to their extinction in the future. Thus our present study on flood hazard mapping of ONP is very crucial. The main objective of this study is to develop a flood hazard map for the ONP utilizing RS and GIS techniques, with the application of the AHP model. By identifying various flood controlling factors within the study area and employing AHP modeling, the aim is to provide a valuable tool for policymakers and planners to mitigate flood hazards in this globally important park effectively. The present study is the first attempt to categorize the study area in different flood susceptible zones using AHP integrated with RS-GIS. No study on flood susceptibility has been carried out earlier in this rich national park.

6.2 Materials and methods

6.2.1 Description of the study area

ONP is situated in Darrang and Sonitpur districts of Assam and is very popular for its diverse flora, fauna, and esthetic beauty. The geographical extension of the park is between 26°28′49.0891″N to 26°35′43.6644″N latitude and 92°14′50.1272″E to 92°24′56.1071″E longitude. It is often known as "mini Kaziranga" due to its landscape being similar to the Kaziranga National Park and its rich population of one-horned *Rhinoceros*. The park is surrounded by the river Brahmaputra on the south, its tributary Dhansiri on the west, and Panchnoi on the east. The northern part of the park is bounded by Nalbari and villages of Darrang district. Due to the close proximity of the park to these three rivers, Orang forms a vital part of the Indo-Burma biodiversity hotspot and is among the attractive parks of Assam that covers an area of 78.08 km^2. The terrain of the region is flat that shows general slopping from north to south. The altitude ranges between 48 and 90 m above the mean sea level. Both young and old alluvial soil are found with varying humus contents and the texture varies from sandy loam to silty loam in nature (Talukdar & Sharma, 1995). The park experiences an average annual RF of 1910 mm (Sarma, Mipun, Talukdar, Kumar, & Basumatary, 2011). A map of the study area is shown in Fig. 6.1.

6.2.2 Sources of data

Various secondary data were used in the present study. These data sources include satellite imagery, digital elevation model (DEM), and climatic data (RF). The Landsat sentinel 2 A image taken on January 19, 2023, was downloaded from US Geological Survey (USGS) Earth Explorer (https://earthexplorer.usgs.gov) website

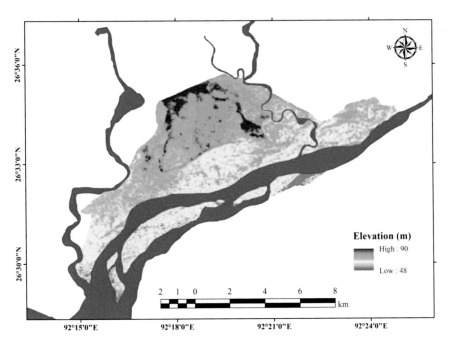

FIGURE 6.1

Study area map.

with 5% cloud cover, which offers satellite images with significant spatial and spectral resolution from different time intervals across the world. Landsat sentinel 2 A images have 10 m spatial resolution and are capable of monitoring the features of the earth's surface such as land use and cover, vegetation health, and drainage characteristics, all of which directly impact flood events in any region. The sentinel 2-A satellite image was georeferenced to the World Geodetic System 1984 datum and Universal Transverse Mercator Zone 46 N coordinate system. Additionally, SRTM DEM data with a 30 m spatial resolution was accessed through the USGS Earth Explorer (https://earthexplorer.usgs.gov). DEM data provide detailed terrain information, and such data can significantly contribute to flood hazard mapping (Muhadi, Abdullah, Bejo, Mahadi, & Mijic, 2020). RF data (0.25° × 0.25° gridded datasets) were obtained from the India Meteorological Department for the year 2022 (Pai et al., 2014). Such high-resolution RF data are very essential for flood modeling and mapping.

6.2.3 Preparation of flood controlling factors

Based on the literature review, personal observation, and discussion with local residents, the flood controlling factors were determined. Accordingly, elevation

(E), slope (S), land use and land cover (LULC), proximity to river, drainage density (DD), normalized difference vegetation index (NDVI), topographic wetness index (TWI), RF, and flow accumulation (FA) were considered as important flood controlling factors in ONP. Table 6.1 depicts the details of mode of preparation of each factor.

6.2.4 The analytical hierarchy process model

The AHP, introduced by Saaty (1980), is a prominent decision-making approach using multiple criteria evaluation. It offers a structured approach to systematically compare and weigh different criteria by employing pairwise comparisons. Relative importance of one criterion over another can be easily accessed by the decision-makers, producing a pairwise comparison matrix (Malczewski, 2000). A standardized scale ranging from 1 to 9 is used in this method where 1 denotes equal importance, 3 moderate importance, 5 strong importance, 7 very strong importance, and 9 extreme importance. Its ability to facilitate group decision-making has been the main compelling aspect of its wide adoption, global recognition, and application. The AHP is a straightforward and adaptable method that is widely used in flood analysis, as it delivers precise outcomes. It can play a vital role in managing flood by helping decision-makers prioritize and make necessary choices regarding various strategies of flood hazard reduction, allocation of resources, and response planning. The model can be used to study various factors and the complex relationship between them in flood hazard assessment. For each controlling factors, the rank values were assigned on the basis of their relative importance on a scale from 1 to 9. Depending upon the importance of the flood-causing factors, pairwise and normalized pairwise comparison matrices were prepared. Weight values of each flood-causing factor were computed, and consistency ratio (CR) was calculated (Saaty & Vargas, 2000; Saaty, 1977, 1980, 1990, 2008). As nine flood-causing factors were considered in the present study, the value of random consistency index is taken as 1.45. The methodological flowchart is shown in Fig. 6.2.

$$\text{Consistancy ratio (CR)} = \frac{\text{CI}}{\text{RI}}$$

where

$$\text{CI(consistency index)} = \frac{\lambda - n}{n - 1}$$

RI = random index (1.45 for nine factors)
n = number of flood-causing factors
λ = average value of the consistency vector

Table 6.1 Methods used in preparation of each factor.

Factors	Mode of preparation of each factor
E	E plays a critical role in influencing the occurrence of floods (Shah & Shah, 2023a,b). Generally areas at lower Es are more susceptible to flooding.
	The spatial E map was created using DEM data and was categorized into five classes in ArcGIS 10.8.2 software.
S	S is a crucial factor in understanding and predicting flood events.
	The S map was developed from SRTM DEM using the surface tool in ArcGIS 10.8.2 software and was classified into five groups.
PR	PR increases the susceptibility to flooding compared to areas located further away (Glenn et al., 2012).
	The PR map is prepared using the Euclidean distance tool in ArcGIS 10.8.2 software (Arora, Pandey, Siddiqui, Hong, & Mishra, 2021; Shadmehri Toosi, Calbimonte, Nouri, & Alaghmand, 2019).
DD	The amount of runoff within a basin is directly and markedly influenced by the river density. Regions characterized by higher river density are at a heightened risk of experiencing floods.
	The DD map was prepared using ArcGIS 10.8.2 software and was grouped into five classes using quantile classification tool.
LULC	LULC patterns have a profound influence on a region's hydrology and can significantly impact the frequency severity of flood
	The LULC classification was performed using supervised classification method (maximum likelihood algorithm). The map was classified into five classes such as moist deciduous forest, swampy land, water bodies, sandbars, and dry grassland.
NDVI	The NDVI is a vegetation health indicator which assesses the health of vegetation. It plays a crucial role in understanding floods dynamics. It ranges from -1 to $+1$ (Khosravi, Nohani, Maroufinia, & Pourghasemi, 2016).
	Here, NDVI values were calculated using Landsat sentinel 2 A image based on following formula
	$NDVI = (NIR - VIS)/(NIR + VIS)$, where NIR = near infrared band, VIS = visible band
RF	RF is one of the primary triggers of flooding. Seasonal RF patterns can lead to prolonged and widespread flooding in certain regions.
	Spatial map of RF pattern was developed using interpolated method by the inverse distance weighting tool in ArcGIS 10.8.2 (Shadmehri Toosi et al., 2019).
TWI	TWI quantifies the prosperity of an area to accumulate water. A higher TWI value indicates areas with a greater to retain water, making them more vulnerable to flooding.
	Here, S and DEM data were applied to compute TWI using following formula $TWI = \ln(F)/(10\ S)$
FA	FA is essential for predicting flood hazard. Areas with high FA are more likely to experience flood (Vojtek & Vojteková, 2019).
	FA raster map was prepared by DEM data using the hydrology tool in ArcGIS 10.8.2 software.

DD, *Drainage density;* DEM, *digital elevation model;* E, *elevation;* FA, *flow accumulation;* LULC, *land use and land cover;* NDVI, *normalized difference vegetation index;* PR, *proximity to river;* RF, *rainfall;* S, *slope;* TWI, *topographic wetness index.*

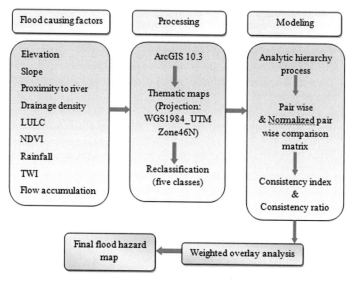

FIGURE 6.2

Methodological flowchart.

6.3 Results

Our research focused on identifying and examining nine flood-causing factors employing rigorous statistical methods for a comprehensive assessment. All the nine factors were subdivided into five classes, as shown in Table 6.2.

Thematic maps were prepared for all the factors and were classified in five different classes showing their spatial distribution. One of the most crucial factors causing flooding in a particular area is the elevation of that region. Generally, water flows from a higher elevation toward a lower elevation. Floods tend to be common on flat terrains with smaller altitude values due to their inverse relationship with the probability of occurrence and thus serve as an accurate indication of vulnerability due to floods (Hammami et al., 2019). The park ranges from 48 to 90 m in elevation. Low elevated regions such as water bodies adjacent to others are covered with water. The red represents the highest elevation from a western perspective of the park, while the dark blue demonstrates the lowest elevation toward the south-eastern end of the park (Fig. 6.3A).

The aspect of the slope is essential in indicating surface zonation, and the rate and duration of the water flow largely depend on the slope. The water moves slower, piles up for a longer time, and accumulates longer on a flatter surface, leading to flood (Rimba, Setiawati, Sambah, & Miura, 2017). The slope of the study area was from 0° to 24.64°. The areas with the slope within 0°−0.052° and 0.053°−0.097° indicates water bodies and flat lands with a very gentle slope,

Table 6.2 Classes of flood-causing factors and weight.

Factor	Class	Hazard class rating	Weight (%)
Elevation (m)	48–56.89	5	13
	56.9–62	4	
	62.01–66.94	3	
	66.95–73.86	2	
	73.87–90	1	
Slope (degree)	0–0.052	5	11
	0.053–0.097	4	
	0.098–9.06	3	
	9.07–16.35	2	
	16.35–24.64	1	
Proximity to river (m)	0–190.8	5	19
	190.9–500.4	4	
	500.5–1131	3	
	1132–2060	2	
	2061–3752	1	
Drainage density (m/km)	0–44.57	1	13
	44.58–125.6	2	
	125.7–210.7	3	
	210.8–305.9	4	
	306–516.6	5	
LULC (level)	Moist deciduous forest	1	12
	Water bodies	5	
	Swampy land	3	
	Sandbars	4	
	Dry grassland	2	
NDVI (level)	−0.145 to 0.0267	5	7
	0.0268–0.161	4	
	0.162–0.266	3	
	0.267–0368	2	
	0.369–0.54	1	
Rainfall deviation (mm/year)	1770–1839	3	8
	1840–1894	2	
	1895–1941	1	
	1942–1998	1	
	1999–2111	2	
TWI (level)	−16.52 to −7	1	8
	−6.999 to −3.417	2	
	−3.416 to −0.1418	3	
	−0.1417 to 2.724	4	
	2.725–9.583	5	
Flow accumulation	0–2.056.8	1	10
	2056.9–8432.8	2	
	8432.9–18,511	3	
	18,512–28,795	4	
	28,796–52,448	5	

LULC, Land use and land cover; NDVI, normalized difference vegetation index; TWI, topographic wetness index.

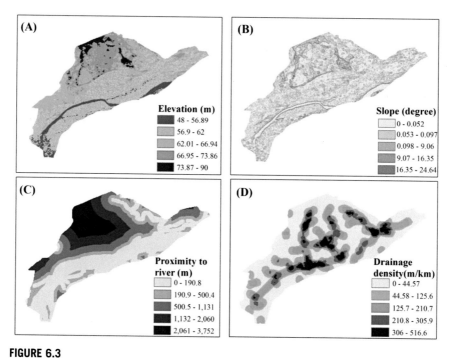

FIGURE 6.3

Flood-causing factors: (A) elevation, (B) slope, (C) proximity to river, and (D) drainage density.

which are suitable for occurrence of floods compared to areas with a moderate to steep slope (Fig. 6.3B). Flood inundation is more intense in areas closer to a stream confluence compared to those remote, since at a point of a stream confluence, the channel often carries a combined load and flow of two or more upstream tributaries. When a river overflows, the flow is more than the drainage capacity of the river, and therefore, it increases the depth of the water around river margins. This flood will not affect only the closest river point, but the surrounding areas will also experience water flooding and risks of flood (Chakraborty & Mukhopadhyay, 2019). Classification of the study area was done into five classes ranging from 0 to 190.8, 190.9 to 500.4, 500.5 to 1131, 1132 to 2060, and 2061 to 3752 m. The areas grouped as very high and high flood risk are located within 0−190.8 and 190.9−500.4 m, respectively, as depicted in Fig. 6.3C. Additionally, DD is another major element that significantly adds to the flood occurrence in a place (Onuşluel Gül, 2013). DD per unit area has a direct impact on runoff occurring from the basin. Areas of a high river concentration are susceptible to flooding (Shah & Shah, 2023a,b). This analysis revealed that the minimum DD was between 0 and 44.57 m/km, while the maximum was a high of 516.6 m/km, as shown in Fig. 6.3D. Flood susceptibility is higher in an area with high DD than in an area with a low DD. The impact of various LULC types on susceptibility to flooding in any region can be quite substantial. The interrelationship of various hydrological parameters such as runoff, infiltration, and RF abstraction is controlled by the LULC. For example, forests and natural trees enhance water penetration and seepage. Significant parts covered by water, sand, and marshlands have an immense influence on flooding. The LULC classification of the park is categorized as moist deciduous forest (19.4 km^2), swampy land (33.15 km^2), water bodies (10.83 km^2), sandbars (4.20 km^2), and dry grassland (10.5 km^2). The dominant LULC in the study area is swamps (42.45%), as depicted in Fig. 6.4A.

The presence of vegetation across any surface also causes reduction in speed of flow of water while increasing soil water infiltration associated with that surface (Zhao et al., 2019). Vegetative cover is helpful in reducing the magnitude and swiftness of flood. A high positive value indicates high density, thus implying that the area is covered with dense forests, while a low positive value indicates low density, thus implying that the area has grasslands or shrubs. Flood susceptibility goes down as NDVI increases. NDVI varied between 0.145 and 0.54 for the study area. A spatial distribution map indicates high frequency of very low and

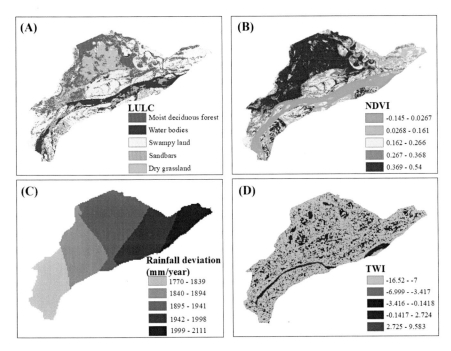

FIGURE 6.4

Flood-causing factors: (A) land use and land cover, (B) normalized difference vegetation index, (C) rainfall deviation, and (D) topographic wetness index.

low index zones within the southern, south-eastern, and southwestern regions portrayed as green and light green coloration, as shown in Fig. 6.4B. The probability of flood in an area increases with the increasing RF within a specific time period. The RF value was found to be in between 1770 and 2111 mm/year. RF deviation in the study area has been categorized into five classes: 1770−1839, 1840−1894, 1895−1941, 1942−1998, and 1999−2111 mm/year (Fig. 6.4C). The amount of soil moisture in an area can be forecast using TWI, which describes the tendency of accumulated water in that particular area. The TWI index clearly depicts how a slope affects hydrological processes. TWI explains the water accumulation trend at a particular region, and the local slope shows the influence of gravitational forces on the flow of water (Fernández & Lutz, 2010; Pourali, Arrowsmith, Chrisman, Matkan, & Mitchell, 2016). A spatial distribution map of TWI was prepared and grouped into five classes such as −16.52 to −7, −6.999 to −3.417, −3.416 to −0.1418, −0.1417 to 2.724, and 2.725 to 9.583. The pink color depicted the highest values, which is shown in Fig. 6.4D. FA is also an essential parameter in determining flood hazards. The values of accumulated flow show regions with a concentrated water flow area, thereby indicating the probability of a flood hazard. The thematic map was grouped into five classes such as 0−2.056.8, 2056.9−8432.8, 8432.9−18,511, 18,512−28,795, and 28,796−52,448 (Fig. 6.5A). The higher score of FA indicates that the area has lower probability flood hazards because of its less concentrated water in that area and vice versa.

Table 6.3 presents the pairwise matrix, while Table 6.4 shows the normalized pairwise matrix of AHP. CR was computed and found to be 0.042 (Table 6.5). According to Saaty (1990), a CR value between 0 and 0.1 is considered to be consistent. Hence, in the present case, the CR value can be acceptable in decision-making. The CI value was found to be 0.061. The factors were prioritized in a systematic approach as per the geographical characteristics of the study area,

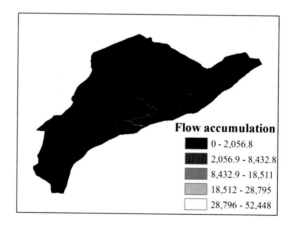

Flow accumulation
- 0 - 2,056.8
- 2,056.9 - 8,432.8
- 8,432.9 - 18,511
- 18,512 - 28,795
- 28,796 - 52,448

FIGURE 6.5

Flood-causing factors: flow accumulation.

Table 6.3 Pairwise comparison matrix.

Factors	E	S	PR	DD	LULC	NDVI	RF	TWI	FA
E	1	1	1	2	1	2	1	1	1
S	1	1	0.50	1	1	2	2	1	1
PR	1	2	1	2	2	3	2	3	2
DD	1	1	0.50	1	1	3	1	3	1
LULC	1	1	0.50	1	1	2	2	2	1
NDVI	1	0.50	0.33	0.30	0.50	1	1	1	1
RF	1	0.50	0.50	1	0.50	1	1	1	1
TWI	1	1	0.33	0.30	0.50	1	1	1	1
FA	1	1	0.50	1	1	1	1	1	1
Sum	9	9	5.16	9.60	8.50	16	12	14	10

DD, Drainage density; E, elevation; FA, flow accumulation; LULC, land use and land cover; NDVI, normalized difference vegetation index; PR, proximity to river; RF, rainfall; S, slope; TWI, topographic wetness index.

Table 6.4 Normalized pairwise comparison matrix.

Factors	E	S	PR	DD	LULC	NDVI	RF	TWI	FA
E	0.11	0.11	0.19	0.21	0.12	0.13	0.08	0.07	0.10
S	0.11	0.11	0.10	0.10	0.12	0.13	0.17	0.07	0.10
PR	0.11	0.22	0.19	0.21	0.24	0.19	0.17	0.21	0.20
DD	0.11	0.11	0.10	0.10	0.12	0.19	0.08	0.21	0.10
LULC	0.11	0.11	0.10	0.10	0.12	0.13	0.17	0.14	0.10
NDVI	0.11	0.06	0.06	0.03	0.06	0.06	0.08	0.07	0.10
RF	0.11	0.06	0.10	0.10	0.06	0.06	0.08	0.07	0.10
TWI	0.11	0.11	0.06	0.03	0.06	0.06	0.08	0.07	0.10
FA	0.11	0.11	0.10	0.10	0.12	0.06	0.08	0.07	0.10

DD, Drainage density; E, elevation; FA, flow accumulation; LULC, land use and land cover; NDVI, normalized difference vegetation index; PR, proximity to river; RF, rainfall; S, slope; TWI, topographic wetness index.

background information, and previous studies (Rahmati, Haghizadeh, & Stefanidis, 2016; Samanta, Bhunia, Shit, & Pourghasemi, 2018; Ullah and Zhang, 2020). The final weight for each factor is shown in Table 6.2.

A total of nine raster images (reclassified into 30 m spatial resolution) were used in the AHP model in order to identify the flood hazard zones. Flood hazard indices (FHIs) using the weight database were computed. The final thematic map of ONP was classified into four classes by weighted overlay tool in ArcGIS 10.8.2, namely, low (1−2), moderate (3), high (4), and very high (5) hazard

Table 6.5 Calculation of λ_{max}, consistency index, and consistency ratio.

Factors	E	S	PR	DD	LULC	NDVI	RF	TWI	FA	Weighted sum value	Ratio
E	1.03	0.92	1.56	2.07	0.99	1.10	0.65	0.60	0.77	9.70	9.39
S	1.03	0.92	0.78	1.04	0.99	1.10	1.31	0.60	0.77	8.53	9.33
PR	1.03	1.83	1.56	2.07	1.97	1.65	1.31	1.81	1.54	14.78	9.47
DD	1.03	0.92	0.78	1.04	0.99	1.65	0.65	1.81	0.77	9.64	9.29
LULC	1.03	0.92	0.78	1.04	0.99	1.10	1.31	1.21	0.77	9.14	9.26
NDVI	1.03	0.46	0.52	0.31	0.49	0.55	0.65	0.60	0.77	5.39	9.81
RF	1.03	0.46	0.78	1.04	0.49	0.55	0.65	0.60	0.77	6.38	9.74
TWI	1.03	0.92	0.52	0.31	0.49	0.55	0.65	0.60	0.77	5.85	9.67
FA	1.03	0.92	0.78	1.04	0.99	0.55	0.65	0.60	0.77	7.33	9.53
Total											85.49

$\lambda_{max} = 85.48/9$
$= 9.498$

$CI = (9.49 - 9)/8$
$= 0.061$

$CR = 0.061/1.45$
$= 0.042$

CI, Consistency index; CR, consistency ratio; DD, drainage density; E, elevation; FA, flow accumulation; LULC, land use and land cover; NDVI, normalized difference vegetation index; PR, proximity to river; RF, rainfall; S, slope; TWI, topographic wetness index.

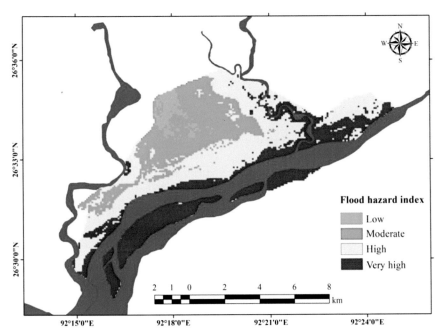

FIGURE 6.6

Flood hazard map.

Table 6.6 Statistics of flood hazard classes.

Sl. no.	Hazard class	Flood hazard indices	Area (%)
1	Low	1–2	4.14
2	Moderate	3	20.60
3	High	4	39.74
4	Very high	5	35.52

zones, as portrayed in Fig. 6.6 and Table 6.6. Higher FHI indicates high chances for flood events, while low FHI corresponds to scarce flood cases in this case.

Our study revealed that 35.52% of the land area of ONP falls under a very high hazard risk zone with an FHI of 5% and 39.74% under high and 20.60% under moderate hazard risk zone with FHI of 4 and 3, respectively. Only 4.14% of the total area is under low risk to flooding. By classifying the ONP into different hazard zones, our finding provides vital information, which will be crucial for management and planning purposes to mitigate the impacts of flooding on the park's ecosystem and wildlife. The study investigates various factors contributing to flood risk, such as elevation, river density, slope gradient, forest cover, and the dynamic nature of river systems like the Brahmaputra and its tributaries. Understanding these factors can help in developing effective strategies for flood management and conservation efforts within the ONP. By linking flood risk to factors such as river dynamics and forest cover, our research may underscore the importance of conservation measures in mitigating the impacts of flooding on biodiversity and habitat loss.

6.4 Discussion

Flood hazard assessment forms part and parcel of flood management and mitigation that seeks to minimize the dangers associated with flooding. ONP located at the bank of Brahmaputra River has always been at risk of being submerged under water due to a series of occurrences of flooding throughout its existence. In cases of the flood hazard assessment, many parameters can be positive or negative depending on their properties in relation to the flood risk. Even though there is no general rule for deciding these factors, previous researchers have investigated different flood controlling elements considering the specificity of the site, data accessibility, and knowledge acquired from prior investigations (Rahmati et al., 2016; Samanta et al., 2018; Wang et al., 2011; Ullah & Zhang, 2020). As per our findings, more than 35% of the total area of the national park is under a high hazard risk zone. Very high and high hazard risk areas are mainly located along the adjoining parts of riverbanks. These areas fall toward the southern, south-eastern, and southwestern part of ONP. The high risk zone is found to have low elevation, higher river density, and slope gradient and is closely located to the active river channels. The low forest cover and the presence of large swamps along with the dynamic nature of river Brahmaputra and its tributaries (Dhansiri and Panchnoi) are mainly responsible for flooding in this area. The river Brahmaputra is also responsible for degradation of areas of Kaziranga and Dibru-Saikhowa national parks of Assam, as reported by Areendran et al. (2020) and Shah and Shah (2023a,b), respectively.

Being a national park, ONP is a habitat to thousands of plants and animals including some threatened and endemic organisms. The Royal Bengal tiger is the national animal of India, and more than 75% of world's tiger population is found in India. In 2016 the National Tiger Conservation Authority declares the park as a tiger reserve to boost the conservation of tigers in this region. The park is the 49th tiger reserve among 53 in India and one of the highly tiger-dense areas, which was reported to have a tiger population of 21 by Jhala and Qureshi (2019). According to the spatial distribution map of tiger prepared by the Assam forest department (Sandeep, 2020), some of the tigers are found in areas that according to our findings fall within high and moderate risk zones. The riverine islands which are very high risk to flood are also utilized by the tigers and other animals such as elephants and ungulates from time to time, as reported by Borah, Firoz, and Sarma (2010). The findings highlight the critical importance of preserving the habitat within ONP for the conservation of its diverse and endangered wildlife, particularly the Royal Bengal tiger and the critically endangered Pygmy hog. With India hosting a significant portion of the world's tiger population, the designation of ONP as a tiger reserve in 2016 underscores its significance as a stronghold for tiger conservation efforts. The spatial risk zonation map developed in the study offers practical implications for wildlife conservationists, policymakers, and local authorities. By identifying high and moderate flood risk zones within the park, this map serves as a valuable tool for decision-making regarding habitat management and wildlife protection strategies. One practical implication is the need for targeted conservation measures to mitigate the impacts of flooding on wildlife populations, particularly those occupying high-risk zones. This may include the relocation of animals to less flood-prone areas during periods of heightened risk, thereby reducing the potential for habitat loss and population declines. Furthermore, the zonation map can inform land use planning and development policies within and around ONP. By delineating flood hazard risk zones, authorities can implement regulations to minimize human encroachment and infrastructure development in vulnerable areas, thereby reducing the potential for habitat fragmentation and degradation. The findings also underscore the importance of collaborative efforts among government agencies, nongovernmental organizations, and local communities in implementing effective conservation strategies. Community-based initiatives such as habitat restoration, antipoaching patrols, and public awareness campaigns can complement formal conservation measures and contribute to the long-term sustainability of ONP's biodiversity. Moreover, the study emphasizes the role of scientific research in informing evidence-based conservation practices. By conducting the first spatial risk zonation of ONP (Fig. 6.6), our research have provided valuable insights into the complex interactions between natural hazards and wildlife habitat suitability, laying the groundwork for future studies and conservation efforts in the region. The practical implications of the study findings extend beyond scientific research, offering actionable insights for wildlife conservation and land management practices in ONP. By leveraging the spatial risk zonation map, stakeholders can implement targeted conservation strategies to safeguard the park's rich biodiversity and ensure its long-term viability as a protected area.

6.5 Limitations of the study

1. One of the primary limitations of our study lies in the selection of flood-causing factors for the susceptibility analysis. While we considered nine factors based on available data, there is a lack of specific guidelines or standards for determining which factors to include. This limitation suggests

that our analysis may not encompass all relevant variables that could influence flood susceptibility. Consequently, there may be additional factors that were not accounted for in our model, potentially impacting the accuracy and comprehensiveness of our flood hazard map.

2. The AHP method utilized for assigning weights to the selected factors inherently involves a degree of subjectivity. The weights assigned to each factor depend on the knowledge and expertise of the researcher, as well as the availability and quality of data. This subjectivity introduces the possibility of error, as different researchers may prioritize factors differently or interpret their significance in varying ways. As a result, the accuracy of our flood hazard map may be influenced by the subjective decisions made during the weighting process, affecting the reliability of our results.

3. Another limitation stems from the reclassification of thematic maps, which can affect the resolution and accuracy of the final flood hazard map. The process of reclassifying data involves simplifying and categorizing information, which may lead to loss of detail or nuance. This loss of resolution could potentially obscure subtle variations in flood susceptibility within the study area, resulting in a less precise representation of flood risk. Therefore the reclassification process introduces a source of uncertainty that may compromise the reliability of our findings.

4. The lack of sufficient financial support represents a significant practical limitation of our study. Limited resources may have constrained our ability to collect comprehensive data, employ more sophisticated analytical techniques, or conduct field validations to enhance the robustness of our results. Additionally, financial constraints have hindered efforts to acquire specialized software or access additional datasets that could have enriched our analysis. Consequently, the scope and depth of our study may have been constrained by financial limitations, potentially impacting the thoroughness and accuracy of our findings.

While our study provides valuable insights into flood hazard mapping in the ONP, it is important to acknowledge these limitations, which may affect the comprehensiveness, accuracy, and reliability of our results. Future research efforts should aim to address these limitations by adopting more systematic approaches to factor selection, minimizing subjectivity in weight assignment, improving resolution in thematic map reclassification, and securing adequate financial support to enhance the rigor and validity of flood susceptibility analysis.

6.6 Challenges and solutions

1. The response from local residents at the fringes of the ONP regarding previous flood events was unsatisfactory. This lack of information can hinder the understanding of historical flood patterns, which is crucial for effective

flood management and mitigation strategies. Several approaches can be taken to address this challenge. Working closely with local authorities, emergency management agencies and community leaders for long to access any existing data or documentation related to previous flood events can be helpful. Their expertise and knowledge of the area can enhance the understanding of historical flood patterns.

2. Obtaining cloud-free satellite data for the study area proved to be challenging, with only a 5% success rate despite multiple attempts. Cloud cover can obstruct visibility and hinder the interpretation of satellite imagery, impacting the accuracy of flood mapping and analysis. To address this challenge different satellite options with advanced imaging capabilities can be explored, such as high-resolution sensors or synthetic aperture radar technology, which are less affected by cloud cover. Historical weather patterns have to be identified to know about optimal time windows for data collection.

6.7 **Recommendations**

Effective flood management strategies for ONP (mainly in the high flood prone areas) are of utmost importance. Implementing a variety of strategies can mitigate the potential damage to the park's ecosystem and infrastructure.

1. Deployment of geobags along riverbanks: Geobags are large, durable bags filled with sand or soil used to reinforce and stabilize riverbanks. This strategy helps prevent erosion and protects against flooding by strengthening vulnerable areas along the riverbanks. It is a cost-effective and relatively simple solution that can be implemented quickly.

2. Reinforced concrete porcupines: Reinforced concrete porcupines are structures placed along riverbanks to slow down the flow of water and reduce erosion. These structures mimic natural features like rocks and logs, providing a habitat for aquatic species, while also serving a practical purpose in flood management. They are durable and low-maintenance, making them a sustainable option for long-term flood protection.

3. Levee construction: Levees are raised embankments built along rivers to contain floodwaters and prevent them from spilling over into surrounding areas. Constructing levees in high flood-prone areas of ONP can provide a reliable barrier against flooding, protecting both natural habitats and human settlements within the park.

4. Stone spurs: Stone spurs are structures built perpendicular to the riverbank to deflect water flow and reduce erosion. They are often made of large rocks or boulders arranged in a staggered pattern to create barriers that slow down the current. Stone spurs can be strategically placed in vulnerable areas to redirect water away from critical habitats and infrastructure.

5. Geomattresses and geotubes: Geomattresses and geotubes are flexible, erosion-control systems made from synthetic materials. They can be laid on the riverbed or installed along the banks to stabilize soil, prevent erosion, and protect against flooding. These innovative solutions offer versatility and adaptability, making them suitable for various terrain types within ONP.

6. Relocating animals from high-risk zones to safer areas within the park is a proactive measure that can help minimize the loss of wildlife during floods. By identifying and designating moderate and low-risk zones as relocation sites, park authorities can safeguard vulnerable species and reduce the overall impact of flooding on biodiversity. Additionally, the creation of flood diversion zones and storage areas for surplus water can help manage floodwaters more effectively, especially in regions with dense vegetation and sensitive ecosystems. These designated areas serve as buffers, absorbing excess water and reducing the risk of inundation in critical habitats.

6.8 Conclusion

One important area of research for a flood disaster study includes flood hazard mapping. It is important for effective emergency management planning and in making informed land use and protection decisions. In this study, the focus was on flood hazard mapping, a crucial aspect of disaster management, particularly in areas like the ONP. By utilizing the AHP model and considering nine significant flood-causing factors, we were able to construct a comprehensive flood hazard map using ArcGIS 10.8.2 software. This map categorized the park into four risk zones: low, moderate, high, and very high. The findings of our analysis indicate that a substantial portion of the ONP, approximately 35%, falls within the very high hazard risk zone, while an additional 39% is classified as high risk. Notably, the southern, south-eastern, and southwestern regions of the park emerged as particularly vulnerable to flooding, highlighting areas where focused mitigation efforts may be necessary. Furthermore, about 20% of the park's land area was identified as having a moderate hazard risk. ONP is not only a biodiversity hotspot but also a habitat for numerous endangered species; effective flood management strategies are imperative. Our research contributes significantly by providing a detailed flood hazard map that can serve as a valuable resource for various stakeholders, including the Assam Disaster Management Authority, Brahmaputra Board, Assam Forest Department, hydrologists, engineers, environmentalists, and conservationists. By utilizing this map, decision-makers can make informed choices regarding land use, emergency response planning, and infrastructure development within the park. Additionally, it can aid in the formulation of targeted conservation efforts aimed at protecting the delicate ecosystem of ONP and mitigating the adverse impacts of flooding on both wildlife and human communities in the surrounding areas. Additionally, this study underscores the importance of proactive flood management measures in safeguarding natural habitats and enhancing overall resilience to disasters. It highlights the critical role of interdisciplinary collaboration and data-driven approaches in addressing complex environmental challenges, ultimately contributing to the sustainable management of precious ecological resources like the ONP.

Acknowledgment

The authors are grateful to the faculty members of the Department of Geography, Cotton University, Guwahati, Assam for help and support.

References

Ajin, R. S. (2013). Flood hazard assessment of Vamanapuram river basin, Kerala, India: An approach using remote sensing & GIS techniques. *Advances in Applied Science Research*, *4*(3), 263−274.

Ali, S. A., Khatun, R., Ahmad, A., & Ahmad, S. N. (2019). Application of GIS-based analytic hierarchy process and frequency ratio model to flood vulnerable mapping and risk area estimation at Sundarban region, India. *Modeling Earth Systems and Environment*, *5*(3), 1083−1102. Available from https://doi.org/10.1007/s40808-019-00593-z, http://springer.com/journal/40808.

Antoine, J., Fischer, G., & Makowski, M. (1997). Multiple criteria land use analysis. *Applied Mathematics and Computation*, *83*(2−3), 195−215. Available from https://doi.org/10.1016/S0096-3003(96)00190-7.

Areendran, G., Raj, K., Sharma, A., Bora, P. J., Sarmah, A., Sahana, M., & Ranjan, K. (2020). Documenting the land use pattern in the corridor complexes of Kaziranga National Park using high resolution satellite imagery. *Trees, Forests and People*, *2*, 100039. Available from https://doi.org/10.1016/j.tfp.2020.100039.

Argaz, A. (2019). Flood hazard mapping using remote sensing and GIS Tools: A case study of souss watershed. *Journal of Materials and Environmental Sciences*, *10*(2), 170−181.

Armenakis, C., Du, E., Natesan, S., Persad, R., & Zhang, Y. (2017). Flood risk assessment in urban areas based on spatial analytics and social factors. *Geosciences*, *7*(4), 123. Available from https://doi.org/10.3390/geosciences7040123.

Arora, A., Pandey, M., Siddiqui, M. A., Hong, H., & Mishra, V. N. (2021). Spatial flood susceptibility prediction in Middle Ganga Plain: Comparison of frequency ratio and Shannon's entropy models. *Geocarto International*, *36*(18), 2085−2116. Available from https://doi.org/10.1080/10106049.2019.1687594, http://www.tandfonline.com/toc/tgei20/current.

Arya, A. K., & Singh, A. P. (2021). Multi criteria analysis for flood hazard mapping using GIS techniques: A case study of Ghaghara River basin in Uttar Pradesh, India. *Arabian Journal of Geosciences*, *14*(8). Available from https://doi.org/10.1007/s12517-021-06971-1, http://www.springer.com/geosciences/journal/12517?cm_mmc = AD-_-enews-_-PSE1892-_-0.

Bapalu, G. V., & Sinha, R. (2005). GIS in flood hazard mapping: A case study of Kosi River Basin, India. *GIS Development Weekly*, *1*(13), 1−3. Available from https://doi.org/10.13140/RG.2.1.1492.2720.

Bhatt, G. D., Sinha, K., Deka, P. K., & Kumar, A. (2014). Flood hazard and risk assessment in Chamoli District, Uttarakhand using satellite remote sensing and GIS techniques. *International Journal of Innovative Research in Science, Engineering and Technology*, *03*(08), 15348−15356. Available from https://doi.org/10.15680/IJIRSET.2014.0308039.

Borah, J., Ahmed, M., & Sarma, P. (2010). Brahmaputra River islands as potential corridors for dispersing tigers: a case study from Assam, India. *International Journal of Biodiversity and Conservation*, *2*, 350−358.

Chakdar, B., Singha, H., & Choudhury, M. R. (2019). Bird community of Rajiv Gandhi Orang National Park, Assam. *Journal of Asia-Pacific Biodiversity*, *12*(4), 498−507. Available from https://doi.org/10.1016/j.japb.2019.07.003, http://www.journals.elsevier.com/journal-of-asia-pacific-biodiversity/.

Chakraborty, S., & Mukhopadhyay, S. (2019). Assessing flood risk using analytical hierarchy process (AHP) and geographical information system (GIS): Application in Coochbehar district of West Bengal, India. *Natural Hazards*, *99*(1), 247−274. Available from https://doi.org/10.1007/s11069-019-03737-7, http://www.wkap.nl/journalhome.htm/0921-030X.

Choudhury, A. U. (2000). *The birds of Assam*. Gibbon Books & WWF India NE Region.

CRED. (2020). *EM-DAT, The Int disaster database*. Centre for Research on the Epidemiology of Disasters.

Danumah, J. H., Odai, S. N., Saley, B. M., Szarzynski, J., Thiel, M., Kwaku, A., ... Akpa, L. Y. (2016). Flood risk assessment and mapping in Abidjan district using multi-criteria analysis (AHP) model and geoinformation techniques, (cote d'ivoire). *Geoenvironmental Disasters*, *3*(1). Available from https://doi.org/10.1186/s40677-016-0044-y, https://link.springer.com/journal/40677.

Debbarma, A., & Deen, S. (2020). Flood disaster management in Assam. *Shodh Sanchar Bulletin*, *10*(40), 105−109.

Dhar, O. N., & Nandargi, S. (2004). Rainfall distribution over the Arunachal Pradesh Himalayas. *Weather*, *59*(6), 155−157. Available from https://doi.org/10.1256/wea.87.03.

Duan, W., He, B., Takara, K., Luo, P., Nover, D., Yamashiki, Y., & Huang, W. (2014). Anomalous atmospheric events leading to Kyushu's flash floods, July 11−14, 2012. *Natural Hazards*, *73*(3), 1255−1267. Available from https://doi.org/10.1007/s11069-014-1134-3, http://www.wkap.nl/journalhome.htm/0921-030X.

El-Haddad, B. A., Youssef, A. M., Pourghasemi, H. R., Pradhan, B., El-Shater, A. H., & El-Khashab, M. H. (2021). Flood susceptibility prediction using four machine learning techniques and comparison of their performance at Wadi Qena Basin, Egypt. *Natural Hazards*, *105*(1), 83−114. Available from https://doi.org/10.1007/s11069-020-04296-y, http://www.wkap.nl/journalhome.htm/0921-030X.

Fernández, D. S., & Lutz, M. A. (2010). Urban flood hazard zoning in Tucumán Province, Argentina, using GIS and multicriteria decision analysis. *Engineering Geology*, *111*(1−4), 90−98. Available from https://doi.org/10.1016/j.enggeo.2009.12.006.

Fonseca, A. R., Santos, M., & Santos, J. A. (2018). Hydrological and flood hazard assessment using a coupled modelling approach for a mountainous catchment in Portugal.

Stochastic Environmental Research and Risk Assessment, 32(7), 2165−2177. Available from https://doi.org/10.1007/s00477-018-1525-1, http://link.springer-ny.com/link/service/journals/00477/index.htm.

Forkuo, E. K. (2011). Flood hazard mapping using Aster image data with GIS. *International Journal of Geomatics and Geosciences, 4*, 932−950.

United Nations Office for Disaster Risk Reduction (UNDRR). (2017). *Gar Atlas: Unveiling global disaster risk.* UNDRR.

Gil, Y., & Kellerman, A. (1993). A multicriteria model for the location of solid waste transfer stations: The case of Ashdod, Israel. *GeoJournal, 29*(4), 377−384. Available from https://doi.org/10.1007/BF00807540.

Glas, H., Rocabado, I., Huysentruyt, S., Maroy, E., Salazar Cortez, D., Coorevits, K., ... Deruyter, G. (2019). Flood risk mapping worldwide: A flexible methodology and toolbox. *Water, 11*(11), 2371. Available from https://doi.org/10.3390/w11112371.

Glenn, E. P., Morino, K., Nagler, P. L., Murray, R. S., Pearlstein, S., & Hultine, K. R. (2012). Roles of saltcedar (*Tamarix* spp.) and capillary rise in salinizing a non-flooding terrace on a flow-regulated desert river. *Journal of Arid Environments, 79*, 56−65. Available from https://doi.org/10.1016/j.jaridenv.2011.11.025.

United Nations Office for Disaster Risk Reduction (UNDRR). (2019). *Global assessment report on disaster risk reduction.*

Gupta, L., & Dixit, J. (2022). A GIS-based flood risk mapping of Assam, India, using the MCDA-AHP approach at the regional and administrative level. *Geocarto International, 37*(26), 11867−11899. Available from https://doi.org/10.1080/10106049.2022.2060329, http://www.tandfonline.com/toc/tgei20/current.

Hallegatte, S., Green, C., Nicholls, R. J., & Corfee-Morlot, J. (2013). Future flood losses in major coastal cities. *Nature Climate Change, 3*(9), 802−806. Available from https://doi.org/10.1038/nclimate1979.

Hammami, S., Zouhri, L., Souissi, D., Souei, A., Zghibi, A., Marzougui, A., & Dlala, M. (2019). Application of the GIS based multi-criteria decision analysis and analytical hierarchy process (AHP) in the flood susceptibility mapping (Tunisia). *Arabian Journal of Geosciences, 12*(21). Available from https://doi.org/10.1007/s12517-019-4754-9, http://www.springer.com/geosciences/journal/12517?cm_mmc = AD-_-enews-_-PSE1892-_-0.

Hapuarachchi, H. A. P., Wang, Q. J., & Pagano, T. C. (2011). A review of advances in flash flood forecasting. *Hydrological Processes, 25*(18), 2771−2784. Available from https://doi.org/10.1002/hyp.8040.

Hazarika, B. C., & Saikia, P. K. (2010). A study on the behaviour of Great Indian One-horned Rhino (*Rhinoceros unicornis* Linn.) in the Rajiv Gandhi Orang National Park. *NeBIO, 1*(2), 62−74.

Jhala, Y. V., & Qureshi, Q. (2019). *Status of tigers, co-predators and prey in India 2018. Summary report.* National Tiger Conservation Authority, Government of India.

Khaing, Z. M., Zhang, K., Sawano, H., Shrestha, B. B., Sayama, T., Nakamura, K., & Schumann, G. J.-P. (2019). Flood hazard mapping and assessment in data-scarce Nyaungdon area, Myanmar. *PLoS One, 14*(11), e0224558. Available from https://doi.org/10.1371/journal.pone.0224558.

Khosravi, K., Nohani, E., Maroufinia, E., & Pourghasemi, H. R. (2016). A GIS-based flood susceptibility assessment and its mapping in Iran: A comparison between frequency ratio and weights-of-evidence bivariate statistical models with multi-criteria decision-making technique. *Natural Hazards*, *83*(2), 947–987. Available from https://doi.org/10.1007/s11069-016-2357-2, http://www.wkap.nl/journalhome.htm/0921-030X.

Laouacheria, F., Kechida, S., & Chabi, M. (2019). Modelling the impact of design rainfall on the urban drainage system by storm water management model. *Journal of Water and Land Development*, *40*(1), 119–125. Available from https://doi.org/10.2478/jwld-2019-0013, http://versita.com/science/environment/jwld/.

Malczewski, J. (2000). On the use of weighted linear combination method in GIS: Common and best practice approaches. *Transactions in GIS*, *4*(1), 5–22. Available from https://doi.org/10.1111/1467-9671.00035, http://onlinelibrary.wiley.com/journal/10.1111/(ISSN)1467-9671.

Muhadi, N. A., Abdullah, A. F., Bejo, S. K., Mahadi, M. R., & Mijic, A. (2020). The use of LiDAR-derived DEM in flood applications: A review. *Remote Sensing*, *12*(14). Available from https://doi.org/10.3390/rs12142308, https://res.mdpi.com/d_attachment/remotesensing/remotesensing-12-02308/article_deploy/remotesensing-12-02308.pdf.

Onuşluel Gül, G. (2013). Estimating flood exposure potentials in Turkish catchments through index-based flood mapping. *Natural Hazards*, *69*(1), 403–423. Available from https://doi.org/10.1007/s11069-013-0717-8, http://www.wkap.nl/journalhome.htm/0921-030X.

Pai, D. S., Sridhar, L., Rajeevan, M., Sreejith, O. P., Satbhai, N. S., & Mukhopadhyay, B. (2014). Development of a new high spatial resolution (0.25° × 0.25°) long period (1901-2010) daily gridded rainfall data set over India and its comparison with existing data sets over the region. *Mausam*, *65*(1), 1–18. Available from http://www.imd.ernet.in/main_new.htm.

Pohekar, S. D., & Ramachandran, M. (2004). Application of multi-criteria decision making to sustainable energy planning – A review. *Renewable and Sustainable Energy Reviews*, *8*(4), 365–381. Available from https://doi.org/10.1016/j.rser.2003.12.007.

Pourali, S. H., Arrowsmith, C., Chrisman, N., Matkan, A. A., & Mitchell, D. (2016). Topography wetness index application in flood-risk-based land use planning. *Applied Spatial Analysis and Policy*, *9*(1), 39–54. Available from https://doi.org/10.1007/s12061-014-9130-2, http://www.springer.com/geography/human + geography/journal/12061.

Rahmani, A. R., Narayan, G., Rosalind, L., & Sankaran R. (1990). Status of the Bengal Florican in India. In: *Status and ecology of the lesser and Bengal Floricans, with reports on Jerdon's courser and mountain quail. Final report* (pp. 55–78). Bombay Natural History Society.

Rahmati, O., Falah, F., Naghibi, S. A., Biggs, T., Soltani, M., Deo, R. C., … Tien Bui, D. (2019). Land subsidence modelling using tree-based machine learning algorithms. *Science of the Total Environment*, *672*, 239–252. Available from https://doi.org/10.1016/j.scitotenv.2019.03.496, http://www.elsevier.com/locate/scitotenv.

Rahmati, O., Haghizadeh, A., & Stefanidis, S. (2016). Assessing the accuracy of GIS-based analytical hierarchy process for watershed prioritization; Gorganrood River Basin, Iran. *Water Resources Management*, *30*(3), 1131–1150. Available from https://doi.org/10.1007/s11269-015-1215-4, http://www.wkap.nl/journalhome.htm/0920-4741.

Rangari, V. A., Sridhar, V., Umamahesh, N. V., & Patel, A. K. (2019). Floodplain mapping and management of urban catchment using HEC-RAS: A case study of Hyderabad City. *Journal of The Institution of Engineers (India): Series A*, *100*(1), 49–63.

Available from https://doi.org/10.1007/s40030-018-0345-0, http://www.springer.com/engineering/civil + engineering/journal/40030.

Rikalovic, A., Cosic, I., & Lazarevic, D. (2014). GIS based multi-criteria analysis for industrial site selection. *Procedia Engineering*, *69*, 1054−1063. Available from https://doi.org/10.1016/j.proeng.2014.03.090, http://www.sciencedirect.com/science/journal/18777058.

Rimba, A., Setiawati, M., Sambah, A., & Miura, F. (2017). Physical flood vulnerability mapping applying geospatial techniques in Okazaki City, Aichi Prefecture, Japan. *Urban Science*, *1*(1), 7. Available from https://doi.org/10.3390/urbansci1010007.

Saaty, T. L. (1977). A scaling method for priorities in hierarchical structures. *Journal of Mathematical Psychology*, *15*(3), 234−281. Available from https://doi.org/10.1016/0022-2496(77)90033-5.

Saaty, T. L. (1980). *The analytic hierarchy process*. Mc Graw Hill Company.

Saaty, T. L. (1990). How to make a decision: The analytic hierarchy process. *European Journal of Operational Research*, *48*(1), 9−26. Available from https://doi.org/10.1016/0377-2217(90)90057-I.

Saaty, T. L. (2008). Decision making with the analytic hierarchy process. *International Journal of Services Sciences*, *1*(1), 83. Available from https://doi.org/10.1504/IJSSCI.2008.017590.

Saaty, T. L., & Vargas, L. G. (2000). *Models, methods, concepts and applications of the analytic hierarchy process*.

Saki, F., Dehghani, H., Jodeiri Shokri, B., & Bogdanovic, D. (2020). Determination of the most appropriate tools of multi-criteria decision analysis for underground mining method selection—A case study. *Arabian Journal of Geosciences*, *13*(23). Available from https://doi.org/10.1007/s12517-020-06233-6, http://www.springer.com/geosciences/journal/12517?cm_mmc = AD-_-enews-_-PSE1892-_-0.

Saleh, S. K., Aliani, H., & Amoushahi, S. (2020). Application of modeling based on fuzzy logic with multi-criteria method in determining appropriate municipal landfill sites (case study: Kerman City). *Arabian Journal of Geosciences*, *13*(22). Available from https://doi.org/10.1007/s12517-020-06213-w, http://www.springer.com/geosciences/journal/12517?cm_mmc = AD-_-enews-_-PSE1892-_-0.

Samanta, R. K., Bhunia, G. S., Shit, P. K., & Pourghasemi, H. R. (2018). Flood susceptibility mapping using geospatial frequency ratio technique: A case study of Subarnarekha River Basin, India. *Modeling Earth Systems and Environment*, *4*(1), 395−408. Available from https://doi.org/10.1007/s40808-018-0427-z, http://springer.com/journal/40808.

Sandeep, B. V. (2020). *Tiger conservation plan 2021-30 Orang Tiger Reserve core area*. Assam Forest Department, GoI.

Sarma, P. K., Mipun, B. S., Talukdar, B. K, Kumar, R., & Basumatary, A. K. (2011). Evaluation of habitat suitability for rhino (*Rhinoceros unicornis*) in Orang National Park using geo-spatial tools. *ISRN Ecology*, *2011*, 1−9. Available from https://doi.org/10.5402/2011/498258.

Shadmehri Toosi, A., Calbimonte, G. H., Nouri, H., & Alaghmand, S. (2019). River basin-scale flood hazard assessment using a modified multi-criteria decision analysis approach: A case study. *Journal of Hydrology*, *574*, 660−671. Available from https://doi.org/10.1016/j.jhydrol.2019.04.072, http://www.elsevier.com/inca/publications/store/5/0/3/3/4/3.

Shah, R. K., & Shah, R. K. (2023a). Forest cover change detection using remote sensing and GIS in Dibru-Saikhowa National Park, Assam: A spatio-temporal study. *Proceedings of the National Academy of Sciences India Section B: Biological Sciences*, *93*(3), 559−564. Available from https://www.springer.com/journal/40011, https://doi.org/10.1007/s40011-023-01449-4.

Shah, R. K., & Shah, R. K. (2023b). GIS-based flood susceptibility analysis using multi-parametric approach of analytical hierarchy process in Majuli Island, Assam, India. *Sustainable Water Resources Management*, *9*(5). Available from https://doi.org/10.1007/s40899-023-00924-0, https://www.springer.com/journal/40899.

Stefanidis, S., & Stathis, D. (2013). Assessment of flood hazard based on natural and anthropogenic factors using analytic hierarchy process (AHP). *Natural Hazards*, *68*(2), 569−585. Available from https://doi.org/10.1007/s11069-013-0639-5.

Subrahmanyam, V. P. (1988). Hazards of floods and droughts in India. In *Natural and man-made hazards. Proceedings of the symposium* (pp. 337−356). Rimouski, Quebec, 1986. Available from https://doi.org/10.1007/978-94-009-1433-9_24.

Talukdar, B. N., & Sharma. (1995). *Checklist of the birds of Orang Wildlife Sanctuary.*

Tsakiris, G. (2014). Flood risk assessment: Concepts, modelling, applications. *Natural Hazards and Earth System Sciences*, *14*(5), 1361−1369. Available from https://doi.org/10.5194/nhess-14-1361-2014.

Ullah, K., & Zhang, J. (2020). GIS-based flood hazard mapping using relative frequency ratio method: A case study of panjkora river basin, eastern Hindu Kush, Pakistan. *PLoS One*, *15*(3). Available from https://doi.org/10.1371/journal.pone.0229153, https://journals.plos.org/plosone/article/file?id = 10.1371/journal.pone.0229153 &type = printable.

Vojtek, M., & Vojteková, J. (2019). Flood susceptibility mapping on a national scale in Slovakia using the analytical hierarchy process. *Water*, *11*(2), 364. Available from https://doi.org/10.3390/w11020364.

Wang, J. J., Jing, Y. Y., Zhang, C. F., & Zhao, J. H. (2009). Review on multi-criteria decision analysis aid in sustainable energy decision-making. *Renewable and Sustainable Energy Reviews*, *13*(9), 2263−2278. Available from https://doi.org/10.1016/j.rser.2009.06.021.

Wang, Y., Li, Z., Tang, Z., & Zeng, G. (2011). A GIS-based spatial multi-criteria approach for flood risk assessment in the Dongting Lake Region, Hunan, Central China. *Water Resources Management*, *25*(13), 3465−3484. Available from https://doi.org/10.1007/s11269-011-9866-2.

Wiles, Jason J., & Levine, Norman S. (2002). A combined GIS and HEC model for the analysis of the effect of urbanization on flooding; The Swan Creek watershed, Ohio. *Environmental and Engineering Geoscience*, *8*(1), 47−61. Available from https://doi.org/10.2113/gseegeosci.8.1.47.

Youssef, A. M., Pradhan, B., & Sefry, S. A. (2016). Flash flood susceptibility assessment in Jeddah city (Kingdom of Saudi Arabia) using bivariate and multivariate statistical models. *Environmental Earth Sciences*, *75*(1), 1−16. Available from https://doi.org/10.1007/s12665-015-4830-8, http://www.springerlink.com/content/121380/.

Zhang, J., Zhou, C., Xu, K., & Watanabe, M. (2002). Flood disaster monitoring and evaluation in china. *Environmental Hazards*, *4*(2), 33−43. Available from https://doi.org/10.3763/ehaz.2002.0404.

Zhao, G., Xu, Z., Pang, B., Tu, T., Xu, L., & Du, L. (2019). An enhanced inundation method for urban flood hazard mapping at the large catchment scale. *Journal of Hydrology*, *571*, 873−882. Available from https://doi.org/10.1016/j.jhydrol.2019.02.008, http://www.elsevier.com/inca/publications/store/5/0/3/3/4/3.

Zwenzner, H., & Voigt, S. (2009). Improved estimation of flood parameters by combining space based SAR data with very high resolution digital elevation data. *Hydrology and Earth System Sciences*, *13*(5), 567−576. Available from https://doi.org/10.5194/hess-13-567-2009, http://www.hydrol-earth-syst-sci.net/volumes_and_issues.html.

Anatomization of flood risk and vulnerability using the analytical hierarchy process and frequency ratio in Dibrugarh district of Assam, India

Jyoti Saikia[1], Archita Hazarika[2] and Sailajananda Saikia[3]

[1]*Department of Geography, DHSK College, Dibrugarh, Assam, India*
[2]*Centre for Studies in Geography, Dibrugarh University, Dibrugarh, Assam, India*
[3]*Department of Geography, Rajiv Gandhi University, Itanagar, Arunachal Pradesh, India*

7.1 Introduction

Flood is regarded as one of the most prevalent natural disasters throughout the globe (Gómez-Palacios, Torres, & Reinoso, 2017; Khosravi, Nohani, Maroufinia, & Pourghasemi, 2016). The changing trend of land use and land cover (LULC) and also the changing climate are the most vital causes of the increase in floods (Bronstert, 2003; Dang & Kumar, 2017; Kjeldsen, 2010). The precipitation pattern has drastically been altered by climate change, resulting in heavy rainfall in a very short period, which leads to flood events as the permeability capacity of the soil declines. The formation of a flood is largely influenced by geomorphological parameters like drainage size, slope, soil structure, low discharge capacity, and infiltration rate of rivers and dams. Humans have been a major manipulator of the ecosystem since their origin. Land use methods like agricultural use and deforestation are some of the anthropogenic causes that also influence floods (Bera et al., 2022; Fu et al., 2022; Haq et al., 2022; Prasad, Loveson, Chandra, & Kotha, 2022; Ullah et al., 2022). Hence, the awareness of anthropogenic manipulation along with susceptibility and risk analysis is of great prominence and utmost need (Costache et al., 2022; Jalayer, Sharifi, Abbasi-Moghadam, Tariq, & Qin, 2022; Tariq, Yan, Gagnon, Riaz Khan, & Mumtaz, 2023). With an increase in developmental activities in vulnerable areas, exposures to flood disasters are reckoned to increase in the future (IPCC, 2012).

Assessment of flood risk requires the inclusion of numerous hazard and vulnerability criteria, as it threatens the three pillars of sustainability such as social,

Applications of Geospatial Technology and Modeling for River Basin Management.
DOI: https://doi.org/10.1016/B978-0-443-23890-1.00007-4

environmental, and economic in the built-up areas. Felsenstein and Lichter (2014) state that despite flood prevention and risk mitigation research from several perspectives and locations, the sustainability of the built environment is still threatened by extreme flood events. Floods being natural hazards cannot be prevented but can be mitigated (Bertilsson et al., 2019) by building a resilient infrastructure (socioeconomic and political). Hence, it is of utmost need to evaluate the probable damages of the flood by applying both quantitative (Tariq, 2013; Tehrany, Pradhan, Mansor, & Ahmad, 2015) and qualitative (De Risi et al., 2020; Perrone, Inam, Albano, Adamowski, & Sole, 2020) aspects. Floods result in socioeconomic changes and also have a severe impact on the environment and ecosystem. Thus, flood risk analysis, susceptibility mapping, and vulnerability assessment are of great importance for hazard mitigation (Ali, Khatun, Ahmad, & Ahmad, 2019; Esteves, 2013; Feng & Wang, 2011; Haghizadeh, Siahkamari, Haghiabi, & Rahmati, 2017; Khosravi et al., 2016; Kia et al., 2012; Lee, Kang, & Jeon, 2012; Rahmati, Haghizadeh, Pourghasemi, & Noormohamadi, 2016; Schober, Hauer, & Habersack, 2015).

The development of geospatial analysis applied for cause and effect parameters is of great importance to reduce mass movement and also to prevent disaster aftermaths (Aslam et al., 2022; Basin, 2021; Hussain et al., 2022; Islam et al., 2022; Sharifi, Mahdipour, Moradi, & Tariq, 2022). Heavy rainfall, including the low discharge capacity of rivers and dams, is a prominent reason for flood. The occurrence and intensity of floods have increased drastically in recent years due to climate change (Kundzewicz et al., 2018). Modeling of flood susceptibility is a prominent factor of flood hazard research (Kundzewicz et al., 2018; Siddayao, Valdez, & Fernandez, 2014). Today, necessary data regarding flood modeling and mapping can be derived by using remote sensing and geographic information system (GIS) techniques (Nashwan, Shahid, & Wang, 2019; Tariq et al., 2023; Wahla et al., 2022). GIS is also used in different parts of the globe for flood susceptibility assessment, hazard monitoring, and GIS-based frequency ratio (FR) models (Hussain et al., 2022; Khan et al., 2022; Shah, Jianguo, Jahangir, Tariq, & Aslam, 2022; Waqas et al., 2021). Hence, it is proven that the incorporation of geotechnology is of great benefit in obtaining data related to flood-related studies. Geotechnology acts as a key tool for developing multicriteria zoning decision analysis (Nyaupane & Chhetri, 2009; Tariq et al., 2023). Studies also highlight that geostatistical calculations have a higher degree of accuracy than spatial data alone. Logistic regression, analytical hierarchy process (AHP), Shannon's entropy model, artificial neural network (ANN), FR, decision tree, and fuzzy logic are commonly used in flood susceptibility studies (Sharifi et al., 2022; Zamani et al., 2022; Zamani, Sharifi, Felegari, Tariq, & Zhao, 2022). Abdullahi and Pradhan (2018) state that the ANN model of flood is widely used in India and Malaysia for flood susceptibility and modeling. The AHP model used in GIS has been of great interest in mapping flood events in recent times (Aslam et al., 2022; Imran, Ahmad, Sattar, & Tariq, 2022; Pal & Ziaul, 2017; Haq et al., 2022) and is more user-friendly and understandable. Similarly, the bivariate statistics and FR methods are also considered to be effective in assessing natural calamities (Tariq et al., 2021; Yerramilli, 2012).

Frequent floods and tremendous damage are common natural hazards in tropical countries. Approximately 7.5 million hectares of land area in India are affected by floods, which cost nearly INR 1805 billion and 1600 human lives annually (National Disaster Management Authority (NDMA), 2022). López et al. (2020) are of the view that the Brahmaputra catchment area in Northeastern India is one of the most flood-prone zones of the country. National Remote Sensing Centre (NRSC) (2022) states that out of 34 districts in Assam, 17 districts are in severe flood hazard zones. The Rashtriya Barh Ayog (Rashtriya Barh Ayog (RBA), 2021) stated that nearly 3.12 million hectares (almost 40%) area of Assam is prone to floods. Cinderby (2016) stated that to minimize loss from natural disasters, evidence-based management is of high importance. Studies suggest that land use and management practices have the potential to increase flood events and they also have the potential to mitigate the risk by altering land use and management (Juarez-Lucas, Kibler, Ohara, & Sayama, 2016; Shabani, Kumar, & Esmaeili, 2014).

7.2 Materials and methods

7.2.1 Description of the study area

Situated in the easternmost part of the Brahmaputra valley, Dibrugarh extends from $27°0'$ to $27°45'$N latitude and $94°30'$ to $95°30'$E longitudes, as shown in Fig. 7.1. The district covers an area of about 3,381 sq. km. The mean temperature of the district ranges between less than $7°C$ and more than $36°C$ (Choudhury, 2009) with annual rainfall between 2500 and 3500 mm. Subtropical monsoon climate prevails in the region.

7.2.2 Data and sources

For the demarcation of the flood deluge areas of Dibrugarh, the AHP and FR models were taken into consideration. Satellite imageries, rainfall data, soil data, and digital elevation model were used to elicit the indicators of flood susceptibility. Satellite data of the flood season were used for extracting flood zones for the FR model. Table 7.1 highlights the detailed components of the data used in the study.

7.2.3 Data processing and methodology

AHP and FR models were used with the aid of ArcGIS 10.3 to anatomize the flood susceptibility in Dibrugarh. Elevation map, distance from river map, stream density map, average annual rainfall map, LULC maps, slope map, topographic wetness index (TWI) map, and soil map were used as flood susceptibility

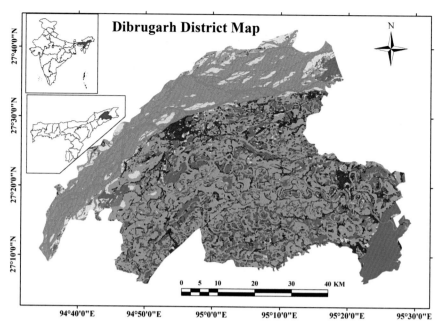

FIGURE 7.1

Location map of the study area. Map lines delineate study areas and do not necessarily depict accepted national boundaries.

Table 7.1 Details of the data used.

Sl. no.	Data types	Data format	Spatial resolution	Sources of data	Derived map(s)
1	DEM	Raster	12.5 m	Alaska Satellite Facility	Elevation, distance from river, stream density
2	Sentinel-2	Raster	10 m	Copernicus Open Access Hub	LULC
3	Rainfall	Raster	0.04° or 4 km	https://chrsdata.eng.uci.edu/	Average annual rainfall distribution
4	Soil	Vector	NA	https://www.fao.org/soils-portal	Soil map (Raster)
5	Landsat 4–5 (TM)	Raster	30 m	USGS Earth Explorer	Flood monitoring during the summer season

DEM, *Digital elevation model;* LULC, *land use and land cover,* TM, *thematic mapper.*

measures in both models. ArcGIS 10.3 and ERDAS Imagine 2014 were used for the preprocessing of the satellite data and LULC. A schematic structure of the methods used in the study is explained in Fig. 7.2.

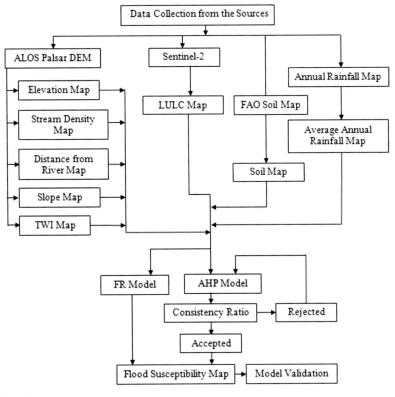

FIGURE 7.2

Methodological flow chart for the flood susceptibility model.

7.2.3.1 Analytical hierarchy process

Lots of studies used the AHP model (Choudhury, Basak, Biswas, & Das, 2022; Msabi & Makonyo, 2021) and the FR model (Chamling, Bera, & Sarkar, 2022; Hasanuzzaman, Adhikary, Bera, & Shit, 2022) in their studies. AHP is a multicriteria decision analysis, whereas FR is a bivariate statistical analysis that helps researchers develop a scale of preference from a set of flood susceptibility indicators (Ayalew & Yamagishi, 2005).

Saaty introduced a pair-wise comparison matrix (PCM) to obtain a consistency ratio (CR) (Singha & Swain, 2016) by applying an intensity score (Saaty, 1980). The CR was defined as the judgment of validating scores assigned to the flood susceptibility indicators, which was obtained as the ratio between the consistency index (CI) and the random index (RI) (Eq. 7.1).

$$CR = \frac{CI}{RI} \qquad (7.1)$$

Table 7.2 Random consistency index (Saaty, 1977).

Factors (n)	1	2	3	4	5	6	7	8	9	10
RI value	0	0	0.58	0.90	1.12	1.24	1.32	1.41	1.45	1.49

RI, *Random index.*

The value of the RI introduced by Saaty (1977) is highlighted in Table 7.2. The scores of the comparison matrix must be reassigned until the CR value is <0.1 to validate the assigned scores. Similarly, the value of CI is calculated as the subtraction of n from λ_{max} and further divided by $n - 1$ (Eq. 7.2).

$$CI = \frac{\lambda_{max} - n}{n - 1} \tag{7.2}$$

The value of λ_{max} is the average ratio between the A3 Matrix and A2 Matrix (Eq. 7.3). The A2 Matrix is the fractions of individual geometric mean and the sum of the geometric mean of the PCM of the indicators (Eq. 7.4). The A3 Matrix is the product of the PCM and the A2 Matrix.

$$\lambda_{max} = \text{Average of } \frac{\text{A3 Matrix}}{\text{A2 Matrix}} \tag{7.3}$$

$$\text{A2 Matrix} = \frac{\sqrt[n]{X_{1i}.X_{2i}.\ldots.X_{ni}}}{\sum \left(\sqrt[n]{X_1.X_2.\ldots.X_n} \right)} \tag{7.4}$$

7.2.3.2 Frequency ratio

The FR model is a statistical analysis where the FR value for each indicator has to be assigned as a weightage value in a GIS environment (Chamling et al., 2022; Hasanuzzaman et al., 2022). The flood susceptibility index (FSI) is the sum of the FR values of the flood indicators (Eq. 7.5). The FR value of each flood indicator is expressed as mentioned in Eq. (7.6).

$$FSI = \sum FR \tag{7.5}$$

$$FR = (\text{NpixE}/\text{NpixT}) / \left(\sum \text{pixE}/ \sum \text{pixE} \right) \tag{7.6}$$

where NpixE is the number of pixels containing flood points in class N, NpixT is the number of pixels in class N, $\sum$pixE is the number of pixels containing flood points, and $\sum$pixT is the number of pixels in the whole study area.

7.3 Indicators of flood susceptibility

Eight flood susceptibility indicators were used to validate the AHP and FR model in the study area, that is, elevation, distance from river, stream density, rainfall,

Table 7.3 Weights assigned to different indicators of flood susceptibility.

Indicators	Weights	Influence (%)	Classes	Class weight
Elevation	0.25	25	<45	5
			45–55	4
			55–75	3
			75–100	2
			>100	1
Distance from river	0.22	22	<0.012	5
			0.012–0.025	4
			0.025–0.045	3
			0.045–0.095	2
			>0.095	1
Stream density	0.16	16	<14	1
			14–44	2
			44–75	3
			75–110	4
			>110	5
Rainfall	0.12	12	<10,500	1
			10,500–11,000	2
			11,000–11,500	3
			11,500–12,000	4
			>12,000	5
LULC	0.09	9	Vegetation	2
			Settlement	3
			Agricultural land	4
			Sand bars	5
			Water body	5
Slope	0.07	7	<1.1	5
			1.1–3.5	4
			3.5–7.4	3
			7.4–16.0	2
			>16.0	1
TWI	0.05	5	<4	1
			4–6	2
			6–8	3
			8–10	4
			>10	5
Soil	0.04	4	Ao76–2/3c	1
			Af48–2ab	3
			Ao79-a	5

LULC, *Land use and land cover;* TWI, *topographic wetness index.*

LULC, slope, TWI, and soil (Choudhury et al., 2022; Hasanuzzaman et al., 2022; Msabi & Makonyo, 2021). Among all the indicators, elevation and distance from river are two major indicators that are responsible for flood in the Dibrugarh District. The weightage values 1—5 were assigned for the different classes of the flood indicators: (1) very less vulnerable for flood; (2) less vulnerable for flood; (3) moderately vulnerable for flood; (4) highly vulnerable for flood; and (5) very highly vulnerable for flood. Table 7.3 explains the class-wise weightage values against the individual indicators of flood susceptibility.

7.3.1 Elevation

Researchers have asserted that elevation plays a great role in the movement of water, flow direction, and depth of floods (Botzen, Aerts, & van den Bergh,

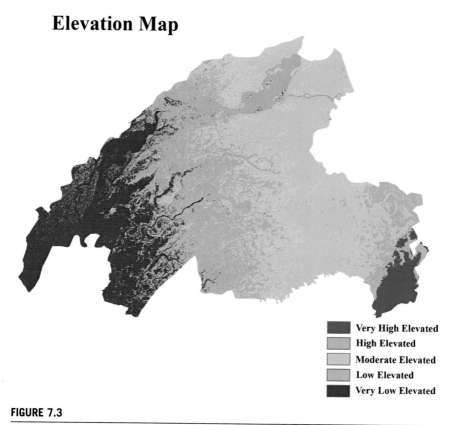

Elevation Map

	Very High Elevated
	High Elevated
	Moderate Elevated
	Low Elevated
	Very Low Elevated

FIGURE 7.3

Elevation map. Map lines delineate study areas and do not necessarily depict accepted national boundaries.

2013; Das, 2019; Msabi & Makonyo, 2021; Stieglitz, Rind, Famiglietti, & Rosenzweig, 1997). Hence, elevation is considered to be a vital factor in flood susceptibility analysis and mapping (Nachappa et al., 2020). Dahri and Abida (2017) stated that low-lying areas experience more flooding than higher areas (Fig. 7.3); a flash flood in the lower region is the result of the continuous flow of water in the higher altitudes (Pham et al., 2020).

7.3.2 Distance from river

The distance from the river is generally deemed to be disproportionately related to the occurrence of floods (Fig. 7.4). Therefore, flash floods can be considered to be a common phenomenon in regions that are near rivers, and on the other hand, regions that are located a greater distance away from the rivers are less vulnerable to flood damage (Souissi et al., 2020).

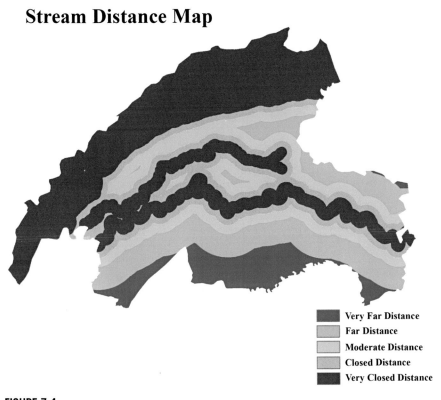

FIGURE 7.4

Distance from the river map. Map lines delineate study areas and do not necessarily depict accepted national boundaries.

7.3.3 Stream density

Greenbaum (1989) stated that drainage density is the ratio of the total drainage length in the grid to the area of the commensurate cell (Fig. 7.5). The occurrence of the flood is largely influenced by the density of the drainage. Kumar, Gopinath, and Seralathan (2007) are of the view that high flood susceptibility areas are usually associated with drainage density of a high percentage, highlighting a higher percentage of surface run-off.

7.3.4 Rainfall

Rainfall is a vital climatic component in the occurrences of flash floods; it is the principal source of surface run-off (Pham et al., 2020). Intense torrential rainfall over a small area results in the increase of surface water flow causing flash

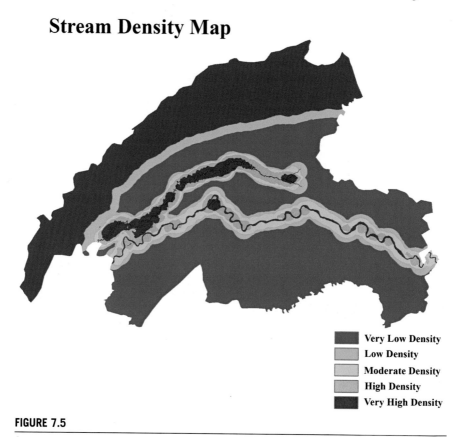

Stream Density Map

Very Low Density
Low Density
Moderate Density
High Density
Very High Density

FIGURE 7.5

Stream density map. Map lines delineate study areas and do not necessarily depict accepted national boundaries.

Rainfall Map

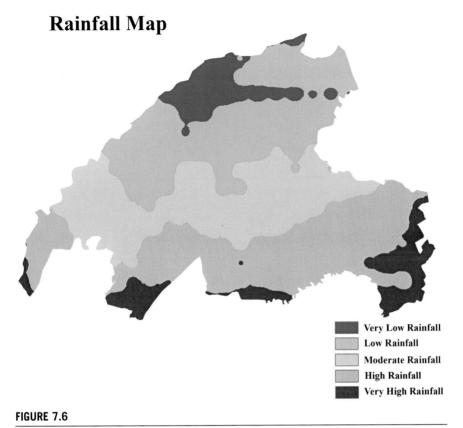

FIGURE 7.6

Rainfall map. Map lines delineate study areas and do not necessarily depict accepted national boundaries.

floods. Hence, events of flash floods are proportionate to the amount of rainfall (Fig. 7.6).

7.3.5 Land use and land cover

The alteration of the landscape in the region is the result of the changing LULC of that region (Vignesh, Anandakumar, Ranjan, & Borah, 2021). The LULC map is a prominent parameter for determining a flood-prone region. The study area was divided into five LULC classes: water body, sand bar, vegetation cover, agricultural land, and settlement (Fig. 7.7). For the determination of a flood-prone area, the LULC map is a vital component, as it plays a consequential role in the hydrological processes such as run-off, evaporation, evapotranspiration, and infiltration (Rahmati et al., 2016; Souissi et al., 2020).

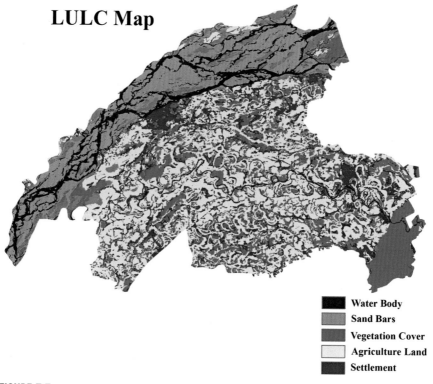

LULC Map

■ Water Body
▨ Sand Bars
▨ Vegetation Cover
☐ Agriculture Land
■ Settlement

FIGURE 7.7

LULC map. *LULC*, Land use and land cover. Map lines delineate study areas and do not necessarily depict accepted national boundaries.

7.3.6 Slope

The slope of a surface has a notable impact on numerous hydrological processes, especially on the process of infiltration and surface water run-off (Msabi & Makonyo, 2021; Paquette & Lowry, 2012). Thus, it acts as a vital factor in mapping flash flood susceptibility (Fig. 7.8). Slopes with steep gradients have lower infiltration capability as compared to gentle gradient slopes; however, the speed of surface run-off is higher in steep slopes. Flash floods in flat regions are the result of excessive flow of surface water, for which flatter regions near higher slopes experience high flash flood events.

7.3.7 Topographic wetness index

For the assessment of topographical events in certain hydrological processes, mainly flood phenomenon, the TWI (Fig. 7.9) acts as a vital measure (Khosravi

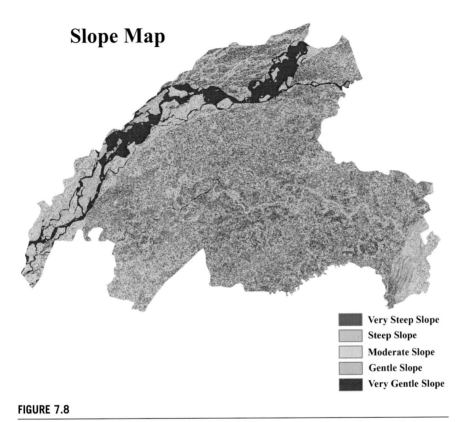

Slope Map

Very Steep Slope
Steep Slope
Moderate Slope
Gentle Slope
Very Gentle Slope

FIGURE 7.8

Slope map. Map lines delineate study areas and do not necessarily depict accepted national boundaries.

et al., 2018; Tehrany et al., 2015). Choudhury et al. (2022) stated in their study that the high value of TWI indicates the soil is well saturated and also has high flood susceptibility.

7.3.8 Soil type

The occurrence of flood can be highly affected by the soil type of the specific regions different soils have different permeability, which dictates the amount of water that the soil can retain; soil type (Fig. 7.10) and texture are the prominent factors for the infiltration of water (Msabi & Makonyo, 2021). According to the Food and Agriculture Organization of the United Nations, there are three types of soils found in the district: Ao76−2/3c type, Af48−2ab type, and Ao79-a type. Ao76−2/3c type represents hilly soil, Af48−2ab type represents old alluviums, and Ao79-a type represents new alluviums.

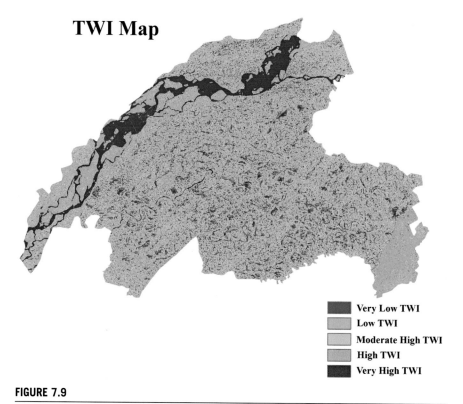

FIGURE 7.9

Topographic wetness index map. Map lines delineate study areas and do not necessarily depict accepted national boundaries.

7.4 Results

7.4.1 Flood susceptibility mapping using the analytical hierarchy process model

The weightage value for each indicator is calculated based on the PCM. The matrix was prepared by assigning each indicator an intensity score (Saaty, 1980), ranging from 1 (both the indicators are equally significant) to 9 (one indicator is extremely more significant than another), as highlighted in Table 7.4. However, the minimum and maximum intensity scores assigned in the study were 1 and 5 (Table 7.5).

The intensity scores were reassigned so that the CR could be of the accepted level, that is, less than 0.1. All the validated scores calculated from the PCM are highlighted in Table 7.6. Among the five indicators, elevation got a maximum weightage value of 25%, followed by distance from river of 22%. Slope, TWI, and soil got very little weightage among all the indicators, that is, 7%, 5%, and 4%, respectively.

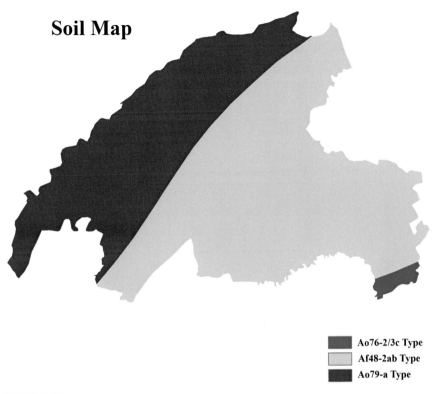

Soil Map

Ao76-2/3c Type
Af48-2ab Type
Ao79-a Type

FIGURE 7.10

Soil map. Map lines delineate study areas and do not necessarily depict accepted national boundaries.

Table 7.4 Saaty's intensity score for the analytical hierarchy process model.

Score	Definition	Explanation
1	Equally significant	Both indicators contribute equally
2	Weak or slight	The transition score
3	Moderately significant	Slightly favor one indicator over another
4	Moderate plus	The transition score
5	Strongly more significant	Strongly favor one indicator over another
6	Strong plus	The transition score
7	Very strong, more significant	An indicator is favored very strongly over another
8	Very strong plus	The transition score
9	Extremely more significant	Favoring one indicator over another is of the highest possible order

Table 7.5 Pair-wise comparison matrix of the flood indicators.

Components	Elevation	DR	SD	Rainfall	LULC	Slope	TWI	Soil	Weights
Elevation	1	3	3	2	2	3	3	3	0.25
DR	0.33	1	2	4	3	4	3	4	0.22
SD	0.33	0.50	1	3	2	2	4	3	0.16
Rainfall	0.50	0.25	0.33	1	3	2	3	3	0.12
LULC	0.50	0.33	0.50	0.33	1	3	2	2	0.09
Slope	0.33	0.25	0.50	0.50	0.33	1	3	3	0.07
TWI	0.33	0.25	0.33	0.33	0.50	0.33	1	2	0.05
Soil	0.33	0.25	0.33	0.33	0.50	0.33	0.50	1	0.04

DR, distance from river; SD, stream density; LULC, land use and land cover; TWI, topographic wetness index.

Table 7.6 Results for validating of the analytical hierarchy process model.

λ_{max}	n	RI	CI	CR	Consistency check
8.85	8	1.41	0.122	0.086	CR < 0.1 (accepted)

CI, Consistency index; CR, consistency ratio; RI, random index.

Flood Susceptibility Map Using AHP Model

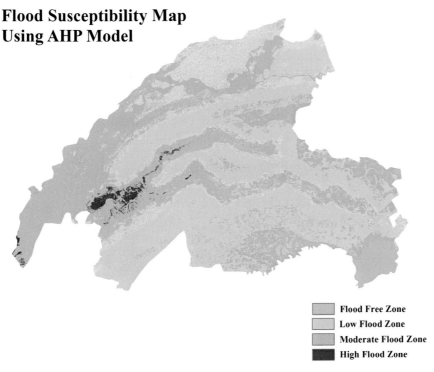

Flood Free Zone
Low Flood Zone
Moderate Flood Zone
High Flood Zone

FIGURE 7.11

Flood susceptibility map the using AHP model. Map lines delineate study areas and do not necessarily depict accepted national boundaries. *AHP,* Analytical hierarchy process.

The output AHP model (Fig. 7.11) shows that only 30.51 sq. km (0.99%) area of the district is under the very high flood susceptibility zone, whereas 1130.95 sq. km (36.60%), 1695.36 sq. km (54.86%), and 235.25 sq. km (7.55%) areas are under high flood susceptibility zone, moderate flood susceptibility zone, and no flood susceptibility zone respectively. The maximum area under the very high flood susceptibility zone was found in the mouth of Burhi Dihing River. The stream density plays a supportive role in having frequent floods in the mentioned part of the study area. The model shows an unexpected result that the area under the Brahmaputra River is not found in the very high flood susceptibility zone.

However, the maximum area under the river is found in the next category, that is, in the high flood susceptibility zone.

7.4.2 Flood susceptibility mapping using frequency ratio model

The output FR model (Fig. 7.12) shows that 855.94 sq. km (27.70%) area of the district is under the very high flood susceptibility zone, whereas 496.67 sq. km (16.07%), 774.13 sq. km (25.05%), and 963.34 sq. km (31.18%) areas are under the high flood susceptibility zone, moderate flood susceptibility zone and no flood susceptibility zone, respectively.

The FR model gives a better result than that of the AHP model covering all the riverine areas under the very high flood susceptibility zone. The model also highlights that the area under the very high flood susceptibility zone in the district is 855.94 sq. km. However, the zone includes the river beds of all the three rivers in the district. The area under the river in the district is 825.53 sq. km. Hence, the actual very high flood susceptibility zone in the land is 30.60 sq. km., which is almost equal to the same zone in the AHP model.

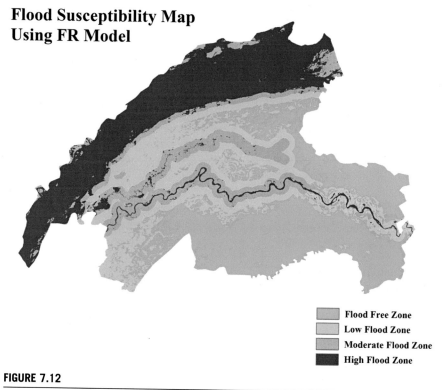

Flood Susceptibility Map Using FR Model

Flood Free Zone
Low Flood Zone
Moderate Flood Zone
High Flood Zone

FIGURE 7.12

Flood susceptibility map using FR model. Map lines delineate study areas and do not necessarily depict accepted national boundaries. *FR*, Frequency ratio.

7.5 Discussion

Choosing the flood susceptibility indicators is very much debatable, and the weightage is assigned to the respective indicators too (Msabi & Makonyo, 2021; Vojtek & Vojteková, 2019). However, field experiences and historical flood events can be of great help to the researcher in the selection of the best indicators. Eight important indicators were considered for the preparation of the flood susceptibility model in the study area. The flood susceptibility maps were generated by using the AHP and FR model integrating eight indicators. The susceptibility zones were categorized as very high flood susceptibility zone, high flood susceptibility zone, moderate flood susceptibility zone, and no flood susceptibility zone. The Brahmaputra River passes through a long narrow valley in Assam and also touches the Dibrugarh district. The tributaries and subtributaries of the Brahmaputra River contribute ample amounts of water every minute. However, the maximum catchment areas of the tributaries of the Brahmaputra River are in the hilly terrain of Arunachal Pradesh, Manipur, Meghalaya, Mizoram, and Nagaland (Bhattachaiyya & Bora, 1997). Only a single tributary (Burhi Dihing River) and a subtributary (Sessa River) of the Brahmaputra River are passing through the Dibrugarh district. During the peak season, the Brahmaputra River carries a large number of sediments and deposits it within the bed and on either side of the bank which is also a factor accelerating flood in the neighboring areas of the river. Similarly, the geological setting of the basin also accelerates river bank erosion, resulting in flood and vice versa (Bordoloi, Nikam, Srivastav, & Sahariah, 2020). The elevation map shows that a few parts of the district are lower elevated compared to that of the bed of the Brahmaputra River. The study shows that a remarkable area in the district is prone to very high flood and high flood. The Dibrugarh town also comes under the category, which is very close to the Brahmaputra River. Dibrugarh Town Protection (DTP) dyke is the only protective measure, which usually comes in the news for the erosion of the Brahmaputra River.

7.6 Conclusion

The study on the anatomization of flood risk and vulnerability in Dibrugarh District of Assam, India, utilizing the AHP and FR has provided valuable insights into the complex dynamics of flood-prone areas. The integration of AHP facilitated the identification and prioritization of key risk factors, considering their interdependencies and relative importance. This methodological approach proved effective in enhancing the accuracy and comprehensiveness of flood risk assessment. Furthermore, the incorporation of FR analysis allowed for a quantitative evaluation of the spatial distribution of vulnerability, aiding in the identification of high-risk zones within the district. The findings of this research are crucial for informing evidence-based policy-making and resource allocation to mitigate the impacts of floods in the region. The study not only contributes to the scientific understanding of flood risk but also provides actionable recommendations for local authorities and communities to enhance their resilience against future flood events. Additionally, the methodology employed in this research can serve as a valuable template for similar studies in other flood-prone regions, fostering a more systematic and comprehensive approach to flood risk assessment and management. This research makes a significant contribution to the field of disaster risk reduction, emphasizing the importance of a multidimensional analysis to better understand and address the intricacies of flood vulnerability in the context of Dibrugarh District and beyond. The present study deployed the AHP and FR models to demarcate the flood-inundated areas of the district. Eight indicators including elevation map, distance from river map, stream density map, average annual rainfall map, LULC map, slope map, TWI map, and soil map were assigned to both the models for comparison and were assigned weightage as the main parameters for flood susceptibility mapping. The weightage value for each indicator was calculated based on the PCM ranging from 1 to 9. For the validation of the models, along with satellite data, 5 years of flood data were also collected. Though the district does not have several rivers for causing floods, the Brahmaputra River is enough, which is very vulnerable to flood disaster. Lateral erosion and accretion is one of the characteristics of the river, which is also a factor of flood disaster. The study also highlights that most of the areas of the district are prone to flood disasters, but the DTP Dyke is the only protective measure for the same. Hence, the dyke must be protected to prevent the district from the flood.

Acknowledgment

The authors offer their sincere gratitude to the DHSK College in Assam, India, and the Rajiv Gandhi University in Arunachal Pradesh, India, for giving them such opportunities to conduct the study smoothly. Authors also express their heartfelt thanks to the scholars who contributed their valuable works in their respective fields.

Conflict of interest

There is no conflict of interest in the present study.

References

Abdullahi, S., & Pradhan, B. (2018). Land use change modeling and the effect of compact city paradigms: Integration of GIS-based cellular automata and weights-of-evidence techniques. *Environmental Earth Sciences, 77*(6). Available from https://doi.org/10.1007/s12665-018-7429-z.

Ali, S. A., Khatun, R., Ahmad, A., & Ahmad, S. N. (2019). Application of GIS-based analytic hierarchy process and frequency ratio model to flood vulnerable mapping and risk area estimation at Sundarban region, India. *Modeling Earth Systems and Environment, 5*(3), 1083−1102. Available from https://doi.org/10.1007/s40808-019-00593-z. Available from, springer.com/journal/40808.

Aslam, M., Ye, D., Tariq, A., Asad, M., Hanif, M., Ndzi, D., ... Jilani, S. F. (2022). Adaptive machine learning based distributed denial-of-services attacks detection and mitigation system for SDN-enabled IoT. *Sensors, 22*(7). Available from https://doi.org/10.3390/s22072697.

Ayalew, L., & Yamagishi, H. (2005). The application of GIS-based logistic regression for landslide susceptibility mapping in the Kakuda-Yahiko Mountains, Central Japan. *Geomorphology, 65*(1−2), 15−31. Available from https://doi.org/10.1016/j.geomorph.2004.06.010.

Basin, I. (2021). *Water availability, use and challenges in Pakistan − Water sector challenges in the Indus Basin and impact of climate change.*

Bera, D., Chatterjee, N. D., Mumtaz, F., Dinda, S., Ghosh, S., Zhao, N., ... Tariq, A. (2022). Integrated influencing mechanism of potential drivers on seasonal variability of LST in Kolkata Municipal Corporation, India. *Land, 11*(9). Available from https://doi.org/10.3390/land11091461.

Bertilsson, L., Wiklund, K., de Moura Tebaldi, I., Rezende, O. M., Veról, A. P., & Miguez, M. G. (2019). Urban flood resilience − A multi-criteria index to integrate flood resilience into urban planning. *Journal of Hydrology, 573*, 970−982. Available from https://doi.org/10.1016/j.jhydrol.2018.06.052.

Bhattachaiyya, N. N., & Bora, A. K. (1997). Floods of the Brahmaputra River in India. *Water International, 22*(4), 222−229. Available from https://doi.org/10.1080/02508069708686709.

Bordoloi, K., Nikam, B. R., Srivastav, S. K., & Sahariah, D. (2020). Assessment of riverbank erosion and erosion probability using geospatial approach: A case study of the Subansiri River, Assam, India. *Applied Geomatics, 12*(3), 265−280. Available from https://doi.org/10.1007/s12518-019-00296-1, http://www.springerlink.com/content/1866-9298/.

Botzen, W. J. W., Aerts, J. C. J. H., & van den Bergh, J. C. J. M. (2013). Individual preferences for reducing flood risk to near zero through elevation. *Mitigation and Adaptation Strategies for Global Change, 18*(2), 229−244. Available from https://doi.org/10.1007/s11027-012-9359-5.

Bronstert, A. (2003). Floods and climate change: Interactions and impacts. *Risk Analysis, 23*(3), 545−557. Available from https://doi.org/10.1111/1539-6924.00335.

Chamling, M., Bera, B., & Sarkar, S. (2022). *Large-scale human intervention and estimation of flood susceptibility applying frequency ratio model* (pp. 161−183). Springer Science and Business Media LLC. Available from 10.1007/978-3-030-94544-2_10.

Choudhury, A. (2009). The hoolock gibbon (*Hoolock hoolock*) in Tinsukia and Dibrugarh districts of Assam. India. *Asian Primates Journal, 1*(2), 24−30.

Choudhury, S., Basak, A., Biswas, S., & Das, J. (2022). *Flash flood susceptibility mapping using GIS-based AHP method* (pp. 119−142). Springer Science and Business Media LLC. Available from 10.1007/978-3-030-94544-2_8.

Cinderby, S. (2016). Co-designing possible flooding solutions: Participatory mapping methods to identify flood management options from a UK borders case study. *GI_Forum, 4* (1), 149−156. Available from https://doi.org/10.1553/giscience2016_01_s149.

Costache, R., Trung Tin, T., Arabameri, A., Crăciun, A., Ajin, R. S., Costache, I., . . . Thai Pham, B. (2022). Flash-flood hazard using deep learning based on H_2O R package and fuzzy-multicriteria decision-making analysis. *Journal of Hydrology, 609*. Available from https://doi.org/10.1016/j.jhydrol.2022.127747.

Dahri, N., & Abida, H. (2017). Monte Carlo simulation-aided analytical hierarchy process (AHP) for flood susceptibility mapping in Gabes Basin (southeastern Tunisia). *Environmental Earth Sciences, 76*(7). Available from https://doi.org/10.1007/s12665-017-6619-4.

Dang, A. T. N., & Kumar, L. (2017). Application of remote sensing and GIS-based hydrological modelling for flood risk analysis: A case study of District 8, Ho Chi Minh city, Vietnam. *Geomatics, Natural Hazards and Risk, 8*(2), 1792−1811. Available from https://doi.org/10.1080/19475705.2017.1388853, http://www.tandfonline.com/toc/tgnh20/current.

Das, S. (2019). Geospatial mapping of flood susceptibility and hydro-geomorphic response to the floods in Ulhas basin, India. *Remote Sensing Applications: Society and Environment, 14*, 60−74. Available from https://doi.org/10.1016/j.rsase.2019.02.006.

De Risi, R., Jalayer, F., De Paola, F., Carozza, S., Yonas, N., Giugni, M., & Gasparini, P. (2020). From flood risk mapping toward reducing vulnerability: The case of Addis Ababa. *Natural Hazards, 100*(1), 387−415. Available from https://doi.org/10.1007/s11069-019-03817-8.

Esteves, L. S. (2013). Consequences to flood management of using different probability distributions to estimate extreme rainfall. *Journal of Environmental Management, 115*, 98−105. Available from https://doi.org/10.1016/j.jenvman.2012.11.013, https://www.sciencedirect.com/journal/journal-of-environmental-management.

Felsenstein, D., & Lichter, M. (2014). Social and economic vulnerability of coastal communities to sea-level rise and extreme flooding. *Natural Hazards, 71*(1), 463−491. Available from https://doi.org/10.1007/s11069-013-0929-y.

Feng, C. C., & Wang, Y. C. (2011). GIScience research challenges for emergency management in Southeast Asia. *Natural Hazards, 59*(1), 597−616. Available from https://doi.org/10.1007/s11069-011-9778-8.

Fu, C., Cheng, L., Qin, S., Tariq, A., Liu, P., Zou, K., & Chang, L. (2022). Timely plastic-mulched cropland extraction method from complex mixed surfaces in arid regions. *Remote Sensing, 14*(16). Available from https://doi.org/10.3390/rs14164051.

Gómez-Palacios, D., Torres, M. A., & Reinoso, E. (2017). Flood mapping through principal component analysis of multitemporal satellite imagery considering the alteration of water spectral properties due to turbidity conditions. *Geomatics, Natural Hazards and Risk, 8*(2), 607−623. Available from https://doi.org/10.1080/19475705.2016.1250115. Available from, http://www.tandfonline.com/toc/tgnh20/current.

Greenbaum, D. (1989). Hydrogeological applications of remote sensing in areas of crystalline basement. In *1989 Proceeding of groundwater exploration and development in crystalline basement aquifers*.

Haghizadeh, A., Siahkamari, S., Haghiabi, A. H., & Rahmati, O. (2017). Forecasting flood-prone areas using Shannon's entropy model. *Journal of Earth System Science, 126*(3). Available from https://doi.org/10.1007/s12040-017-0819-x, http://www.ias.ac.in/article/fulltext/jess/126/03/0039.

Haq, S. M., Tariq, A., Li, Q., Yaqoob, U., Majeed, M., Hassan, M., . . . Aslam, M. (2022). Influence of edaphic properties in determining forest community patterns of the Zabarwan Mountain Range in the Kashmir Himalayas. *Forests*, *13*(8). Available from https://doi.org/10.3390/f13081214.

Hasanuzzaman, M., Adhikary, P. P., Bera, B., & Shit, P. K. (2022). *Flood vulnerability assessment using AHP and frequency ratio techniques* (pp. 91−104). Springer Science and Business Media LLC. Available from 10.1007/978-3-030-94544-2_6.

Hussain, S., Lu, L., Mubeen, M., Nasim, W., Karuppannan, S., Fahad, S., . . . Aslam, M. (2022). Spatiotemporal variation in land use land cover in the response to local climate change using multispectral remote sensing data. *Land*, *11*(5). Available from https://doi.org/10.3390/land11050595.

Imran, M., Ahmad, S., Sattar, A., & Tariq, A. (2022). Mapping sequences and mineral deposits in poorly exposed lithologies of inaccessible regions in Azad Jammu and Kashmir using SVM with ASTER satellite data. *Arabian Journal of Geosciences*, *15* (6). Available from https://doi.org/10.1007/s12517-022-09806-9.

IPCC. (2012). *Managing the risks of extreme events and disasters to advance climate change adaptation: A special report of working groups I and II*. Cambridge University Press.

Islam, F., Riaz, S., Ghaffar, B., Tariq, A., Shah, S. U., Nawaz, M., . . . Aslam, M. (2022). Landslide susceptibility mapping (LSM) of Swat District, Hindu Kush Himalayan region of Pakistan, using GIS-based bivariate modeling. *Frontiers in Environmental Science*, *10*. Available from https://doi.org/10.3389/fenvs.2022.1027423, journal.frontiersin.org/journal/environmental-science.

Jalayer, S., Sharifi, A., Abbasi-Moghadam, D., Tariq, A., & Qin, S. (2022). Modeling and predicting land use land cover spatiotemporal changes: A case study in Chalus watershed, Iran. *IEEE Journal of Selected Topics in Applied Earth Observations and Remote Sensing*, *15*, 5496−5513. Available from https://doi.org/10.1109/jstars.2022.3189528.

Juarez-Lucas, A. M., Kibler, K. M., Ohara, M., & Sayama, T. (2016). Benefits of flood-prone land use and the role of coping capacity, Candaba floodplains, Philippines. *Natural Hazards*, *84*(3), 2243−2264. Available from https://doi.org/10.1007/s11069-016-2551-2.

Khan, A. M., Li, Q., Saqib, Z., Khan, N., Habib, T., Khalid, N., . . . Tariq, A. (2022). MaxEnt modelling and impact of climate change on habitat suitability variations of economically important Chilgoza pine (*Pinus gerardiana* Wall.) in South Asia. *Forests*, *13*(5). Available from https://doi.org/10.3390/f13050715.

Khosravi, K., Nohani, E., Maroufinia, E., & Pourghasemi, H. R. (2016). A GIS-based flood susceptibility assessment and its mapping in Iran: A comparison between frequency ratio and weights-of-evidence bivariate statistical models with multi-criteria decision-making technique. *Natural Hazards*, *83*(2), 947−987. Available from https://doi.org/10.1007/s11069-016-2357-2.

Khosravi, K., Pham, B. T., Chapi, K., Shirzadi, A., Shahabi, H., Revhaug, I., . . . Tien Bui, D. (2018). A comparative assessment of decision trees algorithms for flash flood susceptibility modeling at Haraz watershed, northern Iran. *Science of the Total Environment*, *627*, 744−755. Available from https://doi.org/10.1016/j.scitotenv.2018.01.266.

Kia, M. B., Pirasteh, S., Pradhan, B., Mahmud, A. R., Sulaiman, W. N. A., & Moradi, A. (2012). An artificial neural network model for flood simulation using GIS: Johor River

Basin, Malaysia. *Environmental Earth Sciences*, *67*(1), 251–264. Available from https://doi.org/10.1007/s12665-011-1504-z.

Kjeldsen, T. R. (2010). Modelling the impact of urbanization on flood frequency relationships in the UK. *Hydrology Research*, *41*(5), 391–405. Available from https://doi.org/10.2166/nh.2010.056.

Kumar, P. K. D., Gopinath, G., & Seralathan, P. (2007). Application of remote sensing and GIS for the demarcation of groundwater potential zones of a river basin in Kerala, southwest coast of India. *International Journal of Remote Sensing*, *28*(24), 5583–5601. Available from https://doi.org/10.1080/01431160601086050, https://www.tandfonline.com/loi/tres20.

Kundzewicz, Z. W., Su, B., Wang, Y., Xia, J., Huang, J., Jiang, T., … Rahman, A. U. (2018). Analysis of flood susceptibility and zonation for risk management using frequency ratio model in District Charsadda. Pakistan. *International Journal of Environment and Geoinformatics*, *130*(2), 140–153.

Lee, M.J., Kang, J.E., & Jeon, S. (2012). Application of frequency ratio model and validation for predictive flooded area susceptibility mapping using GIS. In *International geoscience and remote sensing symposium (IGARSS)* (pp. 895–898). South Korea. Available from https://doi.org/10.1109/IGARSS.2012.6351414.

López, P. L., Sultana, T., Kafi, M. A. H., Hossain, M. S., Khan, A. S., & Masud, M. S. (2020). Evaluation of global water resources reanalysis data for estimating flood events in the Brahmaputra River Basin. *Water Resources Management*, *34*(7), 2201–2220. Available from https://doi.org/10.1007/s11269-020-02546-z, http://www.wkap.nl/journalhome.htm/0920-4741.

Msabi, M. M., & Makonyo, M. (2021). Flood susceptibility mapping using GIS and multi-criteria decision analysis: A case of Dodoma region, central Tanzania. *Remote Sensing Applications: Society and Environment*, *21*. Available from https://doi.org/10.1016/j.rsase.2020.100445, http://www.journals.elsevier.com/remote-sensing-applications-society-and-environment/.

Nachappa, T. G., Piralilou, S. T., Gholamnia, K., Ghorbanzadeh, O., Rahmati, O., & Blaschke, T. (2020). Flood susceptibility mapping with machine learning, multi-criteria decision analysis and ensemble using Dempster Shafer Theory. *Journal of hydrology*, *590*.

Nashwan, M. S., Shahid, S., & Wang, X. (2019). Assessment of satellite-based precipitation measurement products over the hot desert climate of Egypt. *Remote Sensing*, *11*(5). Available from https://doi.org/10.3390/rs11050555.

National Disaster Management Authority (NDMA). (2022). Available from https://ndma.gov.in/.

National Remote Sensing Centre (NRSC) (2022). Available from https://www.nrsc.gov.in.

Nyaupane, G. P., & Chhetri, N. (2009). Vulnerability to climate change of nature-based tourism in the Nepalese Himalayas. *Tourism Geographies*, *11*(1), 95–119. Available from https://doi.org/10.1080/14616680802643359.

Pal, Swades, & Ziaul, Sk (2017). Detection of land use and land cover change and land surface temperature in English Bazar urban centre. *The Egyptian Journal of Remote Sensing and Space Science*, *20*(1), 125–145. Available from https://doi.org/10.1016/j.ejrs.2016.11.003.

Paquette, Jessy, & Lowry, John (2012). Flood hazard modelling and risk assessment in the Nadi River Basin, Fiji, using GIS and MCDA. *The South Pacific Journal of Natural and Applied Sciences*, *30*(1), 33. Available from https://doi.org/10.1071/sp12003.

Perrone, A., Inam, A., Albano, R., Adamowski, J., & Sole, A. (2020). A participatory system dynamics modeling approach to facilitate collaborative flood risk management: A case study in the Bradano River (Italy). *Journal of Hydrology, 580*. Available from https://doi.org/10.1016/j.jhydrol.2019.124354.

Pham, B. T., Avand, M., Janizadeh, S., Phong, T. V., Al-Ansari, N., Ho, L. S., ... Prakash, I. (2020). GIS based hybrid computational approaches for flash flood susceptibility assessment. *Water (Switzerland), 12*(3). Available from https://doi.org/10.3390/w12030683, https://res.mdpi.com/d_attachment/water/water-12-00683/article_deploy/water-12-00683.pdf.

Prasad, P., Loveson, V. J., Chandra, P., & Kotha, M. (2022). Evaluation and comparison of the earth observing sensors in land cover/land use studies using machine learning algorithms. *Ecological Informatics, 68*. Available from https://doi.org/10.1016/j.ecoinf.2021.101522.

Rahmati, O., Haghizadeh, A., Pourghasemi, H. R., & Noormohamadi, F. (2016). Gully erosion susceptibility mapping: The role of GIS-based bivariate statistical models and their comparison. *Natural Hazards, 82*(2), 1231−1258. Available from https://doi.org/10.1007/s11069-016-2239-7.

Rashtriya Barh Ayog (RBA). (2021). Available from https://nvli.in/rashtriya-barh-ayog-national-commission-floods-report

Saaty, T. L. (1977). A scaling method for priorities in hierarchical structures. *Journal of Mathematical Psychology, 15*(3), 234−281. Available from https://doi.org/10.1016/0022-2496(77)90033-5.

Saaty, T. L. (1980). *The analytic hierarchy process: Planning, priority setting, resource allocation.* McGraw-Hill International Book Co.

Schober, B., Hauer, C., & Habersack, H. (2015). A novel assessment of the role of Danube floodplains in flood hazard reduction (FEM method). *Natural Hazards, 75*(S1), 33−50. Available from https://doi.org/10.1007/s11069-013-0880-y.

Shabani, F., Kumar, L., & Esmaeili, A. (2014). Improvement to the prediction of the USLE K factor. *Geomorphology, 204*, 229−234. Available from https://doi.org/10.1016/j.geomorph.2013.08.008.

Shah, S. H. I. A., Jianguo, Y., Jahangir, Z., Tariq, A., & Aslam, B. (2022). Integrated geophysical technique for groundwater salinity delineation, an approach to agriculture sustainability for Nankana Sahib Area, Pakistan. *Geomatics, Natural Hazards and Risk, 13*(1), 1043−1064. Available from https://doi.org/10.1080/19475705.2022.2063077, http://www.tandfonline.com/toc/tgnh20/current.

Sharifi, A., Mahdipour, H., Moradi, E., & Tariq, A. (2022). Agricultural field extraction with deep learning algorithm and satellite imagery. *Journal of the Indian Society of Remote Sensing, 50*(2), 417−423. Available from https://doi.org/10.1007/s12524-021-01475-7.

Siddayao, Generino P., Valdez, Sony E., & Fernandez, Proceso L. (2014). Analytic hierarchy process (AHP) in spatial modeling for floodplain risk assessment. *International Journal of Machine Learning and Computing, 4*(5), 450−457. Available from https://doi.org/10.7763/IJMLC.2014.V4.453.

Singha, C., & Swain, K. C. (2016). Land suitability evaluation criteria for agricultural crop selection: A review. *Agricultural Reviews, 37*(2). Available from https://doi.org/10.18805/ar.v37i2.10737.

Souissi, D., Zouhri, L., Hammami, S., Msaddek, M. H., Zghibi, A., & Dlala, M. (2020). GIS-based MCDM−AHP modeling for flood susceptibility mapping of arid areas, southeastern

Tunisia. *Geocarto International, 35*(9), 991−1017. Available from https://doi.org/10.1080/10106049.2019.1566405, http://www.tandfonline.com/toc/tgei20/current.

Stieglitz, Marc, Rind, David, Famiglietti, James, & Rosenzweig, Cynthia (1997). An efficient approach to modeling the topographic control of surface hydrology for regional and global climate modeling. *Journal of Climate, 10*(1), 118−137. Available from https://doi.org/10.1175/1520-0442(1997)010 < 0118:aeatmt > 2.0.co;2.

Tariq, A., Shu, H., Kuriqi, A., Siddiqui, S., Gagnon, A. S., Lu, L., ... Pham, Q. B. (2021). Characterization of the 2014 indus river flood using hydraulic simulations and satellite images. *Remote Sensing, 13*(11). Available from https://doi.org/10.3390/rs13112053, https://www.mdpi.com/2072-4292/13/11/2053/pdf.

Tariq, A., Yan, J., Gagnon, A. S., Riaz Khan, M., & Mumtaz, F. (2023). Mapping of cropland, cropping patterns and crop types by combining optical remote sensing images with decision tree classifier and random forest. *Geo-spatial Information Science, 26*, 302−320.

Tariq, M. A. U. R. (2013). Risk-based flood zoning employing expected annual damages: The Chenab River case study. *Stochastic Environmental Research and Risk Assessment, 27*(8), 1957−1966. Available from https://doi.org/10.1007/s00477-013-0730-1.

Tehrany, M. S., Pradhan, B., Mansor, S., & Ahmad, N. (2015). Flood susceptibility assessment using GIS-based support vector machine model with different kernel types. *Catena, 125*, 91−101. Available from https://doi.org/10.1016/j.catena.2014.10.017.

Ullah, I., Aslam, B., Shah, S. H. I. A., Tariq, A., Qin, S., Majeed, M., & Havenith, H. B. (2022). An integrated approach of machine learning, remote sensing, and GIS data for the landslide susceptibility mapping. *Land, 11*(8). Available from https://doi.org/10.3390/land11081265.

Vignesh, K. S., Anandakumar, I., Ranjan, R., & Borah, D. (2021). Flood vulnerability assessment using an integrated approach of multi-criteria decision-making model and geospatial techniques. *Modeling Earth Systems and Environment, 7*(2), 767−781. Available from https://doi.org/10.1007/s40808-020-00997-2, http://springer.com/journal/40808.

Vojtek, M., & Vojteková, J. (2019). Flood susceptibility mapping on a national scale in Slovakia using the analytical hierarchy process. *Water, 11*(2), 364. Available from https://doi.org/10.3390/w11020364.

Wahla, S. S., Kazmi, J. H., Sharifi, A., Shirazi, S. A., Tariq, A., & Joyell Smith, H. (2022). Assessing spatiotemporal mapping and monitoring of climatic variability using SPEI and RF machine learning models. *Geocarto International, 37*(27), 14963−14982. Available from https://doi.org/10.1080/10106049.2022.2093411, http://www.tandfonline.com/toc/tgei20/current.

Waqas, H., Lu, L., Tariq, A., Li, Q., Baqa, M. F., Xing, J., & Sajjad, A. (2021). Flash flood susceptibility assessment and zonation using an integrating analytic hierarchy process and frequency ratio model for the Chitral District, Khyber Pakhtunkhwa, Pakistan. *Water, 13*(12). Available from https://doi.org/10.3390/w13121650.

Yerramilli, S. (2012). A hybrid approach of integrating HEC-RAS and GIS towards the identification and assessment of flood risk vulnerability in the city of Jackson, MS. *American Journal of Geographic Information System, 1*(1), 7−16. Available from https://doi.org/10.5923/j.ajgis.20120101.02.

Zamani, A., Sharifi, A., Felegari, S., Tariq, A., & Zhao, N. (2022). Agro climatic zoning of saffron culture in Miyaneh city by using WLC method and remote sensing data. *Agriculture, 12*(1). Available from https://doi.org/10.3390/agriculture12010118.

Quantifying land use/land cover dynamics and fragmentation analysis along the stretch of Bhagirathi River in Murshidabad district, West Bengal

8

Anukul Chandra Mandal[1], Raja Majumder[2], Partha Gorai[3] and Gouri Sankar Bhunia[3]

[1]*Department of Geography, Manipur International University, Imphal, Manipur, India*

[2]*Department of Geography, State Aided College Teacher Dinabandhu Mahavidyalaya, Bangaon, West Bengal, India*

[3]*Department of Geography, Seacom Skills University, Kendradangal, Saturia, West Bengal, India*

8.1 Introduction

Social and economic growth frequently results in profound changes to the surface landscape, and changes in land use can have an unintended impact on the quantity, shape, and configuration of river systems (Julian, Wilgruber, de Beurs, Mayer, & Jawarneh, 2015). Since they are being degraded by river controls and increased land use pressure, floodplains are among the most vulnerable ecosystems in worldwide. They are widely grown all over the world and are the main source of income, particularly in South-East Asia (El Bastawesy, Ramadan Ali, Faid, & El Osta, 2013). Erosion and aggradation were first described by Strahler (1956) as responses of systems when the equilibrium state was disturbed by human activity. The intentional modification of river systems to accommodate population growth (e.g., agricultural alterations, industrial development, and urbanization) through restructuring, expanding, occupation, burial, and so forth has been determined as one of the most pervasive and dramatic impacts on river systems since the middle of the 20th century. Though land use/land cover (LULC) change is a worldwide phenomenon, its nature and scope differ from region to region. However, its effects are significant, especially in riverine and vulnerable ecological areas. Consequently, regular monitoring of LULC changes

Applications of Geospatial Technology and Modeling for River Basin Management.
DOI: https://doi.org/10.1016/B978-0-443-23890-1.00008-6

is necessary to evaluate ecosystem health, identify the reasons that set off dynamic operations, and determine the effects of such changes on the environment (Fichera, Modica, & Pollino, 2012). In addition, knowledge of LULC dynamics helps in evaluating environmental alterations and creating efficient land management and planning plans at both the local and national scales (Etefa et al., 2018). Bhagirathi River is one of the most devastating rivers in the Murshidabad district (West Bengal) for its life-threatening nature, particularly during the rainy season. The dynamic nature of Bhagirathi River modifies the floodplains and associated LULC changes (Debnath, Das Pan, Ahmed, & Bhowmik, 2017). During the rainy season, the Bhagirathi River is distinguished by its unusually immense acceleration, unrelenting fluctuations in channel morphology, rapid bed aggradation, and transformation of the bank line (Mandal, Kayet, Chakrabarty, & Rahaman, 2016). The lateral migration of the bank line every year creates large volumes of productive land. During the period between 1978 and 1987, severe bank erosion was observed at Fazilpur and Jangipur barrage. Since 1994, Jalangi Bazar has been severely affected by erosion because of the development of central char land and construction of concrete bund for protecting the Rajsahi town of Bangladesh (Ghosh, 2015). Between 2011 and 2012, severe erosion along the river course started at Maya and the erosion was also rejuvenated in 2015. The agricultural plantation land at the Mithipur area was eroded during this period. Parua (2009) stated that the average annual loss of land is more than 200 ha in the study area. The area endured severe LULC changes due to Bhagirathi River dynamics and associated natural hazard, which has exaggerated the livelihood of its natives. Consequently, increasing human population also causes deforestation, destruction of the natural ecosystem, and urbanization, which contribute to widespread deterioration of the LULC.

Fragmentation, loss, and degradation of the habitat are often regarded as the most important driving elements from (Yang, Damen, & Van Zuidam, 1999) the perspective of biodiversity; as a result, this has been a major study area (Fischer & Lindenmayer, 2007). The growth of infrastructure, an increase in population, the loss of forest cover, and the altering of the terrain for agricultural use are the main causes of landscape fragmentation. Through the estimation of a number of metrics to show fragmentation and spatial allocation, land cover fragmentation analyses are carried out to comprehend the impact of land cover changes within a landscape. The preservation and sustainability of land cover has raised serious concerns due to its interaction with physical, biological, and human-made processes as well as taking into account how anthropogenic activities affect natural and quasinatural ecosystems (De Groot, 2006). In recent years, there has been an increase in interest in using free and open-source software in ecological research (Boyd & Foody, 2011). Open-source software that is freely available provides a number of benefits for research, including the ability to independently study and verify the computational and statistical foundation of the analysis. Free software can also improve biological research and knowledge transmission in underdeveloped nations when access to proprietary alternatives may be limited by cost (Steiniger & Hay, 2009). Globally, LULC mapping and change detection have been heavily reliant on geographic information systems (GISs) and remote sensing. Additionally, developments in remote sensing have boosted the use of satellite imagery, including Landsat data, in studies focused on variations in LULC across a variety of regional and temporal dimensions (Mohamed, Saleh, & Belal, 2014). This study aims to examine the viability of two commonly utilized LULC research methodologies: the study of landscape fragmentation and changes in land cover along the Bhagirathi River channel in the Murshidabad district. Thus using a combination of remote sensing and GIS techniques, the present research assesses the pattern and status of LULC changes over the past 40 years along a 10 km buffer region of the Bhagirathi River in Murshidabad district, as well as the fragmentation analysis.

8.2 Study area

The Bhagirathi River course lies between latitude $23°45'00.46''N-24°48'06.39''N$ and $87°54'13.62''E-88°16'14.84''E$ longitude in the Murshadabad district of West Bengal (India). The river Bhagirathi originates from the Ganga at Nurpur, located at 25 mi away from the South of Farakka. In the Murshidabad district, the river flows through the Moti jheel lake, which is situated north of the Murshidabad town. The second course of Bhagirathi River is eastward, through the path of Kati-Ganga. The third course of Bhagirathi River is along the path of the present "Bishunpur Bill" and "Chaltia Bill" and then toward Haridasmati in the South. The region comes under the tropical monsoon climate. Geologically, the eastern part is more favorable for agriculture than the western part. The entire district is fully encased by an order of quaternary sediments of Bengal delta. The region is sprinkled with several meander scar, swamps, levees, ox-bow lake, and paleo-channels, and these are dispersedly distributed in and along the Bhagirathi River course. The district has a population density of 1334 per sq km (rural population density 1130 per sq km and urban population density 5076 per sq km).

8.3 Data sources and database creation

For studying LULC change of the Bhagirathi River course in the Murshidabad district, we used Landsat Multispectral Scanner (MSS, DOP—October 30, 1977), Landsat (4, 5, 8) Thematic Mapper-TM (November 21, 1990; October 26, 2000; November 21, 2010), and Operational Land Imager-OLI (November 7, 2020) cloud free scenes. Due to their suitable spatial resolution (30 m) and less than 10% cloud cover, the imageries used in this investigation were chosen. It is possible to examine long-term regional and global LULC change using the Landsat data archive since its images are sufficiently consistent with data from older missions (Irons, Dwyer, & Barsi, 2012). No radiometric calibrations or atmospheric adjustments were made because the images had previously been preprocessed. Medium-resolution images were used in addition to supplementary data collection, which comprised topographic maps, ground truth information, and Google images. The reference data points were utilized, as the ground truth data were gathered using the GPS from October 2019 to February 2020 for image processing, image classification, and accuracy evaluation of the findings. The vector layers of the river course were digitized in a polygon shape through QGIS software v2.4.1. A buffer zone was created with 5 km distance for each year separately. The study area was delineated by the clipping and subsetting method in ERDAS Imagine Software v9.0.

8.4 Methods

8.4.1 Land use/land cover classification

By using supervised classification with a maximum likelihood classification technique, LULC maps were created from the training regions (Lillesand & Kiefer, 1994; Tarantino, Novelli, Aquilino, Figorito, & Fratino, 2015). Many researchers have utilized this technique with success. The remaining area of the image is classified using training polygons, which are field verified places (Jensen, 1996). The LULC classes were evaluated by supervised classification using ERDAS IMAGINE v9.0. An algorithm for stratified random sampling was used to identify ground control points in order to assess the accuracy of the various categorized images. 100 ground control sites were selected in total. Changes were separately assessed for every sample point. The magnitude of change for different classified images has been established by a survey of topographic maps of India, survey of published literature, discussions with locals, categorized map released by National Remote Sensing Centre, and ground truth information. Overall accuracy, producer accuracy, user accuracy, and kappa coefficient were used to measure the correctness of all the images (Hassan et al., 2016; Jensen, 1996). The methodological flow chart is represented in Fig. 8.1.

8.4.2 Areal distribution of land use/land cover change analysis

The amount of LULC transformation denotes the form of land use variations in the size range, imitated in the dissimilar sorts of LULC alteration on the total, shimmering the overall circumstances of land use configuration fluctuations. The mathematical model is used to calculate the change as

$$\Delta U = Y_a - Y_b$$

In this equation, Y_a and Y_b show at the beginning and end of the analysis of the land use type area, respectively.

8.4.3 Landscape metric analysis

Using the QGIS LecosS (Landscape Ecology Statistics) plugin (http://plugins.qgis.org/plugins/LecoS/), landscape metrics at the patch and class levels were

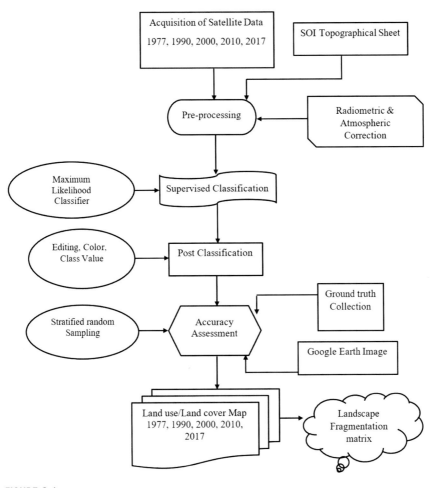

FIGURE 8.1

Flow chart map of study methodology.

found for the LULC classes in the present study. It offers various methods to do landscape analysis and heavily utilizes the scientific python libraries SciPy and Numpy (Oliphant, 2007) to compute both basic and complex landscape measures. As shown in Table 8.1, we compute the landscape proportion, edge density, patch density, and mean patch area (MPA).

The Shannon diversity index (SHDI), Shannon equitability (SHEI), and Simpson diversity index (SIDI) were the three worldwide metrics we used to emphasize the variety in order to characterize the research region (Chelaru, Oiçte, & Mihai, 2014).

The proportionate abundance of each patch of a certain type, multiplied by that percentage, is expressed as the SHDI (Shannon & Weaver, 1949). SHDI is 0 in landscapes where there is only a patch, indicating a lack of variety. The SHDI

Table 8.1 Estimation of the landscape matrix.

Landscape matrix/index	Description
Landscape proportion	The percentage of cells in a certain class out of the total number of cells in the identified LULC image (%).
Edge density	The ratio of the lengths of all edge elements comprising a certain patch type to the overall area of the landscape (ha).
Patch density	Equals the proportion of the overall landscape region taken up by patches of the respective patch type (no./100 ha).
MPA	The area where the average number of cells found over the whole region is represented by a patch (ha).

LULC, *Land use/land cover.*

value rises when there are more patches of various sorts, and the corresponding distribution of each type of patch becomes fair.

$$\text{SHDI} = 1 - \sum_{i=1} Np_i x \ln p_i$$

where "N" is the no. of and cover types p_i = proportional abundance of the ith type

The SIDI has the same characteristics as the Shannon index, except that it produces results between 0 and 1, which are larger when diversity is greater (Simpson, 1949). McGarigal et al. (2002) asserted that the SIDI interpretation is more comprehensible than Shannon's index and that the values of the SIDI are less sensitive to the existence of unusual types. The likelihood that two randomly chosen pixels belong to separate patches is shown by the SIDI values.

$$\text{SIDI} = 1 - \sum_{i=1} Np_i x p_i$$

The distribution of the patches throughout the entire region is displayed by SHEI. It is calculated by dividing SHDI by $\text{SHDI}_{\max}$ (here $\text{SHDI}_{\max} = \ln S$). SHEI approaches zero as the distribution of the various patches gets increasingly uneven, and it is zero when there is just one patch in the landscape (i.e., no diversity). Equitability is predicated on an array between 0 and 1, with 1 representing ideal evenness.

$$\text{SHEI} = \frac{\text{SHDI}}{\text{SHDI}_{\max}} = \frac{\text{SHDI}}{\ln S}$$

8.5 Results

The spatial distribution of LULC characteristics of the Bhagirathi River region within a 5 km surrounding during the period between 1977 and 1990 is represented in Figs. 8.2 and 8.3. The region has been classified into eight major land

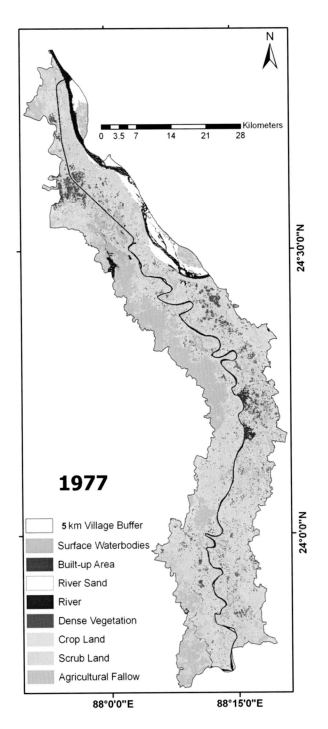

FIGURE 8.2

Land use/land cover map of the Bhagirathi River course (within a 5 km buffer) in the Murshidabad district in 1977 and 1990. Map lines delineate study areas and do not necessarily depict accepted national boundaries.

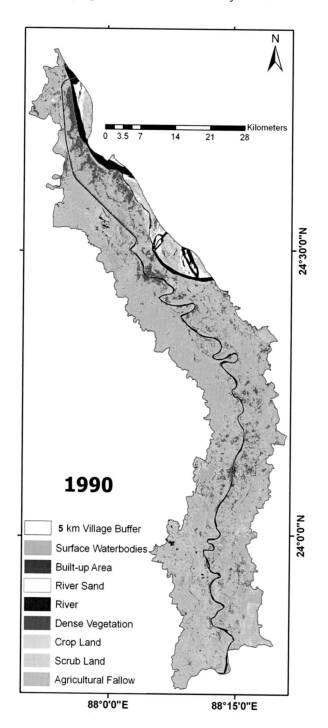

FIGURE 8.3

Land use/land cover map of the Bhagirathi River course (within a 5 km buffer) in the Murshidabad district in 1977 and 1990. Map lines delineate study areas and do not necessarily depict accepted national boundaries.

cover classes, namely, surface waterbodies, built-up area, river sand, river, dense vegetation, crop land, scrubland, and agricultural fallow. The results of the analysis showed that most of the area was covered with the scrubland and agricultural fallow land during the period between 1977 and 1990. The scrubland was mostly observed in the riverine region of the river Ganges and south-west part of the Bhagirathi River. However, these areas gradually decreased and converted into the agricultural fallow land. Dense vegetation was distributed heterogeneously with small patches in the northern part of the study area. Results also showed that vegetation density increases in the central and southern part of the study site during the period between 2010 and 2020 (Figs. 8.4 and 8.5). In 2017 the concentration of vegetation was observed in the east central part of the study site. However, during the period between 1977 and 2020, the spatial distribution of waterbodies was mostly observed in the southern part of the study area. The river sand was found in the north-east of the study area along the Ganga River course and in the small pocket of the entire Bhagirathi River course. The agricultural fallow lands were heterogeneously distributed in the entire study site. Mostly, the western part of the study site was covered with the agricultural fallow land in 2000−20. The outcomes of the study site showed gradual decrease of surface waterbodies during the entire study period. In 1977 the area of surface waterbodies in the study area was calculated as 53.07 ha (2.87%), whereas in 2020, the estimated area of the surface waterbodies was 22.27 ha (1.20%). The area of built-up land in the study area was in an increasing trend. The calculated area of built-up land was estimated to be 3.47 (0.19%), 13.93 (0.75%), 65.81 (3.55%), 144.51 (7.80%), and 245.00 ha (13.23%) in 1977, 1990, 2000, 2010, and 2020, respectively (Table 8.2). The areal distribution of river sand was varied during the period between 1977 and 2020.

The built-up area, crop land, and agricultural fallow land are emphasized as having undergone the most significant changes in terms of the development of the total area of the eight LULC classes throughout the studied period (Table 8.3). The vegetative land of the study area is decreasing in trend from 1977 to 2020. In 1977 the area of the vegetative land was calculated as 292.67 ha (15.81%). In 2020 the estimated area of vegetative area in the study site was calculated as 179.75 ha (9.71%). The crop land in the study area is an increasing trend. The estimated area of the crop land were 145.75, 389.53, 467.36, 638.3, and 795.9 ha in 1977, 1990, 2000, 2010, and 2020, respectively. The areal distribution of scrubland in the study area is varied between 1977 and 2020. However, in 1977 the area of scrubland was estimated as 370.34 ha (20.00%), whereas the estimated area of scrubland in 2020 was 238.28 ha (12.87%). The area of agricultural land in the study area is an decreasing trend. The calculated area of agricultural land was 843.27 (45.54%), 761.03 (41.10%), 685.75 (37.04%), 512.2 (27.66%), and 233.11 ha (12.59%) in 1977, 1990, 2000, 2010, and 2020, respectively.

In the classified map of 1977, overall accuracy and kappa statistics were calculated as 83% and 0.84, respectively. In 1990 the overall accuracy of the LULC class was estimated as 86.00% with kappa statistics of 0.80. In 2000 the estimated

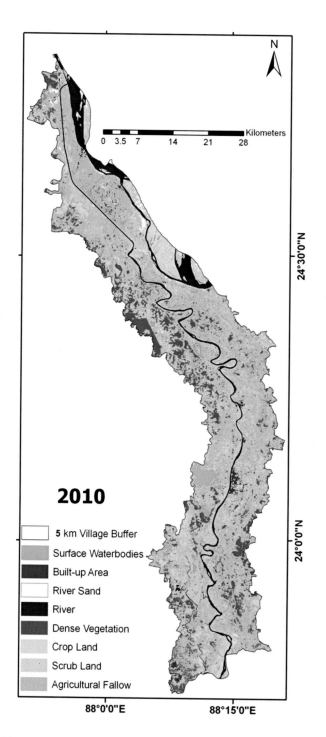

FIGURE 8.4

Land use/land cover map of the Bhagirathi River course (within a 5 km buffer) in the Murshidabad district in 2000 and 2010. Map lines delineate study areas and do not necessarily depict accepted national boundaries.

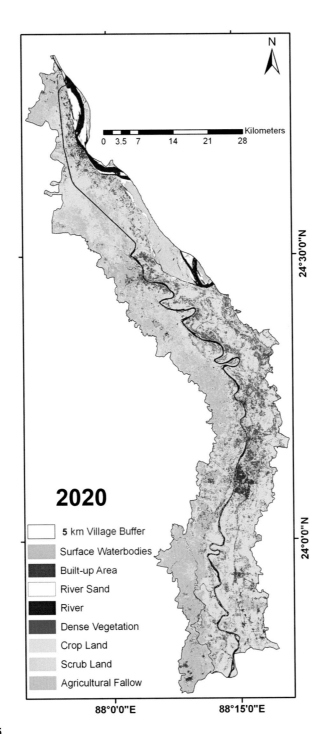

FIGURE 8.5

Land use/land cover map of the Bhagirathi River course (within a 5 km buffer) in the Murshidabad district in 2020. Map lines delineate study areas and do not necessarily depict accepted national boundaries.

Table 8.2 Areal distribution of land use/land cover (LULC) characteristics (in hectares) in the study area.

LULC type	1977	%	1990	%	2000	%	2010	%	2017	%
Surface waterbodies	53.07	2.87	51.73	2.79	45.24	2.44	40.73	2.20	22.27	1.20
Built-up area	3.47	0.19	13.93	0.75	65.81	3.55	144.51	7.80	245.00	13.23
River sand	41.24	2.23	35.80	1.93	42.94	2.32	32.14	1.74	39.87	2.15
River	101.80	5.50	101.66	5.49	99.42	5.37	97.64	5.27	97.52	5.27
Dense vegetation	292.67	15.81	261.40	14.12	239.62	12.94	165.84	8.96	179.75	9.71
Crop land	145.75	7.87	389.53	21.04	467.36	25.24	638.30	34.47	795.90	42.98
Scrubland	370.34	20.00	236.53	12.77	205.46	11.10	220.22	11.89	238.28	12.87
Agricultural fallow	843.27	45.54	761.03	41.10	685.75	37.04	512.20	27.66	233.11	12.59

Table 8.3 Areal changes of land use/land cover (LULC) characteristics (in hectares) during the study period.

LULC type	1977–90	% of change	1990–2000	% of change	2000–10	% of change	2010–17	% of change	1977–2017	% of change
Built-up area	−10.46	−0.57	−51.88	−2.80	−78.70	−4.25	−100.50	−5.43	−241.53	−13.04
Sand	5.44	0.29	−7.15	−0.39	10.81	0.58	−7.74	−0.42	1.36	0.07
Surface waterbodies	1.35	0.07	6.49	0.35	4.52	0.24	18.46	1.00	30.80	1.66
River	0.14	0.01	2.24	0.12	1.78	0.10	0.12	0.01	4.28	0.23
Dense vegetation	31.27	1.69	21.78	1.18	73.79	3.98	−13.91	−0.75	112.93	6.10
Agricultural fallow	82.24	4.44	75.28	4.07	173.55	9.37	279.09	15.07	610.15	32.95
Scrubland	133.80	7.23	31.07	1.68	−14.75	−0.80	−18.06	−0.97	132.06	7.13
Crop land	−243.78	−13.17	−77.83	−4.20	−170.94	−9.23	−157.60	−8.51	−650.15	−35.11

Note: (−) sign indicates positive growth and (+) positive sign indicates negative growth.

overall accuracy was 88.00% with the overall kappa statistics of 0.86. In 2010 the calculated overall accuracy of the LULC class in the study area was recorded as 92.00% with the kappa statistics of 0.91. The overall accuracy and kappa statistics was calculated as 94.00% and 0.93, respectively, in 2020. Using an error/confusion matrix, the final LULC maps for the different time periods (1977 and 2020) are displayed in Figs. 8.2–8.4. This is the method that is frequently used to assess per-pixel categorization (Lu & Weng, 2007). For each classed map, the kappa statistics/index was also calculated to assess the precision of the findings. In the classified map of 1977, overall accuracy and kappa statistics were calculated as 83% and 0.84, respectively. In 1990 the overall accuracy of the LULC class was estimated as 86.00% with kappa statistics of 0.80. In 2000 the estimated overall accuracy was 88.00% with the overall kappa statistics of 0.86. In 2010 the calculated overall accuracy of the LULC class in the study area was recorded as 92.00% with the kappa statistics of 0.91. The overall accuracy and kappa statistics were calculated as 94.00% and 0.93, respectively, in 2020. This was allowed for the following analysis and change detection because of its generally respectable accuracy (Lea & Curtis, 2010) .

The quantity of patch creation and reduction throughout the Bhagirathi riverine area may be used to visualize the continuing scenario of transformation and fragmentation that is presented by the land use class matrices. Overall, there are more patches now than there were previously, which suggests that classes have not been transformed into a continuous horizontal extension. Based on the examination of the estimated landscape metrics, it was discovered that between 1977 and 2020, there was an increase in the level of human interference resulting in a higher fragmentation of uses. The invasion of the different class seeds in the Bhagirathi River regions of Murshidabad district of other LULC classes is explained by the growth and contraction of the areas under various classes. Depending on the viability of the class and its affinity for the surroundings, further expanding the patches also provides the chance for the relevant class to develop in the new area. Moreover, the current land cover is threatened. The study primarily highlights crop land fragmentation and built-up areas, particularly in the post-1990 period (Table 8.4). The amount of scrubland, agricultural fallow land, and dense vegetation patches peaked in 2000, possibly as a result of the increasing economic activity in the reality industry. Surface waterbodies, dense vegetation, and agriculturally fallow land are the ones that are most likely to be established into built-up structures. Crop land due to the need for food, civil infrastructure, and population expansion area decline do direct us to the process of change taking place due to population growing demands.

The edge density serves a similar purpose in the quantitative interpretation of the landscape by making it easier to compare various sized LULC classes within the research region. By examining the data, it is possible to see that the index values for four classes—built-up area, sand, surface waterbody, and crop land—decreased in 1977, which may be an indication of a trend toward landscape homogeneity (Table 8.5). By analyzing the data, it is possible to see that the index

Table 8.4 Estimation of landscape proportion (%) in the study area during the period between 2017 and 2020.

LULC class	1977	1990	2000	2010	2020
Built-up area	0.0038	0.0075	0.01124	0.02424	0.038
Sand	0.06543	0.0329	0.0193	0.01331	0.015
Surface waterbody	0.05423	0.0286	0.0276	0.02548	0.009
River	0.0397	0.0489	0.05379	0.06146	0.042
Dense vegetation	0.0591	0.079	0.14568	0.10085	0.0949
Agricultural fallow	0.47431	0.4506	0.37101	0.278	0.2212
Scrubland	0.3472	0.177	0.1482	0.14637	0.1308
Crop land	0.13923	0.16512	0.2201	0.2668	0.359

LULC, *Land use/land cover.*

Table 8.5 Estimation of edge density in the study area during the period between 1977 and 2020.

LULC class	1977	1990	2000	2010	2020
Built-up area	0.0000357	0.00044	0.000525	0.0023	0.003
Sand	0.00000967	0.0002	0.00024	0.00048	0.0038
Surface waterbody	0.0019	0.00059	0.00041	0.0003	0.0000145
River	0.00057	0.00000391	0.0007	0.00047	0.0008
Dense vegetation	0.0000698	0.00172	0.0054	0.0034	0.0018
Agricultural Fallow	0.1045	0.0215	0.00904	0.0034	0.0000153
Scrubland	0.0121	0.0089	0.0074	0.00404	0.0000135
Crop land	0.0000179	0.00479	0.0064	0.0103	0.0125

LULC, *Land use/land cover.*

values for three classifications (agricultural fallow land, surface waterbody, and scrubland) decreased in 2020, which suggests a trend toward landscape uniformity.

For determining how fragmented a landscape is, edge density and patch density are good markers. According to the figures acquired for the fragmentation, the degree of fragmentation in the research region has increased, leading to a propensity for clustering. In the context of biodiversity and ecological sustainability, the degree of landscape fragmentation is a crucial environmental measure. Information on the extent of landscape fragmentation is also important for regional planning and alternatives on where to build or eliminate infrastructure. Its study of many time series demonstrates the strength and direction of the present trends. ED of the built-up area in the study area showed an increasing trend during the study period (1977 and 2020). The waterbody has revealed an

intriguing correlation that suggests the effects of population growth and urbanization, since the number of waterbody patches has grown between 1977 and 2020
(Table 8.6). Due to the low resolution of image quality in 1977, there were fewer
patches, and in 2000, the sparse rainfall caused many waterbodies to dry out. The
interaction boundary between individual classes and other LULC classes is
another parameter like edge density. The length has demonstrated that there has
been a noticeable decrease of the borders. Therefore the increased number of
recent patches indicates the development of additional soggy regions.

   The amount of isolation they experience inside the LULC class may be determined in large part by calculating mean patch distance. The degree of isolation of
a particular LULC class with ramifications for the system's functional relationships increases as the derived values increase. In the study area, MPAs of built-
up, sand, and crop land LULC classes have been increased progressively during
the period between 1977 and 2020. Moreover, the MPAs of surface waterbody,
river, scrubland, and agricultural fallow have been decreased (Table 8.7).

Table 8.6 Estimation of patch density in the study area during the period
between 1977 and 2020.

LULC class	1977	1990	2000	2010	2020
Built-up area	0.00000011	0.00000106	0.000012	0.000014	0.00007
Sand	0.00000018	0.00000102	0.000012	0.000043	0.000042
Surface waterbody	0.000067	0.000059	0.00000776	0.0000081	0.00000459
River	0.0000102	0.0000061	0.0000052	0.00000016	0.000000901
Dense vegetation	0.00000129	0.00000279	0.0000079	0.000005	0.0000039
Agricultural fallow	0.000269	0.0001271	0.000027	0.0000384	0.00000395
Scrubland	0.000067	0.000036	0.000027	0.00000625	0.00000356
Crop land	0.000031	0.0000204	0.000015	0.00000581	0.0000035

Table 8.7 Estimation of the mean patch area in the study area during the
period between 1977 and 2020.

LULC class	1977	1990	2000	2010	2020
Built-up area	2102.44	2041.91	2661.25	10,558.57	35,352
Sand	5282.09	12,991.88	19,975.78	18,670.09	35,571.04
Surface waterbody	55,520.47	35,079	19,072.89	11,459.48	9208.49
River	681,402.39	251,877.13	125,367.64	103,994.95	41,743.73
Dense vegetation	13,589.59	14,423.96	28,000	21,727.57	18,264.07
Agricultural fallow	57,682.619	33,467.02	13,532.64	12,335.14	10,340.49
Scrubland	23,708.09	10,010.362	9676.752	5514.34	4799.24
Crop land	28,430.64	27,690.832	23,775.26	12,292.743	4488.04

LULC, *Land use/land cover.*

Table 8.8 Estimation of the landscape diversity index in different time periods (1977–2020).

Diversity index	1977	1990	2000	2010	2020
Shannon index	1.538	1.542	1.584	1.598	1.788
Shannon equitability	0.739	0.742	0.762	0.768	0.859
Simpson index	0.711	0.719	0.750	0.752	0.789

These diversity indexes all incorporate assessments of richness and evenness. They grow when there are more different forms of land cover (landscape richness), more evenly distributed land among the different cover types (landscape evenness), or both (Table 8.8). Results of these analysis showed that all these indices were progressively increased during the period between 1977 and 2020.

8.6 Discussion

In this study, the land cover change for the Murshidabad district's Bhagirathi River course area between 1977 and 2020 was estimated. Based on the medium-resolution remote sensing data, the Bhagirathi River basin has been classified into eight major classes (surface waterbodies, built-up area, river sand, river, dense vegetation, cultivated land, scrubland, and agricultural fallow) during the time span of 1977 and 2020. After establishment of Farakka barrage in 1975, the Bhagirathi River course was rejuvenated and river bank erosion and flood both upsurged in various fragments of the Murshidabad district. Rudra (2005) investigated that the main cause of erosion along the river bank was the agony of the people existing along the river front and the consequences instigated by the expansive Padma River. The built-up land mainly lies on the east bank of the Bhagirathi River. The LULC map indicates land predominantly used for agricultural activities, located on the periphery of built-up land and edges of the riverine plain. It may be indicated that an overall increase of built-up areas from 1977 to 2020 demonstrated a population growth in the floodplains (Ardha & Karuniasa, 2023). The vegetated areas of the study area are manifested in the form of agricultural plantation and small to medium scrub. All the vegetative areas are situated adjacent to the "bil" (waterbodies) and multifarious rural inhabitants. The forest areas mainly constitute agricultural plantation and deciduous forest in the study region. Lands with scrub were mainly confined in the Ganges River plain. Marshy lands were mainly concentrated in the south and extreme north of the study region. Erosion along the Bhagirathi River in different spaces is concomitant with the demoralizing flood that occurred in 2000. The part of the earthen embankment in Ajodhyanagar near the Bermhampore town was devastated, and the cultivated land near the Sheikhpura area was infringed by the Bhagirathi River (Ghosh, 2015). Classification measurements for entire 5 km buffer areas around the Bhagirathi rivers showed decrease in agricultural fallow and wet land areas over the years. The increased impact of erosion along the river course could be attributed to this transformation of scrubland and agricultural fallow land into cultivated land. The waterbody is known as an ecologically vulnerable area that contains small and medium forms of wetlands in the course of the Bhagirathi River (Mondal, 2012). Consequently, the adjoining region of the Bhagirathi River course has experienced seasonal inundations because of the low depressions filled up by the seasonal rainfall and overflowing of water during the rainy season. During the period between 1977 and 1990, there were many changes in the scrubland and agricultural land, which may be due to the decrease in the upper reach of the Bhagirathi River between Biswanathpur and Giria (Parua, 2009).

Riverine forests are extremely important to society and the biosphere, yet they are also quite vulnerable. The essential mechanisms that create and preserve riverine vegetation are altered by the hydrologic changes brought forth by river control. Using satellite imagery with improved spatial/spectral resolution can significantly contribute to efforts aimed at better understanding the extent and direction of land cover fragmentation and changing landscapes by helping figure out the underlying causes of such changes (Kastha & Khatun, 2022). Identifying the rate of transition from one class to another and employing more thorough fragmentation indexes incorporating patch-level analysis might assist in determining the patterns for subsequent LULC transformation to assess the endurance of these key riverine systems (Gaur, Sinha, Adhikari, & Ramesh, 2019). How different geographical extents affected the calculation of LULC features and landscape metrics was also examined. The local scale important to conservation managers and land planners revealed a highly substantial correlation between remotely sensed landscape characteristics and riverine biological richness (Nagendra, 2002; Naha, Rico-Ramirez, & Rosolem, 2021). Classification measurements for the entire 5 km buffer areas around the Bhagirathi rivers showed decrease in agricultural fallow and wet land areas over the years. The increased impact of erosion along the river course could be attributed to this transformation of scrubland and agricultural fallow land into cultivated land. This is because in comparison to other urban or industrial land uses, the Bhagirathi River region is mostly characterized by population migration, crop land, and urbanization, all of which sustain a disproportionately high number of fragmented LULC. Riverine landscape structure, which is influenced by the size, form, and nature of the land use patches within the landscape, is a key aspect in assessing the pattern and impacts of numerous natural processes (Gu et al., 2021; Molles, 2006). Understanding the effects of land use changes requires a landscape topology study. Landscape fragmentation is a complicated process that affects an intricate framework and produces a diverse array of spatial patterns. In order to separate ecological services provided by landscapes and allocate them wisely to maintain a balance between growth and development, planners will benefit from the quantitative of pattern measures (Wang, Gamon, Cavender-Bares, Townsend, & Zygielbaum, 2018). The landscape metrics change depending on the LULC classes. Regarding the landscape of the study region, there is a prevalence of some types (built-up area or crop land), which have grown more rapidly than others, placing a significant strain on the local biodiversity and under natural circumstances. Riverine buffers have been used as a conservation measure over the past few decades in an effort to preserve the naturally occurring processes and functions that make up running-water ecosystems and, as a result, preserve local aquatic species. They are hotspots for climate change resilience in the near future because of their biodiversity importance (Das, 2014). The LecoS plugin was used to calculate the landscape metrics, and it was noted that the research region had fragmentation issues because of the high level of human disturbance (Batar, Watanabe, & Kumar, 2017; Olsen, Dale, & Foster, 2007). In the absence of any constraints, QGIS proved to be quite practical and rather intuitive when it came to calculating landscape measures (Turner & Gardner, 2001). Realizing the fragility of these constantly changing systems requires this kind of insight, which makes it necessary to carry out the research at a larger temporal and spatial scale. The study's overall findings demonstrate the value of global indicators in landscape change modeling and the growing significance of this form of LULC analysis for local actors who must make decisions in accordance with the landscape potential of each location.

8.7 Conclusion

The present investigation has limitations despite best attempts. In general, edge length increases as resolution decreases (i.e., as more detail is used to define edges). Edges may look as relatively straight lines at coarser resolutions, but they may also appear as very complicated lines at finer resolutions. Because of this, results generated for edge measurements should not be compared between images of various resolutions. The research was limited to satellite remote sensing data with a 30 m spatial resolution and a 10 year or so temporal resolution. Investigating how size (let us say less than 30 m) affects the estimate of these measures and how well they capture trends would be useful. The outcomes exhibited that most affected LULC class is built-up land and cultivated land. In the local framework, the LULC change pattern influenced by river dynamics unswervingly impacts the livelihood of the floodplain inhabitants. Moreover, in order to make the case for taking into account landscape measurements as possible tools in the creation of land use policy for future growth and development activities, some of the landscape metrics were calculated in order to illustrate their usefulness when paired with the spatial analysis. As a result, the quantitative methods used in this research to evaluate the LULC dynamics as a function of interactions among LULC substances may aid in monitoring the influential and efficient components and in abstracting the LULC scenario of a region, thereby expanding the range of options for decision-making among scientists, planners, and executives to assure environmental sustainability. This strategy will aid in planning for ecological connection in the region that is resilient to climate change.

Acknowledgments

We affirm our earnest gratitude to USGS (http://www.usgs.gov/) for making available of satellite datasets freely to carry out the research work.

References

Ardha, M. J., & Karuniasa, M. (2023). Land cover change and fragmentation analysis in Sintang Regency, West Kalimantan, Indonesia. *IOP Conference Series: Earth and Environmental Science*, *1275*(1), 012003. Available from https://doi.org/10.1088/1755-1315/1275/1/012003.

Batar, A., Watanabe, T., & Kumar, A. (2017). Assessment of land-use/land-cover change and forest fragmentation in the Garhwal Himalayan Region of India. *Environments*, *4*(2), 34. Available from https://doi.org/10.3390/environments4020034.

Boyd, D. S., & Foody, G. M. (2011). An overview of recent remote sensing and GIS based research in ecological informatics. *Ecological Informatics*, *6*(1), 25−36. Available from https://doi.org/10.1016/j.ecoinf.2010.07.007.

Chelaru, D. A., Oiçte, A. M., & Mihai, F. C. (2014). Quantifying the changes in landscape configuration using open source GIS. Case study: Bistrita subcarpathian valley. *International Multidisciplinary Scientific GeoConference Surveying Geology and Mining Ecology Management, SGEM*, *5*(1), 557−564. Available from http://www.sgem.org/.

Das, S. (2014). Ganga − Our endangered heritage. In *Our National River Ganga: Lifeline of millions*. Springer International Publishing. https://doi.org/10.1007/978-3-319-00530-0_2.

Debnath, J., Das (Pan), N., Ahmed, I., & Bhowmik, M. (2017). Channel migration and its impact on land use/land cover using RS and GIS: A study on Khowai River of Tripura, North-East India. *The Egyptian Journal of Remote Sensing and Space Science*, *20*(2), 197−210. Available from https://doi.org/10.1016/j.ejrs.2017.01.009.

De Groot, R. (2006). Function-analysis and valuation as a tool to assess land use conflicts in planning for sustainable, multi-functional landscapes. *Landscape and Urban Planning*, *75*(3−4), 175−186. Available from https://doi.org/10.1016/j.landurbplan.2005.02.016.

Etefa, G., Frankl, A., Lanckriet, S., Biadgilgn, D., Gebreyohannes, Z., Amanuel, Z., ... Nyssen, J. (2018). Changes in land use/cover mapped over 80 years in the Highlands of Northern Ethiopia. *Journal of Geographical Sciences*, *28*(10), 1538−1563. Available from https://doi.org/10.1007/s11442-018-1560-3.

El Bastawesy, M., Ramadan Ali, R., Faid, A., & El Osta, M. (2013). Assessment of water-logging in agricultural megaprojects in the closed drainage basins of the Western Desert of Egypt. *Hydrology and Earth System Sciences*, *17*(4), 1493−1501. Available from https://doi.org/10.5194/hess-17-1493-2013.

Fichera, C. R., Modica, G., & Pollino, M. (2012). Land cover classification and change-detection analysis using multi-temporal remote sensed imagery and landscape metrics. *European Journal of Remote Sensing*, *45*(1), 1−18. Available from https://doi.org/10.5721/EuJRS20124501, http://server-geolab.agr.unifi.it/public/completed/2012_EuJRS_45_001_018_Fichera.pdf.

Fischer, J., & Lindenmayer, D. B. (2007). Landscape modification and habitat fragmentation: A synthesis. *Global Ecology and Biogeography*, *16*(3), 265−280. Available from https://doi.org/10.1111/j.1466-8238.2007.00287.x.

Gaur, T., Sinha, A., Adhikari, B. S., & Ramesh, K. (2019). Dynamics of landscape change in a mountainous river basin: A case study of the Bhagirathi River, western Himalaya. *Applied Ecology and Environmental Research*, *17*(4), 8271−8289. Available from https://doi.org/10.15666/aeer/1704_82718289, http://aloki.hu/pdf/1704_82718289.pdf.

Ghosh, S. (2015). River bank erosion and environmental neo-refugees: A case study of Murshidabad district in West Bengal. India. *International Journal of Current Research*, *7*(5), 16582−16589.

Gu, C., Zhang, Y., Liu, L., Li, L., Li, S., Zhang, B., . . . Rai, M. K. (2021). Qualifying land use and land cover dynamics and their impacts on ecosystem service in central Himalaya transboundary landscape based on Google Earth Engine. *Land*, *10*(2), 173. Available from https://doi.org/10.3390/land10020173.

Hassan, Z., Shabbir, R., Ahmad, S. S., Malik, A. H., Aziz, N., Butt, A., & Erum, S. (2016). Dynamics of land use and land cover change (LULCC) using geospatial techniques: A case study of Islamabad Pakistan. *SpringerPlus*, *5*(1). Available from https://doi.org/10.1186/s40064-016-2414-z, http://www.springerplus.com/archive.

Irons, J. R., Dwyer, J. L., & Barsi, J. A. (2012). The next Landsat satellite: The Landsat data continuity mission. *Remote Sensing of Environment*, *122*, 11−21. Available from https://doi.org/10.1016/j.rse.2011.08.026.

Jensen, J. R. (1996). *Introductory digital image processing: A remote sensing perspective*. Prentice Hall.

Julian, J. P., Wilgruber, N. A., de Beurs, K. M., Mayer, P. M., & Jawarneh, R. N. (2015). Long-term impacts of land cover changes on stream channel loss. *Science of the Total Environment*, *537*, 399−410. Available from https://doi.org/10.1016/j.scitotenv.2015.07.147, http://www.elsevier.com/locate/scitotenv.

Kastha, S., & Khatun, S. (2022). Quantifying and assessing land use and land cover changes around the critical waterbodies—A case study of Bhagirathi-Hooghly floodplain, East India. *Applied Geomatics*, *14*(2), 315−334. Available from https://doi.org/10.1007/s12518-022-00435-1, http://www.springerlink.com/content/1866-9298/.

Mandal, S.P., Kayet, N., Chakrabarty, A., & Rahaman, G. (2016). *Sustainable development and management of ground water resources, its remedial measures for emerging crisis and climate change in West Bengal* (pp. 72−88). Central Ground Water Authority & Central Ground Water Board Eastern Region.

Lea, C., & Curtis, A. C. (2010). Thematic accuracy assessment procedures: National Park Service Vegetation Inventory, version 2.0. Natural Resource Report NPS/2010/NRR—2010/204. *National Park Service, Fort Collins, Colorado*.

Lillesand, T. M., & Kiefer, R. W. (1994). Remote sensing and image interpretation (3rd ed. p. 9780471305750). Hoboken: John Wiley and Sons, Inc.

Lu, D. S., & Weng, Q. H. (2007). A survey of image classification methods and techniques for improving classification performance. *International Journal of Remote Sensing*, *28*, 823−870.

McGarigal, K. S. A., Cushman, Neel, M. C., & Ene, E (2002). FRAGSTATS: spatial pattern analysis program for categorical maps. Computer software program produced by the authors at the University of Massachusetts, Amherst. Available from http://www.umass.edu/landeco/research/fragstats/fragstats.html.

Mohamed, E. S., Saleh, A. M., & Belal, A. A. (2014). Sustainability indicators for agricultural land use based on GIS spatial modeling in North of Sinai-Egypt. *Egyptian Journal of Remote Sensing and Space Science*, *17*(1), 1−15. Available from https://doi.org/10.1016/j.ejrs.2014.05.001, http://www.elsevier.com/wps/find/journaldescription.cws_home/723780/description#description.

Molles, M. (2006). *Ecology: Concepts and applications*. McGraw Hill.

Mondal, D. (2012). Urban landuse change assessment using RS and GIS: A case study of Berhampore town and its surroundings. *Geo-Analyst*, *2*(1), 1−6.

Nagendra, H. (2002). Opposite trends in response for the Shannon and Simpson indices of landscape diversity. *Applied Geography*, *22*(2), 175−186. Available from https://doi.org/10.1016/s0143-6228(02)00002-4.

Naha, S., Rico-Ramirez, M. A., & Rosolem, R. (2021). Quantifying the impacts of land cover change on hydrological responses in the Mahanadi river basin in India. *Hydrology and Earth System Sciences*, *25*(12), 6339−6357. Available from https://doi.org/10.5194/hess-25-6339-2021, http://www.hydrol-earth-syst-sci.net/volumes_and_issues.html.

Oliphant, E. (2007). Voices and apparitions in Jules Bastien-Lepage's Joan of Arc (exhibition catalogue). In Looking and listening in nineteenth-century France, *Martha Ward, Anne Leonard*, (pp. 42−49). Chicago: Smart Museum of Art.

Olsen, L. M., Dale, V. H., & Foster, T. (2007). Landscape patterns as indicators of ecological change at Fort Benning, Georgia, USA. *Landscape and Urban Planning*, *79*(2), 137−149. Available from https://doi.org/10.1016/j.landurbplan.2006.02.007.

Parua, P. K. (2009). Some facts about Farakka Barrage Project, *The Ganga: Water Use in the Indian-Subcontinent* (pp. 58−86). Dordrecht: Springer.

Rudra, K. (2005). *The encroaching Ganga and social conflicts: The case of West Bengal, India*. Gangapedia (GRBMP).

Shannon, C. E., & Weaver, W. (1949). The Mathematical Theory of Communication. *Urbana, IL: The University of Illinois Press*, 1−117.

Simpson, E. H. (1949). Measurement of diversity. *Nature*, *163*(4148), 688. Available from https://doi.org/10.1038/163688a0.

Steiniger, S., & Hay, G. J. (2009). Free and open source geographic information tools for landscape ecology. *Ecological Informatics*, *4*(4), 183−195. Available from https://doi.org/10.1016/j.ecoinf.2009.07.004.

Strahler, A.N. 1956. The nature of induced erosion and aggradation. In W.L. Thomas, Jr., C.O. Sauer, M. Bates, L. Mumford (Eds). Man's Role in Changing the Face of the Earth; 621−638, Vol. 1; Chicago: University of Chicago Press. 448.

Tarantino, E., Novelli, A., Aquilino, M., Figorito, B., & Fratino, U. (2015). Comparing the MLC and JavaNNS approaches in classifying multi-temporal LANDSAT satellite imagery over an ephemeral river area. *International Journal of Agricultural and Environmental Information Systems*, *6*(4), 83−102. Available from https://doi.org/10.4018/ijaeis.2015100105.

Turner, M.G., & Gardner, R. (2001). *Landscape ecology in theory and practice: Pattern and process*.

Wang, R., Gamon, J. A., Cavender-Bares, J., Townsend, P. A., & Zygielbaum, A. I. (2018). The spatial sensitivity of the spectral diversity-biodiversity relationship: An experimental test in a prairie grassland. *Ecological Applications*, *28*(2), 541−556. Available from https://doi.org/10.1002/eap.1669, http://www.esajournals.org.

Yang, X., Damen, M. C. J., & Van Zuidam, R. A. (1999). Satellite remote sensing and GIS for the analysis of channel migration changes in the active Yellow River Delta, China. *International Journal of Applied Earth Observation and Geoinformation*, *1999*(2), 146−157.

2D hydronumeric flood modeling using open-source software QGIS and freeware BASEMENT (V3) on a section of the Elbe River in Magdeburg, Germany

Shafkat Sharif

Hydroscience and Engineering, Civil and Environmental Engineering, Technische Universität Dresden, Dresden, Germany

9.1 Introduction

Nature and humans both rely on the stability of climate variables for maintaining the status quo. In contrast, global change is forcing the climatic conditions to change abruptly and threatens life as we know it. For instance, climatic models indicate lower annual rainfall in many regions, while extreme rainfall events with higher intensities will be frequent (Salcedo-Sanz et al., 2024). In early 2023, researchers suggested that global average temperature will rise beyond the threshold temperature of 1.5°C set by the United Nation's Paris Agreement much earlier than expected (Diffenbaugh & Barnes, 2023). This presents a unique challenge for food production, human living conditions, and infrastructure, as droughts as well as more extreme flood events are deemed to become more frequent (Rahman et al., 2022). In many parts of the world, these flooding events are already causing urban flood hazards and breaching the existing protection structures. Contemporarily, it is of utmost importance to test and bolster our flood mitigation measures such as dikes and embankments, so that critical habitats and agricultural regions remain protected in the foreseeable future (Haque, Ishtiaque, & Chowdhury, 2020). In the wake of such events, flood modeling has become the go-to strategy to be incorporated in flood risk management all over the world (Jodhani, Patel, & Madhavan, 2023). Currently, substantial research has been done in the field of flood modeling. This led to the development of many commercial tools, databanks, and methods, which are available for numerical flood

Applications of Geospatial Technology and Modeling for River Basin Management.
DOI: https://doi.org/10.1016/B978-0-443-23890-1.00009-8

modeling and analysis. However, to cover the substantial flood-prone areas in the world, hydrologists, researchers, and modelers must have access to open-source and freely available tools and data (Jodhani et al., 2023). Recent flood modeling utilizes geospatial planning, mainly using GIS tools, to build empirical and physical-based flood models. Typically, it involves digital elevation models (DEM) and remotely sensed data such as hydrological and geomorphological variables (Fenglin et al., 2023). All these can be processed using GIS tools coupled with other flood simulation hydronumeric (HN) modeling tools. In that regard, open-source/open-access tools can be valuable assets in this field. Especially since readily accessible datasets are becoming available worldwide, the scope of using open tools to develop simple 2D models increases severalfold (Nkwunonwo, Whitworth, & Baily, 2020).

This chapter deals with simulating a flood event in the outskirts of Germany's Magdeburg (Fig. 9.1). The city is home to one of the largest rivers in Europe, the Elbe River. The reach of the river is through Austria, Czech Republic, Poland, and Germany, and it has one of the largest riverbanks with a catchment area of nearly 150,000 km^2 (Bartl, Schümberg, & Deutsch, 2009). The Elbe in Germany is infamous for its devastating floods. Notably, in 2002 and 2013, it flooded big cities such as Dresden and parts of Magdeburg. In the latter, out of 250,000 residents, 15,000 people were displaced and further 8000 inhabitants were in the highest alert. The town of Schönebeck, 15 km south, flooded completely during the flood of 2002 (Davies, 2013).

This chapter establishes a 2D hydronumerical model to simulate a flood event in the "Alte Elbe," a cut-off channel of the Elbe River near the town of Schönebeck. A 9 km long reach of the Alte Elbe is considered as the modeled area where in the town of Pretziener, a weir is constructed to divert a quarter of the Elbe River discharge through a flood canal. The Pretziener weir started operating in

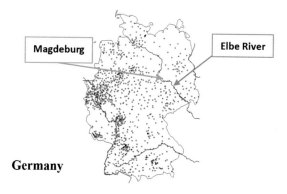

FIGURE 9.1

Map of Germany and its cities (black dots), showing the location of Magdeburg and the Elbe River (plotted using RStudio). Map lines delineate study areas and do not necessarily depict accepted national boundaries.

1875 (Bardua, 2015). In 2002, about 1000 m^3/s of flood water was discharged through the weir, where the capacity is known to be 1200 m^3/s (Bardua, 2015). During the 2013 Elbe River flooding, the inundation depth was 7.45 m, which broke the previous record of 6.72 m. Although the theoretical discharge capacity of the weir was higher, the dike systems protecting the peripheral towns failed during the floods when the weir gates were opened (Davies, 2013). By defining regions and delineating the modeled area with break lines in QGIS, boundary conditions are set up to test the workability of the dikes in the section of the Alte Elbe. The objective is to protect the critical flood-prone areas beyond the left and right bank of the channel. Furthermore, the Pretziener weir is simulated in the model using the simulation software BASEMENT (developed by VAW, ETH Zurich). Outputs such as the water flow velocity, specific discharge, Chezy's Friction coefficient, and water depth are obtained from the simulation of BASEMENT (Vetsch et al., 2022). Afterward, the simulated water depth is compared with a high priority 10-year flood event in the results and discussion section of the chapter. As the risk of increased flooding plagues the Elbe River basin in Germany, it is important to solidify the understanding of the existing flood protection structures. The current study thus aims to develop a dimensionality-based 2D flood simulation developed using BASEMENT and QGIS and then visualized using a QGIS model with plugins such as the TUFLOW viewer. This study is intended to be a demonstration that simple HN models can be built using open tools and readily available datasets. The applied methodology can be reproduced for data-scarce regions around the world where the vulnerability of flood structures and the extent of flooding needs to be tested in a readily and cost-effective manner.

9.2 Rationale of the study

The present study simulates and visualizes the flow characteristics of an Old Elbe River section using a 2D HN model. The section is approximately 9 km long (measured using the QGIS measure tool) and has a maximum floodplain width of 1.2 km (including both sides of the channel). A 2D model is created using the long-term repository of QGIS version 3.16.11 Hannover, using provided maps such as the DEM (2 m by 2 m resolution), land use of the study area, and surrounding dike structures. The model is then simulated using BASEMENT—a software developed by ETH Zurich. This study was initially part of a tertiary-level course work, where personalized boundary conditions and parameters were provided, which was later used in the simulation setup, for instance, differing roughness coefficients for the left and right bank of the river and a lower friction value for the riverbed. Then the model is run until the water reaches a steady state discharge at the outlet of the channel (visualized in QGIS). Additionally, for the current study, a weir in the study area (Pretziener Weir) is included in the modeling and simulation, and furthermore, the floodwater depth is compared with a historical flood event's water depth (the 10-year return period flood).

9.3 **Methodology**
9.3.1 **Delineation of the study area**

Fig. 9.2A shows the section of the Alte Elbe channel that is selected for the flood simulation modeling. Three critical flood-prone areas surround the study area: the villages of Pretzien and Plotzky in the north, which are part of the Schonebeck town, and Bergenfeld, an agricultural land entailing the entire southern region model. Fig. 9.2B is a picture of the Pretziener weir, which falls under the model extent, which protects the villages from receiving flash-flood waters during heavy rainfall events (Maue, 2013).

The Pretziener weir is metal gated with 36 iron barriers; it has a usable flow width of 112.95 m. The water levels of the channel must reach 5.92 m before the weir gates are opened and be below 5.25 m when the gates are closed again (Bardua, 2015). The efficacy of the weir is assessed by incorporating it in the model and simulations.

9.3.2 **Model setup and boundary conditions**

The model for the simulation is built on QGIS 3.16.11 Hannover and subsequently simulated in Basement v3.0. As prepossessing in QGIS is essential for running the simulation further, the available DEM raster file (2 m by 2 m resolution), Dike, and land use vector maps of the study area are added to QGIS for defining the model extent. The corresponding projection coordinates are selected to be EPSG: 25832 ETRS 89/UTM zone 32 N. Using the available layers, especially the DEM and dikes were used to define break lines where the simulated water should be constrained. Later a regions vector file is made to define each area, and a boundary vector file is created to mark the inlet, outlet, and the Pretziener weirs, through which water will flow in the modeled channel. Fig. 9.2A shows the complete model extent, including the boundaries, break lines, and regions constructed in QGIS, whereas Table 9.1 shows the attributes of the regions defined in the model and then used for the model setup.

BASEmesh v2.0 is then used as a QGIS plugin to create a quality mesh layer, which uses the delineated regions from break line polygons, region points, and boundary strings to create a TIN mesh. Here, to indicate the weir in the model, hole-marked fields are selected alongside material ID and maximum elevation area of the model. Outback regions are given the largest areas (1000 m^2), as larger triangular meshes can be present, whereas riverbed, dikes, and the weir are given lower values to enhance flow simulation in those areas. The quality mesh is then combined with the DEM using the Interpolation tool of BASEmesh to create the computational mesh file. Fig. 9.3 shows the mesh created by BASEmesh.

The next step includes setting up the model parameters in BASEMENT, where the BASEPLANE 2D HN model is defined. Notably, after defining the regions (see Table 9.1 for friction values) and strings (three strings for inlet, outlet, and weir),

(A)

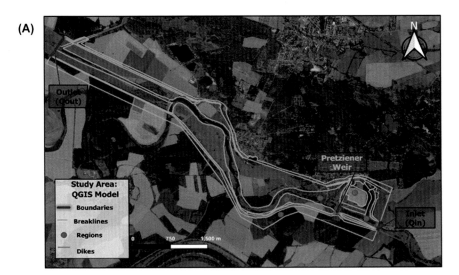

(B)

FIGURE 9.2

(A) QGIS model built for the study area in the Alte Elbe channel section, displaying the boundaries, regions, dikes, and weir; (B) picture of the Pretziener weir. Map lines delineate study areas and do not necessarily depict accepted national boundaries.

From Maue, T. (2013). Pretziener Wehr. Available from https://www.flickr.com/photos/erwinrommel/
8947356710 (CC BY-SA 2.0).

Table 9.1 Table showing the denoted regions, material ID, maximum elevated area, holes, and the associated Strickler's friction coefficient.

Region ID	Material ID	Maximum elevated area (m²)	Location	Hole	Strickler's friction coefficient (m$^{1/3}$/s)
1	3	150	Riverbed	0	38
2	3	150	Riverbed	0	38
3	1	500	Right bank	0	22
4	1	500	Right bank	0	22
5	3	50	Riverbed	0	38
6	4	20	Dike	0	20
7	4	20	Dike	0	20
8	4	20	Dike	0	20
9	4	20	Dike	0	20
10	4	20	Dike	0	20
11	4	20	Dike	0	20
12	2	500	Left bank	0	24
13	1	500	Right bank	0	22
14	4	20	Dike	0	20
15	4	20	Dike	0	20
16	5	1000	Outback	0	20
17	5	1000	Outback	0	20
18	6	15	Weir	1	20
19	4	20	Dike	0	20
20	3	50	Riverbed	0	38
21	2	300	Left bank	0	24
22	2	300	Left bank	0	24
23	2	300	Left bank	0	24

hydraulic parameters are set. **400 m³/s** of flow is defined for uniform inflow, and a uniform slope of **0.0022%** is set for the entire model. For the weir, a linked boundary condition is set alongside the standard boundaries. **0.75** default Poleni factor is kept for the weir with a gate height of **51 m** above the mean sea level. The last step in BASEMENT includes setting up the simulation, with 8 hours (28,800 seconds) of simulation time, requiring output in the results after every minute (60 seconds). The results obtained are then visualized using QGIS temporal animations and TUFLOW QGIS plugin (TUFLOW, 2023). The TUFLOW viewer tool is later used for exporting maps and obtaining timeseries and cross-section graphs of the output from the QGIS interface.

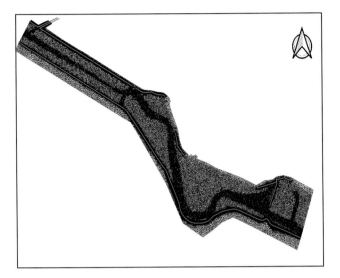

FIGURE 9.3

Map showing the triangular mesh of the computational mesh created by the interpolation tool.

9.4 **Results**

9.4.1 **Flow velocity**

The results obtained from the simulation lead to water entering the forelands and flooding it. The forelands flood up to 1 hour, and when the water reaches a certain depth, it crosses the weir in the first hour. The flow velocity is higher in the inlet, while the rest of the forelands experience velocities between 0.7 and 2.0 m/s. Fig. 9.4 shows the flooding in the forelands along with the flow velocity.

In the following hours, the simulation shows quick progression of water crossing the weir, where the velocity can be visually inspected to be around the range of 0.7–1.4 m/s, and only in some patches of the main channel, the velocity can be seen to reach above 2 m/s. Fig. 9.5A and B shows the maps of flow velocity at the 3rd and 6th hours, respectively. The flow can be seen to reach a steady state just before the 6th hour.

9.4.2 **Inundation depth**

The output results of water depth are mapped using the TUFLOW viewer. Before the Pretziener weir, it is seen that the Alte Elbe channel floods the highest. Fig. 9.6A shows the extent of flooding in the simulated modeled section. The highest water depth is up to 5.05 m, where after the weir, the water levels reach

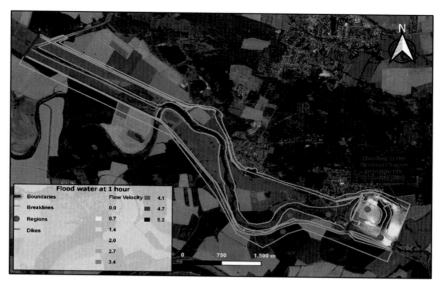

FIGURE 9.4

Map showing the forelands before the Pretziener weir being flooded up to 1 hour in the simulated period. Map lines delineate study areas and do not necessarily depict accepted national boundaries.

up to 4 m in the channel and 3 m or less in the left and right banks of the Alte Elbe. The forelands, however, sees a water depth of 2−3 m in the elevated regions up to 5 m in the channels (LWH, n.d.).

Fig. 9.6B shows the 10-year event (HQ10), which is the highest probability flood event data available in the LHW website. While the flooded areas are similar to the simulated flow, the depth of water is seen to be higher in most of the regions. Notably, the channel beyond the weir inundates from 4 to 5 m up to two-thirds of the modeled area.

9.4.3 Specific discharge and Chezy's friction coefficient

The specific discharge map exported from TUFLOW shows the streamlines of the modeled area. Fig. 9.7A displays the streamlines of specific discharge at 6-hour simulation time, after reaching steady state conditions. The specific discharge is found to be the highest (6.06−7.57 m/s) at the middle of the channel, south of the Plotkzy village during this time. Specific discharge values of other simulated times are discussed in the next section.

Fig. 9.7B shows the results of the Chezy's friction coefficient. After reaching steady state, the squared Chezy's friction values are found to be the highest (above 200) in the bed of the river channel, while lower coefficient values (50−151) can be seen in the rest of the regions.

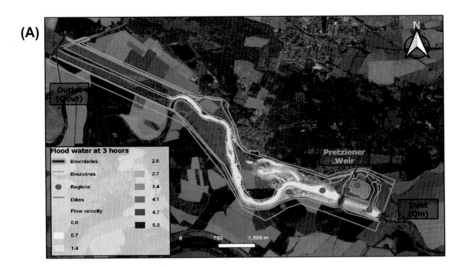

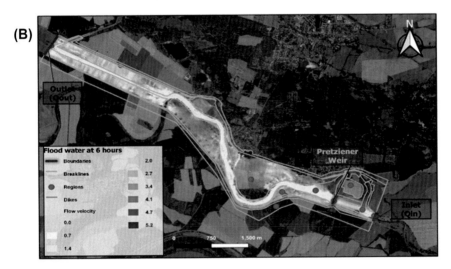

FIGURE 9.5

Evolution of flow and associated velocities for (A) 3rd hour and (B) 6th hour. Map lines delineate study areas and do not necessarily depict accepted national boundaries.

9.5 Discussion

9.5.1 Flow velocity and impact of the weir

Since the inlet boundary is drawn from left to right, the right side of the inlet showed the highest flow velocity in the first hour of simulation. **A** shows the

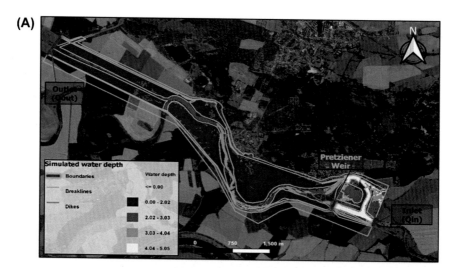

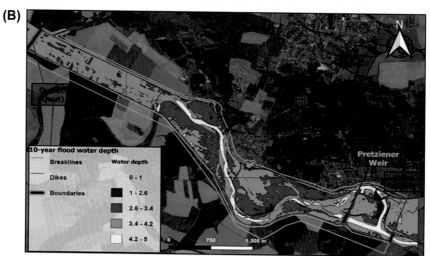

FIGURE 9.6

(A) Output results of water depth at the end of the simulation period (8 hours); (B) water depth map obtained from LHW for a 10-year high probability flood event. Map lines delineate study areas and do not necessarily depict accepted national boundaries.

Adapted from LWH. (n.d.). Hochwassergefahrenkarten—LHW LSA. http://www.geofachdatenserver.de.

cross-sectional velocity profile for the inlet, plotted from TUFLOW viewer's cross-section profile tool. The flow velocity reaches just above 5 m/s at the right side of the inlet boundary. Afterward, the velocity becomes stable and lowers as more water enters the forelands and increases the depth. The outlet and weir receive lower flow velocities as friction and differing elevations slows the flow.

(A)

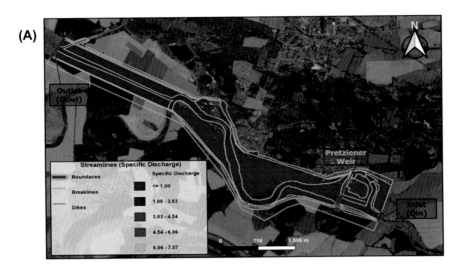

(B)

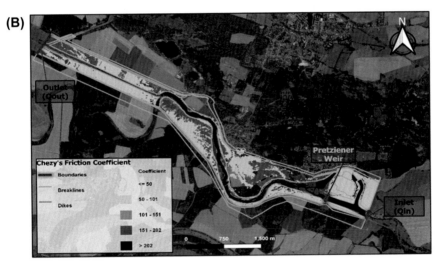

FIGURE 9.7

Maps showing the (A) specific discharge and (B) Chezy's friction coefficient values after reaching steady state conditions (at 6 hours). Map lines delineate study areas and do not necessarily depict accepted national boundaries.

B and **C** show cross-section profiles along the output and weir, respectively. After the forelands are inundated and water levels reach the weir gate height of 51 m asl, the weir sees a lower velocity and specific discharge values compared to the forelands as expected. The entire cross-section is seen to have a fluctuation between a velocity of 1.05−0.95, indicating water flows almost uniformly. This can be called an actual representation of the weir's behavior during flood.

However, in reality, the weir only opens when floodwater depth is higher in the forelands (Bardua, 2015). A height of 52.5 m asl and higher would be an actual representation. However, when simulated at that level, the outback regions received large volumes of water, which breached the 3–5 m high dike systems present in the area. After the water reaches the outlet, higher velocities are seen at the middle of the cross-section in **B**. This is likely because the water at river channel flows faster with lower riverbed friction. Because the outlet of the channel is kept very large (nearly 475 m), most of the flooded water spread across the outlet and a large range of velocities can be seen. In reality, if the outlet is only the river channel, then the corresponding velocities would be higher and with greater chances of inundation.

9.5.2 Inundation depth, discharge, and historical flooding

From the map of inundations in Fig. 9.6A and B of the previous section, we can assume that the simulated water depth for a lower discharge of 400 m³/s can reliably mimic the flooded zones between the break lines, compared to the 10-year flood event. In terms of the depth of inundations, the simulated depths are lower in the left and right banks, but closely resembles the levels of the high probability 10-year event. From the flooding depth, velocity, specific discharge, and friction results, several correlations can be inferred. The timeseries line graph of the entire simulation period shows how water levels at the inlet rise gradually. This is likely because as water starts to log up in the channel, the steady state conditions start to reach over time, and velocity and specific discharge become more uniform and adjusted to the slope and friction of the regions. From the second hour of simulation, we start seeing a constant water depth of 3 m at the right corner of the inlet. As specific discharge is the discharge per unit width of the channel, each unit width of the weir outflow gates discharges a maximum of 3.25 m/s through the weir and fluctuates up to 2.5 m/s. **B** shows the weir's maximum specific discharge and velocity. This implies that when water in the forelands reaches 51 m asl, discharge per unit width at the weir falls by 2 m/s. By looking at Fig. 9.6A, we can see an inundation depth of 3–4 m beyond the weir, which can be because of the lower discharge rate as opposed to the HQ10, the 10-year flood event. Fig. 9.6B shows higher levels of 4–5 m in the channel depth. Overall, specific discharge values correlate heavily with the inundation depth and Chezy's friction coefficients. The higher squared Chezy's friction values as seen in Fig. 9.7B in the previous section also mean less friction for the riverbed (as specified by the Strickler's friction values of the BASEMENT model setup; see Table 9.1). By looking at the 10-year flood event, the simulation also matches real-time data, as real friction values will be relatively similar. This indicates that discharge values may be the only key factor in the change in inundation depth from the simulation and real time. It is noteworthy to mention that HQ50 and HQ100 discharge values (50- and 100-year events) will be much higher, 4530 and 4920 m³/s, respectively (Hübner & Schwandt, 2018; WSV, 2023). This could mean certain breach of dike structures and much larger inundation depth compared to the simulated flood event and the 10-year event.

9.5.3 Limitations of the model

From the flood hazard map in the LHW website (LWH, n.d.), it can be seen that in the southern east of the Plozky village, there is a small pond, surrounding which there is a chance of inundations through leakages or breach of the dikes. Fig. 9.8A shows a section of the flood hazard map and it indicates the pond. The green area surrounding it is known to flood. This could mean that the lower elevations surrounding the pond makes the pond into a larger flood retention basin. Similarly in the modeled area, the pond which falls in the region of the outback (between the break lines) also shows flooding (Fig. 9.8B). This is the only noticeable breach of water in the model. The water entry point timeseries (Fig. 9.8C)

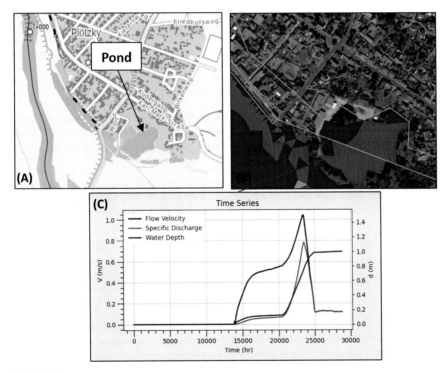

FIGURE 9.8

(A) Section of the flood hazard map obtained from the LHW website, (B) section from the simulated water depth map with the pond, and (C) timeseries at the point where water enters the pond. Map lines delineate study areas and do not necessarily depict accepted national boundaries.

shows that a low velocity of about 1.05 m/s is observed around 6-hour of simulation time just when the model reaches steady state.

A subsequent water depth of 1 m is also seen from the timeseries graph at that point. This also coincides with the inundations observed from the 10-year flood event. This means that future models should adopt the pond as part of a natural retention basin, and it should be kept within the break lines. Other issues include inconsistency in the DEM, where roads and highways obstruct actual elevation measures. A higher resolution DEM would remove these cavities in the elevation and produce better results for the simulation. This may also ensure that water will not breach or leak the dikes in certain areas, as it does not do so in reality with discharges such as 400 m³/s. According to Bardua (2015), the dike systems should be able to hold 1200 m³/s of water during flooding events, given that the weir is properly operated (Bardua, 2015). Lastly, the Poleni factor for the weir is changed to mimic a metal-gated weir (0.64), but later, the default value of 0.75 gave a faster flow and hence kept unchanged for final outputs.

9.5.4 Future directions and knowledge gaps

Improving flood preparedness, in the face of unpredicted weather events, is a top priority around the world (Shoma Mitkari, Suryakar, Mali, Shaikh, & Mullani, 2023). Especially with advances in AI and machine learning, authorities and flood researchers are doubling down on creating intelligent early warning systems and flood detection for the future. The ultimate goal is to reduce the impact of flooding on people, infrastructure, and the environment (Shoma Mitkari et al., 2023). Similarly, testing the existing flood protection infrastructure for flood events can add an extra layer of early warning and be a vital tool for future mitigation planning. The current study used open-source software, freeware, and open data to model a flood event that has approximately one-third of the discharge of extreme flood events in the area. Even at lower capacity than the design, the model showed vulnerability of the dikes to protect nearby villages and demonstrated potential floodwater breaches. Easily replicable and frequent studies should be made using open-access tools for other vulnerable areas around the world. Pertinently, these studies should be made reusable and open for everyone as flood mitigation is of global importance (Schymanski & Schymanski, 2023). These tools are valuable, in particular, because they are cost-effective and can be validated by others using a variety of commercial tools and observation data. Although developed nations have publicly available hydrological data, elevation maps, and land use and land change maps required for these studies, it is difficult to obtain the required resources for the rest of the world. Oftentimes, high resolution data is not released or unavailable altogether. One of the key components of this study is the availability of the DEM, land use map for the study area, and websites in Germany having flood inundation maps. In other regions of the world, these data must be publicly available for scientists to do these sorts of validation of the flood protection structures under different scenarios. Only then these modeling studies can be made readily, and reasonable measures can be taken by responsible authorities to enhance flood infrastructures.

Even though proper calibration and validation was not done for the current HN model, inundation maps could be compared to observation values of the 10-year flood event due to data availability. However, this is a challenge for data-scarce regions. Usually, models made with limited data have very high uncertainty (Jafarzadegan & Merwade, 2017). This is another reason why basic input data such as DEM, river bathymetry, land use, and so forth need to be made available. A lack of computational power and proper data can be a challenge in conventional modeling approaches, but this gap can be closed using surrogate methods. By using global geomorphic DEM/Digital Terrain Models (DTM), which are made publicly available in the last decade, floodplain maps with higher certainty can be created (Jafarzadegan & Merwade, 2017). This study is a demonstration that flood models can be built using vector files of land use maps and existing flood structures alongside a DEM. The vector maps can be made using geospatial tools such as QGIS, and coupling them with freeware simulation tools like BASEMENT V3.0, QGIS-based models can be built with few complications. These can be used as preliminary flood models, in data-scarce/developing regions, and validated using historical observation data or other hydraulic modeling tools. More saliently, benefit−cost analysis studies for flood infrastructure are a crucial part of flood risk management (Johnson et al., 2020). For instance, Germany alone is estimated to incur losses of 529 million to nearly 9 billion euros due to flooding (Sairam et al., 2021). Cost-effective 2D HN models coupled with other multicriteria analyses can help estimate flood damage costs in different land use types such as commercial, agricultural, and residential. These open source and open data studies have the potential to save billions of euros in flood damages by testing the resiliency of the flood infrastructures and developing early economic assessments (Sairam et al., 2021). Governments and responsible entities all over the world should invest in these freely available tools and advocate for open hydrological data to generate more peer-reviewed flood risk analysis studies.

9.6 Conclusion

Although there are limitations with the 2D HN model, it can be said that the simulation outputs show realistic results. The simulation is run for 8 hours and reaches a steady state just before the 6th hour. For a discharge rate of 400 m^3/s, it can be said that the dike structures and elevated zones are sufficient to protect Pretziener and Plotzky villages, as well as the Bergenfeld agricultural area. Incorporating the Pretziener weir in the model made the study more realistic, and enhancement to the weir boundary conditions (such as changing the Polemi factor to 0.64 for metal gates) can be further studied in detail. Also, comparison to actual high probability flood events showed significant correlation visually, but if

(A)

```
},                        C./02002/02002/02002/02002/02002, DOCUMENTS/20 DESIGN/Watershed Management 11,
  "HYDRAULICS": {
    "BOUNDARY": {
      "LINKED": [
        {
          "name": "Weir",
          "poleni_factor": 0.75,
          "string_name": "WEIR_IN",
          "string_name_downstream": "WEIR_OUT",
          "type": "weir_linked_constant",
          "weir_elevation": 51.0
        }
      ],
      "STANDARD": [
        {
          "discharge": 400.0,
          "name": "Inlet",
          "slope": 0.0022,
          "string_name": "INLET",
          "type": "uniform_in"
        },
        {
          "name": "Outlet",
          "slope": 0.0022,
          "string_name": "OUTLET",
          "type": "uniform_out"
        }
      ]
    },
    "FRICTION": {
      "default_friction": 20.0,
      "regions": [
        {
          "friction": 24.0,
          "region_name": "Leftbank"
        },
        {
          "friction": 28.0,
          "region_name": "Riverbed"
        },
        {
          "friction": 22.0,
          "region_name": "Rightbank"
        }
      ],
      "type": "strickler"
    },
    "INITIAL": {
      "type": "dry"
    },
    "PARAMETER": {
      "CFL": 0.6,
      "fluid_density": 1000.0,
      "max_time_step": 100.0,
      "minimum_water_depth": 0.0005,
      "safe_mode": "on"
    }
  }
},
"PHYSICAL_PROPERTIES": {
  "gravity": 9.81
}
},
"simulation_name": "RUNFILE"
```

FIGURE 9.9

The images represent the input file from the BASEMENT (V3) software. The parameters used, output variables measured, and other inputs can be replicated keeping these inputs as an example. (A) Hydraulic conditions setup; (B) Geometry definitions and boundaries setup; (C) Output file setup.

(B)

```
{
  "SETUP": {
    "DOMAIN": {
      "BASEPLANE_2D": {
        "GEOMETRY": {
          "INTERPOLATION": {
            "method": "weighted"
          },
          "REGIONDEF": [
            {
              "index": [
                3
              ],
              "name": "Riverbed"
            },
            {
              "index": [
                1
              ],
              "name": "Rightbank"
            },
            {
              "index": [
                2
              ],
              "name": "Leftbank"
            }
          ],
          "STRINGDEF": [
            {
              "name": "INLET",
              "upstream_direction": "right"
            },
            {
              "name": "OUTLET",
              "upstream_direction": "right"
            },
            {
              "name": "WEIR_IN",
              "upstream_direction": "right"
            },
            {
              "name": "WEIR_OUT",
              "upstream_direction": "right"
            }
          ],
          "mesh_file": "C:/Users/Shafkat Sharif/Documents/TU Dresden/Watershed Management II/Assignment/Basement/Trial CM.2dm"
        },
        "HYDRAULICS": {
          "BOUNDARY": {
            "LINKED": [
              {
                "name": "Weir",
                "poleni_factor": 0.75,
                "string_name": "WEIR_IN",
```

(C)

```
{
  "SIMULATION": {
    "OUTPUT": [
      "friction_chezy",
      "flow_velocity",
      "spec_discharge",
      "water_depth"
    ],
    "TIME": {
      "end": 28800.0,
      "out": 60.0,
      "start": 0.0
    },
    "TIMESTEP": {
      "init": 0.002,
      "minimum": 0.0001
    }
  }
}
```

FIGURE 9.9

(Continued).

higher resolution/consistent DEMs can be used, then the results can accurately mimic real conditions. The model results should be taken with a pinch of salt, as no station data or probabilistic approaches are taken into consideration while simulating, and only a uniform discharge value is provided for simulation. However, similar studies in flood-prone areas can be made to assess the effectiveness of flood protection structures. This study is low-cost and can be achieved using open-source software such as QGIS and freeware BASEMENT. 2D HN modeling like this can be done remotely from anywhere, given that DEM, land use, and/or protection structures are correctly prepared as model layers.

9.6.1 Setup file and used parameters on BASEMENT

Fig. 9.9A and B shows the input parameters of the HN BASEMENT model built for the study. The selected output variables can also be seen in Fig. 9.9C.

Acknowledgments

The current study was part of a master's course called "Watershed Management II" of TU Dresden. The author is grateful toward Lars Backhaus, the course coordinator and research associate at the Institute of Hydraulic Engineering and Technical HydromechanIcs (IWD) of TU Dresden. Mr. Backhaus helped develop the assignment and provide valuable training, input parameters, and the required land use maps and digital elevation model.

References

Bardua, S. (2015). Das Pretziener Wehr an der Elbe - Band 17 Historische Wahrzeichen der Ingenieurbaukunst (1st ed.). Bundesingenieurkammer. Available from https://d-nb.info/1073136450.

Bartl, S., Schümberg, S., & Deutsch, M. (2009). Revising time series of the Elbe River discharge for flood frequency determination at gauge Dresden. *Natural Hazards and Earth System Sciences*, 9(6), 1805–1814. Available from https://doi.org/10.5194/nhess-9-1805-2009.

Davies, R. (2013). The Elbe in Magdeburg, Germany. FloodList. https://floodlist.com/europe/magdeburg-germany

Diffenbaugh, N. S., & Barnes, E. A. (2023). Data-driven predictions of the time remaining until critical global warming thresholds are reached. *Proceedings of the National Academy of Sciences*, 120(6). Available from https://doi.org/10.1073/pnas.2207183120, e2207183120, https://doi.org/10.1073/pnas.2207183120.

Fenglin, W., Ahmad, I., Zelenakova, M., Fenta, A., Dar, M. A., Teka, A. H., … Damtie, M. (2023). Exploratory regression modeling for flood susceptibility mapping in the GIS environment. *Scientific Reports*, 13(1), 247. Available from https://doi.org/10.1038/s41598-023-27447-0, https://doi.org/10.1038/s41598-023-27447-0.

Haque, A., Ishtiaque, A., & Chowdhury, A. (2020). Water, flood management and water security under a changing climate. In *Proceedings from the seventh international conference on water and flood management*. Available from https://doi.org/10.1007/978-3-030-47786-8.

Hübner, G., & Schwandt, D. (2018). Extreme low flow and water quality − A long-term view on the River Elbe. *Erdkunde*, *72*(3), 235−252. Available from https://www.jstor.org/stable/26503557.

Jafarzadegan, K., & Merwade, V. (2017). A DEM-based approach for large-scale floodplain mapping in ungauged watersheds. *Journal of Hydrology*, *550*, 650−662. Available from https://doi.org/10.1016/j.jhydrol.2017.04.053, https://www.sciencedirect.com/science/article/pii/S0022169417302731.

Jodhani, K. H., Patel, D., & Madhavan, N. (2023). A review on analysis of flood modelling using different numerical models. *Materials Today Proceedings*, *80*, 3867−3876. Available from https://doi.org/10.1016/j.matpr.2021.07.405, https://www.sciencedirect.com/science/article/pii/S221478532105269X.

Johnson, K. A., Wing, O. E. J., Bates, P. D., Fargione, J., Kroeger, T., Larson, W. D., … … Smith, A. M. (2020). A benefit−cost analysis of floodplain land acquisition for US flood damage reduction. *Nature Sustainability*, *3*(1), 56−62. Available from https://doi.org/10.1038/s41893-019-0437-5, https://doi.org/10.1038/s41893-019-0437-5.

LWH. (n.d.). *Hochwassergefahrenkarten—LHW LSA*. http://www.geofachdatenserver.de.

Maue, T. (2013). Pretziener Wehr. Available from https://www.flickr.com/photos/erwin-rommel/8947356710

Nkwunonwo, U. C., Whitworth, M., & Baily, B. (2020). A review of the current status of flood modelling for urban flood risk management in the developing countries. *Scientific African*, *7*, e00269. Available from https://doi.org/10.1016/j.sciaf.2020.e00269, https://www.sciencedirect.com/science/article/pii/S2468227620300077.

Rahman, A., Jahan, S., Yildirim, G., Alim, M. A., Haque, M. M., Rahman, M. M., & Kausher, A. H. M. (2022). A review and analysis of water research, development, and management in Bangladesh. *Water (Switzerland)*, *14*(12). Available from https://doi.org/10.3390/w14121834, https://www.mdpi.com/2073-4441/14/12/1834/pdf?version = 1654596916.

Sairam, N., Brill, F., Sieg, T., Farrag, M., Kellermann, P., Nguyen, V. D., … Kreibich, H. (2021). Process-based flood risk assessment for Germany. *Earth's Future*, *9*(10), e2021EF002259. Available from https://doi.org/10.1029/2021EF002259.

Salcedo-Sanz, S., Pérez-Aracil, J., Ascenso, G., Del Ser, J., Casillas-Pérez, D., Kadow, C., … Castelletti, A. (2024). Analysis, characterization, prediction, and attribution of extreme atmospheric events with machine learning and deep learning techniques: A review. *Theoretical and Applied Climatology*, *155*(1), 1−44. Available from https://doi.org/10.1007/s00704-023-04571-5, https://doi.org/10.1007/s00704-023-04571-5.

Schymanski, E. L., & Schymanski, S. J. (2023). Water science must be open science. *Nature Water*, *1*(1), 4−6. Available from https://doi.org/10.1038/s44221-022-00014-z, https://doi.org/10.1038/s44221-022-00014-z.

Shoma Mitkari, M. S., Suryakar, H., Mali, Y., Shaikh, Y., & Mullani, Z. (2023). Smart flood alert system using IoT. *International Research Journal of Modernization in Engineering Technology and Science*, *05*(05). Available from https://www.irjmets.com/uploadedfiles/paper/issue_5_may_2023/39881/final/fin_irjmets1684917924.pdf, https://www.doi.org/10.56726/IRJMETS39881.

TUFLOW QGIS Plugin. (2023). TUFLOW. https://wiki.tuflow.com/w/index.php?title = TUFLOW_QGIS_Plugin&oldid = 34893.

Vetsch, D., Siviglia, A., Bacigaluppi, P., Bürgler, M., Caponi, F., Conde, D., ... Weberndorfer, M. (2022). *System manuals of BASEMENT, version 3.2.0. Laboratory of hydraulics, glaciology and hydrology (VAW)*. ETH Zurich.

WSV - Wasserstraßen- und Schifffahrtsamt Elbe. (2023). *Levels in the Elbe area: Barby, Elbe*. Undine - Information Platform on Hydrological Extreme Events (High Water, Low Water). https://undine.bafg.de/elbe/pegel/elbe_pegel_barby.html.

Drought monitoring using the Water Scalar Index

10

Sweety Singh[1], Hemlata Patel[1], Swapnil S. Vyas[2] and Bimal Bhattacharya[3]

[1]Department of Geography, SP Pune University, Pune, Maharashtra, India
[2]Department of Geography (Geoinformatics), SP Pune University, Pune, Maharashtra, India
[3]Space Applications Centre, ISRO, Ahmedabad, Gujarat, India

10.1 Introduction

Drought events in a region possibly may be explained by the nature of water deficiency in that region. The failure in precipitation is one of the major causes of water deficiency worldwide (Heim, 2002). In India, the summer monsoon (June to September) contributes to 75%–90% of the annual precipitation in the country and is closely linked with its agricultural economy and the ever-increasing population. Due to the high spatial variability of rainfall in India, smaller regions may suffer drought conditions even if overall monsoon performance is above normal. Similarly, the poor performance of the annual southwest monsoon over the whole country results in flood conditions in a few smaller regions, causing several deaths and economic losses. The complexity and uncertainty of drought analysis lie in its creeping nature and large-scale impacts. However, drought analysis is very essential for the study of water management, adaptation planning, agricultural water management, and natural hazard risk assessment. The extent of population and economy affected by drought events is comparatively higher than any other natural hazards (Wilhite, 2000). Drought is defined in many ways based on its nature, different disciplines, and its approach. In general, drought is defined as the temporary water deficit compared to its long-term average condition (Heim, 2002). Further drought is classified into four categories, namely, meteorological, hydrological, agricultural, and socioeconomic drought (Wilhite, 2000; Tsakiris et al., 2007). These four categories are interrelated to each other, mostly considered as the continuation of the meteorological drought ultimately resulting in water deficit in other sectors (Smakhtin & Schipper, 2008). Though agricultural and hydrological droughts occur at different time scales, they are the subsequent result of the meteorological drought (lack of precipitation), and the socioeconomic drought is a consequence of the preceding three types of droughts (Carrão, Russo, Sepulcre-Canto, & Barbosa, 2016). The effects of the drought events are more prominent in the areas with high spatial variability in the seasonal rainfall. A study carried out by Wilhite (2000) stated that an annual economic loss of

Applications of Geospatial Technology and Modeling for River Basin Management.
DOI: https://doi.org/10.1016/B978-0-443-23890-1.00010-4

239

approximately 6–8 billion dollars globally is caused by drought events alone. It is referred to as one of the costliest natural disasters, affecting more people than any other form of natural disaster. The worst affected sector by droughts is agricultural production, particularly rain-fed agriculture. It is obvious to generalize that with the decrease in agricultural share in the country's economy, the vulnerability and risk associated with droughts also reduced.

It has been observed that there is a substantial reduction in the production of major crops following major drought conditions, leading to potential threats to food security and other socioeconomic implications. The second sustainable development goals (SDGs) out of the 17 SDGs established by the United Nations in 2015 aims to achieve zero hunger by the year 2030 (Nakai, 2018). The increasing food insecurity has become a global issue as the number of hungry population continues to increase in certain parts of the world. In addition to population increase and poverty, drought has always been one of the primary causes of food insecurity around the globe (Miyan, 2015). Thus, taking into consideration the seriousness and pervasiveness of drought, it becomes very important to address the severity of drought, with a perspective to improve the living conditions of the millions, particularly inhabiting the monsoon region. Clearly, reliable predictions of the occurrence of extremes of interannual variation of monsoon rainfall, particularly nonoccurrence of droughts, can contribute toward sustained growth of different sectors. Along with the current techniques of drought assessment, additional methodologies for drought monitoring and prediction, especially regarding agricultural drought onset, progression, and impact on crops, need to be studied at the regional scale.

In recent studies carried out by Tabari, Abghari, and Hosseinzadeh Talaee (2012) and Samantaray, Ramadas, and Panda (2022), it was observed that for drought risk assessment, the climate cannot be presumed to be stationary. Moreover, the drastic changing patterns and intensity of rainfall are not sufficient to address the drought risks, even if a temporal trend is not observed in the rainfall pattern. The impacts of drought and its adverse effect on the population in the Asian subcontinent and Pacific island countries have been studied and reported by several researchers. Most of the studies highlight that global climate variability is associated with El Nino and La Nina effects, causing recurrent dry and wet conditions in South African and South-Asian countries (Mukherjee, Pal, Manna, Saha, & Das, 2023). The strong El Nino effects cause extreme dryness, which can be associated with drought events. In India, the region of western Gujarat has experienced several repeated spells of deficit rainfall between the years 1969 and 2018, causing moderate to severe drought conditions (Vishwakarma & Goswami, 2022). The repeated pattern of dry and wet spells has alarmed the government to investigate the socioeconomic issues related to water scarcity. Several drought monitoring studies have been conducted to understand the effects of temporal distribution of rainfall at the regional level. However, the current monitoring system lags in understanding the severity at a specific location (Samantaray et al., 2022; Singh, Saini, & Bhardwaj, 2021). With the advancement of remote sensing technology, historical indices from remote sensing are being overpowered by the new indices developed using real-time satellite data. The Normalized Difference Moisture Index proposed by Wang and Qu (2007) is one of its kind drought index, which uses an infrared band and two shortwave bands at different wavelengths to monitor drought by estimating the soil and vegetation water content. A brief discussion on the various drought indices used at different spatial scales to characterize various aspects of droughts is done in the following paragraphs.

10.1.1 Global scenario for drought monitoring

The dry subhumid to semiarid lands are most prone to drought conditions, accounting for a total coverage of 23.90% globally (Vishwakarma & Goswami, 2022; Kuruppu & Willie, 2015). In the United States, the drought research is led by the NDMC—National Drought Mitigation. The U. S. Drought Monitor produces a map that shows the drought across the United States; the key indicators used for the composition of these maps are Palmers Drought Severity Index (PDSI), Standardized Precipitation Index (SPI), present normal precipitation, USGS daily stream flow percentile, and Climate Prediction Center soil moisture. In the United States of America, near-real-time drought monitoring is provided with the 1 km spatial resolution maps of the Vegetation Drought Response Index developed by Brown, Diuk-Wasser, Andreadis, and Fish (2008). In Africa, the SPI approach is used for monitoring the effects of agricultural drought in the subregions, especially in the Horn of Africa, which is more prone to droughts and floods. Droughts and floods cause severe impacts on the socioeconomic sector of the region. It uses the SPI approach to monitor the effects of agricultural droughts in the subregion. The other remote sensing-based indicator used for better accounting for soil background and atmospheric aerosol effects is the Enhanced Vegetation Index (Kogan, 2000), an enhancement of the Normalized Difference Vegetation Index (NDVI). However, the index cannot be fully expressed as a ratio index because of the soil adjustment factor. The spatial variance of drought occurrence is monitored using the Vegetation Temperature Condition Index (Wang, Li, Gong, & Song, 2001) at a regional level for a specific period of the year; the index cannot be used for country-level monitoring, as it is site- and time-specific. The occurrence of dryness/wetness in China has increased rapidly while there is a gradual shift in wet areas from coast to inland (Song, 2000). In addition, significant change in the trends of SPI and PDSI was found in the European continent during the last century (Ionita & Nagavciuc, 2021). It is expected that the projected climate change over the world will increase flood and drought events over monsoon regions, in general, and Asian monsoon regions, in particular (Turner & Annamalai, 2012).

10.1.2 Regional scenario for drought monitoring

In India, 33% of the total land has an arid landscape. The high variability in rainfall and wide variation in the physiography of the country are the reasons for recurrent drought phenomena in India (Vishwakarma & Goswami, 2022). In India, both meteorological and agricultural drought monitoring and assessment has been carried out by the India Meteorological Department (IMD) based on the percent of rainfall departure, aridity index, and SPI, respectively (Ray, Sesha Sai, & Chattopadhyay, 2015). In addition, IMD issues the weekly Agromet Advisory Services bulletin under the Gramin Krishi Mausam Sewa project (Singh et al., 2023). Several country-wide drought indices have been examined with the long-term time series (1875–1987) of the subdivision rainfall data over India by Sikka (2004). The subdivision-wise rainfall analysis has shown a decreasing trend over the drought-affected area in northwest India, parts of the central Peninsula, and southern parts of the Indian Peninsula (Guhathakurta, Menon, Inkane, Krishnan, & Sable, 2017). In the recent period, there have also been a few attempts to understand the characteristics and impacts of socioeconomic droughts in India (Gautam & Bana, 2014; Dhangar et al., 2019; Adhav, Chandel, Bhandari, Ponnusamy, & Ram, 2021). These studies have predominantly assessed the local response perspective to identify tailor-made strategies toward droughts and their socioeconomic impacts on agriculture, livestock, health, migration, and overall economy. However, human drivers, such as increasing water demand, inappropriate irrigation methods, and poor water management can further worsen the impacts of droughts. As discussed in the previous paragraphs, based on precipitation studies, several attempts have been made either to improvise existing indices or to develop altogether new drought indices. PDSI, based on the soil balance equation, was extensively used for most studies conducted in temperate latitudes for drought assessment. However, considering the dependency of Indian agriculture on summer monsoon rainfall, new indices need to be used that take into consideration the interseasonal variations in tropical monsoon lands. Additionally, various aspects of drought propagation could be represented by the newly added satellite data that contributed significantly to our understanding of this phenomenon. An attempt of synoptic assessment of agricultural drought over the semiarid tract of India (Gujarat, Maharashtra, and Karnataka), using normalized satellite data during summer monsoon season, has been shown in this research study.

10.1.3 Objective of the study

The Land Surface Water Index (LSWI) is found to be a good indicator for drought monitoring than NDVI in terms of detecting vegetation liquid water content (Ceccato, Gobron, Flasse, Pinty, & Tarantola, 2002) and is less sensitive to atmospheric scattering effects (Gao, 1995). The Shortwave Infrared (SWIR) band shows more sensitivity toward canopy moisture and soil moisture, making it useful for agriculture drought monitoring. The LSWI is successfully tested to detect and monitor surface moisture and the vegetation moisture condition over a larger canopy area; the LSWI is preferred over the NDVI, as it shows improved and quick response to detect drought conditions. The water scalar formulated has a range from 0 to 1, depending on the vegetation type and the leaf water content.

A higher value of the water scalar refers to high leaf water content and high vegetation fraction cover vice versa. The water scalar shows a decreasing trend in the value with increasing water stress. The water scalar derived in this study can be used as a surrogate for the water limiting factor and hence to detect agricultural drought. With the above discussion in the background, the present study aims at evaluating drought conditions, particularly agricultural drought, in Gujarat, Maharashtra, and Karnataka from 2009 to 2013 using a Water Scalar Index derived from the LSWI. In order to achieve this aim, the following objectives were formulated: To evaluate the usefulness of the Water Scalar Index in characterizing agricultural drought based on lag response of water supply and agricultural vigor; characterizing agricultural droughts in different districts of Gujarat, Maharashtra, and Karnataka using the Water Scalar Index; and comparing dekadal variation in drought conditions from 2009 to 2013

10.1.4 Limitations of the study

While this research provides valuable insights into agricultural drought assessment, certain limitations of the study should be acknowledged. The present study relies on remote sensing data, which may have inherent limitations, including the potential for atmospheric interference. Even though the satellite-based observations offer extensive spatial coverage, they inherently lack the ground-level precision that rain gage station data could provide. While comprehensive verification and cross-validation with rain gage stations are challenging for every pixel, a more strategic and representative validation approach is recommended. Therefore, it is suggested that future research endeavors must consider integrating rain gage station data to enhance the accuracy of the agricultural drought assessment. The study uses water scalar derived from LSWI to monitor and characterize drought. The water scalar performs better than the LSWI for drought detection. Water scalar is further compared and validated with the precipitation data in the corresponding 2 dekads. This is a preliminary study and can be further improved by incorporating the NDVI data for lag response in drought detection. A Normalized Difference Drought Index (NDDI), which is a combination of both LSWI and NDVI, is more sensitive for drought monitoring. The NDDI can be modified by replacing LSWI with water scalar that would amplify the anomalies and become more responsive to drought monitoring.

10.2 Data used and methodology

10.2.1 Study area

Indian subcontinent is divided on an administrative level into 29 states and 7 union territories. The administrative units are further divided into smaller units called districts. In all, there are approximately 686 districts in India, out of which

195 entirely or partly experience drought or drought-like conditions (Vyas & Bhattacharya, 2020). These districts are highly concentrated in states like Gujarat, Rajasthan, Haryana, Maharashtra, Karnataka, and Andhra Pradesh in the western and central parts of India (Venkateswarlu, Raju, Rao, & Rama Rao, 2014). The present study focuses on three most drought-affected states of Gujarat, Maharashtra, and Karnataka in the western region of India (Fig. 10.1). These three states are mostly agrarian, and the areas under cultivation are 50%, 59.6%, and 54% in the states of Gujarat, Maharashtra, and Karnataka, respectively. In this region, the agriculture is majorly dependent on the monsoon rainfall (ranging from 300 to 5000 mm). These three states are highly vulnerable to drought (Reddy et al., 2015; Udmale, Ichikawa, Manandhar, Ishidaira, & Kiem, 2014). The three states cover approximately 80% of the semiarid tract of India.

The data from the Indian Satellite INSAT 3A CCD was obtained for the study. The CCD payload consists of red, IR, and SWIR bands with 1000 m spatial and 10-bit radiometric resolution. The INSAT 3A CCD NIR and SWIR band data for 5 years (2009−13) is used for calculation of LSWI and water scalar in this study. Daily rainfall (K1-VHR-IMR) data from June to September or 5 years (2009−13) was obtained from Indian geostationary satellite, Kalpana-1. The Kalpana-1 very high-resolution radiometer has three bands—visible bands with 2 km resolution

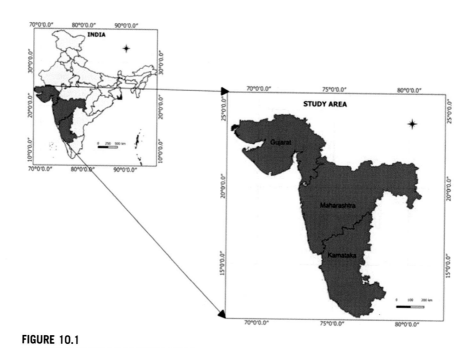

FIGURE 10.1

The study focuses on three drought-prone states in India, viz., Gujarat, Maharashtra, and Karnataka. The three states cover a major portion of semi-arid tracts in the Indian subcontinent.

and water vapor and thermal bands with 8 km resolution, respectively. The daily rainfall product available from the Kalpana-1 satellite was downloaded from the MOSDAC data portal provided by Space Applications Centre, ISRO. An agricultural mask with 1 km resolution was derived using the Satellite Pour Observation de la Terre (SPOT) VGT (vegetation) data to retain the agricultural area within the study area (Agrawal, Joshi, Shukla, & Roy, 2003).

10.2.2 Development of water scalar for drought monitoring

The NDWI provides a good reference to the available water content in the plants. It is considered a good alternative for monitoring crop water stress. The new water scalar (Wscalar) is based on the LSWI derived from INSAT 3A CCD. The maximum and minimum values of LSWI across the 5 years (2009−13) are considered to calculate the water scalar. Comparisons at local scale and over the district level were made, using other recognized drought indices over selected study areas. The following study was designed to monitor the water stress in the crop over the study areas in 2009−13 during the crop growing season.

10.2.2.1 Computation of Land Surface Water Index

The LSWI is calculated using the reflectance of NIR and moisture-sensitive SWIR bands (Chandrasekar, Sesha Sai, Roy, & Dwevedi, 2010). The change in the water content in the vegetation canopies is well captured by SWIR and NIR bands used for calculating LSWI (Zeng et al., 2022). The SWIR band is sensitive to the surface wetness of top soil at the beginning of the cropping season, and with the progression of the crop, it becomes more sensitive to the water content in the canopy (Sesha Sai et al., 2016). The satellite-derived LSWI has been used to monitor drought at regional and local scales (Zhang, Hong, Qin, & Zhu, 2013). The INSAT 3 A data at 0700 GMT in SWIR (1550−1640 nm) and NIR band regions were used to compute LSWI. The LSWI was computed using the following equation:

$$LSWI = \frac{([NIR]_i - [SWIR]_{min})}{(NIR + SWIR)}$$

where i represents the 10-day period,

The 10-day daily LSWI was formulated from INSAT 3A CCD images of 0700GMT using the maximum value composite function. A set of 3 files (1−10, 11−20, 21 to end of month) were generated for each month. The agricultural area within the study area was extracted from the agricultural mask developed using SPOT VGT (vegetation) at a spatial resolution of 1000 m (Agrawal et al., 2003).

10.2.2.2 Computation of Wscalar

The Wscalar was formulated using the LSWI, which was computed on the 10-day interval to filter out and remove cloud contamination. Conceptually, the Wscalar enhances the variations of a surface water index in response to weather

Table 10.1 Classification of droughts into various categories based on water scalar levels.

Water Scalar Index	Severity of droughts
0–0.2	Severe
0.2–0.4	Moderate
0.4–0.6	Mild
0.6–1.0	Normal conditions

fluctuations, also reducing the impact of environmental response (of climate, soils, vegetation type, and topography) Swapnil Dr (2017). Pixel-level Wscalar is calculated from the filtered LSWI data at the 10-day interval using the following formula:

$$\text{Water Scalar(Wscalar)} = \frac{([LSWI]_i - [LSWI]_{min})}{([LSWI]_{max} - [LSWI]_{min})}$$

where Wscalar is a pixel-wise normalization of pixel-wise LSWI, and $LSWI_{max}$ and $LSWI_{min}$ were determined from 5-year cloud-filtered LSWI data from INSAT 3A CCD datasets by evaluating temporal maximum and minimum values, respectively, at a given pixel. Maximum ($LSWI_{max}$) and minimum ($LSWI_{min}$) LSWI vary from 0 (very dry conditions) to 1 (wet, favorable conditions). The $LSWI_{max}$ corresponds to maximum land surface wetness, and $LSWI_{min}$ represents minimum surface wetness. The former one occurs when both soil and canopy were at the peak. The latter one occurs either during the fallow period or at crop harvest. The $LSWI_i$ is cloud-filtered dekadal LSWI. The $LSWI_{max}$ and $LSWI_{min}$ are temporal maximum and minimum LSWI, respectively, at a given pixel. The symbol "i" represents the 10-day period. The Wscalar above 0.6 means close to the normal situation, while Wscalar below 0.3 indicates a severe drought. Thus, various levels of water scalar were used to classify droughts into four categories according to their severity, as shown in Table 10.1.

The drought and non-drought comparison was finally made based on the area-weighted Wscalar generated for each district based on the fraction of available agricultural area in respective categories.

10.2.2.3 Data processing using Harmonic ANalysis of Time Series

The Harmonic ANalysis of Time Series (HANTS) algorithm, described by Menenti et al. (2016) was used for the time series reconstruction including temporal interpolation and outliers' removal in the NDWI generated from INSAT. HANTS estimated the phase and amplitude values in the observed signal and reconstructed the signal from the information of direct climatological and phonological relevance. The HANTS algorithm is widely used for the reconstruction of NDVI time series (Zhou, Jia, & Menenti, 2015). In this study, 5-year data was analyzed separately to address the contamination in NDWI data. The HANTS model was iteratively used to filter out the noise and filtered to remove the high frequency and keep the lower frequency components.

10.3 **Results and discussions**

The present study attempts to characterize agricultural drought using optical remote sensing data from geostationary satellites. The data processing and analysis was carried out for the three states of Gujarat, Karnataka, and Maharashtra constituting the majority of semiarid tract of India. A total of 83 districts were studied, where Gujarat comprised 25 districts, Maharashtra 31 districts, and Karnataka 21 districts. Four districts were not considered, namely, Kheda, Mumbai, Gondia, and Wardha, as they were depicting some error after computation of the Water Scalar Index. These 83 districts were assigned number codes for ease of understanding, as shown in Table 10.2.

The dekadal (10-day period) district mean Wscalar was derived and compared with mean rainfall of all the districts of Gujarat, Maharashtra, and Karnataka. The cumulative rainfall of previous two dekads was used instead of relating weekly rainfall with concurrent weekly Water scalar levels. As the Wscalar represents surface wetness, which is the resultant effect of surface soil moisture, crop vigor, and canopy water status (Nigam, Bhattacharya, & Pandya, 2023), the lag response of water supply on overall agricultural vigor was considered. Prior to the dekadal analysis, seasonal analysis (June to September season) was carried out to associate the rainfall amounts with varying levels of the Water Scalar Index from 2009 to 2013.

10.3.1 **Seasonal analysis**

The seasonal (June to September) average rainfall was computed for all the 5 years from 2009 to 2013. Similarly, the seasonal water scalar was computed by averaging dekadal Wscalar values.

10.3.1.1 *Comparison of seasonal rainfall and Wscalar*

District-wise seasonal rainfall and water scalar for the year 2009 is represented in Fig. 10.2.

The above figure shows the seasonal average of rainfall and levels of Wscalar of all 83 districts in the study area comprising Gujarat, Maharashtra, and Karnataka. It is visible that rainfall is very less in all the districts, rainfall is less than 80 cm, whereas Wscalar is less than 0.6 in all the districts, which shows drought conditions in all the districts. Banas Kanta, Bangalore, Buldana, Gulbarga, Hingoli, Sangli, Surendranagar, the Dangs, Uttar Kannada, Wardha, and Gondiya (12 districts) suffered severe drought. In total, 47 districts experienced moderate drought, whereas 25 districts suffered mild

Table 10.2 List of districts and their number codes.

Number	District name	Number	District name	Number	District name
1	Ahmedabad	22	Chandrapur	43	Kodagu	64	Rajkot
2	Ahmednagar	23	Chikmagalur	44	Kolar	65	Ratnagiri
3	Akola	24	Chitradurga	45	Kolhapur	66	Sabar Kantha
4	Amravati	25	Dakshina Kannada	46	Koppal	67	Sangli
5	Amreli	26	Davanagere	47	Latur	68	Satara
6	Anand	27	Dharwad	48	Mehsana	69	Shimoga
7	Aurangabad	28	Dhule	49	Mysore	70	Sindhudurg
8	Bagalkot	29	Dohad	50	Nagpur	71	Solapur
9	Banas Kantha	30	Gadag	51	Nanded	72	Surat
10	Bengaluru	31	Gadchiroli	52	Nandurbar	73	Surendranagar
11	Bengaluru Rural	32	Gandhinagar	53	Narmada	74	Thane
12	Belgaum	33	Gondiya	54	Nashik	75	The Dangs
13	Bellary	34	Gulbarga	55	Navsari	76	Tumkur
14	Bhandara	35	Hassan	56	Osmanabad	77	Udupi
15	Bharuch	36	Haveri	57	Panch Mahals	78	Uttara Kannada
16	Bhavnagar	37	Hingoli	58	Parbhani	79	Vadodara
17	Bid	38	Jalgaon	59	Patan	80	Valsad
18	Bidar	39	Jalna	60	Porbandar	81	Wardha
19	Bijapur	40	Jamnagar	61	Pune	82	Washim
20	Buldhana	41	Junagadh	62	Raichur	83	Yavatmal
21	Chamarajanagar	42	Kachchh	63	Raigarh		

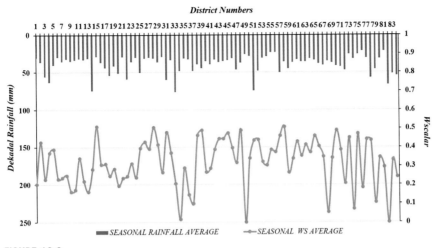

FIGURE 10.2

District-wise comparison of Wscalar and dekadal rainfall for the year 2009. The figure describes the relation between the cumulative rainfall of previous two dekads rainfall and the Wscalar for the year 2009.

drought. With the seasonal average, it is observed that all districts fall below 0.6 category of drought, which shows that all three states experienced drought in year 2009, which was because there was very little rainfall in almost all the districts (below 100 cm); hence, 2009 was drought year. Fig. 10.3 represents the relationship between seasonal average of rainfall and Wscalar for all the districts of the study area in 2010. Many of the districts experienced a normal condition (above 0.6), whereas Chikmagalur, Dakshin Kannada, Gadchiroli, Kodagu, the Dangs, and Udupi (6 districts) faced a severe drought condition; 46 districts experienced moderate drought conditions and 25 districts faced mild drought conditions. Rainfall was quite less in many districts, whereas some of the districts received normal rainfall, but overall 2010 was a non-drought year.

Fig. 10.4 represents the seasonal rainfall as well as Wscalar average for the year 2011 of all the districts of the study area. Many districts were observed under normal conditions. Dakshin Kannada, Gadchiroli, Kodagu, the Dangs, Udupi, and Uttar Kannada (6 districts) faced the severe drought conditions in year 2011. 18 districts faced moderate drought conditions, whereas 50 districts faced mild drought conditions. Many districts faced drought conditions because they received less rainfall, even if the overall state rainfall was near-normal.

Fig. 10.5 represents seasonal average of rainfall and Wscalar for the year 2012 of the study area. It is observed that in the year 2012, every district received very less amount of rainfall. Gadchiroli, Kodagu, Kolhapur, Raigarh, Sindhudurg, the Dangs, Udupi, Uttar Kannada, and Valsad (9 districts) were the worst affected

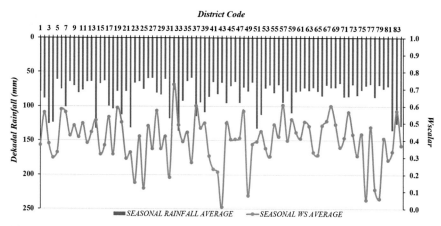

FIGURE 10.3

District-wise comparison of Wscalar and dekadal rainfall for the year 2010. The figure describes the relation between the cumulative rainfall of previous two dekads rainfall and the Wscalar for the year 2010.

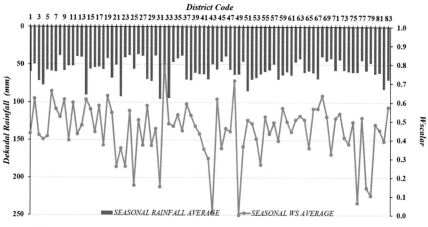

FIGURE 10.4

District-wise comparison of Wscalar and dekadal rainfall for the year 2011. The figure describes the relation between the cumulative rainfall of previous two dekads rainfall and the Wscalar for the year 2011.

districts as these districts faced severe drought conditions with a very low amount of rainfall. 52 districts faced moderate drought conditions, whereas 19 districts faced mild drought conditions. Drought conditions were due to very less rainfall over the study area; hence, 2012 was drought year.

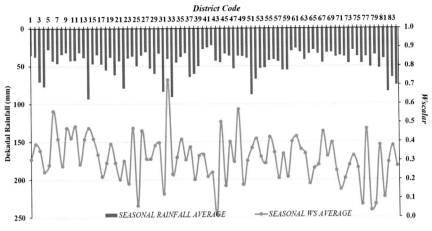

FIGURE 10.5

District-wise comparison of Wscalar and dekadal rainfall for the year 2012. The figure describes the relation between the cumulative rainfall of previous two dekads rainfall and the Wscalar for the year 2012.

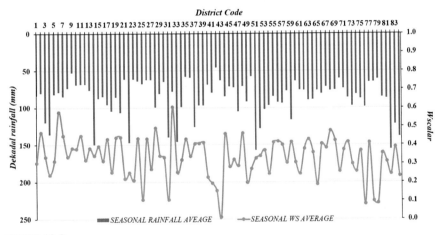

FIGURE 10.6

District-wise comparison of Wscalar and dekadal rainfall for the year 2013. The figure describes the relation between the cumulative rainfall of previous two dekads rainfall and the Wscalar for the year 2013.

   The association of seasonal rainfall and Wscalar averages for various districts of Gujarat, Maharashtra, and Karnataka for 2013 are represented in Fig. 10.6. It is observed that very few districts fall under the normal condition of the study area. Most of the districts received less rainfall and resulted in severe drought conditions in the early summer monsoon season. Seasonal rainfall in 2013 shows that very less amount of rainfall was received, and this resulted in low water scalar. Seasonal average shows that the Wscalar is very less in year 2013, which indicates that there was drought in many of the districts of the study area.

10.3.2 Drought monitoring and characterization using water scalar

Drought monitoring over a spatial scale at the dekadal interval was performed to study the progression of drought. The effect of water availability for crops in the form of rainfall and its effects were mapped in form of the Water Scalar Index for the two drought years 2009 and 2012 and normal years 2010, 2011, and 2013. The provided maps offer a comprehensive depiction of area-weighted water scalar levels based on the fraction of available agricultural pixels in different categories. These maps illustrate the dekad-wise progression of drought conditions across each district in the study region. Before this analysis, an agricultural mask was applied, so that the characterization of agricultural droughts can be effectively done. It can be seen that in the year 2009, the values of the Water Scalar Index predominantly fell below 0.6, indicating severe water stress conditions. The year 2009 was a typical drought year where the rainfall deficiency stood at 21.8%, as declared by the India Meteorological Department. The Water Scalar Index has consistently captured these water stress conditions throughout the entire summer monsoon season across the study region, as shown in Fig. 10.7.

Beyond highlighting water stress, the Water Scalar Index adeptly portrays normal and surplus water levels, as evident from the dekadal maps of 2010. As per the reports of the India Meteorological Department (2010), the southwest monsoon of 2010 was delayed by about 1 week across the west coast, following the formation of a very severe cyclonic storm Phet over the Arabian Sea in the first

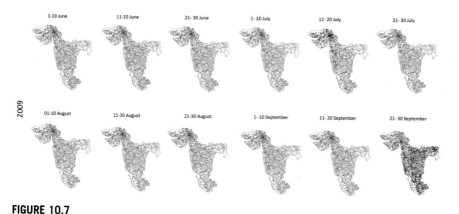

FIGURE 10.7

Wscalar for drought monitoring and characterization for the year 2009. The spatial distribution of drought intensity is explained using the Wscalar index.

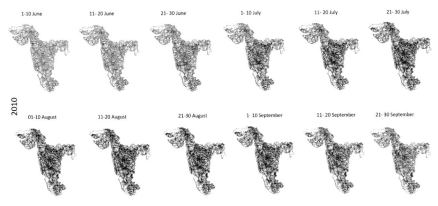

FIGURE 10.8

Wscalar for drought monitoring and characterization for the year 2010. The spatial distribution of drought intensity is explained using the Wscalar index.

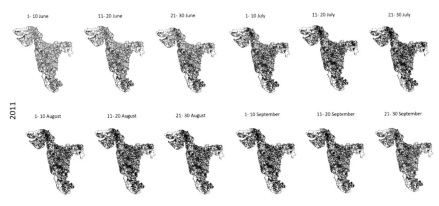

FIGURE 10.9

Wscalar for drought monitoring and characterization for the year 2011. The spatial distribution of drought intensity is explained using the Wscalar index.

week of June. This initial water scarcity can be easily traced from the water scalar maps during the early summer monsoon season. However, as the monsoon was activated by the middle of June, water stress conditions transformed into water surplus conditions in most of the districts, signifying a year with near-normal to above-normal rainfall, as shown in Fig. 10.8.

In the year 2011, the southwest monsoon contributed 101% precipitation of its long-period average, with the South Peninsula homogeneous region receiving 100% of its normal rainfall (Guhathakurta, Rajeevan, Sikka, & Tyagi, 2015). The Water Scalar Index effectively depicts these excess water conditions, as shown in Fig. 10.9, in almost all the districts of the study area.

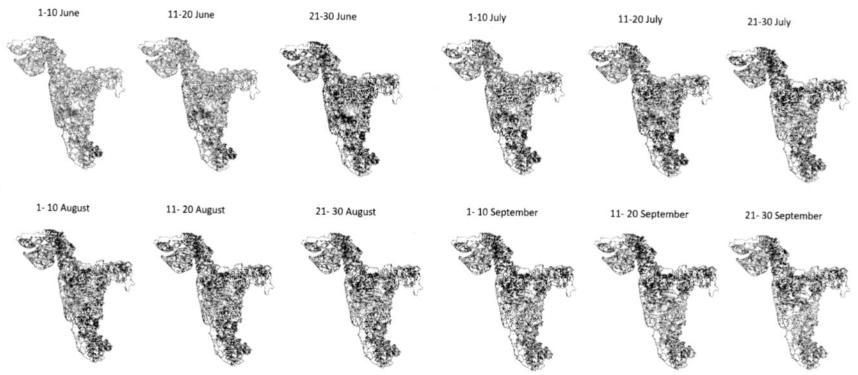

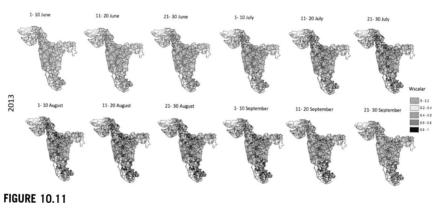

FIGURE 10.10

Wscalar for drought monitoring and characterization for the year 2012. The spatial distribution of drought intensity is explained using the Wscalar index.

FIGURE 10.11

Wscalar for drought monitoring and characterization for the year 2013. The spatial distribution of drought intensity is explained using the Wscalar index.

The year 2012 witnessed drought conditions in most of the districts in Gujarat, followed by 80% of districts in Maharashtra and 50% of districts in Karnataka. Although officially declared as drought on a regional level, pixel-level Water Scalar Index analysis reveals spatiotemporal variations in water scarcity throughout the early mid-season of the southwest monsoon, as shown in Fig. 10.10.

As per the reports of the India Meteorological Department, the pace of advance of southwest monsoon 2013 had been the fastest during the period 1941−2013 (Pai, Sridhar, & Ramesh Kumar, 2016). Despite this unusual rapidity, there was a drastic reduction in rainfall during the postonset period in the majority of districts in Gujarat state, owing to suppressed convection over the North Arabian Sea. This deficiency of rainfall is very well captured by the water scalar maps of 2013, as shown in Fig. 10.11, where almost all districts of Gujarat have Water Scalar Index values below 0.4 in the early summer monsoon season. As the southwest monsoon sets in the entire country, normal rainfall prevails across the entire country and in most districts of the study region.

10.4 Conclusions

This research study has successfully identified and characterized drought events that occurred in Gujarat, Maharashtra, and Karnataka from 2009 to 2013 using a Wscalar derived from LSWI. The efficiency of Wscalar in providing a thorough understanding of drought severity makes it a robust index for regional drought assessment. The two drought years, of 2009 and 2012, were meticulously delineated using the Water Scalar Index. The water scalar performed well in the non-drought years of 2010, 2011, and 2013 showing a surplus of moisture availability. A distinct feature of the Wscalar is its pixel-wise normalization of LSWI, enabling a more detailed and location-specific evaluation of drought conditions. The study can be further improved by combining the Wscalar with other drought indices or ground data to improve drought monitoring. By using remote sensing technologies, the study contributes to hazard management at a much finer spatial scale laying the foundation for the development of early warning indicators for drought events. This investigation holds significant promise for fostering resilience in the SAT of India in the face of recurring drought events. As climate variability continues to pose challenges to agricultural systems, the insights gained from this study serve as a guide for further research studies for minimizing the adverse effects of drought on food security and rural livelihoods.

References

Adhav, C. A., Senthil, R., Chandel, B. S., Bhandari, G., Ponnusamy, K., & Ram, H. (2021). Socio-economic vulnerability to climate change—Index development and mapping for districts in Maharashtra, India. *SSRN Electronic Journal*, 1556—5068. Available from https://doi.org/10.2139/ssrn.3854297.

Agrawal, S., Joshi, P. K., Shukla, Y., & Roy, P. S. (2003). Spot vegetation multi temporal data for classifying vegetation in south central Asia. *Current Science*, *84*(11), 1440—1448.

Brown, H., Diuk-Wasser, M., Andreadis, T., & Fish, D. (2008). Remotely-sensed vegetation indices identify mosquito clusters of West Nile virus vectors in an urban landscape in the Northeastern United States. *Vector-Borne and Zoonotic Diseases*, *8*(2), 197—206. Available from https://doi.org/10.1089/vbz.2007.0154.

Carrão, H., Russo, S., Sepulcre-Canto, G., & Barbosa, P. (2016). An empirical standardized soil moisture index for agricultural drought assessment from remotely sensed data. *International Journal of Applied Earth Observation and Geoinformation*, *48*, 74—84. Available from http://www.elsevier.com/locate/jag, https://doi.org/10.1016/j.jag.2015.06.011.

Ceccato, P., Gobron, N., Flasse, S., Pinty, B., & Tarantola, S. (2002). Designing a spectral index to estimate vegetation water content from remote sensing data: Part 1. *Remote Sensing of Environment, 82*(2−3), 188−197. Available from https://doi.org/10.1016/s0034-4257(02)00037-8.

Chandrasekar, K., Sesha Sai, M. V. R., Roy, P. S., & Dwevedi, R. S. (2010). Land Surface Water Index (LSWI) response to rainfall and NDVI using the MODIS vegetation index product. *International Journal of Remote Sensing, 31*(15), 3987−4005. Available from https://www.tandfonline.com/loi/tres20, https://doi.org/10.1080/01431160802575653.

Dhangar, N., Vyas, S., Guhathakurta, P., Mukim, S., Tidke, N., Balasubramanian, R., & Chattopadhyay, N. (2019). Drought monitoring over India using multi-scalar standardized precipitation evapotranspiration index. *Mausam, 70*(4), 833−840. Available from https://metnet.imd.gov.in/mausamdocs/570415.pdf.

Gao, B. C. (1995). A normalized difference water index for remote sensing of vegetation liquid water from space. *Proceedings of SPIE − The International Society for Optical Engineering, 2480*, 225−236. Available from https://doi.org/10.1117/12.210877, http://spie.org/x1848.xml.

Gautam, R. C., & Bana, R. S. (2014). Drought in India: Its impact and mitigation strategies—A review. *India Indian Journal of Agronomy, 59*(2), 179−190. Available from http://www.indianjournals.com/ijor.aspx?target = ijor:ija&volume = 59&issue = 2&article = 001&type = pdf.

Guhathakurta, P., Menon, P., Inkane, P. M., Krishnan, U., & Sable, S. T. (2017). Trends and variability of meteorological drought over the districts of India using standardized precipitation index. *Journal of Earth System Science, 126*(8). Available from http://www.ias.ac.in/article/fulltext/jess/126/08/0120, https://doi.org/10.1007/s12040-017-0896-x.

Guhathakurta, P., Rajeevan, M., Sikka, D. R., & Tyagi, A. (2015). Observed changes in southwest monsoon rainfall over India during 1901−2011. *International Journal of Climatology, 35*(8), 1881−1898. Available from http://onlinelibrary.wiley.com/journal/10.1002/(ISSN)1097-0088, https://doi.org/10.1002/joc.4095.

Heim, R. R. (2002). A review of twentieth-century drought indices used in the United States. *Bulletin of the American Meteorological Society, 83*(8), 1149−1166. Available from https://doi.org/10.1175/1520-0477-83.8.1149.

Ionita, M., & Nagavciuc, V. (2021). Changes in drought features at the European level over the last 120 years. *atural Hazards and Earth System Sciences, 21*(5), 1685−1701. Available from http://www.nat-hazards-earth-syst-sci.net/volumes_and_issues.html, https://doi.org/10.5194/nhess-21-1685-2021.

Kogan, F.N. (2000). Contribution of remote sensing to drought early warning. In *Early warning systems for drought preparedness and drought management* (pp. 75−87).

Kuruppu, N., & Willie, R. (2015). Barriers to reducing climate enhanced disaster risks in least developed country-small islands through anticipatory adaptation. *Weather and Climate Extremes, 7*, 72−83. Available from http://www.journals.elsevier.com/weather-and-climate-extremes/, https://doi.org/10.1016/j.wace.2014.06.001.

Menenti, M., Malamiri, H. R. G., Shang, H., Alfieri, S. M., Maffei, C., & Jia, L. (2016). Observing the response of terrestrial vegetation to climate variability across a range of time scales by time series analysis of land surface temperature. *Remote Sensing and Digital Image Processing, 20*. Available from http://www.springer.com/series/6477, https://doi.org/10.1007/978-3-319-47037-5-14.

Miyan, M. A. (2015). Droughts in Asian least developed countries: Vulnerability and sustainability. *Weather and Climate Extremes*, *7*, 8−23. Available from http://www.journals.elsevier.com/weather-and-climate-extremes/, https://doi.org/10.1016/j.wace.2014.06.003.

Mukherjee, S., Pal, J., Manna, S., Saha, A., & Das, D. (2023). *El-Niño Southern Oscillation and its effects* (pp. 207−228). Elsevier BV. Available from http://doi.org/10.1016/b978-0-323-99714-0.00013-3.

Nakai, J. (2018). Food and Agriculture Organization of the United Nations and the sustainable development goals. *Sustainable Development*, *22*, 1−450.

Nigam, R., Bhattacharya, B. K., & Pandya, M. R. (2023). Satellite agromet products and their adaptation for advisory services to Indian farming community. *Journal of Agrometeorology*, *25*(1), 42−50. Available from https://journal.agrimetassociation.org/index.php/jam/article/download/2084/1444, https://doi.org/10.54386/jam.v25i1.2084.

Pai, D. S., Sridhar, L., & Ramesh Kumar, M. R. (2016). Active and break events of Indian summer monsoon during 1901−2014. *Climate Dynamics*, *46*(11−12), 3921−3939. Available from http://link.springer.de/link/service/journals/00382/index.htm, https://doi.org/10.1007/s00382-015-2813-9.

Ray, S. S., Sesha Sai, M. V. R., & Chattopadhyay, N. (2015). *Agricultural drought assessment: Operational approaches in India with special emphasis on 2012. High-impact weather events over the SAARC region* (pp. 349−364). India: Springer International Publishing. Available from http://doi.org/10.1007/978-3-319-10217-7-24.

Reddy, S. S., Prashanth, R., Yashodha Devi, B. K., Chugh, N., Kaur, A., & Thomas, N. (2015). Prevalence of oral mucosal lesions among chewing tobacco users: A cross-sectional study. *Indian Journal of Dental Research*, *26*(5), 537−541,. Available from http://www.ijdr.in, https://doi.org/10.4103/0970-9290.172083.

Samantaray, A. K., Ramadas, M., & Panda, R. K. (2022). Changes in drought characteristics based on rainfall pattern drought index and the CMIP6 multi-model ensemble. *Agricultural Water Management*, *266*, 18732283. Available from http://www.journals.elsevier.com/agricultural-water-management/, https://doi.org/10.1016/j.agwat.2022.107568.

Sesha Sai, M. V. R., Murthy, C. S., Chandrasekar, K., Jeyaseelan, A. T., Diwakar, P. G., & Dadhwal, V. K. (2016). Agricultural drought assessment & monitoring. *MAUSAM*, *67*(1), 131−142. Available from https://doi.org/10.54302/mausam.v67i1.1155.

Sikka, D.R. (2004). Monsoon drought a natural hazard in India. In *Resource conservation and food security: An Indian experience* (vol. 1).

Singh, K. K., Ghosh, K., Bhan, S. C., Singh, P., Vishnoi, L., Balasubramanian, R., . . . Goroshi, S., & Singh, R. (2023). Decision support system for digitally climate informed services to farmers in India. *Journal of Agrometeorology*, *25*(2), 205−214. Available from https://journal.agrimetassociation.org/index.php/jam/article/view/2094/1478, https://doi.org/10.54386/jam.v25i2.2094.

Singh, O., Saini, D., & Bhardwaj, P. (2021). Characterization of meteorological drought over a dryland ecosystem in north western India. *Natural Hazards*, *109*(1), 785−826. Available from http://www.wkap.nl/journalhome.htm/0921-030X, https://doi.org/10.1007/s11069-021-04857-9.

Smakhtin, V. U., & Schipper, E. L. F. (2008). Droughts: The impact of semantics and perceptions. *Water Policy*, *10*(2), 131−143. Available from https://doi.org/10.2166/wp.2008.036.

Song, J. (2000). Changes in dryness/wetness in China during the last 529 years. *International Journal of Climatology*, *20*(9), 1003−1016. Available from http://onlinelibrary.wiley.com/journal/10.1002/(ISSN)1097-0088, https://doi.org/10.1002/1097-0088(200007)20:9 < 1003::AID-JOC529 > 3.0.CO;2-S.

Swapnil, V. (2017). *Regional agricultural drought characterization using remote sensing based observations from geostationary satellites* (Ph.D. thesis). Nirma University, Ahmadabad, Unpublished content. http://hdl.handle.net/10603/188037.

Tabari, H., Abghari, H., & Hosseinzadeh Talaee, P. (2012). Temporal trends and spatial characteristics of drought and rainfall in arid and semiarid regions of Iran. *Hydrological Processes*, 26(22), 3351–3361. Available from https://doi.org/10.1002/hyp.8460.

Tsakiris, G., Loukas, A., Pangalou, D., Vangelis, H., Tigkas, D., Rossi, G., & Cancelliere, A. (2007). Drought characterization. In Iglesias A., Moneo M., & López-Francos A. (Eds.). Drought management guidelines technical annex. Zaragoza : CIHEAM / EC MEDA Water, 58, 85–102.

Turner, A. G., & Annamalai, H. (2012). Climate change and the South Asian summer monsoon. *Nature Climate Change*, 2(8), 587–595. Available from https://doi.org/10.1038/nclimate1495.

Udmale, P., Ichikawa, Y., Manandhar, S., Ishidaira, H., & Kiem, A. S. (2014). Farmers' perception of drought impacts, local adaptation and administrative mitigation measures in Maharashtra State, India. *International Journal of Disaster Risk Reduction*, 10, 250–269. Available from https://doi.org/10.1016/j.ijdrr.2014.09.011, http://www.journals.elsevier.com/international-journal-of-disaster-risk-reduction/.

Venkateswarlu, B., Raju, B. M. K., Rao, K. V., & Rama Rao, C. A. (2014). Revisiting drought-prone districts in India. *Economic and Political Weekly, India Economic and Political Weekly* (25), 71–75. Available from http://www.epw.in.

Vishwakarma, A., & Goswami, A. (2022). The dynamics of meteorological droughts over a semi-arid terrain in western India: A last five decadal hydro-climatic evaluation. *Groundwater for Sustainable Development*, 16, 100703. Available from https://doi.org/10.1016/j.gsd.2021.100703.

Vyas, S. S., & Bhattacharya, B. K. (2020). Agricultural drought early warning from geostationary meteorological satellites: Concept and demonstration over semi-arid tract in India. *Environmental Monitoring and Assessment*, 192(5). Available from https://link.springer.com/journal/10661, https://doi.org/10.1007/s10661-020-08272-8.

Wang, L., & Qu, J. J. (2007). NMDI: A normalized multi-band drought index for monitoring soil and vegetation moisture with satellite remote sensing. *Geophysical Research Letters*, 34(20). Available from https://doi.org/10.1029/2007GL031021.

Wang, P. X., Li, X. W., Gong, J. Y., & Song, C. (2001). Vegetation temperature condition index and its application for drought monitoring. *International Geoscience and Remote Sensing Symposium (IGARSS)*, 1, 141–143.

Wilhite, D.A. (2000). *Drought as a natural hazard: Concepts and definitions*.

Zeng, Y., Hao, D., Huete, A., Dechant, B., Berry, J., Chen, J. M., ... Chen, M. (2022). Optical vegetation indices for monitoring terrestrial ecosystems globally. *Nature Reviews Earth and Environment*, 3(7), 477–493. Available from http://nature.com/natrevearthenviron/, https://doi.org/10.1038/s43017-022-00298-5.

Zhang, N., Hong, Y., Qin, Q., & Zhu, L. (2013). Evaluation of the visible and shortwave infrared drought index in China. *International Journal of Disaster Risk Science*, 4(2), 68–76. Available from http://www.springer.com/earth + sciences + and + geography/natural + hazards/journal/13753, https://doi.org/10.1007/s13753-013-0008-8.

Zhou, J., Jia, L., & Menenti, M. (2015). Reconstruction of global MODIS NDVI time series: Performance of Harmonic ANalysis of Time Series (HANTS). *Remote Sensing of Environment*, 163, 217–228. Available from http://www.elsevier.com/inca/publications/store/5/0/5/7/3/3, https://doi.org/10.1016/j.rse.2015.03.018.

Edaphology

Agricultural drought monitoring using Google Earth Engine: a study of Paschim Medinipur district, West Bengal

Pulakesh Pradhan and Sribas Patra

Department of Geography, Ravenshaw University, Cuttack, Odisha, India

11.1 Introduction

Drought is a natural climatological phenomenon (Sreekesh, Kaur, & Naik, 2019) characterized by the deficiency in the precipitation (Kogan, 1995; Mckee, Doesken, & Kleist, 1993; Sreekesh et al., 2019) and abnormally dry weather conditions (Alahacoon, Edirisinghe, & Ranagalage, 2021; Eklund, Mohr, & Dinc, 2024; Kogan, 1995) with water scarcity (Senamaw, Addisu, & Suryabhagavan, 2021). Agricultural drought refers to a period of insufficient soil moisture for crop growth (Ji & Peters, 2003; Roy, Hazra, & Chanda, 2023; Senapati, Raha, Das, & Gayen, 2021) that negatively affects crop production (Eklund et al., 2024; Mishra & Singh, 2010; Pan et al., 2023). There are several other driving factors that lead to the agricultural drought like excessive evapotranspiration due to rising temperatures (Anderson, Norman, Mecikalski, Otkin, & Kustas, 2007; Vicente-Serrano, Beguería, & López-Moreno, 2010), climatic variabilities and changes in climatic patters (Senapati et al., 2021), deforestation and degradation (Tran et al., 2023), weakening of Indian monsoon (Chattopadhyay, Malathi, Tidke, Attri, & Ray, 2020), global warming (Cai et al., 2024), poor water management practices (Alahacoon et al., 2021), and so forth. Identifying agricultural drought in a timely and accurate manner is critical for assessing crop conditions, managing irrigation water, and mitigating farming risks (Moran, Clarke, Inoue, & Vidal, 1994). It is helpful for developing agricultural water strategies (Zhang, Huang, & Li, 2024), ensuring food security (Becker-Reshef, Vermote, Lindeman, & Justice, 2010), and livelihood (Seshasai et al., 2016), essential for selecting appropriate drought management practices for farmers, government bodies, media, and insurance companies (Sreekesh et al., 2019). Agricultural drought monitoring through remote sensing and climate datasets (Cai et al., 2023; Mistry & Suryanarayana, 2023) has become an important approach for assessing crop–climate relationships

(Alahacoon et al., 2021; Ji & Peters, 2003; Kogan, 1995; Mishra & Singh, 2010; Thavorntam & Tantemsapya, 2013). Multiple studies have used vegetation indices like Normalized Difference Vegetation Index (NDVI), Temperature Condition Index (TCI), and Vegetation Condition Index (VCI) (Alahacoon et al., 2021; Thavorntam & Tantemsapya, 2013), drought indices like Standardized Precipitation Index (SPI), Palmer Drought Severity Index (PDSI), and so forth (Bashit, Ristianti, & Ulfiana, 2022), and climatic indices like Normalized Difference Drought Index and Rainfall Anomaly Index (Alahacoon et al., 2021) to access the agricultural drought patterns and crop responses (Chattopadhyay et al., 2020; Hasan & Abdullah, 2022).

11.1.1 Drought indices

For the drought monitoring, precipitation-based SPI (Mckee et al., 1993), PDSI incorporating moisture supply and demand, and Standardized Precipitation Evapotranspiration Index (SPEI) combining precipitation and potential evapotranspiration (PET) (Vicente-Serrano et al., 2010) are widely utilized. Satellite-derived actual evapotranspiration (AET) provides direct crop water use estimates (González & Valdés, 2006). Moisture stress is quantified by the Evaporative Stress Index (Anderson et al., 2007), evaporative fraction, and Crop Water Deficit Index. Soil moisture downscaling techniques also improve drought (Sreekesh et al., 2019; Zhang et al., 2024).

11.1.2 Vegetation indices

Vegetation indices from satellite optical and thermal sensors, particularly NDVI, have been widely used to monitor agricultural drought impacts on crop condition and yields (Amin et al., 2020; Becker-Reshef et al., 2010; Zhang et al., 2022). Decreasing crop NDVI during key phenological stages signals agricultural drought onset and stress (Ji & Peters, 2003; Kogan, 1995). VCI and TCI incorporate historical NDVI and land surface temperature (LST) for agricultural drought detection (Dong, Li, Yuan, You, & Chen, 2017; Kogan, 1995). Combining NDVI and LST, the Vegetation Health Index (VHI) indicates overall crop status (Geng, Zhang, Gu, He, & Zheng, 2024; Kogan, 1995; Possega, Ojeda, & Gámiz-Fortis, 2023). Microwave indices including Vegetation Optical Depth and Vegetation Water Content provide complementary information on vegetation and soil moisture dynamics (Liu et al., 2011; Sims, Niyogi, & Raman, 2002). Radiometric indices such as Normalized Multiband Drought Index are also applied for agricultural drought monitoring (Wang et al., 2022).

11.1.3 Agricultural drought studies

Senapati et al. (2021) carried out a detailed assessment of the agricultural drought susceptibility in the Purulia and Bankura districts, which are located in the western part of West Bengal. Using MODIS and SAR data (Kloos, Yuan, Castelli, & Menzel, 2021; Roy et al., 2023), the monthly agricultural droughts were examined, which occurred in the red and lateritic zones of West Bengal throughout the monsoon seasons. They determined that the severe to extreme drought conditions that resulted in crop failures occurred in three acute drought years: 2005, 2010, and 2015. Using a variety of remote sensing and GIS tools (Kaur, Rishi, & Chaudhary, 2022), another study attempted to determine the spatiotemporal variance of agricultural drought. In India, Chattopadhyay et al. (2020) used a composite drought index to track agricultural drought. The survey was conducted in five Indian states, Andhra Pradesh, Chhattisgarh, Haryana, Maharashtra, and Telangana, over the course of three consecutive years: 2014, 2015, and 2016. A different study examined the nation's current operational and research projects, as well as methods for assessing droughts. The spatiotemporal patterns of agricultural drought in West Bengal, India, are better understood thanks to these studies.

11.1.4 Approach in the present study

This study utilizes several key vegetation and drought indices that were derived from optical, thermal, and microwave satellite data to analyze agricultural drought patterns and crop responses over a 20-year period in the major farming region. Climate-based drought indices such as the SPI (Mckee et al., 1993) and the SPEI (Vicente-Serrano et al., 2010) can identify shortfalls in precipitation when compared to average seasonal norms (Kumar & Chu, 2024; Liu et al., 2016). These indicate drought caused by lack of rainfall recharge for soils. The TerraClimate Dataset is a comprehensive resource that provides several key indicators for analyzing drought conditions. These include AET, which measures the actual transfer of water from the land to the atmosphere, offering insights into the water cycle and plant growth potential. The climate water deficit measures the difference between potential and AET, serving as an indicator of environmental water stress. Soil moisture is another critical indicator, reflecting the water available in the soil for plant uptake, with changes in its levels often signaling the onset of drought conditions. Additionally, the dataset includes minimum and maximum temperature, as temperature extremes can intensify drought conditions by increasing evaporation rates and water demand. Lastly, the PDSI provides a standardized measure of drought severity, incorporating both precipitation and temperature. Together, these indicators offer a holistic view of drought conditions, enabling more accurate predictions and facilitating effective water resource management strategies (Tanarhte, De Vries, Zittis, & Chfadi, 2024). Changes in vegetation during the pre- and postmonsoon period were quantified by analyzing satellite vegetation indices like NDVI, VCI, and VHI. Climatic influence was analyzed and qualified through LST and TCI, and the water condition was analyzed through the NDWI (Fig. 11.1).

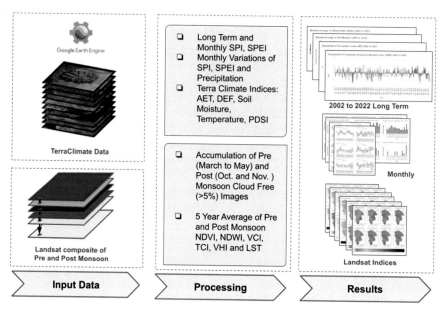

FIGURE 11.1

Methodology chart of the study.

11.2 Study area

The study area, Paschim Medinipur district, is located in the state of West Bengal, India (Fig. 11.2). The district is located in the southern part of West Bengal adjacent to the Bay of Bengal. It is bounded by the Purba Medinipur district to the east, Jhargram district to the west, Bankura and Purulia districts to the north, and the state of Odisha to the south. Paschim Medinipur has a subtropical climate influenced by the southwest monsoon. Average annual rainfall is around 1400 mm, of which 80% is received during the summer monsoon months of June to September (Jana, Sit, & Chanda, 2021; Panda, Upadhyay, Jha, & Sharma, 2020). Temperatures range from 12°C in winter to over 40°C prior to monsoon onset, with high humidity levels. The primary river systems are the Kangsabati, Rasulpur, and Haldi, which are tributaries of the Hooghly River. Canals from these rivers provide irrigation to some areas, although much of the district depends on rain-fed agriculture. Paschim Medinipur is predominantly rural and agrarian, with over 80% of the population engaged in agriculture and related activities. Rice is the major crop during the kharif (summer monsoon) season. Oil seeds such as groundnut and sesame, pulses like arhar, and vegetables including okra, brinjal, and cucurbits are also cultivated across the district (Sen & Bhakat, 2021). The main rabi (winter) crops are sunflower, mustard, potatoes, and vegetables. Forested areas are mainly found in the western region.

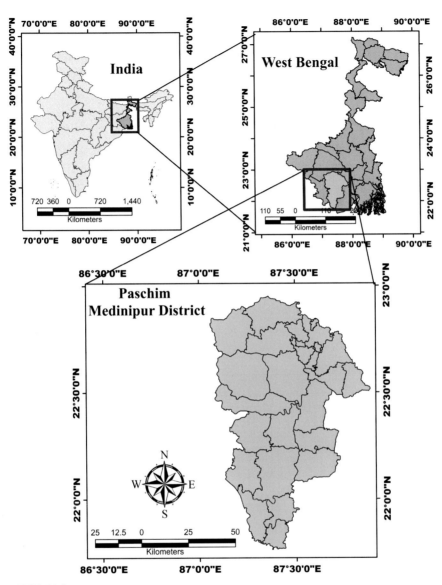

FIGURE 11.2

Location map of the study area. Map lines delineate study areas and do not necessarily depict accepted national boundaries.

11.2.1 Drought vulnerability

High rainfall variability coupled with limited irrigation infrastructure leaves the region prone to frequent agricultural droughts. Kharif rice is highly vulnerable to monsoon rainfall deficits. Recurrent droughts lead to low farm incomes, food insecurity, indebtedness, and distress migration. Integrating satellite and climate datasets can delineate drought patterns, associated crop responses, and vulnerable zones to inform adaptation strategies (Li et al., 2024; Xiao et al., 2024).

11.3 Datasets

TerraClimate and Landsat series datasets were used in this investigation (Ghaleb et al., 2015; Hazaymeh & Hassan, 2016). Landsat data were gathered seasonally, particularly before and after the monsoon. Premonsoon months were taken from March to May, while postmonsoon months were taken from October to November. Datasets with 5-year intervals were gathered from 2002 to 2022. The datasets were obtained from the US Geological Survey via the Google Earth Engine (GEE) platform (Gorelick et al., 2017). The Landsat datasets were collected with a 5% cloud cover. This study employed a total of 83 Landsat 5 and 118 Landsat 8 images. TerraClimate datasets were also gathered from 2002 to 2022 using the GEE platform. Monthly average precipitation data, as well as AET, climate water deficit, soil moisture, minimum and maximum temperature, and the PDSI, were collected (Table 11.1). The images were processed in the GEE platform.

11.3.1 Methodology

In the present study, multiple indices were used for determination of agriculture drought in Paschim Medinipur district of West Bengal. This study made use of

Table 11.1 Different data sources for the study.

Datasets	Name of the data	Duration	Months	Numbers of images
Terra-climate datasets	Precipitation	2002–22	Jan–Dec	4638.3 m spatial resolution
	AET	2002–22	Jan–Dec	
	Climate water deficit	2002–22	Jan–Dec	
	Soil moisture	2002–22	Jan–Dec	
	Minimum and maximum temperature	2002–22	Jan–Dec	
	PDSI	2002–22	Jan–Dec	
Satellite datasets	Landsat 5 TM	2003–07	March–May and Oct–Nov	47
		2008–12		36
	Landsat 8 TIRS/OLI	2013–17		58
		2018–22		60

the following indices: LST, NDVI, Normalized Difference Water Index (NDWI), SPI, SPEI, VCI, TCI, VHI, and so on. The study's whole methodology is displayed in Fig. 11.4.

11.3.1.1 Standardized Precipitation Index

The SPI is one of the drought indices that the WMO has approved for use globally, which was put forth by Mckee et al. (1993). The SPI is a multiscalar index that solely relies on precipitation. This method gained a lot of acceptance and is now extensively used to research drought for applications in various sectors because it is straightforward and takes various timescales into account (Fig. 11.3).

Eqs. 11.1−11.6 represent the procedure of calculating the SPI. For the precipitation value of x over a given period, the distribution of probability density function of Γ is as follows:

$$F(x) = \frac{1}{\beta^\gamma \Gamma(\gamma_0)} x^{\gamma-1} e^{-x/\beta}, x > 0 \tag{11.1}$$

where β and y are scale and shape parameters of the distribution function ($^\Gamma$), respectively.

The probability of all precipitation events (x) that are smaller than the x_0 for a given year is:

$$F(x < x_0) = \int_0^\infty f(x)dx \tag{11.2}$$

and

$$F(x = 0) = m/n \tag{11.3}$$

where m is the number of days with no precipitation ($x = 0$) and n is the total number of days.

The normal standardizing of the Γ probability function is done through substitution of the result of probability value in the standardized normal distribution function:

$$f(x < x_0) = \frac{1}{\sqrt{2\pi}} \int_0^\infty e^{-x^2/2} dx \tag{11.4}$$

By solving the abovementioned equation, the SPI would be as follows:

$$SPI = S \frac{t - (c_2 t + c_1) + c_0}{[(d_3 + d_2)t + d_1]t + 1} \tag{11.5}$$

where

$$t = \sqrt{\ln \frac{1}{F^2}} \tag{11.6}$$

```
// SPI
function getGammaDistParams(col){
    // Shape/rate-based gamma distribution as defined in Wikipedia: Get the gamma distribution parameters. Input = the collection
    var col_no0 = col.map(function(img){return img.updateMask(img.neq(0))});// Masks pixels with 0 values
    var N=col_no0.count(); // Reduces an image collection by calculating the number of images with a valid mask at each pixel across the stack of all matching bands. Bands
    var sum_log=col_no0.map(function(img){return img.log()}).sum();
    var A = (col_no0.mean().log()).subtract(sum_log.divide(N));
    var iA4 = ee.Image.constant(4).multiply(A);
    var i1 = ee.Image.constant(1);
    var i3 = ee.Image.constant(3);
    var alpha = (i1.divide(iA4)).multiply(i1.add((i1.add(iA4.divide(i3))).sqrt()));
    var beta = col_no0.mean().divide(alpha);
    var param = alpha.addBands(beta).select([0,1],["alpha","beta"]);
    return param;
}
// Get q, the probability of zeros. Input = the collection
function getQ(col){ // Get the probability of a zero CHECK that the probability is on Pi and not Pr
    var PiNot0 = col.map(function(img){return img.updateMask(img.neq(0))}); // Mask all null values
    var N = col.count();
    var m = N.subtract(PiNot0.count()); // calculating the number of images with 0 at each pixel
    var q = ee.Image(m.divide(N));
    return q.select([0], ['q']);
}
function getHxSPI(img){
    var i1 = ee.Image.constant(1);
    var coef = img.divide(beta);
    var Gx = ee.Image(alpha.gammainc(coef)); // CDF for shape/scale based gamma functin. Definition found in wikipedia
    // Note: gammainc(x,a) is the regularized lower incomplete gamma function, which is equal to the lower incomplete gamma divided by gamma(x). So, in your particular cas
    var Hx = Gx.addBands(q.add((i1.subtract(q)).multiply(Gx))).select([0, 1], ['Gx', 'Hx']);
    //Calculate SPI
    var c0 = ee.Image.constant(2.515517);
    var c1 = ee.Image.constant(0.802853);
    var c2 = ee.Image.constant(0.010328);
    var d1 = ee.Image.constant(1.432788);
    var d2 = ee.Image.constant(0.189269);
    var d3 = ee.Image.constant(0.001308);

    var in1 = ee.Image.constant(-1);
    var t1 = ee.Image(((i1.divide(Hx.pow(2))).log()).pow(0.5)).select([0], ["t1"]);
    var t2 = ee.Image(((i1.divide(i1.subtract(Hx)).pow(2)).log()).pow(0.5)).select([0], ["t2"]);
    var t = t1.where(Hx.select('Hx').gt(.5), t2).select([0], ['t']);
    var div = t.subtract((c0.add(c1.multiply(t)).add(c2.multiply(t.pow(2)))).divide(i1.add(d1.multiply(t)).add(d2.multiply(t.pow(2))).add(d3.multiply(t.pow(3)))));
    var SPI = in1.multiply(div).where(Hx.select('Hx').lte(.5), div).select([0], ['SPI']);

    return Hx.addBands(SPI).select([0, 1, 2],['Gx', 'Hx', 'SPI'])
        .set({'time': ee.Date.fromYMD(img.get('year'), img.get('month'), 1)});//,
}
// *** getLoglogisticDistParam: Obtain parameters of the log logistic distribution
// Input:  the collection holding Di = Pi-ETi data over the chosen window
function getLoglogisticDistParam(col){
var N = col.count();
// Calculate probability-weighted moments
// Convert the entire collection to an array-based image, and arraySort it
var array = col.toArray();
var DiSorted = array.arraySlice(1, 0).arraySort().arrayProject([0]);
/// START CHANGE///
var i = ee.Image.constant(ee.Array(ee.List.sequence(1, col.size())));
/// END CHANGE///
var Fi = ((i.subtract(ee.Image.constant(.35))).divide(N));
var FiDi = DiSorted.addBands(Fi).select([0,1], ['DiSorted', 'Fi']);
var pwm0 = ee.Image(((ee.Image.constant(1).subtract(Fi)).pow(0).multiply(DiSorted)))
    .arrayAccum(0).arraySlice(0, N.subtract(1)).divide(N);
var pwm1 = ee.Image(((ee.Image.constant(1).subtract(Fi)).pow(1).multiply(DiSorted)))
    .arrayAccum(0).arraySlice(0, N.subtract(1)).divide(N);
var pwm2 = ee.Image(((ee.Image.constant(1).subtract(Fi)).pow(2).multiply(DiSorted)))
    .arrayAccum(0).arraySlice(0, N.subtract(1)).divide(N);
var pwm = pwm0.addBands(pwm1).addBands(pwm2);
// Calculate alpha, beta, gamma
var i1 = ee.Image.constant(1);
var i2 = ee.Image.constant(2);
var i6 = ee.Image.constant(6);
var beta = (i2.multiply(pwm1)).subtract(pwm0))
    .divide((i6.multiply(pwm1)).subtract(pwm0).subtract(i6.multiply(pwm2)));
var gamF1 = (i1.add(i1.divide(beta))).gamma();
var gamF2 = (i1.subtract(i1.divide(beta))).gamma();
var alpha = ((pwm0.subtract(i2.multiply(pwm1))).multiply(beta)) // // Error in the Serano paper (multiply(beta) and NOT divide(beta); according to
    .divide((gamF1).multiply(gamF2));
var gamm = pwm0.subtract((alpha).multiply(gamF1).multiply(gamF2)); // Error in the Serano paper according to the source paper Singh 1993
var param = beta.addBands(alpha).addBands(gamm).select([0,1,2],['beta','alpha','gamma']);
return param;
}
//*** getFxSPEI: obtain Fx and SPEI
/// START CHANGE ///
function getFxSPEI(img){
    var i1 = ee.Image.constant(1);
    var in1 = ee.Image.constant(-1);
    var in2 = ee.Image.constant(-2);

    var Fx = (i1.add((alpha.divide(img.subtract(gamm))).pow(beta))).pow(-1).arrayFlatten([['0']]);
    var Px = i1.subtract(Fx);
    var W1 = (in2.multiply((Px).log())).sqrt();
    var W2 = (in2.multiply((i1.subtract(Px)).log())).sqrt();

    var W = W1.where(Px.gt(ee.Image.constant(.5)), W2);
    var C0 = ee.Image.constant(2.515517);
    var C1 = ee.Image.constant(0.802853);
    var C2 = ee.Image.constant(0.010328);
    var d1 = ee.Image.constant(1.432788);
    var d2 = ee.Image.constant(0.189269);
    var d3 = ee.Image.constant(0.001308);

    var SPEI00 = W.subtract((C0.add(C1.multiply(W)).add(C2.multiply(W.pow(2))))
            .divide(i1.add(d1.multiply(W)).add(d2.multiply(W.pow(2))).add(d3.multiply(W.pow(3)))));

    var SPEI0 = SPEI00.where(Px.gt(ee.Image.constant(.5)), in1.multiply(SPEI00));

    var SPEI = Fx.addBands(SPEI0).select([0,1], ['Fx', 'SPEI'])
        .set({'time': ee.Date.fromYMD(img.get('year'), img.get('month'), 1)});
    return SPEI;//
}
```

FIGURE 11.3

Google Earth Engine script for Standardized Precipitation Index and Standardized
Precipitation Evapotranspiration Index.

$$c_0 = 2.515517; c_1 = 0.802853; c_2 = 0.010328$$
$$d_1 = 1.432788; d_2 = 0.189269; d_3 = 0.001308$$

$c_0, c_1, c_2, d_1, d_2,$ and d_3 are constant parameters of SPI. The sign of the SPI is identified by coefficient S that is based on the value of F. For $F > 0.5$, the S equals 1 and for $F \leq 0.5$, the S equals -1.

11.3.1.2 Standardized Precipitation Evapotranspiration Index

The SPEI, which added the PET as an input, was proposed by Vicente-Serrano et al. (2010), based on the SPI method with a basic climatic water balance. PDSI's sensitivity to atmospheric water demand and SPI's multiscale characteristic are both present in SPEI (Fig. 11.3). Compared to other approaches, it is more favorable because of these benefits and its relative simplicity. According to Vicente-Serrano et al. (2010), the SPEI computation process is shown in Eqs. 11.7–11.14. To estimate the PET, a variety of techniques are available (such as Thornthwaite approach, Penman–Monteith method, Hargreaves method, etc.). The following would be the moisture-deficit for the month after calculating the PET using any of the different methods:

$$D_i = P_i - \text{PET}_i \tag{11.7}$$

The standardization is followed by selection of the probability density function of a three-parameter loglogistic distributed variable as follows:

$$f(x) = \frac{\beta}{\alpha} \left(\frac{x-\gamma}{\alpha} \right)^{\beta-1} \left(1 + \left(\frac{x-\gamma}{\alpha} \right)^{\beta} \right)^{-2} \tag{11.8}$$

in which $\alpha, \beta,$ and y are scale, shape, and origin parameters, respectively; with D values in the range of $y > D < \infty$. These parameters are obtained through Eqs. 11.9–11.11:

$$\beta = \frac{2w_1 - w_0}{6w_1 - w_0 - 6w_2} \tag{11.9}$$

$$\alpha = \frac{(w_0 - 2w_1)\beta}{\Gamma(1 + 1/\beta)\Gamma(1 - 1/\beta)} \tag{11.10}$$

$$\gamma = w_0 - \alpha\Gamma\left(1 + \frac{1}{\beta}\right)\Gamma\left(1 - \frac{1}{\beta}\right) \tag{11.11}$$

where $\Gamma(\beta)$ is the gamma function of β.

The probability density function of D based on the given distribution of three-parameter log-logistic is as follows:

$$F(x) = \left[a + \left(\frac{\alpha}{x-\gamma} \right)^{\beta} \right]^{-1} \tag{11.12}$$

Then, the SPEI is calculated as the standardized values of $F(x)$ as follows:

$$\text{SPEI} = W - \frac{C_0 + C_1 W + C_2 W^2}{1 + d_1 W + d_2 W^2 + d_3 W^3} \tag{11.13}$$

Given P is the probability of exceeding a given D value ($P = 1 - F(x)$), for $P \leq 0.5$, the W is:

$$W = -2\ln(P) \tag{11.14}$$

And, if $P > 0.5$, then P is replaced by $1 - P$ and the sign of the resultant *SPEI* is reversed.

Six constant parameters of the SPEI equation are as follows:

$$c_0 = 2.515517; c_1 = 0.802853; c_2 = 0.010328$$
$$d_1 = 1.432788; d_2 = 0.189269; d_3 = 0.001308$$

11.3.1.3 Normalized Difference Vegetation Index

NDVI as a standard method for assessing the healthy and green vegetation by calculating the difference between near-infrared and red light (Senamaw et al., 2021; Sruthi, Mohammed Aslam, 2015), as shown in Eq. 11.15.

$$\text{NDVI} = \frac{(\text{NIR} - R)}{(\text{NIR} + R)} \tag{11.15}$$

where NIR = near infrared band, R = Red band

According to Thavorntam and Tantemsapya (2013), the NDVI is a potent indicator for tracking the amount of vegetation covering large regions as well as for identifying the frequency and duration of droughts. It gives an indication of the quantity and health of plants at the surface of the ground. The amount of photosynthetic activity in the observed vegetation is correlated with the NDVI magnitude. Higher NDVI scores often correspond to higher levels of vegetation vigor.

11.3.1.4 Vegetation Condition Index

The VCI exhibited the changeable moisture conditions and indicated vegetation dynamics. Maximum levels of this measure show healthy, undisturbed vegetation. The index is very useful for measuring vegetation stress and examining how plant responds (Dutta et al., 2015) Eq. 11.16.

$$\text{VCI} = \frac{100 \times (\text{NDVI} - \text{NDVI}_{\min})}{\text{NDVI}_{\max} - \text{NDVI}_{\min}} \tag{11.16}$$

The proportion of the VCI value is expressed in terms of $1-100$. While readings ranging from 50% to 35% indicate a drought state and values below 35% indicate a severe drought condition, the range between 50% and 100% indicates above average vegetation condition.

11.3.1.5 Temperature Condition Index

TCI is used to assess the stress that high temperatures and heavy moisture have on vegetation. It is calculated by Eq. 11.17 (Alahacoon et al., 2021)

$$TCI = \frac{100 \times LST_{max} - LST}{LST_{max} - LST_{min}} \tag{11.17}$$

A TCI rating of 50% denotes normal conditions or the absence of drought. TCI levels below 50% indicate different degrees of drought severity, whereas TCI values above 50% reflect the ideal/above normal situation.

11.3.1.6 Vegetation Health Index

Analysis of the drought VHI is one of the important remote sensing indices. Both the NDVI and the LST can be used to calculate the VHI by Eq. 11.18 (Alahacoon et al., 2021).

$$VHI = (0.5 \times VCI) + (0.5 \times TCI) \tag{11.18}$$

The VHI value is distributed between 0 and 100. Drought is indicated by low VHI values, whereas robust vegetation is indicated by high values.

11.3.1.7 Normalized Difference Water Index

When assessing drought, NDWI with SWIR can be used in addition to NDVI, especially early in the drought season. It is calculated following Eq. 11.19

$$NDVI = \frac{(NIR - SWIR)}{(NIR + SWIR)} \tag{11.19}$$

where NIR is the near-infrared band, and SWIR is the short wave infrared band of the satellite imageries.

11.3.1.8 Land surface temperature

The LST of the Paschim Medinipur district was determined using Landsat TM 5 imagery for the years 2002 to 2012 and Landsat 8 OLI (Operational Land Imager) for the years 2013 to 2022 with less than 5% cloud cover. While Landsat 8 OLI imagery includes 11 bands, including bands 10 and 11, which are thermal bands (Eqs. 11.20−11.25), Landsat TM 5 imagery only provides 7 bands. The steps involved in converting digital values into spectral radiance for temperature calculation from TM and OLI images are as follows:

Step 1: Landsat TM 5 band 6 digital values to spectral radiances conversion.

The following formula was used to convert band 6 digital values into radiance values ($L\lambda$) (Ara, Alif, & Islam, 2021)

$$L\lambda = \frac{LMAX\lambda - LMIN\lambda}{QCALMAX - QCALMIN} \times (QCAL - QCALMIN) + LMIN\lambda \tag{11.20}$$

Here, $L\lambda$ is the atmospherically corrected cell value as the radiance, QCAL is the digital image value, LMINλ is the spectral radiance scaled to QCALMIN, LMAXλ

is the spectral radiance scaled to QCALMAX, and QCALMIN is the minimum quantization calibration. The radiance pixel value (usually 1) and QCALMAX are the maximum values of quantized calibrated pixels (usually 255). Landsat 8 OLI spectral radiances were converted from band 10 and 11 digital data.

The following formula was used to convert the Landsat OLI band 10 and 11 digital data.

$$L\lambda = ML \times QCAL + AL \tag{11.21}$$

where $L\lambda$ is the spectral radiance at the top of the atmosphere, ML denotes the radiance multiband X, AL denotes the radiance add band X, QCAL denotes the quantized and calibrated standard product pixel value, and X denotes the band number. The band-specific multiplicative rescaling factor ML and the band-specific additive rescaling factor AL are obtained from the metadata file (MTL file).

Step 2: At-satellite brightness spectral radiance temperature conversion— emissivity modifications were added to radiant temperature based on the land cover nature following Ara et al. (2021).

$$T = \frac{K_2}{Ln\left(\frac{K_1}{L\lambda} + 1\right)} - 273 \cdot 15 \tag{11.22}$$

where T is the at-satellite brightness temperature in Kelvin (K), $L\lambda$ is the at-satellite radiance in W/(m^2 sr μm), and K_1 and K_2 are the thermal calibration constants in W/(m^2 sr μm), respectively. For Landsat-5 TM, $K_1 = 607.76$ and $K_2 = 1260.56$ for band 6, and for Landsat 8 OLI, K_1 for band 10 and 11 is 774.8853 and 480.8883 respectively, and K_2 for band 10 and 11 is 1321.0789 and 1201.1442, respectively. The metadata file provided the values for K_2 and K_1. For a better understanding, the thermal constant values for Landsat TM and Landsat OLI are converted from K to °C using the equation 0°C = 273.15K.

Step 3: Emissions from the ground surface are measured (E).

The temperature values derived above are compared to a black body. As a result, spectral emissivity (E) adjustments are required. These can be done according to the land cover type or by calculating the emissivity values for each pixel from the proportion of vegetation (P_V) data (Shahfahad et al., 2021).

$$E = 0.004 * P_V + 0.986 \tag{11.23}$$

where the P_V can be calculated as:

$$P_V = \left\{ \frac{(NDVI_{max} - NDVI_{min})}{(NDVI_{max} - NDVI_{min})} \right\}^2 \tag{11.24}$$

Step 4: LST is calculated using the equation below (Govind & Ramesh, 2020).

$$\frac{BT}{1} + W * \left(\frac{BT}{P}\right) * Ln(E) \tag{11.25}$$

where BT is the brightness temperature at the satellite, W is the wavelength of emitted radiance, $P = h^*c/s$ (1.438 10−2 m K), h is the Planck's constant (6.626 10−34 Js), s is the Boltzmann constant (1.38 10−23 J/K), and the light velocity is 2.998 10 8 m/s.

11.4 **Results and discussion**

The entire analysis has been segmented in three parts; the first section focuses on analysis of precipitation with SPI and SPEI indices calculated on the GEE platform. The second section discusses the drought condition with available drought monitoring indices in TerraClimate datasets, and in the last section, the various indices based on Landsat datasets on pre- and postmonsoon seasons are analyzed to quantify the variation on vegetation and climate through indices like NDVI, VHI, VCI, TCI, LST, and NDWI.

11.4.1 **Distribution of precipitation over the time period**

Precipitation plays a crucial role in the analysis of drought conditions. It is the primary source of water for agricultural activities, and any significant deviations in its patterns can lead to agricultural droughts. Understanding precipitation trends is therefore crucial for predicting and managing drought conditions. In this present study, the **TerraClimate** data was analyzed from January to December (Fig. 11.4A). We observed substantial fluctuations in precipitation, particularly in January, February, November, and December. During these months, some years had almost no precipitation, indicating potential periods of agricultural drought. The monthly detailed analysis is given below:

January: The average precipitation is **8.63 mm**. The standard deviation is high at **11.45 mm**, indicating the spread from **almost no rainfall** in many dry years (2003, 2006, 2007, 2011, 2018, and 2010) to a **33 mm** moderately wet outlier in 2012 (Fig. 11.5). Most Januarys (over two-thirds) see rainfall of 0−15 mm. However, the high variation poses farming challenges. **February**: The **13.14 mm** average precipitation also has substantial fluctuation between 0 mm droughts (2005, 2006, 2009) and a very wet 66 mm year (2019), with a standard deviation of **16.89 mm**. The majority of years record modest 0−15 mm rain, but extremes disrupt crop planning. **March**: The **18.85 mm** average includes a huge 0−64 mm swing between extreme dry and wet Marches. With a **16.55 mm** standard deviation, most years see rainfall group between 15 and 30 mm. However, unpredictable large shifts in moisture early in the growing season complicate decisions. **April**: Despite averaging **45.63 mm** precipitation, April has varied enormously between basically none up to a flooded 150 mm (2020) due to a high **34.24 mm** standard deviation. Most Aprils receive 25−100 mm, but such

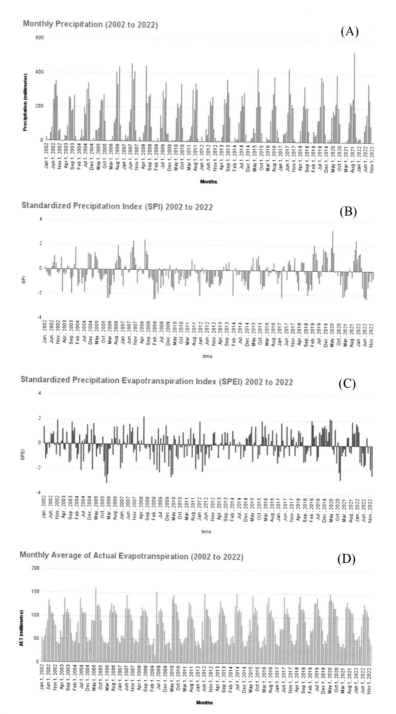

FIGURE 11.4

Timeseries for (A) Precipitation, (B) Standardized Precipitation Index, (C) Standardized Precipitation Evapotranspiration Index and (D) Actual Evapotranspiration from 2002 to 2022.

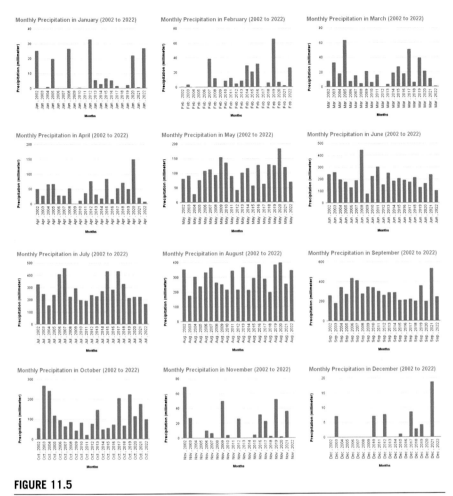

FIGURE 11.5

Monthly precipitation from 2002 to 2022.

volatility hampers crop growth. **May**: The **101.17 mm** average precipitation extends from 29 mm (2004) to over double that in very wet Mays. With a **37.56 mm** standard deviation, farmers must prepare both for half or double the benchmarks. Total rain of 50–150 mm is typical. **June**: Although averaging **204.24 mm**, June drenching swings wildly from just 78 mm (2009) to almost triple that due to a **77.37 mm** standard deviation. Most Junes average 125–300 mm. Planning amidst such variance is demanding.

July: Despite averaging in a hot **277.14 mm**, July precipitation has reached just 156 mm or over triple, reflecting its **90.35** mm standard deviation. Most rainfall is spread across a damp 200–400 mm historically. However, sensitivity is acute. **August**: The **295.87 mm** precipitation average sees volatility between just 176 mm to nearly double with a **69.93** mm standard deviation. Most Augusts cluster within 200–375 mm. However, reliability is tough with shifts from damp to soaked conditions. **September**: The average precipitation is **297.05 mm**, but its **88.14 mm** standard deviation means it fluctuates between 185 and 533 mm. Most years see 200–375 mm total moisture. Yet early monsoons or extended dry periods pressurize farmers. **October**: Its **113.31 mm** precipitation mean ranges between just 22 and 269 mm, thanks to its **70.74 mm** standard deviation. Most rainfall is 50–200 mm. However, such instability right before harvests is problematic. **November**: Despite averaging **16.51 mm** precipitation, its **20.98 mm** standard deviation means many Novembers record zero rain, while the wet outlier saw nearly 70 mm. Most years cluster around 0–25 mm. However, moisture before the winter freeze is precious. **December**: With just **2.76 mm** average precipitation, Decembers tend to be bone dry with negligible **4.79 mm** standard deviation. Most see no moisture as winter sets in. Yet an 18 mm outlier poses challenges calculating frost and snow loads.

11.4.2 Temporal pattern of Standardized Precipitation Index and Standardized Precipitation Evapotranspiration Index

The SPI is a critical tool in the analysis of drought conditions. It operates on various timescales, making it a versatile index for drought monitoring. On shorter timescales, the SPI is closely tied to soil moisture levels, providing insights into **immediate drought conditions**. On longer timescales, it relates to groundwater and reservoir storage, offering a broader view of water availability. The SPI quantifies anomalies in accumulated precipitation over a specified period (e.g., 1, 3, and 12 months), making it an effective indicator for detecting and characterizing the onset, intensity, and duration of meteorological droughts (Dimyati et al., 2024). In this present study, the 1 month SPI was calculated to identify drought conditions. The time series given below represents the monthly SPI value from January 2002 to December 2022 (Fig. 11.4B). The precipitation dataset was used from the TerraClimate dataset for this analysis. Overall, the SPI value was positive in most months of the years 2005, 2007, 2019, 2021, and so forth and negative in most months of the years 2004, 2005. 2006, 2009, 2012, 2014. 2017, and 2022. The detailed monthly explanation is given in the next part.

The SPEI builds upon the SPI by incorporating PET into its calculations. This inclusion of PET allows the SPEI to account for the effects of temperature variability on drought conditions, in addition to precipitation. Therefore, the **SPEI provides a more comprehensive understanding of drought conditions** as it considers both water supply (precipitation) and demand (evapotranspiration). Like the SPI, the SPEI can also measure the severity of drought according to its intensity and duration and can identify the onset and end of drought episodes. In this study, 1 months SPEI was calculated to identify potential months with drought conditions.

Overall, the time series represents the drought condition from 2002 to 2022 (Fig. 11.4C). In the following months, the extreme positive SPEI was found, indicating flood or wet conditions (June 2008, April 2020, May 2020, February 2019, December 2015, September 2021, November 2002, January 2012, and December 2021). Extreme negative means dry conditions were found in the months of January 2006, December 2020, March 2010, December 2005, February 2006, April 2010, December 2022, June 2009, March 2012, November 2020, May 2004, February 2005, and April 2009. The detailed Monthly explanation is given below.

January: In January, the years with the driest conditions were 2006, 2009, and 2014, indicating severe drought with significantly less than median precipitation. The wettest years were 2003, 2020, and 2022, which experienced higher than normal precipitation signaling very wet conditions. The year's closest to normal with SPI and SPEI values near the median were 2002, 2004, and 2013 (Fig. 11.6).

February: For February, the driest years were 2006, 2009, and 2017, with very low SPI and SPEI values pointing to extreme droughts. On the other hand, 2007, 2016, and 2019 were the wettest years, with high index values showing greater than average precipitation. February in the years 2008, 2014, and 2020 saw near median precipitation and normal conditions.

March: In terms of March, the years facing severe droughts and markedly less than normal rains were 2006, 2010, and 2012 based on the very low SPI and SPEI numbers. The opposite very wet years were 2005, 2017, and 2019 when the indices rose considerably beyond the median. Median levels aligning with normal precipitation were registered in 2008, 2014, and 2020.

April: April in 2009, 2010, and 2022 represented years of intense drought and the lowest SPI and SPEI readings. There was higher than typical precipitation in 2005, 2015, and 2020, the wettest years. 2004, 2007, and 2018 exhibited near normal April precipitation patterns and index numbers.

May: May 2004, 2012, and 2022 had dramatically less than normal rains pointing squarely toward severe droughts. Excess precipitation well beyond the median happened in 2005, 2019, and 2020, the wettest years. Normal median levels appeared in 2014, 2015, and 2016 for May (Fig. 11.6).

June: In June, acute dryness linked to rainfall shortages prevailed in 2009, 2012, and 2022, as seen through low SPI and SPEI metrics. The years 2008, 2011, and 2020 were instead the wettest owing to copious above average June showers. Normal precipitation similar to the median occurred in the years 2014, 2015, and 2016 (Fig. 11.6).

July: In July, the years facing severe drought with less than normal rains were 2004, 2012, and 2022 based on very low SPI and SPEI values. The opposite very wet years were 2007, 2008, and 2015, which saw higher than typical precipitation. Somewhere in between, with median July precipitation, were the years 2003, 2013, and 2016.

August: For August, acute dry spells linked to low rainfall happened in 2004, 2010, and 2012, as evident from the low index numbers. Excessive precipitation characterized 2002, 2007, and 2020, the wettest years. Normal August rainfall aligning with median index values occurred in 2020, 2018, and 2014.

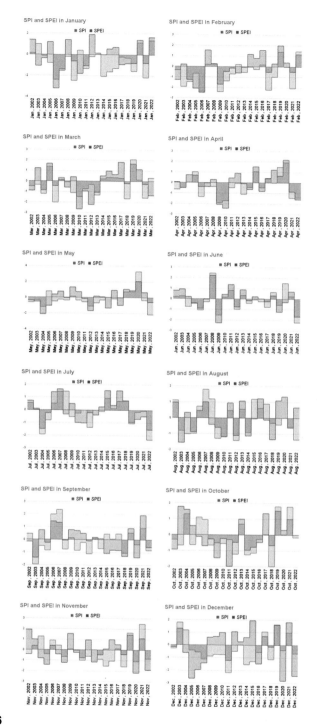

FIGURE 11.6

Monthly Standardized Precipitation Index (SPI) and Standardized Precipitation Evapotranspiration Index (SPEI) form 2002 to 2022.

September: September 2003, 2012, and 2022 faced extreme drought conditions owing to significantly less than average rains according to the SPI and SPEI. In contrast, the years 2006, 2007, and 2021 were very wet with ample above normal September showers. Near media precipitation fell in 2013, 2014, and 2016.

October; In October 2009, 2011, and 2018, SPI and SPEI metrics pointed strongly to intense droughts. The opposite situation of excess rainfall marked 2004, 2006, and 2019 as the wettest years. Normal precipitation similar to median levels happened in 2016, 2020, and 2022.

November: For November, 2005, 2015, and 2022 were the driest years as large rainfall deficiencies led to acute drought scenarios based on very low index numbers. 2002, 2006, and 2021 instead had the highest precipitation, classifying them as the wettest years. Average rainfall consistent with normal November patterns occurred in 2010, 2013, and 2017.

December: December 2011, 2014, and 2022 experienced severely dry conditions, with SPI and SPEI values indicating significant lack of precipitation. The wettest Decembers were 2003, 2019, and 2021 with rain surplus driving index values upward. Near media levels signifying normal rainfall prevailed in 2002, 2013, and 2018.

11.4.3 Distribution of actual evapotranspiration over the time

AET: It represents the actual amount of water that is evaporated and transpired, which is crucial for understanding water deficits during times of precipitation deficits. Fig. 11.4D shows monthly AET values in millimeters from January 2002 to December 2022. Over this two decades, AET levels followed a seasonal pattern, with the highest levels occurring during the summer months of June, July, and August and the lowest levels in the winter months of December, January, and February. Overall, June had the peak monthly AET averages over 130 mm in most years, exceedingly even 150 mm in some summers. In contrast, December and January AET averages were typically below 45 mm. The data indicates strong variability from year to year as well, with some years seeing relatively high evapotranspiration (e.g., the summer of 2020), while other years had lower levels (e.g., the low AET in the spring of 2009). There does not appear to be an obvious long-term trend up or down in average monthly AET over the time period. Instead, monthly AET seems to fluctuate considerably between years while retaining a consistent seasonal cycle peaking in mid-summer.

11.4.4 Distribution of climate water deficit over the time

It estimates how much additional water plants need. When water deficit is outside the normal range of plant adaptations, it restricts plant growth and other ecological processes. Fig. 11.7A shows monthly climate water deficit values in millimeters from January 2002 to December 2022. Deficit refers to the lack of water compared to the needs of plants. Over the two decades, deficit

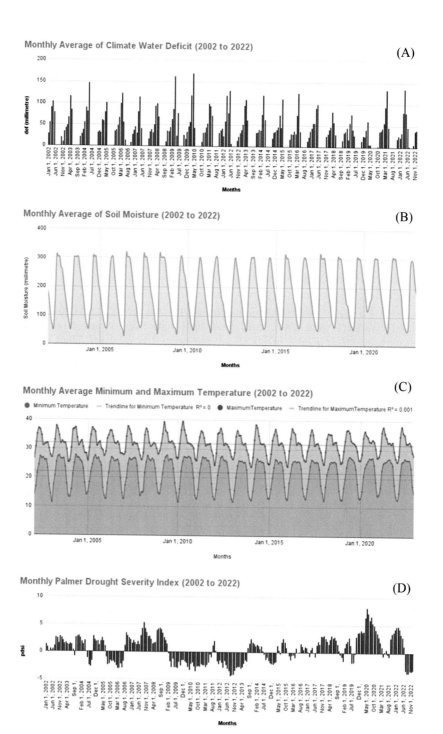

FIGURE 11.7

Timeseries for (A) Crop Water Deficit, (B) Soil Moisture, (C) Temperature, and (D) Palmer Drought Severity Index from 2002 to 2022.

followed a consistent seasonal pattern, with zero deficit occurring during the peak growing season in the summer months of June, July, and August when water demand is highest. During the winter months of December, January, and February, deficit levels were typically highest, exceeding 30 mm in most years. This reflects the lower water needs of plants in winter. The data shows considerable variability between years as well. Some years experienced prolonged zero deficits spanning spring and fall (e.g., 2003), while other years saw deficits persisting into the late spring and reemerging in early fall (e.g., 2009). The highest deficits appeared in the transitional seasons when water demand was rising or falling. There is no evident long-term trend over the 21-year period. Instead, monthly climate water deficit fluctuates year-to-year, while retaining a distinct seasonal cycle marked by zero summer deficit and peaks in the winter.

11.4.5 Distribution of soil moisture

Soil moisture: Reduced soil water levels are typically associated with soil water stress for vegetation, which constitutes a major constraint on the physiological functioning of natural and cultivated ecosystems and can thus lead to large impacts on agricultural production. Fig. 11.7B shows monthly soil moisture values in millimeters in the top 1 m of soil from January 2002 to December 2022. Over the 21-year record, soil moisture followed a consistent seasonal cycle, with the lowest levels occurring during the summer growing season and the highest moisture levels in the winter. June, July, and August showed very consistent soil moisture around 300 mm in most years, indicating field capacity conditions where the soil is fully saturated. In contrast, soil moisture declined sharply in the spring, reaching a low of around 50 mm on average in May just before the start of summer. Between years, soil moisture shows large variability during the transitional seasons, with some years retaining higher moisture later into spring (e.g., 2009), while other years dry down faster (e.g., 2004). On average, the data indicates a typical seasonal fluctuation between fully saturated winter soil moisture levels around 300 mm declining to roughly 50 mm just prior to summer, before rewetting in the fall. There is no obvious multiyear trend over the 21-year period. Rather, monthly soil moisture appears to vary year-to-year while retaining a distinct annual cycle.

11.4.6 Distribution of minimum and maximum temperature

They are important for understanding the thermal conditions of an area, which can influence evapotranspiration rates and thus the availability of water.

11.4.6.1 Minimum temperatures

The monthly minimum temperature data shows the lowest near-surface air temperatures in degrees Celsius from January 2002 to December 2022 (Fig. 11.7C).

Over this period, minimum temperatures followed a consistent seasonal cycle, reaching their lowest levels during winter months. The winter months of December, January, and February saw average minimums of 12°C−15°C. Minimum temperatures then steadily increased into the spring and summer months, typically peaking in July and August around 25°C−27°C. There was, however, considerable year-to-year variability in seasonal minimums overlaid on this pattern. Some years saw unusually cold winter minimums under 12°C, while other years had relatively warm winter lows of 15°C or greater. Overall there was no clear long-term trend toward warmer or colder minimum temperatures over the 21-year record. Rather, monthly minimum temperatures varied from year to year, while retaining a distinct seasonal cycle with winter lows and summer peaks.

11.4.6.2 Maximum temperatures

The monthly maximum temperature data shows the highest near-surface air temperatures in degrees Celsius over the same 2002−22 period. These maximum temperatures revealed a similar seasonal pattern as the minimums, with the hottest temperatures occurring during the summer and the coolest in the winter. On average, maximum temperatures peaked in the summer months of June, July, and August around 35°C−37°C. In contrast, winter maximum temperatures dropped to seasonal lows averaging around 25°C−28°C in December, January, and February (Fig. 11.7C). As with the minimums, however, substantial year-to-year variability is evident in the data. Some years saw significantly hotter summers or cooler winters. Overall, there is no clear long-term warming or cooling trend in maximum temperatures over the 21-year record when averaged on a monthly basis. Instead, monthly maximum temperatures appear to shift up and down year-to-year around a consistent seasonal cycle with summer peaks and winter troughs.

11.4.7 Distribution of Palmer Drought Severity Index over the study period

It is widely used for drought calculation and has the advantage of representing the actual drought characteristics while considering land and vegetation variability.

The PDSI data provides a measurement of relative drought and moisture conditions over the period from January 2002 to December 2022 (Fig. 11.7D). The index ranges from extreme drought indicated by large negative values to very wet moist periods shown by large positive values. Over the 21-year record, PDSI followed a fluctuating pattern, with some extended droughts punctuated by very moist wet spells. The most severe drought stretch occurred during the 2006−12 period, when PDSI remained almost continuously negative, dropping below −4 at times to indicate extreme drought. In contrast, brief wet periods emerged in some years like 2008–09 and 2019–20 when PDSI rose above +4. There was also large variability from month-to-month overlaid on multiyear wet/dry cycles. On average, late spring through early fall months appeared to be more prone to drought, while late fall into winter was climatologically wetter. However, seasonal impacts were modulated by variability on interannual scales. Overall, there was no clear long-term trend toward progressively drier or wetter conditions based on monthly PDSI over the period of record.

11.4.8 Correlation of drought indices

The provided correlation matrix displays relationships between nine drought indices (Fig. 11.8). The SPEI has strong positive correlations with several indices, indicating similar directional movements. There are also moderate and

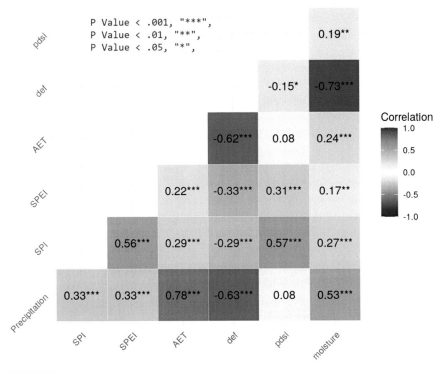

FIGURE 11.8

Correlation between drought indices.

weak correlations among other indices. Notably, as SPEI increases, the climate water deficit decreases, showing a negative correlation. The matrix's color coding visually represents these correlations, with asterisks indicating statistical significance. This matrix aids in understanding complex relationships between drought indices, informing drought risk assessments and management strategies. It emphasizes SPEI as a comprehensive drought measure but also indicates the need for a multiindex approach in drought analysis due to varying correlations with SPEI.

11.4.9 Impacts assessment of drought condition through different indices

Multiple satellite-based environmental indices are studied to understand agricultural drought patterns and farming potential in the region over 2002 to 2022 in the 5-year interval. The greenness, moisture stress, and temperature levels before and after monsoon are evaluated using indicators like NDVI, VHI, NDWI, LST, VCI, and TCI. Their trends showcase the climate-related challenges for regional agriculture as well as emerging resilience capacities. Steadily rising NDVI, NDWI, and VCI denote lowering crop moisture deficits over time. Cooling peak summertime temperatures are also positive for farming. Detailed analysis about the indicators is given below in Table 11.2.

11.4.9.1 Normalized Difference Vegetation Index

The NDVI values showed steady positive trends signaling gradual improvements in agricultural landscape health over the timeframes (Fig. 11.9). Premonsoon NDVI which reflects moisture adequacy for crops in the lead up to rains rose from 0.39 in 2003−07 to 0.57 in 2018−22. Postmonsoon NDVI, an indicator of rainwater utilization potential and agricultural productivity, increased from 0.51 to 0.62 over the same periods (Fig. 11.9). The rising cropland photosynthetic activity indicates progressive enhancements in vegetation condition and moisture availability despite climate risks. Possible factors enabling effective agricultural drought mitigation could be favorable shifts in precipitation patterns, expanded irrigation infrastructure, uptake of water conservation technologies, as well as cultivation of heat/drought resilient varieties.

11.4.9.2 Normalized Difference Water Index

The postmonsoon NDWI showed a transition from negative figures indicating agricultural droughts and moisture stress in early years to positive values in recent periods denoting adequate water content in croplands (Fig. 11.10). The NDWI rise from −0.42 in 2003−07 to 0.25 in 2018−22 after rains indicates wetter vegetation, improved rainwater retention in agricultural soils, and better hydrological

Table 11.2 Average of indices during pre- and postmonsoon over the time intervals.

Mean	2003–07 Pre	2003–07 Post	2008–12 Pre	2008–12 Post	2013–17 Pre	2013–17 Post	2018–22 Pre	2018–22 Post
NDVI	0.39	0.51	0.52	0.53	0.52	0.54	0.57	0.62
NDWI	−0.42	−0.49	−0.50	−0.45	−0.45	0.19	0.12	0.25
LST (°C)	31.13	29.78	30.72	30.08	31.10	30.29	30.98	30.26
VCI	0.46	0.55	0.68	0.73	0.51	0.53	0.55	0.56
TCI	0.51	0.42	0.48	0.41	0.52	0.46	0.54	0.57
VHI	0.48	0.49	0.58	0.57	0.52	0.49	0.54	0.56

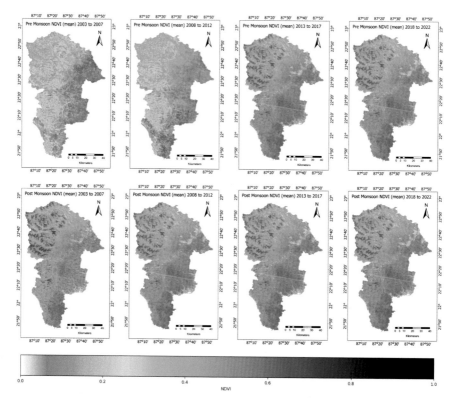

FIGURE 11.9

Normalized Difference Vegetation Index (NDVI) in pre- and postmonsoon.

accessibility for crops after the monsoon season (Fig. 11.10). This signals that the core monsoonal climate is now more effective in replenishing soil moisture and supporting agriculture in regions prone to moisture deficits. Contributing factors could include favorable spatial-temporal rainfall variations, adoption of soil/water conservation techniques, as well as expansion in the irrigated area enabled by groundwater access. Relief from higher evapotranspiration demands owing to tempering peak warming also plays a role.

11.4.9.3 Vegetation Condition Index

The VCI, indicating agricultural moisture stress and heat damage, showed consistent rising trajectories both before and after monsoons over the assessment time frames. Premonsoon VCI rose from 0.46 in 2003−07 to 0.55 in 2018−22. Postmonsoon VCI increased further from 0.55 to 0.56 (Fig. 11.11). The steady

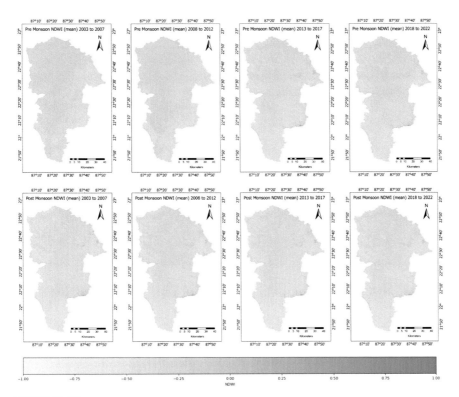

FIGURE 11.10

Normalized Difference Water Index (NDWI) in pre- and postmonsoon.

trends denote progressively lower moisture deficits and warming impacts on regional croplands in recent periods. Possible reasons include enhanced rainwater conservation in soils, expansion of irrigation infrastructure, shift toward heat resilient crop varieties, and so forth alongside the overarching rainfall improvements. Tempering of peak temperatures also plays a key role in alleviating farm-level climate risks. Marginally higher postmonsoon VCI signals effective utilization of monsoonal precipitation for agriculture. More importantly, narrowing gaps between pre- and postmonsoon VCI showcase strengthened capabilities of croplands to maintain robust health during drier periods.

11.4.9.4 Temperature Condition Index

\The postmonsoon TCI demonstrated a progressive increasing trend, rising from 0.42 to 0.57 over the assessment time periods (Fig. 11.12). This direction aligns with the cooling peak temperature patterns, indicating that agricultural landscape thermal

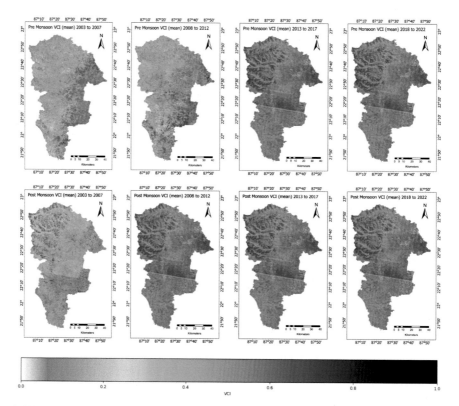

FIGURE 11.11

Vegetation Condition Index (VCI) in pre- and postmonsoon.

dynamics are turning more conducive, especially after rains. However, premonsoon TCI shows interyear fluctuations rather than any distinct trend. This points toward variability in heat extremes building up to the monsoon, a persisting area of concern for regional farming. Yet, the postmonsoon TCI gains showcase agriculture's enhanced capacities to offset periodic heat spikes in the lead up to rains by capitalizing on subsequent cooler temperatures after precipitation. Strengthening moisture retention abilities during rainy months also buffer premonsoon stresses.

11.4.9.5 Vegetation Health Index

The postmonsoon VHI fluctuated within a narrow band between 0.49 and 0.57 over the period (Fig. 11.13). This relative stability indicates average agricultural landscape health after monsoons, with typical vegetation conditions during the main cropping season neither improving drastically nor deteriorating markedly.

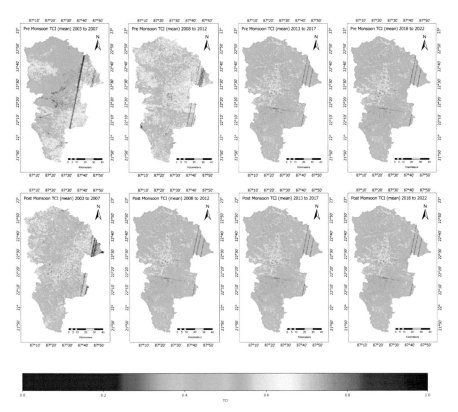

FIGURE 11.12

Temperature Condition Index (TCI) in pre- and postmonsoon.

However, the underlying contributors, viz., VCI and TCI showed independent positive movements when analyzed separately. The rising VCI denotes lowering crop moisture stress, while the TCI indicates relieving heat extremes, both conducive factors for rain-fed agriculture. Although balanced out in the aggregated VHI, these trends point toward reduced environmental constraints and expanded production potential. The steady VHI trajectory aligns with and substantiates the observed individual VCI and TCI shifts. In summation, typical postmonsoon VHI reassures stable cropland health outcomes, enabled by emerging warmth and soil moisture sufficiency that help counteract climate risks after rains.

11.4.9.6 Land surface temperature

The postmonsoon LST demonstrated a cooling trend over the time periods, dropping from 31.13°C in 2003−07 down to 30.26°C in 2018−22 (Fig. 11.14).

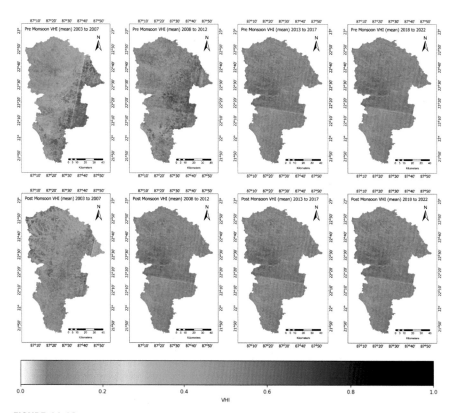

FIGURE 11.13

Vegetation Health Index (VHI) in pre- and postmonsoon.

This near 1°C fall in peak summertime warmth has favorable impacts for regional agriculture. Lower temperatures reduce evaporative moisture losses from soils and crops, preserving soil hydration in rain-fed croplands. Cooler peaks also minimize exposure to heat extremes during flowering and fruiting—critical crop growth stages vulnerable to warming. Together with the NDWI, reduced agricultural LST signifies stabilized hydrological cycling. Although premonsoon heat variability persists, the postmonsoon cooling enables crops to offset periodic stresses. It allows rainfall to translate into more effective moisture gains. Lower thermal extremes coupled with soil hydration maintenance ensure reduced agricultural drought risks. Consequently, heat stress-related farm income losses are averted, while also expanding regional climate resilience.

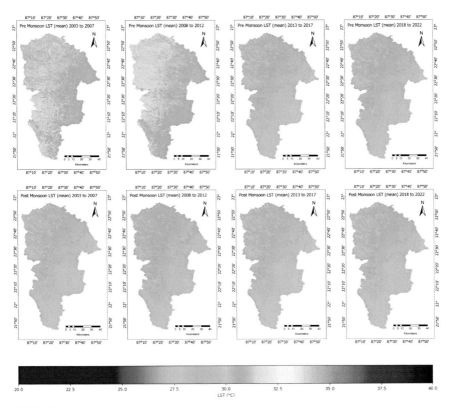

FIGURE 11.14

Land Surface Temperature (LST) in pre- and postmonsoon.

11.5 Conclusions

The analysis of Paschim Medinipur's agricultural drought patterns reveals several crucial conclusions regarding the region's climate risks and emergence of resilience. Precipitation is hugely variable, fluctuating wildly across years, seasons, and even months. Although the summer monsoon from June to September brings the bulk of annual rainfall, deviations from average norms are very common. This uncertainty around the quantity and timing of moisture availability leaves regional agriculture highly vulnerable to frequent drought occurrences, especially during critical crop establishment and harvesting periods. Multiple satellite-derived drought indicators confirm the prevalence of intense dry spells when below normal rains coincide with excessive heat. The years 2004, 2006, and 2009−12 experienced particularly devastating agricultural droughts as evidenced by extremely low values in SPI, SPEI, and PDSI metrics coupled with vegetation moisture stress. Crop yields and harvested area shrunk drastically during these acute events, which also triggered substantial distress migration. In contrast, brief intervening periods like 2008–09 and selective years in the early 2010s and 2020 offered relative moisture sufficiency, pointing to multiyear cycling between opposite extremes. Responses in key vegetation indices like NDVI and VCI also trace these precipitation fluctuations, with moisture adequacy supporting productive croplands, while deficits severely impair crop conditions and farming livelihoods. However, overlaid on this precipitation variability are early indications that regional agriculture is developing capacities to better cope with and offset climatic uncertainty. The gradual long-term rises observed in postmonsoon NDVI, NDWI, and VCI denote enhanced vegetation health, soil moisture levels, and lowering crop water deficits in recent years aided by expanding irrigation infrastructure. Sizeable areas have adopted microirrigation and other water saving techniques. The concurrent cooling of peak summertime temperatures further alleviates heat extremes during critical flowering and yield formation stages. Together, these synergistic developments are equipping farms to capitalize on periodic wet spells, while withstanding frequent dry periods. In summary, while climate risks continue to disrupt Paschim Medinipur's rain-dependent agriculture, strategic interventions supporting emerging moisture conservation capacities can protect and enhance rural livelihoods. The momentum toward resilience evident through remote sensing indicators needs focused investments into adaptive irrigation techniques, heat-tolerant varieties, and associated ecosystem-based farm solutions to help decisively overcome precipitation variability. This will enable regional food and economic security imperatives to be sustainably achieved despite inevitable climate uncertainty ahead.

Acknowledgment

The authors extend their gratitude to the USGS and University of California Merced for providing free data and to Google Earth Engine for their complimentary computational resources. Your support has been invaluable to our study.

Appendix

1. Results tables: https://docs.google.com/spreadsheets/d/
   1Z12l486Uir09cyGme40RmutAwJhskAcYwn0hic4OyXM/edit?usp = sharing
2. Google Earth Engine Scripts:
   a. Landsat 5: https://code.earthengine.google.com/
      8c01e98cc86dfd0a183c1e3cc293174f

b. Landsat 8: https://code.earthengine.google.com/
e58c5bc76fcf57ad3cb27ca9ee61a160
c. SPI: https://code.earthengine.google.com/
aa4eebe621c1173afb5c22425cb7c015
d. SPEI: https://code.earthengine.google.com/
7a125e4c51b66219527b01a026e7231f
e. TerraClimate Timeseries: https://code.earthengine.google.com/
50cb4631f068cd742f8eb105b14d84b5

References

Alahacoon, N., Edirisinghe, M., & Ranagalage, M. (2021). Satellite-based meteorological and agricultural drought monitoring for agricultural sustainability in Sri Lanka. *Sustainability*, *13*(6). Available from https://doi.org/10.3390/su13063427.

Amin, M., Khan, M. R., Hassan, S. S., Khan, A. A., Imran, M., Goheer, M. A., ... Perveen, A. (2020). Monitoring agricultural drought using geospatial techniques: A case study of thal region of Punjab, Pakistan. *Journal of Water and Climate Change*, *11*(1S), 203−216. Available from https://doi.org/10.2166/wcc.2020.232.

Anderson, M. C., Norman, J. M., Mecikalski, J. R., Otkin, J. A., & Kustas, W. P. (2007). A climatological study of evapotranspiration and moisture stress across the continental United States based on thermal remote sensing: 1. Model formulation. *Journal of Geophysical Research Atmospheres*, *112*(10). Available from https://doi.org/10.1029/2006JD007506, http://onlinelibrary.wiley.com/journal/10.1002/(ISSN)2169-8996.

Ara, S., Alif, M. A. U. J., & Islam, K. M. A. (2021). Impact of tourism on LULC and LST in a coastal island of Bangladesh: A geospatial approach on St. Martin's Island of Bay of Bengal. *Journal of the Indian Society of Remote Sensing*, *49*(10), 2329−2345. Available from https://doi.org/10.1007/s12524-021-01389-4, http://www.springerlink.com/content/0255-660X.

Becker-Reshef, I., Vermote, E., Lindeman, M., & Justice, C. (2010). A generalized regression-based model for forecasting winter wheat yields in Kansas and Ukraine using MODIS data. *Remote Sensing of Environment*, *114*(6), 1312−1323. Available from https://doi.org/10.1016/j.rse.2010.01.010.

Cai, S., Zuo, D., Wang, H., Han, Y., Xu, Z., Wang, G., & Yang, H. (2024). Improvement of drought assessment capability based on optimal weighting methods and a new threshold classification scheme. *Journal of Hydrology*, *631*, 130758. Available from https://doi.org/10.1016/j.jhydrol.2024.130758, https://www.sciencedirect.com/science/article/pii/S0022169424001525.

Cai, S., Zuo, D., Wang, H., Xu, Z., Wang, G. Q., & Yang, H. (2023). Assessment of agricultural drought based on multi-source remote sensing data in a major grain producing area of Northwest China. *Agricultural Water Management*, *278*. Available from https://doi.org/10.1016/j.agwat.2023.108142.

Chattopadhyay, N., Malathi, K., Tidke, N., Attri, S. D., & Ray, K. (2020). Monitoring agricultural drought using combined drought index in India. *Journal of Earth System Science*, *129*(1). Available from https://doi.org/10.1007/s12040-020-01417-w.

Dimyati, M., Rustanto, A., Shiddiq, I. P. A., Indratmoko, S., Siswanto., Dimyati, R. D., ... Auni, R. (2024). Spatiotemporal relation of satellite-based meteorological to agricultural drought in the downstream Citarum watershed, Indonesia. *Environmental and Sustainability Indicators*, 100339. Available from https://doi.org/10.1016/j.indic.2024.100339, https://www.sciencedirect.com/science/article/pii/S2665972724000072.

Dong, H., Li, J., Yuan, Y., You, L., & Chen, C. (2017). A component-based system for agricultural drought monitoring by remote sensing. *PLoS One*, *12*(12). Available from https://doi.org/10.1371/journal.pone.0188687.

Dutta, D., Kundu, A., Patel, N. R., Saha, S. K., & Siddiqui, A. R. (2015). Assessment of agricultural drought in Rajasthan (India) using remote sensing derived Vegetation Condition Index (VCI) and Standardized Precipitation Index (SPI). *The Egyptian Journal of Remote Sensing and Space Science*, *18*(1), 53−63. Available from https://doi.org/10.1016/j.ejrs.2015.03.006.

Eklund, L., Mohr, B., & Dinc, P. (2024). Cropland abandonment in the context of drought, economic restructuring, and migration in northeast Syria. *Environmental Research Letters*, *19*(1). Available from https://doi.org/10.1088/1748-9326/ad1723.

Geng, G., Zhang, B., Gu, Q., He, Z., & Zheng, R. (2024). Drought propagation characteristics across China: Time, probability, and threshold. *Journal of Hydrology*, *631*, 130805. Available from https://doi.org/10.1016/j.jhydrol.2024.130805, https://www.sciencedirect.com/science/article/pii/S0022169424001999.

Ghaleb, F., Mario, M., & Sandra, A. N. (2015). Regional landsat-based drought monitoring from 1982 to 2014. *Climate*, *3*(3), 563−577. Available from https://doi.org/10.3390/cli3030563.

González, J., & Valdés, J. B. (2006). New drought frequency index: Definition and comparative performance analysis. *Water Resources Research*, *42*(11). Available from https://doi.org/10.1029/2005WR004308.

Gorelick, N., Hancher, M., Dixon, M., Ilyushchenko, S., Thau, D., & Moore, R. (2017). Google Earth Engine: Planetary-scale geospatial analysis for everyone. *Remote Sensing of Environment*, *202*, 18−27. Available from https://doi.org/10.1016/j.rse.2017.06.031.

Govind, N. R., & Ramesh, H. (2020). Exploring the relationship between LST and land cover of Bengaluru by concentric ring approach. *Environmental Monitoring and Assessment*, *192*(10). Available from https://doi.org/10.1007/s10661-020-08601-x, https://link.springer.com/journal/10661.

Hasan, I. F., & Abdullah, R. (2022). Agricultural drought characteristics analysis using copula. *Water Resources Management*, *36*(15), 5915−5930. Available from https://doi.org/10.1007/s11269-022-03331-w, https://www.springer.com/journal/11269.

Hazaymeh, K., & Hassan, Q. (2016). Remote sensing of agricultural drought monitoring: A state of art review. *AIMS Environmental Science*, *3*, 604−630. Available from https://doi.org/10.3934/environsci.2016.4.604.

Jana, A., Sit, G., & Chanda, A. (2021). Ichthyofaunal diversity of river Kapaleswari at Paschim Medinipur district of West Bengal, India. *Flora and Fauna*, *27*(1). Available from https://doi.org/10.33451/florafauna.v27i1pp113-124.

Ji, L., & Peters, A. J. (2003). Assessing vegetation response to drought in the northern Great Plains using vegetation and drought indices. *Remote Sensing of Environment*, *87*(1), 85−98. Available from https://doi.org/10.1016/S0034-4257(03)00174-3, http://www.elsevier.com/inca/publications/store/5/0/5/7/3/3.

Kaur, L., Rishi, M. S., & Chaudhary, B. S. (2022). Assessment of meteorological and agricultural droughts using remote sensing and their impact on groundwater in an agriculturally productive part of Northwest India. *Agricultural Water Management*, *274*. Available from https://doi.org/10.1016/j.agwat.2022.107956.

Kloos, S., Yuan, Y., Castelli, M., & Menzel, A. (2021). Agricultural drought detection with modis based vegetation health indices in southeast Germany. *Remote Sensing*, *13* (19). Available from https://doi.org/10.3390/rs13193907.

Kogan, F. N. (1995). Application of vegetation index and brightness temperature for drought detection. *Advances in Space Research*, *15*(11), 91–100. Available from https://doi.org/10.1016/0273-1177(95)00079-t.

Kumar, V., & Chu, H.-J. (2024). Seasonal drought severity identification using a modified multivariate index: a case study of Indo-Gangetic Plains in India. *Journal of Hydrology*, *629*. Available from https://doi.org/10.1016/j.jhydrol.2024.130632.

Liu, Y. Y., Parinussa, R. M., Dorigo, W. A., De Jeu, R. A. M., Wagner, W., Van Dijk, A. I. J. M., ... Evans, J. P. (2011). Developing an improved soil moisture dataset by blending passive and active microwave satellite-based retrievals. *Hydrology and Earth System Sciences*, *15*(2), 425–436. Available from https://doi.org/10.5194/hess-15-425-2011.

Liu, X., Zhu, X., Pan, Y., Li, S., Liu, Y., & Ma, Y. (2016). Agricultural drought monitoring: Progress, challenges, and prospects. *Journal of Geographical Sciences*, *26*(6), 750–767. Available from https://doi.org/10.1007/s11442-016-1297-9.

Li, H., Yin, Y., Zhou, J., & Li, F. (2024). Improved agricultural drought monitoring with an Integrated Drought Condition Index in Xinjiang, China. *Water (Switzerland)*, *16*(2). Available from https://doi.org/10.3390/w16020325.

Mckee, T.B., Doesken, N.J., & Kleist, J. (1993). The relationship of drought frequency and duration to time scales. In *Eighth conference on applied climatology* (pp. 17–22).

Mishra, A. K., & Singh, V. P. (2010). A review of drought concepts. *Journal of Hydrology*, *391*(1–2), 202–216. Available from https://doi.org/10.1016/j.jhydrol.2010.07.012.

Mistry, P., & Suryanarayana, T. (2023). Assessment & monitoring of agricultural drought indices using remote sensing techniques and their inter-comparisonassessment & monitoring of agricultural drought indices using remote sensing techniques and their inter-comparison. *Ecological Perspective*, *3*, 1–8. Available from https://doi.org/10.53463/ecopers.20230171.

Moran, M. S., Clarke, T. R., Inoue, Y., & Vidal, A. (1994). Estimating crop water deficit using the relation between surface-air temperature and spectral vegetation index. *Remote Sensing of Environment*, *49*(3), 246–263. Available from https://doi.org/10.1016/0034-4257(94)90020-5.

Panda, K. P., Upadhyay, A., Jha, M. K., & Sharma, S. P. (2020). Mapping of laterite zones using 2D electrical resistivity tomography survey in parts of Paschim Medinipur, West Bengal, India: An approach for artificial groundwater recharge. *Journal of Earth System Science.*, *129*(1). Available from https://doi.org/10.1007/s12040-020-01390-4, https://rd.springer.com/journal/12040.

Pan, Y., Zhu, Y., Lü, H., Yagci, A. L., Fu, X., Liu, E., ... Liu, R. (2023). Accuracy of agricultural drought indices and analysis of agricultural drought characteristics in China between 2000 and 2019. *Agricultural Water Management*, *283*. Available from https://doi.org/10.1016/j.agwat.2023.108305.

Possega, M., Ojeda, M. G.-V., & Gámiz-Fortis, S. R. (2023). Multi-scale analysis of agricultural drought propagation on the iberian peninsula using non-parametric indices. *Water (Switzerland)*, *15*(11). Available from https://doi.org/10.3390/w15112032.

Bashit, N., Ristianti, N. S., & Ulfiana, D. (2022). Drought assessment using remote sensing and geographic information systems (GIS) techniques (Case Study: Klaten Regency, Indonesia). *International Journal of Geoinformatics*, *18*(5), 115−127. Available from https://doi.org/10.52939/ijg.v18i5.2393.

Roy, S., Hazra, S., & Chanda, A. (2023). Assessment of wet season agricultural droughts using monthly MODIS and SAR data in the red and lateritic zone of West Bengal, India. *Spatial Information Research*, *31*(2), 195−210. Available from https://doi.org/10.1007/s41324-022-00485-y.

Senamaw, A., Addisu, S., & Suryabhagavan, K. V. (2021). Mapping the spatial and temporal variation of agricultural and meteorological drought using geospatial techniques, Ethiopia. *Environmental Systems Research*, *10*(1). Available from https://doi.org/10.1186/s40068-020-00204-2.

Senapati, U., Raha, S., Das, T. K., & Gayen, S. K. (2021). *A composite assessment of agricultural drought susceptibility using analytic hierarchy process: Case study of western region of West Bengal. Agriculture, food and nutrition security: A study of availability and sustainability in India* (pp. 15−40). Springer International Publishing, https://link.springer.com/book/10.1007/978-3-030-69333-6. doi: 10.1007/978-3-030-69333-6_2.

Sen, U. K., & Bhakat, R. (2021). Quantitative evaluation of biological spectrum and phenological pattern of vegetation of a sacred grove of West Midnapore District, Eastern India. *Asian Journal of Forestry.*, *5*, 83−100. Available from https://doi.org/10.13057/asianjfor/r050206.

Seshasai, M., Murthy, C., Chandrasekar, K., Jeyaseelan, A., Diwakar, P., & Dadhwal, V. (2016). Agricultural drought: Assessment & monitoirng. *Mausam*, *67*. Available from https://doi.org/10.54302/mausam.v67i1.1155.

Shahfahad, M., Rihan, M. W., Naikoo, M. A., Ali, T. M., & Usmani, A. R. (2021). Urban heat island dynamics in response to land-use/land-cover change in the Coastal City of Mumbai. *Journal of the Indian Society of Remote Sensing*, *49*(9), 2227−2247. Available from https://doi.org/10.1007/s12524-021-01394-7, http://www.springerlink.com/content/0255-660X.

Sims, A. P., Niyogi, D. D. S., & Raman, S. (2002). Adopting drought indices for estimating soil moisture: A North Carolina case study. *Geophysical Research Letters*, *29*(8). Available from https://doi.org/10.1029/2001GL013343, http://onlinelibrary.wiley.com/journal/10.1002/(ISSN)1944-8007/issues?year = 2012.

Sreekesh, S., Kaur, N., & Naik, S.S. R. (2019). Agricultural drought and soil moisture analysis using satellite image based indices. In *International archives of the photogrammetry, remote sensing and spatial information sciences − ISPRS archives* (pp. 507−514). International Society for Photogrammetry and Remote Sensing India. http://www.isprs.org/proceedings/XXXVIII/4-W15/ 16821750 3 42 10.5194/isprs-archives-XLII-3-W6-507-2019.

Sruthi, S., & Mohammed Aslam, M.A. (2015). Agricultural drought analysis using the NDVI and land surface temperature data; a case study of Raichur District. In *International conference on water resources, coastal and ocean engineering (ICWRCOE'15)* (vol. 4, pp. 1258−1264). 2214-241X. https://www.sciencedirect.com/science/article/pii/S2214241X15001650. doi: 10.1016/j.aqpro.2015.02.164.

Tanarhte, M., De Vries, A. J., Zittis, G., & Chfadi, T. (2024). Severe droughts in North Africa: A review of drivers, impacts and management. *Earth-Science Reviews*, *250*, 104701. Available from https://doi.org/10.1016/j.earscirev.2024.104701, https://www.sciencedirect.com/science/article/pii/S001282522400028X.

Thavorntam, W., & Tantemsapya, N. (2013). Vegetation greenness modeling in response to climate change for Northeast Thailand. *Journal of Geographical Sciences*, *23*(6), 1052−1068. Available from https://doi.org/10.1007/s11442-013-1062-2.

Tran, T. V., Bruce, D., Huang, C.-Y., Tran, D. X., Myint, S. W., & Nguyen, D. B. (2023). Decadal assessment of agricultural drought in the context of land use land cover change using MODIS multivariate spectral index time-series data. *GIScience and Remote Sensing*, *60*(1). Available from https://doi.org/10.1080/15481603.2022.2163070.

Vicente-Serrano, S. M., Beguería, S., & López-Moreno, J. I. (2010). A multiscalar drought index sensitive to global warming: The standardized precipitation evapotranspiration index. *Journal of Climate*, *23*(7), 1696−1718. Available from https://doi.org/10.1175/2009JCLI2909.1Spain, http://journals.ametsoc.org/doi/pdf/10.1175/2009JCLI2909.1.

Wang, S., Chu, H., Gong, C., Wang, P., Wu, F., & Zhao, C. (2022). The effects of COVID-19 lockdown on air pollutant concentrations across China: A Google Earth Engine-based analysis. *International Journal of Environmental Research and Public Health*, *19*(24). Available from https://doi.org/10.3390/ijerph192417056.

Xiao, X., Ming, W., Luo, X., Yang, L., Li, M., Yang, P., ... Li, Y. (2024). Leveraging multisource data for accurate agricultural drought monitoring: A hybrid deep learning model. *Agricultural Water Management*, *293*. Available from https://doi.org/10.1016/j.agwat.2024.108692, https://doi.org/10.1016/j.agwat.2024.108692.

Zhang, Z., Guo, H., Feng, K., Wang, F., Zhang, W., & Liu, J. (2024). Analysis of agricultural drought evolution characteristics and driving factors in Inner Mongolia inland river basin based on three-dimensional recognition. *Water*, *16*(3). Available from https://doi.org/10.3390/w16030440, 2073-4441.

Zhang, L., Huang, W., & Li, S. (2024). Multivariate time series convolutional neural networks for long-term agricultural drought prediction under global warming. *Agricultural Water Management*, *292*. Available from https://doi.org/10.1016/j.agwat.2024.108683, https://doi.org/10.1016/j.agwat.2024.108683.

Zhang, Y., Wu, Z., Singh, V. P., Jin, J., Zhou, Y., Xu, S., & Li, L. (2022). Agricultural drought assessment in a typical plain region based on coupled hydrology−crop growth model and remote sensing data. *Remote Sensing*, *14*(23). Available from https://doi.org/10.3390/rs14235994.

Geospatial and statistical techniques of regionalization for river basin management

Muhammad Azam[1], Daye Kim[2], Maeng Seung Jin[2], Seungwook Lee[3] and Jung Yongbae[4]

[1]*Department of Structures and Environmental Engineering, Faculty of Agricultural Engineering & Technology, PMAS Arid Agriculture University Rawalpindi, Rawalpindi, Pakistan*
[2]*Department of Agricultural and Rural Engineering, Chungbuk National University, Cheongju-si, Chungcheongbuk-do, South Korea*
[3]*Chungbuk Research Institute, Choengju-si, Chungcheongbuk-do, South Korea*
[4]*K-water Seomjingang Dam Office, Deajeon, Republic of Korea*

12.1 Introduction

Definitions of drought can be divided into operational and conceptual categories. Conceptual definitions are those that are written in general words to define the parameters of the idea of drought. Drought has been defined in a previous study (Wilhite, 1987) as "a prolonged scarcity of rainfall, particularly throughout a growing season." Drought is a persistent state of water deficiency, which has been a complex task to address in the field of water management resources. It is one of the most severe and poorly understood natural disasters. Drought is a natural and multifaceted disaster with three distinct types that are commonly used, namely, agricultural, meteorological, and hydrological droughts (Kao & Govindaraju, 2010). Meteorological drought is closely attributed to weather conditions and is typically the result of insufficient precipitation over an area. Agricultural drought occurs when there is insufficient water available to the plant roots for its growth. A long-term decrease in precipitation causes a scarcity of surface and ground water availability, which is referred as hydrological drought (Van Loon, 2015).

Various studies indicate that in addition to precipitation, several factors such as temperature, elevation, evaporation, and evapotranspiration have an essential role to play in the occurrence of droughts (Mishra & Singh, 2010). The effect of rising temperatures prompted the highest evapotranspiration in the Mediterranean, resulting in drought in the region (Wilhite, 1987). As a result, the combined behavior of factors such as climatic, physiographic, and hydrological features

Applications of Geospatial Technology and Modeling for River Basin Management.
DOI: https://doi.org/10.1016/B978-0-443-23890-1.00012-8

may be characterized, and drought can be correlated with variables that are not frequently independent. Multivariate strategies in regional modeling analysis are the accumulation of procedures connected with hydrometeorological, climatic, and physiographic variables that should be strongly correlated. The primary objective of multivariate analysis is to evaluate and investigate the similarities and differences between individual datasets using dimension reduced subspace. This task can be accomplished using traditional techniques, which include factor analysis and principal component analysis. Both of the techniques form a combination of linear dimensionality of the features (Kilmer, 2010). Since the assessment of multivariate sets of data and the significant number of drought variables make these datasets challenging to interpret, appropriate modeling of the parameters must be undertaken to limit the number of variables to minimal data. As a result, the initial data screening will provide a framework for the categorization of multidimensional variables in order to undertake regional frequency analysis (Um, Yun, Jeong, & Heo, 2011).

Pakistan faces significant water crisis and management complications. The country ranks fifth in the globe in terms of population. Furthermore, climate change is causing a rise in the number of flood and drought events, thus causing major disruptions in water planning. Pakistan includes several catastrophic droughts with long durations and devastating consequences. Pakistan has a diverse topography from the lowest to highest altitude. The lowest altitude in the south and the highest altitude in the north receive the annual maximum rainfall of approximately 2000 and 250 mm, respectively. Rainfall tends to increase with the increase in altitude. Therefore, a greater diversity of the rainfall makes it difficult to identify the homogeneous regions (Krakauer, Lakhankar, & Dars, 2019). The coastal regions are located in the south and the largest glaciers in the north of the county. As a result, the greatest diversity exists in terms of climatic variation, which leads to extreme weather events. The majority of the land is in the arid zone, with half of it experiencing extreme aridity (Adnan, Ullah, Gao, Khosa, & Wang, 2017). The appropriate delineation of homogeneous regions facilitates parametric estimation of regional drought frequency analysis. This has greater applicability in regional drought risk management and planning.

Preprocessing of multivariate datasets can ideally be performed using dimensional reduction approaches (clustering algorithms). There are many clustering algorithms based on fuzzy logic and principal components analysis for the delineation of regions. For example, fuzzy C-means were used along with L-moments approach in the Canadian provinces to form homogeneous clusters for hydrological sites (Sadri & Burn, 2011). In addition, entropy-based cluster analysis and modified Andrews curve have been applied for the cluster analysis and L-moment based technique for the construction of homogenous regions and regional frequency analysis of drought (Kaluba, Verbist, Cornelis, & Van Ranst, 2017; Modarres, 2010; Zulfatman & Rahmat, 2009). Furthermore, hierarchical clustering and other partitioning techniques used to obtain the data lead to division of variables into the homogeneous clusters by distance measurements (Euclidean distance). Principal component analysis (PCA) derives the small number of components that can be responsible for the variability that exists in a large number of datasets. This procedure, commonly known as data reduction, is usually used by researchers where they do not deal with the large datasets, whereas FA accounts for the highly correlation that exists among the variables. To determine spatiotemporal patterns of the climatic variables, PCA has been used widely by many researchers for this purpose (Pett, Lackey, & Sullivan, 2003). The PCA technique was used in water quality assessment for drinking purposes in northern Pakistan, especially in Gilgit Baltistan, to evaluate the major components contributing to water quality deterioration (Fatima et al., 2022).

Drought, on the other hand, is a multidimensional association of numerous associated random variables such as drought length and intensity; hence, univariate drought analysis cannot describe the greatest variability of the drought situation (Mirabbasi, Fakheri-Fard, & Dinpashoh, 2012). This is because of the fact that findings of different studies suggest that the drought phenomena were influenced mainly due to the topographic and climatic variables and the precipitation pattern was influenced highly by monsoon due to high intensity of rainfall in a very short period (Chang & Kwon, 2007). Therefore, this chapter mainly focused on the multivariate nature of drought. The various studies have been conducted using multidimensional analysis of drought. Many are focused on stochastic methods to simulate drought time-space variability (Rossi, Benedini, Tsakiris, & Giakoumakis, 1992). While the others are also focused on classifying the subregional patterns of the winter precipitation using PCA (Santos, Fragoso, & Santos, 2017; Sarfaraz, 2014). Furthermore, we identified five homogenous climatic zones in Pakistan using a single variable of drought episodes taken from the Reconnaissance Drought Index (RDI) and Standardized Precipitation Index (SPI) datasets (Ullah, Akbar, & Khan, 2020). Drought regionalization was performed in Pakistan over the period of 1902 to 2015 using Standardized Precipitation and Evapotranspiration Index (SPEI) using cluster analysis and tree edge removal techniques (Jamro, Dars, Ansari, & Krakauer, 2019).

The SPEI drought has been selected in this study because it is capable of distinguishing the influence of rising temperature on drought severity (Ahmed, Shahid, Harun, & Wang, 2016). Furthermore, it is suggested that SPEI is more beneficial in analyzing the relationship between climate change and drought variation than SPI. Based on 6-month SPEI, PCA has been used with VARIMAX rotation to identify spatially adjacent and homogenous areas (Wang, Zhu, Xu, & Liu, 2015). It was concluded that the reduction in precipitation and a rise in temperature caused increased atmospheric water demands, which drove the recent droughts. The majority of the research has focused on at-site (local) multivariate analysis of topological, meteorological, and drought variables. The main objective of the study can be expressed as (1) to identify the appropriate number of clusters using multivariate clustering algorithm with the cluster validation indices and (2) to illustrate the spatial and temporal variation of drought and identify bivariate homogeneous regions using bivariate discordancy and heterogeneity measures.

12.2 Limitations of the study

The lack of long-term records limits drought frequency analysis, which influences the robustness of statistical forecasts of drought quantiles. To mitigate this issue of data shortage, the regionalization is an approach that is frequently used to assist the gathering of identical records in one region. The L-moments approach has several theoretical advantages in modeling a wide variety of distributions. Hence, the L-moment approach is most recommended in regional analysis because it can consider correlations between samples, despite the presence of imbalanced datasets. The performance evaluation regionalization using the L-moment technique is most commonly checked by heterogeneity measures in the univariate framework. However, the issue of the univariate technique is that it represents each variable as its own homogeneous region. As a result, the univariate heterogeneity tests provide only a restricted assessment of drought at ungauged locations and are incapable of representing the bivariate occurrences of drought duration and severity. Therefore, L-moment-based multivariate approaches to assess the homogeneity of a region with several features provide an alternative solution (Serfling & Xiao, 2007). The multivariate test is more useful than the univariate homogeneity test because it can account for first-order error and the connection among the variables (Wilhite, 1987). Therefore, this study adopted the multivariate regionalization approach.

12.3 Materials and methods

12.3.1 Standardized precipitation evapotranspiration index

In this study, the SPEI has been used to evaluate the severity and duration of the drought on a time scale of 12 months. The monthly precipitation data from 41 meteorological stations were obtained from Pakistan Metrological Department (PMD) from 1990 to 2020 (30 Years). In addition to this, the monthly precipitation, monthly maximum temperature, and monthly lowest temperature were collected and preprocessed. The SPEI was calculated according to the methodology proposed in a previous study (Vicente-Serrano, Beguería, & López-Moreno, 2010). Its dry and wet conditions are assessed by subtracting precipitation (P) from potential evapotranspiration (PET). The theory with full description and calculations are available in the previous literature (Beguería, Vicente-Serrano, Reig, & Latorre, 2014). The following equation was used to determine the climate-- water balance:

$$D_i = P_i - PET_i \qquad (12.1)$$

monthly precipitation is represented by P_i, the values aggregated on different timescales in that month (mm) is indicated by D_i, and the potential amount of transpiration in that month (mm) is represented by PET. In this study, the

Hargreaves approach was used to estimate the PET. Precipitation, minimum and maximum temperatures, and other meteorological data are all taken into account in this approach. The SPEI drought severity classes are shown, which have been grouped from the extreme wet to extreme drought (Alam et al., 2017; Dikshit, Pradhan, & Huete, 2021).

Drought characteristics were derived using the idea of the run theory analysis (Yevjevich, 1969). The run theory model is used to calculate droughts parameters by identifying dry length and severity based on predefined threshold limit (Mishra & Singh, 2011; Yevjevich & Jeng, 1969). A value of -1 before and succeeding high values, which indicates the time of drought (precipitation deficiency), was used to determine the duration of any drought. Drought severity is calculated following the equation $S = \sum_{i=1}^{D} |-I_i|$; here, i indicates the month of drought commencing (Yevjevich & Jeng, 1969). The cumulative SPEI during the drought period is used to determine the amount of drought severity.

12.3.2 Principle component analysis

The purpose of PCA is to reduce the data to a smaller number of dimensions without disturbing the balance of the data. The PCA method plays an important role in expressing the key information of the variables (Hotelling, 1933).

The variables used as input for the PCA analysis in the given table shall be highly correlated. It is recommended to assess and measure all the variables in the same unit because the principal components depend on the metrological variables. Those variables are passed through the standardization procedure to normalize all of the variables. A statistical correlation method is used as the foundation for the performance of further analysis.

PCA is performed in three phases: (1) variable standardization and correlation of matrix; (2) loading matrix estimation using PCA; and (3) estimation of eigenvalues. The following procedure was followed to standardize all of the variables.

$$X = \frac{xijxi}{Sj} \tag{12.2}$$

In this case here, i and j represent the observation numbers (up to N) of the variables (1 to V). X represents the matrix of the standard parameter, whereas xij signifies the ith observation of the jth variable; and Sj denotes the standardization of jth variable.

12.3.3 Hierarchical clustering on principal components

PCA, tree-based clustering (HC), and segmented grouping are combined in HCPC (particularly K-means) are used (Argüelles, Benavides, & Fernández, 2014). PCA is performed on the different attributes. Hierarchical clustering was then applied to PCA using Ward criteria to classify individuals into homogeneous groups. The HCPC method is used to pair the variables on the factorial map that are closest together until the desired level of clustering is reached. In this study, the HCPC clustering technique is used for the analysis, which is contrary to factor analysis. HCPC implements an objective clustering to PCA data, leading to improved clustering results. Secondly, the inclusion of the mixed technique (Ward's classification technique blended with the K-means algorithm) strengthens the stability of the final clusters. As a consequence, the HCPC approach assists in reducing subjective adjustment in cluster analysis.

12.3.4 Cluster validation

The initially the formed clusters were validated using four different cluster validation indices. The R package was used for cluster validation. The validation indices include Silhouette width (Rousseeuw, 1987), Dunne index (Dunn, 1974), Calinski and Harabasz index (Caliñski & Harabasz, 1974), and Davies–Bouldine Index.

12.3.5 Bivariate L-moments

After the cluster analysis using the clustering algorithm, the cluster was tested using the univariate L-moment based discordancy and homogeneity tests. The mathematical calculations of univariate L-moment methods are available in the previous literature (Hosking & Wallis, 1993; Hosking, 1990). The bivariate L-moment algorithm is similar in structure to that of the univariate L-moment (Argüelles et al., 2014). However, it is able to capture the multiple variables at the same time (Serfling & Xiao, 2007). For $j = 1, 2$, assume $X(j)$ to be a random variable with the distribution $F(j)$. Multivariate L-moments, like covariance representations of L-moments of order $K \geq 1$, are matrices $\wedge K$ with L-comoment components defined as:

$$\lambda_{k[ij]} = \mathrm{Cov}(X^{(i)}, P_{K-1}(F_j(X^{(j)}))), i, j = 1, 2 \text{ and } k = 2, 3, \dots \tag{12.3}$$

Here, P_K is known as shifted Legendre polynomial. It should be observed that the components $[ij]$ and $[ji]$ are not always equal. In particular, the following are the first L-comoment:

$$\lambda_{2[12]} = 2\mathrm{Cov}\left(X^{(1)}, \left(F_2(X^{(2)})\right)\right) \tag{12.4}$$

$$\lambda_{3[12]} = 6\mathrm{Cov}\left(X^{(1)}, \left(F_2(X^{(2)}) - \frac{1}{2}\right)^2\right)(7)\lambda_{4[12]} = 6\mathrm{Cov}\left(X^{(1)}, \left(F_2(X^{(2)}) - \frac{1}{2}\right)^3\right)$$

$$- 3\left(F_2(X^{(2)}) - \frac{1}{2}\right) + 1 \tag{12.5}$$

These are the L-covariance, L-coskewness, and L-cokurtosis, within this sequence. The following are the L-comoment coefficients:

$$\tau_{k[12]} = \frac{\lambda_{k[12]}}{\lambda_2^1}, k \geq 3; \tau_{2[12]} = \frac{\lambda_{2[12]}}{\lambda_1^1} \tag{12.6}$$

Here, $\lambda_{\kappa}^{(j)} = \lambda_{k[ij]}$ which is the conventional *kth* L-moment of variables $X(j)$, where $j = 1, 2$, defined by (Serfling & Xiao, 2007). So, the L-moment coefficients matrix is expressed as;

$$\underline{v}^* = \left(\tau_{[ij]}\right) i, j = 1, 2, = \begin{pmatrix} \tau_{k[11]} & \tau_{k[12]} \\ \tau_{k[21]} & \tau_{k[22]} \end{pmatrix} \tag{12.7}$$

Essentially, L-comoments are structurally analogous to univariate L-moments and can preserve their appealing qualities (Chebana & Ouarda, 2007).

12.3.5.1 Bivariate discordancy test

Chebana and Ouarda (2007) modified the discordancy test following the multivariate analysis (Serfling & Xiao, 2007). Discordancy of site i among the number of sites N can used to filter the data as a preliminary step. Discordancy of the region was tested using two variables such as duration and severity. For each site of the region, we propose $U_i^t = [\underline{v}_2^{*(i)} \underline{v}_3^{*(i)} \underline{v}_4^{*(i)}]$, a matrix of site i which consists of three L-moment matrices $\Lambda_2^{(i)} \Lambda_3^{(i)} \Lambda_4^{(i)}$ defined as in the following Eq. (12.7)

$$\overline{U} = N^{-1} \sum_{i=1}^{N} Ui, i = 1, \dots, N. \tag{12.8}$$

The discordancy of a site shall be measured through the equation below

$$D_i = \frac{1}{3} \left(U_i - \overline{U}\right)^T S^{-1} \left(U_i - \overline{U}\right)^T \tag{12.9}$$

As a result, if the value D_i of the site is bigger than the critical value of the region, it will consider as discordant site. The critical discordancy value if the constant c for a large is usually taken as $x1 - 0.5^{\frac{3}{3}} = 2.6049$. Here, in this case, $x_{1-\alpha^d}$ is the quantile of an order αchi-square distribution of the d degree of freedom. Criteria proposed in Chebana and Ouarda (2007) as $(D_i > 2.6049)$ is regarded as a criterion for evaluating the regional discordancy.

12.3.5.2 Bivariate test for homogeneity

Drought duration and severity, as stated in the previous section, are used to assess the homogeneity of the region (Chebana & Ouarda, 2007). The bivariate homogeneity test is a multivariate analog of the statistic presented in literature (Hosking & Wallis, 1993). Let the statistic be defined as follows:

$$V_{|.|} = \left(\left(\sum_{i=1}^{N} n_i \right)^{-1} \sum_{i=1}^{N} n_i \left| \Lambda_2^{(i)} - \overline{\Lambda_2} \right|^2 \right)^{1/2} \tag{12.10}$$

The L-covariance matrix of the . norm operator $\Lambda_2^{(i)}$ for site i with record length $n_i, i = 1, ., N$

$$\overline{\Lambda_2} = \left(\sum_{i=1}^{N} n_i \right)^{-1} \sum_{i=1}^{N} n_i \Lambda_2^{(i)} \tag{12.11}$$

To evaluate what is envisaged, we should compare the actual value to its simulated bivariate homogeneous locations. This technique is based on 500 simulations for each severity and duration variable fitting the kappa distribution. To generate a joint distribution, the bivariate Gumbel copula distribution is employed (Chebana & Ouarda, 2007). Then, we compare the mean and standard deviation of the 500 simulated bivariate data to the observed value of V.

$$H_{|.|} = \frac{V_{|.|} - \mu_{V\text{sim}}}{\sigma_{V\text{sim}}} \tag{12.12}$$

where $\mu_{V\text{sim}}$ and $\sigma_{V\text{sim}}$ are the mean and standard deviation of the N_{sim} values of V of simulated areas, respectively. Bivariate heterogeneity measures, like univariate heterogeneity measures, include criteria for determining whether a region is homogenous or not. The criteria used to evaluate the homogeneity of the region is described in previous literature. Bivariate discordancy and homogeneity assessments were used for quantile estimation by expanding the technique to the bivariate framework. The detailed methodology is provided in the flow chart, as shown in Fig. 12.1.

12.3.6 Study area

Pakistan has a total area of 796,096 km^2 and is located in the range of 23°39 N−37°01 N and 60°49E−77°40E on the equator, as shown in Fig. 12.1. It has a tropical to subtropical climate due to versatile geographical location. During the summer and winter seasons, rainfall mainly occurs due to two primary

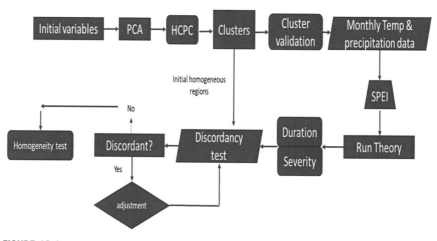

FIGURE 12.1

Study flow chart.

meteorological phenomena: (1) the western disturbances and (2) westerly and South Asian summer monsoon. During the monsoon and winter seasons, these systems of rainfall pouring about 45% and 31% of the annual rainfall intensity, respectively. During the summer and winter, the mean temperature varies from 20°C to 12°C and 35°C to 19 °C, respectively. The average and mean annual precipitation ranges from 30 to 400 mm from low (south) to high (north) latitudes, with a maximum of 900−1800 mm in the semi mountainous areas in the north (Hanif, Khan, & Adnan, 2013). The monsoon season alleviates water stress during the Kharif season and meets crop water needs throughout the Rabi season. The desert area encompasses 75% of land in Pakistan, with the greatest concentration in the southern half of the country, making it more prone to drought. Forty-one rainfall stations of PMD served as the basis for this study area, as shown in Fig. 12.2.

12.4 Results

12.4.1 Selected Parameters

The parameters and variables selected for the cluster analysis include different topographical and hydrometrological variables that have a direct effect on drought

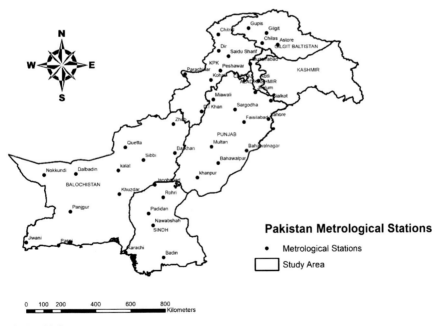

FIGURE 12.2

Study area.

Table 12.1 Summary statistics for all the variables used for the classification of clusters.

Variables	Mean	SD	Max	Min	Med	CV
Lat	30.88	3.36	36.16	24.63	31.35	10.90
Long	70.19	3.54	74.90	61.80	71.43	5.04
EL	706.32	702.92	2168.00	9.00	327.00	99.52
MAP	437.18	406.08	1734.64	34.91	261.88	92.89
MDMXT	27.63	9.66	37.45	-3.45	30.26	34.96
MDMNT	14.44	7.79	23.92	-6.82	16.60	53.97
ET_o	163.55	409.39	19.28	140.10	99.11	60.60
MRH	478.04	146.68	860.14	222.50	447.67	30.68

Table 12.2 Drought characteristics with SPEI-12-month time scale.

Time scale	Total drought events	Drought duration			Drought severity		
		Mean	Max	Min	Mean	Max	Min
SPEI-12	2717	5.611	33	1	8.740	54.935	1.003

across Pakistan. The statistical description of selected parameters is presented in Table 12.1. Murree and Nawabshah recorded the highest mean annual precipitation and highest mean daily maximum temperature, respectively.

The major attributes used for the analysis are as follows: (1) mean daily maximum temperature; (2) latitude (LAT); (3) longitude (Long); (4) elevation level (EL); (5) mean daily minimum temperature; (6) evapotranspiration (ET_o), and (7) mean relative humidity (MRH). The literature review showed that multivariate statistical techniques can be used for at-site regionalization of variables (Burn, 1988).

12.4.2 Temporal variation of drought

The temporal analysis of drought is accomplished by drawing the SPEI plots for 41 stations at the timescales of 12 months. The drought characteristics based on run theory analysis have been presented in Table 12.2. The total recorded number of drought events from 1981 to 2018 was approximately 2717. The basic statistics of drought duration and severity are presented in Table 12.2. The temporal fluctuation of drought has been analyzed across at the all stations from 1981 to 2018 on a time scale of 12 months. However, data for only three stations, Sibbi, Dalbadin, and Bahawalnagar, has been shown in Figs. 12.3−12.5, respectively.

A longer range, such as 12 months, revealed more frequent and longer-lasting droughts. The results showed that severest droughts have been observed from the years 2000 to 2004.

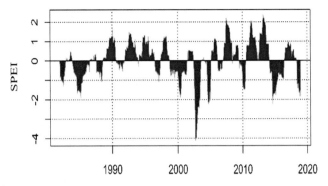

FIGURE 12.3

SPEI-12 drought analysis at Sibbi stations.

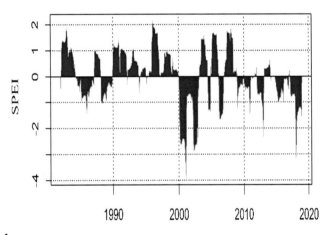

FIGURE 12.4

SPEI-12 drought analysis at Dalbadin stations.

12.4.3 Principle component analysis

Each of the eight variables presented in Table 12.1 was normalized using Eq. (12.2) to eliminate the influence of units. Variables were standardized in such a way that they have a standard deviation of 1 and a mean of 0. The intercorrelation matrix of the standardized variables was computed by employing Eq. (12.5), as shown in Fig. 12.6. The result of eight attributes for the intercorrelation matrix showed a correlation coefficient of 0.50 between MAP and LONG and 0.55 between LAT and EL, 0.54 was observed between MAP and LAT, as shown in Fig. 12.6 while MDMXT and MDMNT showed the strongest correlation with ET_o of about 0.8 and 0.79. On the other hand, EL with MAP, LONG, and RH shows a moderate correlation of about 0.16 to 0.3.

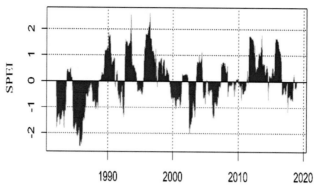

FIGURE 12.5

SPEI-12 drought analysis at Bahawalnagar stations.

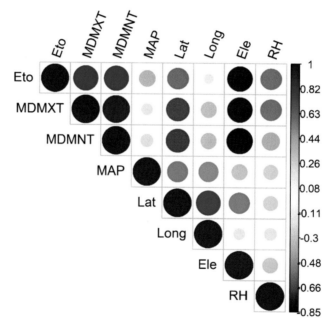

FIGURE 12.6

Correlation plot for all the selected variables.

The characteristics such as MDMXT, MDMNT, and RH have a relative correlation link with other parameters like EL, ET_o, MAP, LAT, and LONG.

PCA applied on a correlation matrix showed that 95.2% of the variance has been explained in first four PCs. For instance, 57.9% for the first PC, 18.6% for the second PC, 10.6% for the third PC, and 8.1% for the fourth PC, as shown in

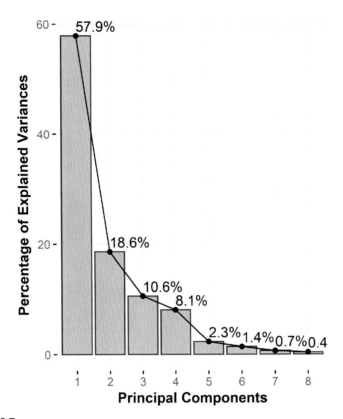

FIGURE 12.7

Percentage of variances explained by each principal component.

Fig. 12.7. We have chosen the first four PCs for the cluster analysis because their eigenvalues were close to or greater than one (first constituent 4.91, second 1.32, third 0.80, and fourth 0.57). Based on the correlation matrix result, Eq. (12.6) has been used to assemble the fundamental element matrix.

The contribution of characteristics in the form of % is presented in Figs. 12.8–12.11. The most important attributes are those which influence considerably the first and second major components. However, features that belong solely to the final dimensions or do not influence to any of the basic components are regarded as less significant in interpreting the variation in the dataset. The expected average contribution can be seen by the red dotted lines on the plots. If the contribution of the features is uniform, the predicted average participation value is (1/ number of attributes) $\times$ 100. In our study, the estimated average contribution is 12.5%. A value higher than this red dashed line is considered as higher contribution. It can be observed that MDMXT, LAT, EL, MDMNT, and ET_o contributed the most to the first principal component, Long and MAP contributed the most to

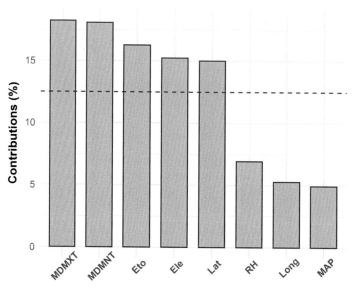

FIGURE 12.8

Percentage of contribution of all the variables for the first principal component.

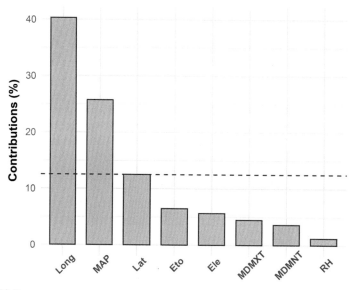

FIGURE 12.9

Percentage of contribution of all the variables for the second principal component.

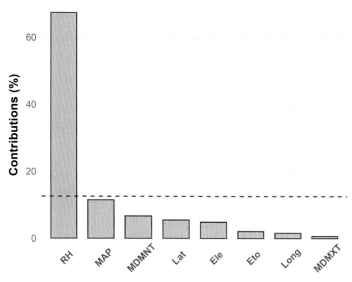

FIGURE 12.10

Percentage of contribution of all the variables for the third principal component.

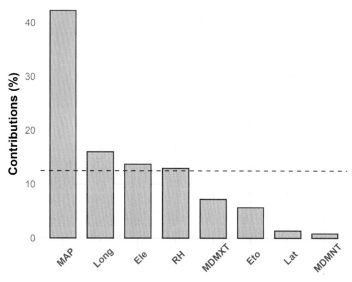

FIGURE 12.11

Percentage of contribution of all the variables for the fourth principal component.

the second principal component, RH contributed the most to the third principal component, and MAP and LONG contributed the most to the fourth principal component. Due to the limited contribution to the first three main components, the RH is considered as the least relevant feature in capturing the dispersion of the dataset.

12.4.4 Cluster analysis

Clusters were formulated by applying the HCPC approach. The initially formed clusters were mapped on Pakistan's geographical space. Cluster dendrogram is presented for all the stations showing the cluster vise location of each station, as shown in Fig. 12.12. For the construction of homogenous regions, we adopted a multistage technique. In the first level, agglomerative hierarchical clustering is used to locate subjective homogeneous groups of meteorological stations using a mix of Ward's linkage and the Euclidian distance approach. The site characteristics of the selected meteorological stations across Pakistan were employed, and the outcome has been presented in the form of dendrogram. Clusters revealed five subjective homogeneous groupings. Some meteorological stations were transferred from one category to another based on geographical attachment, climatic conditions, and homogeneity metrics in order to produce adjusted areas. Dendrogram is constructed on the basis of Ward's method using the variance of clusters. However, because the visual plotting method is subjective and may result in erroneous cluster estimations, cluster validation indices were used to aid in determining the optimal number of clusters.

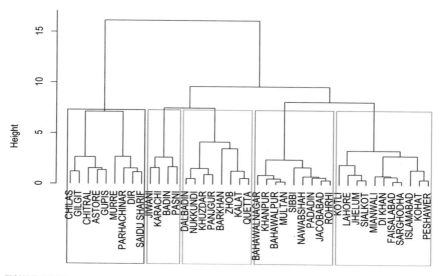

FIGURE 12.12

Cluster dendrogram of all the metrological stations using Ward linkage, Euclidean Distance.

12.4.5 Cluster validation indices

Cluster validation was performed using commonly used cluster validation indices. The appropriate number of clusters identified using Dunne index, Silhouette width, Calinski and Harabasz index, and Davies—Bouldine validation index is shown in Figs. 12.13—12.16, respectively. It is evident from the cluster validation indices that Pakistan can be divided into five clusters, which can be considered as an optimum number of clusters.

12.4.6 Regionalization of drought across Pakistan

Clustering algorithms yield statistically diverse zones in most of the cases. Therefore, preliminary clusters computed through clustering algorithm should go through alteration in order to become homogeneous. It is unlikely that the regions delineated by the HCPC algorithm are statistically homogeneous. Therefore, it requires a regional modification process to improve the homogeneity of the regions. Many methods for reconfiguring the initial groups shaped by cluster analysis were presented in the previous literature (Hosking & Wallis, 1993). The regions defined by the HCPC procedure and validated using different indices did not undergo substantial changes. Since the number of clusters is suitably picked using the HCPC technique, only small adjustments of discordant sites have been made. Under this analysis, SPEI for 12-month time scale serves as the basis for the measurement of discordancy and homogeneity of the regions at the 12-month time scale of SPEI. In the univariate framework, only one station, namely, "Bahawalnagar" is appeared to

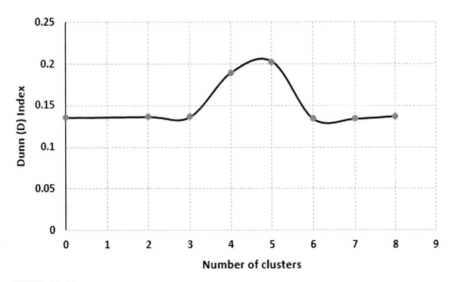

FIGURE 12.13

Variation in value of the Dunne index from cluster 1 to 9.

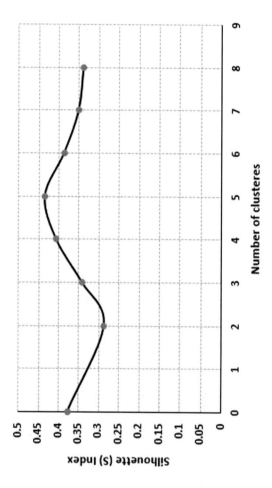

FIGURE 12.14

Variation in value of the silhouette width from cluster 1 to 9.

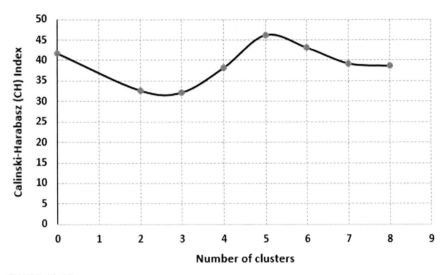

FIGURE 12.15

Variation in value of the Calinski and Harabasz index from cluster 1 to 9.

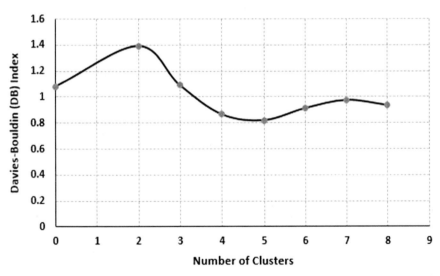

FIGURE 12.16

Variation in value of the Davies—Bouldine Index from cluster 1 to 9.

be discordant in cluster 4, as shown in Table 12.3. The D_i (discordancy) reading revealed that the Bahawalnagar station exceeds the critical discordant value (2.33). Results showed that Bhawalnagar station needed to be adjusted according to suggested guideline in previous literature (Hosking & Wallis, 1993).

Table 12.3 The results of discordancy measures for SPEI-12. In bold, $D_{i,D}$ and $D_{i,S}$ indicate discordancy in the case of drought duration and severity at univariate framework, respectively and $D_{i,DS}$ indicates the discordancy measure in bivariate framework.

Cluster 1 (critical value $C_u = 2.33$, $C_b = 2.60$)							
Site	$D_{i,D}$	$D_{i,S}$	$D_{i,DS}$	Site	$D_{i,D}$	$D_{i,S}$	$D_{i,DS}$
Chilas	1.07	0.83	0.91	Gupis	0.29	0.53	0.45
Gilgit	1.96	1.70	1.80	Muree	0.54	1.70	1.51
Chitral	1.02	0.23	0.89	Parachinar	0.59	1.41	1.21
Astore	0.16	0.83	0.65	Dir	1.75	0.43	0.98
Saidu Sharif	1.61	1.34	1.12				
Cluster 2 (critical value $C_u = 3.00$, $C_b = 2.60$)							
Jiwani	1.00	1.00	1.00	Badin	1.00	1.00	1.00
Karachi	1.00	1.00	1.00	Pasni	1.00	1.00	1.00
Cluster 3 (critical value $C_u = 2.14$, $C_b = 2.60$)							
Dalbandin	0.46	0.37	0.41	Barkhan	0.24	0.58	0.41
Nokkundi	1.87	1.88	1.69	Zhob	1.08	0.51	0.98
Khuzdar	0.81	1.83	1.21	Kalat	2.09	1.36	1.61
Panjgur	0.12	1.03	0.79	Quetta	1.34	0.46	0.91
Cluster 4 (critical value $C_u = 2.33$, $C_b = 2.60$)							
Bahawalnagar	2.49	2.62	2.51	Nawab shah	1.09	0.71	0.97
Bahawalpur	1.08	1.34	1.21	Padidian	1.37	0.45	1.21
Khanpur	1.38	1.06	1.37	Jacobabad	0.47	0.53	0.62
Mulatan	0.20	0.14	0.24	Rohri	0.22	0.05	0.221
Sibbi	0.69	2.10	1.58				
Cluster 5 (critical value $C_u = 2.63$, $C_b = 2.60$)							
Kotli	0.81	0.76	0.79	Faisalabad	0.16	1.35	1.15
Lahore	0.86	0.51	0.75	Sargodha	0.92	1.22	0.68
Jhelum	0.24	0.38	0.21	Islamabad	2.05	1.76	1.45
Sialkot	0.87	1.17	0.98	Kohat	2.13	1.75	1.28
Miawali	0.98	0.68	0.74	Peshawar	1.25	1.03	1.11
D.I. khan	0.71	0.39	0.69				

12.4.7 Construction of homogenous regions

The D_i values for all meteorological stations were calculated after the adjustment of Bahawalnagar (discordant station) in the second stage to verify corrected areas for any discordant station(s), as shown in Table 12.4. After shifting and correcting, the Bahawalnagar station in cluster 4 was moved to cluster 5, while Panjgur in cluster 3 was moved to cluster 2 to meet the discordancy measure statistics.

The revised clusters after the adjustments of stations showed that none of the stations is discordant. The corrected clusters also improved the size of cluster 2, which previously had the least number of stations.

12.4.8 Evaluation of homogeneous clusters

A single value heterogeneity statistical measure (H_1) was computed for each region and is provided in Table 12.5. These numbers show that the five zones are homogenous (i.e., H_1 less than 1), except region 5 at the bivariate framework, as it has a H_1 value greater than 1 (acceptably homogeneous). Results based on physiographic

Table 12.4 The final clusters after adjustment.

Cluster 1 ($C_u = 2.33$, $C_b = 2.60$)

Site	$D_{i,D}$	$D_{i,S}$	$D_{i,DS}$	Site	$D_{i,D}$	$D_{i,S}$	$D_{i,DS}$
Chilas	1.07	0.83	1.31	Gupis	0.29	0.53	0.34
Gilgit	1.96	1.70	1.65	Muree	0.54	1.70	1.32
Chitral	1.02	0.23	0.81	Parachinar	0.59	1.41	0.93
Astore	0.16	0.83	0.67	Dir	1.75	0.43	1.12
Saidu Sharif	1.61	1.34	0.95				

Cluster 2 ($C_u = 2.33$, $C_b = 2.60$)

Jiwani	1.29	1.10	1.02	Badin	1.02	0.16	0.74
Karachi	1.22	1.12	1.15	Pasni	0.73	1.29	1.09
Panjgur	0.74	1.33	0.85				

Cluster 3 ($C_u = 1.92$, $C_b = 2.60$)

Dalbandin	0.42	0.35	0.25	Zhob	0.95	0.85	0.74
Nokkundi	1.64	1.60	1.23	Kalat	1.79	1.45	1.21
Khuzdar	0.69	1.68	0.96	Quetta	1.27	0.58	1.11
Barkhan	0.25	0.49	0.39				

Cluster 4 ($C_u = 2.14$, $C_b = 2.60$)

Bahawalpur	0.95	1.22	1.07	Padidian	1.70	1.85	1.45
Khanpur	1.36	1.15	1.20	Jacobabad	1.14	0.45	0.95
Mulatan	0.58	0.31	0.62	Rohri	0.50	0.21	0.40
Sibbi	0.84	2.13	1.12	Nawab shah	0.93	0.67	0.50

Cluster 5 ($C_u = 2.76$, $C_b = 2.60$)

Kotli	0.87	1.18	0.95	Faisalabad	0.08	0.30	0.52
Lahore	0.97	0.57	0.69	Sargodha	1.00	1.61	1.05
Jhelum	0.28	0.66	0.74	Islamabad	1.96	1.96	1.25
Sialkot	0.97	1.65	1.17	Kohat	2.09	1.97	1.51
Miawali	1.1	0.57	0.85	Peshawar	0.74	0.92	0.57
D.I khan	0.66	0.31	0.69	Bahawalnagar	1.29	1.11	1.01

Table 12.5 Heterogeneity value of the final adjusted homogeneous regions.

Scale (months)	Region	Droughts events	Stations	D	S	DS
SPEI-12	1	558	9	−0.44(H[a])	−0.76(H[a])	−0.88(H[a])
	2	386	5	0.67(H[a])	0.16(H[a])	−0.11(H[a])
	3	424	7	−0.64 (H[a])	−1.08 (H[a])	−1.30(H[a])
	4	521	8	−1.73(H[a])	−0.49(H[a])	−0.42(H[a])
	5	820	12	0.64(H[a])	0.94(H[a])	1.2(A.H.[b])

[a]H = Homogeneous region.
[b]A.H. = Acceptably homogeneous region.

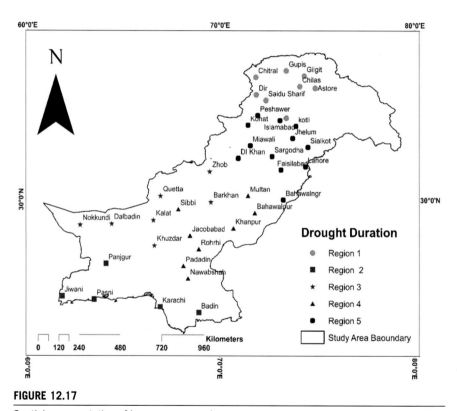

FIGURE 12.17

Spatial representation of homogenous regions.

and climatic characteristics of the Pakistan showed that the country can be divided into five homogeneous regions, as shown in Fig. 12.17. It has greater applicability for the river basin management and water resource planning in Pakistan.

12.5 **Discussion**

The topography of Pakistan has extreme geographical variations; therefore, the climate varies greatly from north to south. The northern and western mountainous regions of the country are greatly influenced by the diversities of climate. The majority of the area is extremely vulnerable to the fluctuations of precipitation. The majority of the higher latitude regions are prone to the flood, whereas the southern part is subjected to the drought phenomena. The monsoon season leads to almost 60% of the rainfall from June to September (Hanif et al., 2013). The average relative humidity of Pakistan is 70%, which varies from 54% to 80% in January and August. The relative humidity is an essential parameter to be used in investigating drought phenomena due to its high influence on drought. RH has been used for the precipitation characterization (Dinpashoh, Fakheri-Fard, Moghaddam, Jahanbakhsh, & Mirnia, 2004).

It can be observed that the frequency of the drought during wet and dry periods tends to be shorter at the smaller time scale. The SPEI for a shorter time scale is commonly known as an agriculture drought index (Ji & Peters, 2003; Sims, Niyogi, & Raman, 2002). It reflects the water content of the plant and soil conditions (Mckee, Doesken, & Kleist, 1993). The result of SPEI analysis on the time scale of 12 months showed that the drought duration lasting for longer periods was less frequent. River water flows can also be monitored using the SPEI over longer periods, like 12 months, which is regarded as the hydrological drought index (Mckee et al., 1993).

Movement of the stations between the regions showed that stations may frequently be adjusted to improve the physical structure and minimize regional variability (Hosking & Wallis, 1993). During the adjustment process, it may be advantageous to add, update, or even delete a site or a few sites from one region to another. In this study, various discordant stations have been shifted from one region to another depending on geographical proximity and meteorological conditions. Because of differences in the first three regions, the majority of changes have been made in these groups. The first region contains the world's third biggest glacier, the second has the highest annual average rainfall, and the third has the lowest annual average rainfall.

12.6 **Challenges and solutions**

Pakistan is experiencing severe droughts as a result of climate change; global warming has raised the country's temperature to an alarming stage. The frequency of various severe occurrences such as floods, droughts, and heat waves is rapidly rising (Panagoulia, Panagoulia, & Dimou, 1998). Severe droughts are commonly occurring phenomena (i.e., occurring around four out of every 10 years in Pakistan). Pakistan faced its worst drought since its inception in 1947 and 1998, which continued until 2002 (Jamro et al., 2019). Therefore, water resource management in the presence of climate extremes (i.e., floods and drought) is the biggest challenge in Pakistan. This study provides the solution by dividing the statistically homogeneous regions according to different geographical and meteorological variables.

12.7 Conclusions

The regionalization of the drought based on physiographic and climate variables is the main objective of this study. The regional water resource management and planning require the statistically homogeneous regions. The complicated hydroclimatic and topographical characteristics of Pakistan make it a difficult task. The SPEI truncation level approach was adopted to collect the drought attributes across 41 meteorological stations. The following are the main points concluded in the study.

1. The SPEI drought index at the time scale of 12 month for all the stations revealed that severe droughts have been observed from the year 2000 to 2004.
2. The PCA approach application on dependent variables revealed that all of the selected attributes strongly contribute to the leading principal components (95.2% of the variance explained) due to the existence of the correlation between the variables.
3. The mixed/hybrid nature of the HCPC clustering method is helpful in the identification of robust homogeneous regions on the basis of hydroclimatic and topographical features of the region. Furthermore, cluster validation indices indicated that Pakistan can hydrologically be divided into five homogeneous regions.
4. The clusters formed using the HCPC cluster algorithm were further tested for the drought attributes (drought duration and severity) using the univariate and bivariate framework. Analysis revealed that only one station (Bahawalnagar) is discordant in the univariate framework of drought duration and severity.
5. The several revisions of the process for the delineation of homogenous regions with the shifting of discordant station from one region to another revealed that Bahawalnagar station can ideally be fitted in region 5 to make all regions hydrologically homogeneous. It is noted that bivariate homogeneity and discordancy measure is better able to capture the correlation between the variables (drought duration and severity combined) as compared to univariate homogeneity and discordancy measure (either duration or severity separately).

The hydrologically homogeneous regions delineated in this study are expected to serve as the basis for regional water resource planning and river basin management in Pakistan.

Acknowledgments

This work was supported by Korea Institute of Planning and Evaluation for Technology in Food, Agriculture and Forestry (IPET) through Agricultural Foundation and Disaster Response Technology Development Program, funded by Ministry of Agriculture, Food and Rural Affairs (MAFRA) (RS-2022-IP322083).

References

Adnan, S., Ullah, K., Gao, S., Khosa, A. H., & Wang, Z. (2017). Shifting of agro-climatic zones, their drought vulnerability, and precipitation and temperature trends in Pakistan. *International Journal of Climatology*, *37*, 529−543. Available from https://doi.org/10.1002/joc.5019, http://onlinelibrary.wiley.com/journal/10.1002/(ISSN)1097-0088.

Ahmed, K., Shahid, S., Harun, Sb, & Wang, Xj (2016). Characterization of seasonal droughts in Balochistan Province, Pakistan. *Stochastic Environmental Research and Risk Assessment*, *30*(2), 747−762. Available from https://doi.org/10.1007/s00477-015-1117-2, http://link.springer-ny.com/link/service/journals/00477/index.htm.

Alam, N. M., Sharma, G. C., Moreira, E., Jana, C., Mishra, P. K., Sharma, N. K., & Mandal, D. (2017). Evaluation of drought using SPEI drought class transitions and log-linear models for different agro-ecological regions of India. *Physics and Chemistry of the Earth*, *100*, 31−43. Available from https://doi.org/10.1016/j.pce.2017.02.008, http://www.journals.elsevier.com/physics-and-chemistry-of-the-earth/.

Argüelles, M., Benavides, C., & Fernández, I. (2014). A new approach to the identification of regional clusters: Hierarchical clustering on principal components. *Applied Economics*, *46*(21), 2511−2519. Available from https://doi.org/10.1080/00036846.2014.904491, http://www.tandf.co.uk/journals/titles/14664283.asp.

Beguería, S., Vicente-Serrano, S. M., Reig, F., & Latorre, B. (2014). Standardized precipitation evapotranspiration index (SPEI) revisited: Parameter fitting, evapotranspiration models, tools, datasets and drought monitoring. *International Journal of Climatology*, *34*(10), 3001−3023. Available from https://doi.org/10.1002/joc.3887, http://onlinelibrary.wiley.com/journal/10.1002/(ISSN)1097-0088.

Burn, D. H. (1988). Delineation of groups for regional flood frequency analysis. *Journal of Hydrology*, *104*(1-4), 345−361. Available from https://doi.org/10.1016/0022-1694(88)90174-6.

Caliñski, T., & Harabasz, J. (1974). A dendrite method Foe cluster analysis. *Communications in Statistics*, *3*(1), 1−27. Available from https://doi.org/10.1080/03610927408827101.

Chang, H., & Kwon, W. T. (2007). Spatial variations of summer precipitation trends in South Korea. *Environmental Research Letters*, *2*(4). Available from https://doi.org/10.1088/1748-9326/2/4/045012, http://www.iop.org/EJ/journal/1748-9326.

Chebana, F., & Ouarda, T. B. M. J. (2007). Multivariate L-moment homogeneity test. *Water Resources Research*, *43*(8). Available from https://doi.org/10.1029/2006WR005639.

Dikshit, A., Pradhan, B., & Huete, A. (2021). An improved SPEI drought forecasting approach using the long short-term memory neural network. *Journal of Environmental Management*, *283*, 111979. Available from https://doi.org/10.1016/j.jenvman.2021.111979.

Dinpashoh, Y., Fakheri-Fard, A., Moghaddam, M., Jahanbakhsh, S., & Mirnia, M. (2004). Selection of variables for the purpose of regionalization of Iran's precipitation climate using multivariate methods. *Journal of Hydrology*, *297*(1–4), 109–123. Available from https://doi.org/10.1016/j.jhydrol.2004.04.009.

Dunn, J. C. (1974). Well-separated clusters and optimal fuzzy partitions. *Journal of Cybernetics*, *4*(1), 95–104. Available from https://doi.org/10.1080/01969727408546059.

Fatima, S. U., Khan, M. A., Siddiqui, F., Mahmood, N., Salman, N., Alamgir, A., & Shaukat, S. S. (2022). Geospatial assessment of water quality using principal components analysis (PCA) and water quality index (WQI) in Basho Valley, Gilgit Baltistan (Northern Areas of Pakistan. *Environmental Monitoring and Assessment*, *194*(3). Available from https://doi.org/10.1007/s10661-022-09845-5, https://link.springer.com/journal/10661.

Hanif, M., Khan, A. H., & Adnan, S. (2013). Latitudinal precipitation characteristics and trends in Pakistan. *Journal of Hydrology*, *492*, 266–272. Available from https://doi.org/10.1016/j.jhydrol.2013.03.040.

Hosking, J. R. M. (1990). L-Moments: Analysis and estimation of distributions using linear combinations of order statistics. *Journal of the Royal Statistical Society Series B*.

Hosking, J. R. M., & Wallis, J. R. (1993). Some statistics useful in regional frequency analysis. *Water Resources Research*, *29*(2), 271–281. Available from https://doi.org/10.1029/92wr01980.

Hotelling, H. (1933). Analysis of a complex of statistical variables into principal components. *Journal of Educational Psychology*, *24*(6), 417–441. Available from https://doi.org/10.1037/h0071325.

Jamro, S., Dars, G. H., Ansari, K., & Krakauer, N. Y. (2019). Spatio-temporal variability of drought in Pakistan using standardized precipitation evapotranspiration index. *Applied Sciences (Switzerland)*, *9*(21). Available from https://doi.org/10.3390/app9214588, https://res.mdpi.com/d_attachment/applsci/applsci-09-04588/article_deploy/applsci-09-04588-v2.pdf.

Ji, L., & Peters, A. J. (2003). Assessing vegetation response to drought in the northern Great Plains using vegetation and drought indices. *Remote Sensing of Environment*, *87*(1), 85–98. Available from https://doi.org/10.1016/S0034-4257(03)00174-3, http://www.elsevier.com/inca/publications/store/5/0/5/7/3/3.

Kaluba, P., Verbist, K. M. J., Cornelis, W. M., & Van Ranst, E. (2017). Spatial mapping of drought in Zambia using regional frequency analysis. *Hydrological Sciences Journal*, *62*(11), 1825–1839. Available from https://doi.org/10.1080/02626667.2017.1343475, http://www.tandfonline.com/loi/thsj20.

Kao, S. C., & Govindaraju, R. S. (2010). A copula-based joint deficit index for droughts. *Journal of Hydrology*, *380*(1–2), 121–134. Available from https://doi.org/10.1016/j.jhydrol.2009.10.029.

Kilmer, P. D. (2010). Review article: Review article. *Journalism*, *11*(3), 369–373. Available from https://doi.org/10.1177/1461444810365020.

Krakauer, N. Y., Lakhankar, T., & Dars, G. H. (2019). Precipitation trends over the Indus basin. *Climate*, *7*(10). Available from https://doi.org/10.3390/cli7100116, https://res.mdpi.com/d_attachment/climate/climate-07-00116/article_deploy/climate-07-00116-v2.pdf.

Mckee, T.B., Doesken, N.J., & Kleist, J. (1993). The relationship of drought frequency and duration to time scales. In *AMS 28th conference on applied climatology* (pp. 179–184).

Mirabbasi, R., Fakheri-Fard, A., & Dinpashoh, Y. (2012). Bivariate drought frequency analysis using the copula method. *Theoretical and Applied Climatology*, *108*(1−2), 191−206. Available from https://doi.org/10.1007/s00704-011-0524-7, http://link.springer.de/link/service/journals/00704/.

Mishra, A. K., & Singh, V. P. (2010). A review of drought concepts. *Journal of Hydrology*, *391*(1-2), 202−216. Available from https://doi.org/10.1016/j.jhydrol.2010.07.012.

Mishra, A. K., & Singh, V. P. (2011). Drought modeling − A review. *Journal of Hydrology*, *403*(1−2), 157−175. Available from https://doi.org/10.1016/j.jhydrol.2011.03.049.

Modarres, R. (2010). Regional dry spells frequency analysis by L-moment and multivariate analysis. *Water Resources Management*, *24*(10), 2365−2380. Available from https://doi.org/10.1007/s11269-009-9556-5.

Panagoulia, D.G., Panagoulia, D., Dimou, G. (1998). *Definitions and effects of droughts definition and effects of droughts.*

Pett, M., Lackey, N., & Sullivan, J. (2003). *An overview of factor analysis* (pp. 2−12). SAGE Publications. Available from http://doi.org/10.4135/9781412984898.n1.

Rossi, G., Benedini, M., Tsakiris, G., & Giakoumakis, S. (1992). On regional drought estimation and analysis. *Water Resources Management*, *6*(4), 249−277. Available from https://doi.org/10.1007/BF00872280.

Rousseeuw, P. J. (1987). Silhouettes: A graphical aid to the interpretation and validation of cluster analysis. *Journal of Computational and Applied Mathematics*, *20*(C), 53−65. Available from https://doi.org/10.1016/0377-0427(87)90125-7.

Sadri, S., & Burn, D. H. (2011). A fuzzy *C*-means approach for regionalization using a bivariate homogeneity and discordancy approach. *Journal of Hydrology*, *401*(3−4), 231−239. Available from https://doi.org/10.1016/j.jhydrol.2011.02.027.

Santos, M., Fragoso, M., & Santos, J. A. (2017). Regionalization and susceptibility assessment to daily precipitation extremes in mainland Portugal. *Applied Geography*, *86*, 128−138. Available from https://doi.org/10.1016/j.apgeog.2017.06.020, http://www.elsevier.com/inca/publications/store/3/0/3/9/0/index.htt.

Sarfaraz, S. (2014). The sub-regional classification of Pakistan's winter precipitation based on principal components analysis. *Pakistan Journal of Meteorology*, *10*, 57−66.

Serfling, R., & Xiao, P. (2007). A contribution to multivariate L-moments: L-comoment matrices. *Journal of Multivariate Analysis*, *98*(9), 1765−1781. Available from https://doi.org/10.1016/j.jmva.2007.01.008.

Sims, A. P., Niyogi, D. D. S., & Raman, S. (2002). Adopting drought indices for estimating soil moisture: A North Carolina case study. *Geophysical Research Letters*, *29*(8). Available from https://doi.org/10.1029/2001GL013343, http://onlinelibrary.wiley.com/journal/10.1002/(ISSN)1944-8007/issues?year = 2012.

Ullah, H., Akbar, M., & Khan, F. (2020). Droughts' projections in homogeneous climatic regions using Standardized Precipitation Index in Pakistan. *Theoretical and Applied Climatology*, *140*(1−2), 787−803. Available from https://doi.org/10.1007/s00704-020-03109-3, https://rd.springer.com/journal/volumesAndIssues/704.

Um, M. J., Yun, H., Jeong, C. S., & Heo, J. H. (2011). Factor analysis and multiple regression between topography and precipitation on Jeju Island, Korea. *Journal of Hydrology*, *410*(3−4), 189−203. Available from https://doi.org/10.1016/j.jhydrol.2011.09.016.

Van Loon, A. F. (2015). Hydrological drought explained. *WIREs Water*, *2*(4), 359−392. Available from https://doi.org/10.1002/wat2.1085.

Vicente-Serrano, S. M., Beguería, S., & López-Moreno, J. I. (2010). A multiscalar drought index sensitive to global warming: The standardized precipitation evapotranspiration index. *Journal of Climate*, *23*(7), 1696–1718. Available from https://doi.org/10.1175/2009jcli2909.1.

Wang, W., Zhu, Y., Xu, R., & Liu, J. (2015). Drought severity change in China during 1961–2012 indicated by SPI and SPEI. *Natural Hazards*, *75*(3), 2437–2451. Available from https://doi.org/10.1007/s11069-014-1436-5, http://www.wkap.nl/journalhome.htm/0921-030X.

Wilhite. (1987). *The role of definitions understanding: The drought phenomenon: The role of definitions* (pp. 11–27). Westview Press.

Yevjevich, V. (1969). An objective approach to definitions and investigations of continental hydrologic droughts. *Journal of Hydrology*, *7*, 90110–90113. Available from https://doi.org/10.1016/0022-1694(69).

Yevjevich, V.M., & Jeng, R.I. S. (1969). *Properties of non-homogensous hydrologic series.*

Zulfatman., & Rahmat, M. F. (2009). Application of self-tuning fuzzy pid controller on industrial hydraulic actuator using system identification approach. *International Journal on Smart Sensing and Intelligent Systems*, *2*(2), 246–261. Available from https://doi.org/10.21307/ijssis-2017-349.

Analysis of watershed attributes for water resource management using geospatial technology: a case study of Haringmuri watershed

13

Sandip Chand[1] and Tapas Mistri[2]
[1]Indian Institute of Technology Kharagpur, Kharagpur, West Bengal, India
[2]The University of Burdwan, Burdwan, West Bengal, India

13.1 Introduction

A watershed represents natural hydrological units encompassing a particular area of land surface, directing rainfall runoff to a defined drainage point, be it a drain, channel, stream, or river. While terms like region, basin, catchment, and watershed are commonly used interchangeably, they have distinct technical meanings (Omernik, Griffith, Hughes, Glover, & Weber, 2017; Shao, Fu, Li, Altan, & Cheng, 2019). The size of a watershed is determined by the stream it occupies and holds practical significance in development programs (Ikram et al., 2024; Javed, Khanday, & Rais, 2011). Attributes of watersheds hold importance for effective water resource management (Gebre, Kibru, Tesfaye, & Taye, 2015; Girma, Abraham, & Muluneh, 2020). In order to manage natural resources like land and water and mitigating the effects of natural disaster for sustainable development, a watershed is a perfect unit for analysis and measurement (Rao, Latha, Kumar, & Krishna, 2010). Along with a hydrological unit, a watershed is a social, political, and ecological entity, playing an important role in social and economic security, ensuring food, and helping rural communities with various life-supporting services (Mengistu & Assefa, 2020). Morphometric assessment, when coupled with geomorphology and geology, aids in providing a primary hydrological diagnosis, enabling the prediction of the approximate behavior of a watershed (Chakrabortty, Ghosh, Pal, Das, & Malik, 2018; Soni, 2017). The physiographic features of a river basin, such as its size, form, slope, and the length and width of its streams, are all related to its hydrological response (Jha, Chowdhury, Chowdary, & Peiffer, 2007). Therefore, morphometric analysis is deemed an

Applications of Geospatial Technology and Modeling for River Basin Management.
DOI: https://doi.org/10.1016/B978-0-443-23890-1.00013-X

essential initial step toward a fundamental understanding of watershed dynamics. The hydrologic and geomorphic processes within the watershed contribute to land surface process formation and development (Tiwari & Rai, 2015). Drainage characteristics of river basins and subbasins globally have been studied conventionally using methods like those of Horton, (1932), Horton (1945), Strahler (1957), Strahler (1964). Morphometric analysis seeks precise measurements of the stream network's observable attributes within the drainage basin. These physiographic characteristics such as size, shape, slope, drainage density, and size and length of contributories can be linked to a range of hydrological phenomena (Das, 2016; Sarkar, 2012; Wani & Garg, 2010). Drainage analysis lacks completeness without a systematic approach to the development of the drainage basin (Rai, Mohan, Mishra, Ahmad, & Mishra, 2017). Drainage lines not only elucidate 3D geometry of an area, but it also recounts its evolutionary process (Suhail et al., 2024). Morphometric analysis incorporates numerous linear, areal, and relief parameters (Borgohain, Khajuria, Garg, Koti, & Bhardwaj, 2023; Gebre et al., 2015) that collectively describe a complex network of morphometric diversity (Al-Saady, Al-Suhail, Al-Tawash, & Othman, 2016; Benzougagh et al., 2022). This analysis aids in identifying microscale landscape units with unique characteristics, impacting land use patterns (Jasrotia, Kumar, & Aasim, 2011). Morphometric parameter evaluation involves the analysis of various drainage parameters such as stream ordering, watershed area and perimeter measurement, stream length, stream length ratio, bifurcation ratio, drainage density, and relief ratio (Gebre et al., 2015). Quantitative analysis of drainage systems is crucial for understanding watershed characteristics, with applications in hydrological investigations, basin management, groundwater management, groundwater potential assessment, and environmental assessment (Arabameri, Rezaei, Cerda, Lombardo, & Rodrigo-Comino, 2019; Fenta, Yasuda, Shimizu, Haregeweyn, & Woldearegay, 2017; Ghosh, Adhikary, Bera, Bhunia, & Shit, 2022; Mink et al., 2014). The Geological Survey of India record is considered a fundamental aspect in describing the geology of the study area (Das, Mondal, Das, & Ghosh, 2011). A low drainage density value suggests highly permeable subsoil and dense vegetative cover in the basin (Rao & Latha, 2019). The circularity ratio value indicates a strongly elongated basin with highly permeable homogeneous geological materials.

The emerging physical developments like townships, industrial units, and academic campuses heavily rely on nearby natural water resources (Tiwari & Rai, 2015). Conflicts among stakeholders arise due to these developments. Meeting the Millennium Development Goal for adequate availability of freshwater standards necessitates ensuring surface and ground water availability through pragmatic and location-specific solutions at macro- and microwatershed levels (Dass et al., 2021; Hall, Ranganathan, & Raj, 2017). Watersheds function as natural systems, collecting, storing, and releasing water (Grigg, 2016; Li et al., 2018). The quality and quantity of surface runoff and groundwater depend on the climate, geographical and geological profile, and vegetation. Integrated watershed planning at macro and micro levels is crucial for sustainable ecosystem development and human habitation (Jadhav & Babar, 2013). Traditional Indian settlements were developed in harmony with the existing water resources, a practice overlooked in modern planning, resulting in water scarcity and quality degradation (Krishan, 2011; Kumar, Gurunath, Madhusudan Rao, & Ajitha, 2020; Mascarenhas, 2021). Scientific studies on urban disasters highlight the need for planning in accordance with watersheds (Huang, Shen, & Mardin, 2018; Tezer, Lutfi Sen, Aksehirli, Cetin, & Onur, 2012). While the existing studies have highlighted the importance of geospatial technology in understanding watershed dynamics, there is a need for more nuanced investigations into the integration of diverse watershed attributes (Dakin Kuiper et al., 2024; Daniel et al., 2010; Strager et al., 2010). Further research is required to develop standardized protocols for the application of geospatial technology in assessing watershed attributes, with a focus on improving accuracy, scalability, and adaptability across different geographic regions and hydrological settings. Addressing these gaps will contribute to the refinement of water resource management strategies, facilitating more sustainable and effective decision-making processes. The major objectives of the present study are as follows:

1. To find out the watershed attributes of the region.
2. To evaluate the land based on the soil parameters.
3. To find out the major land use land cover of the study area and to evaluate the water resource management through watershed attributes of the region.

13.2 Materials and methods

13.2.1 Study area

The Bankura district in West Bengal is located in its western part, and the study area, the "Haringmuri watershed," is positioned in the central part of Bankura district, as shown in Fig. 13.1. The Haringmuri River serves as the left bank tributary of the Dwarakeswar River. Originating from the Tilboni hill in Purulia district, the Dwarakeswar River enters Bankura district near Chhatna. The Haringmuri River, on the other hand, originates from the elevated region of 110 m in the southeastern part of Bankura-I block, merging with the Dwarakeswar River near Mukundapur Village in the Bishnupur block. The study area spans latitudinally from $23°7'10''$N to $23°16'10''$N and longitudinally from $87°14'31''$E to $87°26'38''$ E. Encompassing a total area of 215.19 sq. km, it includes the northwestern part of Bishnupur block, the southern part of Sonamukhi block, the southern part of Barjora block, the eastern part of Bankura-I, and the northern part of Onda block. This region is characterized by a plateau fringe, and the slope generally extends from northwest to southeast.

13.2.2 Methodology

The labor-intensive process of manually extracting drainage networks and assigning stream orders from Survey of India's topo maps and georeferenced satellite

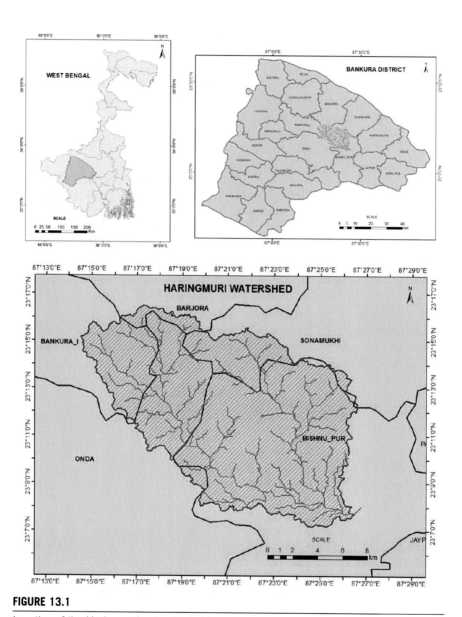

FIGURE 13.1

Location of the Haringmuri watershed with administrative boundary.

image over a large region is time-consuming and tiresome. To address this challenge, an automated extraction technique has been employed to assess morphometric parameters within a basin. This involves the extraction of the stream networks and basin boundary from Haringmuri watershed. The process utilizes ASTER digital elevation model (DEM) in combination with georeferenced SOI toposheets dated 1972 (73M/7 and 73M/8, at a 1:50,000 scale) utilizing ERDAS Imagine-9.0 and ARC GIS-10.4 software. The resulting stream network and basin are then projected onto a regional coordinate system (WGS1984, UTM zone 44N). Various morphometric parameters have been identified, as depicted in Table 13.2.

13.2.2.1 Extraction of the Haringmuri River watershed

The Haringmuri watershed is automatically delineated utilizing ASTER DEM data with 15 m spatial resolution, incorporating georeferenced Survey of India's topo maps. Employing various geoprocessing techniques in ArcGIS-10.4, the contributing basin area is extracted. The essential input parameters for this extraction are DEM and pour point. The pour point refers to a location indicated by the user within cells with the highest accumulation of flow (Lindsay, Rothwell, & Davies, 2008). The outcome of this procedure generates a polygon delineating the watershed boundary based on flow direction (Suwandana, Kawamura, Sakuno, & Kustiyanto, 2012).

13.2.2.2 Extraction of the Haringmuri drainage network

The drainage network within the Haringmuri watershed is derived through a sequence of geoprocessing tools in ArcGIS-10.4. This methodology lays the foundation for constructing a drainage network grid, incorporating stream orders based on Strahler's (1964) classification system. According to this method, the stream which has no tributary is the first-order stream; the stream which has a tributary of only first-order stream is second-order stream, and so forth (Strahler, 1964). This approach necessitates the DEM and a minimum upstream area as parameters for input model (Chorowicz, Ichoku, Riazanoff, Kim, & Cervelle, 1992; O'Callaghan & Mark, 1984). The resulting drainage network is then smoothed using a tool for creating smooth lines in ArcGIS-10.4. To assess the morphometry of the basin, various parameters are analyzed using standard mathematical formula outlined in Table 13.1. Additionally, the slope and aspect maps of Haringmuri Watershed are derived from the ASTER DEM with the help of spatial analyst module in the ArcGIS-10.4.

13.3 Result and discussion

13.3.1 Morphometric parameters

The quantitative analysis of the drainage system's morphometry is a crucial component in the characterization of watersheds (Strahler, 1964). Morphometric parameters can be broadly categorized into three types: linear, areal, and relief parameters.

Table 13.1 Linear parameters of Haringmuri watershed.

SI. no	Stream order (Su)	Stream number (Nu)	Stream length (Lu)	Bifurcation ratio (Rb)	Mean bifurcation ratio (Rbm)	Mean stream length (Lum)	Stream length ratio (RL)
1	1st	103	117.95		3.395	1.15	
2	2nd	22	61.11	4.68		2.78	2.417
3	3rd	5	13.66	4.4		2.73	0.982
4	4th	2	28.22	2.5		14.11	5.168
5	5th	1	3.29	2		3.29	0.233
	Total	133	224.23	13.58		24.06	8.800

13.3.1.1 Linear parameters

13.3.1.1.1 Stream order and stream number

The initial phase of morphometric analysis for a drainage basin involves stream ordering. The stream ordering of the Haringmuri River is determined using Strahler's (1964) method. In the current investigation, the Haringmuri watershed is classified as a fifth-order watershed, comprising a total of 133 streams of various orders across an area of 215.19 km², as shown in Fig. 13.2. The majority of streams in the study area are of lower orders. Specifically, there are 103 first-order streams, 22 second-order streams, 5 third-order streams, 2 fourth-order streams, and 1 indicating a fifth-order stream, as shown in Fig. 13.2. Higher numbers of stream segment in a specific order suggest ongoing topographic erosion. The drainage pattern of the stream network in the basin predominantly exhibits a dendritic type, indicating a uniform texture and an absence of structural control. This pattern resembles a tree-like or fern-like configuration, with branches mostly intersecting at the acute angle. Understanding the properties of the stream network is crucial for studying water resources in the area.

13.3.1.1.2 Bifurcation ratio

The "bifurcation ratio" (Rb) denotes the ratio of the number of streams of a particular order to the number of streams in the next higher order (Schumm, 1956). Typically, bifurcation ratios fall within the range of 3.0−5.0 for basins where the drainage pattern has not been distorted by geological structures (Strahler, 1964). Strahler's work in 1957 (Strahler, 1957) illustrated that the bifurcation ratio exhibits a limited range of variation, emphasizing the dominance of regional or environmental factors. For the study area, the mean bifurcation ratio value is calculated as 3.395, as shown in Table 13.1, indicating that the geological structures have a relatively minor impact on the drainage pattern.

13.3.1.1.3 Mean bifurcation ratio

Mean bifurcation ratio (Rbm) is the average of bifurcation ratios of all order streams. The calculated mean bifurcation ratio of the Haringmuri watershed is 3.395.

13.3.1.1.4 Stream length

Stream length calculations are based on the methodology proposed by Horton (1945), considering the total length within a specific order. This parameter holds considerable hydrological significance for the basin, as it provides insights into the characteristics of surface runoff. Longer stream lengths typically indicate smoother gradients, while smaller stream lengths are suggestive of places with finer textures and steeper slopes. The overall length of stream segments in the Haringmuri watershed is maximum for first-order streams and diminishes with the increasing order. The number of streams of different orders and lengths from the mouth to the dividing ridgeline are estimated using geographic information

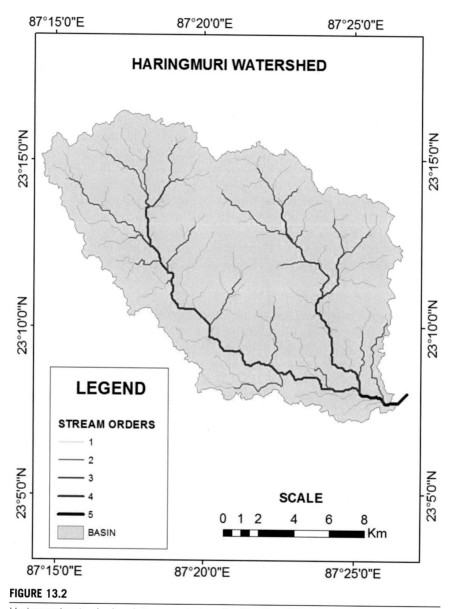

FIGURE 13.2

Haringmuri watershed and stream ordering after Strahler, (1952).

system (GIS) software. The overall stream length for the Haringmuri watershed is 224.23 km, distributed across different orders as follows: 117.95 km in the first order, 66.11 km in second order, 13.66 km in third order, 28.22 km in fourth order, and 3.29 km in fifth order, as presented in Table 13.1.

13.3.1.1.5 Mean stream length

The mean stream length (Lsm) serves as an indicator of the typical size of components within a drainage network and its contributing surfaces, as outlined by Strahler in 1964. This metric is computed by dividing the total stream length of order "u" by the number of stream segments in that order (refer to Table 13.2). For the Haringmuri watershed, Lsm values range from 1.15 to 14.11 km, with an average Lsm value of 4.82 km. It is significant that for every stream order in the basin, the Lsm value is higher than the lower order and lower than the next higher order. Different basins have an impact on the variety in Lsm values, which is closely correlated with each basin's terrain and size. Lsm is a distinctive trait linked to the dimensions of a drainage network and the surfaces that correspond to it, as highlighted by Strahler (1964). The mean stream length (in km) for the Haringmuri watershed across various stream orders is presented in Table 13.1.

13.3.1.1.6 Stream length ratio

According to Horton (1945), there is a strong association between surface flow and discharge and the stream length ratio, which is defined as the ratio of the mean stream length of a particular order to the mean stream length of the next lower order. More specifically, for every pair of orders, the RL is determined. In some cases, stream length ratios may follow the expected pattern, such as decreasing as stream order increases, as seen in many natural river networks. However, variations in stream length ratios may occur due to geological factors, land use changes, or anthropogenic alterations of the landscape.

13.3.1.1.7 Basin length

The longest dimension of a basin to its principal drainage channel is known as basin length. The basin length of the Haringmuri watershed is 24.42 km.

13.3.1.1.8 Length of overland flow

It represents the distance water covers on the surface before converging into defined stream channels (Horton, 1945). This aspect has an inverse relationship with the average slope of the channel and closely aligns with the length of sheet flow. It is roughly equivalent to half of the reciprocal of drainage density (Horton, 1945). The Lg of this region is 0.481 (Table 13.2).

13.3.1.2 Areal parameters

Various widely used methods have been employed to calculate the geometry of basin shape in order to determine it.

13.3.1.2.1 Form factor

Horton (1945) introduced a quantitative measure for expressing the form of a drainage basin through the form factor ratio (Rf). This dimensionless ratio is derived by dividing the basin area by the square of the basin length. Basin shape

Table 13.2 Morphometric parameters of Haringmuri watershed.

Sl. no.	Morphometric parameters	Formula	Result
	Linear		
1	Stream order (u)	Hierarchical rank	1st to 5th
2	Bifurcation ratio (Rb)	$Rb = Nu/Nu + 1$ where Nu = total no. of stream segment of order "u" $Nu + 1$ = total no. of stream segment of the next higher order	
3	Mean bifurcation ratio (Rbm)	Rbm = Average of bifurcation ratios of all order	3.395
4	Stream length (Lu)	Length of the stream	224.23
5	Mean stream length (Lsm)	$Lsm = Lu/Nu$ where, Lu = total stream length of the order "u" Nu = Total no. of stream segments of the order "u"	
6	Stream length ratio (RL)	$RL = Lu/Lu - 1$ where, Lu = total stream length of the order "u" $Lu - 1$ = total stream length of its next lower order	
7	Basin length (Lb)	Longest dimension of the basin	24.42 km
8	Length of overland flow (Lg)	$Lg = 1/2D$ where D = drainage density	0.481
	Areal		
1	Form factor (Rf)	$Rf = A/Lb^2$ where A = area of the basin (km^2) Lb^2 = square basin length	0.36
2	Shape factor (Rs)	$Rs = Lb^2/A$ where A = area of the basin (km^2) Lb^2 = square basin length	2.77
3	Circulatory ratio (Rc)	$Rc = 4 \times Pi \times A/P^2$ where Pi = "Pi" value, i.e., 3.14 A = basin area (km^2) P^2 = square of the perimeter (km)	0.359
4	Elongation ratio (Re)	$Re = 2\sqrt{(A/Pi)}/Lb$ where A = basin area (km^2) Pi = "Pi" value, i.e., 3.14 Lb = basin length	0.6778
5	Stream frequency (Fs)	$Fs = Nu/A$ where Nu = total number of stream segments of all order A = basin area (km^2)	0.618/sq km.

(Continued)

Table 13.2 Morphometric parameters of Haringmuri watershed. *Continued*

Sl. no.	Morphometric parameters	Formula	Result
	Linear		
6	Drainage density (Dd)	$Dd = Lu/A$ where Lu = total stream length of all order A = Area of the basin (in km^2)	1.04 km/ sq km.
7	Drainage texture (Dt)	$Dt = N/P$ where N = total no. of stream segments of all order P = basin perimeter	1.533
	Relief		
1	Watershed relief (H)	H = highest elevation − lowest elevation	69 m
2	Relief ratio (Rh)	$Rh = H/Lb$ where H = total relief (relative relief) of the basin (km) Lb = basin length	0.0028

can be characterized using simple dimensionless ratios involving fundamental measurements such as area, perimeter, and length (Singh, Bhatt, & Prasad, 2003). A watershed that is exactly round would have a form factor value of 0.7854. As the value of the form factor decreases, the watershed becomes more elongated. Watersheds with higher form factors exhibit elevated crest flows of shorter duration, whereas lengthened watersheds with minimal form factors sustain flow for longer periods. For the Haringmuri watershed, the form factor value is calculated as 0.36, indicating a lower value and consequently an elongated shape. The presence of a low form factor and elongated basin indicates a longer flatter flow peak. Managing flood flows in such elongated basins is generally more manageable than in circular basins.

13.3.1.2.2 Shape factor

The shape factor is the ratio of the square of the basin length to the area of the basin and is inversely related to the form factor (Rf) (Horton, 1945). In the current study, the shape factor is determined to be 2.77, indicating an elongated basin shape.

13.3.1.2.3 Circularity ratio

This value signifies the resemblance of the watershed's area to the area of a circle with the same perimeter as the watershed. Values approaching 1 suggest a circular watershed, implying a more uniform infiltration and an extended duration for excess water to be discharged from the watershed. The circularity ratio of the Haringmuri watershed is calculated as 0.359.

13.3.1.2.4 Elongation ratio

Schumm (1956) introduced the concept of an elongation ratio (Re), which is measured as the proportion of the diameter of a circle with the same area as the basin to the maximum basin length. The Re values can be from 0 (representing a highly elongated shape) to unity (1.0, representing a circular shape). Therefore, a higher elongation ratio implies a more circular shape for the basin. Re values which are close to 1 are characteristic of regions with very low relief, while values ranging from 0.6 to 0.8 are typically related with high relief and steep slopes (Strahler, 1964). These values can be categorized as follows:

The Haringmuri watershed shows a low value of elongation ratio, 0.6778, which represents the elongated basin. A circular basin demonstrates higher efficiency in runoff discharge compared to an elongated basin.

13.3.1.2.5 Drainage frequency

The stream frequency (Fs), also known as channel frequency or drainage frequency, for the entire basin is recorded as $0.618/km^2$, as depicted in Table 13.2. This frequency is primarily influenced by the lithology of the basin and serves as an indicator of the drainage network's texture. It serves as an index for different stages of landscape evolution, being dependent on factors such as rock structure, infiltration capacity, vegetation cover, relief, the nature and quantity of rainfall, and the permeability of subsurface materials. The stream frequency of the Haringmuri watershed suggests the presence of good vegetation, a moderate relief, high infiltration capacity, and delayed peak discharges due to a low runoff rate. There is a positive correlation with drainage density and stream frequency. Lower stream frequency and drainage density in a basin result in slower runoff, thereby reducing the likelihood of flooding in basins with low to moderate drainage density and stream frequency.

13.3.1.2.6 Drainage density

Drainage density (Dd) plays a crucial role in determining the travel time of water. The measurement of Dd serves as a valuable numerical indicator of landscape dissection and the potential for runoff. Whereas a low drainage density denotes a poorly drained basin with a slower hydrologic reaction, a high drainage density denotes a well divided drainage basin with a relatively quick hydrological response to a rainfall event. Dd is the product of several interrelated elements that control surface runoff and, in turn, affect the drainage basin's water and sediment production. The value of Dd is varied based on properties of soil and rock, vegetation, climate, and relief. In the case of Haringmuri watershed, the moderate Dd value ($1.04 km/km^2$) suggests that the basin possesses permeable subsurface materials, a robust vegetation cover, and a moderate relief. These factors contribute to increased water infiltration and groundwater aquifer recharge. The relationship between drainage density and mean annual flood is closely linked, indicating that the basin as a whole is less susceptible to flooding.

13.3.1.2.7 Drainage texture

Drainage texture (Dt) is one of several morphometric parameters used to analyze the characteristics of drainage basins. Drainage texture refers to the relative fineness or coarseness of a drainage network within a specific watershed. It essentially describes the complexity and density of streams present in the region. It is a ratio between the number of stream segments and perimeter of the watershed. The drainage texture value of the Haringmuri watershed is 1.533. Finer drainage textures suggest faster and higher water runoff, while coarser textures suggest slower infiltration and potentially higher potential for flooding. Finer textures can be associated with higher susceptibility to erosion, while coarser textures might indicate lower erosion rates. Drainage texture can influence the formation of various landforms and the distribution of plant and animal life within the basin.

13.3.1.3 Relief aspects

Relief refers to the disparity in elevation between two designated reference points. The relief measurement of a region serves as an indicator of the potential energy within its drainage system. In regions with high relief, there is the capacity to transfer elevated energy into the drainage system. The relative relief within a region is essentially the variation in elevation between its highest and lowest points.

13.3.1.3.1 Watershed relief

The elevation difference between the watershed's discharge point and the highest point on the water divide line is known as the watershed relief.

$H = $ (*Elevation of the highest point of the watershed—Elevation of the watershed outlet*)

The variation in elevation between the furthest point and the discharge point is determined from the existing DEM. The highest point in the watershed is situated at an elevation of 110 m above the mean sea level (msl), while the lowest relief at the outlet is recorded at 41 m above msl. Consequently, the total relief calculated for the watershed amounts to 69 m.

13.3.1.3.2 Relief ratio

The horizontal distance parallel to the principal drainage line along the basin's longest dimension is known as the relief ratio (Schumm, 1956). It represents the ratio of basin relief to basin length. This ratio measures a river basin's overall steepness and indicates the strength of erosion processes that are taking place on the basin's slopes. The size of the drainage basin and the drainage area typically have an inverse relationship with the relief ratio. It is computed as 0.0028 for the current study, as shown in Table 13.2.

13.3.1.4 Aspect and slope

The aspect map serves as a crucial parameter for comprehending the sun's impact on the local climate of an area. Generally, the west-facing slopes tend to be warmer than the east-facing slopes, particularly in the afternoon (Burnett, Meyer, & McFadden, 2008; Pelletier et al., 2018). Aspect plays a significant role in influencing the distribution of vegetation. Derived from ASTER DEM, the aspect map of the Haringmuri watershed denotes the compass direction, with 0 representing true north and 90 indicating an east-facing slope (see Fig. 13.3). The results show that in comparison to west-facing slopes, the majority of the slopes in the Haringmuri watershed are east-facing, with comparatively higher soil moisture content and moderate vegetation.

The highest elevation in the Haringmuri watershed, reaching 102 m above the sea level, is located in its northwestern part. This area produces higher runoff, reducing the probability of infiltration of rainfall. The slope map is classified into six classes based on percentages: 0%−8% (flat or almost flat), 8%−15% (gentle sloping), 15%−25% (moderately steep), 25%−45% (steep), and >45% (very steep), as shown in Fig. 13.4. The majority of the Haringmuri watershed is categorized as a gentle slope, considered "excellent" for groundwater management due to the favorable conditions for infiltration on a nearly flat terrain. Because of its slightly undulating topography, which promotes maximal percolation or partial runoff, a moderately sloping area is also considered good. Because they have considerable surface runoff and little soil infiltration, "steep" areas are thought to be good places to build dams for water harvesting or infiltration ponds to replenish groundwater. One important factor that directly affects how runoff responses and soil infiltration rate are balanced in a terrain is slope. Higher slope regions tend to produce more runoff, leading to less soil infiltration, which significantly shapes the development of aquifers.

13.3.2 Land use/land cover mapping

The land use/land cover (LULC) map of the watershed, depicted in Fig. 13.5, encompasses water bodies, forested areas, settlements, croplands, fallow lands, and degraded forest lands. The majority of the watershed is characterized with a combination of forested and cultivated lands. Changes in LULC patterns are crucial considerations for evaluating water resource conditions. Water resources face significant challenges due to the effects of climate change and land use practices that require a lot of water.

The alterations in land use patterns, along with their assessment, provide insights into the utilization of land resources through human activities, especially in agriculture and urbanization. Hydrological inferences drawn from land use patterns contribute to the understanding of evolving scenarios of water demand across various activities, such as agricultural use, domestic needs, and industrialization. Moreover, they aid in comprehending rainwater infiltration in the

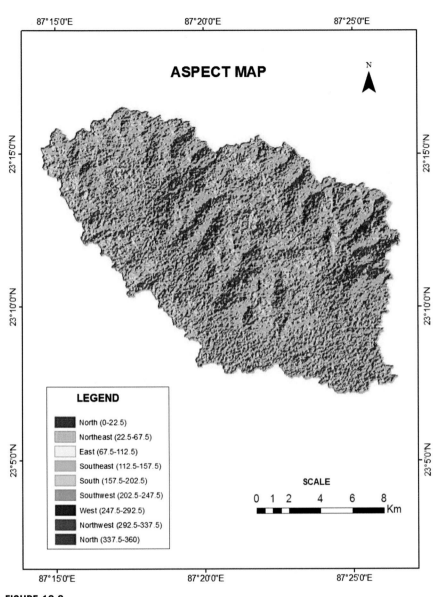

FIGURE 13.3

Aspect map of the Haringmuri watershed.

watershed, groundwater recharge, and surface runoff rates. Changes in land use patterns are integral to hydrological monitoring, modeling, and overall natural resource management. It is essential to analyze changes in land use for hydrologic processes in order to predict future requirements. This encompasses shifts in

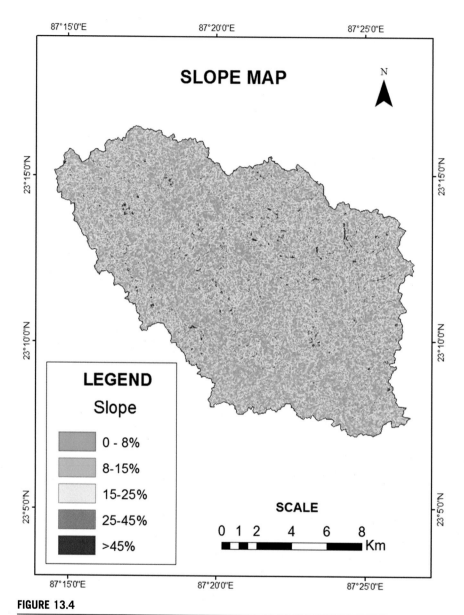

FIGURE 13.4

Slope map of the Haringmuri watershed.

water demand resulting from evolving land use activities, including the irrigation in agricultural land shown in Fig. 13.6 and urbanization and alterations in supply of water due to changes in hydrological processes like surface runoff, infiltration, and groundwater recharge.

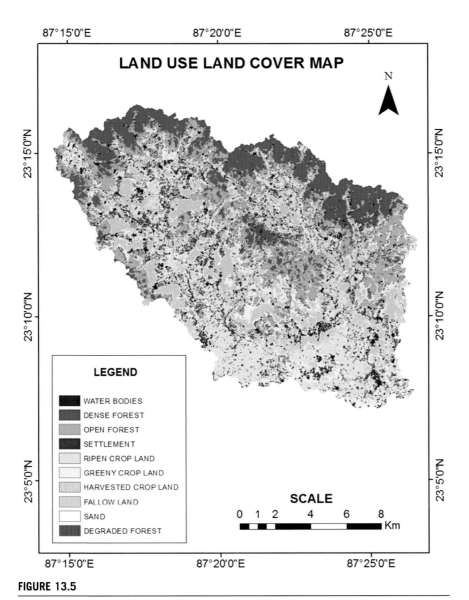

FIGURE 13.5

Land use land cover map of the Haringmuri watershed.

Numerous researchers have emphasized the importance of land use maps as crucial inputs for understanding and managing hydrological conditions within a watershed. The definition of the LULC pattern in the microwatershed reveals that a significant portion comprises forested (Fig. 13.7) and cultivated lands (Fig. 13.8), indirectly supporting the potential for future watershed development and management.

FIGURE 13.6

Agricultural field during the dry season in the Pandua village.

13.3.3 Soil type classification

The soils in the studied area can be broadly categorized into three main types.

13.3.3.1 Lateritic soil

The northern, eastern, and western parts of the region are characterized by later-itic soil, notably in regions encompassing Patharmara, Icharia, Layekbandh villages, and the northeastern part of the study area. This soil type primarily originates from laterite, containing quartz, pebbles, and other rock fragments. It is predominantly loam to sandy loam with a reddish-brown color. Iron concretions are present, and the soil is characterized as dry and completely bleached. The water table varies, reaching 27 feet and above in summer and 8−15 feet during the rainy season (Fig. 13.9).

13.3.3.2 Old alluvial soil

This soil type is located in the southwestern and eastern-central parts of the area. The old alluvial soil belongs mainly to the Vindhya family and is derived from the alluvium transported from the Chhotanagpur Plateau. The soil is predominantly brown to gray-brown to olive-brown in color, with a texture ranging from clay-loam to clay, and in certain areas, sandy clay loam.

FIGURE 13.7

Forest of *Shorea robusta* (Shal Tree) in the Kastha Sanga village.

FIGURE 13.8

Agricultural field during the monsoon period in the Pandua village.

FIGURE 13.9

Laterite soil in the Patharmara village.

13.3.3.3 New alluvial soil

This soil is situated along the Haringmuri River Valley and is primarily formed on recent alluvium of newer origin. The color of the soil ranges from brown to gray-brown, and its texture varies from sandy clay-loam to clay-loam.

13.3.4 Effects of watershed attribute analysis in the water resource management

The utilization of GIS and DEM for the examination of morphological attributes in a watershed is crucial for a comprehensive hydrological analysis of any terrain. This analysis indirectly contributes to maintaining the hydrogeological conditions within the watershed. The quantitative assessment of watershed attributes proves highly valuable in tasks such as watershed demarcation, water and soil conservation, and overall management of the watershed. An examination of the attributes within the Haringmuri watershed revealed that the area possesses a moderate relief and an elongated shape. The selection of runoff harvesting and artificial recharge sites for groundwater development is based on a detailed topographic map. Drainage analysis, when coupled with remote sensing and GIS tools, significantly aids in identifying suitable artificial recharge sites. Comparative indices for the permeability of rock surfaces are produced by the analysis of drainage parameters. Integration of this data with other hydrological characteristics enhances the efficacy of water harvesting and silting recharge strategies, contributing to a more robust groundwater management plan for a particular region. The classification of the drainage pattern within the studied watershed is dendritic, possibly influenced by structural control and lithological homogeneity. Groundwater potential zones are best identified in regions with a moderate drainage density. One important factor impacting infiltration and runoff production is the microwatershed's slope, with a notable inverse relationship between slope and infiltration; gentler slopes are associated with higher infiltration rates.

13.4 Recommended management measures

Throughout the whole Haringmuri watershed, some measures have taken for water resource management. Many small- and medium-sized traditional handmade ponds have been constructed by the local villagers. Two medium-sized check dams have been constructed near Bansh Kopa and Dapanjuri. Two large-sized check dams are constructed near Pandua and Sanatanpur village (Fig. 13.10). However, on the basis of our analysis, the following management measures are recommended.

13.4.1 Surface water storage

13.4.1.1 Large ponds

Large ponds are water reservoirs constructed by communities or households, filled through rainfall, groundwater infiltration, or surface runoff. These ponds serve as valuable resources for farmers, assisting them in coping with dry periods. Notably, these ponds are characterized by the absence of concrete embankments.

13.4.2 Artificial recharge techniques and designs

A diverse range of methods are employed to replenish groundwater reservoirs. Just as the hydrogeological framework exhibits variations, the techniques for artificial recharge also differ significantly. These artificial recharge methods are classified as follows:

- Direct surface methods like stream augmentation, percolation tanks, ditch, and furrow systems.
- Direct subsurface methods include borehole flooding, dug well recharge, injection well or recharge well, and recharge pit and shaft.
- Indirect techniques like induced recharge from surface water sources.

13.4.2.1 Ground water dams or subsurface dykes

A groundwater dam is an underground barrier constructed across a stream, slowing down the natural flow of groundwater within the systems and storing water beneath the ground surface to fulfill the demand in specific periods. The primary objective of a groundwater dam is to impede the groundwater outflow from the subbasin, thereby enhancing storage in the aquifer. The water levels upstream of the groundwater dam increase, saturating areas of the aquifer that would otherwise remain dry. The underground dam offers several benefits. As the aquifer is a

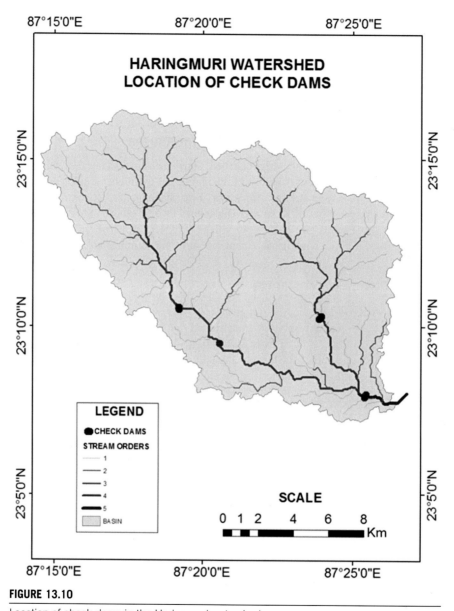

FIGURE 13.10

Location of check dams in the Haringmuri watershed.

reservoir for water, there is no land submergence, allowing the use of land above the reservoir even after dam construction. There is no loss through evaporation from the reservoir. Siltation in the reservoir is eliminated, and the risk of potential disasters, such as dam collapse, can be avoided.

13.5 Limitations of the study

The timeframe available for an in-depth study was limited. As a geographer, it is not feasible to thoroughly examine all the characteristic components and mechanisms of watersheds and water resources, which require an in-depth interdisciplinary investigation.

13.6 Conclusions

Prior to embarking on any artificial recharge initiative, it is imperative to assess the accessibility of water sources for recharging the groundwater reservoir. Hydrological investigations are essential for determining the source water availability in the watershed where artificial recharge structures are planned. In situ precipitation is generally available at most locations, but its adequacy for causing artificial recharge may vary. Runoff that goes unutilized outside the watershed may be stored using simple recharge facilities at appropriate sites. In-depth hydrological studies can help to determine the surplus monsoon runoff that can be utilized as a source of water for artificial recharge. The findings from this study are applicable for conducting an analysis of site suitability for water and soil conservation structures. Within the GIS domain, these parameters are then merged with additional data on hydrology, landforms, geology, land use and cover, water levels, and soil characteristics. This integration aids in making informed decisions about identifying suitable sites for the establishment of soil and water conservation structures, particularly for groundwater development and management (Altaf, Meraj, & Romshoo, 2013). The analysis of watershed attributes using geospatial technology in the case study of Haringmuri Watershed provides valuable insights for effective water resource management. The integration of geospatial tools and techniques has proven to be instrumental in understanding the complex dynamics of the watershed, enabling a comprehensive assessment of its characteristics. Through the examination of various watershed attributes, including LULC, soil types, slope, and hydrological parameters, a holistic understanding of the Haringmuri Watershed has been achieved. This knowledge is essential for making informed decisions and implementing sustainable water resource management practices. The study highlights the significance of accurate and up-to-date spatial information in delineating watershed boundaries and identifying critical areas prone to erosion, sedimentation, and other environmental challenges. Such information is crucial for developing targeted strategies to address issues related to water quality and quantity within the watershed. Additionally, the geospatial analysis provides a robust foundation for creating hydrological models and simulations, aiding in the prediction of water flow patterns and the identification of potential flood-prone zones. This predictive capability is essential for early warning systems and emergency preparedness. Furthermore, the study underscores the importance of stakeholder involvement and community engagement in water resource management initiatives. The geospatial technology employed facilitates effective communication of the findings to diverse stakeholders, fostering collaboration and collective decision-making for sustainable development. The insights gained from this study contribute to the broader understanding of watershed dynamics and can serve as a model for similar regions facing water management challenges. Implementing the recommendations derived from this analysis can lead to more resilient and sustainable water resource management practices, ensuring the long-term health and vitality of the Haringmuri Watershed and similar ecosystems.

References

Al-Saady, Y. I., Al-Suhail, Q. A., Al-Tawash, B. S., & Othman, A. A. (2016). Drainage network extraction and morphometric analysis using remote sensing and GIS mapping techniques (Lesser Zab River Basin, Iraq and Iran. *Environmental Earth Sciences*, *75*(18). Available from: http://www.springerlink.com/content/121380/, https://doi.org/10.1007/s12665-016-6038-y.

Altaf, F., Meraj, G., & Romshoo, S. A. (2013). Morphometric analysis to infer hydrological behaviour of Lidder Watershed, Western Himalaya, India. *Geography Journal*, *2013*, 1−14. Available from https://doi.org/10.1155/2013/178021.

Arabameri, A., Rezaei, K., Cerda, A., Lombardo, L., & Rodrigo-Comino, J. (2019). GIS-based groundwater potential mapping in Shahroud plain, Iran. A comparison among statistical (bivariate and multivariate), data mining and MCDM approaches. *Science of the Total Environment*, *658*, 160−177. Available from: http://www.elsevier.com/locate/scitotenv, https://doi.org/10.1016/j.scitotenv.2018.12.115.

Benzougagh, B., Meshram, S. G., Dridri, A., Boudad, L., Baamar, B., Sadkaoui, D., & Khedher, K. M. (2022). Identification of critical watershed at risk of soil erosion using morphometric and geographic information system analysis. *Applied Water Science*, *12*(1). Available from: http://www.springer.com/earth + sciences + and + geography/hydrogeology/journal/13201, https://doi.org/10.1007/s13201-021-01532-z.

Borgohain, A., Khajuria, V., Garg, V., Koti, S. R., & Bhardwaj, A. (2023). Comparison of geomorphological parameters detected using MERIT and FABDEM products. *Environmental Sciences Proceedings*, *12*(1), 59.

Burnett, B. N., Meyer, G. A., & McFadden, L. D. (2008). Aspect-related microclimatic influences on slope forms and processes northeastern Arizona. *Journal of Geophysical Research: Earth Surface*, *113*(3). Available from: http://onlinelibrary.wiley.com/journal/10.1002/(ISSN)2169-9011, https://doi.org/10.1029/2007JF000789.

Chakrabortty, R., Ghosh, S., Pal, S. C., Das, B., & Malik, S. (2018). Morphometric analysis for hydrological assessment using remote sensing and GIS technique: A case study of Dwarkeswar River Basin of Bankura district, West Bengal. *Asian Journal of Research in Social Sciences and Humanities*, *8*(4), 113. Available from https://doi.org/10.5958/2249-7315.2018.00074.6.

Chorowicz, J., Ichoku, C., Riazanoff, S., Kim, Y., & Cervelle, B. (1992). A combined algorithm for automated drainage network extraction. *Water Resources Research*, *28*(5), 1293−1302. Available from https://doi.org/10.1029/91WR03098.

Dakin Kuiper, S., Coops, N. C., White, J. C., Hinch, S. G., Tompalski, P., & Stackhouse, L. A. (2024). Enhanced watershed status evaluation: Towards an integrated framework to assess fish habitat in forested watersheds using airborne laser scanning data. *Forest Ecology and Management*, *555*. Available from: https://www.sciencedirect.com/science/journal/03781127. https://doi.org/10.1016/j.foreco.2024.121720.

Daniel, E. B., Camp, J. V., le Boeuf, E. J., Penrod, J. R., Abkowitz, M. D., & Dobbins, J. P. (2010). Watershed modeling using GIS technology: A critical review. *Journal of Spatial Hydrology*, *10*(2), 13−28. Available from: http://www.spatialhydrology.com/journal/PaperFall2010/Watershed%20Modeling%20Using%20GIS%20Technology.pdf.

Das, A., Mondal, M., Das, B., & Ghosh, A. R. (2011). Analysis of drainage morphometry and watershed prioritization in Bandu Watershed, Purulia, West Bengal through remote sensing and GIS technology—A case study. *International journal of Geomatics and Geosciences*, *2*(4), 995−1013.

Das, P. (2016). Morphometric diversity on Kuya River Basin. *International Research Journal of Earth Science*, *4*, 17−28.

Dass, B., Abhishek., Sen, S., Bamola, V., Sharma, A., & Sen, D. (2021). Assessment of spring flows in Indian Himalayan micro-watersheds—A hydro-geological approach. *Journal of Hydrology*, *598*, 126354. Available from https://doi.org/10.1016/j.jhydrol.2021.126354.

Fenta, A. A., Yasuda, H., Shimizu, K., Haregeweyn, N., & Woldearegay, K. (2017). Quantitative analysis and implications of drainage morphometry of the Agula watershed in the semi-arid northern Ethiopia. *Applied Water Science*, *7*(7), 3825−3840. Available from: https://www.springer.com/journal/13201, https://doi.org/10.1007/s13201-017-0534-4.

Gebre, T., Kibru, T., Tesfaye, S., & Taye, G. (2015). Analysis of watershed attributes for water resources management using GIS: The case of Chelekot micro-watershed, Tigray, Ethiopia. *Journal of Geographic Information System*, *07*(02), 177−190. Available from https://doi.org/10.4236/jgis.2015.72015.

Ghosh, A., Adhikary, P. P., Bera, B., Bhunia, G. S., & Shit, P. K. (2022). Assessment of groundwater potential zone using MCDA and AHP techniques: Case study from a tropical river basin of India. *Applied Water Science*, *12*(3). Available from: http://www.springer.com/earth + sciences + and + geography/hydrogeology/journal/13201, https://doi.org/10.1007/s13201-021-01548-5.

Girma, R., Abraham, T., & Muluneh, A. (2020). Quantitative evaluation of watershed attributes for water resources management in the Rift Valley Lakes Basin, Ethiopia: A case from Tikur Wuha river watershed. *Applied Water Science*, *10*(8). Available from: https://www.springer.com/journal/13201, https://doi.org/10.1007/s13201-020-01281-5.

Grigg, N. S. (2016). *Watersheds as social-ecological systems* (pp. 139−149). Springer Nature. Available from http://doi.org/10.1057/978-1-137-57615-6_7.

Hall, R., Ranganathan, S., & Raj, G. C. (2017). A general micro-level modeling approach to analyzing interconnected SDGs: Achieving SDG 6 and more through multiple-use water services (MUS. *Sustainability*, *9*(2), 314. Available from https://doi.org/10.3390/su9020314.

Horton, R. E. (1932). Drainage-basin characteristics. *Transactions, American Geophysical Union, 13*(1), 350−361.

Horton, R. E. (1945). Erosion development of streams and their drainage basins: Hydrophysical approach to quantitative morphology. *Bulletin of the Geological Society of America, 56*, 275−370.

Huang, G., Shen, Z., & Mardin, R. (2018). *Overview of urban planning and water-related disaster management* (pp. 1−10). Springer Nature. Available from http://doi.org/10.1007/978-3-319-90173-2_1.

Ikram, R. M. A., Meshram, S. G., Hasan, M. A., Cao, X., Alvandi, E., Meshram, C., & Islam, S. (2024). The application of multi-attribute decision making methods in integrated watershed management. *Stochastic Environmental Research and Risk Assessment, 38*(1), 297−313. Available from: https://www.springer.com/journal/477, https://doi.org/10.1007/s00477-023-02557-3.

Jadhav, S. I., & Babar, M. (2013). Morphometric analysis with reference to hydrogeological repercussion on Domri River sub-basin of Sindphana River Basin, Maharashtra, India. *Journal of Geosciences and Geomatics, 1*, 29−35.

Jasrotia, A. S., Kumar., & Aasim, M. (2011). Morphometric analysis and hydrogeomorphology for delineating groundwater potential zones of Western Doon Valley. *International Journal of Geomatics and Geosciences, 2*(4), 1078−1096.

Javed, A., Khanday, M. Y., & Rais, S. (2011). Watershed prioritization using morphometric and land use/land cover parameters: A remote sensing and GIS based approach. *Journal of the Geological Society of India, 78*(1), 63−75. Available from https://doi.org/10.1007/s12594-011-0068-6.

Jha, M. K., Chowdhury, A., Chowdary, V. M., & Peiffer, S. (2007). Groundwater management and development by integrated remote sensing and geographic information systems: Prospects and constraints. *Water Resources Management, 21*(2), 427−467. Available from https://doi.org/10.1007/s11269-006-9024-4.

Krishan, S. (2011). Water harvesting traditions and the social milieu in India: A second look. *Economic and Political Weekly, 46*(26−27), 87−95.

Kumar, K., Gurunath, S., Madhusudan Rao, Y., & Ajitha, M. (2020). Formulation, characterization and evaluation to establish the bioavailability of gastroretentive mucoadhesive dosage of atenolol in human subjects with possible in-vitro-in-vivo correlation. *Iranian Journal of Pharmaceutical Sciences, 16*(4), 53−70. Available from: http://www.ijps.ir/article_242597_41ca4c4dcf30b4293f82be4b3cd32eac.pdf.

Li, X., Cheng, G., Lin, H., Cai, X., Fang, M., Ge, Y., ... Li, W. (2018). Watershed system model: The essentials to model complex human-nature system at the River Basin Scale. *Journal of Geophysical Research: Atmospheres, 123*(6), 3019−3034. Available from: http://onlinelibrary.wiley.com/journal/10.1002/(ISSN)2169-8996, https://doi.org/10.1002/2017JD028154.

Lindsay, J. B., Rothwell, J. J., & Davies, H. (2008). Mapping outlet points used for watershed delineation onto DEM-derived stream networks. *Water Resources Research, 44*(8). Available from https://doi.org/10.1029/2007WR006507.

Mascarenhas, P. V. (2021). Water culture connection. *Ekistics and the New Habitat, 80*(2), 33−39. Available from https://doi.org/10.53910/26531313-e2020802528.

Mengistu, F., & Assefa, E. (2020). Towards sustaining watershed management practices in Ethiopia: A synthesis of local perception, community participation, adoption and livelihoods. *Environmental Science and Policy, 112*, 414−430. Available from: http://www.elsevier.com/wps/find/journaldescription.cws_home/601264/description#description. https://doi.org/10.1016/j.envsci.2020.06.019.

Mink, S., López-Martínez, J., Maestro, A., Garrote, J., Ortega, J. A., Serrano, E., ... Schmid, T. (2014). Insights into deglaciation of the largest ice-free area in the South Shetland Islands (Antarctica) from quantitative analysis of the drainage system. *Geomorphology*, *225*(C), 4−24. Available from: http://www.elsevier.com/inca/publications/store/5/0/3/3/3/4/, https://doi.org/10.1016/j.geomorph.2014.03.028.

Omernik, J. M., Griffith, G. E., Hughes, R. M., Glover, J. B., & Weber, M. H. (2017). How misapplication of the hydrologic unit framework diminishes the meaning of watersheds. *Environmental Management*, *60*(1), 1−11. Available from: http://link.springer.de/link/service/journals/00267/index.htm, https://doi.org/10.1007/s00267-017-0854-z.

O'Callaghan, J. F., & Mark, D. M. (1984). The extraction of drainage networks from digital elevation data. *Computer Vision, Graphics, & Image Processing*, *28*(3), 323−344. Available from https://doi.org/10.1016/S0734-189X(84)80011-0.

Pelletier, J. D., Barron-Gafford, G. A., Gutiérrez-Jurado, H., Hinckley, E. L. S., Istanbulluoglu, E., McGuire, L. A., ... Tucker, G. E. (2018). Which way do you lean? Using slope aspect variations to understand Critical Zone processes and feedbacks. *Earth Surface Processes and Landforms*, *43*(5), 1133−1154. Available from: http://onlinelibrary.wiley.com/journal/10.1002/(ISSN)1096-9837. https://doi.org/10.1002/esp.4306.

Rai, P. K., Mohan, K., Mishra, S., Ahmad, A., & Mishra, V. N. (2017). A GIS-based approach in drainage morphometric analysis of Kanhar River Basin, India. *Applied Water Science*, *7*(1), 217−232. Available from: https://www.springer.com/journal/13201, https://doi.org/10.1007/s13201-014-0238-y.

Rao, K. N., & Latha, P. S. (2019). Groundwater quality assessment using water quality index with a special focus on vulnerable tribal region of Eastern Ghats hard rock terrain, Southern India. *Journal of Geosciences*, *12*(8). Available from: http://www.springer.com/geosciences/journal/12517?cm_mmc = AD-_-enews-_-PSE1892-_-0, https://doi.org/10.1007/s12517-019-4440-y.

Rao, N. K., Latha, S. P., Kumar, A. P., & Krishna, H. M. (2010). Morphometric analysis of Gostani river basin in Andhra Pradesh State, India using spatial information technology. *International Journal of Geomatics and Geosciences*, *1*(2), 179−187.

Sarkar, A. (2012). *Practical geography: A systematic approach.* Orient Black Swan Publcatio.

Schumm, S. A. (1956). Evolution of drainage systems and slopes in badlands at Perth Amboy, New Jersey. *Bulletin of the Geological Society of America*, *67*(5), 597−646. Available from: http://gsabulletin.gsapubs.org/content/by/year. https://doi.org/10.1130/0016-7606(1956)67[597:EODSAS]2.0.CO;2.

Shao, Z., Fu, H., Li, D., Altan, O., & Cheng, T. (2019). Remote sensing monitoring of multiscale watersheds impermeability for urban hydrological evaluation. *Remote Sensing of Environment*, *232*, 111338. Available from https://doi.org/10.1016/j.rse.2019.111338.

Singh, R. K., Bhatt, C. M., & Prasad, V. H. (2003). Morphological study of a watershed using remote sensing and GIS techniques. *Hydrology Journal*, *26*, 55−66.

Soni, S. (2017). Assessment of morphometric characteristics of Chakrar watershed in Madhya Pradesh India using geospatial technique. *Applied Water Science*, *7*(5), 2089−2102. Available from: https://www.springer.com/journal/13201, https://doi.org/10.1007/s13201-016-0395-2.

Strager, M. P., Fletcher, J. J., Strager, J. M., Yuill, C. B., Eli, R. N., Todd Petty, J., & Lamont, S. J. (2010). Watershed analysis with GIS: The watershed characterization and modeling system software application. *Computers and Geosciences*, *36*(7), 970−976. Available from https://doi.org/10.1016/j.cageo.2010.01.003.

Strahler, A. N. (1952). Hypsometric (area-altitude) analysis of erosional topography. *Geological Society of America Bulletin, 63(11)*, 1117–1142.

Strahler, A. N. (1957). Quantitative analysis of watershed geomorphology. *Eos, Transactions American Geophysical Union, 38*(6), 913–920. Available from https://doi.org/10.1029/TR038i006p00913.

Strahler, A. N. (1964). *Quantative geomorphology of drainage basins and channel networks.* McGraw Hill Book Company.

Suhail, H. A., Yang, R., Nie, Q., Zhang, X., Pu, Y., & Wu, X. (2024). Increasing uplift rate since ∼10 Ma in the eastern Tibetan Plateau from river profile inversion. *Journal of Earth System Science, 133*(1). Available from https://doi.org/10.1007/s12040-024-02263-w.

Suwandana, E., Kawamura, K., Sakuno, Y., & Kustiyanto, E. (2012). Thematic information content assessment of the ASTER GDEM: A case study of watershed delineation in West Java, Indonesia. *Remote Sensing Letters, 3*(5), 423–432. Available from https://doi.org/10.1080/01431161.2011.593580.

Tezer, A., Lutfi Sen, O., Aksehirli, L., Cetin, N. I., & Onur, A. C. T. (2012). Integrated planning for the resilience of urban riverine ecosystems: The Istanbul-Omerli Watershed case. *Ecohydrology & Hydrobiology, 12*(2), 153–163. Available from https://doi.org/10.2478/v10104-012-0015-1.

Tiwari, A. V., & Rai, A. R. B. (2015). *Integrating micro watershed management for the evolution of physical development plan.*

Wani, S. P., & Garg, K. K. (2010). *Watershed management concept and principle.*

Ecological and economical services of coastal landscape: a case study on Ramnagar-I and II Blocks, Purba Medinipur, West Bengal, India

14

Amrit Kamila[1] and Jatisankar Bandyopadhyay[2]

[1]*Coastal Observatory and Outreach Centre (COOC), Vidyasagar University, Midnapore, West Bengal, India*
[2]*Remote Sensing and GIS, Vidyasagar University, Midnapore, West Bengal, India*

14.1 Introduction

Landscape ecology is a constraint related to environmental complexity in addition to geographical variability and the implications of measure. Landscape service perception contains the service-related trajectories of ecological, landscape, and environmental facilities, which underscore the connection between actual frameworks (environments or scenes) and human qualities. Facilities must be designed for the continuation and convenience of humanity (Daily, 1997; De Groot, 2006; Groot, 2002). There has been debate and formation regarding the methodical system of services ever since the phrase was first used in the literature, such as the separation of functions and processes from services and their explanation (Bastian, Krönert, & Lipský, 2006; Wallace, 2007). Coral reefs, seagrass beds, open water, mangroves, and sand are just a few of the habitat types that are altered in tropical marine ecosystems, which frequently exist as energetic, spatially diverse seascapes that are connected by numerous biological, physical, and chemical processes. Water movement, including tides and currents, facilitates the exchange of nutrients, chemical pollutants, pathogens, sediments, and organisms between the activities of the viewing area. The dynamic movement of organisms also unifies environmental zones along the marine landscape (Gillanders, Elsdon, & Roughan, 2011; Sale et al., 2010). Incredibly portable species might bond together during their everyday scavenging exercises, including flowing developments, as well as during bigger scope ventures for multiplication and relocation of occasional home (Kramer & Chapman, 1999; Zeller, 1998). The money saved

Applications of Geospatial Technology and Modeling for River Basin Management.
DOI: https://doi.org/10.1016/B978-0-443-23890-1.00014-1

on repairing damage can be used to cover the costs of creating and maintaining vegetation strips. Converts in commodities and services put additional strain on people's welfare (Daily, 1997; Groot, 2002).

In general, the values that these learnings produce are very appropriate, depending on the landscape and the revolution that is expected to occur in that area. Therefore, these exercise values must be included to the nonuse values in a more thorough evaluation of the landscapes' total monetary value. There cannot be any assumption made about the value categories' equilibrium anywhere. Integrating spatial technologies into the teaching of landscape ecology makes it possible to acquire the theoretical and functional frameworks necessary to attempt these global environmental applications at regular spatial scales (Wu, 2006). Recently, a comprehensive set of concepts, terminology, and research tools to examine the link between the spatial configuration of the land surface and ecological measures has been developed (Turner, 2005). The study estimates the overall value based on the subdepletion of ecosystem services to standardize the cost of the whole ecosystem. The valuation of environmental units is thought to be founded on scene biological system administrations.

When assessing the landscape, the efficacy of various land uses, including wetlands, fishing areas, agricultural fields, and other productive lands, is taken into consideration. Recent scientific achievements in remote sensing, acoustic telemetry, geographic information systems, and spatial statistics have enabled us to capture, manage, and analyze the data needed for connectivity research, which follows a spatially precise approach at an appropriately large spatial scale (Crooks & Sanjayan, 2006).

It is well known that the main difficulty involved in taking into account the diversity of biodiversity services in economic choices lies in the fact that many of these services are not valued in the market (Heywood & Watson, 1995). The economic worth of biodiversity and market valuation differ from one another. The services provided by natural products that are valued even if they are not priced in the market must first be documented before they may, to the maximum degree feasible, be made profitable in order to bridge these valuation gaps. It was not until 1967 that survival values were recognized as possible benefits from natural goods. They were often anonymous before this date (Krutilla, 1967; Tilman, 1997).

14.2 Physical setup of the study area

The study area is the Ramnagar-I and II blocks of the Purba Medinipur district in West Bengal. It is a large coastal area that is a part of the Purba Medinipur district and is situated in the exciting southwest of the state on the Bay of Bengal coastal plain. This emerging coastal plain, consisting of sand and mud deposited by fluvial and eolian processes, is also part of the east-central Kanthi coastal plain (Fig. 14.1).

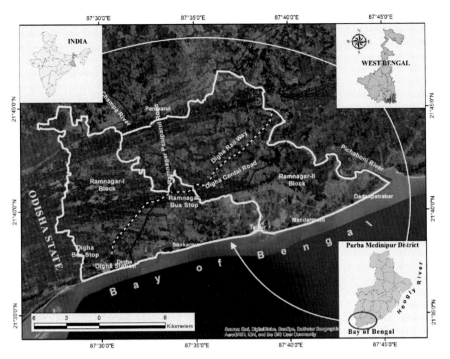

FIGURE 14.1

Location map of the study area.

The extensive beaches and networks of coves, mangroves, mudflats, frequent coastal dunes, and long stretches of sandbanks are the main features of this coastal region. Climate variability in the study area is greater between the monsoon season and the premonsoon season. Temperatures change from at least of 9°C in winter to a high of 38°C in summer. Relative mugginess ranges from 90% to 96% in most months. The study area does not have extensive forest land and grass land. On the other hand, the gricultural land is the leading plants of this study area. Trees such as casuarina, eucalyptus, and *Acacia auriculiformis* have been grown in this area, while coconut, banana, bamboo, and mango are native floral species.

14.3 **Materials and methods**

Several researchers have developed numerous methods of landscape valuation assessment. In this context, the study separates the procedures into three sorts, such as market-based strategies, uncovered inclination methods, and expressed inclination procedures, extending from integral market methods to nonmarket procedures. To account for the diversity of floral species, this study considers the transect method to assess the species community at the micro level. Transects has been drawn at three areas close Jaldah Mohana, close Tajpur, and close Ancient Digha based on the proximity of compact floral variances. Vegetation classification based on arithmetic operation of images was done considering seasonal data of two periods like premonsoon and postmonsoon. The Normalized Difference Vegetation Index (NDVI) technique is used to classify vegetation types. The NDVI proportion is calculated by separating the contrast between the near infrared (NIR) and red band by the sum of the NIR and red groups for each picture pixel (Rouse, Haas, Schell, & Deering, 1974). Such parameters (organic carbon, oxygen production, ground water recharge capacity, and fire wood) are considered for the estimation of the ecological as well as economical services of the coastal vegetated surface. On the other hand, carbon storage, nitrogen emission, methane emission, and ground water recharge capacity are measured to evaluate the valuation of the agricultural landscapes. In the case of different fish firm practices, carbon sequestration, nitrogen emission, and methane emission are estimated to achieve the ecological/economical services of these plots.

The market appraising technique used to calculate the economic profits of traded properties is constructed on the quantities purchased at diverse values and the quantities provided at diverse values. Revealed preference approaches are considered on goods and services that are clearly related to the ecological value of the landscape and are addressed in a variety of ways, such as the link between travel costs and market statistics. These data can be mined for hypothetical values. The stated preference techniques used so far fall short of reproducing all the values that landscape ecology can offer, by requiring no utilization or inactive utilization of natural benefits. It is evident that individuals are willing to compensate such benefits.

14.4 Results and discussions

The coastal vegetated surface is divided into three types, such as vegetation of the low lying coastal plain surface (30 km²), vegetated beach ridge surface with sloping flats and older natural levees (48 km²), and ridge crest vegetation (53 km²) based on the NDVI techniques (Fig. 14.2). The NDVI value ranges from -1.0 to $+1.0$, but negative NDVI (values approaching -1) indicates the deep-water part of the surface. A value close to zero (-0.1 to 0.1) indicates desert rocks, sands, and snow zones. Low positive values represent brambles and grasslands (around $0.2-0.4$), while high values represent calm and tropical rainforest (values move in the direction of 1). The taxonomy was isolated using the spectral responses of the different species groups within each grid.

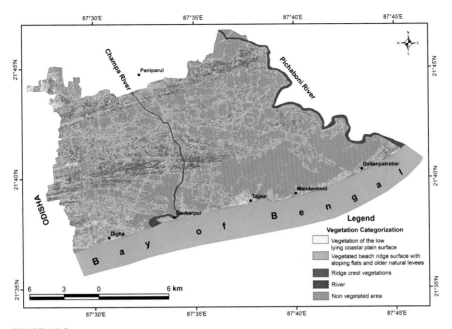

FIGURE 14.2

Classification of coastal vegetations within the different topographic units.

14.4.1 Valuation of the vegetated area

Sixteen sampling sites have been selected to estimate the above ground biomass (AGB) and below ground biomass (BGB) of the study area, which almost absolutely covers the entire vegetation cover (Fig. 14.3). According to their connotation, the presence of tree height conspires against each other to understand the equality of vegetation (Fig. 14.4). The normal tallness of the trees is continuously superior in Grid-1 to Grid-7. These characteristics appearances that the considerable range is partial abundant category of vegetation and the occurrence of vegetation is well progression. On the other hand, Grid No. 8 to 16 shows mixed or bush type vegetation because presence of tree and height of tree are not segregate of the region.

14.4.1.1 Sampling strategy for biomass estimation...

As per the Kyoto Protocol norms, the grid method and random sampling techniques are mostly accepted by the Clean Development Mechanism for cost-effective base line and afforestation/reforestation projects. So, the recent study comprises 16 grids with 15 m × 15 m grid size, which were arbitrarily placed to determine indicator parameters such as tree diameter at breast height (DBH) or height with allometric functions to calculate the biomass and simplifying the

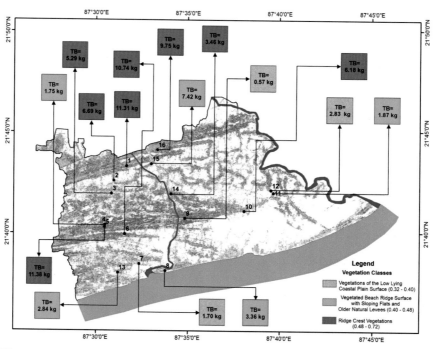

FIGURE 14.3

Location of grid and amount of total biomass of the coastal vegetated surface.

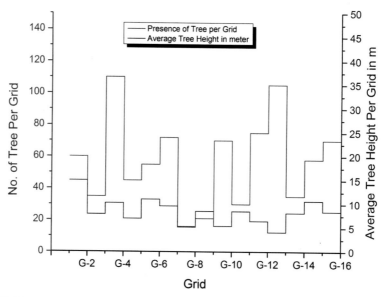

FIGURE 14.4

Grid-wise comparison between the number of trees and height.

value of a hectare and the estimate of the endeavor per quarter. Ensuing perceptions were calculated to gage the ground biomass drag over the plain soil.

Tree biomass was estimated based on DBH and tree height. DBH can be determined by calculating the girth breast height (GBH) of the tree, approximately 1.3 m from the ground level. GBH of trees more than 10 cm wide was calculated directly with a meter (Table 14.1). Tree height was calculated according to the following method proposed by Pearson, Brown, and Ravindranath (2005).

14.4.1.1.1 Above ground biomass

The AGB includes the shoots, branches, leaves, flowers, and fruits of the whole tree. This is treated with the method reported by Zanne et al. (2009), so the AGB of trees includes all above-ground parts of the plant. The GBH of a tree is directly measured by measuring adhesive tape, and the height of the tree is calculated by operating the Abney Level instrument. It is a commonly utilized gadget for measuring tree tallness based on trigonometric standards, and it gives the correct point of rise and fall. Allometric circumstances for biomass estimation is the most section of data on stem width at breast height DBH (m), add up to tree figure H (m), and wood thickness (kg/m^3). The component of the AGB similar to the allometric condition is the kilogram (kg). AGB is measured using the following equation:

$$\text{AGB } (\text{kg/tree}) = \text{Volume of tree } (m^3) \times \text{Wood thickness } (\text{kg}/m^3)$$
$$= \pi r2H \ (m^3) \times \text{Wood Thickness } (\text{kg/m3}) = (\text{GBH})2/\ 4\pi \times H \times \text{Wood thickness } (\text{kg}/m^3)$$
$$\text{where, } r = \text{Span of the tree } (m) = \text{GBH}/2\pi H = \text{Stature of the tree } (m)$$

If the wood thickness of the tree variety was not available, a standard average of 0.6 g/cm^3 was used (Warran & Patwardhan, 2001; Zanne et al., 2009).

14.4.1.1.2 Below ground biomass

The BGB includes the available root biomass and excludes fine roots having <2 mm diameter. To determine the BGB, the AGB is multiplied with a root shoot ratio coefficient of 0.26 (Ravindranath & Ostwald, 2007). The nondestructive method is used to estimate the carbon stock of a major tree. The BGB is calculated by using the following equation: BGB (kg/tree) or (ton/tree) = AGB (kg/tree) or (ton/tree) $\times$ 0.26 (Hangarge, Kulkarni, Gaikwad, Mahajan, & Chaudhari, 2012; MacDicken, 1997).

14.4.1.1.3 Total biomass

The total biomass of the trees is calculated by the sum of AGB and BGB of trees (MacDicken, 1997; Sheikh, Kumar, Bussman, & Todaria, 2011). Hence, the total biomass (kg/tree) or (ton/tree) = AGB + BGB, as shown in Table 14.2.

14.4.1.2 Carbon and oxygen estimation

Generally, for any floral types of species 50% of its biomass is restrained as carbon (Pearson et al., 2005), that is, carbon capacity breaks even with 50%

Table 14.1 Grid-wise tree sampling for biomass estimation (November, 2018).

Grid no.	No. of trees present in the grid	Longitude	Latitude	Average diameter of the tree in cm	Average elevation angle (e) in degree	Distance of the tree from the instrument (D) in m
1	60	87°31′40.65″E	21°43′19.56″N	45.72	35	14.99
2	35	87°30′58.59″E	21°42′35.71″N	63.50	15	7.89
3	110	87°30′51.23″E	21°41′57.10″N	27.94	32	10.25
4	45	87°30′29.19″E	21°40′24.86″N	30.48	30	6.95
5	55	87°30′28.59″E	21°40′17.56″N	55.88	38	11.02
6	72	87°31′34.22″E	21°39′57.31″N	50.80	35	9.61
7	16	87°32′21.58″E	21°38′28.98″N	58.42	20	5.19
8	21	87°33′46.43″E	21°38′8.49″N	55.88	21	8.52
9	70	87°34′48.85″E	21°40′43.75″N	17.78	23	5.38
10	30	87°38′3.59″E	21°41′4.78″N	63.50	27	8.49
11	75	87°39′39.20″E	21°41′58.19″N	25.40	18	6.45
12	105	87°39′31.30″E	21°42′6.34″N	33.02	19	4.12
13	35	87°31′11.92″E	21°38′2.53″N	40.64	20	8.16
14	58	87°34′5.07″E	21°41′55.78″N	30.48	31	10.67
15	70	87°33′0.94″E	21°43′24.84″N	45.72	33	8.43
16	150	87°33′21.33″E	21°44′6.65″N	38.10	38	7.45

(Tan e × D) + h (eye height). Each grid is considered 15 m × 15 m (225 m^2). Eye height (h) = 1.5 m (59 in.).

Table 14.2 Calculation of vegetation biomass (November, 2018).

Grid no.	No. of trees present in the grid	Average GBH (cm)	Average tree height (m)	Above ground biomass (ton/tree)	Below ground biomass (ton/tree)	Total biomass (ton/tree)	Total biomass (ton/grid)	Organic carbon (ton/grid)
1	60	45.72	14.99	0.1496	0.0388	0.1884	11.30	5.655
2	35	63.50	7.89	0.1519	0.0394	0.1913	6.69	3.345
3	110	27.94	10.25	0.0382	0.0099	0.0481	5.29	2.645
4	45	30.48	6.95	0.0308	0.0080	0.0388	1.75	0.875
5	55	55.88	11.02	0.1642	0.0427	0.2069	11.38	5.690
6	72	50.80	9.61	0.1184	0.0307	0.1491	10.74	5.37
7	16	58.42	5.19	0.0845	0.0219	0.1064	1.70	0.850
8	21	55.88	8.52	0.1270	0.0330	0.1600	3.36	1.680
9	70	17.78	5.38	0.0081	0.0021	0.0102	0.57	0.285
10	30	63.50	8.49	0.1634	0.0424	0.2058	6.18	3.090
11	75	25.40	6.45	0.0198	0.0051	0.0249	1.87	0.935
12	105	33.02	4.12	0.0214	0.0055	0.0269	2.83	1.415
13	35	40.64	8.16	0.0643	0.0167	0.0810	2.84	1.420
14	58	30.48	10.67	0.0473	0.0123	0.0596	3.46	1.730
15	70	45.72	8.43	0.0841	0.0218	0.1059	7.42	3.710
16	150	38.10	7.45	0.0516	0.0134	0.0650	9.75	4.875

Each grid is considered 15 m × 15 m (225 m^2).

Biomass. A growing logs plant produces oxygen in proportion to the retrained carbon in the biomass. However, 1 oxygen molecule is created when 1 molecule is charged, that is, in case 1 kg of carbon ties to the biomass, 2.67 kg of oxygen is produced (Quora, 2017).

14.4.1.3 Fire wood estimation

After calculating the whole biomass and natural carbon of each framework, this information are coordinates over the three plant species to gauge the prudent administrations of the vegetation. To calculate the economic value of vegetation, the value of fire wood in each grid is estimated (Table 14.3). For the calculation of the amount of firewood for a given tree, the volume of that tree must be measured by integrating the radius and height of the tree (Blozan, 2006).

Volume can be calculated directly from felled plants or trunks, but this is habitually predicted from measurements such as the smallest span or length of the piece (Tables 14.4 and 14.5). Direct volume measurement is usually done by cutting the wood into smaller pieces that are not defined as cylinders. Eucalyptus are usually fairly large trees, 40−60 ft tall and 40−45 in. wide. The stem of the tree is usually straight and makes up half of the total height. The wood is bought back

Table 14.3 Estimation of fire wood value for each grid.

Grid No.	No. of tree present in the grid	Average diameter of the tree in cm	Height of the tree in m	Volume $(\pi r^2 h)$	Fuel wood in kg/ tree	Fire wood in kg/ 4 year	Fire Wood in kg/ year
1	60	45.72	14.99	2.46	61.76	3705.83	926.46
2	35	63.5	7.89	2.50	62.71	2194.89	548.72
3	110	27.94	10.25	0.63	15.77	1734.96	433.74
4	45	30.48	6.95	0.51	12.73	572.73	143.18
5	55	55.88	11.02	2.70	67.83	3730.59	932.65
6	72	50.8	9.61	1.95	48.88	3519.69	879.92
7	16	58.42	5.19	1.39	34.91	558.64	139.66
8	21	55.88	8.52	2.09	52.44	1101.27	275.32
9	70	17.78	5.38	0.13	3.35	234.67	58.67
10	30	63.5	8.49	2.69	67.48	2024.40	506.10
11	75	25.4	6.45	0.33	8.20	615.19	153.80
12	105	33.02	4.12	0.35	8.85	929.74	232.43
13	35	40.64	8.16	1.06	26.57	929.79	232.45
14	58	30.48	10.67	0.78	19.54	1133.30	283.32
15	70	45.72	8.43	1.38	34.73	2431.41	607.85
16	150	38.1	7.45	0.85	21.32	3197.56	799.39

Each grid is considered 15 m × 15 m (225 m²).

Table 14.4 Grid-wise biomass and fire wood calculation of coastal vegetated surface.

Grid no.	Vegetation types	Total biomass in ton/ 225 m² grid/year	Organic carbon retained in ton/225 m² grid/ year	Oxygen production in ton/225 m² grid/ year	Total fuel wood production in ton/225 m² grid/ year
1, 2, 3, 5, 6, 10, 14, 16	Ridge crest vegetations (53,000,000 m² or 53.00 km²)	8.1 ton	4.05 ton	10.81 ton	0.66 ton (663.79 kg)
4, 8, 11, 13, 15	Vegetated beach ridge surface with sloping flats and older natural levees (47,594,800 m² or 47.59 km²)	3.51 ton	1.724 ton	4.60 ton	0.28 ton (282.52 kg)
7, 9, 12	Vegetation of the low lying coastal plain surface (29,544,600 m² or 29.54 km²)	1.7 ton	0.85 ton	2.27 ton	0.14 ton (143.59 kg)

Table 14.5 Ecological and economical services of coastal vegetated surface/ ha/year.

Ecological factor	Ridge crest vegetations	Vegetated beach ridge surface with sloping flats and older natural levees	Vegetation of the low lying coastal plain surface
Organic carbon retained (ton/ha/year)	180 ton (180,000 kg)	76.6 ton (76,600 kg)	37.77 ton (37,770 kg)
Oxygen production (ton/ha/year)	480.6 ton (480,600 kg)	204.58 ton (204,580 kg)	100.86 ton (100,860 kg)
Ground water recharge (L/ha/year)	0.27 million gallon (1033,355 L)		
Wood fuel (ton/ha/year)	29.50 ton (29,500 kg)	12.56 ton (12,560 kg)	6.38 ton (6380 kg)

(1 ton = 1000 kg); (1 million gallon = 3785,411.8 L); (1 ha = 10,000 sq. m)

Table 14.6 Estimation of ecological and economical valuation in Rs./ha/year.

Ecological and economical parameters	Ridge crest vegetations (ha/year)	Vegetated beach ridge surface with sloping flats and older natural levees (ha/year)	Vegetation of the low lying coastal plain surface (ha/year)
Market value of organic carbon (Rs. 30/kg)	Rs. 5400,000	Rs. 2298,000	Rs. 1121,100
Market value of oxygen production (Rs. 493/kg)	Rs. 236,935,800	Rs. 100,857,940	Rs. 49,723,980
Market value of ground water (Rs. 6/L)	Rs. 6200,130	Rs. 6200,130	Rs. 6200,130
Total ecological value (Rs./year)	Rs. 248,535,930	Rs. 109,356,070	Rs. 57,045,210
Market value of wood fuel (Rs. 5/kg)	Rs. 147,500	Rs. 62,800	Rs. 31,900

at Rs. 5–6/kg. The development frequency is 300–400 kg/tree in 4–5 years. (Agropedia, 2017).

Ramnagar-I and II Blocks received an estimated 118.92 mm (119 mm) of annual rainfall in 2014, according to IMD (premonsoon 132.4 mm from March to May, monsoon 1157.8 mm from June to September, postmonsoon 89.0 mm from October to December, and winter 47.8 mm). A typical yearly boost is thought to be around 830 million gallons for each square mile each year (Hampton, 1963). Vegetated land has a groundwater recharge limit of 0.27 million gallons/ha/year (1033,355 L/ha/year). Lastly, the ecological and economic principles of the three plants types of the study area are calculated. However, ecological and economic services of landscapes are converted into rupees/ha/year. However, the market prices of carbon, oxygen, water, and fire wood are Rs. 30/kg (Indiamart, 2018c), Rs. 493/kg (Sigma-Aldrich, 2018), Rs. 6/L (Indiamart, 2018d), and Rs. 5/kg (Agropedia, 2017), respectively. Edge peak vegetations offer more natural types of assistance and higher monetary worth than other vegetation classifications (Table 14.6).

14.4.2 Valuation of agricultural lands

In this study, 20 sampling sites of single and double cropping cultivated lands are selected for the estimation of the valuation of the agricultural land (Fig. 14.5). For this study, a repeated respondent survey was carried out in the months of

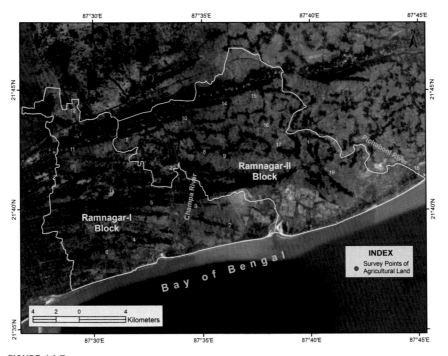

FIGURE 14.5

Location of the agriculture sampling sites.

June and December each year for 4 years, from 2014 to 2018. The region's cropping land is used to compute the quintal yield of paddy, which is then converted to a market value expressed in rupees. Be that as it may, the yearly efficiency potential of twofold editing is Rs. 14,659/bigha/year, and that of single editing is Rs. 6800/bigha/year (Table 14.7).

After the calculation of financial estimate of the agricultural land/ha/year, the biological services of agricultural lands are assessed afterward through diverse steps. In humid regions, the average carbon storage capacity of rice fields with agroforestry practices is projected to be approximately 50 tons per hectare annually (Montagnini & Nair, 2004). The size of the paddy field region is 15,000 ha, of which almost two thirds are flooded by groundwater. The yearly precipitation in the region is roughly 1500 mm. Ground water energize from paddy areas assimilates around 42% (443 million m^3 per year) of the full energize to the shallow aquifer (Elhassan, Goto, & Mizutani, 2003). Therefore, the water recharge capacity of a paddy field is calculated to be 248,179.2 L of drinking water per hectare in a year. The methane emission (CH_4) from the rice field in the damp phase is 41.8 (418) and 15 g/m^2/year (150 kg/ha/year) in the irrigated phase. Thus, the average methane emission from rice farming is 284 kg/ha/year, and correspondingly, the nitrogen oxide (N_2O) emanation from the rice ground is 250 kg/

Table 14.7 Calculation of economic value for agricultural production per/year in INR.

Sl. no.	Farmer's name	Production of paddy per Bigha in Quintal	Prices of paddy per Quintal (Rs.)	Production of paddy per Bigha (Rs.)	Investment for production per Bigha (Rs.)	Net profit per Bigha (Rs.)	Annual profit of double and single cropping land (Rs.)
1	Swapan Dhara	Boro = 13 + Amon = 10	1600	20,800	8500	12,300	Profit of double cropping land (Boro) Rs. 8292 per year
			1400	14,000	6000	8000	
2	Satyan Khara	Boro = 13 + Amon = 8	1500	19,500	10,000	9500	Profit of double cropping land (Amon) Rs. 6367 per year
			1300	10,400	5000	5400	
3	Ganesh Kamilya	Boro = 10 + Amon = 9	1500	15,000	8000	7000	Annual profit of double cropping land Rs. 14,659 per year
			1400	12,600	7000	5600	
4	Purna Khatua	Boro = 13 + Amon = 8	1400	18,200	8000	10,200	
			1300	10,400	6000	4400	
5	Gopal Chanda	Boro = 12 + Amon = 10	1500	18,000	12,000	6000	
			1350	13,500	7000	6500	
6	Sanatan Bodhuk	Boro = 12 + Amon = 8	1400	16,800	9000	7800	
			1300	10,400	5000	5400	
7	Manas Tripathi	Boro = 14 + Amon = 10	1500	21,000	10,000	11,000	
			1400	14,000	6000	8000	
8	Nimai Dinda	Boro = 12 + Amon = 9	1500	18,000	14,000	4000	
			1400	12,600	7000	5600	
9	Basanta Bag	Boro = 10 + Amon = 8	1400	14,000	7000	7000	
			1300	10,400	5000	5400	
10	Babu Bag	Boro = 12 + Amon = 10	1600	19,200	9000	10,200	
			1400	14,000	6000	8000	
11	NakulBera	Boro = 10 + Amon = 8	1600	16,000	8500	7500	
			1450	11,600	6000	5600	
12	Shakti Nayek	Boro = 10 + Amon = 10	1500	15,000	8000	7000	
			1350	13,500	5000	8500	

No.	Name						Annul profit of single cropping land Rs. 6800 per year
13	Ranjit Patra	Amon = 8	1400	11,200	6000	5200	
14	Bimal Nag	Amon = 12	1450	17,400	8000	9400	
15	Swapan Jana	Amon = 10	1400	14,000	6000	8000	
16	Anupam Pani	Amon = 8	1450	11,600	5000	6600	
17	Probodh Mondal	Amon = 10	1400	14,000	6500	7500	
18	Sukumar Patra	Amon = 8	1500	12,000	5000	7000	
19	Rathikanta Giri	Amon = 7	1500	10,500	5000	5500	
20	Kalyan Sardar	Amon = 8	1400	11,200	6000	5200	

(1 Decimal = 436 F²); (1 Decimal = 40.5 m²); (1 Hectare (ha) = 10, 000 m²); (1 Acre = 100 Decimal); (1 Acre = 0.405 ha).

Table 14.8 Ecological and economical valuation of agricultural land in Rs./ha/year.

Ecological input	Single or double cropping paddy field	Market value per kg	Total economic value in rupees
Carbon storage kg/ha/year	50,000 kg (50 ton)	Rs. 30/kg	Rs. 1500,000
Nitrogen emission kg/ha/year	250 kg (0.25 ton)	Rs. 230/kg	Rs. 57,500
Methane emission kg/ha/year	284 kg (0.284 ton)	Rs. 3900/kg	Rs. 1107,600
Ground water recharge liter /ha/year	248,179.2 L	Rs. 6/L	Rs. 1489,075.2
Total ecological value		Rs. 4154,175.2	
Economical value		Rs. 52,982.72	

ha/year (Purkait, Sengupta, De, & Chakrabarty, 2005; Tang, Wang, Li, Wang, & Qu, 2018). The biological services are exchanged within the costs of market price. The market price of nitrogen and methane is Rs. 230/kg (Indiamart, 2018a) and Rs. 3,900/kg (Indiamart, 2018b), respectively, and the price of carbon and drinking water is, as of now, being evaluated over. Agronomic lands have high capacity of carbon and groundwater, which reflects the high financial strength of the region compared to other environmental supervisions (Table 14.8). Therefore, agricultural land provides many more ecological services than economic production.

14.4.3 Valuation of fishery lands

In the coastal area of Digha−Sankarpur, such fishing activities (commercial fishing, subsistence fishing in coastal wetlands, fishing in village ponds, and abandoned fishing) take place in the backshore region. In order to understand the annual fish production from such activities, 43 sampling sites were considered for a repeat survey (Table 14.9). The location of sampling points from different fishing zones are delineated using mapping techniques to better understand the area (Fig. 14.6). Annual profits from fishing were estimated based on the simulations of repeated respondent surveys.

The financial values of the commercial angling, town lake angling, and subsistence angling within the coastal wetlands are calculated as Rs. 1400,000, Rs. 1104,841, and 123,621/ha/year, respectively (Table 14.10). The commercial fishing ground has much greater economic value than fishing in village ponds and subsistence fishing in coastal wetlands.

Commercial and subsistence fishing plots are mainly carried out in coastal wetland areas. To categorize sensitive locations of fishing areas, a thematic map was developed for the identification of abandoned, commercial fishing, and subsistence fishing areas, respectively: 18.02, 10.84, and 4.09 km^2 out of 53.50 km^2

Table 14.9 Sampling sites of different fish firm plots (March, 2018).

Sl. no.	Longitude	Latitude	Village name	Name of the fishing activities
1	87°44′46.946″E	21°41′39.381″N	Dakshin Purusotyampur	Open marine fishing (Soula Mouth)
2	87°38′31.096″E	21°39′24.142″N	Mandarmoni	Open marine fishing (Jaldha)
3	87°38′36.757″E	21°39′17.904″N	Mandarmoni	Open marine fishing (Jaldha)
4	87°32′37.717″E	21°37′56.017″N	Gangadharpur	Open marine fishing (Digha Mohana)
5	87°29′7.943″E	21°36′38.432″N	Duttapur	Open marine fishing (Udaipur)
6	87°34′14.369″E	21°38′27.663″N	Digha	Open marine fishing (Sankarpur)
7	87°43′9.847″E	21°40′8.941″N	Dadanpatrabar	Open marine fishing (Dadanpatrabar)
8	87°44′44.582″E	21°41′30.648″N	Dakshin Purusotyampur	Commercial fishing (Sankar mandi)
9	87°40′51.053″E	21°40′11.810″N	Sonamuhi	Commercial fishing (Astik Das)
10	87°42′29.182″E	21°41′48.979″N	Dakshin Purusotyampur	Commercial fishing
11	87°43′3.408″E	21°41′6.364″N	Dakshin Purusotyampur	Commercial fishing
12	87°42′0.451″E	21°40′6.705″N	Dadanpatrabar	Commercial fishing
13	87°39′49.023″E	21°39′21.473″N	Silampur	Commercial fishing
14	87°37′38.173″E	21°40′22.649″N	Junbani	Commercial fishing
15	87°37′7.203″E	21°39′45.902″N	Jaldha	Commercial fishing
16	87°36′22.230″E	21°40′15.293″N	Dublabari	Commercial fishing
17	87°33′34.121″E	21°38′27.798″N	Purba Mukundapur	Commercial fishing
18	87°33′41.822″E	21°40′6.411″N	Deulbatta	Commercial fishing
19	87°32′49.200″E	21°38′19.646″N	Maitrapur	Commercial fishing
20	87°31′41.604″E	21°38′3.739″N	Ghersai	Commercial fishing
21	87°42′2.679″E	21°40′32.119″N	Dadanpatrabar	Subsistence fishing in the coastal wetlands (Joyanta Bar)
22	87°42′20.130″E	21°41′27.215″N	Dakshin Purusotyampur	Subsistence fishing in the coastal wetlands
23	87°42′59.661″E	21°41′25.784″N	Dakshin Purusotyampur	Subsistence fishing in the coastal wetlands
24	87°42′55.554″E	21°40′26.389″N	Dadanpatrabar	Subsistence fishing in the coastal wetlands
25	87°42′7.039″E	21°41′14.335″N	Mania	Subsistence fishing in the coastal wetlands

(Continued)

Table 14.9 Sampling sites of different fish firm plots (March, 2018). *Continued*

Sl. no.	Longitude	Latitude	Village name	Name of the fishing activities
26	87°40′13.067″E	21°40′29.252″N	Baichibonia	Subsistence fishing in the coastal wetlands
27	87°38′36.550″E	21°40′46.427″N	Patharmuha	Subsistence fishing in the coastal wetlands
28	87°38′24.742″E	21°40′2.773″N	Silampur	Subsistence fishing in the coastal wetlands
29	87°37′59.843″E	21°40′16.609″N	Junbani	Subsistence fishing in the coastal wetlands
30	87°33′56.582″E	21°38′55.257″N	Purba Mukundapur	Subsistence fishing in the coastal wetlands
31	87°35′59.819″E	21°38′35.154″N	Chandanpur	Village pond (Anita Majhi)
32	87°35′18.733″E	21°39′4.283″N	Tengramari	Village pond
33	87°29′43.063″E	21°38′13.462″N	Saripur	Village pond
34	87°30′29.267″E	21°38′14.735″N	Dakshinsimulia	Village pond
35	87°32′55.583″E	21°38′37.959″N	Purba Mukundapur	Village pond
36	87°30′29.374″E	21°40′16.748″N	Gangpura	Village pond
37	87°29′45.565″E	21°43′5.296″N	Chandapur	Village pond
38	87°37′27.615″E	21°41′12.725″N	Satilapur	Village pond
39	87°41′23.773″E	21°40′50.462″N	Rania	Village pond
40	87°40′20.113″E	21°42′36.043″N	Purbba Bar	Village pond
41	87°37′25.561″E	21°42′59.573″N	Sherpur	Village pond
42	87°36′17.109″E	21°44′27.960″N	Kadua	Village pond
43	87°34′40.592″E	21°42′58.937″N	Kanpur	Village pond

(Fig. 14.7) in the regions. On the other hand, open marine fishing activities mainly take place at six key locations of Ramnagar-I and II blocks. A tedious respondent survey was conducted within the crew members of every boat to know the fish catch amount of per operation.

However, fishermen in this area typically carry out 36 operations per year, so the annual fish catch is calculated and converted into market value at each location (Table 14.11). The annual profits of sea trawling in the Sankarpur area are much higher than that in other locations, and low profits can be seen in the Jaldha Mohana area. The zonation of the open marine angling divisions (e.g., Udaipur Division; Digha Mohana; Sankarpur Segment; Jaldha Mohana; Dadanpatrabar Division; and Soula Waterway Mouth) is mapped to understand the distance from the harbor to the seaward side (Fig. 14.8).

Coastal wetlands provide good ecological services and play an important role in regulating the coastal environment. Several key ecological services are

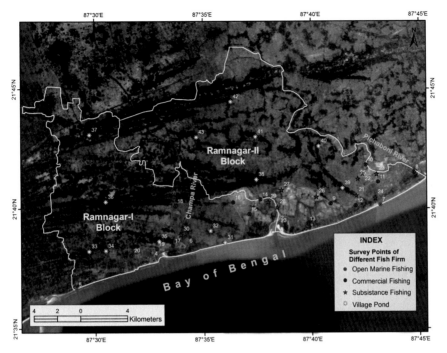

FIGURE 14.6

Sampling location fish firm plots.

identified in this particular area. Wetlands or swampy areas sequester an average of 317 g carbon/m^2/year (Climate Citizen, 2012). In contrast, the amount of methane that brackish water fluxes get is 13.8 g/m^2/year (Holm et al., 2016) and coastal wetlands get 1500 kg nitrogen/ha/year (Waquoit Bay, 2018). All biological services are changed over into showcase values to evaluate the environmental value of coastal wetlands in this range (Table 14.12).

14.4.4 Ecological and economical valuation of coastal landscape units

Ecological and economical services changed over to the showcase cost for the valuation of distinctive scene units, like, vegetated arrive, coastal wetland and agrarian arrive. Carbon sequestration in vegetated lands is very high compared to other landscape units, while coastal wetlands have a greater nitrogen uptake capacity than the other two (Table 14.13). The overall biomass of edge peak vegetation is exceptionally high throughout the study area, and commercial fishing is financially much more vital than other financial exercises within the presently considered region.

Table 14.10 Estimation of yearly production of different fish firm plots.

Fishing ground	Investment for fishing in Rs. (ha/year)	Production of fish in 1 year (quintal)	Market value of captured fish per kg (Rs.)	Total production in 1 year (Rs.)	Annual profit of fishery (Rs.)
Commercial fishing	5600,000	175	400 kg.	7000,000	1400,000
Village pond fishing	185,159	86	150/kg	1290,00	1104,841
Subsistence fishing in the coastal wetlands	246,879	988	375/kg	370,500	123,621

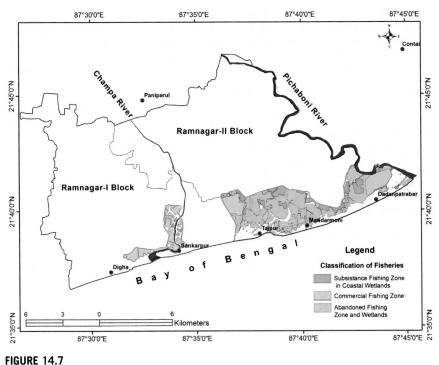

FIGURE 14.7

Spatial distribution of different fishery sectors.

Table 14.11 Estimation of yearly production of open marine fishing.

Fishing sector	Profit for one operation in Rs.	Profit for one year in Rs. (36 operation/year)
Udaipur Sector	8000	288,000
Digha Mohana	85,000	3060,000
Sankarpur Sector	420,000	15,120,000
Jaldha Mohana	3500	126,000
Dadanpatrabar Sector	60,000	2160,000
Soula River Mouth	90,000	3240,000

Assessment of ecological services shows a higher estimated value of the diverse vegetation zones of the coastal belt; on the other hand, economic services show a higher estimated value that can benefit commercial fish farms and open marine fisheries. However, the study also demonstrates that the region's agricultural rice fields have the lowest value. Based on the current findings, the researchers recommend the use of ridge crest vegetations to provide environmental services and forest resource exploitation, mainly cashew nut processing, to restore the sustainability of this special

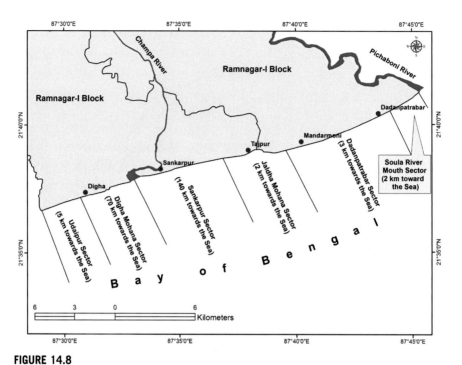

FIGURE 14.8

Seaward extension of the open marine fishing.

Table 14.12 Ecological valuation of fish firm plots in Rs./ha/year.

Ecological parameters	Depressional wetland, brackish water and swamps area	Market value in Rs. per kg	Total economic value in Rs.
Carbon sequestration kg/ha/year	3,170 kg	30/kg	95,100
Nitrogen receive kg/ha/year	1500 kg	230/kg	345,000
Methane emission kg/ha/year	138 kg	3900/kg	538,200
Total ecological value			978,300

(1 ha = 10,000 sq. m/0.01 sq. km); (1 sq. km = 100 ha).

coastal area. Similarly, for other coastal landscape units such as the vegetated side edge surface with inclining pads, more established natural levees, vegetation of the low-lying beach front plain surface, agricultural land (paddy fields), commercial fishing, village pond and subsistence fishing in the coastal wetlands should also be recommended, which can help restore the entire low-lying coastal plain area (Table 14.14). The ecological and economical value of subsistence fishing in the coastal wetlands is very much lower, which needs effective recommendations.

Table 14.13 Ecological and economical valuation of coastal landscapes units/ha/year.

Valuation of landscape units (per/ha/year)	Ecological services in Rs.	Economic services in Rs.	Total value in Rs.
Ridge crest vegetations	248,535,930	147,500	248,683,430
Vegetated beach ridge surface with sloping flats and older natural levees	109,356,070	62,800	109,418,870
Vegetation of the low lying coastal plain surface	57,045,210	31,900	57,077,110
Agricultural land (paddy field)	4154,175.2	52,983	4207,157.92
Village pond	978,300	1104,841	2083,141
Subsistence fishing in the coastal wetlands	978,300	123,621	1101,921
Commercial fishing	978,300	1400,000	2378,300

Table 14.14 Recommended actual uses of landscape valuation for habitats and society.

Valuation of landscape units (per/ha/year)	Total value in Rs.	Exiting use	Recommended use	Remarks
Ridge crest vegetations	248,683,430	Multiple use (settlement, infrastructures, forestry and pastures)	Providing environmental services and options for harvesting forest-based resources	Cashew nut processing may be practiced
Vegetated beach ridge surface with sloping flats and older natural levees	109,418,870	Low height scrubs and small trees with settlements	Orchards with settlements to be developed	Slope control measures by increasing wilderness
Vegetation of the low lying coastal plain surface	57,077,110	Pastures and strip plantations	Nature park development	Flood-prone areas with seasonal use by livestock grazing
Agricultural land (paddy field)	4207,158	Food crop cultivations	Low lying rice and salt-tolerant rice with seasonal practices of cash crops	Mixed farming practices are needed and scope for ground water recharging

(Continued)

Table 14.14 Recommended actual uses of landscape valuation for habitats and society. *Continued*

Valuation of landscape units (per/ha/year)	Total value in Rs.	Exiting use	Recommended use	Remarks
Commercial fishing	2378,300	Fisheries for commercial gains	Mangrove dominated agro-forestry with fishing	Rotational practices with rain water harvesting
Village pond	2083,141	Fresh water fishing	Forestry with fresh water fishing and rain water harvesting	Ground water recharging sites
Subsistence fishing in the coastal wetlands	1101,921	Traditional fish farming practices	Other forms of aquaculture (crabs, crustaceans, and algae)	Mangrove conservations and salt processing

14.5 Conclusion

This study shows that diverse landscape components, such as vegetated lands, coastal wetlands, and agricultural lands, are carefully considered to appraise their services within the framework of financial and environmental valuation. The total biomass concentration of vegetation is exceptionally high at the edge peak portion and at the same time low at the low lying coastal plain surface, whereas the groundwater recharge capacity (infiltration rate) is exceptionally very high in vegetated land (1033,355 L/ha/year). Comparatively, ecological services of different landscapes /ha/year show that the carbon sequestration capacity of vegetated soils is very high compared to other landscape units. Then again, coastal wetlands have a more prominent nitrogen take-up limit than the other two landscape units. With respect to environmental services and efficiency, vegetated land accounts for the foremost ecological services and commercial fisheries have greater production capacity than other landscape units.

References

Agropedia. (2017). *Eucalyptus plantation information guide*. Retrieved from http://agropedia.iitk.ac.in/content/eucalyptus-plantation-information-guide.

Bastian, O., Krönert, R., & Lipský, Z. (2006). Landscape diagnosis on different space and time scales — A challenge for landscape planning. *Landscape Ecology, 21*, 359–374.

Blozan, W. (2006). Tree measuring guidelines of the eastern native tree society. *Bulletin of the Eastern Native Tree Society, 1*(1), 3–10.

Climate Citizen. (2012). *Freshwater wetlands are important carbon sinks says scientific study.* Retrieved from https://takvera.blogspot.com/2012/01/freshwater-wetlands-are-important.html.

Crooks, K. R., & Sanjayan, M. (Eds.), (2006). *Connectivity conservation* (Vol. 14). Cambridge University Press.

Daily G. C. (Ed.) (1997). Introduction: what are ecosystem services. *Nature's services: Societal dependence on natural ecosystems,* 1(1) Washington, DC: Island Press.

De Groot, R. (2006). Function-analysis and valuation as a tool to assess land use conflicts in planning for sustainable, multi-functional landscapes. *Landscape and Urban Planning, 75*(3–4), 175–186.

Elhassan, A. M., Goto, A., & Mizutani, M. (2003). Effect of conjunctive use of water for paddy field irrigation on groundwater budget in an alluvial fan. *Agricultural Engineering International: CIGR Journal.*

Gillanders, B.M., Elsdon, T.S., & Roughan, M. (2011). 7.06 Connectivity of estuaries. Treatise on estuarine and coastal science.

Groot, D. E. (2002). A typology for classification, description and valuation of ecosystem functions, goods and services. *Ecological Economics: the Journal of the International Society for Ecological Economics, 41,* 393–408.

Hampton, E.R. (1963). *Ground water in the coastal dune area near Florence, Oregon (No. 1539-K).* US Govt. Print. Off.

Hangarge, L. M., Kulkarni, D. K., Gaikwad, V. B., Mahajan, D. M., & Chaudhari, N. (2012). Carbon sequestration potential of tree species in Somjaichi Rai (Sacred grove) at Nandghur village, in Bhor region of Pune District, Maharashtra State, India. *Annals of Biological Research, 3*(7), 3426–3429.

Heywood, V. H., & Watson, R. T. (1995). *Global biodiversity assessment* (Vol. 1140). Cambridge: Cambridge University Press.

Holm, G. O., Perez, B. C., McWhorter, D. E., Krauss, K. W., Johnson, D. J., Raynie, R. C., & Killebrew, C. J. (2016). Ecosystem level methane fluxes from tidal freshwater and brackish marshes of the Mississippi River Delta: Implications for coastal wetland carbon projects. *Wetlands, 36,* 401–413.

Indiamart. (2018a). *Natural, industrial & medical gasses. Nitrogen gas.* Retrieved from https://www.indiamart.com/proddctail/nitrogen-gas-9317363991.html.

Indiamart. (2018b). *Natural, industrial & medical gasses. Methane gas.* Retrieved from https://www.indiamart.com/proddetail/methane-gas-11691715730.html.

Indiamart. (2018c). *Natural, industrial & medical gasses. Carbon dioxide gas.* Retrieved from https://www.indiamart.com/proddetail/carbon-dioxide-gas-11178548797.html.

Indiamart. (2018d). *Juices, soups & soft drinks, drinking water. Mineral water.* Retrieved from https://www.indiamart.com/proddetail/1-liter-mineral-water-10983801555.html.

Kramer, D. L., & Chapman, M. R. (1999). Implications of fish home range size and relocation for marine reserve function. *Environmental Biology of Fishes, 55,* 65–79.

Krutilla, J. V. (1967). Conservation reconsidered. *The American Economic Review, 57*(4), 777–786.

MacDicken, K. (1997). A Guide to Monitoring Carbon Storage in Forestry and Agroforestry Projects. Arlington, VA: Winrock International.

Montagnini, F., & Nair, P. R. (2004). *Carbon sequestration: an underexploited environmental benefit of agroforestry systems,* . *New vistas in agroforestry: A compendium for 1st World Congress of Agroforestry* (2004, pp. 281–295). Springer Netherlands.

Pearson, T.R., Brown, S., & Ravindranath, N.H. (2005). *Integrating carbon benefit estimates into GEF projects: guidelines.* Capacity Development and Adaptation Group, Global Environment Facility, United Nations Development Programme, New York.

Purkait, N.N., Sengupta, M.K., De, S., & Chakrabarty, D.K. (2005). *Methane emission from the rice fields of West Bengal over a decade.* 89.60. Fe; 92.60. Sz; 92.70. Cp.

Quora. (2017). *How much carbon dioxide we exhale and how much oxygen is given out by plants or trees?* Retrieved from https://www.quora.com/How-much-carbon-dioxide-we-exhale-and-how-muchoxygen-is-given-out-by-plants-or-trees.

Ravindranath, N. H., & Ostwald, M. (2007). *Carbon inventory methods: Handbook for greenhouse gas inventory, carbon mitigation and roundwood production projects* (Vol. 29). Springer Science & Business Media.

Rouse, J. W., Haas, R. H., Schell, J. A., & Deering, D. W. (1974). Monitoring vegetation systems in the great plains with ERTS. *NASA Special Publications, 351*(1), 309.

Sale, P. F., Van Lavieren, H., Lagman, M. A., Atema, J., Butler, M., Fauvelot, C., & Stewart, H. L. (2010). Preserving reef connectivity: A handbook for marine protected area managers. Connectivity Working Group. *Coral Reef Targeted Research & Capacity Building for Management Program, UNU-INWEH, 88.*

Sheikh, M. A., Kumar, M., Bussman, R. W., & Todaria, N. P. (2011). Forest carbon stocks and fluxes in physiographic zones of India. *Carbon Balance and Management, 6*(1), 1–10.

Sigma-Aldrich. (2018). *00476-Oxygen. Oxygen gas.* Retrieved from https://www.sigmaaldrich.com/catalog/product/sial/00476?lang = en&region = IN&cm_sp = Insite-_-prodRecCold_xviews-_-prodRecCold5-1.

Tang, J., Wang, J., Li, Z., Wang, S., & Qu, Y. (2018). Effects of irrigation regime and nitrogen fertilizer management on CH_4, N_2O and CO_2 emissions from saline–alkaline paddy fields in Northeast China. *Sustainability, 10*(2), 475.

Tilman, D. (1997). Community invasibility, recruitment limitation, and grassland biodiversity. *Ecology, 78*(1), 81–92.

Turner, M. G. (2005). Landscape ecology: What is the state of the science? *Annual Review of Ecology, Evolution, and Systematics, 36,* 319–344.

Wallace, K. J. (2007). Classification of ecosystem services: problems and solutions. *Biological Conservation, 139*(3–4), 235–246.

Waquoit Bay. (2018). *Bringing wetlands to market. Phase 1 – Nitrogen & coastal blue carbon (2011–2015).* Retrieved from https://www.waquoitbayreserve.org/research-monitoring/salt-marsh-carbon-project/.

Warran, A., & Patwardhan, A. (2001). *Carbon sequestration potential of trees in and around Pune city.* Case *Study* Department of Environmental Sciences, University of *Pune.*

Wu, J. (2006). Landscape ecology, cross-disciplinarity, and sustainability science. *Landscape Ecology, 21,* 1–4.

Zanne, A.E., Lopez-Gonzalez, G., Coomes, D.A., Ilic, J., Jansen, S., Lewis, S.L., & Chave, J. (2009). Global wood density database.

Zeller, D. C. (1998). Spawning aggregations: patterns of movement of the coral trout *Plectropomus leopardus* (Serranidae) as determined by ultrasonic telemetry. *Marine Ecology Progress Series, 162,* 253–263.

Comparative study of remote sensing-derived indices for meteorological and agricultural drought monitoring: a review

Argha Ghosh[1], Momsona Mondal[2], Debolina Sarkar[2] and Manoj Kumar Nanda[2]

[1]*Agricultural Meteorology, Odisha University of Agriculture and Technology, Bhubaneswar, Odisha, India*

[2]*Agricultural Meteorology and Physics, Bidhan Chandra Krishi Viswavidyalaya, Mohanpur, West Bengal, India*

15.1 Introduction

Global climate change, acknowledged as both the most significant problem and the most intricate challenge facing humanity, has garnered widespread attention from the public and governments worldwide. The frequency and intensity of extreme climate events, such as droughts, have shown a significant increase since the 1970s. These events, marked by abnormality, unpredictability, and heightened sensitivity to climate change, stand as the primary contributors to terrestrial ecosystem instability, profoundly impacting the sustainable development of ecosystems and the global economy (Stocker et al., 2013). The Earth system model predicts a further escalation of the risk of global drought in the 21st century (Dai, 2011). The evolving climate, characterized by increased spatiotemporal variability (Huang et al., 2019), necessitates a renewed focus on quantifying extreme events, considering their implications for agriculture, infrastructure, and the economy. Among the multitude of adverse effects of drought, its direct and substantial influence on agriculture stands out. Agriculture, intricately linked to national food security and social stability, faces severe constraints imposed by climate and weather conditions (Dai, 2011). Consequently, research on agricultural drought has gained prominence among governments and scholars globally. In agricultural countries like India, the heightened risk of future extremes is attributed to greater variability in monsoon rainfall, increased extraction of groundwater for rice and wheat production, and the looming threat of food demand from a growing population (Zhang, Obringer, Wei, Chen, & Niyogi, 2017). As a result, addressing how to react to and mitigate drought and its impact has become an urgent scientific issue. Statistical data reveals

Applications of Geospatial Technology and Modeling for River Basin Management.
DOI: https://doi.org/10.1016/B978-0-443-23890-1.00015-3

that economic losses due to global meteorological disasters account for 85%, with drought alone contributing to over 50% of these losses (Dai, 2011). The global concern for climate change has prompted significant attention to the quantitative analysis of drought. Over the years, various indices have been employed for drought assessment. Notably, drought indices based solely on rainfall data, such as the Standardized Precipitation Index (SPI) (McKee, Doesken, & Kleist, 1993), have proven to be simple to compute and more effective (Olukayode Oladipo, 1985). SPI, recognized by the World Meteorological Organization as a standard drought monitoring index (Hayes, Svoboda, Wardlow, Anderson, & Kogan, 2012), is one of two indices, the other being the Aridity Anomaly Index, used by the Indian Meteorological Department. In recent decades, satellite-based remote sensing (RS) has provided high spatial and temporal resolution observations of the Earth, making it invaluable for large-area drought monitoring. Satellite-derived vegetation indices have been developed to monitor drought at various scales, and researchers continue to refine drought monitoring tools by integrating climate, satellite, and environmental data. Hybrid drought indices have emerged, incorporating multiple data sources, and recent satellite-based instruments have enabled the estimation of key variables related to drought, including land surface temperature (LST), evapotranspiration, soil moisture, and precipitation. Microwave and radar instruments aboard satellites are increasingly utilized for soil moisture and precipitation estimation. Given the complex nature of drought events in their development and termination (Parry, Wilby, Prudhomme, & Wood, 2016), regular observations of key variables, facilitated by RS, have become crucial. Daily to weekly observations, such as daily rainfall from the Global Precipitation Measurement mission and weekly/bi-weekly vegetation condition indices from moderate resolution imaging spectroradiometer, Landsat, and Sentinel-2, offer valuable insights. This frequent observation, coupled with enhanced spatial coverage, has filled data gaps in traditionally data-sparse regions, particularly in developing countries. The integration of RS into drought monitoring not only enhances our understanding of drought dynamics but also contributes to more effective and informed drought management strategies on a global scale. The present chapter aims to review and discuss about several RS indices for monitoring meteorological and agricultural drought. A comparative study was made among the RS-derived indices. Further, the conventional and RS-based drought monitoring indices were compared in this chapter. Application of the machine learning (ML) algorithms for prediction of drought was also discussed. Finally, this chapter identified some research gaps in monitoring drought using RS indices.

15.2 Drought scenario of India

The observable increase in drought intensity and frequency in recent decades has raised concerns about its societal impacts in India. The escalating drought conditions have induced shifts in agricultural patterns, posing a persistent threat to food security. Several studies reported the variability in drought across the country (Sharma & Goyal, 2020). A rise in the area under moderate drought, designating 1987 as one of the severe drought years, was noted. Additionally, years such as 1985, 1990, 1993, 1997, 1999, 2000, 2001, 2004, 2006, and 2010 have been identified as drought years (Zhang et al., 2017). Regional variations in drought intensity are also notable, emphasizing hydrological drought over the Indo-Gangetic plains. Their analysis reveals that 40%−60% of the study region is affected by mild hydrological drought, while 2%−7% faces extreme hydrological drought, impacting groundwater storage levels negatively. Addressing these shifts is crucial for effective drought management and mitigation strategies in the region. Numerous studies have explored the general attributes of droughts in India (Mallya, Mishra, Niyogi, Tripathi, & Govindaraju, 2015), alongside detailed investigations into specific drought episodes (Thomas, Nayak, & Ghosh, 2015). Considering India's dense population and extensive agricultural landscape, there exists a crucial necessity to enhance our understanding of drought risk in the region. Comprehensive drought characterization becomes pivotal for activities such as early warning systems for droughts and analysis of drought risk, facilitating improved readiness and contingency planning (Hayes et al., 2012). This knowledge contributes significantly to developing effective strategies for managing the impact of droughts in a country heavily reliant on agriculture.

15.3 Conventional methods of drought monitoring

Traditional drought monitoring systems heavily rely on rainfall-based assessments. Various drought indices, derived from meteorological data, play a pivotal role in evaluating and quantifying drought conditions. These indices include the SPI, Palmer Drought Severity Index (PDSI), Crop Moisture Index (CMI), Reclamation Drought Index, among others. Each index offers a unique perspective on drought severity and provides valuable information for understanding and managing drought-related challenges. The diversity of these indices allows for a comprehensive analysis of meteorological conditions, aiding in the development of effective strategies for drought monitoring and mitigation. However, the accuracy of these methods across diverse countries has been questioned (Kogan, 1990). In the station-level rainfall-based drought monitoring, difficulty lies in spatial representation of rainfall data. Station-level rainfall very often does not represent larger regions like districts. This lacuna can be overcome using RS-based gridded rainfall data. Gridded data represent the whole region as a satellite gives a holistic view. Another shortcoming of the traditional system of drought monitoring is the vast time scale. The conventional system does not account for short-term or day-to-day distribution of rainfall, which might have a strong influence on crop performance. The traditional system monitors drought in 1-month to 12-month scale like SPI monitors drought effectively in 1-month and 3-month scale. This vast time span does not represent the distribution pattern of rainfall. It is not just the rainfall amount, the rainfall distribution at the critical stage of a crop is also crucial. If a crop encounters moisture stress at the critical phenological phase, the growth and yield of the crop will be hampered. The conventional system fails to represent this holistic feature very often. Therefore, monitoring vegetation for accurate representation of drought is vital. Some other limitations of the station-level precipitation data comprise accuracy of measurements, the number of gauging stations and length of the record. Sometimes, the current rainfall data is not available, and most of the time, there are biases in observation. With the use of satellite data, these shortcomings can be overcome as satellite data are timely, accurate, and unbiased.

15.4 Remote sensing for drought monitoring

RS proves highly valuable in comprehending the spatiotemporal changes in land cover concerning fundamental physical properties like surface radiance and emissivity data (Orhan, Ekercin, & Dadaser-Celik, 2014). The effectiveness of RS for drought assessment and disaster reduction hinges on the reliability of a sustained data supply for operational users (Dalezios, Blanta, & Spyropoulos, 2012). Meteorological satellites, including geostationary types like METEOSAT and geosynchronous types like NOAA/AVHRR, play a role in operational drought monitoring and assessment (Dalezios et al., 2012; Zhou et al., 2014). Similarly, environmental satellites such as the LANDSAT series, SPOT, IKONOS, and WORLD VIEW are utilized for studying land cover, disasters, vegetation, and so forth. Despite their infrequent coverage, these satellites contribute to land use classification, qualitative aspects of drought, and, to a lesser extent, quantitative assessments (Peled, Dutra, Viterbo, & Angert, 2010). RS is a valuable tool for detecting meteorological and agricultural drought by providing timely and spatially explicit information on various environmental variables. Elevated LSTs are often associated with drought conditions. RS, particularly thermal infrared sensors, can measure LST, providing information about the temperature of the Earth's surface. Increased LST can indicate reduced soil moisture and heightened drought stress. RS can estimate evapotranspiration rates, which represent the combined loss of water from both the soil surface and plant leaves. During drought, evapotranspiration rates may decrease due to limited water availability. Instruments like thermal infrared sensors are particularly useful for assessing changes in evapotranspiration. Drought indices are quantitative measures developed to evaluate the severity of drought events based on precipitation, soil moisture, and temperature anomalies. These indices establish a standardized framework for comprehensively understanding and quantifying the impact of drought conditions on ecosystems, agriculture, and water resources. Table 15.1 demonstrates the formula of some popular and useful RS-based drought monitoring indices that are being discussed in this chapter. The salient features and major application of the indices are summarized in Table 15.2.

Table 15.1 Details of the remote sensing based indices applicable for drought monitoring.

Name	Formula
PCI (Du et al., 2013)	$\frac{P - P_{mean}}{P_{mean}} \times 100$ P is the actual precipitation for a specific period (e.g., monthly, seasonal, and annual). P mean is the mean or average precipitation over a reference period.
TCI (Tsiros, Domenikiotis, Spiliotopoulos, & Dalezios, 2004)	$\frac{T - T_{mean}}{T_{mean}} \times 100$ T is the actual temperature for a specific period (e.g., monthly, seasonal, and annual). T mean is the mean or average temperature over a reference period.
SMCI (Souza, Ribeiro Neto, & Souza, 2021)	$\frac{Current soil moisture - longterm average soil moisture}{Long term standard deviation of soil moisture}$
NDVI (Rouse, Haas, Schell, & Deering, 1974)	$\frac{R_{NIR} - R_R}{R_{NIR} + R_R}$ R_{NIR} = Reflectance in near infrared (NIR); R_R = reflectance in red band (R).
VCI (Quiring & Ganesh, 2010)	$\frac{Current\ NDVI - Minimum\ NDVI}{Maximum\ NDVI - Minimum\ NDVI} \times 100$
VTCI (Wang, Li, Gong, & Song, 2001)	$\frac{T_b - T_{min}}{T_{max} + T_{min}}$ T_b is the observed or calculated brightness temperature. T_{min} is the historical or long-term minimum brightness temperature for the corresponding period. T_{max} is the historical or long-term maximum brightness temperature for the corresponding period.
VHI (Bento, Gouveia, DaCamara, & Trigo, 2018)	$VHI = 0.5 \times (NDVI + TCI)$ NDVI is the Normalized Difference Vegetation Index. TCI is the Temperature Condition Index.
NDWI-I (Gao, 1996)	$\frac{R_{NIR} - R_{SWIR}}{R_{NIR} + R_{SWIR}}$ R_{NIR} = reflectance in NIR; R_{SWIR} = reflectance in SWIR.
NDWI-II (Teng et al., 2021)	$\frac{R_G - R_{NIR}}{R_G + R_{NIR}}$ R_{NIR} = reflectance in NIR; R_G = reflectance in the green band (G).
NDDI (Dobri et al., 2021)	$\frac{R_{NIR} - R_{MIR}}{R_{NIR} + R_{MIR}}$ R_{NIR} = reflectance in NIR; R_{MIR} = reflectance in mid-wave infrared band.

NDDI, *Normalized Difference Drought Index;* NDVI, *Normalized Difference Vegetation Index;* NDWI-I, *Normalized Difference Water Index;* NDWI-II, *Normalized Difference Water Index;* NIR, *near infrared;* PCI, *Precipitation Condition Index;* SMCI, *Soil Moisture Condition Index;* SWIR, *shortwave infrared band;* TCI, *Temperature Condition Index;* VCI, *Vegetation Condition Index;* VHI, *Vegetation Health Index;* VTCI, *Vegetation Temperature Condition Index.*

Table 15.2 Comparison among remote sensing indices used for drought monitoring.

Sl no	Indices	Characteristics	Interpretation	Application
1	PCI	Focuses on precipitation patterns and amounts	Higher PCI values indicate wetter conditions, while lower values suggest drier conditions	Reflects the adequacy and distribution of rainfall over a certain period. Typically used to assess drought conditions or water availability for agricultural purposes.
2	TCI	Measures temperature anomalies or deviations from the normal temperature range	Higher TCI values may indicate unusually warm conditions, while lower values may indicate unusually cold conditions.	Used to monitor heat stress or cold stress in ecosystems or agricultural areas. Reflects the impact of temperature extremes on various biological processes
3	SMCI	Calculated based on factors such as rainfall, evapotranspiration, and soil characteristics	Higher SMCI values typically indicate wetter soil conditions, while lower values suggest drier soil conditions.	Indicates the level of moisture content in the soil. Reflects the availability of water for plant growth and other soil-dependent processes
4	VCI	Derived from satellite imagery or ground-based observations	Higher VCI values indicate healthier and more vigorous vegetation, while lower values may indicate stress or decline in vegetation health	Impact of environmental factors such as temperature, precipitation, and soil moisture on vegetation growth and greenness
5	NDVI	Calculated based on the difference between NIR and red light reflectance from the Earth's surface	Higher NDVI values typically indicate denser, healthier vegetation cover, while lower values may indicate sparse or stressed vegetation	Vegetation dynamics, crop health, land cover changes, and biomass production
6	VTCI	Derived from thermal infrared imagery, comparing the temperature of vegetation pixels to a reference surface temperature	Higher VTCI values suggest cooler vegetation, which may indicate healthier vegetation with adequate water availability. Lower values may indicate warmer vegetation, suggesting stress due to factors like water scarcity or disease	Monitoring agricultural fields, detecting irrigation efficiency, assessing water stress in vegetation, and identifying areas prone to drought or other environmental stresses
7	VHI	Combines NDVI and VTCI into a single index, often using weighted averages or other statistical methods	Higher VHI values generally indicate healthier vegetation, considering both the vigor (NDVI) and thermal condition (VTCI) of the vegetation cover. Lower values may indicate stress or decline in vegetation health	Monitoring large-scale vegetation conditions, assessing drought impacts, evaluating crop health, and informing agricultural management decisions

#	Index	Description	Notes / Application
8	NDWI	Compares NIR and green band reflectance	Higher NDWI values typically indicate the presence of water, while lower values suggest drier conditions	Monitoring water bodies, wetlands, flood extent, and changes in vegetation water content, especially in areas affected by drought or inundation
9	NDDI	Compares NIR and SWIR reflectance	Negative NDDI values indicate vegetation stress or drought conditions, while positive values suggest healthy vegetation	Vegetation health, assess drought severity, and identify areas vulnerable to water stress in agriculture and ecosystems
10	NMDI	Involves a combination of visible, NIR, and SWIR bands	Negative NMDI values indicate drought stress in vegetation, while positive values indicate healthy conditions	Comprehensive assessment of drought conditions by considering multiple spectral bands, useful for monitoring large-scale drought events and assessing vegetation response
11	SDCI	Involves scaling and aggregating multiple vegetation indices, often with statistical methods	Lower SDCI values indicate more severe drought conditions, while higher values suggest less severe drought or normal conditions	Composite measure of drought severity, useful for drought monitoring, risk assessment, and drought impact analysis
12	SWASI	Quantifies changes in shortwave radiation scattering due to vegetation stress or moisture deficit.	Higher SWASI values indicate greater vegetation stress or moisture deficiency, often associated with drought conditions.	Vegetation health, drought impact on ecosystems, and water resource management strategies
13	MIDI	Integrates microwave measurements of soil moisture, vegetation water content, and other relevant parameters.	Lower MIDI values indicate drier conditions, while higher values suggest wetter conditions.	Soil moisture dynamics, vegetation water stress, and drought severity, useful for agricultural planning, hydrological modeling, and drought early warning systems.

MIDI, *Microwave Integrated Drought Index*; NDDI, *Normalized Difference Drought Index*; NDVI, *Normalized Difference Vegetation Index*; NDWI, *Normalized Difference Water Index*; NIR, *near-infrared*; NMDI, *Normalized Multiband Drought Index*; PCI, *Precipitation Condition Index*; SDCI, *Scaled Drought Condition Index*; SMCI, *Soil Moisture Condition Index*; SWASI, *Shortwave Angle Slope Index*; SWIR, *shortwave infrared*; VCI, *Vegetation Condition Index*; VHI, *Vegetation Health Index*; VTCI, *Vegetation Temperature Condition Index*.

15.4.1 Precipitation Condition Index

The Precipitation Condition Index (PCI) assesses deviations in precipitation patterns from historical norms. It provides a numerical representation of the impact of precipitation anomalies on drought severity, with negative values indicating drier-than-normal conditions. PCI operates on a scale ranging from 0 (extremely unfavorable) to 100 (optimal conditions). In instances of meteorological drought, the PCI tends to approach or reach values close to 0, indicating the severity of the drought event. Conversely, during flood, the PCI is close to 100. This two-sided scale provides a standardized and easily interpretable measure, facilitating the characterization of meteorological droughts and flooding events based on observed PCI values. The PCI serves as a quantitative tool for measuring both precipitation deficit and surplus to identify drought events. PCI contributes to the systematic assessment of precipitation variations, aiding in the timely identification of drought events and enabling proactive responses to water scarcity challenges. Additionally, the PCI finds practical application in agricultural planning by providing farmers with crucial insights to adapt strategies based on expected precipitation conditions. The initiation of the PCI was spurred by a critical need for a unified metric to appraise the repercussions of precipitation anomalies on drought scenarios. Researchers orchestrated the development of PCI to furnish a dependable and consistent measure (Hayes, Svoboda, Wilhite, & Vanyarkho, 1999; McKee et al., 1993). This endeavor aimed to streamline the evaluation of precipitation anomalies, allowing for seamless cross-regional and temporal comparisons.

15.4.2 Temperature Condition Index

The Temperature Condition Index (TCI) quantifies temperature anomalies to assess their contribution to drought severity. Negative values indicate warmer-than-normal conditions, contributing to increased evaporation and potential drought stress. TCI operates on a scale ranging from 0 (extremely unfavorable) to 100 (optimal conditions), serving as a comprehensive measure for characterizing the impact of temperature conditions on vegetation health. This standardized scale provides an easily interpretable metric that encapsulates the diverse responses of vegetation to temperature variations. By offering a quantifiable range, TCI facilitates a systematic assessment of stress factors induced by temperature fluctuations and excess wetness, enabling a nuanced understanding of vegetation well-being. TCI evaluates the role of temperature anomalies in exacerbating drought conditions. This index helps to assess heat-related stress on ecosystems, agriculture, and water resources. By providing a standardized measure for estimating the impact of temperature conditions on vegetation health, TCI aids in gauging stress levels across diverse vegetation types. Its application extends to monitoring and understanding the stress experienced by vegetation in response to temperature fluctuations and excessive wetness. The genesis of TCI unfolded from a pressing necessity for a standardized gage in evaluating temperature anomalies during drought occurrences. This venture played a pivotal role in enriching our comprehension of climatic factors shaping drought severity (Dai, 2011; Sheffield & Wood, 2008).

15.4.3 Soil Moisture Condition Index

The Soil Moisture Condition Index (SMCI) evaluates soil moisture anomalies to quantify drought conditions. It reflects the relative state of soil moisture compared to historical averages, with negative values indicating drier-than-normal soil conditions. The SMCI spans a scale from 0 to 100, representing changes in soil moisture conditions ranging from extremely unfavorable to optimal. During meteorological droughts characterized by markedly low soil moisture levels, the SMCI tends to approximate or reach values close to 0. In contrast, under moisture saturation conditions, the SMCI tends toward values close to 100. This numerical scale provides a standardized metric for characterizing the relative dryness or wetness of soil, offering an easily interpretable measure to gage soil moisture conditions. SMCI plays a crucial role in understanding the impact of drought on soil moisture, influencing agricultural productivity, water availability, and ecosystem health. Its practical applications extend to supporting land use planning and facilitating drought risk assessments. By providing insights into soil moisture conditions, the SMCI assists in making informed decisions related to land management practices and water resource planning. The index's versatility makes it a valuable tool for a range of sectors that are sensitive to variations in soil moisture levels. The emergence of the SMCI germinated from the imperative of establishing a consistent measure for appraising soil moisture anomalies during drought occurrences. Leveraging the insights, the SMCI was meticulously fashioned to mirror the relative dryness or wetness of soil in comparison to historical benchmarks (Otkin et al., 2016; Sheffield, Wood, & Roderick, 2012). Navigating a terrain from 0 to 100, the SMCI emerges as a standardized metric adept at characterizing soil moisture conditions. Its versatility extends to unraveling the impacts of drought on agricultural productivity, water resource availability, and the overall health of ecosystems. The SMCI lends crucial support to land use planning endeavors and serves as an invaluable compass for informed decision-making concerning soil moisture management amid the climatic patterns.

15.4.4 Normalized Difference Vegetation Index

The Normalized Difference Vegetation Index (NDVI) is a widely used RS technique that assesses vegetation health and density based on the reflectance of different wavelengths of light. It is a numerical indicator that uses RS to assess and measure the health or density of vegetation in a particular area. The NDVI is calculated using the reflectance of near-infrared and red light. The resulting NDVI values range from -1 to 1, with higher values indicating healthier and denser vegetation. This formula leverages the fact that healthy vegetation strongly absorbs visible light and reflects infrared light. NDVI's simplicity and effectiveness make it valuable in assessing vegetation dynamics, land cover changes, and environmental monitoring. NDVI can successfully be applied to assess crop health and detect stress, to optimize irrigation and fertilization based on vegetation vigor, and to monitor changes in vegetation over time. The sensitivity of near infrared (NIR) radiation to vegetation and water content helps to detect the moisture scarcity in the crop fields. Before the introduction of the NDVI, simpler vegetation indices were used, but they often lacked the sensitivity and accuracy needed for comprehensive vegetation assessment. NDVI's ability to account for variations in atmospheric conditions and provide a standardized measure of vegetation health has made it a widely adopted tool in the field of RS (Al-Quraishi, Gaznayee, & Crespi, 2021; Bushra et al., 2019; Hu et al., 2019). The climate change impacts, focusing on rainfall fluctuations in northern Iraq from 1980 to 2010, were investigated (Al-Hedny & Muhaimeed, 2020) using the SPI and NDVI derived from moderate resolution imaging spectroradiometer time series data.

15.4.5 Vegetation Condition Index

The Vegetation Condition Index (VCI) is a quantitative measure used to assess the health and vigor of vegetation based on RS data. It combines various spectral indices to evaluate the physiological status of plants, reflecting their response to environmental stressors such as drought or disease (Bouras et al., 2021). The development of the VCI can be attributed to the growing need for accurate and timely information on vegetation health, especially in the face of changing environmental conditions, climate variability, and land use changes. Monitoring vegetation conditions is crucial for early detection of stress, disease, or other factors that can affect plant productivity. This information is vital for decision-making processes related to land management, resource allocation, and sustainable development. The VCI provides a valuable tool for monitoring and managing ecosystems by indicating the relative well-being of vegetation over time. The concept of VCI revolves around the analysis of spectral reflectance patterns emitted by vegetation. By interpreting these patterns, the VCI captures the dynamic changes in vegetation health, offering insights into ecological conditions. High VCI values suggest robust vegetation, while lower values indicate stress or degradation. This index aids in early detection of environmental threats, supporting timely intervention and sustainable land management practices. The VCI helps to monitor the agricultural drought (Dutta, Kundu, Patel, Saha, & Siddiqui, 2015). The VCI is commonly used to assess the impact of drought on vegetation health (Pei et al., 2018). Lower VCI values may indicate reduced moisture availability and stress on plant growth. In agriculture, the VCI is used to monitor the condition of crops, helping farmers and policymakers make informed decisions related to irrigation, pest control, and other agricultural practices to get optimum yield (Pham, Awange, Kuhn, Nguyen, & Bui, 2022). The past spatiotemporal variations in spring vegetative drought conditions from 1981 to 2015 in China were analyzed using the VCI (Liang, Qiu, Yan, Shi, & Geng, 2021). Frequent occurrences of spring vegetative drought, notably moderate drought, showed distinct geo-graphical differences influenced by monsoons. Wavelet analysis revealed short-period oscillations (5−7 years) and long-period oscillations (23−28 years), with dominance based on susceptibility to monsoons.

15.4.6 **Vegetation Temperature Condition Index**

The Vegetation Temperature Condition Index (VTCI) is a quantitative measure used to assess the health and stress level of vegetation based on its temperature. The VTCI provides a comprehensive analysis of the current temperature conditions of vegetation across a given area. The VTCI combines data from RS technologies, such as visible, multispectral, and thermal infrared sensors. By comparing the dif-ference between actual vegetation temperature and its potential maximum and mini-mum temperatures, the VTCI provides valuable insights into the state of vegetation, including its moisture levels and heat stress, which make the index a valuable mea-sure for drought monitoring and assessment and forecasting (Tian, Wang, & Khan, 2016). The index was developed as a response to the growing need for accurate and efficient methods of monitoring vegetation temperature, particularly in urban environments. Urban environments often experience higher temperatures compared to rural areas due to the "heat island" effect. The heat island effect occurs when urban areas, with their abundance of concrete and asphalt, absorb and retain more heat than the surrounding rural areas. This can lead to higher temperatures within cities, impacting the health and vitality of vegetation. Field scale drought monitor-ing is possible by using the VTCI (Zhou et al., 2020).

15.4.7 **Vegetation Health Index**

The Vegetation Health Index (VHI) is a measure of assessing the health and condi-tion of vegetation based on the correlation between two key environmental factors: the NDVI and LST, which have been found to be strongly related to moisture avail-ability and soil water stress (Marengo et al., 2018). It is an important tool for moni-toring and assessing the health and conditions of vegetation on a regional or global scale by considering the relationship between vegetation greenness and temperature. The VHI takes into account both the density and growth status of vegetation to pro-vide information on vegetation coverage and health (Jiang et al., 2021). The VHI can be considered as a proxy characterizing vegetation health or a combine estima-tion of moisture and thermal conditions. This information is especially important for agricultural purposes, as it can help farmers to monitor the health of the crops and make informed decisions regarding irrigation and pest control measures to opti-mize yield and minimize losses. The VHI is successfully applied to monitor the spatial extent of agricultural drought (Sholihah et al., 2016). The effectiveness of the VHI for detecting vegetation drought was examined through an improved algo-rithm based on a multisource dataset (Zeng et al., 2001). The enhanced VHI demonstrates significant improvements in vegetation drought detection. Findings indicate worsening vegetation drought events in high latitudes, revealing distinct responses among various vegetation types. The vulnerability of agriculture to drought in East Java, Indonesia, was analyzed using the VHI from MODIS Data (Kirana, Ariyanto, Ririd, & Amalia, 2020). The drought maps, based on VHI values during this period, predominantly indicated a prevalence of moderate drought in East Java's agricultural regions.

15.4.8 Normalized Difference Water Index (NDWI)

The Normalized Difference Water Index (NDWI) is an indicator used to assess the availability of water resources in a given area. The development of NDWI was driven by the need to accurately monitor and detect water content in vegetation. It is calculated by comparing the reflectance values of specific wavelengths related to water absorption, typically in the green and near-infrared spectral regions. This index is widely used in RS applications to detect and monitor surface water features, such as lakes, rivers, and reservoirs. The NDWI helps to highlight the presence of water by emphasizing the differences in how water and nonwater features reflect light in these specific bands (Li et al., 2013). This helps to identify the suitable areas for crop growing in the drought-prone regions. By utilizing the near-infrared and shortwave infrared reflectance values, the NDWI can also be computed, which is able to monitor the vegetation water content (Chai et al., 2021; Sartori, Sbruzzi, & Fonseca, 2020). Several researchers applied the NDWI for drought monitoring (Al-Quraishi et al., 2021; Amalo, Ma'rufah, & Permatasari, 2018). The drought scenario in West Java during the 2015 El Niño was studied using the NDWI (Amalo et al., 2018). NDWI highlighted widespread drought in West Java, peaking strongly from September to November. The region predominantly experienced strong and moderate drought, indicating the substantial impact of El Niño 2015 on heightened drought conditions and reduced rainfall in West Java. The NDWI proved effective in spatially correlating with rainfall, providing a reliable method for detecting past drought occurrences in the area.

15.4.9 Normalized Difference Drought Index

The Normalized Difference Drought Index (NDDI) is a widely used meteorological index that quantifies the severity of drought conditions by measuring the difference between near-infrared and visible red wavebands in RS data (Sun, Wang, Li, & Wang, 2020). It provides valuable information on vegetation health and can be used to monitor drought severity on a global scale. The concept behind NDDI is to determine the dryness of agricultural land by calculating the difference between near-infrared and visible red wavebands, which are sensitive to changes in vegetation health and water availability (Mujiyo, Nurdianti, & Sutarno, 2023). The NDDI was developed to address the need for a reliable and standardized measure of drought severity. The NDDI helps to overcome the limitations of traditional drought monitoring methods and provide a more accurate and comprehensive assessment of drought conditions. This index takes into account the sensitivity of vegetation to drought conditions, as well as the availability of water in the soil. It can be used to identify regions experiencing severe drought conditions, monitor the progression of drought over time, and inform decision-making in agriculture, water management, and disaster response. This index can also be used in ecological studies to assess the effects of drought on natural vegetation and ecosystems. Drought frequency and severity across Romania's arable lands from 2001 to 2020 was assessed using the NDDI derived from MODIS satellite data (Dobri, Sfîcă, Amihǎesei, Apostol, & Țîmpu, 2021). The NDDI showed a negative correlation with annual rainfall, confirming its reliability as a valuable tool for drought assessment in Romania's arable areas from a climatic perspective.

15.4.10 Normalized Multiband Drought Index

The Normalized Multiband Drought Index (NMDI) is a measure used to assess and monitor drought conditions over a specific area. It is calculated by combining the information from multiple bands of satellite imagery, such as the NDVI and the NDVI, to estimate soil and vegetation moisture (Fayne, Ahamed, Roberts-Pierel, Rumsey, & Kirschbaum, 2019). The index was developed to address the need for a more accurate and sensitive measure of drought severity. The concept behind the NMDI is to utilize satellite imagery and spectral indices to effectively measure and monitor drought conditions (Mirzaee & Mirzakhani Nafchi, 2023). The index takes into account the reflectance in specific bands (0.86, 1.64, and 2.13 μm) to estimate soil moisture and detect vegetation and soil water content. The common approach is to calculate the index by subtracting NDWI from NDVI. This subtraction helps in quantifying the drought conditions by comparing the vegetation vigor and cover with the availability of water. The NMDI is widely used in agriculture, hydrology, and environmental monitoring to assess and monitor drought conditions. It is applied to detect and assess vegetation and soil water content, providing accurate assessments of drought severity. It provides valuable information on the dryness of agricultural land, as well as the effects of drought on vegetation and soil moisture. This index is particularly useful for precise irrigation in agriculture, as it can inform decisions regarding water allocation and management. Drought conditions in Turkey from February 2000 to January 2019 was assessed using MODIS satellite data on the Google Earth Engine (GEE) platform (Aksoy, Gorucu, & Sertel, 2019). Employing VHI, NMDI, and NDDI spatiotemporal distribution of drought severity was analyzed at the country level for different years and months. Results indicated that MODIS-derived drought indices offer valuable geospatial information for country-level drought assessment.

15.4.11 Scaled Drought Condition Index

The Scaled Drought Condition Index (SDCI) is a metric used to quantify and monitor the severity of drought conditions within a specific region. It takes into account

various factors such as precipitation, temperature, vegetation health, and soil moisture to provide a comprehensive assessment of drought intensity. The SDCI is a
composite index that combines multiple RS indices, including the VCI, TCI, and
Standardized Vegetation Index, to create a more robust and reliable measure of
drought conditions. The SDCI is computed to overcome the limitations of using a
single drought index by incorporating multiple RS condition indices. The formula
for calculating the SDCI involves a linear combination of several RS condition
indices (Zhao & Wang, 2021). This allows for a more comprehensive understanding of drought conditions, as it captures various aspects of the environment that
contribute to drought severity. Furthermore, the SDCI is designed to be applicable
in both arid and humid regions, making it a versatile tool for drought monitoring
(Venkatesh et al., 2022; Zhao & Wang, 2021). The optimal SDCI (OSDCI) was
introduced as an enhanced agricultural drought index, addressing limitations in the
existing SDCI (Guo et al., 2019). Utilizing alternative precipitation data (GSMaP),
varied lag time, optimal variable weights, and a revised severity classification, the
OSDCI outperformed three experimental SDCIs in Central Asia. The OSDCI with
the revised classification scheme significantly reduced drought frequency and area
error, effectively capturing both temporal and spatial drought patterns.

15.4.12 Shortwave Angle Slope Index

The Shortwave Angle Slope Index (SWASI) is a meteorological parameter used
to describe the angle at which shortwave radiation intersects the Earth's surface.
The SWASI is based on the relationship between the angle at shortwave infrared
(SWIR) and the difference between the reflectance at NIR and SWIR to quantify
the intersection of shortwave radiation on the Earth's surface. This index takes
into account the relationship between bands, rather than focusing on the absolute
value of reflectance at any one band (Das, Murthy C, & Mvr, 2014). The SWASI
can be calculated by drawing a triangle shape in a two-dimensional space that
relates the wavelength and reflectance values of land covers using three consecutive bands as vertices. The SWASI is calculated by measuring the angle formed
between three consecutive bands—NIR, SWIR1, and SWIR2 of MODIS sensor
bands (Das, Sahay, Seshasai, & Dutta, 2017). By analyzing the angle formed
between three consecutive bands (NIR, SWIR1, and SWIR2), the SWASI provides information on the spectral shape of the land cover, allowing for the detection of moisture content. One of the main applications of SWASI is to estimate
surface moisture and discriminate land cover (Das, Murthy, & Seshasai, 2013).

15.4.13 Microwave Integrated Drought Index

The Microwave Integrated Drought Index (MIDI) is a comprehensive and highly
advanced drought monitoring tool that utilizes satellite-based microwave observations to assess and quantify the moisture content in the soil, vegetation health,

and overall drought conditions over a specific area or region. The MIDI was developed to address the need for a reliable and accurate method to monitor drought conditions, especially meteorological drought, in various regions. Traditional single-parameter drought indices often fail to capture the complexity and nuances of drought occurrence (Zhang, Jia, & Wang, 2019). Moreover, the MIDI takes into account the principle of water balance and utilizes statistical methods to ensure a more comprehensive and accurate assessment of drought. The MIDI combines data from passive microwave sensors to estimate soil moisture levels, which is a critical parameter for monitoring and predicting agricultural droughts. By directly measuring soil moisture using microwave signals, the MIDI offers a more direct and reliable method of drought monitoring compared to traditional approaches that rely on optical RS data. The MIDI takes into account three main variables: TRMM-derived precipitation, advanced microwave scanning radiometer for earth observing system-derived soil moisture, and AMSR-E-derived LST. The formula for calculating the MIDI involves assigning proper weights to the three components: TRMM-derived precipitation, advanced microwave scanning radiometer for earth observing system-derived soil moisture, and AMSR-E-derived LST (Liu et al., 2017). The critical importance of monitoring meteorological drought was emphasized in tropical and subtropical water-limited ecosystems, integral to the global carbon cycle and ecosystem services (Zhang et al., 2019). With an optimal ensemble of satellite precipitation products, the MIDI exhibited enhanced meteorological drought monitoring across diverse bioclimatic zones, showcasing reliability and applicability in all weather conditions. This study highlights the MIDI as a valuable tool for both long-term and near-real-time meteorological drought assessments using operational satellites.

15.5 Comparison of traditional drought indices and remote sensing-based drought indices

Traditional drought indices, relying on meteorological variables, have been the cornerstone of drought assessment for many years. However, recent advancements in RS technology have provided new tools for monitoring drought with improved spatial and temporal resolution. The comparison between traditional and RS-based drought monitoring methods are summarized in Table 15.3.

15.5.1 Traditional drought indices

Traditional drought indices, including PDSI, SPI, and CMI, primarily rely on meteorological variables for their calculations. The PDSI integrates precipitation, temperature, and potential evapotranspiration to assess soil moisture conditions over an extended time scale (Palmer, 1965). In contrast, the SPI focuses solely on

Table 15.3 Comparisons between conventional and remote sensing-based drought monitoring.

Sl no	Characteristics	Conventional monitoring	Remote sensing-based monitoring
1	Data sources	Relies on ground-based data, including meteorological stations, rain gages, and groundwater monitoring wells.	Utilizes satellite, airborne, or drone-based sensors to capture information from a broader spatial scale.
2	Spatial coverage	Limited by the distribution of monitoring stations, resulting in sparse spatial coverage.	Provides broader spatial coverage, allowing for monitoring of large areas or regions with diverse landscapes.
3	Temporal resolution	Data is typically collected at fixed intervals (e.g., daily, monthly), which can limit the ability to capture rapid changes.	Offers higher temporal resolution, with frequent revisits to the same area, enabling the detection of changes over shorter time intervals.
4	Parameter measurement	Primarily relies on meteorological parameters such as precipitation, temperature, and soil moisture measured at specific locations.	Measures a wide range of parameters, including vegetation indices, LST, and soil moisture, providing a more comprehensive view of environmental conditions.
5	Accuracy and precision	Measurements are accurate at specific locations but may lack precision in between monitoring stations.	Provides accurate and precise information at different spatial scales, offering a more detailed understanding of the landscape.
6	Response time	Data collection and dissemination may take time, leading to delays in drought detection and response.	Offers near-real-time data, allowing for quicker identification of drought conditions and timely decision-making.
7	Cost-effectiveness	Infrastructure and maintenance costs associated with ground-based monitoring can be high.	Initial setup costs can be high, but once established, RS can be more cost-effective for large-scale and continuous monitoring.
8	Automation	Often requires manual data collection and analysis.	Allows for automated data acquisition and processing, reducing the need for extensive manual labor.
9	Accessibility	Limited accessibility to remote or difficult-to-reach areas.	Provides data for remote and inaccessible regions, improving coverage in areas where ground-based monitoring is challenging.
10	Integration with other data sources	Integration with other data sources may be limited.	Allows for easy integration with meteorological data, hydrological models, and other datasets, enhancing the overall understanding of drought conditions.

LST, *Land surface temperature.*

precipitation anomalies and is widely employed for drought characterization (McKee et al., 1993). These indices employ straightforward mathematical formulas to quantify the severity of drought. A limitation of traditional indices is their coarse spatial resolution, stemming from their reliance on data from ground-based weather stations. This can result in an inaccurate representation of local variations in drought conditions. Additionally, these indices often operate on monthly or seasonal time scales, providing a broader perspective on drought events but potentially overlooking short-term variations. Traditional indices offer the advantage of simplicity in computation and often come with long historical datasets, facilitating the analysis of past drought events. However, challenges arise from limited spatial coverage, particularly in remote or data-scarce regions, and potential biases in ground-based meteorological data (Sheffield et al., 2012). Despite their advantages, these indices may not capture the full complexity of drought dynamics, and efforts to improve spatial resolution and address data limitations are ongoing in the field of drought monitoring and assessment.

15.5.2 Remote sensing-based drought indices

RS technologies have revolutionized drought monitoring by seamlessly integrating high-resolution satellite data. Crucial Earth observation parameters, such as NDVI and LST, are derived from this data to form the backbone of RS-based drought indices. A prime example is VCI, which amalgamates NDVI and LST to provide a nuanced assessment of both vegetation health and water stress (Kogan, 1995). The effectiveness of these indices is underpinned by the application of complex algorithms, adept at processing and interpreting the extensive satellite datasets, thereby elevating the precision and depth of drought assessments. Satellite-based drought monitoring relies on physical parameters such as soil moisture, vegetation biomass, chlorophyll levels, organic matter, canopy or soil temperature, and surface water storage (Sur et al., 2020). A notable strength of RS-based drought indices lies in their exceptional spatial resolution, allowing for granular monitoring of specific regions, and even individual fields. This capability provides a detailed and localized understanding of drought dynamics. Moreover, the use of satellite data enables a heightened temporal resolution, facilitating more frequent updates. This near-real-time information proves invaluable, especially in sectors such as agriculture and water resource management. The enhanced temporal resolution ensures timely decision-making and response to evolving drought conditions. RS-based indices offer a significant advancement over traditional approaches by addressing their limitations. The integration of satellite data provides a holistic view of land surface conditions, contributing to more accurate and comprehensive drought assessments. However, challenges persist, including reliance on satellite data availability, susceptibility to potential atmospheric interference, and the inherent complexity of the algorithms involved.

15.6 Drought prediction using machine learning algorithms

Drought prediction using RS and geographic information system involves the integration of satellite imagery, meteorological data, and spatial analysis to monitor, assess, and predict drought conditions. The ML algorithms can be successfully implemented to analyze the historical data and identify patterns that may indicate drought conditions. Models can be trained to predict future drought events based on the past trends. ML techniques can be powerful tools for drought prediction, allowing for the analysis of complex datasets and the identification of patterns that may not be apparent through traditional methods.

15.6.1 Random forest

Random forest (RF) is an ensemble learning method that builds multiple decision trees and merges their predictions. It is robust and can handle a mix of categorical and numerical data. RF can be applied to predict drought based on various input features such as meteorological data, soil moisture, and vegetation indices (Elbeltagi et al., 2023). This algorithm gathers the relevant data, including meteorological variables (temperature, precipitation, humidity, wind speed), soil moisture content, vegetation indices (e.g., NDVI), and any other pertinent features, cleans and preprocesses the data, handles the missing values, and normalizes or standardizes the numerical features. RF has a built-in feature importance measure, which can be used to rank the importance of different variables in the prediction (Chen, Li, & Wang, 2012).

15.6.2 Support vector machines

Support vector machine (SVM) is a supervised learning algorithm that can be used for classification and regression tasks. SVM is effective in handling high-dimensional datasets and is suitable for predicting drought conditions based on multiple variables. It encodes the categorical variables. The domain knowledge or statistical methods can be used to select important features for drought prediction. SVM can be effective for drought prediction, especially when dealing with complex, nonlinear relationships in the data (Chiang & Tsai, 2012). It improves the accuracy of prediction (Pham, Yang, Kuo, Tseng, & Yu, 2021). Drought early warning can also be done successfully by applying SVM (Kolachian & Saghafian, 2021).

15.6.3 Neural networks

Deep learning techniques, including neural networks, can be applied for drought prediction. Recurrent neural networks (RNNs) are particularly useful for time-series data, making them suitable for modeling meteorological variables over time. Neural networks can be powerful tools for drought prediction (Dikshit, Pradhan, & Santosh, 2022), especially when dealing with complex, nonlinear relationships in the data. For drought prediction, RNNs or long short-term memory networks can be applied if the data has a temporal component, as these architectures are well-suited for time-series data (Villegas-Ch & García-Ortiz, 2023). The hyperparameters such as learning rate, batch size, and the number of epochs are to be adjusted to optimize the training process. An appropriate loss function (e.g., mean squared error for regression tasks and binary cross-entropy for binary classification) based on the nature of the prediction task is to be chosen. Neural networks can successfully be applied for prediction of precipitation and soil moisture at regional scale (Filipović, Brdar, Mimić, Marko, & Crnojević, 2022).

15.6.4 Gradient boosting

Gradient boosting models, with their ability to handle complex relationships and high predictive accuracy, are well-suited for drought prediction tasks. They can efficiently capture interactions between different variables and adapt to the dynamic nature of environmental data. It is an ensemble learning technique that combines the predictions of weak models to create a strong predictive model. Specifically, models like XGBoost and LightGBM, which are implementations of gradient boosting, are commonly used for their high performance and efficiency. Gradient boosting models inherently perform feature selection as they train, giving more importance to informative features. Gradient boosting can be successfully applied for classification of drought as well as the Standard precipitation Evapotranspiration Index (Danandeh Mehr & Fathollahzadeh Attar, 2021). This algorithm is also effective for drought monitoring (Li, Jia, & Wang, 2023).

15.6.5 K-nearest neighbors

K-nearest neighbors (KNN) is a simple, nonparametric and intuitive ML algorithm that is useful for spatial analysis and can be applied to predict drought conditions based on the similarity of neighboring regions. Further, the missing data can be supplemented by using the KNN approach (Choesang, Ryntathiang, Jacob, Krishnan, & Kokatnoor, 2023). KNN is sensitive to the choice of features and scale of the features. The number of neighbors (K) for the KNN algorithm is to be selected. The choice of K depends on the characteristics of the data and should be determined through experimentation. The future pattern of the climatic variables can be predicted using the KNN algorithm (Golkar Hamzee Yazd, Salehnia, Kolsoumi, & Hoogenboom, 2019). Data mining techniques like KNN helps in drought monitoring and prediction (Fadaei-Kermani, Barani, & Ghaeini-Hessaroeyeh, 2017). Moreover, the KNN technique can forecast hydrological variables, which are important for drought prediction (Azmi, Araghinejad, & Kholghi, 2010).

15.7 Limitations and research gaps

RS-based indices provide valuable information for monitoring agricultural and meteorological droughts. However, there are some limitations to their use in drought monitoring. One limitation is that RS indices may not capture the full complexity of drought conditions. For example, these indices primarily rely on satellite data related to vegetation health and precipitation patterns, but may not take into account other factors such as soil moisture content, crop management practices, or local water availability (Brown, Wardlow, Tadesse, Hayes, & Reed, 2008).

- RS indices may not always provide crop-specific information, making it challenging to assess the impact of drought on specific crops. Different crops may respond differently to drought conditions. Besides, in the Indian scenario, pre- and postmonsoon crops have a varied growing cycle. The sowing and harvesting of the crops are not synchronized. Further, the growing periods of several winter and summer crop are somewhat overlapping. For example, the maturity stage of winter crops such as potato, mustard, and winter vegetables generally coincides with the sowing or germination of summer pulses and vegetables. So at this period, the RS indices will be higher over the winter crop fields and lower over the summer crop fields. Most of the RS-based drought indices are sensitive to the growth and vigor of the crops. Thus, lower values of vegetation indices are sometimes attributed to the initial growing stages of the crop, not the water stress in the crop field. So, the crop and cropping system-specific study is important to overcome the confusion.
- Optical RS indices are vulnerable to meteorological factors, aerosols, water vapor, cloud cover, and land cover. These indices primarily reflect on conditions at the top of the canopy, especially in places with dense vegetation. Microwave RS provides information on live aboveground biomass and canopy density at all weather conditions. Unlike optical sensors, microwave sensors are less affected by air conditions and can penetrate dense foliage. Additionally, microwave sensors have the ability to gather data on vegetation conditions day or night. The integration between optical and microwave RS technologies may yield more reliable indices for drought monitoring (Xue & Su, 2017).
- In order to mitigate the effects of water stress, vegetation reacts to droughts structurally and physiologically (Donohue, Roderick, McVicar, & Farquhar, 2013). Droughts cause the ecosystem as a whole to absorb less CO_2, which raises atmospheric CO_2 concentrations (Van der Molen et al., 2011). Research in this direction can be supported and further enriched by satellite RS observations. The chance to create indicators to help track the effects of drought on the carbon cycle and the terrestrial CO_2 budget is made possible by these new satellite instruments and datasets. The majority of currently used RS-based drought indicators are not effectively adapted or made to track ecological reactions to modifications in the carbon and nitrogen cycles,

despite the fact that these changes can have a significant influence on natural vegetation and the services and functions provided by ecosystems. Gaining more knowledge about the markers associated with the cycling of carbon and nitrogen will help us comprehend how ecosystems react to droughts (AghaKouchak et al., 2015).

- RS indices may not always accurately reflect the ground truth conditions due to cloud cover, sensor limitations, or the presence of other environmental factors that influence the vegetation index values. Despite the development of models and indicators to assess uncertainty, most satellite-based data products lack uncertainty boundaries or estimations. Inaccuracy in the model structure and parameter errors can affect land surface and hydrologic models. Accurate characterization of RS instruments and products is crucial for decision-making. Developing uncertainty products for existing and upcoming satellite datasets will increase the usefulness and acceptance of satellite observations.

- A single vegetation index may not adequately represent the drought conditions for different types of crops or agricultural practices. This limitation highlights the need for complementary data sources and approaches to enhance drought monitoring efforts. Further research is needed to develop reliable mathematical and statistical frameworks for multiindex drought data. The microwave brightness temperature's diurnal variability may be utilized as a proxy for assessing drought in ecosystems or agriculture by serving as an indicator of the total amount of water that is available in the soil and vegetation. The absence of data on the vertical moisture profiles of soil and vegetation is a significant gap in the available datasets. Additional study in the field of soil and vegetation-combined moisture profiles and vegetation density can yield important insights into droughts and water stress.

- RS indices are often sensitive to soil types, and their effectiveness can vary depending on the soil composition in the area of interest. Some RS indices may have limitations in accurately estimating soil moisture levels, which are critical for assessing agricultural drought (Vicente-Serrano et al., 2020). Quantitative soil moisture assessment is still challenging, though, particularly when there is plant present (Dorigo et al., 2012). The effectiveness of certain indices may be influenced by the types of vegetation and land cover present in the area. Changes in land cover, such as deforestation or land-use changes, can impact the applicability of certain indices.

- RS indices may not capture small-scale variations in drought intensity and may not provide real-time or near-real-time monitoring at a fine scale. Further, the spatial and temporal resolution of RS data may not always match the scale at which drought impacts are manifested. Agricultural drought conditions can be highly localized, and coarse-resolution data may not capture the variability within smaller fields. Some satellite platforms have a limited revisit time, meaning they may not capture changes in vegetation conditions frequently enough. This can be a challenge for monitoring rapidly evolving drought situations. Different satellite sensors have varying spectral bands and characteristics. Inconsistencies between sensors may affect the comparability of data over time and hinder the creation of long-term datasets. The study using multiple sensors data can contribute to a great extent to overcome the issues of sensors' capabilities (Jiao, Wang, & McCabe, 2021).

15.8 Contribution of the chapter to the future drought study

Comparison among RS indices refers to the integration of multiple indices derived from satellite data to assess drought conditions. This approach can significantly enhance drought monitoring efforts in several ways. Comparison between RS indices provides a powerful tool for drought monitoring by integrating complementary information from various sources, improving accuracy, enabling early warning systems, enhancing spatial and temporal resolution, identifying vulnerable areas, and supporting long-term monitoring efforts.

- Comparison among RS indices can play a significant role in the comprehensive assessment of drought monitoring by providing a more holistic understanding of drought conditions. Different RS indices capture various aspects of vegetation health, soil moisture, and meteorological conditions. By combining multiple RS indices that capture different aspects of the environment, comparison can improve the accuracy of drought monitoring. Some indices may be more sensitive to early-stage drought indicators, while others may capture longer-term drought impacts. Comparison among these indices can help in detecting drought onset, severity, duration, and recovery phases more effectively. A comparative study allows for a better understanding of the spatial and temporal patterns of drought risk. By integrating information from various indices, it is possible to identify areas that are more vulnerable to drought and assess the potential impact on various sectors such as agriculture, water resources, and ecosystems.
- Combining multiple indices allows for cross-validation, reducing the risk of errors or misinterpretations associated with individual indices. If one index shows a certain pattern, while another contradicts it, comparative analysis helps reconcile the differences and provide a more accurate assessment of drought severity.
- Comparison among RS indices enables the development of early warning systems for drought events by providing more robust indicators and forecasts of drought events, helping authorities and stakeholders to take proactive measures to reduce the impacts on communities and ecosystems. By analyzing trends and anomalies across multiple indices, it becomes possible to detect subtle changes indicative of drought onset or progression. Early detection allows for timely intervention and mitigation efforts. Timely detection of drought conditions is crucial for effective mitigation and response efforts.

- Different indices have varying spatial and temporal resolutions. Integrating these indices can provide a more detailed understanding of drought dynamics at different scales, from regional to local levels, and over different time periods, from seasonal to annual.
- By comparing multiple RS indices, it becomes easier to identify regions that are particularly vulnerable to drought. Areas exhibiting consistent patterns of decline across various indices may indicate regions where drought impacts are most severe, thus guiding targeted relief efforts and resource allocation.
- Comparison among RS indices facilitates long-term monitoring of drought conditions, allowing researchers and policymakers to track changes over time and assess the effectiveness of drought management strategies. By combining historical data from different RS indices, comparison enables the analysis of long-term drought trends and patterns. This can facilitate the identification of climate change impacts on drought occurrence, frequency, and intensity, supporting better-informed decision-making for adaptation and resilience-building strategies.

15.9 Conclusions and future scope of research

A comprehensive approach to drought monitoring involves synergizing the strengths of both traditional and RS-based indices. Traditional indices contribute historical context and serve as a baseline for comparison, while RS data provide detailed, near-real-time insights into current conditions. This combination creates a more robust and nuanced understanding of the evolving nature of drought events. The combination of data from diverse drought-related sources enables the creation of integrated drought indices, producing high-quality spatial and temporal information. This approach involves synthesizing information from various datasets, enhancing the accuracy and comprehensiveness of drought monitoring. As technology continues to advance and satellite missions improve, the collaboration between traditional and RS-based indices is poised to become increasingly vital for comprehensive drought monitoring and effective management. Drought monitoring using RS indices presents a rich field for ongoing and future research, particularly given the increasing frequency and severity of drought events worldwide due to climate change, as listed below,

1. To increase the precision and spatial resolution of drought monitoring systems, researchers can investigate combining data from many RS platforms, including satellites, drones, and ground-based sensors.
2. Development of new indices or improving current ones in order to more accurately capture various aspects of drought conditions.
3. Gaining a better understanding of how drought affects fluctuate across different locations and time periods, research is being done on drought monitoring at numerous geographical and temporal scales.

4. Integration of climate models with data from RS to enhance the forecasting and prediction of drought episodes. Proactive drought management plans and early warning systems may benefit from this.
5. To increase the reliability in the accuracy and dependability of RS-based drought indices, especially in complicated landscapes or with changing climate conditions, extensive validation and uncertainty evaluations should be carried out.
6. Integration of auxiliary data sources to improve the interpretation and precision of RS-based drought monitoring systems, such as soil moisture measurements, groundwater levels, and meteorological data.
7. Formation of frameworks and decision support systems that use data from RS to help policymakers, resource managers, and agricultural stakeholders makes well-informed decisions about adapting to and mitigating the drought.
8. The efficiencies of RS-based drought monitoring indices may be evaluated by concentrating on sensitive regions, such as dry and semiarid areas, areas susceptible to agricultural droughts, or places facing increased drought risk as a result of climate change.
9. Application of RS data for long-term trend analysis and monitoring in order to evaluate how the frequency, severity, and geographical extent of droughts have changed over time and to pinpoint possible causes.
10. To guarantee the applicability and adoption of research findings, capacity building programs should be supported. Stakeholders such as governments, NGOs, and local people should be involved in the creation and execution of RS-based drought monitoring systems.

Through investigating these avenues of inquiry, researchers can enhance the body of knowledge and technology concerning RS-based drought monitoring, so supporting improved preparedness, mitigation, and adaption strategies against the backdrop of climate change.

References

AghaKouchak, A., Farahmand, A., Melton, F. S., Teixeira, J., Anderson, M. C., Wardlow, B. D., & Hain, C. R. (2015). Remote sensing of drought: Progress, challenges and opportunities. *Reviews of Geophysics, 53*(2), 452−480. Available from https://doi.org/10.1002/2014RG000456. Available from, http://onlinelibrary.wiley.com/journal/10.1002/(ISSN)1944-9208.

Aksoy, S., Gorucu, O., & Sertel, E. (2019). Drought monitoring using MODIS derived indices and Google Earth Engine platform. In *Eighth international conference on agro-geoinformatics*. 2019/07/01. Institute of Electrical and Electronics Engineers Inc., Turkey. 9781728121161; 10.1109/Agro-Geoinformatics.2019.8820209. http://ieeexplore.ieee.org/xpl/mostRecentIssue.jsp?punumber = 8811423.

Al-Hedny, S. M., & Muhaimeed, A. S. (2020). *Drought monitoring for northern part of Iraq using temporal NDVI and rainfall indices Springer water* (pp. 301−331). Iraq:

Springer Nature. Available from http://www.springer.com/series/13419, 10.1007/978-3-030-21344-2_13.

Al-Quraishi, A. M. F., Gaznayee, H. A., & Crespi, M. (2021). Drought trend analysis in a semi-arid area of Iraq based on Normalized Difference Vegetation Index, Normalized Difference Water Index and Standardized Precipitation Index. *Journal of Arid Land, 13*(4), 413–430. Available from https://doi.org/10.1007/s40333-021-0062-9. Available from, http://link.springer.com/journal/40333.

Amalo, L. F., Ma'rufah, U., & Permatasari, P. A. (2018). Monitoring 2015 drought in West Java using Normalized Difference Water Index (NDWI. *IOP Conference Series: Earth and Environmental Science, 149*(1), 012007. Available from https://doi.org/10.1088/1755-1315/149/1/012007.

Azmi, M., Araghinejad, S., & Kholghi, M. (2010). Multi model data fusion for hydrological forecasting using K-nearest neighbour method. *Iranian Journal of Science and Technology, Transaction B: Engineering, 34*(1), 81–92.

Bento, V. A., Gouveia, C. M., DaCamara, C. C., & Trigo, I. F. (2018). A climatological assessment of drought impact on vegetation health index. *Agricultural and Forest Meteorology, 259*, 286–295. Available from https://doi.org/10.1016/j.agrformet.2018.050.014. Available from, http://www.elsevier.com/inca/publications/store/5/0/3/2/9/5.

Bouras, E. houssaine, Jarlan, L., Er-Raki, S., Balaghi, R., Amazirh, A., Richard, B., & Khabba, S. (2021). Cereal yield forecasting with satellite drought-based indices, weather data and regional climate indices using machine learning in Morocco. *Remote Sensing, 13*(16), 3101. Available from https://doi.org/10.3390/rs13163101.

Brown, J. F., Wardlow, B. D., Tadesse, T., Hayes, M. J., & Reed, B. C. (2008). The Vegetation Drought Response Index (VegDRI): A new integrated approach for monitoring drought stress in vegetation. *GIScience and Remote Sensing, 45*(1), 16–46. Available from https://doi.org/10.2747/1548-1603.45.1.16.

Bushra, N., Rohli, R. V., Lam, N. S. N., Zou, L., Mostafiz, R. B., & Mihunov, V. (2019). The relationship between the Normalized Difference Vegetation Index and drought indices in the South Central United States. *Natural Hazards, 96*(2), 791–808. Available from https://doi.org/10.1007/s11069-019-03569-5. Available from, http://www.wkap.nl/journalhome.htm/0921-030X.

Chai, L., Jiang, H., Crow, W. T., Liu, S., Zhao, S., Liu, J., & Yang, S. (2021). Estimating corn canopy water content from Normalized Difference Water Index (NDWI): An optimized NDWI-based scheme and its feasibility for retrieving corn VWC. *IEEE Transactions on Geoscience and Remote Sensing, 59*(10), 8168–8181. Available from https://doi.org/10.1109/TGRS.2020.3041039. Available from, https://ieeexplore.ieee.org/servlet/opac?punumber = 36.

Chen, J., Li, M., & Wang, W. (2012). Statistical uncertainty estimation using random forests and its application to drought forecast. *Mathematical Problems in Engineering, 2012*. Available from https://doi.org/10.1155/2012/915053.

Chiang, J. L., & Tsai, Y. S. (2012). Reservoir drought prediction using support vector machines. *Applied Mechanics and Materials, 145*, 455–459, http://www.scientific.net/AMM.145.455 16627482.

Choesang, T., Ryntathiang, S., Jacob, B.A., Krishnan, B., & Kokatnoor, S.A. (2023). Drought prediction—A comparative analysis of supervised machine learning techniques. In *Smart innovation, systems and technologies* (vol. 351, pp. 295–307). Springer

Science and Business Media, Deutschland, GmbH. 10.1007/978-981-99-2468-4_23; 21903026. https://www.springer.com/series/8767.

Dai, A. (2011). Drought under global warming: A review. *Wiley Interdisciplinary Reviews: Climate Change*, 2(1), 45–65. Available from https://doi.org/10.1002/wcc.81. Available from, http://onlinelibrary.wiley.com/journal/10.1002/(ISSN)1757-7799.

Dalezios, N. R., Blanta, A., & Spyropoulos, N. V. (2012). Assessment of remotely sensed drought features in vulnerable agriculture. *Natural Hazards and Earth System Science*, 12(10), 3139–3150. Available from https://doi.org/10.5194/nhess-12-3139-2012.

Danandeh Mehr, A., & Fatollahzadeh Attar, N. (2021). A gradient boosting tree approach for SPEI classification and prediction in Turkey. *Hydrological Sciences Journal*, 66 (11), 1653–1663. Available from https://doi.org/10.1080/02626667.2021.1962884. Available from, http://www.tandfonline.com/loi/thsj20.

Das, P. K., Murthy, S. C., & Seshasai, M. V. R. (2013). Early-season agricultural drought: Detection, assessment and monitoring using Shortwave Angle and Slope Index (SASI) data. *Environmental Monitoring and Assessment*, 185(12), 9889–9902. Available from https://doi.org/10.1007/s10661-013-3299-8. Available from, https://link.springer.com/journal/10661.

Das, P. K., Murthy C, S., & Mvr, S. (2014). Monitoring of seasonal dryness/wetness conditions using shortwave angle slope index for early season agricultural drought assessment. *Geomatics, Natural Hazards and Risk*, 5(3), 232–251. Available from https://doi.org/10.1080/19475705.2013.803267. Available from, http://www.tandfonline.com/toc/tgnh20/current.

Das, P. K., Sahay, B., Seshasai, M. V. R., & Dutta, D. (2017). Generation of improved surface moisture information using angle-based drought index derived from Resourcesat-2 AWiFS for Haryana state, India. *Geomatics, Natural Hazards and Risk*, 8(2), 271–281. Available from https://doi.org/10.1080/19475705.2016.1201149. Available from, http://www.tandfonline.com/toc/tgnh20/current.

Dikshit, A., Pradhan, B., & Santosh, M. (2022). Artificial neural networks in drought prediction in the 21st century – A scientometric analysis. *Applied Soft Computing*, 114, 108080. Available from https://doi.org/10.1016/j.asoc.2021.108080.

Dobri, R. V., Sfîcă, L., Amihăesei, V. A., Apostol, L., & Țîmpu, S. (2021). Drought extent and severity on arable lands in Romania derived from normalized difference drought index (2001–2020). *Remote Sensing*, 13(8), 1478.

Donohue, R. J., Roderick, M. L., McVicar, T. R., & Farquhar, G. D. (2013). Impact of CO2 fertilization on maximum foliage cover across the globe's warm, arid environments. *Geophysical Research Letters*, 40(12), 3031–3035. Available from https://doi.org/10.1002/grl.50563.

Dorigo, W., Jeu, R. de, Chung, D., Parinussa, R., Liu, Y., Wagner, W., & Fernández-Prieto, D. (2012). Evaluating global trends (1988–2010) in harmonized multi-satellite surface soil moisture. *Geophysical Research Letters*, 39(18). Available from https://doi.org/10.1029/2012gl052988.

Du, L., Tian, Q., Yu, T., Meng, Q., Jancso, T., Udvardy, P., & Huang, Y. (2013). A comprehensive drought monitoring method integrating MODIS and TRMM data. *International Journal of Applied Earth Observation and Geoinformation*, 23(1), 245–253. Available from https://doi.org/10.1016/j.jag.2012.090.010. Available from, http://www.elsevier.com/locate/jag.

Dutta, D., Kundu, A., Patel, N. R., Saha, S. K., & Siddiqui, A. R. (2015). Assessment of agricultural drought in Rajasthan (India) using remote sensing derived Vegetation Condition Index (VCI) and Standardized Precipitation Index (SPI. *The Egyptian*

Journal of Remote Sensing and Space Science, *18*(1), 53−63. Available from https://doi.org/10.1016/j.ejrs.2015.030.006.

Elbeltagi, A., Pande, C. B., Kumar, M., Tolche, A. D., Singh, S. K., Kumar, A., & Vishwakarma, D. K. (2023). Prediction of meteorological drought and Standardized Precipitation Index based on the random forest (RF), random tree (RT), and Gaussian process regression (GPR) models. *Environmental Science and Pollution Research*, *30* (15), 43183−43202. Available from https://doi.org/10.1007/s11356-023-25221-3. Available from, https://www.springer.com/journal/11356.

Fadaei-Kermani, E., Barani, G. A., & Ghaeini-Hessaroeyeh, M. (2017). Drought monitoring and prediction using K-nearest neighbor algorithm. *Journal of AI and Data Mining*, *5*(2), 319−325.

Fayne, J. V., Ahamed, A., Roberts-Pierel, J., Rumsey, A. C., & Kirschbaum, D. (2019). Automated satellite-based landslide identification product for Nepal. *Earth Interactions*, *23*(3), 1−21. Available from https://doi.org/10.1175/EI-D-17-0022.1. Available from, http://journals.ametsoc.org/doi/pdf/10.1175/EI-D-17-0022.1.

Filipović, N., Brdar, S., Mimić, G., Marko, O., & Crnojević, V. (2022). Regional soil moisture prediction system based on Long Short-Term Memory network. *Biosystems Engineering*, *213*, 30−38. Available from https://doi.org/10.1016/j.biosystemseng.2021.110.019. Available from, http://www.elsevier.com/inca/publications/store/6/2/2/7/9/5/index.htt.

Gao, B. C. (1996). NDWI − A normalized difference water index for remote sensing of vegetation liquid water from space. *Remote Sensing of Environment.*, *58*(3), 257−266. Available from https://doi.org/10.1016/S0034-4257(96)00067-3.

Golkar Hamzee Yazd, H. R., Salehnia, N., Kolsoumi, S., & Hoogenboom, G. (2019). Prediction of climate variables by comparing the k-nearest neighbor method and MIROC5 outputs in an arid environment. *Climate Research*, *77*(2), 99−114. Available from https://doi.org/10.3354/cr01545. Available from, https://www.int-res.com/articles/cr_oa/c077p099.pdf.

Guo, H., Bao, A., Liu, T., Ndayisaba, F., Jiang, L., Zheng, G., . . . De Maeyer, P. (2019). Determining variable weights for an Optimal Scaled Drought Condition Index (OSDCI): Evaluation in Central Asia. *Remote Sensing of Environment*, *231*, 111220. Available from https://doi.org/10.1016/j.rse.2019.111220.

Hayes, M. J., Svoboda, M. D., Wilhite, D. A., & Vanyarkho, O. V. (1999). Monitoring the 1996 drought using the Standardized Precipitation Index. *Bulletin of the American Meteorological Society*, *80*(3), 429−438. Available from http://ams.allenpress.com, 10.1175/1520-0477(1999)080 < 0429:MTDUTS > 2.0.CO;2.

Hayes, M. J., Svoboda, M. D., Wardlow, B. D., Anderson, M. C., & Kogan, F. (2012). *Drought monitoring: Historical and current perspectives. Remote sensing of drought: Innovative monitoring approaches* (pp. 1−19). CRC Press. Available from http://www.tandfebooks.com/doi/book/10.1201/b11863, 10.1201/b11863.

Hu, X., Ren, H., Tansey, K., Zheng, Y., Ghent, D., Liu, X., & Yan, L. (2019). Agricultural drought monitoring using European Space Agency Sentinel 3A land surface temperature and normalized difference vegetation index imageries. *Agricultural and Forest Meteorology*, *279*, 107707. Available from https://doi.org/10.1016/j.agrformet.2019.107707.

Huang, S., Wang, L., Wang, H., Huang, Q., Leng, G., Fang, W., & Zhang, Y. (2019). Spatio-temporal characteristics of drought structure across China using an integrated drought index. *Agricultural Water Management, 218,* 182–192. Available from https://doi.org/10.1016/j.agwat.2019.030.053. Available from, http://www.journals.elsevier.com/agricultural-water-management/.

Jiang, R., Liang, J., Zhao, Y., Wang, H., Xie, J., Lu, X., & Li, F. (2021). Assessment of vegetation growth and drought conditions using satellite-based vegetation health indices in Jing-Jin-Ji region of China. *Nature Research, China Scientific Reports, 11*(1). Available from https://doi.org/10.1038/s41598-021-93328-z. Available from, http://www.nature.com/srep/index.html.

Jiao, W., Wang, L., & McCabe, M. F. (2021). Multi-sensor remote sensing for drought characterization: Current status, opportunities and a roadmap for the future. *Remote Sensing of Environment, 256.* Available from https://doi.org/10.1016/j.rse.2021.112313. Available from, http://www.elsevier.com/inca/publications/store/5/0/5/7/3/3.

Kirana, A. P., Ariyanto, R., Ririd, A. R. T. H., & Amalia, E. L. (2020). Agricultural drought monitoring based on vegetation health index in East Java Indonesia using MODIS Satellite Data. *IOP Conference Series: Materials Science and Engineering, 732*(1), 012063. Available from https://doi.org/10.1088/1757-899x/732/1/012063.

Kogan, F. N. (1990). Remote sensing of weather impacts on vegetation in non-homogeneous areas. *International Journal of Remote Sensing, 11*(8), 1405–1419. Available from https://doi.org/10.1080/01431169008955102.

Kogan, F. N. (1995). Application of vegetation index and brightness temperature for drought detection. *Advances in Space Research, 15*(11), 91–100. Available from https://doi.org/10.1016/0273-1177(95)00079-T.

Kolachian, R., & Saghafian, B. (2021). Hydrological drought class early warning using support vector machines and rough sets. *Environmental Earth Sciences, 80*(11). Available from https://doi.org/10.1007/s12665-021-09536-3. Available from, https://link.springer.com/journal/12665.

Li, W., Du, Z., Ling, F., Zhou, D., Wang, H., Gui, Y., . . . Zhang, X. (2013). A comparison of land surface water mapping using the normalized difference water index from TM, ETM + and ALI. *Remote Sensing, 5*(11), 5530–5549. Available from https://doi.org/10.3390/rs5115530. Available from, http://www.mdpi.com/2072-4292/5/11/5530/pdf.

Li, X., Jia, H., & Wang, L. (2023). Remote sensing monitoring of drought in southwest China using random forest and eXtreme Gradient Boosting methods. *Remote Sensing, 15*(19), 4840. Available from https://doi.org/10.3390/rs15194840.

Liang, L., Qiu, S., Yan, J., Shi, Y., & Geng, D. (2021). VCI-based analysis on spatiotemporal variations of spring drought in China. *International Journal of Environmental Research and Public Health, 18*(15). Available from https://doi.org/10.3390/ijerph18157967. Available from, https://www.mdpi.com/1660-4601/18/15/7967/pdf.

Liu, L., Liao, J., Chen, X., Zhou, G., Su, Y., Xiang, Z., . . . Shao, H. (2017). The Microwave Temperature Vegetation Drought Index (MTVDI) based on AMSR-E brightness temperatures for long-term drought assessment across China (2003–2010. *Remote Sensing of Environment, 199,* 302–320. Available from https://doi.org/10.1016/j.rse.2017.070.012. Available from, http://www.elsevier.com/inca/publications/store/5/0/5/7/3/3.

Mallya, G., Mishra, V., Niyogi, D., Tripathi, S., & Govindaraju, R. S. (2015). Trends and variability of droughts over the Indian monsoon region. *Weather and Climate Extremes, 12,* 43–68. Available from https://doi.org/10.1016/j.wace.2016.010.002. Available from, http://www.journals.elsevier.com/weather-and-climate-extremes/.

Marengo, J. A., Cunha, A. P., Soares, W. R., Torres, R. R., Alves, L. M., de Barros Brito, S. S., ... Magalhaes, A. R. (2018). *Increase risk of drought in the semiarid lands of northeast Brazil due to regional warming above 4°C. Climate change risks in Brazil* (pp. 181−200). Brazil: Springer International Publishing https://www.springer.com/in/book/9783319928807. Available from 10.1007/978-3-319-92881-4_7.

McKee, T.B., Doesken, N.J., & Kleist, J. (1993). The relationship of drought frequency and duration to time scales. In *Proceedings of the eighth conference on applied climatology* (vol. 17, pp. 179−183).

Mirzaee, S., & Mirzakhani Nafchi, A. (2023). Monitoring spatiotemporal vegetation response to drought using remote sensing data. *Sensors*, *23*(4), 2134. Available from https://doi.org/10.3390/s23042134.

Mujiyo, M., Nurdianti, R., & Sutarno, K. (2023). Agricultural land dryness distribution using the Normalized Difference Drought Index (NDDI) algorithm on Landsat 8 imagery in Eromoko, Indonesia. *Environment and Natural Resources Journal*, *21*(2), 127−139. Available from https://doi.org/10.32526/ennrj/21/202200157. Available from, https://ph02.tci-thaijo.org/index.php/ennrj/article/download/248366/168511.

Olukayode Oladipo, E. (1985). A comparative performance analysis of three meteorological drought indices. *Journal of Climatology*, *5*(6), 655−664. Available from https://doi.org/10.1002/joc.3370050607.

Orhan, O., Ekercin, S., & Dadaser-Celik, F. (2014). Use of Landsat land surface temperature and vegetation indices for monitoring drought in the Salt Lake Basin Area, Turkey. *The Scientific World Journal*, *2014*. Available from https://doi.org/10.1155/2014/142939.

Otkin, J. A., Anderson, M. C., Hain, C., Svoboda, M., Johnson, D., Mueller, R., ... Brown, J. (2016). Assessing the evolution of soil moisture and vegetation conditions during the 2012 United States flash drought. *Agricultural and Forest Meteorology*, *218–219*, 230−242. Available from https://doi.org/10.1016/j.agrformet.2015.120.065. Available from, http://www.elsevier.com/inca/publications/store/5/0/3/2/9/5.

Palmer, W. C. (1965). Meteorological drought. *ResPaper*, *45*, 1−58.

Parry, S., Wilby, L. R., Prudhomme, C., & Wood, J. P. (2016). A systematic assessment of drought termination in the United Kingdom. *Hydrology and Earth System Sciences*, *20*(10), 4265−4281. Available from https://doi.org/10.5194/hess-20-4265-2016. Available from, http://www.hydrol-earth-syst-sci.net/volumes_and_issues.html.

Pei, F., Wu, C., Liu, X., Li, X., Yang, K., Zhou, Y., ... Xia, G. (2018). Monitoring the vegetation activity in China using vegetation health indices. *Agricultural and Forest Meteorology*, *248*, 215−227. Available from https://doi.org/10.1016/j.agrformet.2017.100.001. Available from, http://www.elsevier.com/inca/publications/store/5/0/3/2/9/5.

Peled, E., Dutra, E., Viterbo, P., & Angert, A. (2010). Technical note: Comparing and ranking soil drought indices performance over Europe, through remote-sensing of vegetation. *Hydrology and Earth System Sciences*, *14*(2), 271−277. Available from https://doi.org/10.5194/hess-14-271-2010. Available from, http://www.hydrol-earth-syst-sci.net/volumes_and_issues.html.

Pham, H. T., Awange, J., Kuhn, M., Nguyen, B. V., & Bui, L. K. (2022). Enhancing crop yield prediction utilizing machine learning on satellite-based vegetation health indices. *Sensors*, *22*(3), 719. Available from https://doi.org/10.3390/s22030719.

Pham, Q. B., Yang, T. C., Kuo, C. M., Tseng, H. W., & Yu, P. S. (2021). Coupling singular spectrum analysis with least square support vector machine to improve accuracy of SPI drought forecasting. *Water Resources Management*, *35*(3), 847−868. Available from https://doi.org/10.1007/s11269-020-02746-7. Available from, http://www.wkap.nl/journalhome.htm/0920-4741.

Quiring, S. M., & Ganesh, S. (2010). Evaluating the utility of the Vegetation Condition Index (VCI) for monitoring meteorological drought in Texas. *Agricultural and Forest Meteorology*, *150*(3), 330−339. Available from https://doi.org/10.1016/j.agrformet.2009.110.015.

Rouse Jr., J.W., Haas, R.H., Schell, J.A., & Deering, D.W. (1974). Monitoring vegetation systems in the Great Plains with ERTS. In *Third earth resources technology satellite-1 symposium. Volume I: Technical presentations*. NASA. (1974).

Sartori, J.A., Sbruzzi, J.B., & Fonseca, E.L. (2020). Preliminary analysis for automatic tidal inlets mapping using Google Earth Engine. In *IEEE Latin American GRSS and ISPRS remote sensing conference, LAGIRS 2020 − Proceedings* (pp. 93−97). Institute of Electrical and Electronics Engineers Inc. 10.1109/LAGIRS48042.2020.9165650; 9781728143507. http://ieeexplore.ieee.org/xpl/mostRecentIssue.jsp?punumber = 9162082.

Sharma, A., & Goyal, M. K. (2020). Assessment of drought trend and variability in India using wavelet transform. *Hydrological Sciences Journal*, *65*(9), 1539−1554. Available from https://doi.org/10.1080/02626667.2020.1754422. Available from, http://www.tandfonline.com/loi/thsj20.

Sheffield, J., & Wood, E. F. (2008). Projected changes in drought occurrence under future global warming from multi-model, multi-scenario, IPCC AR4 simulations. *Climate Dynamics*, *31*(1), 79−105. Available from https://doi.org/10.1007/s00382-007-0340-z.

Sheffield, J., Wood, E. F., & Roderick, M. L. (2012). Little change in global drought over the past 60 years. *Nature*, *491*(7424), 435−438. Available from https://doi.org/10.1038/nature11575.

Sholihah, R. I., Trisasongko, B. H., Shiddiq, D., Iman, L. O. S., Kusdaryanto, S., Manijo., & Panuju, D. R. (2016). Identification of agricultural drought extent based on vegetation health indices of Landsat data: Case of Subang and Karawang, Indonesia. *Procedia Environmental Sciences*, *33*, 14−20. Available from https://doi.org/10.1016/j.proenv.2016.030.051.

Souza, A. G. S. S., Ribeiro Neto, A., & Souza, L. Ld (2021). Soil moisture-based index for agricultural drought assessment: SMADI application in Pernambuco State-Brazil. *Remote Sensing of Environment*, *252*. Available from https://doi.org/10.1016/j.rse.2020.112124. Available from, http://www.elsevier.com/inca/publications/store/5/0/5/7/3/3.

Stocker, T. F., Qin, D., Plattner, G. K., Alexander, L. V., Allen, S. K., Bindoff, N. L., ... Forster, E. (2013). *Technical summary* (pp. 33−115). Cambridge University Press.

Sun, X., Wang, M., Li, G., & Wang, Y. (2020). Regional-scale drought monitor using synthesized index based on remote sensing in northeast China. *Open Geosciences*, *12*(1), 163−173. Available from https://doi.org/10.1515/geo-2020-0037. Available from, http://www.degruyter.com/view/j/geo.

Sur, C., Kang, D. H., Lim, K. J., Yang, J. E., Shin, Y., & Jung, Y. (2020). Soil moisture-vegetation-carbon flux relationship under agricultural drought condition using optical multispectral sensor. *Remote Sensing, 12*(9). Available from https://doi.org/10.3390/RS12091359. Available from, https://www.mdpi.com/2072-4292/12/9/1359.

Teng, J., Xia, S., Liu, Y., Yu, X., Duan, H., Xiao, H., & Zhao, C. (2021). Assessing habitat suitability for wintering geese by using Normalized Difference Water Index (NDWI) in a large floodplain wetland, China. *Ecological Indicators, 122*, 107260. Available from https://doi.org/10.1016/j.ecolind.2020.107260.

Thomas, T., Nayak, P. C., & Ghosh, N. C. (2015). Spatiotemporal analysis of drought characteristics in the Bundelkhand region of central India using the Standardized Precipitation Index. *Journal of Hydrologic Engineering, 20*(11). Available from https://doi.org/10.1061/(ASCE)HE.1943-5584.0001189. Available from, https://ascelibrary.org/journal/jhyeff.

Tian, M., Wang, P., & Khan, J. (2016). Drought forecasting with vegetation temperature condition index using ARIMA models in the Guanzhong Plain. *Remote Sensing, 8*(9), 690. Available from https://doi.org/10.3390/rs8090690.

Tsiros, E., Domenikiotis, C., Spiliotopoulos, M., & Dalezios, N.R. (2004). Use of NOAA/AVHRR-based vegetation condition index (VCI) and temperature condition index (TCI) for drought monitoring in Thessaly, Greece. In *EWRA symposium on water resources management: Risks and challenges for the 21st century* (pp. 2−4).

Van der Molen, M. K., Dolman, A. J., Ciais, P., Eglin, T., Gobron, N., Law, B. E., … Wang, G. (2011). Drought and ecosystem carbon cycling. *Agricultural and Forest Meteorology, 151*(7), 765−773. Available from https://doi.org/10.1016/j.agrformet.2011.010.018.

Venkatesh, K., John, R., Chen, J., Xiao, J., Goljani Amirkhiz, R., Giannico, V., & Kussainova, M. (2022). Optimal ranges of social-environmental drivers and their impacts on vegetation dynamics in Kazakhstan. *Science of the Total Environment, 847*, 157562. Available from https://doi.org/10.1016/j.scitotenv.2022.157562.

Vicente-Serrano, S. M., Domínguez-Castro, F., McVicar, T. R., Tomas-Burguera, M., Peña-Gallardo, M., Noguera, I., … El Kenawy, A. (2020). Global characterization of hydrological and meteorological droughts under future climate change: The importance of timescales, vegetation-CO_2 feedbacks and changes to distribution functions. *International Journal of Climatology, 40*(5), 2557−2567. Available from https://doi.org/10.1002/joc.6350. Available from, http://onlinelibrary.wiley.com/journal/10.1002/(ISSN)1097-0088.

Villegas-Ch, W., & García-Ortiz, J. (2023). A long short-term memory-based prototype model for drought prediction. *Electronics, 12*(18), 3956. Available from https://doi.org/10.3390/electronics12183956.

Wang, P.X., Li, X.W., Gong, J.Y., & Song, C. (2001). Vegetation temperature condition index and its application for drought monitoring. In *International geoscience and remote sensing symposium (IGARSS)* (pp. 141−143).

Xue, J., & Su, B. (2017). Significant remote sensing vegetation indices: A review of developments and applications. *Journal of Sensors, 2017*. Available from https://doi.org/10.1155/2017/1353691. Available from, http://www.hindawi.com/journals/js/biblio.html.

Zeng, J., Zhang, R., Qu, Y., Bento, V. A., Zhou, T., Lin, Y., & Wang. (2001). Improving the drought monitoring capability of VHI at the global scale via ensemble indices for various vegetation types from 2001 to 2018. *Weather and Climate Extremes, 35*, 100412.

Zhang, A., Jia, G., & Wang, H. (2019). Improving meteorological drought monitoring capability over tropical and subtropical water-limited ecosystems: Evaluation and ensemble of the Microwave Integrated Drought Index. *Environmental Research Letters*, *14*(4), 044025. Available from https://doi.org/10.1088/1748-9326/ab005e.

Zhang, X., Obringer, R., Wei, C., Chen, N., & Niyogi, D. (2017). Droughts in India from 1981 to 2013 and implications to wheat production. *Scientific Reports*, *7*. Available from https://doi.org/10.1038/srep44552. Available from, http://www.nature.com/srep/index.html.

Zhao, Z., & Wang, K. (2021). Capability of existing drought indices in reflecting agricultural drought in China. *Journal of Geophysical Research: Biogeosciences*, *126*(8). Available from https://doi.org/10.1029/2020JG006064. Available from, https://agupubs.onlinelibrary.wiley.com/loi/21698961.

Zhou, X., Wang, P., Tansey, K., Zhang, S., Li, H., & Wang, L. (2020). Developing a fused vegetation temperature condition index for drought monitoring at field scales using Sentinel-2 and MODIS imagery. *Computers and Electronics in Agriculture*, *168*, 105144. Available from https://doi.org/10.1016/j.compag.2019.105144.

Zhou, Y., Liu, L., Zhou, P., Jin, J., Li, J., & Wu, C. (2014). Identification of drought and frequency analysis of drought characteristics based on palmer drought severity index model. *Nongye Gongcheng Xuebao/Transactions of the Chinese Society of Agricultural Engineering*, *30*(23), 174–184. Available from https://doi.org/10.3969/j.issn.1002-6819.2014.230.022. Available from, http://www.tcsae.org/nygcxb/ch/index.aspx.

Geographic information system-based statistical mapping of socioeconomic vulnerability in the Upper Citarum River, West Java Province, Indonesia

16

Setiawan Hari Harjanto[1], Tanjung Mahdi Ibrahim[2], Abdullah Abdullah[3], Djaenudin Djaenudin[4] and Siswoyo Suhandy[5]

[1]*National Research and Innovation Agency (BRIN), Research Center for Social Welfare, Villages and Connectivity, Jakarta, Indonesia*
[2]*Ministry of Public Works and Housing, Water Resources, Bandung, Indonesia*
[3]*School of Economics and Business, Department of Management, Telkom University, Bandung, Indonesia*
[4]*National Research and Innovation Agency (BRIN), Center for Environmental Research and Clean Technology, Jakarta, Indonesia*
[5]*Indonesian University of Education (UPI), Architecture Study Program, Bandung, Indonesia*

16.1 Introduction

The Citarum watershed is the longest river basin in West Java in Indonesia. The length of this river is ± 3332.97 km, with 105 tributaries flowing and ending in the Java Sea area, Karawang Regency. The Citarum watershed area covers an area of $\pm 682,227$ ha, which is utilized by the Saguling Reservoir, Cirata Reservoir, and Jatiluhur Reservoir for various development activities. This watershed is a raw source of drinking water, agricultural irrigation water, electricity generation, and a raw source of industrial water. The upper reaches of the Citarum River are strategic and support various economic activities in West Java Province and DKI Jakarta. Housing construction causes a lack of water absorption during the rainy season. The high capacity of water overflow means that river flows cannot be accommodated, resulting in flooding (Jurumai, Abidin, Soetrisno, & Wibowo, 2023). Continuous flooding causes socioeconomic vulnerabilities that require solutions. Community vulnerability is a population negatively impacted by social and economic factors (Rahadiati et al., 2019). This vulnerability can be seen in various forms, including high poverty levels, limited access to resources,

Applications of Geospatial Technology and Modeling for River Basin Management.
DOI: https://doi.org/10.1016/B978-0-443-23890-1.00016-5

and reduced capacity to cope with disruptions or disasters (Verschuur, Koks, & Hall, 2020). Socioeconomic vulnerability mapping is essential to effectively assess social and economic conditions (Deroliya et al., 2022). Mapping results will be better using the geographic information system (GIS). This computer-based mapping information system is used to enter, store, retrieve, process, analyze, and produce data. GIS-based statistical mapping approaches can be used to analyze and visualize data related to socioeconomic variables such as population density, poverty levels, infrastructure quality, access to services, and environmental risks. The Citarum River is one of the strategic rivers in West Java Province, which has the most dense population in Indonesia. Flood disasters can cause socioeconomic impacts on society, such as infrastructure damage, trauma, and even loss of life (Svetlana, Radovan, & Ján, 2015). Socioeconomic vulnerability can be seen from various factors, including family livelihood, gender, and age. In addressing socioeconomic vulnerability in the Upper Citarum River, it is important to consider the specific risk factors that contribute to this vulnerability. These risk factors may include limited access to resources, high levels of poverty, and reduced capacity to cope with disasters. Climate change also has an impact on socioeconomic vulnerability in the Upper Citarum River, which is getting worse. Communities at high risk of experiencing vulnerability are communities whose livelihoods as farmers depend on natural conditions. People who work as farmers and their subsistence activities depend directly on weather patterns. Addressing socioeconomic vulnerabilities in the Upper Citarum River is critical to considering the impacts of climate change in the region (Schilling, Hertig, Tramblay, & Scheffran, 2020). Understanding the socioeconomic vulnerability of communities along the upper reaches of the Citarum River with GIS is very important for disaster risk management and sustainable development in the region (Alves, Gersonius, Kapelan, & Sanchez, 2019). Analyzing community socioeconomic vulnerability mapping with GIS is essential for policymakers and stakeholders. They can identify areas that require intervention to increase community resilience. This mapping also contributes to knowledge about the severity of socioeconomic vulnerabilities in river basins to reduce risks and improve community welfare. The upper reaches of the Citarum River are an area that faces various flooding challenges due to changes in land use and uncontrolled urban development (Kuntoro, Putro, Kusuma, & Natasaputra, 2017). These challenges have significant implications for the socioeconomic well-being of communities in this region. This study, which focuses on the Upper Citarum River area, provides insight into the socioeconomic vulnerability of the community so that more targeted interventions are needed. This chapter will explain the demographic profile of the Citarum River upstream, socioeconomic vulnerability, flood disaster mitigation, and the GIS-based early warning system (EWS). The benefit obtained is knowing the actual problems in the Upper Citarum River so that the causes of these problems can be analyzed and practical solutions can be found. The effectiveness of flood disaster mitigation in the area is implemented to reduce the impact. The role of stakeholders and community participation in flood management through the EWS needs to be well-socialized to minimize the effects of flooding. So, the socioeconomic impact of flooding in the upstream Citarum River must be intervened comprehensively to resolve the problems there.

16.2 The rationale of the study

16.2.1 Socioeconomic vulnerability

Physical vulnerability is the condition of an area against the physical impacts of natural disasters such as floods, landslides, and earthquakes (Kappes, Papathoma-Köhle, & Keiler, 2012). In the case of the Citarum River Headwaters, physical vulnerability is a significant concern because of its vulnerability to flooding due to uncontrolled changes in land use for urban development. These changes have altered the characteristics of the river, leading to an increased risk of flooding. Factors for assessing physical vulnerability in the Upper Citarum River include proximity to the river, altitude of the area, frequency of flooding, and severity (Bouaakkaz, El Morjani and Bouchaou, 2023). These factors contribute to the potential for physical damage during a flood event. Steep slopes and unstable ground conditions can also increase the region's vulnerability to landslides threatening settlements. Understanding the physical vulnerability of the Upper Citarum River is very important to mitigate the potential impact of floods and landslides in this region. This mitigation helps implement appropriate measures to protect affected communities. Apart from the physical vulnerability factors mentioned previously, the socioeconomic vulnerability of communities in the Upper Citarum River area also plays a vital role in determining their ability to overcome and recover from flood disasters. Socioeconomic vulnerability is the condition of communities and individuals negatively impacted by natural disasters based on their socioeconomic characteristics, such as income level and access to resources (De Silva & Kawasaki, 2018). Specific population subgroups, such as women and senior citizens, are particularly vulnerable regarding their ability to cope with flooding (Chakraborty, Rus, Henstra, Thistlethwaite, & Scott, 2020). Taking socioeconomic vulnerabilities into account is necessary to ensure that flood mitigation and relief efforts are carried out equitably and meet the needs of all communities. Understanding the socioeconomic vulnerability of communities along the upper reaches of the Citarum River is critical for effective disaster risk management and sustainable development in the region. Mapping and analyzing socioeconomic vulnerabilities is helpful for policymakers in creating targeted programs. The upper reaches of the Citarum River are a critical river basin in Indonesia. This region faces various hydrological, hydraulic, and flooding challenges due to changes in land use caused by uncontrolled urban development activities (Kuntoro et al., 2017). These challenges have significant implications for the socioeconomic well-being of this region. Understanding the socioeconomic vulnerability of communities along the upper reaches of the Citarum River is very important for managing the risk of frequent flood disasters. Therefore, understanding the socioeconomic vulnerabilities of these communities is very important to develop effective strategies for mitigating flood risks and encouraging sustainable development in the Upper Citarum River region (Abdi-Dehkordi, Bozorg-Haddad, Salavitabar, Mohammad-Azari, & Goharian, 2021).

16.2.2 Geographic information system-based risk management

GIS-based mapping techniques can analyze factors such as height, slope, soil type, and land cover in the study area to assess physical vulnerability (Tehrany, Pradhan, Mansor, & Ahmad, 2015). GIS can also examine the factors that caused flood events in the past. This technique can also identify areas most at risk and determine priority mitigation strategies. It is essential to realize that physical factors will impact the socioeconomic vulnerability felt by the community. Socioeconomic vulnerability, namely, the vulnerability of society based on its socioeconomic conditions, is essential in flood vulnerability. Socioeconomic factors, such as income level, education, and resource access, can influence a community's ability to prepare for and recover from flooding. Considering these factors is essential to comprehensively analyze flood vulnerability to develop effective risk reduction strategies completely (Abbas, Amjath-Babu, Kächele, Usman, & Müller, 2015). The social resilience of communities around the Upper Citarum River is essential in determining the community's ability to survive and recover from flood disasters. Factors contributing to community social resilience include socioeconomic status, access to resources and services, level of preparedness, community cohesion, and strong social networks and support systems (Saja, Goonetilleke, Teo, & Ziyath, 2019). GIS-based statistical mapping can collect data on these factors to assess the social resilience of communities in the Upper Citarum River area. This can include collecting data on variables such as income level, educational attainment, access to health facilities, infrastructure availability, and community groups' existence. By mapping and analyzing these variables, we can identify areas of social vulnerability to determine which communities need additional support in building their resilience (Imran, Sumra, Amer Mahmood, & Faisal Sajjad, 2019). The flow of thought for flood disaster management in the Upper Citarum River area can be seen in Fig. 16.1.

Risk management is an action taken to reduce the socioeconomic impact of a disaster (Jia et al., 2022). This step is expected to minimize disaster victims and reduce the losses incurred. Disaster risk management is critical to implement in countries located in areas prone to natural disasters, such as Indonesia. There are several stages in disaster risk management. Starting from the mitigation plan as a preventive stage and then continuing with the early warning stage. Then, there are the evacuation process stages when a disaster occurs and the post-disaster recovery stage to repair the affected infrastructure. Managing flood disasters on the Citarum River requires a recording system supported by spatial data. The river's flow at several points results in the drainage being unable to accommodate excess

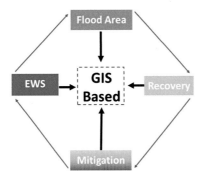

FIGURE 16.1

GIS-based Upper Citarum River risk management.

water because it is damaged and blocked. A system is needed to prevent premature flooding. Therefore, using spatial data analysis theory, it is necessary to provide spatial information supported by GIS to solve problems and meet asset management needs. Spatial data analysis consists of several parameters, which will be assessed by giving values and weights according to classification and then superimposed to produce a map.

16.3 Materials and methods

This book chapter of the book uses a qualitative descriptive approach (Doyle, McCabe, Keogh, Brady, & McCann, 2020). This approach will provide an in-depth understanding of socioeconomic vulnerability and explore contextual factors regarding little-known empirical phenomena (Côté-Boileau, Gaboury, Breton, & Denis, 2020) regarding flooding in the Upper Citarum River. Empirical data regarding the socioeconomic vulnerability of communities around the upper reaches of the Citarum River was also obtained through this approach. Research results were developed by thoroughly interpreting the facts at the research location. The research location was deliberately determined (Marcella-Hood & Marcella, 2023) around the upper reaches of the Citarum River. This location was chosen because it has the highest socioeconomic vulnerability in West Java. The increasing population and development activities make this area increasingly vulnerable to flooding. Vulnerability assessments are critical to determining disaster preparedness, mitigation, response, and recovery. Informants were chosen purposively because they understand the vulnerability around the Citarum River Headwaters. Secondary data sources include data and information from related institutions, both government and NGOs. Data collection techniques used in-depth interviews with selected informants, observation, and document review. The data analysis technique uses a qualitative approach with a descriptive process starting from reducing, presenting, and verifying conclusions (Sundler, Lindberg, Nilsson, & Palmér, 2019). The three components of data analysis, which include reduction, presentation, and discovery, are interconnected during and after data collection. The validity of this research data uses data source triangulation (Jentoft & Olsen, 2019). The GIS data will be analyzed by connecting it with phenomena in the Upper Citarum River. This geographic analysis system is a computer-based information system that can process spatial data to get an overview of the changing patterns of the Citarum River Upstream phenomenon. This GIS can help predict the occurrence of natural disasters and provide a spatial model for constructing the Nanjung Tunnel. The three types of geographic analysis used are buffering, overlay, and network. Buffering analysis is a GIS analysis that produces new regional and territorial boundaries. This analysis is suitable for planning, environmental protection, and zoning mapping. Overlay analysis overlaps two or more spatial data to determine unique land suitability and informal integration. Network analysis analyzes interconnected paths or points (Singh, Gupta, & Singh, 2014). GIS analysis is a collection of methods that can be used to carry out GIS processing.

16.4 Results and discussion

16.4.1 Demographic profile

The Citarum River is an ancient river that flows into Lake Cisanti on the slopes of Mount Wayang. Bandung Regency, West Java Province, Indonesia. This river has the most expansive river basin (DAS) in West Java, with an altitude of ± 1700 m above sea level. The area is approximately 6615 sq. km or 661,500 ha. This river flows downstream about 300 km through Bandung Regency, West Bandung Regency, Cianjur Regency, and Purwakarta Regency. This river flows to the north coast of Java Island, precisely at the end of Karawang Regency. Etymologically, the name Citarum comes from two words, namely, "ci" which means "water" and "tarum," which is a type of plant that was previously used to dye cloth. Some people think the name Citarum is related to the name of the oldest kingdom in West Java, namely, Tarumanagara. People widely use rivers for washing and bathing, as a source of drinking water and as a source of livelihood. People also often use rivers as water reservoirs, means of transportation, recreation, and sources of irrigation for agriculture (Brierley et al., 2023). The Citarum watershed is divided into three subwatersheds: the Upper Citarum subwatershed, the Central Citarum subwatershed, and the Lower Citarum watershed, all united in a hydrological unity. The Upper Citarum subwatershed is the focus of this book because this subwatershed is the primary water catchment area and flow regulator. The condition of the Upper Citarum subwatershed has long been under pressure due to the intervention of human activities that are not environmentally friendly, thus becoming a source of flooding problems (Priya et al., 2023). For example, the Upper Citarum subwatershed can restore its natural function, positively impacting the Central Citarum subwatershed and Lower Citarum subwatershed. Population pressure and increasing economic activity are causing changes in land use in this segment. Indications of this damage are felt by decreasing minimum extreme discharge and increasing maximum powerful discharge. Increasing the water runoff coefficient value causes flooding phenomena in the rainy season and drought in the dry season. The worsening quality of river water is due to many industrial, residential, and livestock activities (Anh, Can, Nhan, Schmalz, & Luu, 2023). This industry dumps its waste directly into the Citarum River and its tributaries without processing it first, adding to the problems in the Upper Citarum subwatershed.

Various efforts have been made to improve the environment of the Upper Citarum subwatershed, including by cutting straightening twists and turns to deal with flooding. They are building wastewater treatment plants in Bojongsoang and Cisirung to deal with domestic and industrial waste pollution. Many dams must be built to overcome erosion and sedimentation. All these efforts have not been able to overcome the problems in the Upper Citarum subwatershed because flooding, landslides, erosion, sedimentation, and river pollution continue to this day. An incomprehensive understanding of the natural character and population of the Upper Citarum subwatershed caused this failure. An image of the risk of flooding can be seen in Fig. 16.2.

Other problems with the Citarum watershed include deforestation in upstream areas (conservation) and land clearing without planning and supervision. Soil erosion in the upstream part causes high levels of sedimentation in the middle and downstream parts, causing river water to overflow beyond capacity, resulting in flooding. The problem in the Citarum River Basin faced by the government and society is flooding. This occurs due to land subsidence, large amounts of land conversion, silting, and narrowing of rivers, which have increased over the last 20 years. Locations where flooding frequently occurs in the Citarum watershed area, which requires special handling, include flooding in the Rancaekek, Dayeuhkolot, Pasteur, Pagarsih, Gedebage, Melong, and Margaasih areas.

16.4.2 Socioeconomic vulnerability

To understand the socioeconomic vulnerability of the community in the Hulu Sungai Citarum area, it is necessary to consider physical factors. This can be achieved by assessing indicators such as income level, access to basic services, education level, and social networks. These indicators provide insight into a community's resources and capacity to prepare for, respond to, and recover from flood events. Other physiographic factors, including elevation, slope, and land cover, contribute to flood vulnerability (Deepak, Rajan, & Jairaj, 2020). A community's vulnerability to flooding is not only determined by physical and socioeconomic factors. Residents' perceptions of risk and sociocultural control variables also play an essential role in shaping vulnerability to flood hazards

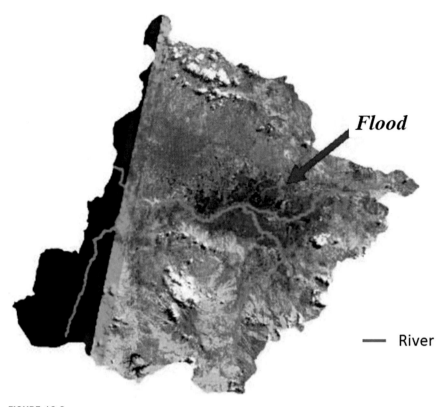

FIGURE 16.2

Map of the flood risk of the Upper Citarum River. Map lines delineate study areas and do not necessarily depict accepted national boundaries.

(Ahmed & Roy, 2015). Therefore, a holistic approach to flood risk management is needed to effectively address vulnerability in the Upper Citarum River community effectively (Chakraborty et al., 2020). Socioeconomic vulnerability assessments must consider physical vulnerability because, initially, there is physical damage due to flooding. A more comprehensive understanding of vulnerability can be achieved by considering socioeconomic factors that influence vulnerability (Thakur & Mohanty, 2023). By examining various vulnerability factors, including physical, socioeconomic, attitudinal, and sociocultural, we can understand the vulnerability of the Upper Citarum River community to flooding. One important group that may be particularly vulnerable during floods is women and older people. These groups often face challenges during floods. They have limited mobility and access to resources, which makes their condition vulnerable. In assessing socioeconomic vulnerability, it is essential to consider the vulnerability of specific groups in society. Socioeconomic vulnerability in the Upper Citarum River, West Java Province, Indonesia, is influenced by several factors (Widianingsih,

Riswanda, & Paskarina, 2019). Physical characteristics that influence socioeconomic vulnerability include population density, need for clean water, and dependence on land. Socioeconomic factors affected by physical vulnerability include income, access to essential services, and education. By adopting a holistic approach, policymakers and stakeholders can develop more effective strategies to address the vulnerability of the Hulu Sungai Citarum community. In conclusion, a holistic flood risk management approach is crucial in overcoming flood hazard vulnerability. The Upper Citarum watershed is mainly in the Bandung administrative area, covering 56% (Belinawati, Soesilo, Herdiansyah, & Aini, 2018). This condition makes this area vulnerable to natural disasters, exacerbating socioeconomic vulnerability. Rural settlements have a reasonably low vulnerability value because their population is in the medium category compared to densely populated urban areas. Based on this analysis, controlling population growth in areas with high vulnerability is necessary. Maintaining the evacuation system also needs to be carried out considering the high number of residents, children, elderly, and women. The social vulnerability parameter has a significant influence because floods occur in residential areas with a high population. According to informants through interviews, this social vulnerability parameter is considered the most important in flood vulnerability. Therefore, social aspects need to be managed well through increasing capacity in dealing with disasters as a form of disaster mitigation effort (Tate, Rahman, Emrich, & Sampson, 2021). Areas that are prone to flooding are Majalaya and Rancaekek Districts, which can be seen in Fig. 16.3.

Baleendah District is one of three subdistricts affected by flooding due to the overflow of the upstream Citarum River. Most of the area is at the bottom of the Bandung Basin. This is what causes Baleendah District to be vulnerable to flood disasters. Social vulnerability parameters include population size, sex ratio, and age group ratio (Hinojos, McPhillips, Stempel, & Grady, 2023). The population is in the highest position because a high population can create obstacles in the evacuation process. Flood disasters can cause physical, economic, social, and ecological impacts, for example, infrastructure damage and loss of life. Socially, flood disasters cause trauma and other psychological effects on flood victims. Apart from Majalaya and Rancaekek Districts, areas prone to flood disasters are Majalaya District, which can be seen in Fig. 16.4.

A high population can also indicate a high possibility of loss of life when a disaster occurs—the ratio of vulnerable age groups, namely, children and elderly age groups. Children and older people are especially vulnerable to disasters because of their limited mobility and difficulty evacuating independently (Li, Xia, Zhou, Deng, & Dong, 2024). Furthermore, the sex ratio describes the population composition between women and men. This condition is taken into account because women are considered more vulnerable when a disaster occurs. In general, women's mobility in developing countries is limited by a culture that requires them to be homemakers, so most of their time is spent at home. So the death rate due to floods is greater, meaning they are more vulnerable than men

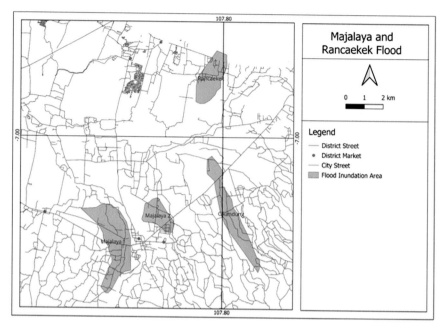

FIGURE 16.3

Flood hazard map in Majalaya and Rancaekek. Map lines delineate study areas and do not necessarily depict accepted national boundaries.

(Krishnan, 2023). Social vulnerability is a description of the condition of the level of social fragility in facing a disaster. Social vulnerability comprises population density parameters and vulnerable groups (Ahmadu, Nukpezah, French, & Menifield, 2024). Vulnerable groups include the sex ratio, poverty ratio, disabled person ratio, and age group ratio (Xue, Huang, Xie, Sun, & Fei, 2024). Population density is a condition that changes every year due to changes in population in an area, either naturally or due to population movement from other regions, and this indicator helps see the population density in a spatial unit. A sex ratio is a number that shows the ratio between the number of males and the number of females in a particular area and time, usually expressed as the number of males per 100 females. The poverty ratio is between poor households and all households in an area. The ratio of disabled people is the ratio of the number and proportion of disabled people to the population as a whole. The age group ratio is the ratio of nonproductive and productive ages in an area. Socioeconomic vulnerability in the Upper Citarum River area is also influenced by rapid urbanization and population growth. Urbanization and population growth have increased the demand for resources and services, putting pressure on limited regional capacity to meet these needs. This can result in inadequate infrastructure, limited access to essential services such as clean water and sanitation, and inadequate health and education facilities. The conversion of forest areas into housing also contributes

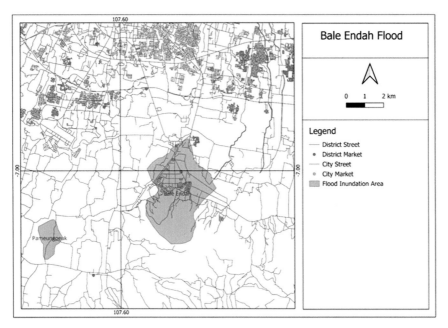

FIGURE 16.4

Flood hazard map in Baleendah. Map lines delineate study areas and do not necessarily depict accepted national boundaries.

to changes in land use, which is the main factor in flooding in the area (Ma, Wang, Zhao, & Jiang, 2023). The vulnerability of the health sector is also felt by the people living around the Hulu Sungai Citarum. This river is in the international spotlight because of its serious pollution level. Pollution causes include industrial, domestic, agricultural, and garbage (Noor et al., 2023). This causes a decline in river water quality, as seen from several indicators. Chemical pollution is caused by the many industries along the river that dispose of liquid waste containing dangerous chemicals. The turbidity of river water is caused by settlements along the river flow, which are the disposition of domestic land. Nutrient increases caused by unsustainable agricultural practices, such as excessive use of fertilizers and pesticides, can pollute rivers. Bacteria and germs are caused by many people throwing away illegal waste. The housing sector experienced more losses due to major floods than the agricultural sector. The highest building structure losses occurred in the residential sector, while cleaning costs were not too high compared to the other two categories. However, the total residential losses from these three sectors reached IDR 600−700 billion for the entire Hulu Sungai Citarum area. The agricultural sector also experienced losses due to flooding. The total amount that can be harvested depends on the depth and duration of the flood. If the entire rice crop cannot be harvested during the flood, the total loss will reach IDR 68 billion. It is hoped that information regarding the

Table 16.1 Losses due to floods in 2022.

No	Damage category	Loss
1.	Building	368,687,000
2.	Fill the house	235,922,000
3.	House cleaning	1346,000
4.	Rice plants	40,507,000

distribution of losses experienced by the housing and agricultural sectors can be the basis for creating a zoning map for the level of flood vulnerability in the Citarum watershed. With this zoning map, it is hoped that river watershed management, risk management, disaster adaptation, and mitigation options can be more anticipatory and targeted. The amount of loss due to flooding can be seen in Table 16.1.

Factors that influence the level of vulnerability are population size, gender, age group, population density, poverty level, people with disabilities, level of dependency, number of family members, population growth, and level of education. The area's vulnerability level is addressed to reduce regional disaster risk. Treatment, in this case, includes increasing public awareness regarding vulnerable locations that require disaster response efforts, conducting outreach regarding flood disasters and mitigation efforts, the community's active role in maintaining environmental cleanliness, and so on. The government and communities in the local area carry out disaster mitigation forms.

16.4.3 Flood disaster mitigation

The definition of disaster management is an event or series of events that threaten and disrupt people's lives caused by natural, nonnatural, and social factors. Disasters result in casualties, environmental damage, property loss, and psychological impacts. Disasters are divided into natural, nonnatural, and social disasters. Causes of natural disasters include earthquakes. Nonnatural disasters are caused by nonnatural events such as disease outbreaks. Social disasters are caused by human behavior, such as the social conflict between groups. Mitigation must be implemented as a disaster management effort to reduce the risk of flood disasters in an area physically and nonphysically (Qi et al., 2021). Physical mitigation is carried out by providing infiltration wells, improving drainage systems, building poles, and so forth. Then, nonphysical mitigation is by preparing flood disaster management policies, such as making regulations related to disaster risk reduction efforts (Mesta, Kerschbaum, Cremen, & Galasso, 2023). Other government efforts include outreach, counseling, education, and increasing public awareness of flood disasters. Accurate weather forecasts are important for society. They are key in helping decision-makers and communities prepare to respond to the impacts of flooding. By understanding the factors contributing to social

FIGURE 16.5

Before the construction of the Nanjung Tunnel.

vulnerability to flooding, such as exposure, sensitivity, and adaptive capacity, policymakers and stakeholders can develop more effective strategies to address and mitigate the impacts of flooding on marginalized communities. The form of physical mitigation carried out by the Indonesian government can be seen in Fig. 16.5 before it was done.

The government has completed the construction of the Nanjung Tunnel in Bandung Regency as part of the Citarum River Flood Control System. The construction of the Nanjung Tunnel in the Hulu Citarum area at Curug Jompong facilitates the flow of the Citarum River downstream so that the duration and extent of flood inundation in the South Bandung area can be reduced. The existence of the Nanjung Tunnel has provided the benefit of reducing waterlogging during the peak of rain that flushed parts of West Java Province. So far, during the rainy season, the large flood discharge of the Citarum River has been blocked by large rocks at Curug Jompong, which is also a cultural site, causing flooding due to backwater. Therefore, the Nanjung Tunnel was built to facilitate water flow from the upper reaches of the Citarum River. The Nanjung Tunnel was built in November 2017 and completed in December 2019. It consists of 2 tunnels, each 230 m long and 8 m in diameter. The two tunnels will reduce flooding in Dayeuhkolot, Baleendah, Andir, and surrounding areas with many families. The Nanjung Tunnel will reduce the total inundation area in Bandung

FIGURE 16.6

After construction of the Nanjung Tunnel.

Regency from 3461 to 2761 ha. Some tunnels will increase the capacity of the Citarum River from 570 to 669 m³/second. Once completed, the tunnel can be seen in Fig. 16.6.

The Greater Bandung area is shaped like a bowl with the lowest basin around the Citarum River. During the rainy season, many residential areas around the riverbanks, including Dayeuhkolot, Bojongsoang, and Rancaekek subdistricts, on the north side of the river, and Baleendah subdistrict on the south side are prone to flooding. The Citarum flood overflowed left and right. The vulnerability to flooding in the valley on the banks of the Citarum River is increasingly chronic because it flows into two rivers at once, not far from the Bojongsoang Bridge. On the east side, it flows into the Cikapundung River, which flows from the direction of Bandung City, and the Cisangkuy River from the Pengalengan plateau. Both estuaries are now to the right of the Bojongsoang Bridge, a distance of less than 1 km. If heavy rain hits Greater Bandung, the Citarum discharge around the Bojongsoang Bridge could double to reach above 1000 m³/second. Added to this is the narrow river body and rapid shallowing due to massive erosion in the upstream section. Water overflow is inevitable. Even though it is located in the highlands, with a normal water level of 655 m above the sea level, the contour of the Citarum bottom body is quite sloping for several tens of kilometers downstream. The rate of water movement downstream is limited. After leaving the traffic jam at Curug Jompong, the Citarum River flows rapidly downstream and enters the Saguling Dam gathering area. Apart from making tunnels, the next step is making retention ponds. Both have the function of collecting water from the surrounding rainwater catchment areas. A retention pond is a pond that functions to temporarily store rainwater by allowing it to seep into the ground, the operation of which can be combined with a pump or sluice gate. The function of retention ponds is to collect rainwater and channel river water into ponds to be stored temporarily and then discharged back into the river or channeled in to the ground so that infiltration occurs. The size, volume, area, and depth of this pond depend on how much land is converted into residential areas. The location before the retention pond was built can be seen in Fig. 16.7.

The retention pond was constructed by four polder ponds spread across Dayeuhkolot and Baleendah subdistricts. The benefit is that the accumulation area, originally 247 ha, has decreased to 32 ha. If the benefits are calculated, it can protect coverage for 5192 families or the equivalent of 15,973 people, including protecting district and provincial roads. The retention pond was built on an area of 4.85 ha with a water catchment area of 149 ha and a storage volume of up to 160,000 m^3. The entire drainage system in the water intake area is closed. All water flows into the retention pond. There are no more sewers that discharge water into Citarum. Embankments also protect the area to control river water

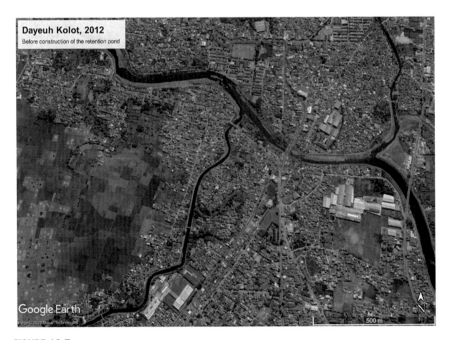

FIGURE 16.7

Before the construction of the retention pond.

runoff during floods. Under normal circumstances, the water retention gate in the pool opens to channel water into the river. However, there is heavy rain when the river water is high, and the floodgates are closed tightly. Water from the rain catchment area in the retention pond is pumped into the river. Thus, 149 ha of the area is safe from shelters. Three water pump units have a total capacity of 500 liters per second. After the retention pond is built, the peak flood discharge during overflow will decrease due to the additional time for water concentration to flow on the surface. Fig. 16.8.

Retention ponds function to hold water and keep it free from puddles. During rains and floods, all water access to the river is closed, and pumps remove the air. Retention ponds can function as water absorption areas into the ground. Groundwater in the area can guarantee its existence. Previously, it had been operating the Cieunteung Retention Pool since early 2018. This retention pool has a pool area of 4.75 ha and a depth of 6.5 m and can accommodate 190,000 m^2 of rain runoff plus household wastewater. With 4 pump units and a total capacity of 12.50 m^3/second, this retention pond can accommodate an area of 91 ha. The construction of tunnels and retention ponds has saved around 3100 houses and some commercial buildings, provincial roads, and article roads in Baleendah. Complementing the retention ponds, new embankments continue to be built on the banks of the Citarum, Cikapundung, and Cisangkuy Rivers. Small rivers that used to flow

FIGURE 16.8

After completion of the construction of the retention pond.

directly in Citarum now have sluice gates that close when the river water rises. The water flow is diverted into polder ponds or retention ponds. The existing retention ponds are still insufficient to control flooding in the Greater Bandung Basin. Bandung alone still needs 30 new polder pools, each with an area of 1000 m^2. No retention ponds are available in Bandung Regency, Bojongsoang, and Dayeuhkolot. Several sections of the Citarum River also require dredging and widening.

16.4.4 Early warning system

Early warning is a series of activities to provide warnings as soon as possible to the public about the possibility of a disaster occurring in a place by authorized institutions. Several stages can be achieved effectively by implementing early flood warnings (Mandal, Dharanirajan, Meena, Jaman, & Rana, 2023). First, timely detection data is processed and monitored to obtain information about the possibility of flooding. This information is then passed on to provide warnings without going through forecasting. Second, forecasting using data on water level height and time of arrival of floods is done so that warnings can be made. Third, warning and outreach are key factors for success in a flood warning system using information obtained from the detection stage. Responsible parties can disseminate this information to minimize the risks it poses. Fourth, response to flood warning issues is very important to achieve the goals of implementing flood warnings. If the purpose of a flood warning is to reduce damage through flood preparedness, then one must respond appropriately by deploying personnel ready to evacuate if a flood occurs.

The time required for water to reach Nanjung is used to determine the arrival time of surface rain runoff in Dayeuhkolot. This area frequently experiences flooding and requires warnings to minimize losses. Because telemetry data on water levels in Dayeuhkolot is not yet available, the data used is downstream data, namely, Nanjung. Determining the time required for water to be used from Rain to Dayeuhkolot, the time required to arrive in the Nanjung area must be reduced by the water travel time from Dayeuhkolot to Nanjung. The time needed from Dayeuhkolot to Nanjung is estimated from the length of the river and the average speed of flooding that occurs (Pyatkova, Chen, Butler, Vojinović, & Djordjević, 2019).

From the measurement results, the length of the river from Dayeuhkolot to Nanjung is around 27 km (27,000 m). The journey of flood water from these two places takes around 5 hours to 14 hours with an average of 9 hours. If calculated from the initial mileage of rain, the travel time for water to reach the Nanjung monitoring post from the Dampit, Ciparay and Cipadung monitoring posts is around 11 hours. So the total travel time for water from the start of the rain is around 19 hours. Through good communication between rainwater monitoring posts, early warning can be provided. Monitoring post officers can warn people living in areas around the flood plain to evacuate and save all their belongings (Ringo, Sabai, & Mahenge, 2024).

Another approach to determining water travel time is graphical analysis by comparing hydrographs at both locations (Gangrade et al., 2023). The comparison of the two water surface hydrographs shows four events of water level rise. The scale of the four major events was then enlarged to see the time lag between the two hydrographs. The time lag that can be calculated is between 5 and 6 hours. This shows that calculations using the previous method are acceptable. Studying rainfall events in the Citarum watershed is unique because it turns out that rainfall in a watershed rarely occurs simultaneously.

The Bandung City Government is trying to reduce the risk of flood disasters in the Citarum watershed with the help of the Flood Early Warning and Early Action System (FEWEAS) application. This application will provide disaster warnings and mitigation easily because it can be downloaded on the App Store and Google Store. This application is managed by four West Java Provincial Government units, namely, Perum Jasa Tirta 2, BBWS, West Java BPBD, and West Java Provincial Government. These four institutions will work together to manage the use of FEWEAS to mitigate the risk of flooding for people living in the Citarum watershed area. FEWEAS Citarum was built with the most advanced system. So it provides accurate weather and flood prediction information for the next 3 days in the Citarum River watershed area. Flood warnings are generated from weather predictions with high-resolution and precise inundation predictions (Schumann, Giustarini, Tarpanelli, Jarihani, & Martinis, 2023).

Predictions of rainfall and climate vulnerability can be used to predict long-term floods and anticipate droughts (Busker, Moel, Hurk, & Aerts, 2023). FEWEAS technology is the second generation of the system implemented in Bengawan Solo. The advantage of this application is that it provides the latest information about predicted events for the next 3 days on a village scale. Namely, there is weather information, an overview of whether the colors blue, white, and red are ready to be alerted, and long-term weather and rice field information regarding planting periods. FEWEAS was created for the Citarum watershed. Hydrological disasters are currently of concern to the West Java government. Moreover, the Citarum River is the hanging place for 25 million West Java residents.

Through this application, it will contribute a lot, especially in reducing disaster risk. The government is trying to reduce vulnerability and increase capacity in dealing with hydrological disasters. The Regional Disaster Management Agency will handle human resource capacity, equipment, management, and technology disasters. Early warning to reduce disaster risk is necessary for communities around rivers. Floods not only cause health problems but also damage transportation routes and infrastructure. However, people's welfare is also disrupted. Therefore, overcoming the risk of loss due to floods cannot be seen only sectorally but must involve all stakeholders, namely, the government, private sector, NGOs, and academics (Awah, Belle, Nyam, & Orimoloye, 2024). With support and involvement from various parties, the intervention carried out by the government will be better.

Apart from early warning through applications, the government collaborates with the private sector and universities to create a Flood Disaster Early Warning System using New Renewable Energy Technology. This solar-powered water sensor system will provide early warning via sirens and lights if the Citarum River water level reaches a certain height. A total of 30 points along the Citarum River have installed a flood EWS. This equipment was installed as an anticipatory step so that people on the riverbanks would be more alert. This tool will send a signal containing information about potential flooding via a radio system. If it rains in the Sapan or Majalaya areas, the EWS can predict how long the water will arrive in downstream areas such as Dayeuhkolot or Baleendah. So, this tool can predict potential flooding within 4 hours. Residents around the river were able to save themselves and their belongings several hours before the flood occurred. The public was not surprised because they had received information beforehand. Anticipatory steps can be more effective so that the negative impacts of flooding can be avoided early on (Xu, Li, Tian, & Chen, 2023).

16.5 Research limitations

This research is only limited to answering research problems. This study cannot be generalized, considering that the characteristics of each flood area are always different. However, the results of this research can be used to strengthen community mitigation in dealing with flood disasters around rivers. Therefore, communities with other characteristics must prepare an assessment of the conditions to be able to apply the results of this research. If the characteristics are the same, then direct research can be applied.

This research has limited observation time, which impacts the research results. The researcher lives in the city of Bandung. Even though he does not live in a flood location, information about the flood area always appears, impacting traffic jams in Bandung. This incident was also widely recorded by the public and uploaded to the media during the flood disaster. This information adds to the understanding of writing about this event. Another limitation is that the research results cannot be used at different times and cannot be used as a measurement standard because the social situation of society is unique and unstable.

16.6 Recommendations

Floods cause damage to facilities and buildings, causing socioeconomic losses to the community. Data has an important role in solving flood problems. The availability of data allows someone to make better decisions in dealing with flood problems. The Citarum River Upstream condition improvement program is based on GIS-based data. This requires ongoing outreach and coordination between local governments and stakeholders to maximize results. Data must always be updated regularly from time to time to determine developments during a certain period. This data can be analyzed by linking certain issues related to flooding. The issue of climate change is relevant for analyzing what is related to the intensity of flooding that occurs so that interventions are carried out more comprehensively.

The presence of humans using land determines whether their activities can trigger larger disasters or prevent disasters from occurring. By knowing the quality of the living environment as a whole river basin, development activities must be wiser in making decisions. The direction of land use intensity in subwatersheds with low to medium typology must begin to be limited, directed at the intensity of space use that allows for good water infiltration. So the risk of water discharge can be reduced and does not cause flooding in downstream areas. Space utilization in subwatersheds with high typologies can still be carried out freely, but must still pay attention to conservation, environmentally friendly and sustainable principles.

The mitigation process and EWS must always pay attention to vulnerable groups in the area. The policies created should highlight the importance of considering social vulnerability in flood risk management and adaptation strategies. Groups vulnerable to flood disasters in the upper reaches of the Citarum River are children, mothers who are pregnant or breastfeeding, people with disabilities, and the elderly. They are a vulnerable group because they cannot prepare to face the risks and threats of flood disasters. This community group tends to feel a greater impact than other communities when floods occur. By considering the vulnerabilities of these groups, policymakers and stakeholders can develop more socioeconomic strategies for addressing and mitigating them. Future research needs to study vulnerable groups in flood disasters in detail because these groups require special policies and programs.

16.7 Conclusions

Socioeconomic vulnerability cannot be separated from physical vulnerability in the Upper Citarum River area. A holistic flood risk management approach is essential in overcoming vulnerability to flood hazards arising from human-led housing development. This approach must consider social, economic, and environmental factors that contribute to vulnerability. Involving all members of society in decision-making and developing adaptive measures is also necessary. Sustainable adaptation strategies are essential to address the increasing vulnerability of rivers to flooding. GIS-based monitoring and assessment of flood impacts is very important in determining mitigation to reduce disaster risk.

The river flow pattern in the Upper Citarum watershed has a radial pattern, where the tributaries radially serve a point. Major flooding is possible in such drainage areas, especially near the confluence points of tributaries. The estuaries of the Citarum River, namely, the Cisangkuy River and the Cikapundung River, intersect perpendicular to the Citarum Main River. Floods will occur if surface water overflows due to rotating water and inundates the surrounding area. River conditions like this require a physical development program that can facilitate river flow by creating water tunnels or building retention ponds. This can reduce the intensity of flooding and reduce risks.

Identifying and understanding social vulnerability to flooding in the Citarum watershed is key to socioeconomic risk management. Integrating various aspects, such as socioeconomic conditions, community characteristics, and the specific vulnerabilities of different groups, can help inform interventions to reduce flood vulnerability and increase community resilience. An EWS is a step taken based on GIS-based data to reduce risk. This kind of approach must continue to be socialized to the public so that people realize the importance of early warning. Developing an effective flood risk management strategy is important by considering various sources of information and expertise.

Acknowledgments

All authors are main contributors who have different expertise. Hari Harjanto Setiawan focuses on social vulnerability, Mahdi Ibrahim Tanjung focuses on writing about geospatial mapping, Abdullah Abdullah, who acts as a research assistant (RA) at the social welfare, village and connectivity research center, will focus on economic vulnerability studies, Djaenudin focuses on environmental management, and Suhandy Siswoyo focuses on housing arrangement around the Hulu Sungai Citarum. These five disciplines complement each other in the writing of this book chapter.

References

Abbas, A., Amjath-Babu, T. S., Kächele, H., Usman, M., & Müller, K. (2015). An overview of flood mitigation strategy and research support in South Asia: Implications for sustainable flood risk management. *International Journal of Sustainable Development & World Ecology*, 23(1), 98−111. Available from https://doi.org/10.1080/13504509.2015.1111954.

Abdi-Dehkordi, M., Bozorg-Haddad, O., Salavitabar, A., Mohammad-Azari, S., & Goharian, E. (2021). Development of flood mitigation strategies toward sustainable development. *Natural Hazards*, 108(3), 2543−2567. Available from https://doi.org/10.1007/s11069-021-04788-5, http://www.wkap.nl/journalhome.htm/0921-030X.

Ahmadu, A. S., Nukpezah, J. A., French, P. E., & Menifield, C. E. (2024). Disasters and social vulnerability determinants of federal subsidiarity assistance. *Public Administration Review*, 84(1), 21−39. Available from https://doi.org/10.1111/puar.13671.

Ahmed, R., & Roy, C. (2015). An integrated approach to environmental management in Bangladesh. In *Global sustainability: Cultural perspectives and challenges for transdisciplinary integrated research* (pp. 185−201). Springer International Publishing, Available from: https://doi.org/10.1007/978-3-319-16477-9. doi:10.1007/978-3-319-16477-9_10.

Alves, A., Gersonius, B., Kapelan, Z., & Sanchez, Z. V. A. (2019). Assessing the co-benefits of green-blue-grey infrastructure for sustainable urban flood risk management. *Journal of Environmental Management, 239*, 244−254. Available from https://doi.org/10.1016/j.jenvman.2019.030.036, https://www.sciencedirect.com/science/article/pii/S030147971930338X.

Anh, N. T., Can, L. D., Nhan, N. T., Schmalz, B., & Luu, T. L. (2023). Influences of key factors on river water quality in urban and rural areas: A review. *Case Studies in Chemical and Environmental Engineering, 8*, 100424. Available from https://doi.org/10.1016/j.cscee.2023.100424, https://www.sciencedirect.com/science/article/pii/S2666016423001299.

Awah, L. S., Belle, J. A., Nyam, Y. S., & Orimoloye, I. R. (2024). A participatory systems dynamic modelling approach to understanding flood systems in a coastal community in Cameroon. *International Journal of Disaster Risk Reduction, 101*, 104236. Available from https://doi.org/10.1016/j.ijdrr.2023.104236, https://www.sciencedirect.com/science/article/pii/S2212420923007161.

Belinawati, R.A. P., Soesilo, T.E. B., Herdiansyah, H., & Aini, I.N. (2018). BOD Pressure in the sustainability of the Citarum River. In *E3S web of conferences*, vol. 52. 22671242EDP Sciences Indonesia. Available from https://doi.org/10.1051/e3sconf/20185200037, http://www.e3s-conferences.org/.

Bouaakkaz, B., Abidine El Morjani, Z. E., & Bouchaou, L. (2023). Social vulnerability assessment to flood hazard in Souss basin, Morocco. *Journal of African Earth Sciences, 198*, 104774. Available from https://doi.org/10.1016/j.jafrearsci.2022.104774, https://www.sciencedirect.com/science/article/pii/S1464343X22003260.

Brierley, G. J., Hikuroa, D., Fuller, I. C., Tunnicliffe, J., Allen, K., Brasington, J., … Measures, R. (2023). Reanimating the strangled rivers of Aotearoa New Zealand. *WIREs Water, 10*(2), e1624. Available from https://doi.org/10.1002/wat2.1624, https://doi.org/10.1002/wat2.1624.

Busker, T., Moel, H. de, Hurk, B. van den, & Aerts, J. C. J. H. (2023). Impact-based seasonal rainfall forecasting to trigger early action for droughts. *Science of the total Environment, 898*, 165506. Available from https://doi.org/10.1016/j.scitotenv.2023.165506, https://www.sciencedirect.com/science/article/pii/S0048969723041293.

Chakraborty, L., Rus, H., Henstra, D., Thistlethwaite, J., & Scott, D. (2020). A place-based socioeconomic status index: Measuring social vulnerability to flood hazards in the context of environmental justice. *International Journal of Disaster Risk Reduction, 43*, 101394. Available from https://doi.org/10.1016/j.ijdrr.2019.101394, https://www.sciencedirect.com/science/article/pii/S2212420919309860.

Côté-Boileau, É., Gaboury, I., Breton, M., & Denis, J. L. (2020). Organizational ethnographic case studies: Toward a new generative in-depth qualitative methodology for health care research? *International Journal of Qualitative Methods, 19*. Available from https://doi.org/10.1177/1609406920926904, http://ijq.sagepub.com/content/by/year.

De Silva, M. M. G. T., & Kawasaki, A. (2018). Socioeconomic vulnerability to disaster risk: A case study of flood and drought impact in a Rural Sri Lankan Community. *Ecological Economics, 152*, 131−140. Available from https://doi.org/10.1016/j.ecolecon.2018.050.010, https://www.sciencedirect.com/science/article/pii/S092180091731604X.

Deepak, S., Rajan, G., & Jairaj, P. G. (2020). Geospatial approach for assessment of vulnerability to flood in local self governments. *Geoenvironmental Disasters, 7*(1). Available from https://doi.org/10.1186/s40677-020-00172-w.

Deroliya, P., Ghosh, M., Mohanty, M. P., Ghosh, S., Durga Rao, K. H. V., & Karmakar, S. (2022). A novel flood risk mapping approach with machine learning considering geomorphic and socio-economic vulnerability dimensions. *Science of the Total Environment*, *851*, 158002. Available from https://doi.org/10.1016/j.scitotenv.2022.158002, https://www.sciencedirect.com/science/article/pii/S0048969722051014.

Doyle, L., McCabe, C., Keogh, B., Brady, A., & McCann, M. (2020). An overview of the qualitative descriptive design within nursing research. *Journal of Research in Nursing*, *25*(5), 443−455. Available from https://doi.org/10.1177/1744987119880234, http://www.sagepub.co.uk/journalsProdDesc.nav?prodId = Journal201720.

Gangrade, S., Ghimire, G. R., Kao, S.-C., Morales-Hernández, M., Tavakoly, A. A., Gutenson, J. L., ... Follum, M. L. (2023). Unraveling the 2021 Central Tennessee flood event using a hierarchical multi-model inundation modeling framework. *Journal of Hydrology*, *625*, 130157. Available from https://doi.org/10.1016/j.jhydrol.2023.130157, https://www.sciencedirect.com/science/article/pii/S0022169423010995.

Hinojos, S., McPhillips, L., Stempel, P., & Grady, C. (2023). Social and environmental vulnerability to flooding: Investigating cross-scale hypotheses. *Applied Geography*, *157*, 103017. Available from https://doi.org/10.1016/j.apgeog.2023.103017, https://www.sciencedirect.com/science/article/pii/S0143622823001480.

Imran, M., Sumra, K., Amer Mahmood, S., & Faisal Sajjad, S. (2019). Mapping flood vulnerability from socioeconomic classes and GI data: Linking socially resilient policies to geographically sustainable neighborhoods using PLS-SEMMapping flood vulnerability from socioeconomic classes and GI data: Linking socially resilient policies to geographically sustainable neighborhoods using PLS-SEM. *International Journal of Disaster Risk Reduction*, *41*, 101288. Available from https://doi.org/10.1016/j.ijdrr.2019.101288, https://www.sciencedirect.com/science/article/pii/S2212420919308842.

Jentoft, N., & Olsen, T. S. (2019). Against the flow in data collection: How data triangulation combined with a 'slow' interview technique enriches data. *Qualitative Social Work*, *18*(2), 179−193. Available from https://doi.org/10.1177/1473325017712581, http://www.sagepub.com/journalsIndex.nav.

Jia, H., Chen, F., Pan, D., Du, E., Wang, L., Wang, N., & Yang, A. (2022). Flood risk management in the Yangtze River basin—Comparison of 1998 and 2020 events. *International Journal of Disaster Risk Reduction*, *68*, 102724. Available from https://doi.org/10.1016/j.ijdrr.2021.102724, https://www.sciencedirect.com/science/article/pii/S2212420921006853.

Jurumai, L. P., Abidin, A. A., Soetrisno, E. T., & Wibowo, D. (2023). Impact of population growth and housing development on the riverine environment: Identifying environmental threat and solution in the Wanggu River, Indonesia. *Ecological Modelling*, *486*, 110540. Available from https://doi.org/10.1016/j.ecolmodel.2023.110540, https://www.sciencedirect.com/science/article/pii/S0304380023002703.

Kappes, M. S., Papathoma-Köhle, M., & Keiler, M. (2012). Assessing physical vulnerability for multi-hazards using an indicator-based methodology. *Applied Geography*, *32*(2), 577−590. Available from https://doi.org/10.1016/j.apgeog.2011.070.002, https://www.sciencedirect.com/science/article/pii/S0143622811001378.

Krishnan, S. (2023). Adaptive capacities for women's mobility during displacement after floods and riverbank erosion in Assam, India. *Climate and Development*, *15*(5), 404−417. Available from https://doi.org/10.1080/17565529.2022.2092052.

Kuntoro, A.A., Putro, A.W., Kusuma, M.S. B., & Natasaputra, S. (2017). The effect of land use change to maximum and minimum discharge in Cikapundung River Basin. In

AIP conference proceedings, vol. 1903. American Institute of Physics Inc. 15517616. 10.1063/1.5011621. http://scitation.aip.org/content/aip/proceeding/aipcp.

Li, Q., Xia, J., Zhou, M., Deng, S., & Dong, B. (2024). Risk assessment of metro tunnel evacuation in devastating urban flooding events. *Tunnelling and Underground Space Technology*, *144*, 105540. Available from https://doi.org/10.1016/j.tust.2023.105540, https://www.sciencedirect.com/science/article/pii/S0886779823005606.

Ma, S., Wang, Liang-Jie, Wang, H.-Y., Zhao, Y.-G., & Jiang, J. (2023). Impacts of land use/land cover and soil property changes on soil erosion in the black soil region, China. *Journal of Environmental Management*, *328*, 117024. Available from https://doi.org/10.1016/j.jenv-man.2022.117024, https://www.sciencedirect.com/science/article/pii/S030147972202597X.

Mandal, K. K., Dharanirajan, K., Meena, M. L., Jaman, T., & Rana, S. (2023). Application of geospatial tools in the assessment of Flood hazard impact on social vulnerability of Malda district, West Bengal, India. *Natural Hazards Research*. Available from https://doi.org/10.1016/j.nhres.2023.110.008, https://www.sciencedirect.com/science/article/pii/S266659212300118X.

Marcella-Hood, M., & Marcella, R. (2023). Purposive and non-purposive information behaviour on Instagram. *Journal of Librarianship and Information Science*, *55*(3), 634−657. Available from https://doi.org/10.1177/09610006221097974, https://journals.sagepub.com/home/lis.

Mesta, C., Kerschbaum, D., Cremen, G., & Galasso, C. (2023). Quantifying the potential benefits of risk-mitigation strategies on present and future seismic losses in Kathmandu Valley, Nepal. *Earthquake Spectra*, *39*(1), 377−401. Available from https://doi.org/10.1177/87552930221134950, https://journals.sagepub.com/home/eqs.

Noor, R., Maqsood, A., Baig, A., Pande, C. B., Mishal Zahra, S., Saad, A., Anwar, M., & Kumar Singh, S. (2023). A comprehensive review on water pollution, South Asia Region: Pakistan. *Urban Climate*, *48*, 101413. Available from https://doi.org/10.1016/j.uclim.2023.101413, https://www.sciencedirect.com/science/article/pii/S221209552300007X.

Priya, A. K., Muruganandam, M., Rajamanickam, S., Sivarethinamohan, S., Krishna Reddy Gaddam, M., Velusamy, P., . . . Muniasamy, S. K. (2023). Impact of climate change and anthropogenic activities on aquatic ecosystem − A review. *Environmental Research*, *238*, 117233. Available from https://doi.org/10.1016/j.envres.2023.117233, https://www.sciencedirect.com/science/article/pii/S0013935123020376.

Pyatkova, K., Chen, A. S., Butler, D., Vojinović, Z., & Djordjević, S. (2019). Assessing the knock-on effects of flooding on road transportation. *Journal of Environmental Management*, *244*, 48−60. Available from https://doi.org/10.1016/j.jenvman.2019.050.013, https://www.sciencedirect.com/science/article/pii/S0301479719306115.

Qi, W., Ma, C., Xu, H., Chen, Z., Zhao, K., & Han, H. (2021). A review on applications of urban flood models in flood mitigation strategies. *Natural Hazards*, *108*(1), 31−62. Available from https://doi.org/10.1007/s11069-021-04715-8, http://www.wkap.nl/journalhome.htm/0921-030X.

Rahadiati, A., Prihanto, Y., Suryanegara, E., Rudiastuti, A. W., Nahib, I., & Nursugi. (2019). Assessment of socioeconomic vulnerability of coastal community in management of floods in Mataram. *IOP Conference Series: Earth and Environmental Science*, *399*(1), 012098. Available from https://doi.org/10.1088/1755-1315/399/1/012098.

Ringo, J., Sabai, S., & Mahenge, A. (2024). Performance of early warning systems in mitigating flood effects. A reviewPerformance of early warning systems in mitigating flood effects. A review. *Journal of African Earth Sciences*, *210*, 105134. Available from https://doi.org/10.1016/j.jafrearsci.2023.105134, https://www.sciencedirect.com/science/article/pii/S1464343X23003072.

Saja, A. M. A., Goonetilleke, A., Teo, M., & Ziyath, A. M. (2019). A critical review of social resilience assessment frameworks in disaster management. *International Journal of Disaster Risk Reduction, 35*, 101096. Available from https://doi.org/10.1016/j.ijdrr.2019.101096, https://www.sciencedirect.com/science/article/pii/S2212420918307945.

Schilling, J., Hertig, E., Tramblay, Y., & Scheffran, J. (2020). Climate change vulnerability, water resources and social implications in North Africa. *Regional Environmental Change, 20*(1), 15. Available from https://doi.org/10.1007/s10113-020-01597-7, https://doi.org/10.1007/s10113-020-01597-7.

Schumann, G., Giustarini, L., Tarpanelli, A., Jarihani, B., & Martinis, S. (2023). Flood modeling and prediction using earth observation data. *Luxembourg Surveys in Geophysics, 44*(5), 1553−1578. Available from https://doi.org/10.1007/s10712-022-09751-y, https://www.springer.com/journal/10712.

Singh, P., Gupta, A., & Singh, M. (2014). Hydrological inferences from watershed analysis for water resource management using remote sensing and GIS techniques. *The Egyptian Journal of Remote Sensing and Space Science, 17*(2), 111−121. Available from https://doi.org/10.1016/j.ejrs.2014.090.003, https://www.sciencedirect.com/science/article/pii/S1110982314000271.

Sundler, A. J., Lindberg, E., Nilsson, C., & Palmér, L. (2019). Qualitative thematic analysis based on descriptive phenomenology. *Nursing Open, 6*(3), 733−739. Available from https://doi.org/10.1002/nop20.275, http://onlinelibrary.wiley.com/journal/10.1002/(ISSN)2054-1058.

Svetlana, D., Radovan, D., & Ján, D. (2015). The economic impact of floods and their importance in different regions of the world with emphasis on Europe. *International Scientific Conference: Business Economics and Management (BEM2015), 34*, 649−655. Available from https://doi.org/10.1016/S2212-5671(15)01681-0, https://www.sciencedirect.com/science/article/pii/S2212567115016810.

Tate, E., Rahman, M. A., Emrich, C. T., & Sampson, C. C. (2021). Flood exposure and social vulnerability in the United States. *Natural Hazards, 106*(1), 435−457. Available from https://doi.org/10.1007/s11069-020-04470-2, http://www.wkap.nl/journalhome.htm/0921-030X.

Tehrany, M. S., Pradhan, B., Mansor, S., & Ahmad, N. (2015). Flood susceptibility assessment using GIS-based support vector machine model with different kernel types. *CATENA, 125*, 91−101. Available from https://doi.org/10.1016/j.catena.2014.100.017, https://www.sciencedirect.com/science/article/pii/S034181621400294X.

Thakur, D. A., & Mohanty, M. P. (2023). A synergistic approach towards understanding flood risks over coastal multi-hazard environments: Appraisal of bivariate flood risk mapping through flood hazard, and socio-economic-cum-physical vulnerability dimensions. *Science of The Total Environment, 901*, 166423. Available from https://doi.org/10.1016/j.scitotenv.2023.166423, https://www.sciencedirect.com/science/article/pii/S0048969723050489.

Verschuur, J., Koks, E. E., & Hall, J. W. (2020). Port disruptions due to natural disasters: Insights into port and logistics resilience. *Transportation Research Part D: Transport and Environment, 85*, 102393. Available from https://doi.org/10.1016/j.trd.2020.102393, https://www.sciencedirect.com/science/article/pii/S1361920920305800.

Widianingsih, I., Riswanda, & Paskarina, C. (2019). *Rural-urban linkage and local government capacity in coping with water crisis: A brief note from Indonesia* (pp. 350–354). Atlantis Press. 2352-5398. https://doi.org/10.2991/icdesa-19.2019.71.

Xu, H., Li, H., Tian, S., & Chen, Y. (2023). Effects of flood risk warnings on preparedness behavior: Evidence from northern China. *International Journal of Disaster Risk Reduction*, *96*, 103971. Available from https://doi.org/10.1016/j.ijdrr.2023.103971, https://www.sciencedirect.com/science/article/pii/S221242092300451X.

Xue, P., Huang, S., Xie, K., Sun, Y., & Fei, L. (2024). Identification of the critical factors in flood vulnerability assessment based on an improved DEMATEL method under uncertain environments. *International Journal of Disaster Risk Reduction*, *100*, 104217. Available from https://doi.org/10.1016/j.ijdrr.2023.104217, https://www.sciencedirect.com/science/article/pii/S2212420923006970.

Assessing the impact of channel migration and bank erosion on the loss of agricultural land and land use land cover change in the lower course of Kankai Mai River—a geospatial approach

17

Uttara Nath[1] and Manas Hudait[2]

[1]University of North Bengal, West Bengal, India
[2]Amity University, Kolkata, West Bengal, India

17.1 Introduction

River bank erosion, lateral channel migration, and accretion are the crucial geomorphological processes, which need attention from scientists all over the world in the last century (Dragićević, Tošić, Stepić, Živković, & Novković, 2013). Bank erosion of any rivers is a recurrent and endemic natural hazard (Debnath, Das Pan, Ahmed, & Bhowmik, 2017). The course of any river manifests dynamic equilibrium conditions, which adjust in time and space. Discharge fluctuation and sediment load changes result in the lateral mobility of the channel, which is reflected in flood plains subjected to periodical and seasonal modifications. This phenomenon makes riverbank erosion the most common natural calamity in flood plains (Guite & Bora, 2016). All over the world, floodplains are studied as one of the most endangered areas, as they are degraded by land use changes related to pressures and river regulations policy (Hazarika, Das, & Borah, 2015). The land use pattern along the river channel changes monotonously due to the lateral erosion of river banks (Simon, Curini, Darby, & Langendoen, 2000); moreover, the deposition of sand on the floodplain degrades the fertility condition of the existing agricultural lands. Bank erosion is a very dangerous problem for a fluvial system, as it produces almost 90% of the total sediment yield from a stream catchment (Olley, Murray, Mackenzie, & Edwards, 1993; Prosser & Winchester, 1996;

Applications of Geospatial Technology and Modeling for River Basin Management.
DOI: https://doi.org/10.1016/B978-0-443-23890-1.00017-7

Wallbrink, Murray, Olley, & Olive, 1998; Wasson, Mazari, Starr, & Clifton, 1998). Bank erosion may represent a dire threat to the function played by agricultural land (De Rose & Basher, 2011; Rosgen, 1996; Sass & Keane, 2012; Simon et al., 2000; Simon, Wolfe, & Molinas, 1991; Thorne, 1999). Bank erosion is a remarkable source of sediment load in the stream channel (Bull, 1997; Evans, Gibson, & Rossell, 2006; Fox et al., 2007; Grove, Croke, & Thompson, 2013; Sekely, Mulla, & Bauer, 2002), which also causes bank failure through both of gravitation forces and hydraulic changes during intense rainfall events, which sometimes cause a large area of land to be flooded rapidly (Fischenich, 1989; Thorne, 1999). The lower course of the Kankai Mai River flowing through the Southeastern portion of Kishanganj District and the NE part of Purnea District of Bihar is very prone to flood and massive bank erosion. The river follows a meandering pattern in its lower course before joining Mahananda and deposits an excessive amount of sand and eroded materials in the channel bed. This deposition has led to the spill of the channel, and as a result after every monsoon season, the main flow of the river changes its position very frequently by repositioning the bank lines. The impact of bank erosion on agricultural land loss due to channel migration has never been investigated in the previous works in the lower flood plain of Kankai Mai River.

This study attempts to identify the channel migration pattern and characteristics of the Kankai Mai River in its lower floodplain and its impact on agricultural land loss and land use land cover (LULC) changes. Multispectral satellite images were used to assess the pattern of channel shifting and severity of agricultural land loss for 231 villages along the banks of the Kankai River over 19 years. Remote sensing (RS) and geographic information system (GIS) technology is used to identify and measure the temporal and spatial pattern of agricultural land loss along with LULC change detection. The satellite images ETM + of Landsat 7 and Landsat 8 OLI were considered for the years 2001, 2013, and 2020 and were collected during the winter season (November and December) when the streamline recedes and the maximum extent of bank erosion is visible. This study is trying to convey the implications of channel migration on agricultural land loss along the bank of the alluvial river and how the evidence of these changes can be shown through LULC Maps. The lower course of the Kankai Mai River has been taken as a case study to explore the intensity and magnitude of the impact of channel migration on agricultural land loss through bank erosion. Before measuring the magnitude of agricultural land loss, it is important to understand the dynamic nature of the Kankai Mai River along the lower course. The channel shifting pattern of River Kankai Mai over the 20 years of time period has been divided into two phases, that is, 2001–13 and 2013–20. For a better understanding of the channel shifting pattern of Kankai River in 2013, both the channels of 2001 and 2013 were overlaid. The same process has been adopted to show the river migration pattern in the phase 2013–20, where the 2013 channel has taken as a baseline. A total of 25 and 29 cross sections were demarcated at the places, where both the banks of 2013 and 2020 shifted away from their former position in 2001 and 2013, respectively. The direction of river bank shifting and the maximum distance between the banks of 2001 and 2013 were measured at the cross sections. These cross sections are the areas where the agricultural land loss happened through the two processes, that is, bank erosion and sand deposition. The total area eroded is mentioned in the tables. In the case of the alluvial rivers like Kankai Mai, these incidents are seasonal in the monsoon period due to floods.

17.2 Study area

River Kankai Mai is one of the important rivers of Nepal and the Indian state of Bihar, which joins Mahananda near Dalkhola. For the current study, the lower course of the Kankai Mai River has been selected, which is flowing through the Kishanganj and Purnea districts of Bihar. The total length of the river that lies in the study area is about 60.29 km, from the Dharhar village in the North to Bhasia village in the south, where Kankai meets the Mahananda River. A total of 231 villages from the watershed of the Kankai River were selected in this study, as these are more prone to bank erosion and the occurrence of channel shifting, as shown in Fig. 17.1.

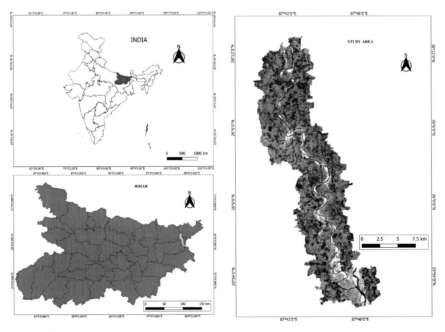

FIGURE 17.1

Location map of the study area.

17.3 Datasets and methodology

17.3.1 Data collection

The datasets, which are used in this study, are collected from different data sources. A vector layer of administrative boundaries (village level) of the year 2001 has been collected from the Socio-Economic Data and Applications Center, NASA. Landsat 8 OLI and Landsat 7 + ETM (Path: 139, Row: 042) satellite images are downloaded from the official website of the US Geological Survey. The details of the images are mentioned in Table 17.1.

17.3.2 Preprocessing

For the current study, the atmospheric correction of the Landsat images was performed, and the study area was clipped using the shapefile of the study area. The image quality was enhanced using a pan-sharpening tool for better image classification. All of the images were kept in the projection system of WGS 84, UTM Zone 45°N. For a better identification of the extended bank line of river Kankai, satellite images were taken in November were selected.

17.3.3 Methodology

The shapefile of the study area has been generated using the village boundary maps downloaded from SEDAC, NASA. A total of 231 villages along the bank of the lower Kankai River were identified and dissolved using the geoprocessing tools of QGIS. Identification of channel shifting is measured by digitizing the channel and the centerline from the satellite images of 2001, 2013, and 2020. To assess the impact of river bank erosion on the loss of agricultural land, the total

Table 17.1 Details of satellite images used in the study.

Characteristics	Landsat mission		
Images	Landsat 7 ETM +	Landsat 8 OLI	Landsat 8 OLI
Data type	Landsat 7 ETM + C2L1	Landsat 8 OLIC2 L1	Landsat 8 OLI C2 L1
Date	December 29, 2001	November 20, 2013	November 23, 2020
Row/path	042/139	042/139	042/139
Bands number	7	11	11
Map projection	UTM	UTM	UTM
Datum	WGS 84	WGS 84	WGS 84
UTM zone	45	45	45

river length of 64.25 and 60.29 km in 2013 and 2020 has been divided into 5 reaches. Each reach of 2013 and 2020 has overlapped with the satellite images of 2001 and 2013, respectively, to identify the agricultural lands, which the river has grasped. The degraded and eroded agricultural lands were then digitized and measured using QGIS software. The LULC classification of the satellite images was done using the maximum likelihood method and divided into a total of six classes, namely, built-up areas, vegetation, agricultural land, fallow land, sand bar, and waterbodies. The overall accuracy and Kappa hat coefficient were generated using SCP Plugin where values of overall accuracy and Kappa coefficient are 91.80%, 86.75%, and 100% and 0.87, 0.81, and 1.00 for the years 2001, 2013, and 2020, respectively, as shown in Fig. 17.2.

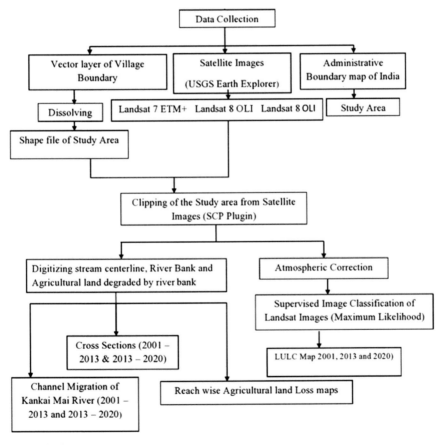

FIGURE 17.2

Methodological workflow of the present study.

17.4 Result and findings

Channel shifting of Kankai Mai River: To identify the channel shifting pattern of Kankai Mai from 2001 to 2020, the period has been divided into two phases, that is, 2001–13 and 2013–20. For a better understanding of the channel shifting pattern of the Kankai River in 2013, both the channels of 2001 and 2013 were overlaid. The same process has been adopted to show the river migration pattern in the phase 2013–20, where the 2013 channel has been taken as a baseline. The pattern of channel shifting of the Kankai River from 2001 to 2013 and from 2013 to 2020 has been identified and mapped (Fig. 17.3).

Further, 25 and 29 cross sections were demarcated at the places (Fig. 17.4), where both the banks of 2013 and 2020 shifted away from their former position in 2001 and 2013, respectively. The direction of river bank shifting and the maximum distance between the banks between 2001 and 2013 were measured at the cross sections. The measurements of the cross sections are listed in Table 17.2. From the table, it can be seen that in the case of the first phase, CS-18 has experienced the maximum lateral migration of both the banks, which is 1644.08 m (Table 17.2) westward of their former position. CS-15 up to CS-19 are showing relatively higher values of distance, which indicates the very dynamic nature of

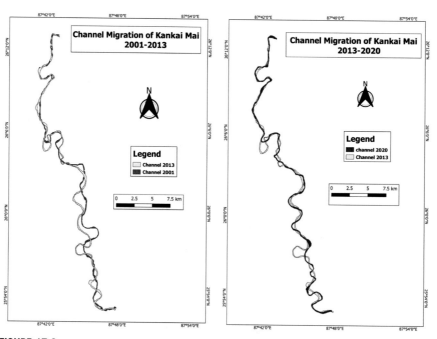

FIGURE 17.3

Migration of river Kankai Mai in its lower course from 2001 to 2020.

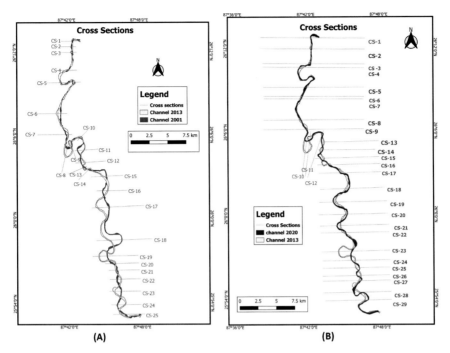

FIGURE 17.4

(A) River cross sections (2001−13) and (B) river cross sections (2013−20).

the channel. In the second phase at CS-23, the channel has shifted at a maximum distance, which is 1681.1 m eastward from its former path in 2013. Also, at the CS-10 (2013−20), the channel has shifted to the maximum (Table 17.3).

Bank erosion and agricultural land loss: The reach wise portion of agricultural land, which went under the river bed or degraded by sand deposition during both the phases, has been analyzed and mapped (Figs. 17.5−17.9). Among the five reaches, the maximum erosion of agricultural land can be seen at reach no 4 and reach no 3 during the phases of 2001−13 and 2013−20, respectively. A total of 628.2 ha of agricultural land has been washed away or degraded by sand deposition in the first phase, whereas it is 433.76 ha during the second phase (Tables 17.4 and 17.5).

Reach 1 covers five cross sections where both the banks of 2013 and 2020 have shifted away from their former position in 2001 and 2013, respectively (Fig. 17.5). The channel between cross sections 4 and 5 has a maximum bend compared to the rest of the cross sections, and as a result, this area has contributed the maximum in degrading agricultural land, mostly through sand deposition and bank erosion. The 108.35 m westward shifting of the channel at CS 4, 229.88 m eastward shifting at CS 5 between the period 2001 and 2013, and 254.47 m

Table 17.2 Cross section across the Kankai River channel (2001–13).

Cross section	Direction	Migration (m)
CS-1	Westward	54.87
CS-2	Eastward	39.87
CS-3	Westward	167.91
CS-4	Westward	108.35
CS-5	Eastward	229.88
CS-6	Westward	347.98
CS-7	Eastward	264.56
CS-8	Southward	996.84
CS-9	Northward	652.18
CS-10	Eastward	143.03
CS-11	Eastward	893.11
CS-12	Eastward	405
CS-13	Westward	140
CS-14	Northward	88.34
CS-15	Westward	658.95
CS-16	Eastward	514.11
CS-17	Westward	1429.71
CS-18	Westward	1644.08
CS-19	Westward	1028.03
CS-20	Eastward	266.37
CS-21	Westward	125.57
CS-22	Eastward	411.2
CS-23	Westward	803
CS-24	Eastward	971.69
CS-25	South-westward	204.98

westward shifting at CS 4 between the period of 2013 and 2020 better capture the volatile situation of the river bank at this area.

Reach 2 covers CS 6, 7, 8, 9, 10, and 11 (CS 2013–20) and 12 (CS 2013–20), where both the banks of 2013 and 2020 shifted away from their former position in 2001 and 2013, respectively. From 2001 to 2013, a total 107.3 ha of agricultural land were degraded. It was 70.51 ha from 2013 to 2020. In this part, cross section 8 and cross section 10 are the spots that have experienced remarkable shifts in 30 years (Fig. 17.6). It is evident from the LULC maps that during the period 2001 to 2013, it was bank erosion that grasped the agricultural land, and after 2013, its sand deposition at that part has contributed to agricultural land degradation.

Reach 3 covers CS 11, 12, 13, 14, 15, and 16 (CS 2001–13) and CS 17, 18, and 19 (CS 2013–20). In this part of the study area, bank erosion is more prominent than sand deposition (Fig. 17.7). From 2001 to 2013, a total of 158.4 ha of

Table 17.3 Cross section across the Kankai River channel (2013−20).

Cross sections	Direction	Migration (m)
CS-1	Westward	28.37
CS-2	Eastward	124.85
CS-3	Eastward	148.22
CS-4	Westward	254.47
CS-5	Westward	38.92
CS-6	Eastward	58.62
CS-7	Eastward	35.67
CS-8	Eastward	68.47
CS-9	Westward	100.37
CS-10	Northward	1415.68
CS-11	Southward	436.07
CS-12	Eastward	124.89
CS-13	Westward	239.17
CS-14	Eastward	261.685
CS-15	Westward	334.09
CS-16	Eastward	520.193
CS-17	Westward	134.25
CS-18	Eastward	242.01
CS-19	Westward	231.67
CS-20	Eastward	224.17
CS-21	Westward	380.84
CS-22	Eastward	153.86
CS-23	Eastward	1681.1
CS-24	Westward	442.96
CS-25	Eastward	388.67
CS-26	Westward	454.71
CS-27	Eastward	178.22
CS-28	Westward	571.25

agricultural land was eroded through this process, and it was 147.41 ha from 2013 to 2020.

Reach 4 and Reach 5 are situated near the confluence of the River Kankai Mai, where the outer lines of the meanders are more defined. Reach 4 (Fig. 17.8) covers CS 17, 18, 19, and 20 (CS 2001−13) and 20, 21, 22, 23, and 24 (CS 2013−20), whereas Reach 5 (Fig. 17.9) covers CS 2, 22, 23, 24, and 25 (CS 2001−13) and 25, 26, 27, 28, and 29 (CS 2013−20).

Here, the outer bank line, which has shifted westward, most of the time has grasped agricultural land through bank erosion. On the other hand, the eastern side is affected by sand deposition. At Reach 4, a total of 264.3 ha of agricultural land was degraded over 30 years compared to 234.77 ha at Reach 5. *LULC*

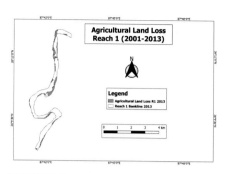
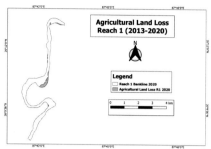

FIGURE 17.5

Loss of agricultural land (Reach 1).

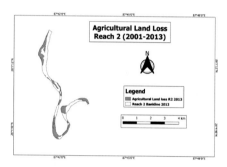
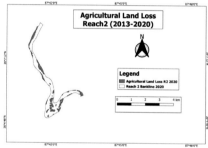

FIGURE 17.6

Loss of agricultural land (Reach 2).

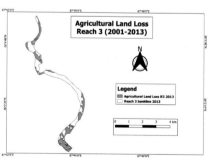
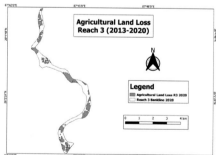

FIGURE 17.7

Loss of agricultural land (Reach 3).

change: LULC change of the study area has been done for the years 2001, 2013, and 2020 (Fig. 17.10). Results show that agricultural land has decreased from 32 sq. km in 2001 to 23.51 sq. km in 2013 and 17.48 sq. km in 2020. Fallow land

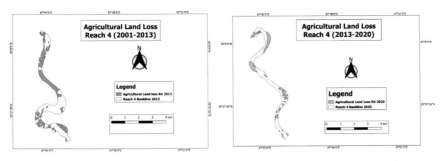

FIGURE 17.8

Loss of agricultural land (Reach 4).

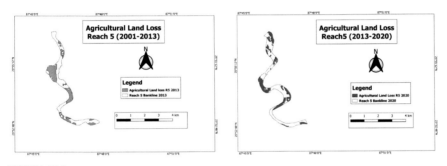

FIGURE 17.9

Loss of agricultural land (Reach 5).

Table 17.4 Loss of agricultural land by bank erosion (2001–13).

Reach	Time period	Agricultural land Loss (ha)
1	2001–13	59.97
2	2001–13	107.3
3	2001–13	158.4
4	2001–13	180
5	2001–13	122.53
Total	–	628.2

has increased from 141.80 sq. km in 2001 to 151.68 sq. km in 2020, but slightly decreased in 2013. Sand bars occupied 12.17 and 13.97 sq. km area in 2001 and 2020, respectively, but covered only a 4.01 sq. km area in 2013. Water bodies showing the almost same pattern covering 10.85 and 13.65 sq. km areas in 2001 and 2020 and 9.92 sq. km area in 2013 (Table 17.6). These patterns indicate that

Table 17.5 Loss of agricultural land by bank erosion (2013–20).

Reach	Time period	Agricultural land loss (ha)
1	2013–20	19.3
2	2013–20	70.51
3	2013–20	147.41
4	2013–20	84.3
5	2013–20	112.24
Total	–	433.76

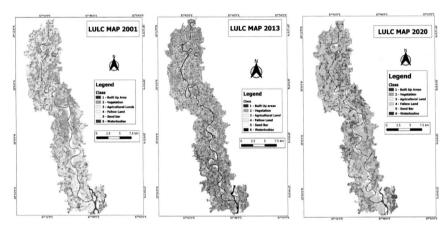

FIGURE 17.10

Land use and land cover map.

Table 17.6 Area under different land use land cove categories.

LULC categories	2001		2013		2020	
	Area (sq. km)	%	Area (sq. km)	%	Area (sq. km)	%
Built-up area	10.33	3.65	43.11	15.23	26.38	9.32
Vegetation	75.89	26.81	103.66	36.63	59.86	21.15
Agricultural land	32	11.35	23.51	8.30	17.48	6.17
Fallow land	141.80	50.09	98.82	34.92	151.68	53.58
Sand bar	12.17	4.29	4.01	1.41	13.97	4.94
Water bodies	10.85	3.83	9.92	3.50	13.65	4.28

between 2001 and 2020, Kankai River was more volatile compared to 2001. As a result, sand deposition in a wide area took place, vegetation washed away, and fallow land covered most of the floodplain. On the other hand, in 2013,

Table 17.7 Accuracy for different land use land cover categories.

LULC categories	SE	PA	UA	Kappa hat
2001				
Built-up area	0.0036	81.69	62.85	0.61
Vegetation	0.0076	89.23	92.78	0.90
Agricultural land	0.0065	100	70.31	0.67
Fallow land	0.0051	96.59	96.92	0.93
Sand bar	0	100	100	1
Water bodies	0.0051	56.37	100	1
2013				
Built-up area	0.010	77.64	58.49	0.53
Vegetation	0.010	100	89.19	0.83
Agricultural land	0.11	60	100	1
Fallow land	0.009	94.88	91.52	0.87
Sand bar	0.002	78.23	100	1
Water bodies	0.009	57.30	100	1
2020				
Built-up area	0.001	100	100	1
Vegetation	0	100	100	1
Agricultural land	0	100	100	1
Fallow land	0	100	100	1
Sand bar	0	100	98.00	1
Water bodies	0	100	100	1

vegetation and agricultural areas can be seen clearly in November as the influence of river water was less.

Assessment of accuracy: Table 17.7 shows the accuracy of the classified maps. The error matrix method has been applied to assess the accuracy of the classified maps. Standard error, producer accuracy, users accuracy, Kappa hat classification, and overall accuracy have been mentioned. The overall accuracy is 92.8077 for the 2001 classified map, 86.76 for the 2013 map, and 100 for the 2020 LULC map. On the other hand, Kappa hat values are 0.87, 0.81, and 1 for the respective areas; thus, we can say that this is good image classification.

17.5 Discussion

It can be seen that the first two reaches are less volatile and less prone to meandering, whereas Reach no 3 to Reach no 5 show active changes of the bank line in both the phases. Arefin, Meshram, and Seker (2021) mentioned that the change detection of LULC due to the shifting of the river banks and cultivated land converted to the sand bar, which can be seen in cross sections 8 (2001−13) and cross section 10 (2013−20). Here, the channel continued to move southward until 2013 and again moved to its former position in 2020, causing agricultural land degradation by the deposition of sand. Serpentine lakes and old riverbeds are clearly visible around the floodplain. The river is characterized by huge sand bars, mid-channel bars, and raised beds, which are indicative of alluvium deposition and prove channel spilling. According to Debnath et al. (2017), the bank erosion frequency of a river is much higher after the monsoon season because there is a decreasing rate in rainfall intensity. Moreover, the recession of the level of water results in a disequilibrium between the level of water and the riverbank. Thus, the loss of cohesive strength of soil leads to the maximum erosion, which is the same factor for bank erosion in the Kankai Mai River. Landsat images of November (after Monsoon) were selected for the study, where the recessed bank lines can be seen in the classified images (Fig. 17.10). The sand deposition and fallow lands along both the banks away from the centerline mark the width of the channel during monsoon season. In some places, embankments are being constructed to prevent bank erosion, but this did not help to secure agricultural lands. Besides this, sand deposition is a huge problem, as we can see nearly 4% area of the flood plain that is sand-covered after the monsoon season in 2001 and 2020 by degrading the quality of fertile lands. At cross Sections 17, 18, and 19, the previous bank line has been shifted nearly 1 km away from its former position in 2013 by degrading the quality of the fertile land area situated between this 1 km migration zone. Mukhopadhayay (2010) found similar results where >1000 ha of land were affected by sand splay in the Ajay River floodplain in the flood years 1973, 1978, 1995, 1999, 2000, 2006, and 2007. In the case of the Kankai Mai River, it is 628.2 and 433.76 ha in the first and second phases, respectively. It is also shown that due to the small amount of slope differences along the lower course, that is, below Illambazar, the siltation rate is very high. This decreases the water accommodating capacity of the river channel, which is pretty similar to the case of the Kankai River floodplain as in 2020 CS-10 and CS-23 were the worst affected regions by the dynamic nature of the channel.

Serious bank stability measures should be taken into consideration to revive the losses. Though concrete structures will not be very fruitful considering the volatile nature of the channel, bank stabilization by using tree roots or indigenous grasses can be used as a better choice. Along with these buildings, sloping terraces along both the banks will reduce the velocity of the water and bank scouring can be reduced. The findings of the study show that sand deposition is a huge problem, as we can see nearly 4% area of the flood plain that is that is sand-covered after the monsoon season in 2001 and 2020 by degrading the quality of fertile lands. At cross sections 17, 18, and 19, the previous bank line has been shifted nearly 1 km away from its former position in 2013 by degrading the quality of the fertile lands, which are situated between this 1 km migration zone. As mentioned above, the cross sections where the rate of channel migration is high are the prone areas of agricultural land loss due to bank erosion and sand deposition. Debnath et al. (2017) found a similar process in their study where they mentioned that settlement and cultivated land have been eroded away and converted into depositional land, which was again used for both cultivation and settlement purposes. However, some portions remain barren due to the presence of a high percentage of sand, which cannot be used either for cultivation or for settlement. Ayman and Ahmed (2009) observed that the bank erosion at a meander in their study area reached about 50 m in only 1 year, which resulted in the destruction of farmers' houses and agricultural lands and imposed a property delineation problem. While visiting the study area, the authors have encountered similar kinds of problems stated by the dwellers living around CS 5 of Reach 1. Some of the agricultural lands submerged in the river and degraded due to sand deposition take a year to several years to recover, as shown in Fig. 17.11. In the meantime, farmers are forced to become migrant laborers or seasonal migrants.

To capture the agricultural land loss and land degradation due to sand deposition, LULC maps of 2001, 2013, and 2020 have been compared. It is evident from the results that agricultural land has decreased from 2001 to 2020 continuously in the study area. On the other hand, fallow land has increased from 141.80 sq. km in 2001 to 151.68 sq. km in 2020 but slightly decreased in 2013. Sand

FIGURE 17.11

Glimpses of bank erosion in the study area.

bars occupied 12.17 and 13.97 sq. km area in 2001 and 2020, respectively, but covered only 4.01 sq. km area in 2013. If the classified image of 2001 is compared with that of 2020, then the development of the fallow area can be easily distinguishable along the lower course of the river. Similar observations by Debnath et al. (2017) in their study area using LULC maps have shown that the cultivated land area is the dominant land use type of the study area, but in this period, from 1975 to 2014, the percentage has gradually been decreased by 4.55% due to bank erosion of the river. Though the LULC maps captured the evidence of the degradation and erosion of the agricultural land, still there is a scope to showcase the areas that are eroded directly by bank erosion and the areas that are degraded by sand deposition over the period.

17.6 Conclusion

River bank erosion and its impact on agricultural land loss along the banks of the Kankai River showed the dynamic nature of the river Kankai concerning spatial and temporal extent. Multispectral satellite images provide a vivid overview of the changes that occurred in the floodplain over 19 years. RS and GIS technology have been very helpful in analyzing the temporal pattern of channel shifting and associated LULC changes in the Kankai River floodplain. The past and present condition of the lower course of the Kankai River shows that the channel has migrated from its former path in a meandering position leaving behind serpentine lakes and old riverbeds. The frequent changes in the river channel lead to severe bank erosion and degraded agricultural and fallow lands. It is evident from the study that agricultural land loss due to channel migration through bank erosion and sand deposition is a very frequent event in the areas of certain climatic conditions. The incidents of bank erosion and sand deposition are well captured through the satellite images, but one should keep in mind that the process of sand deposition and bank erosion starts at the end of monsoon season. Hence, for better capturing the incidents, the satellite images of the months of late October and November are ideal. Along with this, there are fewer atmospheric obstacles. As evident from the study, certain cross sections are prone to more frequent to channel migration and sand deposition. Hence, there is scope to explore the geomorphic characteristics, soil type, and other factors behind it to understand the intensity of land loss and investigate possible volatile areas in other regions with the same characteristics. Moreover, before making any plan for agricultural intervention, infrastructure development, and policy making in the agricultural field, one should pay attention to these findings. While conducting the study, it was observed that there are some specific areas (cross section 8, for example) where the incident of channel shifting is frequent, and these cross sections are the major contributors to grasping the agricultural land or destroying the agricultural land through sand deposit. Hence, when considering any project for developing any infrastructure, opting for welfare schemes for the agricultural communities or planning for any agricultural scheme for these areas, one should consider the volatile nature of the channel and the possible implications. Here, an attempt has been made to capture the changes in agricultural land loss over time through LULC maps, but there are other factors also which are simultaneously responsible for the changes in agricultural land use. So, one can explore in the particular area what are the active contributors to change the agricultural land use pattern over time and for how much contribution each one of these factors is responsible.

References

Arefin, R., Meshram, S. G., & Seker, D. Z. (2021). River channel migration and land-use/land-cover change for Padma River at Bangladesh: A RS- and GIS-based approach. *International Journal of Environmental Science and Technology*. Available from https://doi.org/10.1007/s13762-020-03063-7, https://www.researchgate.net/publication/349453818_River_channel_migration_and_land-useland-cover_change_for_Padma_River_at_Bangladesh_a_RS-_and_GIS-based_approach#fullTextFileContent.

Ayman, A., & Ahmed, F. D. G. (2009). Meandering and bank erosion of the River Nile and its environmental impact on the area between Sohag and El-Minia, Egypt. *Arabian Journal of Geosciences*, *4*, 1−11. Available from https://doi.org/10.1007/s12517-009-0048-y.

Bull, L. J. (1997). Magnitude and variation in the contribution of bank erosion to the suspended sediment load of the River Severn, UK. *Earth Surface Processes and Landforms*, *22*(12), 1109−1123. Available from https://doi.org/10.1002/(SICI)1096-9837, http://onlinelibrary.wiley.com/journal/10.1002/(ISSN)1096-9837.

Debnath, J., Das (Pan), N., Ahmed, I., & Bhowmik, M. (2017). Channel migration and its impact on land use/land cover using RS and GIS: A study on Khowai River of Tripura, North-East India. *The Egyptian Journal of Remote Sensing and Space Science*, *20*(2), 197−210. Available from https://doi.org/10.1016/j.ejrs.2017.01.009.

De Rose, R. C., & Basher, L. R. (2011). Measurement of river bank and cliff erosion from sequential LIDAR and historical aerial photography. *Geomorphology*, *126*(1−2), 132−147. Available from https://doi.org/10.1016/j.geomorph.2010.10.037.

Dragićević, S., Tošić, R., Stepić, M., Živković, N., & Novković, I. (2013). Consequences of the river bank erosion in the southern part of the Pannonian Basin: Case study − Serbia and the Republic of Srpska. *Forum Geografic*, *XII*(1), 5−15. Available from https://doi.org/10.5775/fg.2067-4635.2013.008.i.

Evans, D. J., Gibson, C. E., & Rossell, R. S. (2006). Sediment loads and sources in heavily modified Irish catchments: A move towards informed management strategies. *Geomorphology*, *79*(1−2), 93−113. Available from https://doi.org/10.1016/j.geomorph.2005.09.018.

Fischenich, J.C. (1989). Symposium proceedings headwaters hydrology. In W. Woessmer, & D. F. Potts (Eds.), *American Water Resources Association Channel erosion analysis and control*.

Fox, G. A., Wilson, G. V., Simon, A., Langendoen, E. J., Akay, O., & Fuchs, J. W. (2007). Measuring streambank erosion due to ground water seepage: Correlation to bank pore water pressure, precipitation and stream stage. *Earth Surface Processes and Landforms*, *32*(10), 1558−1573. Available from https://doi.org/10.1002/esp.1490.

Grove, J. R., Croke, J., & Thompson, C. (2013). Quantifying different riverbank erosion processes during an extreme flood event. *Earth Surface Processes and Landforms, 38* (12), 1393−1406. Available from https://doi.org/10.1002/esp.3386.

Guite, L. T. S., & Bora, A. (2016). Impact of river bank erosion on land cover in lower Subansiri river flood plain. *International Journal of Scientific and Research Publications, 6*(5), 480−486. Available from https://www.researchgate.net/publication/305883214_Impact_of_River_Bank_Erosion_on_Land_Cover_in_Lower_Subansiri_River_Flood_Plain#fullTextFileContent.

Hazarika, N., Das, A. K., & Borah, S. B. (2015). Assessing land-use changes driven by river dynamics in chronically flood affected Upper Brahmaputra plains, India, using RS-GIS techniques. *Egyptian Journal of Remote Sensing and Space Science, 18*(1), 107−118. Available from https://doi.org/10.1016/j.ejrs.2015.02.001, http://www.elsevier.com/wps/find/journaldescription.cws_home/723780/description#description.

Mukhopadhayay, S. (2010). A geo-environmental assessment of flood dynamics in lower Ajoy river inducing sand splay problem in Eastern India. *Ethiopian Journal of Environmental Studies and Management, 3.*

Olley, J. M., Murray, A. S., Mackenzie, D. H., & Edwards, K. (1993). Identifying sediment sources in a gullied catchment using natural and anthropogenic radioactivity. *Water Resources Research, 29*(4), 1037−1043. Available from https://doi.org/10.1029/92WR02710.

Prosser, I. P., & Winchester, S. J. (1996). History and processes of gully initiation and development in eastern Australia. *Zeitschrift fur Geomorphologie, Supplementband, 105,* 91−109.

Rosgen, D. (1996). *Applied river morphology.* Wildland Hydrology.

Sass, C. K., & Keane, T. D. (2012). Application of Rosgen's BANCS model for NE Kansas and the development of predictive streambank erosion curves. *Journal of the American Water Resources Association, 48*(4), 774−787. Available from https://doi.org/10.1111/j.1752-1688.2012.00644.x.

Sekely, A. C., Mulla, D. J., & Bauer, D. W. (2002). Streambank slumping and its contribution to the phosphorus and suspended sediment loads of the Blue Earth River, Minnesota. *Journal of Soil and Water Conservation, 57*(5), 243−250.

Simon, A., Curini, A., Darby, S. E., & Langendoen, E. J. (2000). Bank and nearbank processes in incised channel. *Geomorphology, 35*(3), 183−217.

Simon, A., Wolfe, W.J., & Molinas, A. (1991). Mass wasting algorithms in an alluvial channel model. In *Proceedings of the fifth federal inter-agency sedimentation conference* (pp. 22−29).

Thorne, C. R. (1999). *Bank processes and channel evolution in the incised rivers of north-central Mississippi* (pp. 97−121). Chichester.

Wallbrink, P. J., Murray, A. S., Olley, J. M., & Olive, L. J. (1998). Determining sources and transit times of suspended sediment in the Murrumbidgee River, New South Wales, Australia, using fallout 137 Cs and 210 Pb. *Water Resources Research, 34*(4), 879−887. Available from https://doi.org/10.1029/97wr03471.

Wasson, R. J., Mazari, R. K., Starr, B., & Clifton, G. (1998). The recent history of erosion and sedimentation on the Southern Tablelands of southeastern Australia: Sediment flux dominated by channel incision. *Geomorphology, 24*(4), 291−308. Available from https://doi.org/10.1016/S0169-555X(98)00019-1.

Geospatial modeling of potential soil erosion estimation for sustainable soil conservation planning and management

18

Manisha Tikader[1], Debaaditya Mukhopadhyay[2] and Zoheb Islam[3]

[1]*Govt. Bilasa Girls P.G. College, Atal Bihari Vajpayee Vishwavidyalaya, Bilaspur,*
Chhattisgarh, India
[2]*ICFRE—Rain Forest Research Institute, Jorhat, Assam, India*
[3]*Department of Geography, Hari-Har Mahavidyalaya, Bansra, West Bengal, India*

18.1 Introduction

Soil erosion caused by water represents a significant contributor to global land degradation, as highlighted by various studies (Borrelli et al., 2013, 2020; Lal, 2001; Montanarella et al., 2016). This process involves the removal of the nutrient-rich topsoil, where a substantial amount of organic matter and nutrients is concentrated (Pimentel et al., 1995; Zheng, 2005). The Global Assessment of Land Degradation (GLADA) project by the United Nations Environment Programme (UNEP) revealed that approximately 1.1 billion hectares of land worldwide has experienced degradation due to soil erosion, with Asia, Africa, Latin America, and the Caribbean being particularly affected (Bai, Dent, Olsson, & Schaepman, 2008). A potential loss of 74 Pg of soil organic carbon between 1850 and 2005, predominantly occurring in agricultural land and grassland was estimated, attributing it to human-induced accelerated soil erosion (Naipal et al., 2018). This has resulted in a substantial net primary productivity loss of 55% in drought-prone areas (Zika & Erb, 2009). Numerous studies on regional and global scales have reported varying soil erosion rates, with Africa and Asia experiencing the highest rates, primarily attributed to factors such as high rainfall erosivity, population pressure, and the conversion of natural vegetation to cropland and urban areas (Borrelli et al., 2020; Fenta et al., 2020, 2021; Foley et al., 2005; Hansen et al., 2013; Obalum et al., 2012; Panagos et al., 2017; Yang, Kanae, Oki, Koike, & Musiake, 2003; Zhang, Drake, & Wainwright, 2002). These changes in land use have led to not only soil erosion but also significant biodiversity loss and a decline in ecosystem services (Fenta et al., 2020; Li, Tan, & Hao, 2019;

Mertz, Ravnborg, Lövei, Nielsen, & Konijnendijk, 2007). A recent projection by Borrelli et al. (2020) anticipates a global increase of 30%−66% in water-induced soil erosion by 2070, particularly in the Global South. This underscores the need for comprehensive research to enhance our understanding of the key factors influencing soil erosion across diverse geographical regions. To help with the development, execution, and evaluation of mitigation measures, more thorough data on land cover and management techniques is specifically needed (Bouma, 2002).

Out of India's 328 million hectares of land, over 175 million hectares is experiencing severe soil erosion as a result of widespread deforestation and poor land management techniques. Among the most adversely affected regions are the Shiwaliks and the Lesser Himalayas (Dhruva Narayana, 1987). The steep slopes of the Himalayas, reduced forest cover, and heightened seismic activity contribute significantly to soil erosion and sedimentation in river systems (Jain, Kumar, & Varghese, 2001). Changes in the hydrology of the basin are evident, driven by soil erosion and floods within the transboundary Himalayan River basins (Nibanupudiand & Rawat, 2012). The social, economic, and political trends exacerbate these issues (Pimentel, 2000). The main factors contributing to land degradation in the Himalayan region are, according to the International Centre for Integrated Mountain Development (ICIMOD, 1994), the deterioration of pasture and forest areas, as well as natural disasters like mudslides, landslides, terrace collapses, and soil loss from steep slopes. The High Himalayas have received little attention despite the fact that the central Himalayas have been the subject of several research on soil erosion, with a special emphasis on the middle mountain area (Ghimire, Higaki, & Bhattarai, 2013; Kayastha, Dhital, & De Smedt, 2013; Paudel & Andersen, 2010; Uddin, Matin, & Maharjan, 2018). Water erosion has caused the loss of 601 Mt of soil and significant amounts of nitrogen, phosphorus, potassium, manganese, zinc, calcium, and magnesium in the northeastern Himalayan area alone (2011). In different physiographic locations, the topography, slope, land use practices, and population pressure all influence the rates of soil erosion. There are notable variations in sediment losses throughout land use and land cover classifications that are impacted by geology, vegetation, slope, and soil properties. For example, agricultural practices hasten soil erosion, which results in the buildup of soil sediments in water bodies and the deterioration of stream water quality (Yusof, Jamil, Aini, & Manaf, 2018). On the other hand, soil loss in areas covered by vegetation and forests is often negligible (Patric, 1976; Rai & Sharma, 1998). Consequently, in order to estimate erosion in regions that are more sensitive than those that experience less erosion, it is imperative to understand the state of soil erosion across various land covers. The effective use of management techniques to lessen soil erosion is facilitated by the identification of the size and geographic distribution of soil erosion risk regions. In order to forecast future soil loss, several studies examine the possibility of soil loss in the context of climate change (Hateffard et al., 2021; Panagos et al., 2021).

Soil erosion management, integral to sustainable soil management (FAO, 2017), involves measures outlined by the Voluntary Guidelines for Sustainable

Soil Management (VGSSM) and the World Overview of Conservation Approaches (FAO, 2017). VGSSM categorizes erosion control into four groups. The first targets land use changes affecting soil susceptibility to erosion, causing 30%−40% loss of original soil organic carbon during forest or grassland conversion (Guo & Gifford, 2002; Li et al., 2018; Poeplau et al., 2011; Wei, Shao, Gale, & Li, 2014). The second and third groups emphasize safeguarding the soil surface and reducing hillslope runoff through methods like plant cover, mulching, minimum tillage, and agro-ecological approaches (FAO, 2017). Physical barriers like terraces, strip cropping, and grass strips are employed to diminish runoff velocity and depth (FAO, 2017). Minimizing soil particle export and contaminants involves practices like riparian buffers, check dams, and sediment ponds (Dawa & Arjune, 2021). No-till, widely used, reduces soil loss and runoff but exhibits varied effects in different climates (Mhazo, Chivenge, & Chaplot, 2016; Pittelkow et al., 2015; Sun, Zeng, Shi, Pan, & Huang, 2015). The impact of no-till on soil organic carbon remains debated (Mitasova, Hofierka, Zlocha, & Iverson, 1996; Powlson et al., 2014). Success depends on local context and societal acceptance (CGIAR, 2013). Mulching effectively reduces sediment concentration and runoff (Prosdocimi, Tarolli, & Cerdà, 2016). Shelterbelts, reducing wind erosion, may impact adjacent crop yields (Kowalchuk & de Jong, 1995). Terraces, though effective, face abandonment due to structural issues (Wei et al., 2016). Controlling gullies involves grass or forest cover, check dams, and stone bunds (Valentin, Poesen, & Li, 2005), but adoption is hindered by construction and maintenance challenges (Valentin et al., 2005). Regional specificity, nutrient management, and societal considerations are critical for erosion control success.

The study of soil erosion is greatly aided by remote sensing (RS) and geographic information systems (GISs), particularly in areas like the Himalayas that are severely affected by land degradation. The Himalayan landscape, characterized by steep slopes, fragile soils, and erosive rainfall patterns, encounters escalated soil erosion due to deforestation and inappropriate land use. This process diminishes soil fertility, adversely affecting agricultural suitability (Bai et al., 2008; Oldeman, 1994). Large tracts of land, about 73.27 million hectares, in India are impacted by river erosion. The yearly rate of soil erosion is 16.4 t/ha/year, meaning that 5.3 billion tons of soil is lost annually (Narayana & Babu, 1983; Pandey, Chowdary, Mal, & Billib, 2008). The Indian Himalayas, constituting 16.4% of India's total area, are particularly susceptible to soil erosion and landslides due to deforestation and unsustainable agricultural practices (Garde & Kothyari, 1986; Mandal & Sharda, 2011). Understanding soil erosion processes in the Himalayan region is crucial for effective conservation. To address soil and water conservation, the Indian government undertakes watershed management initiatives such as the Integrated Watershed Management Programme. RS and GIS offer valuable tools for studying soil erosion, providing accurate spatial information on land cover, soil types, and terrain characteristics. High-resolution RS data, coupled with digital elevation models (DEMs), facilitates precise terrain parameter retrieval, enabling the estimation of soil erosion risk areas. These geographical data layers are integrated and analyzed using GIS software, which helps with planning conservation measures in watersheds and estimating the danger of erosion (Pandey et al., 2008).

This chapter aims to comprehensively address the geospatial modeling of potential soil erosion estimation for sustainable soil conservation planning and management. It introduces the significance of soil erosion and conservation, highlighting challenges in management. The objectives include elucidating soil erosion processes, detailing geospatial data and technology, evaluating the existing and advanced models, and addressing challenges and future directions in soil conservation planning. This chapter emphasizes the role of geospatial modeling in understanding, predicting, and mitigating soil erosion for sustainable land management.

18.2 Understanding soil erosion processes

The English term erosion comes from the Latin verb "erodere" meaning "to eat away" (rodere implies "to gnaw") or "to excavate." While surface wash and precipitation erosion were referred to as ablation (latin abletino—to take away), the term erosion was first used in geology to describe the development of hollows by water as well as the wearing away of solid mass by the action of river water (Penck, 1894). The issue of river erosion and how it affects surface modeling had gained widespread recognition by the end of the 1800s. Several terms were used to characterize geomorphological processes brought about by wind and water in addition to erosion and ablation. Abrasion (Latin: abradere, to scrape off), denudation (Latin: denudare, to strip), corrasion (Latin: corradere, to scrape together), and corrosion (Latin: corrodere, to eat to pieces) were some of them. However, there is a great deal of ambiguity because these sentences have been utilized and switched around by various writers in different situations. While wind damage is referred to as deflation and abrasion, many authors now use the word erosion to describe any kind of soil or earth's surface deterioration caused by water (Morgan, 2009).

18.2.1 Classification of erosive agents

Water, glaciers, snow, wind, debris, plants, animals, and humans are examples of erosive agents. Because certain phrases are well-established, there may be opposition to a new, standardized vocabulary. These terms can be categorized as follows: main agent erosion. Water erosion, also known as aquatic or hydric erosion, is commonly confused with fluvial erosion, which is more specifically defined as river erosion, erosion caused by glaciers. One aspect of the geomorphic action of snow, known as nivation, was described by Matthes (1900) as snow or nival erosion. It is best to refer to erosion as wind or aeolian. There is the distinct category of erosion brought on by people, plants, and animals. Based on the pedosphere and lithosphere components, there are two groupings. Surface phenomena are included in the first group, and subsurface phenomena are included in the second. There are four types of exomorphic (surface) precipitation erosion: multiform, rock, gully, and sheet erosion. Surface runoff and rainfall are the main causes of sheet erosion. The soil mantle eventually thins due to sheet erosion, exposing a sizable portion of the underlying rock and mineral substrata. Gully formation occurs due to erosion brought on by increased water buildup or the slow inclination of hills and gullies of all shapes and sizes (Zachar, 2011). A basic idea in geomorphology, multiform erosion, captures the many mechanisms by which landscapes change over geological timeframes. The complex interactions between different elements such as water, wind, ice, gravity, and life produce the many landforms that make up the surface of the Earth. Features like valleys and river deltas are shaped by water erosion, which is fueled by variables like river flow and rainfall (Hopley, Smithers, & Parnell, 2007). Sand dunes and other unique landforms are formed by wind erosion, which carries loose particles away from the source (Al-Hamdan et al., 2015). As a result of ice movement, glacial erosion leaves its mark on the terrain, forming fjords and valleys (Nachtergaele & Poesen, 1999). The terrain is further altered by gravity-driven processes like landslides, while biological processes like plant root activity and burrowing increase the complexity of geomorphic development (Brundrett, 2021). In addition, chemical weathering processes are essential for the disintegration of rocks and the creation of distinctive landforms (Herndon, 2012). Thus, multiform erosion provides a thorough framework for comprehending the dynamic and interrelated processes that sculpt the many landscapes on Earth. The way that water erodes soil is affected in different ways by rock fragments or mineral particles with a diameter of two millimeters or more that are located close to the soil's surface or in its topmost layer. The effects of rock fragments, both directly and indirectly, have been documented in the literature. Examples of direct effects on soil erosion include catching spilled material or shielding the soil surface from soil surface separation brought on by raindrop splash and runoff. The following are the main indirect impacts that rock pieces have, despite the fact that there are many others: (1) physical properties of the fine earth, such as porosity and organic matter content, that influence subprocesses of soil erosion; (2) physical soil degradation of the uppermost layer, such as compaction and surface sealing; (3) hydrological processes, such as infiltration and percolation, that influence runoff generation and discharge; (4) discharge (such as percolation and infiltration); and (5) runoff hydraulics (Poesen, Torri, & Bunte, 1994).

18.2.2 Factors influencing soil erosion

Factors influencing soil erosion in river basins include climatic variation, human impact, land use practices, and topographic characteristics (Kumar, Devrani, Kumar, Chatterjee, & Deshmukh, 2022). Soil erosion rates are affected by factors such as rainfall erosivity, soil erodibility, slope length and steepness, cover and management, and support practice. These factors have been studied using models like the revised universal soil loss equation (RUSLE) and techniques like RS and GIS (Roy & Mitra, 2022). The studies have identified areas with high soil erosion probability, categorized soil erosion rates, and proposed soil conservation practices. The analysis has shown that undulating western parts of river basins are more prone to soil erosion, while lower courses of watersheds have lower anticipated soil loss (Ghosh & Maiti, 2021). Logistic regression, decision tree, and random forest models have been used to predict soil erosion probability with varying levels of accuracy. The findings of these studies can be valuable for policymakers in planning and implementing measures to regulate and minimize soil erosion rates in river basins (Gayen & Haque, 2022; Raskar, Gaikwad, Kadekar, & Umrikar, 2019).

Environmental variables impact the types and rates of erosion that take place in a particular site. There are four main components to these conditions — land cover and usage, soil, climate, and geography (Nunes, de Almeida, & Coelho, 2011). Though in slightly different ways, these factors affect erosion caused by wind and water. The foundation for erosion processes, technology for predicting erosion, and erosion management is provided by each. For a large portion of the Earth's land area, the primary geomorphic process is water erosion (Finlayson & Montgomery, 2003). The processes of soil particle dissociation, entrainment, transport, and deposition are combined into water erosion and sedimentation. The main forces behind these processes are shear stresses produced by surface runoff across the ground surface and raindrop contact (Droppo et al., 2015). The forces that raindrop impact and surface runoff apply to the soil in proportion to the soil's resistance to separation result in water erosion. Sediment is the term used to describe soil particles that are moving. Sedimentation results from a watershed's outflow (sediment yield) or from hillslope erosion (soil loss) (Aksoy & Kavvas, 2005). Estimates of the forces applied to the soil to cause water erosion are provided by climate-determined rainfall erosivity (Nearing, Foster, Lane, & Finkner, 1989; Nearing, Yin, Borrelli, & Polyakov, 2017). The characteristics of soil control how easily it erodes naturally. The soil is subject to different pressures depending on its topography, vegetation, and soil surface structure. Additionally, the erodibility of the soil is affected by the presence of biological components and soil management practices. Climate refers to the state of the weather and the changes in it over time (Wang, Zheng, Römkens, & Darboux, 2013). Erosion is impacted by climate both directly and indirectly. Precipitation is the single most significant climatic factor affecting water erosion. Raindrops striking the soil and water passing over it cause rainfall erosion.

Soil, as opposed to solid rock, refers to a loose material that covers the earth's surface. Soil serves various purposes, acting as a nourishing medium for plant growth, a construction material supporting structures and roadways, while simultaneously posing challenges through erosion, which can lead to reservoir sedimentation, water channel obstruction, contamination, and the transport of pollutants detrimental to water quality in streams and lakes (Greiner, Keller, Grêt-Regamey, & Papritz, 2017). Raindrop impact can promote soil compaction and dispersion of tiny soil particles, resulting in the formation of a thick layer near the soil surface. This layer, known as a crust when dry and a seal when wet, decreases infiltration and increases runoff, both of which exacerbate erosion (Williams, Pagliai, & Stoops, 2018). Soils differ in their susceptibility to erosion by raindrop impact versus their susceptibility to erosion by surface runoff. In comparison to a high silt soil that is quickly eroded by surface runoff, a clay soil can be more resistant to erosion by flow. The physical qualities of the soil drive erosion, and second, the physical and chemical properties of the extracted colloids and their relationship to soil attributes. Even under the same climatic and environmental variables such as rainfall, vegetation, and terrain, soils have widely diverse erosive qualities. The Iredell and Davidson soils are typical instances of such variances; the Iredell is very erosive, whereas the Davidson is not (Musgrave & Free, 1937). The Davidson's B-horizon in the field is always friable, granular, porous, and permeable to water, whereas the Iredell's B-horizon is notably plastic, sticky, and impenetrable. Because the Davidson has a larger clay percentage than the Iredell, the variations in erosiveness cannot be explained by texture; the mechanical composition of the two soils should favor more erosion in the Davidson (Norton, Shainberg, Cihacek, & Edwards, 1998).

Topography is the term used to describe the geometry of the ground surface. The most important geometric parameters are the length and steepness of the slope, the shape in the profile view, and the form in the plan view (Evans & Cox, 1999). Uniform slopes, characterized by a consistent steepness throughout, represent the most basic type of slope (Ribolzi et al., 2011). Over regular slopes, erosion increases due to the accumulation of runoff down the slope. More than every other factor combined, land use affects erosion. Land use includes both the overall use of the land and its administration. For instance, land utilization can be the total absence of human activity, maintaining a pristine and undisturbed environment. Another potential land use is forestland, where disturbance is often confined to logging and reseeding. Rangelands are another type of land use in which cattle are allowed to graze. In contrast to where extremely strong grazing eliminates a significant quantity of vegetation, if grazing is correctly regulated, vegetation might be considerably greater. The kind of vegetation is governed mostly by temperature and soil, while the amount of vegetation is determined by agricultural and grazing land management. Management also influences soil disturbance, which impacts erosion as well as vegetation. During urban expansion, landfill filling, or mining, all vegetation is often eliminated, leaving the soil completely exposed to erosion (Toy, Foster, & Renard, 2002).

18.2.3 Soil erosion measurement and estimation techniques

Soil erosion assessment plays a pivotal role in understanding landscape dynamics, with both traditional and modern methods offering insights into sediment transport processes. Estimating suspended load, the finer particles carried by water currents, traditionally involves grab samples taken from the stream, filtered, dried, and weighed for sediment concentration measurement (Hudson, 1993). Depth-integrating samplers, like the fish-shaped frame with a glass bottle, provide a single sample accounting for variations in sediment concentration at different points in the stream (Hudson, 1993). Point-integrating samplers, fixed in the stream, collect continuous samples at various depths. Pumping samplers automate the process, pumping samples into bottles at preset intervals based on depth or flow conditions (Lane, Renard, Foster, & Laflen, 1992). Estimating bedload, comprising larger particles rolling along the streambed, involves direct measurements where debris is removed, weighed, and analyzed (Hudson, 1993). Bedload samplers, which are lowered to the streambed and raised after a predetermined time, offer an alternative method, but they present challenges, such as altering hydraulic conditions and instability on the streambed. Radioactive tracers mimic bedload movement, resembling silt in weight, shape, and size (Hudson, 1993). Empirical estimations, though less common, attempt to compute bedload based on suspended sediment concentration and material texture (Einstein, 1950).

Traditional soil erosion measurement methods encounter challenges, such as the potential lack of representativeness in samples due to spatial heterogeneity and the disturbance of flow dynamics by depth-integrating samplers during deployment (Li et al., 2018, 2019). Point-integrating samplers and pumping samplers face variations in sediment concentration over time, while direct measurements for bedload estimation are labor-intensive and may not comprehensively capture the entire bedload (Perks, 2014). Bedload samplers, despite altering hydraulic conditions, may sink into scouring streambeds, and the use of radioactive tracers raises environmental concerns. Although empirical estimations are theoretically sound, they grapple with the complexity of achieving accurate bedload measurements (Rachlewicz, Zwoliński, Kociuba, & Stawska, 2017). The choice between traditional and modern methods hinges on the specific characteristics of the study site and research objectives, underscoring the necessity for a tailored approach in soil erosion measurement. The evolution of soil erosion measurement methods has introduced a spectrum of approaches to address the complexities of sediment transport; nevertheless, each method inherently possesses limitations that researchers must carefully consider when selecting the most appropriate technique for their specific study context.

Soil erosion measurement and estimation techniques using GIS and RS have been widely studied in various regions. The RUSLE model, combined with GIS and RS, is commonly used for estimating soil erosion rates (Madhukar, Hari, Srivalli, & Neelima, 2023). The RUSLE model integrates factors such as rainfall erosivity, soil erodibility, slope length and steepness, cover management factor, and conservation practices (Ghosh et al., 2023). These factors are derived from geospatial datasets, including rainfall records, soil characteristics, topographic data, and land use/land cover maps. By applying the RUSLE model, researchers can estimate and map soil erosion rates, providing valuable insights into areas most susceptible to erosion (Wondrade, 2023). The integration of GIS and RS techniques allows for the identification of erosion-prone areas and the implementation of soil conservation measures. This approach can be used by policymakers and land managers to prioritize and implement appropriate soil conservation strategies (Ghosh et al., 2023).

18.3 Geo-spatial data and technology

18.3.1 Remote sensing in soil erosion assessment

The indispensability of RS for natural resource inventory in the realm of natural resource management is underscored by its environmental advantages (Lillesand & Kiefer, 1994). Among the widely utilized remotely sensed data for soil erosion modeling, the Landsat series, initiated in 1972, holds prominence, enabling cost-effective and accessible monitoring of impacted areas over an extended temporal scope. The Landsat sensor's primary strength lies in its multi-temporal capabilities (Jong & Leij, 1999). However, a notable limitation of the sensor is its inherent low spectral resolution. In addition to physical deteriorations like crusting, hard setting, and compaction, visual degradation like sheets, rills, or gullies may be used to measure total erosion over time (Baartman, Temme, Schoorl, Braakhekke, & Veldkamp, 2012; Omuto & Shrestha, 2007). On the other hand, enhanced spectral and radiometric properties of the recently launched Landsat 8 sensor make it appropriate for mapping soil erosion at the local and regional levels (Dhakal, Yoneda, Kato, & Kaneko, 2002; Dwivedi, Kumar, & Tewari, 1997; Pickup & Nelson, 1984). The high resolution visible and high resolution visible and infrared sensors on the SPOT series satellites have demonstrated superiority over Landsat TM observations in identifying eroded regions. These sensors detect reflected radiance in three bands at a spatial resolution of 20 m (Bocco & Valenzuela, 1988; Dwivedi et al., 1997). These sensors are important, but they cannot identify and map minor erosion structures like sheet erosion or rills. Additionally, studies demonstrate that satellites like SPOT-5 and QuickBird provide high-quality data for potential application in mapping soil erosion (Vrieling, 2006). Nevertheless, there are still gaps in our knowledge on soil erosion. For the purpose of mapping erosion across wide areas, such high resolution data (IKONOS and QuickBird) are quite costly to gather (Vrieling, de Jong, Sterk, & Rodrigues, 2008). This new sensor should be used in future studies, even if it means contrasting its capabilities with those of the Landsat 8 sensor. While the two sensors have been compared for measuring biomass of plants and grasses (Shoko, Mutanga, Dube, & Slotow, 2018; Sibanda, Sebata, Mufandaedza, & Mawanza, 2016), it is not yet possible to determine if soil erosion modeling would yield results similar to these findings until they are confirmed. A review of studies comparing the two sensors leads one to believe that they can yield almost identical findings when used in soil erosion models (Sepuru & Dube, 2018).

RS techniques have been used in several studies to assess soil erosion in river basins in the Indian context. The RUSLE model, integrated with RS and GIS, has been employed to estimate annual soil loss and identify erosion-prone areas in different river basins. For example, Ghosh et al. (2023) used GIS and RS techniques with the RUSLE model to estimate soil loss in the Mayurakshi river basin in eastern India. Similarly, Prakash, Baiju, Varghese, and Anish (2022) utilized RUSLE methodology in conjunction with RS and GIS to estimate soil loss and identify erosion-prone zones in the Chalakkudy River Basin in southern India. Devrani focused on the north flowing Cratonic (NFC) rivers in the northern Peninsular region and employed the RUSLE model to assess soil erosion and identify erosion-prone areas in different NFC river basins (Kumar et al., 2022). These studies demonstrate the effectiveness of RS and GIS techniques in assessing soil erosion in river basins in the Indian context.

18.3.2 Geographical information systems in soil erosion modeling

An international issue that primarily impacts natural resources and agricultural production is soil loss from erosion. Soil erosion is perhaps one of the riskiest conditions. GISs and RS have become more popular tools for natural resource management and disaster study in recent years. This inquiry requires a large amount of geographical data, which GIS can process efficiently. For this reason, many researchers use GIS as their main tool for estimating soil erosion at all scales (Ali & Hagos, 2016; Belasri & Lakhouili, 2016; Fistikoglu & Harmancioglu, 2002; Parveen & Kumar, 2012; Trinh, Vu, & Do, 2015). In the field of environmental research, the incorporation of GIS into soil erosion models is a significant development. This article presents a perceptive examination of many modern soil erosion models, with an emphasis on the use of GIS in conjunction with the LISEM soil erosion model. Using GIS makes sense since it can handle the regional variability that comes with soil erosion and runoff processes, which calls for the usage of cell sizes that can account for spatial variations (Wischmeier & Smith, 1978). GIS is indispensable when handling the enormous amount of data needed for several cells, which exceeds the capacity of human input. A variety of advantages are provided by GIS, including the ability to quickly create altered input maps that depict various land use patterns or conservation scenarios, simulate large catchments with lots of pixels for more detail, and provide a visual representation of the results in map formats. These features highlight the effectiveness and adaptability of GIS in soil erosion modeling (Flacke, Auerswald, & Neufang, 1990). The USLE is the main tool used in soil loss estimate, and it is particularly well-executed when data on soil erosion-causing factors are stored using the AK-INFO system. Three different techniques of computation are used by the USLE; one way just uses USLE, while the other two methods provide different estimates of soil loss using land information system loss equation (LISLE). The GIS-centric technique offers the vital benefit of quick information delivery on soil loss estimations across various areas of the examined region, in addition to facilitating accurate estimation. With several applications in environmental management and conservation planning, this combination of GIS and soil erosion models provides a comprehensive and effective method of comprehending and reducing soil erosion (De Roo, Hazelhoff, & Burrough, 1989).

The factors that most affect soil erosion rates in different places are the topographical factor (LS), cropping management (C), rainfall erosivity factor (R), practice support factor (P), and soil erodibility factor (K). It is evident that a combination of GIS tools and USLE data sources may be used to calculate soil erosion. Researchers have developed a number of methods to assess soil loss, such as the Soil and Water Assessment Tool (SWAT), the Water Erosion Prediction Project (WEPP), the USLE, the RUSLE, and others. Although there are certain disadvantages owing to its high input data requirements, USLE is the most commonly utilized of them for researching soil erosion by water because of how simple it is to use (Lufafa, Tenywa, Isabirye, Majaliwa, & Woomer, 2003; Parveen & Kumar, 2012; Tiwari, Risse, & Nearing, 2000). The USLE strategy forecasts the long-term average yearly rate of erosion on a field using data on crop system, geography, soil type, rainfall patterns, and management practices. In the midwest of the United States, the USLE was first created to calculate soil erosion on gently sloping agricultural land (Gitas, Douros, Minakou, Silleos, & Karydas, 2009; K.G. Renard, 1997; P. Renard & De Marsily, 1997; Renard et al., 1978). The RUSLE, which was developed lately and has many improvements in identifying input components based on the most recent database updates in the United States, is similar in structure to the USLE (Pham, 2008).

18.3.3 Digital elevation models

Utilizing an expanded set of points delineating bare soil through X, Y, and Z coordinates, a digital elevation model (DEM) is formulated, presenting options for raster or random arrangement. The DHM model serves as an alternative to "digital height," and "DEM" provides a nuanced representation. In contrast, a digital terrain model (DTM) offers comprehensive information, encompassing object positions. Despite the incorporation of additional data and spatial considerations, achieving precise DEM determination becomes challenging, particularly for locations atop visible surfaces such as foliage and building summits. The transformation of a digital surface model to a DEM necessitates the elimination of extraneous features, ensuring alignment with bare soil specifications (Jacobsen, 2003). Word used to refer to any format of digital topography (including bathymetric) data as well as the method(s) used to implicitly interpret heights between measurements. It usually refers to heights of bare earth free of vegetation and structures, but it can also apply to man-made structures like road embankments. Elevations of hydrological features, including rivers and lakes, are usually connected to a surface that is devoid of water (Maune, Schmidt, & Küchler, 2001). DEM is the most important part of DTM. Elevation models and other geographic and natural elements, including rivers and other break lines, are included in the umbrella notion. Topographical data, such as aspect, slope, curvature, visibility, and so on, may also be supplied (Li & Chen, 2005). The quantitative study of terrain formations, or geomorphometry, combines elements from several fields, such as computer science, mathematics, statistics, engineering, and earth sciences. The notions of terrain analysis currently include digital geomorphometry based on the usage of DEMs (Wilson & Gallant, 2000). Although geomorphometry dates back to the 1800s and the pioneers of academic geography, the last 35 years have seen a revolution in the field due to advances in computing power, particularly the mass production of digital elevation maps (Pike 1973, 2002; Yang et al., 2003).

Topographic properties have been classified based on their characteristics and spatial extent. Secondary attributes (e.g., indices for topographic wetness and radiation) are based on two or more primary attributes and may be physically or empirically derived indices. In general, primary attributes (e.g., slope, aspect, and curvature) derived directly from the DEM are distinguished from secondary attributes (Wilson & Gallant, 2000). The primary characteristics are frequently computed using numerical estimates of the first and second partial derivatives of the DEM heights in the lattice axis directions. Derivatives that are local (constrained) and global (unconstrained) must be differentiated. The explicit computation window size is given in terms of aspect and slope, which are local derivatives. On the other hand, global derivatives depend on extensive and unexpected interactions among DEM heights. For instance, the delineation of automated drainage basins is dependent upon the repeated analysis of flow direction data; the extent of the final result is not known until the research is completed (Jenson & Domingue, 1988).

18.4 Geospatial models for soil erosion estimation

Soil erosion models employ mathematical formulations to depict the processes of soil particle detachment, transport, and deposition on the land surface. These models are grounded in an understanding of the physical laws governing surface runoff generation, including detachment capacity and sedimentation across the landscape (Kumar, 2018). These models include soil, topography, soil, land use/cover, and management strategies. They express mathematical correlations between soil erosion causes and processes. Through the clarification of the complex interactions among soil erosion variables and processes, these models evaluate the influence of various land use/cover and management strategies on soil erosion rates. They may also be used to model how agricultural patterns and soil conservation measures affect soil erosion rates, which can help determine the best management techniques for a given terrain. Three main types of erosion models have been produced during the last 30 years: conceptual, empirical, and physically based models (Table 18.1).

Table 18.1 Commonly used soil erosion models.

Sl. no.	Model types	Model spatial scale	Temporal scale	References
1	Empirical	USLE plot/hillslope	Event/annual	Wischmeier and Smith (1965, 1978)
		MUSLE hillslope/ catchment	Annual	Williams (1975)
		RUSLE hillslope/ watershed	Annual	Renard et al. (1978)
2	Conceptual	MMF hillslope/ watershed	Annual	Morgan and Morgan (1984)
		SWAT watershed/ basin	Continuous	Arnold et al. (1998)
		AGNPS small watershed	Event-based	Young, Behzadi, Wang, and Chotai (1987)
3	Physical-based	WEPP hillslope/ watershed	Distributed, event-based, continuous	Nearing et al. (1989)
		KINEROS hillslope/ small watershed	Event-based	Smith, Goodrich, and Quinton (1995)
		ANSWERS small watershed	Distributed, event-based	Beasley, Carpenter, and Jennings (1982)

MUSLE, *Modified universal soil loss equation;* RUSLE, *revised universal soil loss equation;* SWAT, *Soil and Water Assessment Tool;* USLE, *universal soil loss equation;* WEPP, *Water Erosion Prediction Project.* MMF, *maximum-minimum-factor;* AGNPS, *annualized agricultural non-point source pollution model;* KINEROS, *KINematic runoff and EROSion model;* ANSWERS, *areal non-point source watershed environmental response simulation.*

Ecosystem services and agricultural output may suffer from the gradual depletion of soil resources brought on by erosion processes. However, agriculture itself may be a major contributor to soil erosion due to disruption of soil structure, which has the power to alter the long-term evolution of landscapes (Baartman et al., 2012). It is well known that farming on hillslopes causes the highest rates of runoff and soil loss; in particular, erosion risk is often increased when native vegetation is converted to agriculture (Cerdan et al., 2010; Kosmas et al., 1997; Maetens et al., 2012). A variety of soil and water conservation structures, such as various types of terraces designed to shorten and diminish the slope's steepness, are occasionally used in steep slope farming techniques. However, these structures often simply serve to modify or worsen patterns of soil erosion, particularly when poorly designed or maintained (Tarolli, Preti, & Romano, 2014). This section explains several instruments and models for managing and assessing soil erosion, as well as various model parameters (Table 18.2).

Using statistical analysis, empirical models determine correlations between soil erosion rates and variables. Because of their simplicity, they are extensively

Table 18.2 Using geographic information system (GIS) and remote sensing to determine input parameters for soil erosion models.

Sl. no.	Soil erosion factors	Factor elements	Thematic maps	Remote sensing/ancillary data	Methods of analysis
1	Vegetation cover (C)	Crop types, cropping pattern, natural/perennial land cover types	Land use/land cover map	Multidate, remote sensing data at appropriate sensor/scale	Visual analysis, digital supervised/unsupervised classification, spectral indices
2	Land management practices (P)	Field management practices—bunding, terracing, contouring, field size, etc.	Detail land use/land cover types showing management practices	High spatial resolution remote sensing data	Visual analysis/based on farmers' interview
3	Soil erodibility (K)	Soil texture (sand, silt, and clay percentage), organic matter, soil structure, and soil permeability	Soil map	Remote sensing data at appropriate scale	Visual analysis for physiographic soil analysis
4	Topographic (L S)	Elevation, slope, drainage network	Slope map, drainage map	DEM—CartoDEM, ASTER, and SRTM	Terrain analysis based on DEM using GIS/image analysis software, Survey of India toposheet, stereo satellite data, microwave satellite data
5	Climate (R)	Rainfall, rain intensity, rainy days	Location-based rain erosivity	Weather station	Point interpolation (Thiessen map) using GIS

Developed from Aiello, A., Adamo, M., & Canora, F. (2015). Remote sensing and GIS to assess soil erosion with RUSLE3D and USPED at river basin scale in southern Italy. Catena, 131, 174–185. doi: 10.1016/j.catena.2015.04.003; http://www.elsevier.com/inca/publications/store/5/2/4/6/0/9 and Barman, B. K., Rao, K. S., Sonowal, K., Prasad, N. S. R., & Sahoo, U. K. (2020). Soil erosion assessment using revised universal soil loss equation model and geo-spatial technology: A case study of upper Turial river basin. AIMS Geosciences, 6(4), 525–545.

utilized and site-specific. They do not, however, have any understanding of the actual erosion processes. Empirical models such as the USLE have been widely applied in the Indian setting, where heterogeneous landscapes are prevalent (Wischmeier & Smith, 1965, 1978). Based on rainfall erosivity, soil erodibility, slope length, slope steepness, plant cover, and management techniques, the USLE forecasts soil erosion. Conceptual models—also called semiempirical models—offer a more thorough comprehension of erosion causes. Water and sediment continuity equations in spatially lumped forms are used in these models, such as the RUSLE. Conceptual models are essential in India because watersheds are greatly impacted by changes in land use. For example, RUSLE evaluates soil erosion in a difficult terrain with changing land use/cover conditions (K.G. Renard, 1997; P. Renard & De Marsily, 1997). Fundamental mechanisms of erosion are described by sophisticated physical process-based models. They take into account the effects of rainfall, soil particle separation, and runoff movement. Given the intricate topography of the Himalayan area, these models provide a sophisticated knowledge. One tool that simulates water, sediment, nutrient, and pesticide transport at the watershed scale is the SWAT (Setegn, Srinivasan, & Dargahi, 2008). Empirical models like USLE are effective for quick assessments, especially in regions with limited data. However, in scenarios requiring detailed insights into erosion processes, conceptual models like RUSLE provide a more comprehensive understanding. While conceptual models offer a middle ground, physical process-based models, such as SWAT, deliver a detailed representation of erosion processes. In regions like the Himalayas, where terrain dynamics are complex, physical process-based models are crucial for accurate predictions. Empirical models' simplicity is advantageous, but they may lack accuracy. Physical process-based models, accounting for intricate processes, are more suitable for regions like the Himalayas, where diverse topography influences erosion.

18.4.1 Universal soil loss equation

One well-known erosion model that is used to forecast soil erosion at the field or plot level in a variety of crop management systems is the USLE. The USLE model was created empirically by carefully examining more than 10,000 plot-years' worth of runoff and soil loss data from small-scale plots (Wischmeier & Smith, 1978). It is represented mathematically as:

$$A = R \cdot K \cdot L \cdot S \cdot C \cdot P$$

Here,
    A represents the average annual soil loss (t/ha/year), with
    R as the rainfall erosivity index,
    K as the soil erodibility factor,
    L as the slope length factor,
    S as the slope steepness factor,
    C as the vegetation cover factor, and

P as the management practice factor in various land uses/land covers.

The modified USLE (MUSLE) and the RUSLE were developed as a result of modifications made to the USLE model to improve its computational precision. The USLE model has been used extensively in soil conservation planning for more than 30 years (Biswas & Pani, 2015; Renard & De Marsily, 1997; Renard et al., 1978).

18.4.1.1 Modified universal soil loss equation

Sediment loss at the watershed/catchment scale is explicitly addressed by the MUSLE model, which is a derivative of the USLE (Zhang, Degroote, Wolter, & Sugumaran, 2009). Williams (1975) proposed adjustments by substituting a runoff rate component for the rainfall erosivity (R) factor. This is how the MUSLE equation is written:

$$SYe = Xe \cdot K \cdot L \cdot S \cdot C \cdot P$$

where

SYe denotes the rainfall event sediment yield (metric tons), and
Xe is defined as:

$$Xe = 11.8 \cdot (Qe \cdot qp)^{0.56}$$

Here,

Qe represents the surface runoff amount (mm/ha), and
qp is the peak runoff rate (m^3/s) obtained during the rain event.

18.4.1.2 Revised universal soil loss equation

The RUSLE model is an improved version of the USLE that keeps the basic architecture but makes substantial changes to the methods used to determine each erosion component (Renard, 1997; Renard et al., 1978). Enhancements include improved calculation of soil erodibility and rainfall erosivity factors, season-dependent algorithms, updated slope length and steepness, and a new method for figuring out crop cover and management variables. As the most popular model for calculating the yearly soil erosion loss over a long period of time worldwide, the RUSLE model estimates both rill and interrill erosion. The model supports soil erosion risk assessment and conservation planning since it can be adjusted to a variety of land cover situations, such as croplands, rangelands, and forest lands (Millward & Mersey, 1999). The RUSLE model integrates key factors to estimate soil erosion, including rainfall—runoff erosivity (R), soil erodibility (K), slope length (L), steepness (S), cover management (C), and practice management (P). These factors collectively assess the potential for soil loss by considering rainfall, soil properties, slope characteristics, land cover, and conservation practices. The RUSLE model provides a concise evaluation of average annual soil loss in a specific area (Barman, Rao, Sonowal, Prasad, & Sahoo, 2020).

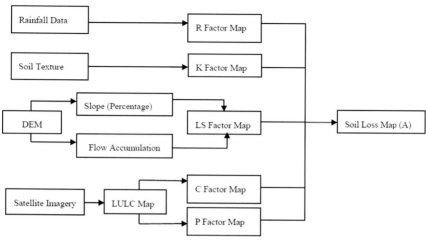

FIGURE 18.1

Flow chart illustrating of the steps involved in simulating soil erosion using revised universal soil loss equation.

The core equation of the RUSLE model is expressed as:

$$A = R \cdot K \cdot LS \cdot C \cdot P$$

Here,

In this case, A denotes the average annual soil loss (t/ha/year), R stands for rainfall erosivity, K for soil erodibility, LS for topography, C for land cover, and P for soil conservation or preventive practices (Fig. 18.1).

18.4.2 **RUSLE-3D**

The RUSLE-3D model is a development of RUSLE that makes a change to the topographic factor (LS) while maintaining similarities in most other variables (Covelli, Cimorelli, Pagliuca, Molino, & Pianese, 2020). By substituting the upslope contributing area for the slope length, DEMs allow for integration with GISs. The potential average soil loss (A) is computed by the model as:

$$A(r) = R \cdot K \cdot LS(r) \cdot C \cdot P$$

Here,

In this case, A(r) represents the average annual soil loss of a grid cell at position r.

18.4.3 **Unit stream power-based erosion deposition**

Unit stream power-based erosion deposition (USPED) is the name of a two-dimensional soil erosion model. In the one-dimensional RUSLE model, soil erosion and deposition are mostly reliant on rainfall detachment capacity; in the

USPED model, however, the primary factor influencing soil erosion and deposition is the sediment transport capacity of surface runoff. The actual quantity of erosion will be much decreased if soil particles have already been separated by rain, but there is insufficient runoff to transfer the soil particles due to the form of the terrain or the influence of plants. According to the USPED method, a land pixel's capacity for transporting sediment is determined by the total amount of water that passes through it. The USLE model includes variables such as rainfall erosivity, soil erodibility, slope length, slope steepness, plant cover, and management techniques. It was developed experimentally from substantial plot-level runoff and soil loss data analysis. Rainfall erosivity index (R), soil erodibility factor (K), slope length factor (L), slope steepness factor (S), plant cover factor (C), and management practice factor (P) are among the elements included in its prediction equation (Kumar, 2018).

18.4.4 Water Erosion Prediction Project

In August 1985, SDA Agricultural Research Service staff and their partners launched the WEPP with the goal of developing next-generation water erosion prediction technology for federal action agencies engaged in environmental planning and assessment, soil and water conservation, and planning and assessment. The USLE was the widely used soil erosion prediction tool at the earlier times and later on was detailed with the circumstances behind the USLE's publication (Flanagan, Gilley, & Franti, 2007). The RUSLE model maintains the basic structure while refining algorithms for erosion factor calculations. Improvements include considerations for seasonal variations in rainfall erosivity and soil erodibility, revised slope factors, and a novel method for crop cover and management factors. The RUSLE equation includes the following variables: land cover (C), rainfall erosivity (R), soil erodibility (K), topographic factor (LS), and soil conservation measures (P) (Kumar, 2018). WEPP uses steady-state sediment continuity models to study rill and interrill dynamics, providing insights at the event, daily, monthly, and yearly scales.

18.4.5 European soil erosion model

For both individual fields and small catchments, the dynamic distributed European soil erosion model (EUROSEM) can model the movement, erosion, and deposition of sediment over the land surface induced by rill and interrill processes in a single storm. The output of the model includes the storm hydrograph, storm sediment graph, total runoff, and total soil loss. The effects of plant cover on rainfall energy and interceptions, the effects of rock fragment (stoniness) on infiltration, flow velocity, splash erosion, and the changes in rill channel shape and size due to erosion and deposition are all explicitly simulated by EUROSEM, in contrast to other erosion models (Smets, Borselli, Poesen, & Torri, 2011).

18.4.6 Soil and Water Assessment Tool

EUROSEM's computation of the water balance takes into consideration the hydrologic balance of the watershed. According to Arnold, Srinivasan, Muttiah, and Williams (1998), it calculates hydrologic components such runoff, stream flow, and evapotranspiration. The components of transpiration and evaporation are calculated independently. Using a modified version of the SCS-CN algorithm, SWAT forecasts runoff. The peak runoff is calculated using the modified rational formula. It calculates the elements of the landscape's water balance throughout a 24-hour period. The SWAT model requires data on soil, terrain, land cover, and daily weather. It is a water balancing tool that was created by the USDA and simulates transpiration, evapotranspiration, streamflow, and runoff. Its formula takes into account variables such as soil erodibility (K), land cover (C), topographic factor (LS), rainfall erosivity (R), and soil conservation methods (P) (Mosbahi, Benabdallah, & Boussema, 2013).

18.4.7 Agricultural Policy Environmental eXtender

A distributed model called Agricultural Policy Environmental eXtender (APEX), an outgrowth of EPIC, simulates surface runoff, sedimentation, and nutrient loss at the watershed level. It uses soil, geography, and land use/cover to split the watershed into homogenous land sections. Weather modeling, hydrology, erosion, sedimentation, nutrient cycling, crop development, tillage techniques, and plant environmental management are all included in APEX's extensive component set. In challenging terrains like the Himalayan region, characterized by dynamic landscapes and limited data, the choice of erosion models becomes critical. While models like RUSLE, SCS-CN-based models, SWAT, APEX, and WEPP find applications, challenges arise due to the rugged terrain and data scarcity. The modified SWAT-VSA is a promising method since it may be used to estimate surface runoff and runoff source regions in the Himalayan environment with little hydrological data. The dearth of studies in this field emphasizes the need for more research on model validation and calibration to increase applicability (Kumar, 2018).

The suitability of advanced geospatial models for soil erosion estimation is contingent upon specific landscape characteristics and the goals of erosion assessment. USPED adopts a two-dimensional approach, focusing on sediment transport capacity over rainfall detachment, making it suitable for terrains where runoff dynamics significantly influence erosion. Designed to advance prediction technologies for federal agencies involved in environmental planning, WEPP is an excellent water erosion predictor. The EUROSEM dynamically models the movement, erosion, and deposition of sediment caused by rill and interill processes. It produces comprehensive results that are particularly useful for small catchments. The physically grounded, geographically dispersed SWAT model is appropriate for watershed-scale evaluations, taking into account the daily time scale transit of water, sediment, nutrients, and pesticides. The model's adaptability is exemplified by its capacity to compute a wide range of hydrologic parameters, rendering it appropriate for a variety of terrain types. The Himalayan region's rugged terrain necessitates models capable of addressing its spatial and temporal variations. RUSLE and its 3D version find relevance due to their ability to consider terrain convexity/concavity. SCS-CN-based models like SWAT, SWAT-VSA, APEX, and WEPP, though semidistributed, are widely used for their adaptability to hilly landscapes. The choice among these models should align with the specific requirements of the study area, considering factors such as terrain, land cover, and the scale of erosion assessment (Pan et al., 2021).

18.5 Challenges and future directions

Soil erosion stands as a formidable challenge in India, further exacerbated by the anticipated impacts of climate change, including heightened erosion risks due to intense storms and deforestation. The Indian Himalayas, marked by rugged terrain and diverse ecosystems, face distinctive challenges that demand customized soil conservation strategies for effective mitigation. Tackling these challenges and steering toward sustainable solutions is imperative for the ecological resilience of the region. The changing climate dynamics in India predict an increase in high-intensity storms, intensifying both wind and water erosion. This necessitates adaptive strategies capable of withstanding and mitigating the impacts of extreme weather events. Additionally, the denudation of forest cover emerges as a significant contributor to soil erosion in the Himalayas and river basins. Preserving and restoring forest ecosystems is paramount for maintaining soil stability and preventing erosion in vulnerable regions (Bhattacharyya et al., 2015). The detrimental impacts of soil erosion on river basins necessitate a comprehensive understanding of the processes involved, and geospatial models emerge as invaluable tools in unraveling the complexities of this phenomenon. Empirical models, exemplified by the USLE, play a crucial role in quickly assessing soil erosion rates, especially in regions characterized by heterogeneous landscapes. In river basins, the USLE, with its variables such as rainfall erosivity, soil erodibility, slope length, slope steepness, plant cover, and management practices, provides a pragmatic approach to estimating soil loss (Ghimire et al., 2013). However, the simplicity of empirical models may limit their accuracy in capturing the intricate dynamics of erosion processes in riverine environments. Conceptual models, such as the RUSLE, offer a more comprehensive understanding of erosion causes in river basins. RUSLE's incorporation of water and sediment continuity equations in spatially lumped forms enhances its capacity to evaluate soil erosion in regions where changes in land use significantly impact watersheds (Thomas, Joseph, & Thrivikramji, 2018). The adaptability of RUSLE to assess soil erosion in challenging terrains, such as those found in river basins, underscores its utility in capturing the nuanced interactions between land use changes and erosion processes. Sophisticated physical process-based models, exemplified by the SWAT, contribute to a detailed representation of erosion processes in river basins. Considering the intricate topography of riverine landscapes, models like SWAT become imperative for accurate predictions. The model's ability to simulate water, sediment, nutrient, and pesticide transport at the watershed scale aligns with the need for a holistic understanding of soil erosion in river basins (Mosbahi et al., 2013). In the specific context of river basins, models like RUSLE-3D and USPED address the unique challenges posed by the dynamic nature of riverine landscapes. RUSLE-3D, by modifying the topographic factor, integrates DEMs into GISs, enhancing its suitability for river basins. USPED, with its focus on sediment transport capacity over rainfall detachment, acknowledges the influence of runoff dynamics in terrains where rivers play a significant role in erosion processes (Aiello, Adamo, & Canora, 2015). The EUROSEM and the WEPP stand out as comprehensive models capable of simulating rill and interrill processes in river basins. EUROSEM's explicit consideration of various factors, including plant cover effects and changes in rill channel shape, provides a holistic view of erosion dynamics in riverine environments (Lahiguera et al., 2021). WEPP, developed to advance water erosion prediction technologies, proves beneficial in federal environmental planning, soil and water conservation, and assessment within river basins. In rugged terrains like the Himalayas, the choice among models becomes critical. RUSLE, SCS-CN-based models, SWAT, APEX, and WEPP find applications, but challenges arise due to the unique topography and limited data. The modified SWAT-VSA emerges as a promising method for estimating surface runoff in such environments, showcasing the need for continual research to enhance model applicability (Senanayake, Pradhan, Huete, & Brennan, 2020).

Watershed development, recognized as a holistic approach, faces challenges in effective implementation. Issues such as inadequate public participation, inappropriate institutional arrangements, and the necessity for tailored interventions based on regional rainfall patterns hinder the success of these programs (Darghouth, Ward, Gambarelli, Styger, & Roux, 2008). Institutional bottlenecks further impede sustainability, emphasizing the need for involving elected institutions, fostering local leadership, and establishing effective linkages with various sectors. Watershed benefits exhibit disparities based on specific annual rainfall ranges, with more pronounced advantages in regions with 700−1000 mm annual rainfall. Bridging the gap between low and high rainfall regions necessitates targeted technological interventions and a nuanced understanding of socio-economic disparities (Bhattacharyya et al., 2016). Looking toward the future, a holistic approach to watershed development should extend beyond agricultural lands, encompassing all lands. Matching the productive capacity of soils and landscapes with appropriate policies and technologies ensures sustainable soil and water conservation. Community engagement emerges as central to the success of watershed programs, requiring a predisposition for collective action within communities and involvement of elected institutions. Educational initiatives play a vital role in promoting awareness and enhancing the adaptive capacity of communities to embrace soil conservation practices and cope with climate change impacts (Fazey et al., 2010). Given that soil erosion is caused by both physical and socioeconomic forces, an integrated strategy is needed. Watershed systems are interrelated, and policies should integrate water management techniques and regulations for complete landscape-level conservation. In addition to protecting the environment, diversification techniques such as conservation agriculture, animal rearing, integrated nutrient management, tree shelterbelts, conservation agriculture, and stress on forage crop and grazing management also help to revitalize rural communities (Sauer et al., 2021). The selection of geospatial models for soil erosion estimation in river basins should align with the specific characteristics of the study area. While empirical models offer simplicity and efficiency, conceptual and physical process-based models provide a more nuanced understanding of erosion processes in dynamic riverine landscapes. Specialized models designed for river basins and comprehensive approaches like EUROSEM and WEPP contribute to a holistic understanding, emphasizing the importance of choosing models that suit the spatial and temporal variations inherent in river basin environments. Continued research and refinement of these models will undoubtedly contribute to more accurate assessments and effective soil conservation strategies in river basins. Mitigating soil erosion challenges in the Indian context, particularly in the river basins, necessitates a multipronged approach, integrating sustainable practices, community involvement, educational initiatives, and responsive policies for the long-term health of the region's soil and water resources.

18.6 Conclusion

This review's thorough examination of soil erosion and mitigation techniques emphasizes the necessity and relevance of implementing long-term soil conservation practices, particularly in environmentally sensitive areas such as river basins. To successfully address soil erosion concerns, a multifaceted strategy is required due to the complex interplay of causes such as climate change, deforestation, and poor land management. The incorporation of adaptive techniques capable of withstanding the expected consequences of climate change holds the key to future soil conservation possibilities. The changing climate dynamics, which foresee an increase in high-intensity storms, emphasize the importance of robust conservation strategies that can withstand extreme weather occurrences. Customized soil conservation strategies, tailored to the unique challenges posed by the rugged terrain and diverse ecosystems of the Indian Himalayas, are imperative for ecological resilience.

Conserving and rebuilding forest ecosystems appears as an important component in reducing soil erosion concerns. Denudation of forest cover contributes greatly to soil erosion; hence, concentrated efforts must be aimed toward conserving and rebuilding these critical biological buffers. Furthermore, watershed development, which is recognized as a comprehensive strategy, necessitates successful execution through community participation, educational activities, and responsive legislation. Watershed programs must bridge socioeconomic divides, refine institutional frameworks, and build local leadership in order to be successful. Mitigation techniques should apply to all lands in the watershed system, not only agricultural fields. A holistic approach that integrates water management techniques and regulations at the landscape level ensures comprehensive soil and water conservation. The utilization of terrain analysis and DEMs as a global strategy further advances comprehension of eroding landscapes. The investigation into renowned geographic models like the USLE and its revised version, the RUSLE, extends beyond the Indian Himalayas to showcase their global resonance. Comparative studies of sophisticated models such as USPED, WEPP, EUROSEM, and SWAT provide valuable insights applicable in diverse global contexts. Diversification techniques, such as conservation agriculture, animal rearing, integrated nutrient management, and stress on forage crop and grazing management, not only protect the environment but also contribute to the revitalization of rural communities in the Himalayan region.

Looking toward the future, the success of soil erosion mitigation hinges on continued community involvement, education, and the integration of adaptive policies. The promotion of collective action within communities, involvement of elected institutions, and awareness-building initiatives are vital components of a sustainable strategy. Ongoing advancements in technology, including RS and GIS, can further enhance our understanding of soil erosion processes, aiding in the development and implementation of effective conservation plans. In conclusion, the road ahead involves a concerted effort to harmonize ecological health with sustainable development, recognizing the intricate web of factors influencing soil erosion. By embracing a holistic and adaptive approach, informed by scientific insights and community participation, we can navigate the challenges of soil erosion in the Indian context and ensure the long-term well-being of the region's soil and water resources.

References

Aiello, A., Adamo, M., & Canora, F. (2015). Remote sensing and GIS to assess soil erosion with RUSLE3D and USPED at river basin scale in southern Italy. *Catena, 131*, 174−185. Available from https://doi.org/10.1016/j.catena.2015.04.003, http://www.elsevier.com/inca/publications/store/5/2/4/6/0/9.

Aksoy, H., & Kavvas, M. L. (2005). A review of hillslope and watershed scale erosion and sediment transport models. *Catena, 64*(2−3), 247−271. Available from https://doi.org/10.1016/j.catena.2005.08.008.

Al-Hamdan, O. Z., Hernandez, M., Pierson, F. B., Nearing, M. A., Williams, C. J., Stone, J. J., ... Weltz, M. A. (2015). Rangeland hydrology and erosion model (RHEM) enhancements for applications on disturbed rangelands. *Hydrological Processes*, *29*(3), 445−457. Available from https://doi.org/10.1002/hyp.10167, http://onlinelibrary.wiley. com/journal/10.1002/(ISSN)1099-1085.

Ali, S. A., & Hagos, H. (2016). Estimation of soil erosion using USLE and GIS in Awassa Catchment, Rift valley, Central Ethiopia. *Geoderma Regional*, *7*(2), 159−166. Available from https://doi.org/10.1016/j.geodrs.2016.03.005, http://www.journals.elsevier.com/geoderma-regional/.

Arnold, J. G., Srinivasan, R., Muttiah, R. S., & Williams, J. R. (1998). Large area hydrologic modeling and assessment Part I: Model development 1. *JAWRA Journal of the American Water Resources Association*, *34*(1), 73−89. Available from https://doi.org/10.1111/j.1752-1688.1998.tb05961.x.

Baartman, J. E. M., Temme, A. J. A. M., Schoorl, J. M., Braakhekke, M. H. A., & Veldkamp, T. A. (2012). Did tillage erosion play a role in millennial scale landscape development? *Earth Surface Processes and Landforms*, *37*(15), 1615−1626. Available from https://doi.org/10.1002/esp.3262.

Bai, Z.G., Dent, D.L., Olsson, L., & Schaepman, M.E. (2008). *Global assessment of land degradation and improvement: 1. Identification by remote sensing*. ISRIC-World Soil Information.

Barman, B. K., Rao, K. S., Sonowal, K., Prasad, N. S. R., & Sahoo, U. K. (2020). Soil erosion assessment using revised universal soil loss equation model and geo-spatial technology: A case study of upper Tuirial river basin. *AIMS Geosciences*, *6*(4), 525−545.

Beasley, T. M., Carpenter, R., & Jennings, C. D. (1982). Plutonium, 241Am and 137Cs ratios, inventories and vertical profiles in Washington and Oregon continental shelf sediments. *Geochimica et Cosmochimica Acta*, *46*(10), 1931−1946. Available from https://doi.org/10.1016/0016-7037(82)90131-4.

Belasri, A., & Lakhouili, A. (2016). Estimation of soil erosion risk using the universal soil loss equation (USLE) and geo-information technology in Oued El Makhazine Watershed, Morocco. *Journal of Geographic Information System*, *08*(01), 98−107. Available from https://doi.org/10.4236/jgis.2016.81010.

Bhattacharyya, R., Ghosh, B. N., Mishra, P. K., Mandal, B., Rao, C. S., Sarkar, D., . . . Franzluebbers, A. J. (2015). Soil degradation in india: Challenges and potential solutions. *Sustainability (Switzerland)*, *7*(4), 3528−3570. Available from https://doi.org/10.3390/su7043528, http://www.mdpi.com/2071-1050/7/4/3528/pdf.

Bhattacharyya, R., Ghosh, B., Dogra, P., Mishra, P., Santra, P., Kumar, S., . . . Parmar, B. (2016). Soil conservation issues in India. *Sustainability*, *8*(6), 565. Available from https://doi.org/10.3390/su8060565.

Biswas, S. S., & Pani, P. (2015). Estimation of soil erosion using RUSLE and GIS techniques: A case study of Barakar River basin, Jharkhand, India. *Modeling Earth Systems and Environment*, *1*(4). Available from https://doi.org/10.1007/s40808-015-0040-3, http://springer.com/journal/40808.

Bocco, G., & Valenzuela, C. R. (1988). Integration of GIS and image processing in soil erosion studies using ILWIS. *ITC Journal*, *4*, 309−319.

Borrelli, P., Robinson, D. A., Fleischer, L. R., Lugato, E., Ballabio, C., Alewell, C., . . . Bagarello, V. (2013). An assessment of the global impact of 21st-century land use change on soil erosion. *Nature Communications*, *8*.

Borrelli, P., Robinson, D. A., Panagos, P., Lugato, E., Yang, J. E., Alewell, C., . . . Ballabio, C. (2020). Land use and climate change impacts on global soil erosion by water (2015-2070). *Proceedings of the National Academy of Sciences of the United States of America*, *117*(36), 21994−22001. Available from https://doi.org/10.1073/pnas.2001403117, https://www.pnas.org/content/pnas/117/36/21994.full.pdf.

Bouma, J. (2002). Netherlands Land quality indicators of sustainable land management across scales. *Agriculture, Ecosystems and Environment*, *88*(2), 129−136. Available from https://doi.org/10.1016/S0167-8809(01)00248-1.

Brundrett, M. C. (2021). One biodiversity hotspot to rule them all: Southwestern Australia—An extraordinary evolutionary centre for plant functional and taxonomic diversity. *Australia Journal of the Royal Society of Western Australia*, *104*, 91−117. Available from http://www.royalsocietyofwa.com/139/journal.

Cerdan, O., Govers, G., Le Bissonnais, Y., Van Oost, K., Poesen, J., Saby, N., ... Dostal, T. (2010). Rates and spatial variations of soil erosion in Europe: A study based on erosion plot data. *Geomorphology*, *122*(1−2), 167−177. Available from https://doi.org/10.1016/j.geomorph.2010.06.011.

CGIAR. (2013). *The Nebraska declaration on conservation agriculture.* https://ispc.cgiar.org/sites/default/files/ISPC_StrategyTrends_ConservationAgriculture_NebraskaDeclaration.pdf

Covelli, C., Cimorelli, L., Pagliuca, D. N., Molino, B., & Pianese, D. (2020). Assessment of erosion in river Basins: A distributed model to estimate the sediment production over watersheds by a 3-dimensional LS factor in RUSLE model. *Hydrology*, *7*(1). Available from https://doi.org/10.3390/hydrology7010013, https://res.mdpi.com/d_attachment/hydrology/hydrology-07-00013/article_deploy/hydrology-07-00013.pdf.

Darghouth, S., Ward, C., Gambarelli, G., Styger, E., & Roux, J. (2008). *Watershed management approaches, policies, and operations: Lessons for scaling up.*

Dawa, D., & Arjune, V. (2021). Identifying potential erosion-prone areas in the Indian Himalayan Region using the revised universal soil loss equation (RUSLE). *Asian Journal of Water, Environment and Pollution*, *18*(1), 15−23. Available from https://doi.org/10.3233/AJW210003, http://www.iospress.nl/loadtop/load.php?isbn = 09729860.

Dhakal, G., Yoneda, T., Kato, M., & Kaneko, K. (2002). Slake durability and mineralogical properties of some pyroclastic and sedimentary rocks. *Engineering Geology*, *65*(1), 31−45. Available from https://doi.org/10.1016/S0013-7952(01)00101-6.

Dhruva Narayana, V. V. (1987). Downstream impacts of soil conservation in the Himalayan Region. *Mountain Research and Development*, *7*(3), 287. Available from https://doi.org/10.2307/3673207.

Droppo, I. G., D'Andrea, L., Krishnappan, B. G., Jaskot, C., Trapp, B., Basuvaraj, M., & Liss, S. N. (2015). Fine-sediment dynamics: towards an improved understanding of sediment erosion and transport. *Journal of Soils and Sediments*, *15*(2), 467−479. Available from https://doi.org/10.1007/s11368-014-1004-3, http://www.springerlink.com/content/1439-0108.

Dwivedi, R. S., Kumar, A. B., & Tewari, K. N. (1997). The utility of multi-sensor data for mapping eroded lands. *International Journal of Remote Sensing*, *18*(11), 2303−2318. Available from https://doi.org/10.1080/014311697217620.

Einstein, H.A. (1950). *The bed-load function for sediment transportation in open channel flows (No. 1026).* US Department ofAgriculture.

Evans, I.S., & Cox, N.J. (1999). *Relations between land surface properties: Altitude, slope and curvature.* In: S. Hergarten & H. J. Neugebauer (Eds.), *Process modelling and landform evolution. Lecture notes in earth sciences, vol 78.* Springer Science and

Business Media LLC, Berlin, Heidelberg. Available from https://doi.org/10.1007/bfb0009718.

FAO. (2017). *Voluntary guidelines for sustainable soil management.* Food and Agriculture Organization of the United Nations. Unpublished content. http://www.fao.org/3/a-bl813e.pdf

Fazey, I., Gamarra, J. G. P., Fischer, J., Reed, M. S., Stringer, L. C., & Christie, M. (2010). Adaptation strategies for reducing vulnerability to future environmental change. *Frontiers in Ecology and the Environment, 8*(8), 414−422. Available from https://doi.org/10.1890/080215, http://www.esajournals.org/doi/pdf/10.1890/080215.

Fenta, A. A., Tsunekawa, A., Haregeweyn, N., Tsubo, M., Yasuda, H., Kawai, T., ... Sultan, D. (2021). Agroecology-based soil erosion assessment for better conservation planning in Ethiopian river basins. *Environmental Research, 195.* Available from https://doi.org/10.1016/j.envres.2021.110786, http://www.elsevier.com/inca/publications/store/6/2/2/8/2/1/index.htt.

Fenta, A. A., Tsunekawa, A., Haregeweyn, N., Poesen, J., Tsubo, M., Borrelli, P., ... Kurosaki, Y. (2020). Land susceptibility to water and wind erosion risks in the East Africa region. *Science of The Total Environment, 703,* 135016. Available from https://doi.org/10.1016/j.scitotenv.2019.135016.

Finlayson, D. P., & Montgomery, D. R. (2003). Modeling large-scale fluvial erosion in geographic information systems. *Geomorphology, 53*(1−2), 147−164. Available from https://doi.org/10.1016/S0169-555X(02)00351-3, http://www.elsevier.com/inca/publications/store/5/0/3/3/3/4/.

Fistikoglu, O., & Harmancioglu, N. B. (2002). Integration of GIS with USLE in assessment of soil erosion. *Water Resources Management, 16*(6), 447−467. Available from https://doi.org/10.1023/A:1022282125760.

Flacke, W., Auerswald, K., & Neufang, L. (1990). Combining a modified Universal Soil Loss Equation with a digital terrain model for computing high resolution maps of soil loss resulting from rain wash. *Catena, 17*(4-5), 383−397. Available from https://doi.org/10.1016/0341-8162(90)90040-K.

Flanagan, D. C., Gilley, J. E., & Franti, T. G. (2007). Water Erosion Prediction Project (WEPP): Development history, model capabilities, and future enhancements. *Transactions of the ASABE, 50*(5), 1603−1612.

Foley, J. A., DeFries, R., Asner, G. P., Barford, C., Bonan, G., Carpenter, S. R., ... Snyder, P. K. (2005). Global consequences of land use. *science, 309*(5734), 570−574.

Garde, & Kothyari, U.C. (1986). *Erosion in Indian catchments.*

Gayen, A., & Haque, S. M. (2022). Soil erodibility assessment of laterite dominant sub-basin watersheds in the humid tropical region of India. *CATENA, 213,* 106161. Available from https://doi.org/10.1016/j.catena.2022.106161.

Ghimire, S. K., Higaki, D., & Bhattarai, T. P. (2013). Estimation of soil erosion rates and eroded sediment in a degraded catchment of the Siwalik Hills, Nepal. *Land, 2*(3), 370−391. Available from https://doi.org/10.3390/land2030370, http://www.mdpi.com/2073-445X/2/3/370/pdf.

Ghosh, A., & Maiti, R. (2021). Soil erosion susceptibility assessment using logistic regression, decision tree and random forest: Study on the Mayurakshi river basin of Eastern India. *Environmental Earth Sciences, 80*(8). Available from https://doi.org/10.1007/s12665-021-09631-5, https://link.springer.com/journal/12665.

Ghosh, A., Rakshit, S., Tikle, S., Das, S., Chatterjee, U., Pande, C. B., ... Mattar, M. A. (2023). Integration of GIS and remote sensing with RUSLE model for estimation of soil erosion. *Land, 12*(1). Available from https://doi.org/10.3390/land12010116, http://www.mdpi.com/journal/land/.

Gitas, I. Z., Douros, K., Minakou, C., Silleos, G. N., & Karydas, C. G. (2009). Multitemporal soil erosion risk assessment in N. Chalkidiki using a modified USLE raster model. *EARSel Eproceedings, 8*, 40–52.

Greiner, L., Keller, A., Grêt-Regamey, A., & Papritz, A. (2017). Soil function assessment: review of methods for quantifying the contributions of soils to ecosystem services. *Land Use Policy, 69*, 224–237. Available from https://doi.org/10.1016/j.landusepol.2017.06.025, http://www.elsevier.com/inca/publications/store/3/0/4/5/1/.

Guo, L. B., & Gifford, R. M. (2002). Soil carbon stocks and land use change: A meta analysis. *Global Change Biology, 8*(4), 345–360. Available from https://doi.org/10.1046/j.1354-1013.2002.00486.x.

Hansen, M. C., Potapov, P. V., Moore, R., Hancher, M., Turubanova, S. A., Tyukavina, A., ... Townshend, J. R. G. (2013). High-resolution global maps of 21st-century forest cover change. *Science (New York, N.Y.), 342*(6160), 850–853. Available from https://doi.org/10.1126/science.1244693, http://www.sciencemag.org/content/342/6160/850.full.pdf.

Hateffard, F., Mohammed, S., Alsafadi, K., Enaruvbe, G. O., Heidari, A., Abdo, H. G., & Rodrigo-Comino, J. (2021). CMIP5 climate projections and RUSLE-based soil erosion assessment in the central part of Iran. *Scientific Reports, 11*(1). Available from https://doi.org/10.1038/s41598-021-86618-z, http://www.nature.com/srep/index.html.

Herndon, E.M. (2012). *Biogeochemistry of manganese contamination in a temperate forested watershed* [Doctoral Dissertation, The Pennsylvania State University]. PennState University Libraries. Available from https://etda.libraries.psu.edu/files/final_submissions/7965

Hopley, D., Smithers, S. G., & Parnell, K. E. (2007). *The geomorphology of the Great Barrier Reef: Development, diversity, and change* (pp. 1–532). Cambridge University Press. Available from http://doi.org/10.1017/CBO9780511535543, https://doi.org/10.1017/CBO9780511535543.

Hudson, N. (1993). Field measurement of soil erosion and runoff. *Food & Agriculture Org, 68*.

ICIMOD. (1994). *Proceedings of the international symposium on mountain environment and development*. International Centre for Integrated Mountain Development Constraints and Opportunities.

Jacobsen, K. (2003). DEM generation from satellite data. *EARSeL Ghent, 4*.

Jain, S. K., Kumar, S., & Varghese, J. (2001). Estimation of soil erosion for a Himalayan watershed using GIS technique. *Water Resources Management, 15*(1), 41–54. Available from https://doi.org/10.1023/A:1012246029263.

Jenson, S. K., & Domingue, J. O. (1988). Extracting topographic structure from digital elevation data for geographic information system analysis. *Photogrammetric Engineering and Remote Sensing, 54*(11), 1593–1600.

Jong, D., & Leij. (1999). Specific contributions of phonological abilities to early reading acquisition: Results from a Dutch latent variable longitudinal study. *Journal of Educational Psychology, 91*(3).

Kayastha, P., Dhital, M. R., & De Smedt, F. (2013). Application of the analytical hierarchy process (AHP) for landslide susceptibility mapping: A case study from the Tinau watershed, west Nepal. *Computers & Geosciences, 52*, 398–408. Available from https://doi.org/10.1016/j.cageo.2012.11.003.

Kosmas, C., Danalatos, N., Cammeraat, L. H., Chabart, M., Diamantopoulos, J., Farand, R., ... Vacca, A. (1997). The effect of land use on runoff and soil erosion rates under Mediterranean conditions. *Catena*, *29*(1), 45−59. Available from https://doi.org/10.1016/S0341-8162(96)00062-8.

Kowalchuk, T. E., & de Jong, E. (1995). Shelterbelts and their effect on crop yield. *Canadian Journal of Soil Science*, *75*(4), 543−550. Available from https://doi.org/10.4141/cjss95-077.

Kumar, R., Devrani, R., Kumar, R., Chatterjee, S., & Deshmukh, B. (2022). Assessment of soil erosion in the north flowing cratonic river basins. In *Peninsular India EGU general assembly conference abstracts* (pp. 22−539).

Kumar, S. (2018). *Geospatial approach in modeling soil erosion processes in predicting soil erosion and nutrient loss in hilly and mountainous landscape. Remote sensing of northwest himalayan ecosystems* (pp. 355−380). Springer Singapore. Available from https://link.springer.com/book/10.1007/978-981-13-2128-3, https://doi.org/10.1007/978-981-13-2128-3_17.

Lahiguera, A., Salvador, M.P., Soto, M.S., Marqués, A., Pons, V., Llinares, J., & Borselli. (2021). *Identification of areas vulnerable to soil erosion in a drainage sub-basin of the Júcar river in La Casella (Alzira, Valencia) in a Climate Change scenario using the EuroSEM prediction model.* pp. 533-536.

Lal, R. (2001). Soil degradation by erosion. *Land Degradation and Development*, *12*(6), 519−539. Available from https://doi.org/10.1002/ldr.472.

Lane, L. J., Renard, K. G., Foster, G. R., & Laflen, J. M. (1992). Development and application of modern soil erosion prediction technology—The usda experience. *Australian Journal of Soil Research*, *30*(6), 893−912. Available from https://doi.org/10.1071/SR9920893.

Li, J., & Chen, W. (2005). A rule-based method for mapping Canada's wetlands using optical, radar and DEM data. *International Journal of Remote Sensing*, *26*(22), 5051−5069. Available from https://doi.org/10.1080/01431160500166516, https://www.tandfonline.com/loi/tres20.

Li, W., Ciais, P., Guenet, B., Peng, S., Chang, J., Chaplot, V., ... Yue, C. (2018). Temporal response of soil organic carbon after grassland-related land-use change. *Global Change Biology*, *24*(10), 4731−4746. Available from https://doi.org/10.1111/gcb.14328, http://onlinelibrary.wiley.com/journal/10.1111/(ISSN)1365-2486.

Li, Y., Tan, M., & Hao, H. (2019). The impact of global cropland changes on terrestrial ecosystem services value, 1992−2015. *Journal of Geographical Sciences*, *29*(3), 323−333. Available from https://doi.org/10.1007/s11442-019-1600-7, http://www.springer.com/sgw/cda/frontpage/0,11855,1-40391-70-66229542-0,00.html?changeHeader = true/.

Lillesand, T.M., & Kiefer, R.W. (1994). *Remote sensing and image interpretation* (3rd Edition, p. 750). Hoboken: John Wiley and Sons, Inc.

Lufafa, A., Tenywa, M. M., Isabirye, M., Majaliwa, M. J. G., & Woomer, P. L. (2003). Prediction of soil erosion in a Lake Victoria basin catchment using a GIS-based Universal Soil Loss model. *Agricultural Systems*, *76*(3), 883−894. Available from https://doi.org/10.1016/S0308-521X(02)00012-4.

Madhukar, A., Hari, N., Srivalli, R., & Neelima, T. L. (2023). Spatial estimation of soil erosion using RUSLE model: A case study of Sangareddy Telangna State, India. *International Journal of Plant & Soil Science*, *35*(18), 490−498. Available from https://doi.org/10.9734/ijpss/2023/v35i183314.

Maetens, W., Vanmaercke, M., Poesen, J., Jankauskas, B., Jankauskiene, G., & Ionita, I. (2012). Effects of land use on annual runoff and soil loss in Europe and the Mediterranean: A meta-analysis of plot data. *Progress in Physical Geography*, *36*(5), 599–653. Available from https://doi.org/10.1177/0309133312451303, https://journals.sagepub.com/home/PPG.

Mandal, D., & Sharda, V. N. (2011). Assessment of permissible soil loss in India employing a quantitative bio-physical model. *Current Science*, *100*(3), 383–390. Available from http://www.ias.ac.in/currsci/10feb2011/383.pdf.

Matthes, F.E. (1900). Glacial sculpture of the Bighorn Mountains, Wyoming. In *US Geological Survey twenty-first annual report Part. II* (pp. 167–190).

Maune, S., Schmidt, C., & Küchler, T. (2001). Messung der ergebnisqualität als ansatz zum qualitätsmanagement in der hals-nasen-ohrenheilkunde. *Laryngo-Rhino-Otologie*, *80*(2), 101–108. Available from https://doi.org/10.1055/s-2001-11852.

Mertz, O., Ravnborg, H. M., Lövei, G. L., Nielsen, I., & Konijnendijk, C. C. (2007). Ecosystem services and biodiversity in developing countries. *Biodiversity and Conservation*, *16*(10), 2729–2737. Available from https://doi.org/10.1007/s10531-007-9216-0.

Mhazo, N., Chivenge, P., & Chaplot, V. (2016). Tillage impact on soil erosion by water: Discrepancies due to climate and soil characteristics. *Agriculture, Ecosystems and Environment*, *230*, 231–241. Available from https://doi.org/10.1016/j.agee.2016.04.033, http://www.elsevier.com/inca/publications/store/5/0/3/2/9/8.

Millward, A. A., & Mersey, J. E. (1999). Adapting the RUSLE to model soil erosion potential in a mountainous tropical watershed. *Catena*, *38*(2), 109–129. Available from https://doi.org/10.1016/S0341-8162(99)00067-3.

Mitasova, H., Hofierka, J., Zlocha, M., & Iverson, L. R. (1996). Modelling topographic potential for erosion and deposition using GIS. *International Journal of Geographical Information Systems*, *10*(5), 629–641. Available from https://doi.org/10.1080/02693799608902101.

Montanarella, L., Pennock, D.J., McKenzie, N., Badraoui, M., Chude, V., Baptista, I., . . . Vargas, C.R. (2016).

Morgan, J. P., & Morgan, K. G. (1984). Stimulus-specific patterns of intracellular calcium levels in smooth muscle of ferret portal vein. *The Journal of Physiology*, *351*(1), 155–167. Available from https://doi.org/10.1113/jphysiol.1984.sp015239.

Morgan, R. P. C. (2009). *Soil erosion and conservation*. John Wiley & Sons.

Mosbahi, M., Benabdallah, S., & Boussema, M. R. (2013). Assessment of soil erosion risk using SWAT model. *Arabian Journal of Geosciences*, *6*(10), 4011–4019. Available from https://doi.org/10.1007/s12517-012-0658-7.

Musgrave, G. W., & Free, G. R. (1937). Preliminary report on a determination of comparative infiltration-rates on some major soil-types. *Eos, Transactions American Geophysical Union*, *18*(2), 345–349. Available from https://doi.org/10.1029/TR018i002p00345.

Nachtergaele, J., & Poesen, J. (1999). Assessment of soil losses by ephemeral gully erosion using high-altitude (stereo) aerial photographs. *Earth Surface Processes and Landforms*, *24*(8), 693–706. Available from https://doi.org/10.1002/(SICI)1096-9837 (199908)24:8 < 693::AID-ESP992 > 3.0.CO;2-7.

Naipal, V., Ciais, P., Wang, Y., Lauerwald, R., Guenet, B., & Van Oost, K. (2018). Global soil organic carbon removal by water erosion under climate change and land use change during AD-1850-2005. *Biogeosciences*, *15*(14), 4459–4480. Available from https://doi.org/10.5194/bg-15-4459-2018, http://www.biogeosciences.net/volumes_an-d_issues.html.

Narayana, D. V. V., & Babu, R. (1983). Estimation of soil erosion in India. *Journal of Irrigation and Drainage Engineering, 109*(4), 419−434. Available from https://doi.org/10.1061/(ASCE)0733-9437(1983)109:4(419).

Nearing, M. A., Foster, G. R., Lane, L. J., & Finkner, S. C. (1989). A process-based soil erosion model for USDA-Water Erosion Prediction Project technology. *Transactions of the ASAE, 32*(5), 1587−1593. Available from https://doi.org/10.13031/2013.31195.

Nearing, M. A., Yin, Sq, Borrelli, P., & Polyakov, V. O. (2017). Rainfall erosivity: An historical review. *Catena, 157*, 357−362. Available from https://doi.org/10.1016/j.catena.2017.06.004, http://www.elsevier.com/inca/publications/store/5/2/4/6/0/9.

Nibanupudiand, H. K., & Rawat, P. (2012). Environmental concerns for drr in hindu-kush himalaya region. *Ecosystem Approach to Disaster Risk Reduction, 85*.

Norton, D., Shainberg, I., Cihacek, L., & Edwards, J. H. (1998). *Erosion and soil chemical properties. Soil quality and soil erosion* (pp. 39−55). CRC Press.

Nunes, A. N., de Almeida, A. C., & Coelho, C. O. A. (2011). Impacts of land use and cover type on runoff and soil erosion in a marginal area of Portugal. *Applied Geography, 31*(2), 687−699. Available from https://doi.org/10.1016/j.apgeog.2010.12.006.

Obalum, S. E., Buri, M. M., Nwite, J. C., Hermansah., Watanabe, Y., Igwe, C. A., & Wakatsuki, T. (2012). Soil degradation-induced decline in productivity of sub-saharan african soils: The prospects of looking downwards the lowlands with the sawah eco-technology. *Applied and Environmental Soil Science, 2012*. Available from https://doi.org/10.1155/2012/673926.

Oldeman. (1994). The global extent of soil degradation. *Soil Resilience and Sustainable Land Use, 9*.

Omuto, C. T., & Shrestha, D. P. (2007). Remote sensing techniques for rapid detection of soil physical degradation. *International Journal of Remote Sensing, 28*(21), 4785−4805. Available from https://doi.org/10.1080/01431160701260357, https://www.tandfonline.com/loi/tres20.

Pan, F., Feng, Q., McGehee, R., Engel, B. A., Flanagan, D. C., & Chen, J. (2021). A framework for automated and spatially-distributed modeling with the Agricultural Policy Environmental eXtender (APEX) model. *Environmental Modelling and Software, 144*. Available from https://doi.org/10.1016/j.envsoft.2021.105147, http://www.elsevier.com/inca/publications/store/4/2/2/9/2/1.

Panagos, P., Ballabio, C., Himics, M., Scarpa, S., Matthews, F., Bogonos, M., . . . Borrelli, P. (2021). Projections of soil loss by water erosion in Europe by 2050. *Environmental Science and Policy, 124*, 380−392. Available from https://doi.org/10.1016/j.envsci.2021.07.012, http://www.elsevier.com/wps/find/journaldescription.cws_home/601264/description#description.

Panagos, P., Borrelli, P., Meusburger, K., Yu, B., Klik, A., Lim, K. J., . . . Ballabio, C. (2017). Global rainfall erosivity assessment based on high-temporal resolution rainfall records. *Scientific Reports, 7*(1). Available from https://doi.org/10.1038/s41598-017-04282-8, http://www.nature.com/srep/index.html.

Pandey, A., Chowdary, V. M., Mal, B. C., & Billib, M. (2008). Runoff and sediment yield modeling from a small agricultural watershed in India using the WEPP model. *Journal of Hydrology, 348*(3–4), 305–319. Available from https://doi.org/10.1016/j.jhydrol.2007.10.010.

Parveen, R., & Kumar, U. (2012). Integrated approach of universal soil loss equation (USLE) and geographical information system (GIS) for soil loss risk assessment in Upper South Koel Basin, Jharkhand. *Journal of Geographic Information System, 04*(06), 588–596. Available from https://doi.org/10.4236/jgis.2012.46061.

Patric, J. H. (1976). Soil erosion in the eastern forest. *Journal of Forestry, 129*(10), 671–677.

Paudel, K. P., & Andersen, P. (2010). Assessing rangeland degradation using multi temporal satellite images and grazing pressure surface model in Upper Mustang, Trans Himalaya, Nepal. *Remote Sensing of Environment, 114*(8), 1845–1855. Available from https://doi.org/10.1016/j.rse.2010.03.011.

Penck, A.,& Engelhorn, J. (1894). Morphologie der erdoberfläche.

Perks, M.T. (2014). Suspended sediment sampling. In *Geomorphological techniques* (Online ed.).

Pham, T.H. (2008). Soil erosion risk modeling within upland landscapes using remotely sensed data and the RUSLE model (A case study in Huong Tra district, Thua Thien Hue province, Vietnam). In *International symposium on geoinformatics for spatial infrastructure development in earth and allied sciences.*

Pickup, G., & Nelson, D. J. (1984). Use of landsat radiance parameters to distinguish soil erosion, stability, and deposition in arid Central Australia. *Remote Sensing of Environment, 16*(3), 195–209. Available from https://doi.org/10.1016/0034-4257(84)90064-6.

Pike, R.J. (2002). A bibliography of terrain modeling (geomorphometry), the quantitative representation of topography. In *USGS open file report* (pp. 2–465).

Pike, S. (1973). Destination image analysis—A review of 142 papers from. *Tourism Management, 23*(5), 541–549.

Pimentel, D. (2000). Soil erosion and the threat to food security and the environment. *Ecosystem Health, 6*(4), 221–226. Available from https://doi.org/10.1046/j.1526-0992.2000.006004221.x.

Pimentel, D., Harvey, C., Resosudarmo, P., Sinclair, K., Kurz, D., McNair, M., ... Blair, R. (1995). Environmental and economic costs of soil erosion and conservation benefits. *Science (New York, N.Y.), 267*(5201), 1117–1123. Available from https://doi.org/10.1126/science.267.5201.1117, http://www.sciencemag.org.

Pittelkow, C. M., Linquist, B. A., Lundy, M. E., Liang, X., van Groenigen, K. J., Lee, J., ... van Kessel, C. (2015). When does no-till yield more? A global meta-analysis. *Field Crops Research, 183*, 156–168. Available from https://doi.org/10.1016/j.fcr.2015.07.020, http://www.elsevier.com/inca/publications/store/5/0/3/3/0/8.

Poeplau, C., Don, A., Vesterdal, L., Leifeld, J., Van Wesemael, B., Schumacher, J., & Gensior, A. (2011). Temporal dynamics of soil organic carbon after land-use change in the temperate zone – Carbon response functions as a model approach. *Global Change Biology, 17*(7), 2415–2427. Available from https://doi.org/10.1111/j.1365-2486.2011.02408.x, http://onlinelibrary.wiley.com/journal/10.1111/(ISSN)1365-2486.

Poesen, J. W., Torri, D., & Bunte, K. (1994). Effects of rock fragments on soil erosion by water at different spatial scales: A review. *Catena, 23*(1–2), 141–166. Available from https://doi.org/10.1016/0341-8162(94)90058-2.

Powlson, D. S., Stirling, C. M., Jat, M. L., Gerard, B. G., Palm, C. A., Sanchez, P. A., & Cassman, K. G. (2014). Limited potential of no-till agriculture for climate change mitigation. *Nature Climate Change*, *4*(8), 678−683. Available from https://doi.org/10.1038/nclimate2292, http://www.nature.com/nclimate/index.html.

Prakash, S., Baiju, K. R., Varghese, A., & Anish, A. U. (2022). Annual soil loss estimation in a tropical river basin of Southern India using RUSLE model and AHP techniques. *Journal of Geosciences Research*, *7*(2). Available from https://doi.org/10.56153/g19088-021-0068-8.

Prosdocimi, M., Tarolli, P., & Cerdà, A. (2016). Mulching practices for reducing soil water erosion: A review. *Earth-Science Reviews*, *161*, 191−203. Available from https://doi.org/10.1016/j.earscirev.2016.08.006, http://www.sciencedirect.com/science/journal/00128252.

Rachlewicz, G., Zwoliński, Z., Kociuba, W., & Stawska, M. (2017). Field testing of three bedload samplers' efficiency in a gravel-bed river, Spitsbergen. *Geomorphology*, *287*, 90−100. Available from https://doi.org/10.1016/j.geomorph.2016.06.001, http://www.elsevier.com/inca/publications/store/5/0/3/3/3/4/.

Rai, S. C., & Sharma, E. (1998). Comparative assessment of runoff characteristics under different land use patterns within a Himalayan watershed. *Hydrological Processes*, *12*(13−14), 2235−2248. Available from http://onlinelibrary.wiley.com/journal/10.1002/(ISSN)1099-1085, https://doi.org/10.1002/(sici)1099-1085(19981030)12:13/14 < 2235::aid-hyp732 > 3.0.co;2-5.

Raskar, T., Gaikwad, H., Kadekar, O., & Umrikar, B. (2019). Impact assessment of water harvesting structures in micro-watersheds of Nira River Basin, Maharashtra, India. *Hydrospatial Analysis*, *3*(2), 72−89. Available from https://doi.org/10.21523/gcj3.19030203.

Renard, K.G., Foster, G.R., Weesies, G.A., McCool, D.K., & Yoder, D.C. (1978). *Predicting soil erosion by water: A guide to conservation planning with the revised universal soil loss equation (RUSLE)* (pp. 1−69). U.S. Department of Agriculture, Agriculture Handbook. Agric. Handb. No. 537. doi: 10.1029/TR039i002p00285.

Renard, K. G. (1997). Predicting soil erosion by water: a guide to conservation planning with the Revised Universal Soil Loss Equation (RUSLE). US Department of Agriculture, Agricultural Research Service.

Renard, P., & De Marsily, G. (1997). Calculating equivalent permeability: A review. *Advances in Water Resources*, *20*(5−6), 253−278. Available from https://doi.org/10.1016/s0309-1708(96)00050-4, http://www.elsevier.com/inca/publications/store/4/2/2/9/1/3/index.htt.

Ribolzi, O., Patin, J., Bresson, L. M., Latsachack, K. O., Mouche, E., Sengtaheuanghoung, O., ... Valentin, C. (2011). Impact of slope gradient on soil surface features and infiltration on steep slopes in northern Laos. *Geomorphology*, *127*(1−2), 53−63. Available from https://doi.org/10.1016/j.geomorph.2010.12.004.

De Roo, A. P. J., Hazelhoff, L., & Burrough, P. A. (1989). Soil erosion modelling using 'answers' and geographical information systems. *Earth Surface Processes and Landforms*, *14*(6), 517−532. Available from https://doi.org/10.1002/esp.3290140608.

Roy, D., & Mitra, R. (2022). Estimation of soil erosion in the Balason River Basin using RUSLE modelling of the Darjeeling Himalayan region, India. *Research Square, India*. https://www.researchsquare.com/browse. Available from https://doi.org/10.21203/rs.3.rs-1766179/v1.

Sauer, T. J., Dold, C., Ashworth, A. J., Nieman, C. C., Hernandez-Ramirez, G., Philipp, D., ... Chendev, Y. G. (2021). *Agroforestry practices for soil conservation and resilient agriculture Agroforestry and Ecosystem Services* (pp. 19−48). Springer International Publishing. Available from https://link.springer.com/book/10.1007/978-3-030-80060-4, 10.1007/978-3-030-80060-4_2.

Senanayake, S., Pradhan, B., Huete, A., & Brennan, J. (2020). A review on assessing and mapping soil erosion hazard using geo-informatics technology for farming system management. *Remote Sensing*, *12*(24), 4063. Available from https://doi.org/10.3390/rs12244063.

Sepuru, T. K., & Dube, T. (2018). An appraisal on the progress of remote sensing applications in soil erosion mapping and monitoring. *Remote Sensing Applications: Society and Environment.*, *9*, 1−9. Available from https://doi.org/10.1016/j.rsase.2017.10.005, http://www.journals.elsevier.com/remote-sensing-applications-society-and-environment/.

Setegn, S. G., Srinivasan, R., & Dargahi, B. (2008). Hydrological modelling in the Lake Tana Basin, Ethiopia using SWAT model. *The Open Hydrology Journal*, *2*(1), 49−62. Available from https://doi.org/10.2174/1874378100802010049.

Shoko, C., Mutanga, O., Dube, T., & Slotow, R. (2018). Characterizing the spatio-temporal variations of C3 and C4 dominated grasslands aboveground biomass in the Drakensberg, South Africa. *International Journal of Applied Earth Observation and Geoinformation*, *68*, 51−60. Available from https://doi.org/10.1016/j.jag.2018.02.006, http://www.elsevier.com/locate/jag.

Sibanda, P., Sebata, A., Mufandaedza, E., & Mawanza, M. (2016). Effect of short-duration overnight cattle kraaling on grass production in a southern African savanna. *African Journal of Range and Forage Science*, *33*(4), 217−223. Available from https://doi.org/10.2989/10220119.2016.1243580, http://www.tandfonline.com/toc/tarf20/current.

Smets, T., Borselli, L., Poesen, J., & Torri, D. (2011). Evaluation of the EUROSEM model for predicting the effects of erosion-control blankets on runoff and interrill soil erosion by water. *Geotextiles and Geomembranes*, *29*(3), 285−297. Available from https://doi.org/10.1016/j.geotexmem.2011.01.012.

Smith, R. E., Goodrich, D. C., & Quinton, J. N. (1995). Dynamic, distributed simulation of watershed erosion: the KINEROS2 and EUROSEM models. *Journal of Soil & Water Conservation*, *50*(5), 517−520.

Sun, Y., Zeng, Y., Shi, Q., Pan, X., & Huang, S. (2015). No-tillage controls on runoff: A meta-analysis. *Soil and Tillage Research*, *153*, 1−6. Available from https://doi.org/10.1016/j.still.2015.04.007, http://www.elsevier.com/inca/publications/store/5/0/3/3/1/8.

Tarolli, P., Preti, F., & Romano, N. (2014). Terraced landscapes: From an old best practice to a potential hazard for soil degradation due to land abandonment. *Anthropocene*, *6*, 10−25. Available from https://doi.org/10.1016/j.ancene.2014.03.002, http://www.journals.elsevier.com/anthropocene/.

Thomas, J., Joseph, S., & Thrivikramji, K. P. (2018). Assessment of soil erosion in a tropical mountain river basin of the southern Western Ghats, India using RUSLE and GIS. *Geoscience Frontiers*, *9*(3), 893−906. Available from https://doi.org/10.1016/j.gsf.2017.05.011, https://www.sciencedirect.com/journal/geoscience-frontiers.

Tiwari, A. K., Risse, L. M., & Nearing, M. A. (2000). Evaluation of WEPP and its comparison with USLE and RUSLE. *Transactions of the American Society of Agricultural Engineers.*, *43*(5), 1129−1135.

Toy, T. J., Foster, G. R., & Renard, K. G. (2002). *Soil erosion: Processes, prediction, measurement, and control*. John Wiley & Sons.

Trinh, L. H., Vu, D. T., & Do, N. H. (2015). Evaluation of soil erosion risk using remote sensing and GIS data (a case study: Lang Chanh District, Thanh Hoa Province, Vietnam). *Vestnik OrelGAU*, *3*(54), 57–64. Available from https://doi.org/10.15217/issn1990-3618.2015.3.57.

Uddin, K., Matin, M. A., & Maharjan, S. (2018). Assessment of land cover change and its impact on changes in soil erosion risk in Nepal. *Sustainability*, *10*(12), 4715. Available from https://doi.org/10.3390/su10124715.

Valentin, C., Poesen, J., & Li, Y. (2005). Laos Gully erosion: Impacts, factors and control. *Catena*, *63*(2–3), 132–153. Available from https://doi.org/10.1016/j.catena.2005.06.001.

Vrieling, A. (2006). Satellite remote sensing for water erosion assessment: A review. *Catena*, *65*(1), 2–18. Available from https://doi.org/10.1016/j.catena.2005.10.005.

Vrieling, A., de Jong, S. M., Sterk, G., & Rodrigues, S. C. (2008). Timing of erosion and satellite data: A multi-resolution approach to soil erosion risk mapping. *International Journal of Applied Earth Observation and Geoinformation*, *10*(3), 267–281. Available from https://doi.org/10.1016/j.jag.2007.10.009, http://www.elsevier.com/locate/jag.

Wang, B., Zheng, F., Römkens, M. J. M., & Darboux, F. (2013). Soil erodibility for water erosion: A perspective and Chinese experiences. *Geomorphology*, *187*, 1–10. Available from https://doi.org/10.1016/j.geomorph.2013.01.018.

Wei, W., Chen, D., Wang, L., Daryanto, S., Chen, L., Yu, Y., . . . Feng, T. (2016). Global synthesis of the classifications, distributions, benefits and issues of terracing. *Earth-Science Reviews*, *159*, 388–403. Available from https://doi.org/10.1016/j.earscirev.2016.06.010, http://www.sciencedirect.com/science/journal/00128252.

Wei, X., Shao, M., Gale, W., & Li, L. (2014). Global pattern of soil carbon losses due to the conversion of forests to agricultural land. *Scientific Reports*, *4*. Available from https://doi.org/10.1038/srep04062, http://www.nature.com/srep/index.html.

Williams, A. J., Pagliai, M., & Stoops, G. (2018). *Physical and biological surface crusts and seals* (pp. 539–574). Elsevier BV. Available from https://doi.org/10.1016/b978-0-444-63522-8.00019-x.

Williams, J. R. (1975). Sediment routing for agricultural watersheds. *JAWRA Journal of the American Water Resources Association*, *11*(5), 965–974. Available from https://doi.org/10.1111/j.1752-1688.1975.tb01817.x.

Wilson, J.P., & Gallant, J.C. (2000). Digital terrain analysis. In *Terrain analysis: Principles and applications*, vol. 6.

Wischmeier, W.H., & Smith, D.D. (1965). *Predicting rainfall-erosion losses from cropland east of the Rocky Mountains: Guide for selection of practices for soil and water conservation*. Agricultural Research Service.

Wischmeier, W.H., & Smith, D.D. (1978). *Predicting rainfall erosion losses: A guide to conservation planning (No. 537)*. Department of Agriculture, Science and Education Administration.

Wondrade, N. (2023). Integrated use of GIS, RS and USLE model for LULC change analysis and soil erosion risk mapping in the Lake Hawassa Watershed, Southern Ethiopia. *Geocarto International*, *38*(1). Available from https://doi.org/10.1080/10106049.2023.2210106, http://www.tandfonline.com/toc/tgei20/current.

Yang, D., Kanae, S., Oki, T., Koike, T., & Musiake, K. (2003). Global potential soil erosion with reference to land use and climate changes. *Hydrological Processes*, *17*(14), 2913–2928. Available from https://doi.org/10.1002/hyp.1441.

Young, P., Behzadi, M. A., Wang, C. L., & Chotai, A. (1987). Direct digital and adaptive control by input-output state variable feedback pole assignment. *International Journal of Control*, *46*(6), 1867–1881. Available from https://doi.org/10.1080/00207178708934021.

Yusof, F.M., Jamil, N.R., Aini, N., & Manaf, L.A. (2018). Land use change and soil loss risk assessment by using geographical information system (GIS): A case study of the lower part of Perak River. In *Proceedings of the IOP conference series: Earth and environmental science*. IOP Publishing LTD.

Zachar, D. (2011). *Soil erosion*. Elsevier.

Zhang, X., Drake, N., & Wainwright, J. (2002). Scaling land surface parameters for global-scale soil erosion estimation. *Water Resources Research*, *38*(9). Available from https://doi.org/10.1029/2001wr000356, http://agupubs.onlinelibrary.wiley.com/hub/journal/10.1002/(ISSN)1944-7973/.

Zhang, Y., Degroote, J., Wolter, C., & Sugumaran, R. (2009). Integration of modified universal soil loss equation (MUSLE) into a GIS framework to assess soil erosion risk. *Land Degradation and Development*, *20*(1), 84–91. Available from https://doi.org/10.1002/ldr.893, http://www3.interscience.wiley.com/cgi-bin/fulltext/121583424/PDFSTART, United States.

Zheng, F. L. (2005). Effects of accelerated soil erosion on soil nutrient loss after deforestation on the Loess Plateau. Soil Science Society of China. *China Pedosphere*, *15*(6), 707–715. Available from http://pedosphere.issas.ac.cn.

Zika, M., & Erb, K. H. (2009). The global loss of net primary production resulting from human-induced soil degradation in drylands. *Ecological Economics*, *69*(2), 310–318. Available from https://doi.org/10.1016/j.ecolecon.2009.06.014.

Assessment of riverbank erosion and its prediction using geospatial and machine learning techniques

19

Md Naimur Rahman[1,2,3], Md Mushfiqus Saleheen[4], Hamza EL Fadili[5] and Md Nazirul Islam Sarker[6]

[1]*Department of Geography, Hong Kong Baptist University, Kowloon, Hong Kong, P.R. China*
[2]*David C Lam Institute for East-West Studies, Hong Kong Baptist University, Kowloon, Hong Kong, P.R. China*
[3]*Department of Development Studies, Daffodil International University, Dhaka, Bangladesh*
[4]*Department of Geography and Environmental Science, Begum Rokeya University, Rangpur, Bangladesh*
[5]*Laboratory of Spectroscopy, Molecular Modeling, Materials, Nanomaterials, Water and Environment, Materials for Environment Team, ENSAM, Mohammed V University in Rabat, Rabat, Morocco*
[6]*Miyan Research Institute, International University of Business Agriculture and Technology, Dhaka, Bangladesh*

19.1 Introduction

One of Bangladesh's three principal rivers, the Jamuna, plays a crucial role in the development of the deltaic terrain. However, it is clear that the river is rapidly losing land along its banks and suffering from significant bank erosion, especially along the Brahmaputra–Jamuna River (Khan & Islam, 2003). While this erosion is an inevitable event, anthropogenic actions have disturbed the river's natural geomorphological dynamics. Extraction of sand from mines, construction of structures on riverbanks, artificial cutoffs, and banks revetted artificially, as well as the creation of reservoirs and changes in land use are examples of disturbances to river stability that need adjustment (Gocić et al., 2020). Riverbank erosion remains a serious threat in morphological terms, which requires further research in order to manage and better understand natural processes as well as human activities in order to ensure long-term sustainability for these vital waterways. Bangladesh presents unique challenges when it comes to flooding patterns; therefore, it is crucial that its people observe and comprehend these fluid patterns that

Applications of Geospatial Technology and Modeling for River Basin Management.
DOI: https://doi.org/10.1016/B978-0-443-23890-1.00019-0

vary in terms of frequency, intensity, and depth. Bangladesh experiences annual flood damage of 21% of its total land surface area, which covers an approximate 31,000-km expanse, each year. Assuming an equal distribution of people, 21% are at high risk from flooding events. Flood waters have the ability to affect up to 60% of an entire nation, meaning up to 70 million people could become victims. Three years, in particular 1987, 1998 and 1999, are an eloquent reminder that Bangladesh endured three catastrophic floods that inflicted untold destruction and immense suffering on its people (Dewan, Nishigaki, & Komatsu, 2003; Mirza, 2003). Bangladesh has seen long-term and frequent riverbank erosion over recent years (Malak, Hossain, Quader, Akter, & Islam, 2021), most commonly in areas along its outer edges. Riverbank erosion tends to occur most frequently on banks that lie adjacent to rivers (Malak et al., 2021). Erosion severity depends on the characteristics of materials used on riverbanks (Uwadiae Oyegun, Olanrewaju, & Mark, 2023). At Jamuna River, in-depth assessments were carried out to understand its bank materials' properties—an important step toward developing predictive tools (Uddin, Shrestha, & Alam, 2011). Therefore, it is vitally important to evaluate both historical patterns of spatial-temporal change and future trends of erosion of riverbanks in an area. Satellite remote sensing has proven to be an efficient means of conducting extensive analyses on fluvial channel dynamics across large geographic regions.

Remote sensing via satellite has proven to be extremely valuable in understanding the dynamics of fluvial channels over large geographic areas, as it allows scientists to measure movement patterns of rivers as well as identify braided channels on terraces and identify old, braided streams that have formed. Many studies have employed geospatial techniques, for instance, by superimposing historic channel maps across various river systems to detect significant channel transformations (Desai, Shukla, & Tandon, 2023; Khatun, Rahaman, Garai, Das, & Tiwari, 2022). Bangladesh remote sensing data have proven to be instrumental in monitoring changes to river channels caused by human impact, as well as actions associated with patterns of land use (Halder & Mowla Chowdhury, 2023; Naimur Rahman, Mushfiqus Saleheen, Shozib, & Towfiqul Islam, 2023). Satellite images reveal a disturbing reality: annual river infiltration onto 6700 ha of arable land is impacting 800,000 lives (DMB, 2008). This highlights the vital significance of satellite remote sensing not only in understanding fluvial systems but also in dealing with their major problems. Studies conducted to date on erosion patterns along the Jamuna River have provided insights into its erosion at several locations (Aktar, 2013; Hassan, Ratna, Hassan, & Tamanna, 2017; Pahlowan & Hossain, 2015), but few examined erosion and accretion trends and predictions for Ulipur upazila. To address these gaps, the purpose of this chapter is to investigate temporally and spatially degrading processes on Jamuna River using artificial neural network (ANN) models as predictors. This study unveils erosion patterns not just caused by erosion but also their variations over time. An in-depth investigation of channel dynamics within Jamuna River was carried out using time series data collected via multispectral satellite images. This study presents us with a great opportunity to advance our understanding of predictive analysis models that accurately forecast river morphological features such as channel movement, erosion patterns, sedimentation rates, and channel processes. This research marks a substantial step in our ability to comprehend the geomorphological changes taking place on Jamuna river that is continuously shifting. This has implications for local residents, political leaders, and environmental scientists committed to sustainable river management and disaster preparedness. Focusing on Ulipur upazilas, this study could highlight erosion's effects on granular scale and accumulation processes and offer specific solutions to residents living there. Furthermore, its aim is to provide an in-depth knowledge of riverine dynamics not often covered in studies of fluvial geomorphology.

19.2 The rationale of the study

Bangladesh has already experienced numerous natural disasters that have proven its susceptibility to disasters, with climate change acting as an aggravating factor (Sarker et al., 2022). Bangladesh is plagued annually by natural disasters such as tropical storm surges, cyclones, flooding and coastal erosion, as well as drought (Sarker, Wu, Alam, & Shouse, 2019). Each of these events causes irreparable harm to both property and lives and can impede development activities. Erosion of riverbanks is an ongoing yet silent tragedy that is the leading source of loss in Bangladesh (Islam, 2017; Mamun, Islam, Alam, Chandra Pal, & Alam, 2022). Rising water levels upstream, sudden floods, sedimentation on riverbeds, and unplanned interventions are some of the leading contributors to erosion in riverbanks. Riverbank erosion is one of the most unpredictable annual disasters that threaten households' properties, shelter, and other essential social and physical assets (Tran et al., 2023). As they move into new communities and establish themselves there, these migrants often lose contact with both family members and social networks that existed before moving. Erosion of riverbanks has an enormously detrimental impact on socioeconomic factors, including family shelters and standing crops, income loss, decreased family ties, social isolation, and eventually homelessness in Bangladesh (Siddik, Zaman, Islam, Hridoy, & Akhtar, 2017). Erosion accounts for half of all homelessness cases in Bangladesh. Jamuna River in Bangladesh has experienced rapid bank erosion over recent decades, creating major socioeconomic and environmental concerns for ecosystems and communities. Predictive models with the capability of accurately forecasting trends in erosion and accretion are integral components of long-term planning and building resilience (Saadon, Abdullah, Muhammad, Ariffin, & Julien, 2021), but their use with Jamuna River remains understudied, creating an information void both academics and local authorities need in order to plan effectively and build resilience. Unfortunately, however, their application in relation to Jamuna remains underresearched, creating gaps in both research as well as applications of findings. Early warning programs, community engagement efforts, and innovative erosion control methods are key components of improving river management tactics and methods. Therefore, this study seeks to fill this gap by providing information from research in terms of Jamuna River management strategies and methods. This chapter evaluates the erosion of the Jamuna River and its potential future changes within Ulipur upazila using supervised and unsupervised classification and the ANN model.

19.3 Materials and methods

19.3.1 Study area

As part of studying the Jamuna River, an extraordinary geological event occurred in 1787 that featured both tectonic shifts and flood waters, leading to extraordinary alterations of Brahmaputra River flow and an extraordinary flood event. This alteration resulted in Jamuna river being created as its main route when entering Bangladesh from India. Jamuna river today follows an approximate path starting in Kurigram district before merging with Padma river near Goalundo Ghat in Bangladesh. Moreover, the study location includes Ulipur upazila, which falls under the Kurigram district (Fig. 19.1). Within this defined region, the climate is characterized as tropical monsoon, with the study area experiencing the seasonal dynamics that accompany this climatic pattern.

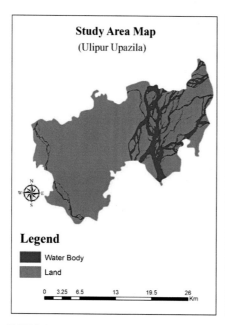

FIGURE 19.1

Study area of Jamuna River in Ulipur Upazila. Map lines delineate study areas and do not necessarily depict accepted national boundaries.

19.3.2 **Data and methods**

Satellite imagery was employed in this study to investigate erosion patterns and predictive modeling. The dataset was sourced from the United States Geological Survey via their Earth Explorer platform (https://earthexplorer.usgs.gov/). The selected temporal range for the satellite images included data from the years 2003, 2013, and 2022. This deliberate choice of a 10-year interval between the years enabled us to effectively capture and visualize the transformative changes that occurred over this period (Fig. 19.2).

It is important to note that at the time of conducting this research, satellite images for the year 2023 were not available for analysis. As a result, the most recent imagery accessible for our investigation was from the year 2022. Additional information pertaining to the satellite imagery employed in this research can be found in Table 19.1.

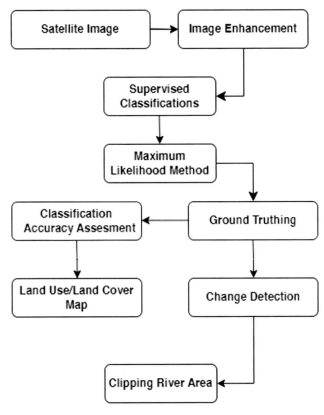

FIGURE 19.2

Methodological flowchart on changes of riverbank erosion.

Table 19.1 Detailed description of satellite imageries.

Area	Year	Landsat	Date of acquisition	Sensor	Path and row
Ulipur area	2003	Landsat-7	November 26, 2003	ETM	138/42
	2013	Landsat-7	November 21, 2013	TM	
	2022	Landsat-8	November 22, 2022	ETM	138/42

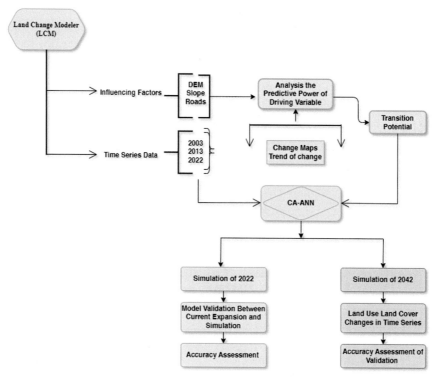

FIGURE 19.3

Methodological flowchart on prediction of riverbank erosion.

Within this study, we utilized both supervised and unsupervised image classification techniques to methodically monitor riverbank erosion over the period from 2003 to 2022. This comprehensive analysis facilitated the projection of future erosion dynamics, primarily relying on the predictive capabilities of an ANN, with a specific focus on the year 2042 (Fig. 19.3).

The basis for our predictive model was drawn from the classified imagery of 2003 and 2022. Subsequently, geospatial analysis was executed using ArcGIS 10.8, allowing us to quantify the extent of riverbank erosion over the specified period. To anticipate the forthcoming spatiotemporal changes in riverbank erosion, QGIS 2.18.15 software was employed, enabling a holistic and forward-looking assessment of this critical environmental phenomenon.

19.4 **Results and discussion**

19.4.1 **Major findings**

Table 19.2 illustrates a decline in the total river area, with the most significant reduction projected for 2042 at 5433.82 ha. Nevertheless, it is noteworthy that an increase was observed in 2013, amounting to 7083 ha, as compared to the initial measurement of 6038 ha in 2003.

Table 19.3 and Fig. 19.4 present a detailed account of the spatiotemporal changes in erosion activity observed along the Jamuna river in the Ulipur upazila, spanning different time intervals from 2003 to 2042. Therefore, the data presented in Tables 19.2 and 19.3, along with the corresponding figures, provide a nuanced understanding of the geomorphological changes occurring along the Jamuna River in Ulipur Upazila. These changes are crucial in understanding the river's dynamics and the implications for local land use and conservation efforts. Table 21.3 categorizes the land area into three distinct classifications, unchanged area, erosion, and accretion, with each category denoting the area in hectares (ha) that corresponds to these classifications. The unchanged area category provides insights into the land that remains consistent along the riverbanks. In the decade from 2003 to 2013, this area amounted to 2937.38 ha. Nonetheless, a noticeable decrease in the area that remained unchanged can be observed in the following timeframe, spanning from 2013 to 2022, with a landmass of 2850.76 ha. Most significantly, from 2022 to 2042, the unchanged area saw a substantial decrease to 2218.20 ha (Fig. 19.5). This diminishing trend may suggest an escalated rate of

Table 19.2 Total area of river channel of Jamuna River in Ulipur Upazila.

2003	2013	2022	2042
6038.9	7083.24	5500.82	5433.82

Table 19.3 Temporal changes of erosion activity of Jamuna river in Ulipur upazila (area in ha).

	2003_13	2013_22	2003_22	2022_42
Unchanged area	2937.381186	2850.761006	2218.203038	5368.2178
Erosion	3101.527928	4232.483488	3820.706076	132.603805
Accretion	4145.863309	2650.060599	3282.618568	65.599109

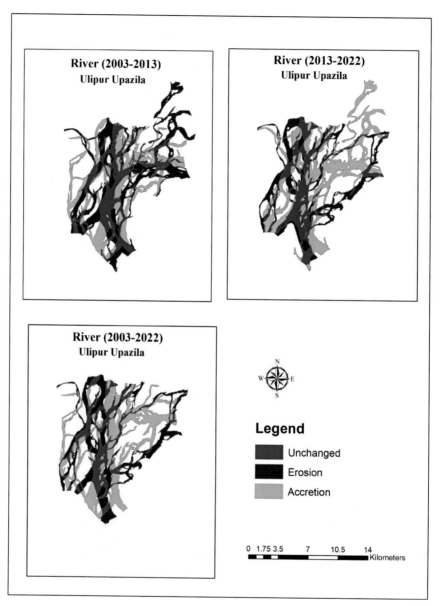

FIGURE 19.4

Spatial pattern of erosion activity of Jamuna river in Ulipur upazila.

erosion and alterations in the region's geomorphology over time. Hence, Table 19.3 and Fig. 19.4, which detail the spatiotemporal changes in erosion, reveal a fluctuating pattern of land stability and loss. The decrease in unchanged

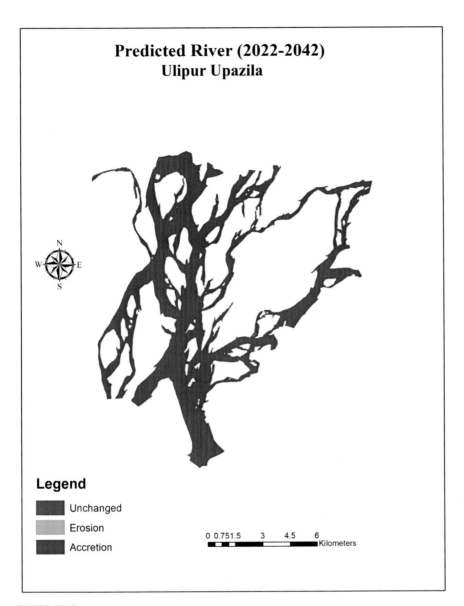

FIGURE 19.5

Predicted spatial pattern of erosion activity of Jamuna river in Ulipur upazila.

area over the years points to an increasing rate of change along the riverbanks, highlighting the need for adaptive management strategies to mitigate the effects of erosion (Tsakiris & Loucks, 2023). The sharp increase in erosion between 2013 and 2022, followed by a drastic reduction, underscores the variable nature of erosion processes, which could be linked to episodic events or changes in river management practices.

Erosion, as a crucial factor shaping the riverbanks, exhibited distinct patterns. Between 2003 and 2013, the area subject to erosion was measured at 3101.53 ha. This erosion activity expanded significantly in the subsequent decade, from 2013 to 2022, with an affected area of 4232.48 ha. Intriguingly, there was a marked reduction in erosion from 2022 to 2042, with only 132.60 ha impacted. The sharp increase in erosion from 2013 to 2022 is a notable concern, potentially pointing to an increased vulnerability of the land to erosional forces. Shifts may influence the subsequent decline in erosion in river dynamics, land management practices, or changes in climate conditions (Debnath et al., 2023; Rowland et al., 2023). **Conversely**, accretion, the process of sediment buildup along riverbanks, displayed noteworthy variations. The spatial pattern includes the highest accretion that happened in the northeast part of the study area. The accretion area was increased by 4145.86 ha from 2003 to 2023. From 2013 to 2022, it saw a remarkable surge in accretion activities, and it is projected to reach 3282.62 ha in 2042 (Fig. 19.5 and Table 19.2). The decrease between 2013 and 2022 may suggest less accumulation of land or sediment on riverbanks, while its subsequent increase may be linked to shifts in river dynamics, shifting sediment transport patterns, or land management practices that aim to promote growth (Hasanuzzaman, Bera, Islam, & Shit, 2023; Langhorst & Pavelsky, 2023). Therefore, this study offers an important and thorough examination of the interactions among erosion, accretion, and other geomorphological processes occurring along the Jamuna River in Ulipur Upazila. The findings underscored the complex and ever-evolving character of geomorphological processes, thus highlighting their ongoing monitoring and investigation to understand and manage them for the benefit of local community and environment. Exploring the intricate dynamics of riverbank erosion and accretion along the Jamuna River, this study was designed with specific goals in mind to analyze past patterns while anticipating future ones. Our analyses directly support these goals, yielding knowledge that not only meets but surpasses initial goals.

19.4.2 Assess historical spatiotemporal patterns of erosion and accumulation

The findings of this study highlight an increase in riverbank erosion over the study period (2003−22), which was accompanied by significant spatial variation (Malak et al., 2021; Uddin et al., 2011). The initial objective was to perform an exhaustive analysis of patterns of accumulation and erosion for over 20 years using machine learning and geospatial technologies. With these methods, we identified the patterns with incredible precision by highlighting areas with significant loss or accumulation. Comparing these results to global landscape, such as that of Mississippi and Yellow Rivers (Tomsett & Leyland, 2019), provides an accurate depiction. Xu et al. (2020) discussed issues associated with managing river dynamics that are affected by human and natural factors, validating our methodology, while showing why universal erosion management strategies must be reconsidered in an innovative fashion.

19.4.3 Prediction of future changes in accumulation and erosion patterns

ANN methods of modeling were employed in this research to accurately predict where changes may take place in order to educate environmental scientists, decision-makers, and local communities on sustainable river management practices (Ahmed et al., 2024). Furthermore, scientifically backed approaches provided to communities will assist them in adjusting or decreasing agricultural systems' impacts on livelihoods as well as biodiversity (Rentier & Cammeraat, 2022). Identification and forecast of factors contributing to erosion enables researchers to create measures designed to protect ecosystems, rivers, and nearby communities against further erosive forces. Comparing erosion patterns along the Jamuna River with those from other major river systems allows us to contextualize our findings in a wider geomorphological framework and more thoroughly study patterns of erosion. Geospatial and machine learning methodologies were employed in our research to explore how global processes could be managed effectively, offering Jamuna River residents new perspectives. Geospatial techniques and machine learning were an integral component to meeting our goal of employing modern methods to study and predict erosion (Nguyen, Chen, Lin, & Seeboonruang, 2021), setting the bar high for future researchers. Our erosion research on riverbanks set this bar very high. All research was intended to increase knowledge about accumulation and erosion processes on riverbanks from both local and global perspectives. Our findings not only met the stated objectives, but also sparked global discussions on sustainable river management by emphasizing technological advancement's significant role in environmental research (Tomsett & Leyland, 2019). With its innovative methodology, this study offers both theoretical and practical information to aid river management professionals worldwide. Concerning our research findings on Jamuna River erosion patterns, it is crucial that they are set in their proper geographical and scientific context. While decreasing erosion rates and anticipated riverbank stabilization by 2042 are positive signs for geomorphology within an area, they also raise serious concerns regarding current practices for water management due to their inability to respond quickly enough to changes that impact the environment. Examining Jamuna river for this study confirmed general findings and demonstrated its resilience and adaptability systems. Effective management requires understanding global as well as local influences. As one example, expected reductions in erosion rates, accumulation rates, erosion rates, and natural drainage rates along the Jamuna by 2042 may be attributable to positive local actions, such as embankments being built or land use laws being changed, climate change-driven hydrological patterns shifting, as well as changes triggered by long-term sustainability issues arising from climate change requiring further investigation or shifts occurring over time. Research done here goes well beyond geomorphological changes; erosion reduction could provide immediate socioeconomic advantages to communities living along the Jamuna River, including better agricultural methods, lower displacement rates, and greater water security akin to that seen on European Rhine River (Siddik et al., 2017). Comparing our results to others across various locations reveals that although river management is a worldwide challenge, its solutions must take account of regional conditions (Ahmed et al., 2024; Mallick, Bandyopadhyay, & Halder, 2023; Vianna et al., 2020). Utilizing machine learning and geospatial methodologies as discussed during our research provides us with an opportunity to develop predictive models, which could assist with effective river-management strategy.

19.5 Limitations of the study

One major drawback of this research is its reliance on satellite images from 2003 to 2022 only; there is an absence of imagery for the year 2023. This gap in data may introduce some uncertainty, particularly when attempting to project future erosion trends to 2042. The absence of 2023 data may hinder the accuracy of predictions for that year.

19.6 Recommendations and practical implications

19.6.1 Advanced monitoring systems for early warning

This research seeks to facilitate the creation of complex monitoring alarm and warning systems using predictive models. These systems alert local authorities and communities about imminent erosion events or significant river dynamics changes which require evacuation or strengthening riverbanks; it quickly transforms research conducted to assess risks into practical tools to increase community security as well as emergency measures.

19.6.2 Community engagement and education programs

Educational and community outreach initiatives must focus on fostering sustainable land usage and protecting riverbanks. To do so, these efforts could include disseminating information regarding erosion causes and effects as well as natural barriers such as vegetation covers to combat it actively by residents themselves. Such an approach fosters both an awareness of their surroundings as well as strengthening individual abilities to adapt and rebuild them more quickly.

19.6.3 **Innovative erosion control technologies**

Modern erosion control technologies, such as bioengineering techniques, should be adopted to combine conventional engineering techniques with natural processes. Bioengineering methods like installing vegetative buffers, building living shorelines, or using geotextiles made from natural materials may offer alternatives to rigid engineering solutions, while simultaneously stabilizing riverbanks, increasing biodiversity, and creating habitats for wildlife in the area.

19.6.4 **Water resource management and flood mitigation strategies**

The findings of this study will help to develop more complete strategies for water resource management and flood prevention. Gaining an understanding of what causes erosion of riverbanks can improve floodplain management efficiency and facilitate the construction of barriers that work in harmony with natural processes that govern river flow, such as creating barriers that resemble natural floodplain barriers in harmony with their flow pattern. Taking this comprehensive approach ensures that any methods used to control flooding will not inadvertently increase erosion or interfere with ecosystem services that protect biodiversity.

19.6.5 **Policy formulation and regulatory frameworks**

Policy formulation and the creation of regulations will be facilitated with the intention of decreasing human activities' effects on riverbank erosion, such as restrictions on mining sand, developing adjacent to riverbanks, or agricultural operations that threaten stabilization efforts. The findings of this study could also be used to create guidelines requiring erosion prediction models in development planning procedures, thus ensuring decisions made regarding land use reflect the long-term effects of river dynamics.

19.6.6 **Cross-sectoral collaboration for integrated river basin management**

Cross-sectoral collaboration must be facilitated for the implementation of an integrated river basin management method. By gathering stakeholders from different industries, such as urban development, agriculture, water resources management, and environment, to work on coordinated management methods of managing riverbanks, using water efficiently as well as developing land in the basin, this collaboration model ensures that erosion management efforts are consistent with larger socioeconomic and environmental goals.

19.7 Conclusions

This research explores the complex spatiotemporal dynamics of erosion along the Jamuna River, in an effort to provide guidance for future actions. This study employs geospatial methods to assess erosion patterns, while using an ANN model to predict future events. Many interesting findings result from this research. As predicted for 2042, an apparent expansion of the Jamuna River's spatial region can be expected due to the shifting geomorphological environment that characterizes it. Study findings indicate that between 2003 and 2022 was the peak period of Jamuna River erosion in its northeastern region. Erosion activity was recorded across various directions, with the greatest frequency occurring between 2013 and 2022 when covering over 4232.48 ha. By 2042, it is expected that both erosion rates and accretion rates will have dramatically decreased to reach only 65% of what they were in 2003. These findings demonstrate the complex and dynamic patterns of accretion and erosion along the Jamuna River, making significant contributions toward our understanding of this critical environmental phenomenon. Findings, particularly related to areas not experiencing spatial change, highlight the necessity of continuously monitoring river dynamics and exercising effective management controls on them in order to ensure sustainable development of this vital watercourse.

References

Ahmed, I. A., Talukdar, S., Baig, M. R. I., Shahfahad., Ramana, G. V., & Rahman, A. (2024). Quantifying soil erosion and influential factors in Guwahati's urban watershed using statistical analysis, machine and deep learning. *Remote Sensing Applications: Society and Environment, 33*. Available from https://doi.org/10.1016/j.rsase.2023.101088, http://www.journals.elsevier.com/remote-sensing-applications-society-and-environment/.

Aktar, M. N. (2013). Historical trend of riverbank erosion along the braided River Jamuna. *International Journal of Sciences: Basic and Applied Research, 11*, 173−180.

Debnath, J., Sahariah, D., Lahon, D., Nath, N., Chand, K., Meraj, G., ... Farooq, M. (2023). Assessing the impacts of current and future changes of the planforms of river Brahmaputra on its land use-land cover. *Geoscience Frontiers, 14*(4), 101557. Available from https://doi.org/10.1016/j.gsf.2023.101557.

Desai, N. P., Shukla, S. H., & Tandon, A. (2023). Exploring geomorphic features in lower sabarmati river basin as signatures of palaeoenvironmental change, Gujarat: A remote sensing approach. *Journal of the Geological Society of India, 99*(2), 233−238. Available from https://doi.org/10.1007/s12594-023-2290-4, https://www.springer.com/journal/12594.

Dewan, A. M., Nishigaki, M., & Komatsu, M. (2003). Floods in Bangladesh: A comparative hydrological investigation on two catastrophic events. *Journal of Faculty of Environmental Science and Technology, 8*, 53−62.

Gocić, M., Dragićević, S., Radivojević, A., Martić Bursać, N., Stričević, L., & Đorđević, M. (2020). Changes in soil erosion intensity caused by land use and demographic changes in the Jablanica River Basin, Serbia. *Agriculture*, *10*(8), 345. Available from https://doi.org/10.3390/agriculture10080345.

Halder, A., & Mowla Chowdhury, R. (2023). Evaluation of the river Padma morphological transition in the central Bangladesh using GIS and remote sensing techniques. *International Journal of River Basin Management*, *21*(1), 21−35. Available from https://doi.org/10.1080/15715124.2021.1879095, http://www.tandfonline.com/loi/trbm20.

Hasanuzzaman, M., Bera, B., Islam, A., & Shit, P. K. (2023). Estimation and prediction of riverbank erosion and accretion rate using DSAS, BEHI, and REBVI models: Evidence from the lower Ganga River in India. *Natural Hazards*, *118*(2), 1163−1190. Available from https://doi.org/10.1007/s11069-023-06044-4, https://www.springer.com/journal/11069.

Hassan, M. A., Ratna, S. J., Hassan, M., & Tamanna, S. (2017). Remote sensing and GIS for the spatio-temporal change analysis of the east and the west river bank erosion and accretion of Jamuna River (1995−2015), Bangladesh. *Journal of Geoscience and Environment Protection*, *05*(09), 79−92. Available from https://doi.org/10.4236/gep.2017.59006.

Islam, M. N. (2017). Community-based responses to flood and river erosion hazards in the active Ganges floodplain of Bangladesh. *Science and technology in disaster risk reduction in Asia: Potentials and challenges* (pp. 301−325). Elsevier. Available from http://www.sciencedirect.com/science/book/9780128127117, 10.1016/B978-0-12-812711-7.00018-3.

Khan, N. I., & Islam, A. (2003). Quantification of erosion patterns in the Brahmaputra-Jamuna River using geographical information system and remote sensing techniques. *Hydrological Processes*, *17*(5), 959−966. Available from https://doi.org/10.1002/hyp.1173.

Khatun, M., Rahaman, S. M., Garai, S., Das, P., & Tiwari, S. (2022). Assessing river bank erosion in the Ganges using remote sensing and GIS. *Advances in geographic information science* (pp. 499−512). Springer Science and Business Media Deutschland GmbH. Available from http://link.springer.com/bookseries/7712, 10.1007/978-3-030-75197-5_22.

Langhorst, T., & Pavelsky, T. (2023). Global observations of riverbank erosion and accretion from Landsat imagery. *Journal of Geophysical Research: Earth Surface*, *128*(2). Available from https://doi.org/10.1029/2022JF006774, http://agupubs.onlinelibrary.wiley.com/agu/jgr/journal/10.1002/(ISSN)2169-9011/.

Siddik, Md. A., Zaman, A. K. M. M., Islam, Md. R., Hridoy, S. K., & Akhtar, Mt. P. (2017). Socio-economic impacts of river bank erosion: A case study on coastal island of Bangladesh. *The Journal of NOAMI*, *34*, 73−84. Available from https://doi.org/10.1046/j.1365-3091.2000.00290.x, https://doi.org/10.1046/j.1365-3091.2000.00290.x.

Malak, M. A., Hossain, N. J., Quader, M. A., Akter, T., & Islam, M. N. (2021). Climate change-induced natural hazard: Population displacement, settlement relocation, and livelihood change due to riverbank erosion in Bangladesh. *Springer climate* (pp. 193−210). Springer Science and Business Media B.V. Available from https://rd.springer.com/bookseries/11741, 10.1007/978-3-030-71950-0_6.

Mallick, R. H., Bandyopadhyay, J., & Halder, B. (2023). Impact assessment of river bank erosion in the lower part of Mahanadi River using geospatial sciences. *Sustainable Horizons*, *8*, 100075. Available from https://doi.org/10.1016/j.horiz.2023.100075.

Mamun, A. A., Islam, A. R. M. T., Alam, E., Chandra Pal, S., & Alam, G. M. M. (2022). Assessing riverbank erosion and livelihood resilience using traditional approaches in northern Bangladesh. *Sustainability*, *14*(4), 2348. Available from https://doi.org/10.3390/su14042348.

Mirza, M. M. Q. (2003). Three recent extreme floods in Bangladesh: A hydrometeorological analysis. *Natural Hazards*, *28*(1), 35−64. Available from https://doi.org/10.1023/A:1021169731325.

Naimur Rahman, M., Mushfiqus Saleheen, M., Shozib, S. H., & Towfiqul Islam, A. R. M. (2023). *Monitoring and prediction of spatiotemporal land-use/land-cover change using Markov chain cellular automata model in Barisal, Bangladesh* (pp. 113−124). Springer Science and Business Media LLC. Available from https://doi.org/10.1007/978-3-031-21587-2_8.

DMB. (2008). *National plan for disaster management*. DMB.

Nguyen, K. A., Chen, W., Lin, B. S., & Seeboonruang, U. (2021). Comparison of ensemble machine learning methods for soil erosion pin measurements. *ISPRS International Journal of Geo-Information*, *10*(1). Available from https://doi.org/10.3390/ijgi10010042, http://www.mdpi.com/journal/ijgi.

Pahlowan, E. U., & Hossain, A. T. M. S. (2015). Jamuna river erosional hazards, accretion & annual water discharge—A remote sensing & GIS approach. *International Archives of the Photogrammetry, Remote Sensing and Spatial Information Sciences − ISPRS Archives*, *40*(7), 831−835. Available from https://doi.org/10.5194/isprsarchives-XL-7-W3-831-2015, http://www.isprs.org/proceedings/XXXVIII/4-W15/.

Rentier, E. S., & Cammeraat, L. H. (2022). The environmental impacts of river sand mining. *Science of the Total Environment*, *838*, 155877. Available from https://doi.org/10.1016/j.scitotenv.2022.155877.

Rowland, J. C., Schwenk, J. P., Shelef, E., Muss, J., Ahrens, D., Stauffer, S., ... Vulis, L. (2023). Scale-dependent influence of permafrost on riverbank erosion rates. *Journal of Geophysical Research: Earth Surface*, *128*(7). Available from https://doi.org/10.1029/2023JF007101, http://agupubs.onlinelibrary.wiley.com/agu/jgr/journal/10.1002/(ISSN)2169-9011/.

Saadon, A., Abdullah, J., Muhammad, N. S., Ariffin, J., & Julien, P. Y. (2021). Predictive models for the estimation of riverbank erosion rates. *CATENA*, *196*, 104917. Available from https://doi.org/10.1016/j.catena.2020.104917.

Sarker, M. N. I., Alam, G. M. M., Firdaus, R. B. R., Biswas, J. C., Islam, A. R. M. T., Raihan, M. L., ... Shaw, R. (2022). Assessment of flood vulnerability of riverine island community using a composite flood vulnerability index. *International Journal of Disaster Risk Reduction*, *82*. Available from https://doi.org/10.1016/j.ijdrr.2022.103306, http://www.journals.elsevier.com/international-journal-of-disaster-risk-reduction/.

Sarker, M. N. I., Wu, M., Alam, G. M. M., & Shouse, R. C. (2019). Livelihood vulnerability of riverine-island dwellers in the face of natural disasters in Bangladesh. *Sustainability (Switzerland)*, *11*(6). Available from https://doi.org/10.3390/su11061623, https://res.mdpi.com/sustainability/sustainability-11-01623/article_deploy/sustainability-11-01623.pdf.

Tomsett, C., & Leyland, J. (2019). Remote sensing of river corridors: A review of current trends and future directions. *River Research and Applications*, *35*(7), 779−803. Available from https://doi.org/10.1002/rra.3479, http://onlinelibrary.wiley.com/journal/10.1002/(ISSN)1535-1467.

Tran, D. D., Thien, N. D., Yuen, K. W., Lau, R. Y. S., Wang, J., & Park, E. (2023). Uncovering the lack of awareness of sand mining impacts on riverbank erosion among Mekong Delta residents: insights from a comprehensive survey. *Scientific Reports*, *13*(1). Available from https://doi.org/10.1038/s41598-023-43114-w, https://www.nature.com/srep/.

Tsakiris, G. P., & Loucks, D. P. (2023). Adaptive water resources management under climate change: An introduction. *Water Resources Management*, *37*(6−7), 2221−2233. Available from https://doi.org/10.1007/s11269-023-03518-9, https://www.springer.com/journal/11269.

Uddin, K., Shrestha, B., & Alam, M.S. (2011). Assessment of morphological changes and vulnerability of river bank erosion alongside the river Jamuna using remote sensing. In *Gi4DM 2011 — GeoInformation for disaster management.*

Uwadiae Oyegun, C., Olanrewaju, L., & Mark, O. (2023). Erosion and accretion along the coastal zone of Nigeria. *World Geomorphological Landscapes*, 243−252. Available from https://doi.org/10.1007/978-3-031-17972-3_17, https://www.springer.com/series/10852.

Vianna, V. F., Fleury, M. P., Menezes, G. B., Coelho, A. T., Bueno, C., da Silva, J. L., & Luz, M. P. (2020). Bioengineering techniques adopted for controlling riverbanks' superficial erosion of the Simplício Hydroelectric Power Plant, Brazil. *Sustainability (Switzerland)*, *12*(19). Available from https://doi.org/10.3390/SU12197886, https://res.mdpi.com/d_attachment/sustainability/sustainability-12-07886/article_deploy/sustainability-12-07886.pdf.

Xu, X., Shrestha, S., Gilani, H., Gumma, M. K., Siddiqui, B. N., & Jain, A. K. (2020). Dynamics and drivers of land use and land cover changes in Bangladesh. *Regional Environmental Change*, *20*(2). Available from https://doi.org/10.1007/s10113-020-01650-5, http://springerlink.metapress.com/app/home/journal.asp?wasp = 64cr5a4mwl drxj984xaw&referrer = parent&backto = browsepublicationsresults,451,542.

Environment and sustainability III

Impact of soil erosion and resultant sediment yield on different stream orders

20

Kaushik Ghosal[1] and Santasmita Das Bhattacharya[2]

[1]*Department of Mining Engineering, Indian Institute of Engineering Science and Technology Shibpur, Howrah, West Bengal, India*
[2]*Department of Geography, Amity Institute of Social Sciences, Amity University, Kolkata, West Bengal, India*

20.1 Introduction

The progressive decrease in a reservoir's storage capacity showed that its life is coming to an end. This storage capacity is diminishing mostly as a result of the massive deposition of silt in it. This massive sedimentation is determined by the sediment production of a river's watershed region located upstream of a reservoir. This sediment accumulation is caused by the huge degraded soil of that watershed region (Ouadja et al., 2021). To calculate the reservoir's life, all characteristics of soil erosion and the quantity of ease of transport of that eroded topsoil in that watershed region must be assessed and estimated. Many scholars have examined soil erosion, sediment deposition, and sedimentation in different reservoirs for its practical value since then. Soil is considered one of the most important resources for sustaining a living. The health and viability of soil nutrients and the ecosystems they support are critical to a state's financial, ecological, and social sustainability. However, natural processes (such as water, wind, tillage, and so on) accelerate the loss of topsoil from the ground surface, resulting in soil erosion and massive soil loss. Worldwide water erosion rates recorded over the previous decade range from around 20 to over 200 Gt/year (FAO & ITPS, 2015). Estimating soil erosion has been a global problem for several decades. Previously, these data were based on direct field measurements. There are various field-based measures that estimate the change in the surface level of a certain landmass or the enormous quantity of lost soil. Due to the expense and time involved with the broad geographical extent, these field-based measurements make estimating soil loss in a big region difficult (Stefanidis, Alexandridis, Chatzichristaki, & Stefanidis, 2021). In contrast to in situ observation, remote sensing allows for the capture of information about an item or phenomena without establishing physical contact with the thing. Soil erosion measures nowadays mostly rely on geographic information system (GIS) and remote sensing-based modeling tools (Kashiwar, Kundu, & Dongarwar, 2022; Stefanidis,

Applications of Geospatial Technology and Modeling for River Basin Management.
DOI: https://doi.org/10.1016/B978-0-443-23890-1.00020-7

Alexandridis, & Ghosal, 2022). Various GIS and remote sensing-based models have been developed to assess soil erosion caused by water activity. The revised universal soil loss equation (RUSLE) is an empirical soil erosion estimation model that is used globally (i.e., in any climatic zone) and has five basic variables for soil loss assessment (rainfall erosivity factor, soil erodibility factor, slope length and stability factor, cover management factor, and conservation practice factor). RUSLE is a well-known model that has been used to assess soil erosion in the upper Bakreshwar river basin.

As a result of the tremendous soil erosion, massive silt has accumulated in the riverbed, primarily at the bottom of the reservoirs downstream of its river. Sedimentation is the accumulation of particles in a reservoir's bed. The hydroelectric plant's safety system will be jeopardized due to excessive silt in a reservoir. Sedimentation even has a negative socioeconomic impact. Sedimentation reduces the reservoir's water carrying capacity, which has a substantial influence on hydroelectric power generation, industrial, and irrigational water use. As a result, the assessment of sedimentation below the reservoir bed as a result of enormous soil erosion is in great demand. Several approaches for estimating sedimentation have been developed, including field-based hydrological surveys, the use of radioactive tracers, and so on. However, due to the significant time and expense requirements of these field-based methodologies, conventional GIS and remote sensing-based models are created. The most widely used GIS and remote sensing-based sediment yield measurement methods are modified universal soil loss equation, Pan-European Soil Erosion Risk Assessment, Normalized Difference Water Index, and sediment delivery ratio (SDR). All other approaches, with the exception of SDR, do not address the topic of sediment transportability inside a meaningful way. In SDR modeling methodologies, various researchers address different characteristics responsible for sediment movement. SDR is exclusively related to the slope of the stream channel by Williams and Berndt (1972). Ferro and Minacapilli (1995) presented a more generic strategy using journey time as the major function. According to Ferro and Minacapilli (1995), the transportability of sediment is determined by the length of the streamflow channel, the kind of landcover available in the surrounding region of flow, as well as the slope of the course. Kothyari and Jain (1997) provided a more complicated strategy for dividing a catchment into time-area segments. Diodato and Grauso (2009) provided a model on hydromorphological functions as well, although data redundancy is possible. Therefore, among all these prior models, the journey time method to determining SDR is the most significant since it is affected by the slope, vegetation, and flow length of a stream. The sediment yield created under the reservoir's bottom is calculated by multiplying the quantity of soil eroded by the quantity of eroded soil carried (i.e., SDR) because the SDR is defined as the ratio of sediment yield to eroded soil. The sediment yield produced by this technology represents the amount of sediment created under the reservoir. Kolli, Opp, and Groll (2021) investigated soil erosion and sediment output in India's Kolleru Lake basin. Kolli et al. (2021) observed the SDR strongly related with the watershed drainage system. Gelagay (2016) investigated soil erosion in the Koga watershed and discovered that stream orders had a significant impact upon this SDR. Gelagay (2016), on the other hand, did not estimate the quantity of sediment output across various stream orders and land use classes. The Bakreshwar river runs seasonally. The dense vegetation cover in forest is not prominent. Agricultural methods are likewise not year-round. As a result, we confront two sorts of issues in this situation. For starters, soil erosion exacerbates the previously listed issues. The second issue is a lack of available water during dry seasons. As a result, our research attempted to determine the pattern of soil erosion, which changes owing to different land use. This research addressed the issue of water availability by determining the pace and amount of silt buildup in the study area's stream channels and the sole accessible reservoir. This study assessed soil erosion considering rainfall, existing soil distribution, topography, plant cover distribution, and land use characteristics. Soil erosion's impact and dispersion across different land use groups have been examined. The obtained sediment yield produced by this soil erosion over various land use categories has been identified. The study's major emphasis is the dispersion of silt buildup across different stream orders. Soil erosion and sediment yield in the Bakreshwar river basin must be assessed since the Bakreshwar dam holds water for irrigation and thermal power generation at the Bakreshwar power station. Several researchers have studied on soil erosion assessment of Bakreshwar river basin for the gradual increasing concern of this region. Ghosh, Mukhopadhyay, and Pal (2015) correlated the runoff of stream with the soil erosion of Bakreshwar river basin. However, this study got very poor correlation among runoff with soil erosion and have not estimated sedimentation in this river basin. Ghosal and Bhattacharya (2021) identified the relationship between the C factor with the land surface temperature (LST) of the upper Bakreshwar river basin. However, there was there was a lack of research on soil erosion-based sediment impact on the Bakreshwar river basin. Our research attempted to identify the types of soil erosion that occur on different land use land covers of the upper Bakreshwar river basin. This study addressed the issue of soil erosion and the rate of sediment deposition along the streams of Upper Bakreshwar basin and in the Bakreshwar reservoir.

20.1.1 Objectives of the study

1. To estimate the average annual soil loss of the upper Bakreshwar river basin.
2. To estimate the SDR of the study area
3. To estimate the sediment yield of the upper Bakreshwar river basin along its different stream orders.

20.2 Description of the study area

The upper Bakreshwar river basin (see Fig. 20.1) stretches between (87°17′15″E, 23°55′55″N) and (87°27′0″E, 23°48′35″N), sharing a brief boundary with the

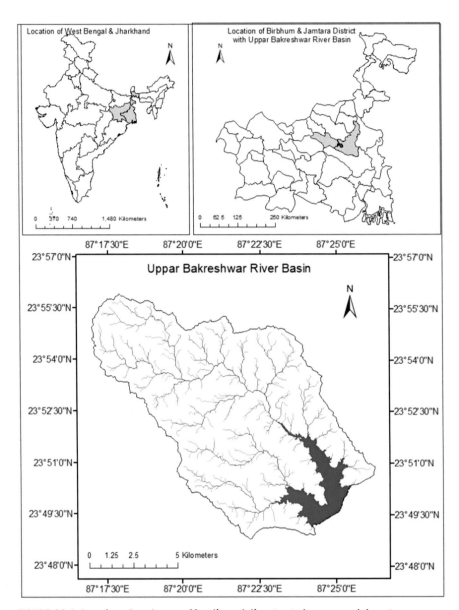

FIGURE 20.1 Location of study area. Map lines delineate study areas and do not necessarily depict accepted national boundaries.

Jamtara, Jharkhand, and occupying Dubrajpur and Rajnagar block of Birbhum, West Bengal. The Bakreshwar River is a fifth tributary (Ghosh et al., 2015) of the Kuya River and a component of the Mayurakshi river system that flows through the Rarh area of Eastern India. The Bakreshwar River began in the Jharkhand district of Jamtara and flows through the Rajnagar and Dubrajpur blocks in the Birbhum, West Bengal. The upper catchment of the Bakreshwar River ends at the Bakreshwar Dam (Nil Nirjane). The upper Bakreshwar River watershed ranges in elevation from 145 to 61 m above the sea level. This research area has a total area of 124.82 km^2. The upper catchment region of the Bakreshwar River is located in the Humid Subtropical climate zone, which has hot summers and moderate winters. The majority of summer rains fall during thunderstorms, which form as a result of the intense surface heating as well as the high sun angle of subtropical climatic zone. The yearly precipitation in this area is between 100 and 115 cm. The average yearly temperature in this region ranges from 10°C to 40°C. The major soil type of this region, which is mostly adjacent to the Santalpargana district, is virtually lateritic with visible granite veins in parts, high medium texture topsoil, and undulated terrain. Along with the clay layer, the old alluvial is discovered. The major soil type in the studied region belongs to the soil series of low dissected plateau. This substantial soil series is represented by four soil classes in this area. There are four of them: loam, sandy clay loam, sandy loam, and silty clay. Alluvial deposits, granite, and weathered rock are the principal soil parent materials in this area. Soil's dominant colors are red and reddish brown. This area's vegetation cover is mostly made up of two types: forested land and cultivated land. In the forested land, species like *Madhuca indica* (Mahua), *Shorea robusta* (Sal), *Butea monosperma* (Palash), *Dalbergia sissoo* (Sissoo), *Agave* spp. (Sisal), *Syzygium cumini* (Jam), *Terminalia arjuna* (Arjun), *Borassus flabellifer* (Palm), and *Phoenix sylvestris* (Date palm) are present.

20.3 **Materials and methods**

The RUSLE model was used to estimate soil erosion. RUSLE is an empirical model that is used all over the world to measure rainfall-induced soil loss from a unit area on an annual time period (Jaiswal & Amin, 2020). Wischmeier and Smith published the USLE in USDA Agricultural Manual 282 in 1965. Renard and others changed the USLE in 1997 to make the factors more precise. The RUSLE was introduced as a refinement of the USLE (as stated by cooper) (Benkobi, Trlica, & Smith, 1994; Cooper, 2011). The rainfall erosivity factor (R), which integrates the kinetic energy created by the impact of rainfall, is one of several elements that have a direct influence on annual soil erosion. Secondly, the soil erodibility factor (K) represents soil strength, indicating the resistance provided by the soil. The third is the slope length factor (L) and slope stiffness factor (S), which both represent the topography of the region, which has a substantial impact on soil erosion. The fourth variable is the cover management factor (C), which depicts the effect of plant cover on soil erosion, and the last parameter is the conservation practice factor (P), which depicts the

Table 20.1 Description of the collected data for this research.

Type of data	Year	Name	Data extracted	Source
Toposheet	2003	73 M/5 Scale- 1:50,000	Contour	Survey of India (SOI), Kolkata
Rainfall data	2020	Rainfall data 4 rain gage stations	Annual average rainfall depth	Indian Meteorological Department (IMD), Pune
Soil data	2007	Soil texture scale- 1:50,000	Grain size distribution, organic content, structural code, and permeability	National Bureau of Soil Survey And Land Use Planning Regional Center, Kolkata
Multispectral satellite data	2020	IRS P6, LISS-4 Resolution- 5m	Three bands (green, red, and infrared)	National Remote Sensing Centre, Hyderabad, Telangana

effect of different human behaviors that have some impact on soil loss (Ghosal & Das Bhattacharya, 2020). Table 20.1 shows the specifics of the data acquired for estimating RUSLE-based soil loss and sediment output. Fig. 20.2 depicts the methods of this investigation.

The methodology of the study has been depicted in Fig. 20.2.

20.3.1 Estimation of rainfall erosivity factor (R) of the upper Bakreshwar river basin

The R factor assesses the influence of kinetic energy caused by rainfall and indicates the amount and rate of runoff that is anticipated to be associated with the rainfall. The R factor fluctuates based on the climatic circumstances. The eastern section of India (Jharkhand, West Bengal, and so on) is predominantly in the humid subtropical climate zone. Several rainfall erosivity factor models have been produced for research in the world's humid subtropical climatic zones (Ghosal & Das Bhattacharya, 2020); however the R factor published by the Korea Institute of Construction Technology (1992) is mostly employed in the eastern region of India (Eq. 20.1).

$$R = 38.5 + 0.35r \tag{20.1}$$

where r denotes the total yearly rainfall in millimeters. R represents the long-term yearly average of the combination of precipitation kinetic energy events expressed MJ mm/ha/year. The approach for calculating the R factor (Fig. 20.3A) was

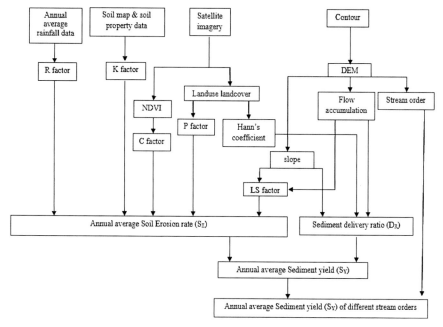

FIGURE 20.2 Methodology flowchart

established in accordance with the local meteorological circumstances (Ghosal & Das Bhattacharya, 2020).

20.3.2 Estimation of soil erodibility factor (*K*) of the upper Bakreshwar river basin

Soil erodibility (*K*) is a measure of the susceptibility of the topsoil to erosion, sediment ease of transport, as well as the amount and runoff for a given rainfall input under normal circumstances. Soil and geological qualities such as bedrock, texture, organic content, structure, permeability, catena, and several other variables influence soil erodibility (Schwab, Fangmeier, & Elliot, 1994). Despite its high matching percentage in the sand and clay fractions, soil erodibility falls when silt content is low (Mhangara, Kakembo, & Lim, 2012). It is determined by the soil parameters used to calculate the *K* factor. The grain size of the soil (percent sand, percent silt, and percent clay), the quantity of moisture content, the amount of organic matter, the permeability rate of soil, structural category of soil, and the color and textural category of the soil are all examples. The developers choose and evaluate various attributes, generating several soil *K* factor models based on their idealization (Ghosal & Das Bhattacharya, 2020). Wischmeier and Smith

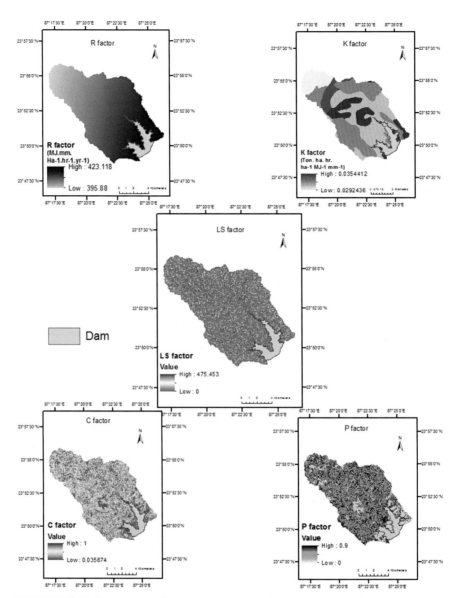

FIGURE 20.3 (A) R factor map, **(B)** *K* factor map, **(C)** *LS* factor map, **(D)** *C* factor map, and **(E)** *P* factor map

(1978) discovered that the erodibility factor of soil is affected not only by particle size but also by the quantity of organic content or carbon content, the chemical bonding or structural category it belongs to, as well as the rate of permeability.

Wischmeier and Smith (1978) created an equation as well as a nomograph depending on it. Many scholars changed the Wischmeier and Smith (1978) K factors formula for greater accuracy in their specific study area after establishing it. For their specific study field, Foster, McCool, Renard, and Moldenhauer (1981), Schwab, Frevert, Edminster, and Barnes (1982), Sharpley and Williams (1990), and Goldman, Jackson, and Bursztynsky (1986) performed numerous regression analyses on the Wischmeier and Smith (1978) equation, and many researchers further employed equations. Rosewell (1993) revised this equation to produce a nomograph. For Indian soil conditions, Rosewell's (1993) solution (Eq. 20.2) is commonly employed. The K factor is calculated in the SI system (t ha hour/ha MJ mm) (Fig. 20.3B).

$$K = \{2.77(10^{-7})(12 - \alpha)M^{1.14}\} + \{4.28(10^{-3})(\beta - 2)\} + \{3.29(10^{-3})(\gamma - 3)\} \quad (20.2)$$

M = (percent silt + percent very fine sand). (100% − percent clay), α = organic matter (percentage), β = structural code, and γ = permeability rating. In this research field, there are five different types of soil (Sarkar, Dutta, & Nayak, 2007). Table 20.2 displays full soil data collected from the Sarkar et al. (2007) study released by NBSS and LUP.

20.3.3 Estimation of slope length and slope steepness factor of the upper Bakreshwar river basin

The LS factor represents the effect of topography on the soil erosion rate by incorporating slope length (L) and slope steepness factors (S) (Fig. 20.3C). The longer the slope, the greater the cumulative runoff. The longer the slope, the greater the cumulative runoff. Furthermore, the higher the slope of the ground, the higher the runoff, which leads to erosion. There are numerous techniques to simulating slope length and stiffness factor. Smith and Wischmeier (1957) invented the LS factor, a nondimensional factor in which L is stated according to slope lengths expressed in meters; nevertheless, using 22 m as the basic slope length and 9% as the principal slope gradient keeps the LS values virtually same. Wischmeier and Smith (1978) updated this model by substituting the sin of the slope angle (θ) for the mean slope gradient percentage (Eq. 20.3).

$$LS = \left(\frac{\lambda}{\Psi}\right)^m .(65.41\sin2\theta + 4.56\sin\theta + 0.065) \quad (20.3)$$

where λ = flow route length (m or feet) = (cell size × flow-accumulation), Ψ = 22.13 for the SI system and 72.6 in English Units (BU), and the LS is defined as the ratio of soil erosion per unit area from the field slope to a 22.13 m length. Angle of slope = m = 0.5, when the % slope is 5 or greater, 0.4 when the slope is 3.5% to 4.5%, 0.3 when the slope is 1% to 3%, and 0.2 when the uniform gradient is less than %.

Table 20.2 Distribution of soil property of the upper Bakreshwar river basin.

Soil group	Sand (%)	Silt (%)	Clay (%)	Organic matter (%) (α)	Structural code (β)	Permeability rating (γ)	M	K factor (t ha/h ha/MJ/mm)
6	48.59	28.60	22.79	0.43	2	3	2979.919	0.029243598
7	59.5	24.89	15.60	0.47	2	2	3561.680	0.032422386
8	73	13	14	0.33	2	2	3698.000	0.034470081
9	42.70	26.79	30.5	0.15	4	4	2415.125	0.0354416
11	10.19	48.70	41.09	0.58	4	5	1734.605	0.030715935

20.3.4 Estimation of the cover management factor (*C*) of the upper Bakreshwar river basin

Soil erosion is affected by plant cover, slope steepness, and length component (Benkobi et al., 1994). Plant cover shields the soil by dispersing raindrop energy before it reaches the soil's surface. The *C* value is determined by the kind of vegetation, the stage of development, and the percentage of cover. In India, there are numerous techniques for determining the *C* factor. Rao (1981) defined the *C* factor depending on the kind of landcover. As a reaction to the necessity for seasonal land use/cover information, an alternate approach was taken here by substituting the *C*-factor with the normalized difference vegetation index (NDVI) (Gitas, 2009; McFarlane, Delroy, & Van, 1991). Morgan (1995) presented a linear equation after seeing that the *C* factor is closely related to the kind of plant cover. According to van der Knijff, Jones, & Montanarella, 2000), the relation among the *C* factor with NDVI is nonlinear (Eq. 20.4). The connection is deteriorating at an exponential rate.

$$C = e^{-}(\alpha(NDVI/\beta - NDVI))$$ (20.4)

The form of the NDVI versus *C* graph is determined by the parameters, $\alpha = 2$ and $\beta = $ of 1 appear to produce satisfactory outcomes (van der Knijff et al., 2000). Higher *C* values (Fig. 20.3D) imply no cover effectiveness and soil loss equivalent to that of a tilled bare plain, whereas lower *C* indicates a strong cover effect with no erosion.

20.3.5 Estimation of the conservation practice factor (*P*) of the upper Bakreshwar river basin

The *P* factor (Fig. 20.3E) is known as the conservation practice factor or support practice factor. It depicts the effect of techniques that reduce the volume and rate of water runoff, thereby reducing erosion. The *P* described as the ratio of soil erosion induced by one support approach to equivalent soil loss produced by up and down gradient, contoured cultivation (Wischmeier & Smith, 1978). Foster et al. (1981) displayed the *P* factor numbers relating to land use classifications (Table 20.3).

Table 20.3 Details of *P* values of various land use and landcover of upper Bakreshwar river basin.

Land use class	Area (ha)	Area (%)	P value
Waterbody	1867.65	14.96	0
Settlement	1567.31	12.56	0.9
Barren land	860.46	6.89	0.9
Agricultural land	5754.84	46.11	0.2
Open scrub	823.53	6.60	0.8
Forest	1607.91	12.88	0.1

The average annual soil loss of the Bakreshwar river catchment region (Fig. 20.4) was calculated using the RUSLE model (Eq. 20.5)

$$S_E = R^* K^* LS^* C^* P$$ (20.5)

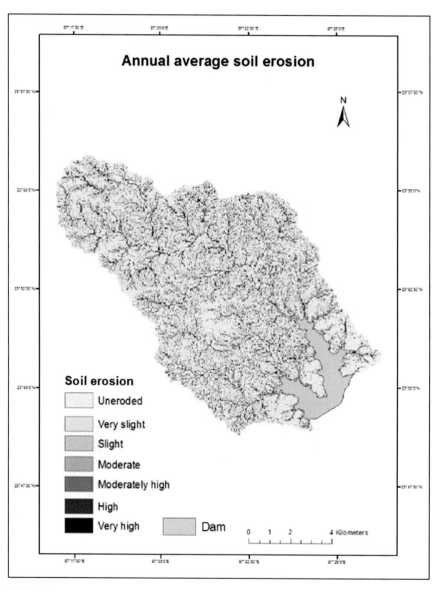

FIGURE 20.4 Average annual soil erosion of the Upper Bakreshwar River basin.

20.3.6 Estimation of sediment delivery ratio and sediment yield of the upper Bakreshwar river basin

The deposition of sediment, or the sediment yield, is measured by the rate of soil erosion in the river basin region, as well as the flow or transport of that eroded soil using slope, flow velocity, the lack of plant cover, and other factors. The preceding study was used to calculate the quantity of soil erosion. SDR estimates the delivery of that eroded soil. The SDR is the percentage of gross erosion transferred from a specific location in a given time interval. A sediment transport rate metric relates the amount of sediment supplied from erosion sources to a downstream of the river to the overall amount of soil eroded above that point in the same region. A portion of the topsoil lost in an overland region accumulates within a basin before reaching its outflow. The SDR is described as the proportion of sediment yield over total soil erosion (SDR). The SDR values for a region are shown to be influenced by the catchment transport system, physiography, sediment sources, eroded material texture, landcover, and so on (Richard, 1993; Walling, 1983; Walling, 1988). Variables like as catchment size, land gradient, and landcover, on the other hand, have mostly been employed as variables in empirical models for SDR (Hadley, Lal, Onstad, & Walling, 1985; Kothyari & Jain, 1997; Maner, 1958; Roehl, 1962; Williams & Berndt, 1972). The main advantages of this model are as follows:

1 The SDR inside a given watershed reflects the catchment's total capacity for storing and transmitting degraded material. It accounts for regions of sediment load that become increasingly essential as catchment area expands, determining the relative importance of sediment source and its deliveries (Lu, Moran, Prosser, & Raupach, 2003).
2 Several physical properties of a watershed, such as drainage basin, gradient, relief-length ratio, rainfall-runoff parameters, land use and landcover distribution, and sediment particle size, all have an impact on it (Mutua, 2006).

SDR (D_r) is the ratio of an area's estimated soil erosion to its cumulative sediment yield (Fig. 20.5A). Ferro and Minacapilli (1995) developed SDR intervals of exponential decline with sediment flow duration are represented in Eq. 20.6:

$$D_r = e^{\left(-\gamma \sum_{i=1}^{m} \left(\frac{l_i}{a_i \cdot s_i^{0.5}}\right)\right)}$$

(20.6)

where γ is a coefficient considered constant for a given catchment. As per Bhattarai and Dutta (2007), γ can be taken as 1 for a small catchment area. "m" is the number of pixels the flow path. ($\frac{l_i}{s_i^{0.5}}$) is the travel time factor. "a_i" is the overland flow time of the concentration coefficient developed by Haan, Barfield, and Hayes (1994). The SDR map shows the sediment fraction delivered to a pixel.

The sediment, as defined by the American Meteorological Society, is total sediment flow from a watershed in relation to the watershed area, given in masses

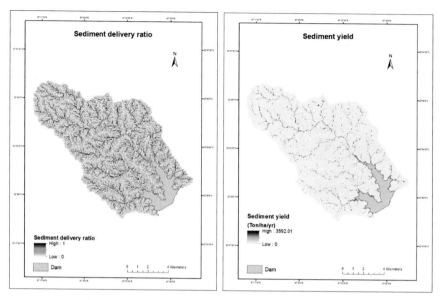

FIGURE 20.5

(A) Sediment delivery ratio and (B) sediment yield of the Upper Bakreshwar River basin.

units per unit time per unit area. The sediment yield demonstrates the accumulated impact of soil erosion as well as the supply of eroded soil for sediment buildup in every pixel (5 m × 5 m) in the research area. This equation was used to compute the sediment yield (Ferro & Minacapilli, 1995):

$$S_y = \sum_{i=1}^{N} D_{r_i} S_{E_i} \tag{20.7}$$

where each pixel's SDR (D_{r_i}) is a nondimensional parameter and annual average soil erosion of each pixel (S_{E_i}) is in (t/ha/year). As a result, the sediment yield generated are in (t/ha/yr). The distribution of sediment yield (Fig. 20.5B) was created by combining both raster outputs.

20.4 Results

20.4.1 Assessment of soil loss vulnerability of the upper Bakreshwar river basin

By combining all R, K, LS, C, and P components, the potential soil loss rate was calculated. Table 20.4 discusses the variability of soil loss risk. According to this table, approximately 42.43% of the area is under soil erosion risk. This catchment area is mostly vulnerable to minor (11.37%) to moderate (11.24%) soil erosion. Table 20.5 shows the influence of soil erosion risk across different land use

Table 20.4 Details of annual average soil erosion of the upper Bakreshwar River basin.

Soil erosion class	Annual average soil erosion (S_E) (t/ha/year)	Area (ha)	Percentage of total gross area (%)
Uneroded	0	6445.45	57.57
Very slight	0–5	1124.22	10.04
Slight	5–10	1273.32	11.37
Moderate	10–20	1258.87	11.24
Moderately high	20–40	1065.98	9.52
High	40–100	873.70	7.80
Very high	>100	440.16	3.93

Table 20.5 Distribution of soil erosion vulnerability over different land use and landcover of the upper Bakreshwar river basin.

Land use/landcover class	Annual average soil erosion rate (t/ha/year)	Area (ha)	Area (%)
Waterbody	3.47	1867.65	14.96
Settlement	30.75	1567.31	12.56
Barren land	16.58	860.46	6.89
Agricultural land	12.95	5754.84	46.11
Open scrub	15.37	823.53	6.60
Forest	7.35	1607.91	12.88

landcover. According to table 20.5, settlement or builtup lands are the most prone to soil loss (30.75 t/ha/year). Aside from habitation, barren terrain is at risk of soil erosion (16.58 t/ha/year). A greater amount of vegetation reduces the risk of soil erosion. As a consequence, the soil erosion rates in open scrub, agricultural areas, and forest regions are 15.37, 12.95, and 7.35 t/acre/year, respectively. Table 20.6 depicts the areal spread of distinct soil erosion classes across different land uses.

The proportion only of soil risk zones over different land uses (Fig. 20.6) demonstrates that urbanization and a lack of plant cover increase the danger of soil loss, which is also increased in soil layers near water bodies.

20.4.2 Assessment sediment yield vulnerability of the upper Bakreshwar river basin

The SDR is mostly determined by the watershed drainage system. The rate of sediment yield (Table 20.7) has been calculated for all land use classifications. Significant sedimentation has been reported (Table 20.7) mostly under river drainage system and water bodies. Aside from the waterbody, the sedimentation

Table 20.6 Distribution of areal coverage of different soil erosion classes over different land uses of the upper Bakreshwar river basin.

Land use/landcover class soil erosion class	Areal coverage (ha)						Total area
	Waterbody	Settlement	Barren land	Agricultural land	Open scrub	Forest	
Uneroded	997.68	782.38	461.11	2804.48	434.27	965.55	6445.45
Very slight	78.08	95.40	43.54	450.04	78.05	379.11	1124.22
Slight	131.78	97.80	77.71	797.04	72.96	96.04	1273.32
Moderate	135.77	128.80	84.60	763.60	88.69	57.41	1258.87
Moderately high	194.36	154.48	79.81	507.82	79.24	50.28	1065.98
High	180.27	193.34	91.29	318.59	48.41	41.81	873.70
Very high	149.72	115.11	22.40	113.30	21.93	17.72	440.16
Total area	1,867.65	1,567.31	860.46	5,754.84	823.53	1,607.91	

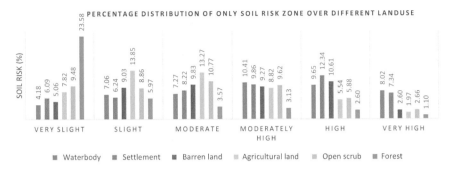

FIGURE 20.6 Soil risk assessment over different land use of the upper Bakreshwar river basin.

Table 20.7 Distribution of sediment yield vulnerability over different land uses and landcover of the upper Bakreshwar river basin.

Land use/landcover class	Area (ha)	Area (%)	Annual average sediment yield (t/ha/year)
Waterbody	1867.65	14.96	39.00
Settlement	1567.31	12.56	25.07
Barren land	860.46	6.89	11.07
Agricultural land	5754.84	46.11	9.36
Open scrub	823.53	6.60	13.40
Forest	1607.91	12.88	5.49

distribution (Fig. 20.7) matches to the pattern of soil erosion over different land use classes.

Because significant sedimentation exists in the waterbody area, the amount of sediment production has been detected across distinct stream orders. Table 20.8 and Fig. 20.8 indicate the average sediment production over five stream orders of the upper Bakreshwar drainage basin region.

This investigation demonstrates that as stream order increases, the sediment deposition rate increases. This pattern demonstrates the vulnerability of Bakreshwar dam's life. The Bakreshwar dam is the sink of all streams of Bakreshwar river. As a result, the sediment accumulation rate beneath this dam is huge, and it is the principal reason of the dam's reduced carrying capacity.

20.5 Discussion

The first objective of the study was to estimate the annual soil loss rate. Using the empirical RUSLE model, this study investigated yearly soil loss of upper

FIGURE 20.7 Soil erosion and sediment yield over land use of the upper Bakreshwar River basin.

Table 20.8 Distribution of the average annual sedimentation rate over different stream orders of the upper Bakreshwar river basin.

Stream order	Area (ha)	Area (%)	Annual average sedimentation rate (t/ha/year)
1	704.88	54.85	26.58
2	309.97	24.12	43.95
3	194.23	15.11	59.87
4	65.01	5.06	108.55
5	11.03	0.86	171.56

Bakreshwar basin (West Bengal, India). The second and third objectives of the study were to estimate the SDR and sediment yield of the study area, respectively. To attain these two objectives, the SDR model has been used. There have been no previous studies that used the RUSLE/SDR model to assess the yearly soil loss and sediment production for this watershed. The results obtained in this study utilizing the RUSLE and SDR methods provided fresh insights for analyzing soil loss in a region dominated with lateritic soil. RUSLE with SDR combinations have been studied in various watersheds. These approaches were used to the Kolleru Lake basin in southeastern India by Kolli et al. (2021). In their research, the most common soil types were black cotton soil (57.6%), sandy clay loams

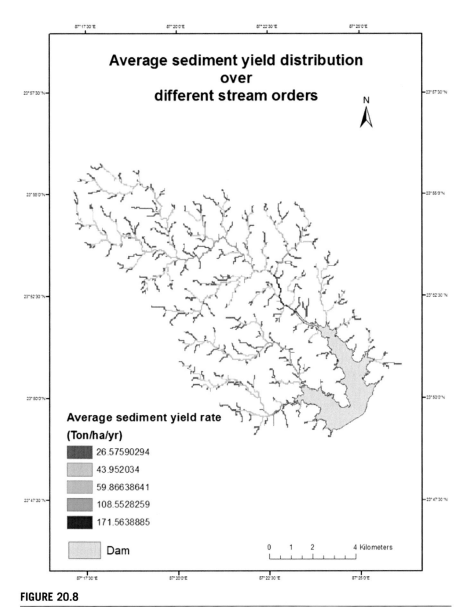

FIGURE 20.8

Average sedimentation rate over different stream orders of the upper Bakreshwar River basin.

(22.3%), and red soils (19.4%). Gelagay (2016) estimated soil erosion and sediment yield in Ethiopia's Koga watershed using the RUSLE with the SDR combination technique. Gelagay (2016) investigated the quantity of sediment output as well as the watershed's various stream orders. In the Nozhian watershed, Ebrahimzadeh, Motagh, and Mahboub (2018) used a similar combination to measure soil erosion and sediment output (western Iran). According to Ebrahimzadeh et al. (2018) and Gelagay (2016), the major cause of soil erosion is a steep slope. Their dirt was not like the soil in the Bakreshwar river catchment. Ghosh et al. (2015) utilized the RUSLE model for predicting soil loss in the Bakreshwar catchment, assuming that runoff was the primary cause of soil loss in the study region despite receiving a poor R^2 value for the aforementioned association. Ghosal and Das Bhattacharya (2020) investigated the effect of the C factor upon LST in the Upper Bakreshwar drainage basin region and discovered a strong relationship between LST and C factor region. They did not, however, determine the region's soil erosion rate or sediment output. The influence of the yearly average soil loss rate on various land use landcover types in the upper Bakreshwar watershed was investigated in this study. We found extensive soil erosion in the settlement area (Fig. 20.7) because the conservation level is high in the settlement area compared to other land use due to presence anthropogenic activities in this area. Aside from settlement areas, the rate of soil erosion is also significant in barren regions due to the lack of plant cover. The rate of soil erosion is lower in the forest zone because the plant cover minimizes the impact of rainfall and the tree roots retain the soil layer in place, and anthropogenic activity is also lower in the forest region. We estimated the sediment production rates of the Upper Bakreshwar area after evaluating soil erosion. Fig. 20.7 depicts the effect of the sedimentation rate on different land uses. We discovered that the sediment substrate concentration takes a similar soil loss pattern across land uses. The sediment buildup rate, on the other hand, is greatest under the waterbody region. The primary objective and goal of this research was to identify the risk of the Bakreshwar dam and its ailing steam from sediment accumulation, as this dam holds water used for generating power (Bakreshwar thermal power plant) and irrigation of significant parts of India's West Bengal and Jharkhand regions. We studied the influence of silt accumulation on the several stream orders of Bakreshwar river to achieve this aim. We discovered that when stream order rises, so does the sediment output rate (Table 20.8 and Fig. 20.8). Because it is the sink for all streams of upper Bakreshwar streams, the Bakreshwar dam is most vulnerable to sediment deposition, which has resulted from the soil erosion across all stream orders of the watershed.

20.6 Conclusion

The study proved that the absence of plant cover increases the risk of soil erosion. The growth in human settlement and builtup area has disturbed the natural soil holding strength, resulting in considerable soil loss. It shows the influence of soil erosion risk across different land use landcover. According to the present study, settlement or builtup lands are the most prone to soil loss (30.75 t/ha/year). Aside from habitation, barren land is also at risk of soil erosion (16.58 t/ha/year). A greater amount of vegetation reduces the risk of soil erosion. As a consequence, the soil erosion rates in open scrub, agricultural areas, and forest regions are 15.37, 12.95, and 7.35 t/acre/year, respectively. The resultant sediment yield from soil erosion has an extensive impact on waterbody structures. River drainage has mostly been blocked as a result of enormous sediment accumulation due to soil erosion. According to the findings of this study, sediment accumulation rises in higher stream order more toward the downstream section. The upper Bakreshwar river basin area's annual average soil loss rate and sediment yield rate are 11.62 and 8.38 t/ha/year, respectively. As a result, over 72.12% of eroded soil accumulates in river bottoms. All river channels go into the Bakreshwar dam. Rise in sediment deposition with channel order indicates a steady deterioration in the Bakreshwar dam's water holding capacity. These findings are of great concern since they have a considerable influence on thermal electricity production and irrigation in most sections of West Bengal and Jharkhand, India. To mitigate this damage, barren regions must be planted with appropriate flora. Deforestation must be avoided because plant roots keep the soil together, and vegetation cover lessens the impact of raindrop kinetic energy, lowering the soil loss. Dredging of upper channel (lower orders) stream beds is required to decrease silt accumulation in stream beds; else, the life of a river and associated downstream reservoirs are gradually endangered. This model will aid in the assessment and monitoring of soil erosion and sedimentation in any catchment region. As a result, this model is critical for reservoir life studies.

Conflict of interest

The authors declare that they have no conflict of interest.

References

Benkobi, L., Trlica, M. J., & Smith, J. L. (1994). Evaluation of a refined surface cover sub-factor for use in RUSLE. *Journal of Range Management, 47*(1), 74−78. Available from https://doi.org/10.2307/4002845, http://uvalde.tamu.edu/jrm/jrmhome.htm.

Bhattarai, R., & Dutta, D. (2007). Estimation of soil erosion and sediment yield using GIS at catchment scale. *Water Resources Management, 21*(10), 1635−1647. Available from https://doi.org/10.1007/s11269-006-9118-z.

Cooper, K. (2011). *Evaluation of the relationships between the RUSLE R-factor and mean annual precipitation.* http://engr.colostate.edu/∼pierre/ce_old/Projects/linkfiles/Cooper%20R-factor-Final.pdf.

Diodato, N., & Grauso, S. (2009). An improved correlation model for sediment delivery ratio assessment. *Environmental Earth Sciences*, *59*(1), 223–231. Available from https://doi.org/10.1007/s12665-009-0020-x.

FAO, & ITPS. (2015). *Status of the World's Soil Resources (SWSR)*.

Ebrahimzadeh, S., Motagh, M., Mahboub, V., et al. (2018). An improved RUSLE/SDR model for the evaluation of soil erosion. *Environmental Earth Sciences*, *77*(12), 454.

Ferro, V., & Minacapilli, M. (1995). Sediment delivery processes at basin scale. *Hydrological Sciences Journal*, *40*(6), 703–717. Available from https://doi.org/10.1080/02626669509491460.

Foster, G. R., McCool, D. K., Renard, K. G., & Moldenhauer, W. C. (1981). Conversion of the universal soil loss equation to SI metric units. *Journal of Soil & Water Conservation*, *36*(6), 355–359. Available from http://www.jswconline.org.

Gelagay, H. S. (2016). RUSLE and SDR model based sediment yield assessment in a GIS and remote sensing environment; a case study of Koga Watershed, Upper Blue Nile Basin, Ethiopia. *Journal of Waste Water Treatment & Analysis*, *7*(2). Available from https://doi.org/10.4172/2157-7587.1000239.

Ghosal, K., & Bhattacharya, S.D. (2021). Identification of the relationship between temporally varying land surface temperature of winter season with the cover management factor of revised universal soil loss equation: A case study from upper Bakreshwar river basin. In Banerjee, T. (Ed.), *Geoinformatics in research & development* (pp. 60–71). South Asian Institute for Advanced Research and Development (SAIARD).

Ghosal, K., & Das Bhattacharya, S. (2020). A review of RUSLE model. *Journal of the Indian Society of Remote Sensing*, *48*(4), 689–707. Available from https://doi.org/10.1007/s12524-019-01097-0.

Ghosh, G., Mukhopadhyay, S., & Pal, S. (2015). Surface runoff and soil erosion dynamics: A case study on Bakreshwar river basin, eastern India. *International Research Journal of Earth Sciences*, *3*(7), 11–22.

Gitas, I. (2009). Multi-temporal soil erosion risk assessment in N. Chalkidiki using a modified USLE raster model. s.l. *EARSeL eProceedings*, 40–52.

Goldman, S.J., Jackson, K., & Bursztynsky, T.A. (1986). *Erosion and sediment control handbook*. 9780070236554.

Haan, C.T., Barfield, B.J., Hayes, J.C. (1994). *Design hydrology and sedimentology for small catchments*.

Hadley, R., Lal, R., Onstad, C., & Walling, D. (1985). Recent developments in erosion and sediment yield studies. In Technical documents in hydrology.

Jaiswal, M. K., & Amin, N. (2020). The impact of land use dynamics on the soil erosion in the Panchnoi River Basin, Northeast India. *Journal of the Geographical Institute Jovan Cvijic SASA*, *70*(1), 1–4. Available from https://doi.org/10.2298/IJGI2001001J, http://www.gi.sanu.ac.rs/zbornik/index.php/zbornik/article/view/186/pdf.

Kashiwar, S. R., Kundu, M. C., & Dongarwar, U. R. (2022). Soil erosion estimation of Bhandara region of Maharashtra, India, by integrated use of RUSLE, remote sensing, and GIS. *Natural Hazards*, *110*(2), 937–959. Available from https://doi.org/10.1007/s11069-021-04974-5, http://www.wkap.nl/journalhome.htm/0921-030X.

Kolli, M. K., Opp, C., & Groll, M. (2021). Estimation of soil erosion and sediment yield concentration across the Kolleru Lake catchment using GIS. *Environmental Earth Sciences*, *80*(4). Available from https://doi.org/10.1007/s12665-021-09443-7.

Korea Institute of Construction Technology. (1992). *The development of selection standard for calculation method of unit sediment yield in river.* KICT.

Kothyari, U. C., & Jain, S. K. (1997). Sediment yield estimation using GIS. *Hydrological Sciences Journal, 42*(6), 833−843. Available from https://doi.org/10.1080/02626669709492082.

Lu, H., Moran, C., Prosser, I., Raupach, M. (2003). *Hillslope erosion and Sediment delivery: A basin wide estimation at medium catchment scale.*

Maner, S. B. (1958). Factors affecting sediment delivery rates in the red hills physiographic area. *Eos, Transactions American Geophysical Union, 39*(4), 669−675. Available from https://doi.org/10.1029/TR039i004p00669.

McFarlane, D., Delroy, N., & Van, S. V. (1991). Water erosion of potato land in Western Australia. *Australian Journal of Soil and Water Conservation*, 33−40.

Mhangara, P., Kakembo, V., & Lim, K. J. (2012). Soil erosion risk assessment of the Keiskamma catchment, South Africa using GIS and remote sensing. *Environmental Earth Sciences, 65*(7), 2087−2102. Available from https://doi.org/10.1007/s12665-011-1190-x.

Morgan, R. (1995). *Soil erosion and conservation* (2nd ed.). Longman.

Mutua, B. (2006). Estimating spatial sediment delivery ratio on a large rural catchment. *Journal of Spatial Hydrology, 6*(1), 64−80.

Ouadja, A., Benfetta, H., Porto, P., Flanagan, D. C., Mihoubi, M. K., Omeir, M. R., ... Talchabhadel, R. (2021). Mapping potential soil erosion using RUSLE, remote sensing, and GIS: A case study in the watershed of Oued El Ardjem, Northwest Algeria. *Arabian Journal of Geosciences, 14*(18). Available from https://doi.org/10.1007/s12517-021-07992-6.

Rao, Y. (1981). Evaluation of cropping management factor in universal soil loss equation under natural rainfall condition of Kharagpur. In *Proceedings of the Southeast Asian regional symposium on problems of soil erosion and sedimentation* (pp. 241−254).

Richard, K. (1993). Sediment delivery and the drainage network. In Channel network hydrology (pp. 222−254).

Roehl, J. (1962). Sediment source areas, delivery ratio sand influencing morphological factors. Symposium of Bari. *International Association of Hydrological Sciences, 59.*

Rosewell, C. (1993). *SOILLOSS − A program to assist in the selection of the management practices to reduce erosion.* Soil Conservation Service of New South Wales.

Sarkar, D., Dutta, D., & Nayak, D. (2007). *Optimising land use of Birbhum district, (West Bengal) soil resource assessment.* NBSS & LUP (ICAR).

Schwab, G. O., Frevert, R. K., Edminster, T. W., & Barnes, K. K. (1982). Soil and water conservation engineering. *Soil Science, 134*(2). Available from https://doi.org/10.1097/00010694-198208000-00013.

Schwab, G., Fangmeier, D., & Elliot , W. (1994). *Soil and water conservation engineering* (fourth ed.).

Sharpley, A., & Williams, J. (1990). EPIC—Erosion/productivity impact calculator: 1. Model documentation. Washington DC: US Department of Agriculture Technical Bulletin No. 1768.

Smith, D. D., & Wischmeier, W. H. (1957). Factors affecting sheet and rill erosion. *Eos, Transactions American Geophysical Union, 38*(6), 889−896. Available from https://doi.org/10.1029/TR038i006p00889.

Stefanidis, S., Alexandridis, V., Chatzichristaki, C., & Stefanidis, P. (2021). Assessing soil loss by water erosion in a typical mediterranean ecosystem of Northern Greece under current and future rainfall erosivity. *Water*, *13*(15). Available from https://doi.org/10.3390/w13152002.

Stefanidis, S., Alexandridis, V., & Ghosal, K. (2022). Assessment of water-induced soil erosion as a threat to natura 2000 protected areas in Crete Island, Greece. *Sustainability*, *14*(5). Available from https://doi.org/10.3390/su14052738.

van der Knijff, J.M, Jones, R., Montanarella, L. (2000). *Soil erosion risk assessment in Europe*. Office for Official Publications of the European Communities.

Walling, D. E. (1983). The sediment delivery problem. *Journal of Hydrology*, *65*(1−3), 209−237. Available from https://doi.org/10.1016/0022-1694(83)90217-2.

Walling, D. E. (1988). Erosion and sediment yield research—Some recent perspectives. *Journal of Hydrology*, *100*(1−3), 113−141. Available from https://doi.org/10.1016/0022-1694(88)90183-7.

Williams, J. R., & Berndt, H. D. (1972). Sediment yield computed with universal equation. *Journal of the Hydraulics Division*, *98*(12), 2087−2098. Available from https://doi.org/10.1061/jyceaj.0003498.

Wischmeier, W., & Smith, D. (1978). *Predicting rainfall erosion losses—A guide to conservation planning*.

Prediction of gully erosion vulnerability using geospatial tools and machine learning algorithms: a critical review

21

Akhilesh Kumar Gupta[1] and Argha Ghosh[2]

[1]*Agricultural Statistics, Odisha University of Agriculture and Technology, Bhubaneswar, Odisha, India*

[2]*Agricultural Meteorology, Odisha University of Agriculture and Technology, Bhubaneswar, Odisha, India*

21.1 Introduction

A gully is a steep-sided channel caused by soil erosion due to heavy and intermittent runoff of water during and immediately following the heavy rains (Poesen, Vandaele, & van Wesemael, 1998). The water flow often develops steep walls and actively eroding head scarp (of on average >0.5 m deep) in these channels, which cannot be generally removed by conventional tillage methods (Thwaites, Brooks, Pietsch, & Spencer, 2022). They are permanent erosional forms and can develop in any type soil or environments; however, arid and semiarid environments are more prone (Aber, Marzolff, & Ries, 2010). The development of gully (Gully erosion) involves two phenomena, that is, waterfall crosion and channel erosion. Gullies usually start with channel erosion and the extension of its head is usually caused by waterfall erosion. The scouring of the bottom and sides of gullies is by channel erosion, which enlarges the depth and width of gullies. The gully erosion, in a very general way, can be defined as the process in which runoff water builds up and frequently recurs in little channels, removing soil from this small region to significant depths in a short amount of time (Poesen, Nachtergaele, Verstraeten, & Valentin, 2003). The study of gully erosion, its impact, and control is crucial, as it is recognized as a key indicator of land degradation, posing threats to agricultural production, economic development, and the ecological environment (Wang et al., 2020). Their control is more difficult because it is extremely difficult to establish trees or any other plant material once erosion has reached the bedrock, and the enormous gullies cannot be fixed (McIvor, Youjun, Daoping, Eyles, & Pu, 2017). The typically unforeseen effects,

Applications of Geospatial Technology and Modeling for River Basin Management.
DOI: https://doi.org/10.1016/B978-0-443-23890-1.00021-9

which include the loss of arable land and an increase in labor costs, are frequently more severe and conspicuous for farmlands. The gully erosion enhances drainage and accelerated aridification processes (Valentin, Poesen, & Li, 2005). Only small valleys contain the majority of the floral biomass and the agricultural fields. The evidence from the parched Negev highlands of southern Israel shows that gully incision erodes loess deposited in the valleys and alluvial deposits (Avni, 2005). It is not only a threat to farmlands, but also in the south-eastern part of Nigeria, it has claimed lives, ruined homes and schools, ripped up highways, and forced people from their homes (Eke & Ogba, 2020). Gully erosion is also a dominant sediment-producing process, which is a major source of pollution for waterways and water supplies. Also, the sediment in stream beds can destroy the habitat of the small stream organisms and disrupt the natural food chain, causing massive declines in fish supply. Several studies have been conducted by various researchers to determine the contribution of gully erosion toward sediment production. In the Chinese Loess Plateau, gully erosion contributes between 60% and 70% of all sediments, and in northwestern highland Ethiopia, a similar proportion (70%) has been reported (Valentin et al., 2005).

The soil erosion, water erosion, or, to be specific, the gully erosion, has a detrimental impact on the environment, agriculture, natural resources, and society. Gully erosion prediction has gained greater importance in recent times for its effective utilization to mitigate the adverse effects of gully erosion on the environment, ecology, and economy. Therefore, accurate prediction and modeling of gully erosion events are crucial for decision-making in soil and environment conservation, flooding prevention, water quality management, sustainable land use planning, and climate change adaptation. Furthermore, the prediction and modeling of gully erosion provide valuable knowledge for mapping, simulating, and accurately assessing the distribution and density of gully systems, as well as their morphology and changes over time in order to effectively manage and mitigate its impacts. The first approach to predict the location of gully heads was the threshold concept applied to geomorphic systems (Patton & Schumm, 1975). Poesen, Torri, and Van Walleghem (2011) presented a detailed review on the different traditional procedures of modeling the soil erosion affected by gullying with respect to three questions; (1) the possibility and location of gullies in a particular terrain; (2) the rate of gully erosion (measured by headcut retreat, channel length, and cross-section); and (3) the relationship between gully formation and other soil erosion processes. Models for gully placement, soil losses, gully headcut retreat, and the relationship between gully erosion and hydrological and other soil erosion processes have been created by several researchers. The prediction of gully erosion is a challenging task owing to its complex nature and interaction between environmental factors and the difficulty of obtaining accurate data on these factors. In recent years, there has been growing interest in developing more efficient and reliable methods for predicting gully erosion (Samani, Rad, Azarakhshi, Rahdari, & Rodrigo-Comino, 2018). Vanmaercke et al. (2021) have extensively listed a summary of studies on gully erosion conducted in Europe and in European Russia, along with empirical models used to forecast gully occurrence and density. Traditional methods have been around for years for assessing erosion vulnerability; however, they often fall short in providing timely and accurate predictions. The advent of geospatial tools, such as remote sensing (RS), geographic information systems (GIS), and statistical modeling coupled with the power of machine learning algorithms, has immense potential to accurately predict gully erosion. This study is an attempt to critically examine the prevalent state-of-the-art statistical and machine-learning models and algorithms used for gully erosion prediction.

21.2 Factors affecting gully erosion (conditioning factors)

Although gully erosion is a natural fluvial erosion process, it is a threshold-dependent process (Valentin et al., 2005) and its phenomenon is driven by a multitude of natural and anthropogenic factors that has led to the initiation and expansion of gullies in recent decades (Jahantigh & Pessarakli, 2011). Gully erosion occurs when a threshold such as flow hydraulics, rainfall (or snowmelt), topography, pedology (or lithology), and land use is exceeded (Poesen et al., 2011). Gullying is a global threat, but the factors that cause gullying can differ from locality to locality. It is very crucial to choose informative conditioning factors as per the local geographic conditions to establish a functional relationship of gullying with conditioning factors with satisfactory performance (Rahmati, Haghizadeh, Pourghasemi, & Noormohamadi, 2016). Various studies used different numbers of conditioning factors as per the local geographic conditions and data availability. The following table summarizes various studies with their number of conditioning factors as per the locality. Table 21.1 presents different conditioning factors used by researchers in various geographical locations for modeling gully erosion.

21.2.1 The slope angle

It plays a significant role in modeling gully erosion, and its impact can be substantial in influencing the initiation, development, and severity of gully erosion. Slope angle is a critical factor in the modeling of gully erosion, affecting the initiation, morphology, and development of gullies. Understanding the relationship between slope angle and erosion processes is essential for effective erosion control and land management strategies. It is often considered alongside other topographic factors, soil characteristics, and land use patterns in comprehensive erosion models. Steeper slope angles generally result in increased runoff velocity. When rainfall or surface water flows over steep slopes, it can lead to the initiation of gully erosion. The higher the slope angle, the greater the potential for concentrated flow and increasing the likelihood of gully formation. There is often a

Table 21.1 Conditioning factors used by researchers in various regions for modeling gully erosion using machine learning techniques.

Authors	Region of the study	No. of factors used	Factors used
Saha et al. (2021)	Phuentsholing, Chukha district, Bhutan	16	Altitude, slope, aspect, clay, sand, silt, CI, relative slope position, TWI, TRI, TPI, valley depth, plan curvature, land use land cover, drainage density, and lithology
Arabameri, Chen, et al. (2020)	Bastam Watershed, Iran	16	Topographical factors: aspect, slope angle, elevation, profile curvature, CI, plan curvature, slope length, TRI, TPI, and SPI Hydrological factors: distance to streams, rainfall, and drainage density Anthropogenic factors: land use/land cover and distance to road
Lei et al. (2020)	Robat Turk Watershed, Iran	12	Altitude, slope, aspect, elevation, plan curvature, profile curvature, distance from river, NDVI, lithology, land use, drainage density, distance from road, and annual mean rainfall
Saha et al. (2020)	Hinglo river basin, India	14	Slope, aspect, elevation, monsoon rainfall, soil type, geology, land use/land cover, NDVI, distance to river, distance to lineament, length of overland flow, TWI, Sediment Transportation Index, and SPI
Arabameri, Nalivan, et al. (2020)	Golestan Dam basin, Iran	16	Slope, slope aspect, plan curvature, profile curvature, drainage density, slope length, soil type, distance from ridge, CI, TPI, TRI, land use/land cover, TWI, NDVI, distance to road, and lithology
Conoscenti et al. (2014)	Watershed of the San Giorgio River, Italy	15	Plan curvature, elevation range, distance from roads, TPI, road network length, flow distance to the river network, land use, slope aspect, elevation, slope angle, profile curvature, SPI, TWI, length–slope factor, and bedrock lithology
Hughes & Prosser (2012)	Murray–Darling Basin, Australia	14	Solum thickness, geology, median soil texture (A-horizon), median soil texture (B-horizon), minimum temperature (coldest period), temperature seasonality, mean annual temperature range, mean annual precipitation, moisture index seasonality, lowest period moisture index, land use, hillslope gradient, groundcover, and hillslope length

CI, Convergence Index; LoF, length of overland flow; NDVI, Normalized Difference Vegetation Index; SPI, Stream Power Index; STI, Sediment Transportation Index; TPI, Topographic Position Index; TRI, Topographic Ruggedness Index; TWI, Topographic Wetness Index.

threshold slope angle, beyond which gully initiation becomes more probable. This threshold varies based on soil characteristics, land cover, and climate conditions (Razavi-Termeh, Sadeghi-Niaraki, & Choi, 2020). Steeper slopes tend to concentrate and accelerate surface water flow. This concentration can lead to the development of well-defined flow paths, which, under certain conditions, can evolve into incised gullies. The increased velocity of water on steeper slopes enhances its erosive power. This can result in more effective sediment transport and contribute to the down-cutting of gullies. Steeper slopes often lead to the formation of deeper and wider gullies. The erosive force of water is amplified on steeper terrain, resulting in more pronounced gully features. Gullies on steep slopes may experience rapid growth, especially during intense rainfall events. The combination of high slope angles and concentrated flow can lead to accelerated erosion and gully expansion. GIS tools are commonly used to model gully erosion, and slope angle data are integrated into these models to assess erosion susceptibility and identify vulnerable areas (Azareh et al., 2019).

21.2.2 Altitude

It is closely linked to temperature changes. As altitude increases, temperatures generally decrease. Temperature influences precipitation patterns, and variations in temperature due to altitude can impact the form and intensity of precipitation events, which, in turn, affect gully erosion. High-altitude areas often serve as headwater regions for river systems. Gully erosion in these areas can significantly contribute to sediment transport downstream. Altitude-related variations in runoff patterns can affect the erosive power of water. Changes in altitude create altitudinal gradients in precipitation, temperature, and vegetation. These gradients can lead to variations in soil moisture, vegetation cover, and erosive forces, influencing gully erosion susceptibility (Mokarram & Zarei, 2021). Altitude is inherently linked to topography. Gully erosion modeling often considers slope and aspect, both of which can be influenced by changes in altitude. Steeper slopes at higher altitudes may enhance erosional processes. Changes in altitude are associated with shifts in vegetation types and land cover. Altitudinal zonation affects the type and density of vegetation, impacting erosion control. Vegetation cover is a crucial factor in gully erosion models. Some gully erosion models may include altitude-related constraints or thresholds based on soil and geological factors (Rahmati, Tahmasebipour, Haghizadeh, Pourghasemi, & Feizizadeh, 2017).

21.2.3 Slope aspect

It refers to the compass direction that a slope faces and can have a significant impact on modeling gully erosion (Ohlmacher & Davis, 2003). Aspect determines the exposure of a slope to solar radiation. South-facing slopes receive more direct sunlight in the Northern Hemisphere, while the opposite is true in the Southern Hemisphere. This influences temperature patterns, affecting evaporation rates and soil moisture content, which are critical factors in gully erosion. The distribution of vegetation on a slope may be altered by aspect. Rainfall or snowfall may vary with aspect due to the orographic effect. Windward slopes (facing the prevailing wind direction) often receive more precipitation, influencing the erosive potential of runoff. The movement of water down a slope is influenced by aspect. North-facing slopes in the Northern Hemisphere, for example, may retain more moisture, affecting runoff patterns and erosion susceptibility. Aspect interacts with slope steepness to influence the overall terrain. Different aspects experience varying rates of drying and wetting. This differential moisture content can impact soil cohesion and erodibility, affecting the likelihood of gully erosion. In cold climates, aspect influences freeze-thaw cycles. North-facing slopes may experience longer periods of freezing, leading to soil expansion and contraction, which can contribute to gully formation (Gayen, Haque, & Saha, 2020).

21.2.4 Soil texture

It is a critical factor in modeling gully erosion, as it influences the ability of the soil to resist erosion and affects the movement of water through the landscape. Soils with coarse textures (sandy soils) generally have larger particles, leading to rapid water infiltration. On the other hand, fine-textured soils (clayey soils) have smaller particles and may be more prone to surface runoff and erosion. Soils with a significant proportion of silt may exhibit intermediate properties between sandy and clayey soils. Silt particles can contribute to soil cohesion and affect erodibility. Sandy soils typically have high infiltration rates, allowing water to penetrate quickly. This may result in less surface runoff and reduced erosion potential. Soil structure, influenced by soil texture, plays a role in erosion resistance. Well-aggregated soils are more resistant to detachment and transport, reducing the likelihood of gully formation. The erodibility of a soil, representing its susceptibility to detachment and transport, is influenced by texture. Fine-textured soils may be more prone to detachment, while coarse-textured soils may be more susceptible to transport. Clayey soils may undergo significant drying and shrinking, leading to the development of cracks and increased surface roughness, which can contribute to gully erosion. Fine-textured soils may impede root penetration, affecting the ability of plants to anchor the soil. This can contribute to increased erosion risk. Agricultural activities can exacerbate the erosion of certain soil textures. Models that simulate water movement through the landscape often incorporate soil texture data to assess runoff and erosion potential (Pal et al., 2020).

21.2.5 Convergence index

It provides a measurement of the divergence and convergence of flow in a cell. Since the erosive strength of flowing water directly affects slope toe erosion and river incision, the CI factor is one of the primary elements regulating slope erosion processes. Additionally, a high CI indicates a significant potential for erosion since it indicates the potential energy available to entrain sediment (Arabameri & Pourghasemi, 2019; Arabameri, Cerda, & Tiefenbacher, 2019). CI is determined by setting a search radius and quantifies the degree to which nearby cells point to the center cell. Up to a scale determined by the predetermined search radius, CI characterizes the overall topography of the terrain (Conoscenti & Rotigliano, 2020). CI is a terrain parameter, indicating the relief structure of the channels, which are convergent areas, and the ridges, which are the divergent areas (Arabameri & Pourghasemi, 2019).

21.2.6 Topographic Ruggedness Index

Topographic Ruggedness Index (TRI) is a measure of terrain heterogeneity, describing the variability in elevation within a specified neighborhood and can have a significant impact on modeling gully erosion. Areas with higher TRI values indicate more rugged and heterogeneous terrain. These areas may experience varied hydrological and erosional processes, influencing the initiation and development of gully erosion. High TRI values can affect the direction and concentration of surface water flow (Mokarram & Zarei, 2021). The ruggedness of the terrain may lead to the formation of preferential flow paths, influencing the pathways through which water moves and potentially contributing to gully erosion. Elevated TRI values can help identify potential erosion hotspots. Steeper and more rugged terrain may be prone to concentrated runoff, which can contribute to gully formation and expansion. Rugged terrain can impact sediment transport dynamics. Higher TRI areas may experience increased sediment transport, influencing the sediment load carried by runoff and its contribution to gully erosion. High TRI values may contribute to the development of a complex gully network. The rugged terrain can create diverse flow patterns, leading to the formation of interconnected gullies. TRI influences the connectivity of different topographic elements. Understanding these connections is crucial for modeling how water and sediment move across the landscape, impacting gully erosion (Hembram, Paul, & Saha, 2020).

21.2.7 Topographic Position Index

The Topographic Position Index (TPI) is a terrain attribute that characterizes the topographic position of a location relative to its surroundings. It is calculated by subtracting the elevation of a point from the mean elevation of the surrounding landscape (Hembram et al., 2020). TPI provides information about whether a location is in a depression, on a ridge, or on a slope. Areas with negative TPI values may indicate depressions or valley bottoms. These locations can be susceptible to gully initiation as they often serve as potential accumulation zones for water, leading to increased erosion potential. Positive TPI values represent ridge or hilltop positions. Gully erosion may not initiate directly on ridges, but these areas can influence the flow of water and sediment downslope, contributing to gully development in adjacent areas. Combining TPI with slope aspect information can provide insights into how gully erosion responds to different aspects. For example, gullies may be more prevalent on certain aspects due to sun exposure and vegetation distribution influenced by topographic position (Arabameri, Cerda, et al., 2020).

21.2.8 Topographic Wetness Index

The Topographic Wetness Index (TWI) is a terrain attribute that characterizes the relative wetness of the landscape based on the upslope contributing area and slope. It is widely used in hydrology and terrain analysis to identify areas that are likely to accumulate water (Conoscenti & Rotigliano, 2020). TWI is derived from the accumulation of flow downslope. Higher TWI values indicate areas with higher water accumulation potential. Gully erosion is often associated with areas of concentrated flow, and TWI can help identify such locations. Areas with higher TWI values generally have better moisture availability. This can influence the initiation of gullies, as the presence of water is a key factor in soil erosion processes. Analyzing the spatial distribution of gullies in relation to TWI values can reveal patterns. Gully erosion may be more prevalent in areas with higher TWI, indicating the importance of moisture conditions in erosion dynamics. Combining TWI with land cover information allows for an assessment of how vegetation and land use practices interact with wetness conditions. Vegetation can both influence and be influenced by the wetness patterns identified by TWI. TWI is related to the potential for runoff. Understanding how TWI influences runoff patterns is crucial for predicting sediment transport, which is a key component of gully erosion. TWI is related to the potential for runoff. Understanding how TWI influences runoff patterns is crucial for predicting sediment transport, which is a key component of gully erosion (Kheir, Wilson, & Deng, 2007).

21.2.9 Plan curvature

Plan curvature is a topographic factor that characterizes the curvature of the land surface in the horizontal plane. It provides information about the shape of the terrain, specifically whether a location is situated on a convex or concave slope. Areas with a concave curvature tend to accumulate water and sediments. In the context of gully erosion, concave slopes may be more prone to water pooling and increased sediment deposition, which can contribute to erosion processes. Areas with a convex curvature tend to promote water divergence. Water may be more likely to flow away from the convex slope, potentially leading to less concentrated runoff and reduced erosive forces. Gullies are often associated with concentrated overland flow (Zabihi et al., 2018). Concave areas may serve as natural pathways for water, leading to increased flow accumulation and potential gully initiation. Sediments may be transported and deposited in concave areas due to the concentration of water flow. This can influence gully morphology and sedimentation patterns. Knowledge of plan curvature can aid in identifying areas vulnerable to gully erosion. Concave areas may require targeted erosion control measures, especially in regions with high precipitation and runoff (Pourghasemi, Sadhasivam, Kariminejad, & Collins, 2020).

21.2.10 Land use and land cover

Its impact on gully erosion is significant and multifaceted. Different land uses and land cover types can influence the susceptibility of an area to gully erosion in various ways. Dense vegetation cover, such as forests or grasslands, plays a crucial role in preventing gully erosion. Vegetation helps in intercepting rainfall, reducing surface runoff, enhancing soil structure, and stabilizing slopes. Land cover with substantial vegetation cover tends to mitigate gully erosion by reducing the erosive force of flowing water and promoting soil conservation. Certain agricultural practices, such as intensive cultivation with minimal groundcover, can increase the risk of gully erosion. Plowing, especially on steep slopes, can expose soil to erosion by water runoff. Implementing conservation practices like contour plowing, cover cropping, and maintaining buffer strips can help reduce gully erosion in agricultural areas. Areas with exposed or disturbed soil, such as construction sites or mining area, are highly susceptible to gully erosion. Lack of vegetative cover increases the vulnerability of these areas. Erosion control measures, such as mulching, revegetation, and stabilization techniques, are essential to prevent gully formation in bare or disturbed soil (Setargie et al., 2023).

21.2.11 Drainage density

Drainage density is a topographic parameter that measures the amount of stream channels or drainage lines within a unit area. It is typically calculated as the total length of channels or drainage lines divided by the total area. The impact of drainage density on gully erosion can be significant, and it is often considered as a key factor in erosion modeling. Areas with high drainage density often indicate a dense network of channels or drainage lines (Zabihi et al., 2018). This can contribute to increased runoff concentration, potentially leading to higher erosion potential. Implementing erosion control measures in areas with high drainage density, such as the construction of check dams, vegetation cover, and contour plowing, can help mitigate gully erosion. Maintaining or enhancing vegetative cover, especially in areas with low drainage density, can help reduce erosion by slowing down and intercepting runoff. There may be a threshold drainage density, beyond which the risk of gully erosion increases significantly. Identifying this threshold can be essential for erosion prediction and management. Adaptive management practices, such as monitoring and adjusting land use practices based on drainage density thresholds, can help prevent gully erosion. High drainage density may enhance hydrological connectivity, allowing runoff to be quickly conveyed through the landscape. This connectivity can increase erosive potential (Amiri & Pourghasemi, 2020).

21.2.12 Stream Power Index

The Stream Power Index (SPI) is a geomorphic index used in the field of hydrology and geomorphology to assess the erosive power of flowing water in a landscape. It considers factors such as slope and drainage area to estimate the potential for erosion. SPI reflects the erosive potential of flowing water based on topographic characteristics. Higher SPI values indicate areas where water is more likely to have greater erosive power, potentially leading to gully erosion. SPI can be integrated into erosion models to identify and prioritize areas with high erosive potential, helping to target erosion control measures more effectively (Conoscenti et al., 2013). SPI incorporates slope and drainage area, emphasizing the influence of topography on erosion. It recognizes that steeper slopes and larger drainage areas contribute to higher erosive forces. When modeling gully erosion, understanding the topographic factors through SPI allows for the identification of areas with elevated erosion risk, aiding in the design of erosion control strategies. SPI allows for spatial analysis of erosive potential across a landscape. This spatial information is valuable for identifying areas prone to gully erosion and understanding the overall erosion dynamics. Incorporating SPI into erosion models aids in the planning of erosion control measures. It helps identify priority areas where interventions are most urgently needed to prevent or mitigate gully erosion (Conforti, Aucelli, Robustelli, & Scarciglia, 2011).

21.3 Application of geospatial tools in gully erosion mapping

21.3.1 Geographic Information System

GIS is fundamental for gully erosion modeling. It allows for the integration, analysis, and visualization of various spatial data layers, such as topography, land use, soil types, and hydrology, which are essential for understanding the factors contributing to gully erosion. GIS integrates digital elevation models (DEMs) to analyze the terrain and calculate slope, aspect, and other topographic parameters influencing gully erosion. It incorporates land cover data to identify areas prone to erosion based on vegetation type and land management practices. Information about soil properties and types is integrated into GIS to assess soil erodibility and susceptibility to erosion (Arabameri, Pradhan, Pourghasemi, Rezaei, & Kerle, 2018). It is used to delineate watersheds and subwatersheds, which are essential for understanding the flow of water and sediment transport contributing to gully erosion. GIS tools help model the flow of water across the landscape, indicating areas where runoff is concentrated and likely to cause erosion. It produces erosion risk maps by combining various factors such as slope, land use, soil types, and precipitation. These maps identify areas with a higher risk of gully erosion (Lei et al., 2020). Temporal changes can be facilitated by GIS integrating satellite imagery. It helps identify areas experiencing land use changes or increased erosion risk over time. GIS-based decision support systems allow researchers and land managers to simulate different scenarios, assessing the impact of land management practices on gully erosion. GIS maps and visualizations can be used to communicate gully erosion risks to the public, raising awareness and promoting community engagement in erosion control initiatives. Web-based GIS applications make gully erosion data accessible to a broader audience, including policymakers, researchers, and the general public. The application of GISs in gully erosion modeling enhances the understanding of spatial relationships and enables informed decision-making for erosion control and land management strategies. Various software packages, including ArcGIS, QGIS, and specialized erosion modeling software, offer spatial analysis tools for processing, modeling, and visualizing gully erosion data (Conoscenti et al., 2013).

21.3.2 Remote sensing

RS plays a crucial role in gully erosion modeling by providing valuable data for monitoring, analyzing, and mapping landscape changes. RS technologies, including satellite and aerial imagery, provide high-resolution data for monitoring land cover changes, detecting erosion features, and assessing the landscape. RS data, especially high-resolution satellite imagery and aerial photographs, can be used to identify and map gullies accurately. Time-series satellite imagery helps monitor changes in land cover and gully morphology over time, facilitating the detection of erosion hotspots. Unmanned aerial vehicles (UAVs) or drones equipped with high-resolution cameras can capture detailed imagery of small areas with high accuracy (Wang et al., 2020). This technology is particularly useful for obtaining up-to-date and fine-scale information for gully erosion modeling. RS technology is effectively used to assess vegetation cover, which is critical factor influencing gully erosion. Changes in vegetation health can indicate areas prone to erosion. RS data, such as stereo satellite imagery, LiDAR, or synthetic aperture radar (SAR), is used to create high-resolution DEMs for accurate topographic mapping, including slope, aspect, and elevation. Additionally, RS data helps delineate watersheds and map surface flow paths, contributing to the understanding of water movement and potential erosion pathways. RS-derived vegetation indices, such as Normalized Difference Vegetation Index (NDVI), are used to monitor vegetation health and identify areas susceptible to erosion during drought conditions. Thermal infrared RS data can be used to estimate soil moisture levels, providing insights into soil conditions and erosion susceptibility. SAR data can be used to monitor soil moisture and surface deformation, aiding in the assessment of sediment transport and erosion risk. RS data is integrated into GISs for spatial analysis, enabling the overlay of various data layers to identify factors contributing to gully erosion. RS data and derived information can be made accessible through web-based platforms, allowing researchers, land managers, and policymakers to access and analyze gully erosion data remotely (Arabameri et al., 2019).

21.4 Methods of gully erosion prediction

Gully erosion modeling can be done in many statistical and machine learning approaches using GISs, machine learning techniques, and RS data analysis. Since machine learning techniques can identify particular types of land deformation phenomena from RS photos, they hold promise for improving the efficiency and accuracy of gully-like landform detection in modeling gully erosion (Shen et al., 2021). The following table represents the authors and the methods employed by them for gully erosion modeling and prediction. It presents the various methodologies used by various researchers in different geographical locations. Table 21.2 presents various statistical and machine-learning methodologies employed by researchers in various geographical locations for modeling gully erosion.

21.4.1 Power function

This function represents the approximation of the slope-area relationship for the incision of gullies (Vandaele, Poesen, Govers, & van Wesemael, 1996). This relationship gives the critical line for incision, meaning no incision occurs below this line.

$$S_{cr} = aA^{-b}$$

where S_{cr} is the critical slope gradient (m/m), A is the drainage area (ha), a is a coefficient, and b is an exponent. Vandaele et al. (1996) established this relationship for the various previous research available for that time period and reported that coefficient a values ranged from 0.0035 to 0.35, while slope b ranged from 0.25 to 0.60.

21.4.2 Multivariate adaptive regression splines

Multivariate adaptive regression splines (MARS) originally developed and discussed by Friedman (1991) have been used to predict gully occurrence in Spain (Gutiérrez, Schnabel, & Felicísimo, 2009) and Italy (Gómez-Gutiérrez, Conoscenti, Angileri, Rotigliano, & Schnabel, 2015). It takes up the form as follows

$$y = f(x) = a_0 + \sum_{m=1}^{M} a_m b_m$$

where y is the value given by a function $f(x)$ that is composed by an initial constant (a_0) and a sum of M terms, each of them including a coefficient of expansion (a_m) and a basic function as $B_m(x) = I(x \in R_m)$. The value of I in the basis function is 1 when an argument is true and if it is false, its value is 0.

Table 21.2 Methodologies employed by researchers for modeling and prediction gully erosion.

Author	Methodological category	Methods	Location
Valentin et al. (2005), Vandaele et al. (1996), and Vandekerckhove et al. (1998, 2000)	Empirical, statistical	Power function ($S = aA^{-b}$) based on geomorphic threshold conditions	Guadalentin basin (Spain), Bragança (Portugal)
			Rambla Chortal (Spain)
Nachtergaele, Poesen, Sidorchuk, & Torri (2002)	Empirical, statistical	Ephemeral gully erosion model	Alentejo region (Portugal)
			Brussels and Leuven (Belgium)
Gutiérrez et al. (2009), Gómez-Gutiérrez et al. (2015)	Statistical	MARS	San Giorgio basin (Italy)
			Mula basin (Spain)
Rahmati et al. (2016)	Statistical	Frequency ratio model	Ilam province, Iran
		WoE model	
Pourghasemi et al. (2017)	Machine learning	Maximum entropy	Aghemam Watershed, Iran
		Support vector machine	
		Artificial neural network	
Arabameri et al. (2018)	Statistical and machine learning	MARS	Shahroud watershed, Iran
		Boosted regression trees	
		Random forest	
Arabameri et al. (2019)	Statistical	WoE model	Bastam watershed, Iran
		Index of entropy	
Arabameri et al. (2018)	Statistical and machine learning	Natural risk factor	Bayazeh Watershed, Iran
		Frequency ratio	
		Binary logistic regression	
		Boosted regression trees	
Pal et al. (2020)	Machine learning	Boosted regression Tree	Paschim Medinipur district in West Bengal, India
		Bagging	
		Ensemble of BRT and bagging	

(Continued)

Table 21.2 Methodologies employed by researchers for modeling and prediction gully erosion. *Continued*

Author	Methodological category	Methods	Location
Roy et al. (2020)	Statistical and machine learning	Boosted Regression Tree	Hinglo river basin, India
		Spatial Logistic Regression	
		MARS	
Yang et al. (2021)	Statistical and machine learning	Gradient-boosted decision trees	Mizhigou watershed, China
		Weight of Evidence Model	
		Random Forest	
		Extreme gradient boosting machine	
Baiddah et al. (2023)	Statistical and machine learning	Linear discriminant analysis	Chichaoua watershed, Morocco
		Classification and regression tree	
		Logistic regression	
		k-nearest neighbors	

MARS, *Multivariate adaptive regression splines;* WoE, *weights-of-evidence.*

These coefficients of expansion are varied in such a way that the model gives the best fit on the basis of a parameter called generalized cross-validation (GCV). The model with the lowest GCV is selected as the best model.

21.4.3 Artificial neural networks

An artificial neural network is a neural network architecture that mimics the functioning of the human brain to learn any process or phenomenon. It has been successfully used for a wide variety of tasks, including classification and forecasting (Zhang, Eddy Patuwo, & Hu, 1998). The gully erosion prediction is basically a classification problem, which means that a trained artificial neural network should be able to predict whether a particular region is susceptible to gully erosion or not on the basis of data of various conditioning factors. Artificial neural networks generally consist of three layers called input layer, hidden layer, and output layer. The input layer is made of nonlinear elements called neurons (alternatively called as nodes), which simply accept the predictor value. Neurons in successive levels take in information from the ones before them. Nodes in the layer above use the outputs of their corresponding layers as inputs. The output layer is the final layer. Hidden layers are those that lie between the input and output layers. Each node in each hidden layer receives input values from the nodes of the previous layer and

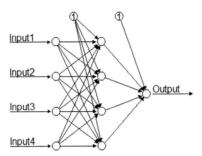

FIGURE 21.1 A fully connected multilayered perception network with one hidden layer.

It is a general presentation of a feed-forward neural network with one hidden layer.

processes it through an activation function or transfer function to produce a transformed output for the nodes in the next layer or output layer as presented in Fig. 21.1 (Gupta, Sarkar, Bhattacharya, & Dhakre, 2022).

For a causal relationship problem, the information given to the neurons of the input layer in an ANN is the predictor variable. The functional relationship estimated by the ANN can be written as $y = f(x_1, x_2, x_p)$, where x_1, x_2, and x_p are predictor variables and y is the response.

The output of jth hidden node in the neural network is given by

$$\text{output}_j = g\left(\theta_j + \sum_{i=1}^{k} w_{ij}x_i\right)$$

where g is a transfer or activation function, θ_j is the bias of the node j, w_{1j}, ..., w_{kj} are weights of node j, and $x_i (i = 1, 2,..., k)$ are the input variables. The final output value y_t in an MLP feed-forward neural network is given by

$$y_t = f\left[\varphi_t + \sum_{y=i}^{n} \alpha_j g\left(\theta_j + \sum_{i=1}^{k} w_{ij}x_i\right)\right]$$

where k is the number of input nodes (independent variables), m is the number of hidden nodes, w_{ij} is the weight attached to the connection between the ith input node and the jth node of the hidden layer, α_j is the weight attached to the connection from the jth hidden node to the output node, and φ_t is the bias of the output node. Additionally, g and f denote the activation functions at the hidden and output layers, respectively. The meaning of the remaining notations is the same as it was before. Thus, the network has an interpretation of the input-output model, with weights associated with each neuron and biases in each layer as free parameters of the model. These networks can model a function of high complexity in which the number of hidden layers and the number of neurons in each layer determine the degree of complexity of the function. In designing an MLP, one must determine the number of input nodes, the number of hidden layers and nodes in each hidden layer, and the number of output nodes (Gupta, Sarkar, Dhakre, & Bhattacharya, 2023).

21.4.4 Number of neurons (nodes) at each layer

The number of input nodes is usually the number of input variables, and the number of output nodes is the number of dependent variable levels considered in the study. In the case of gully erosion, modeling the number of input nodes is equal to the gully conditioning factors used in the study and the output nodes are equal to two (0 and 1), 0 for the absence of a gully and 1 for the presence of gully (Pourghasemi, Yousefi, Kornejady, & Cerdà, 2017). To conduct the complex nonlinear mapping between input and output variables, neural networks need to be able to recognize and capture characteristics and patterns in the data, which is made possible by the hidden layer and nodes in ANN modeling. To date, there is no reliable theoretical foundation for choosing these parameters; instead, the most popular approach for figuring out how many hidden layers and nodes is trial and error (Zhang et al., 1998).

21.4.5 Learning algorithm of the network

The ANN modeling basically consists of training the network and then validating the trained network using the hold-out data that were not used for training. In the training process, the network produces its own output through the learning algorithm and tries to minimize the difference between its own output and the desired output (target value). The minimization of difference is done by adjusting the weights of neurons and biases. This is called the learning process of the network. The most common learning algorithm is backpropagation in terms of gradient descent and momentum (Saha, Sarkar, Thapa, & Roy, 2021).

21.4.6 Activation function

In artificial neural network training, the activation function also plays a very critical role since that determines the relationship between the input and outputs of a node and introduces nonlinearity in the model, which is the core of ANNs. The sigmoid (logistic) function is the most popular activation function in neural networks (Shmueli, Bruce, Yahav, Patel, & Lichtendahl, 2018). Some other options for activation function are rectified linear unit function, hyperbolic tangent function, and identity function.

By applying artificial neural networks in gully erosion modeling and forecasting, we can gain valuable insights into the factors contributing to gully formation and predicting future erosion patterns. These neural networks can analyze various inputs, such as rainfall data, slope gradients, soil types, land cover, and land use patterns, to create a comprehensive model that accurately predicts the occurrence and severity of gully erosion. Some of the studies reported (Arabameri et al., 2020, 2021; Band et al., 2020) in recent years on the application of neural networks in gully erosion modeling and forecasting have shown promising results. These studies have demonstrated the ability of artificial neural networks to capture the complex relationships between environmental variables and gully erosion, leading to more accurate predictions and improved land management strategies to mitigate the impacts of gully erosion.

21.4.7 Logistic regression

Logistic regression is a form of generalized linear models which have been widely used algorithms for solving classification problems in various practical applications. It utilizes a logit transformation to simulate the link between independent variables and a binary dependent variable, which makes it very useful for forecasting whether gully erosion will occur or not (Arabameri, Pradhan, Rezaei, et al., 2018). Its application in gully erosion prediction has been reported by various researchers such as the study conducted by Conoscenti et al. (2014) in the San Giorgio River basin, Arabameri, Pradhan, Rezaei, et al. (2018) in Toroud Watershed, Iran, Azedou, Lahssini, Khattabi, Meliho, and Rifai (2021) in Souss basin, Morocco, and Roy and Saha (2019) in Hinglo River Basin, India. The logistic regression function can be expressed as

$$p(y) = \frac{1}{1 + e^{-(b_0 + \sum_{i=1}^{k} b_i x_i)}}$$

where y is the value of a binary dependent variable (either 0 or 1), x_i are the independent variables, conditioning factors in the case of gully erosion modeling, and b_i's are coefficients. To make it convenient for modeling, it is transformed through logit transformation, so that the equation is linear in its parameters

$$\log\frac{p(y)}{1 - p(y)} = \text{logit} p(y) = b_0 + \sum_{i=1}^{k} b_i x_i$$

Kernel logistic regression is another much more powerful classifier algorithm than logistic regression with advantages like the potential to expand to multiclass classification problems and the ability to forecast an occurrence based on its probability (Lei et al., 2020). They employed kernel logistic regression to predict gully erosion in Robat Turk Watershed, Iran. Spatial logistic regression is another modification of logistic regression to account for the spatial autocorrelation between data. In spatial logistic regression, the spatial structure is included in the LR model by changing some expressions (Roy et al., 2020). They applied this technique to model the gullies in the Hinglo River basin region of eastern India.

21.4.8 Support vector machine

Support vector machines (SVMs) are a powerful machine learning tool first introduced in the mid-1960s by Vladimir Vapnik. The techniques have evolved as a supervised learning algorithm that uses structural risk minimization and statistical learning (Sain & Vapnik, 1996) and can be applied for classification as well as regression problems (Mountrakis, Im, & Ogole, 2011). However, for the gully erosion modeling and prediction, the classification approach of the SVM is applicable. The SVM algorithm attempts to generate a division of the hyperplane between the points of two different classes in the original space of n coordinates (x_i parameters in vector $\mathbf{x}$) (Marjanović, Kovačević, Bajat, & Voženílek, 2011), in which one class indicated above the hyperplane, indicating the presence of gully erosion and other one located below the hyperplane for non-gully erosion (Band et al., 2020). The nonlinear function of an SVM is given by

$$y = f(x) = w^T \phi(x_i) + b_0$$

where $\mathbf{w}$ is the weight vector, b_0 is the bias, and $\Phi(\mathbf{x}_i)$ is the linearly mapped high dimensional feature space. The above equation is optimized with some constraints (Vapnik, 1995) such as

$$\text{minimize} \frac{1}{2} w^T w + C \sum_{i=1}^{m} (\xi_i^+ + \xi_i^-)$$

$$\text{Subject to } y_i - w^T \phi(x_i) - b_0 \leq \epsilon + \xi_i^+$$

$$w^T \phi(x_i) + b_0 - y_i \leq \epsilon + \xi_i^+$$

$$\xi_i^+, \xi_i^- \geq 0$$

where C is a prespecified penalty factor which represents the weight of the loss function, ξ_i^+ and ξ_i^- indicate slack variables within the model, and ε represents the optimized performance of the model (Chowdhuri et al., 2020).

The data processing in the SVM of a nonlinear relationship is done through the kernel function (Naghibi, Moghaddam, Kalantar, Pradhan, & Kisi, 2017). Polynomial kernels and radial basis function, also referred to as Gaussian kernels and sigmoid kernel function, are the most often utilized kernels in SVM classification applications. By expressing the solution for optimal hyperplane weights in the linear form using any of the kernels, the SVM equation can be expressed as

$$y = f(x) = \text{sgn} \sum \alpha_i y_i k(x_i, x) + b_0$$

21.4.9 Decision trees

Decision tree algorithms are widely used machine learning methods for classification as well as regression problems. When it comes to predicting and modeling hazards such as gully erosion, decision tree algorithms are favored due to their interpretability and ease of implementation. These algorithms have been successfully applied in various studies to predict gully erosion and assess erosion susceptibility in different geographical locations. Some sources have compared the performance of decision tree algorithms with other machine learning models, such as random forest and boosted regression trees, in predicting gully erosion potential (Malinowski, Heckrath, Rybicki, & Eltner, 2023). Other studies have utilized decision trees to model discrete properties related to gully erosion, such as the presence or absence of gully headcuts (Arabameri, Cerda, et al., 2020). Additionally, the use of decision tree algorithms in combination with RS data and GIS has shown promising results in predicting gully erosion susceptibility (Saha, Roy, Arabameri, Blaschke, & Bui, 2020). In recent research, Arabameri et al. (2021) proposed a novel ensemble of machine learning algorithms, including decision trees, for predicting maps of gully erosion susceptibility (Iqbal, Perez, & Barthelemy, 2021). These studies and approaches highlight the effectiveness of decision tree algorithms in gully erosion prediction and emphasize their potential for improved accuracy and robustness compared to other machine learning methods. Among the decision trees used for gully erosion prediction are random forest, gradient-boosted regression tree, naïve Bayes tree, and tree ensemble models (Busch, Hardt, Nir, & Schütt, 2021). These algorithms, particularly the gradient-boosted regression tree, have shown promising results in accurately detecting gully erosion-prone zones (Saha et al., 2020).

21.4.10 Random forest

A machine learning approach called random forest builds a large number of decision trees during training and outputs the class that is the mean prediction (regression) of the individual trees or the mode of the classes (classification) (Austin et al., 2022). It is a highly flexible algorithm that is capable of handling both categorical and numerical data, making it suitable for various types of predictions, including gully erosion prediction. Three user-defined parameters are present in an random forest: (1) the number of variables utilized to build each decision tree, which displays the independent tree's power; (2) the total number of trees in the RF; and (3) the minimum number of nodes in each tree (Arabameri, Chen, et al., 2020). The random forest algorithm has been successfully applied in predicting gully erosion, achieving high accuracies and robustness compared to other machine learning approaches (Malinowski et al., 2023). By utilizing a random forest algorithm for gully erosion prediction, researchers and practitioners can harness the power of ensemble learning to improve the accuracy and reliability of their predictions (Yang et al., 2021). In a study conducted by Saha et al. (2020), the Hinglo River basin's gully erosion susceptibility was assessed using the random forest algorithm, and it was discovered that RF had the best simulation effect and the highest prediction accuracy, suggesting that it could be used to assess gully erosion susceptibility in other locations with comparable geological conditions.

To summarize from previous studies, the use of geospatial tools and machine learning algorithms for predicting gully erosion vulnerability has shown promising results. Since machine-learning techniques do not rely on stringent assumptions and are self-adoptive and data-driven, the selection of appropriate machine-learning algorithms and the optimization of model parameters are crucial in ensuring reliable predictions. Researchers should explore the performance of different algorithms and identify the most suitable ones for gully erosion vulnerability assessment. Furthermore, integrating domain knowledge and expert inputs into the modeling process can help in refining the predictions and increasing the interpretability of the models. At the same time, the applicability of statistical models cannot be sidelined, as they provide very important insight to efficiently and parsimoniously develop the architecture of neural networks for particular data.

Despite all the encouraging positive aspects of high-end tools and algorithms for predicting gully erosion vulnerability, it is important to critically review the limitations and potential areas for improvement in this approach. One of the key limitations is the availability and quality of input data, including topographic, soil, land use, and rainfall data. Inaccuracies in these input data can lead to unreliable vulnerability predictions. Therefore, future research should focus on improving data collection methods and integrating multiple data sources to enhance the accuracy of gully erosion vulnerability assessments.

Moreover, there is a need for standardized protocols and guidelines for assessing gully erosion vulnerability using geospatial tools and machine learning algorithms. This would facilitate comparability across studies and enhance the reliability and robustness of vulnerability assessments. In conclusion, while geospatial tools and machine learning algorithms offer great potential for predicting gully erosion vulnerability, addressing the limitations and exploring opportunities for improvement is essential for advancing the accuracy and reliability of these predictions. Future research should focus on enhancing data quality, refining model selection and parameter optimization, and developing standardized protocols for gully erosion vulnerability assessment.

21.5 Future research directions for predicting gully erosion vulnerability

21.5.1 Enhanced data collection and integration

Future research in this area should prioritize the improvement of data collection methods and integration of multiple data sources. This includes exploring advanced RS techniques, such as LiDAR and UAV-based surveys, to capture high-resolution topographic and land use data. Furthermore, efforts should be made to enhance the quality and accessibility of soil and rainfall data through improved monitoring and data-sharing initiatives.

21.5.2 Model selection and parameter optimization

Researchers should continue to explore the performance of different machine learning algorithms, considering advancements in this field, and identify the most suitable ones for gully erosion vulnerability assessments. Additionally, the optimization of model parameters should be a focal point, utilizing techniques such as grid search and cross-validation to enhance the robustness of the predictions.

21.5.3 Integration of domain knowledge

The incorporation of domain knowledge and expert inputs into the modeling process should be further emphasized. Collaboration with geoscientists, hydrologists, and soil experts can provide valuable insights for refining the predictions and increasing the interpretability of the models.

21.5.4 Standardized protocols and guidelines

The development of standardized protocols and guidelines for gully erosion vulnerability assessment using geospatial tools and machine learning algorithms is imperative. This should involve the establishment of best practices for data preprocessing, model training, and validation, ensuring consistency and comparability across studies.

21.5.5 Interdisciplinary collaborations

Encouraging interdisciplinary collaborations between researchers from geospatial science, environmental engineering, and earth sciences can lead to comprehensive advancements in predicting gully erosion vulnerability. Leveraging diverse expertise can facilitate the development of holistic approaches for addressing the complexities of gully erosion dynamics.

By addressing these future scopes of work, the accuracy, reliability, and applicability of geospatial tools and machine learning algorithms in predicting gully erosion vulnerability can be significantly enhanced, furthering the understanding and management of this critical environmental issue.

21.6 Conclusion

Gully erosion is a significant environmental issue that poses threats to landscapes and communities worldwide. To address this problem, researchers have increasingly turned to machine learning techniques for gully erosion prediction. These techniques leverage algorithms and statistical models to analyze various environmental data, such as rainfall patterns, soil composition, topography, and land use factors. The critical review of gully erosion vulnerability prediction using geospatial tools and machine learning algorithms has shown significant potential for improving accuracy and efficiency in identifying areas at risk. The integration of geospatial data and machine learning algorithms has proven to be a valuable approach for assessing gully erosion susceptibility. By training these models on historical gully erosion data, machine learning algorithms can identify patterns and relationships between the environmental variables and gully erosion occurrences. This enables the creation of predictive models that can assess the susceptibility of different areas to gully erosion. By combining different machine learning algorithms, researchers have been able to improve the accuracy of gully erosion predictions and provide valuable insights for land management and planning purposes. However, while these methods offer promising results, there is a need for further research to address the challenges associated with data integration, model validation, and the incorporation of temporal dynamics. Furthermore, the use of genetic algorithms in combination with machine learning models has shown promise in enhancing the prediction accuracy of gully erosion susceptibility maps. These advancements in machine learning have revolutionized gully erosion research by providing more accurate and reliable predictions. Additionally, the application of these techniques in diverse geographical regions and under different environmental conditions remains a crucial area for future investigation. Nonetheless, the advancements in geospatial tools and machine learning algorithms provide a solid foundation for enhancing our understanding of gully erosion vulnerability and for developing effective mitigation strategies.

References

Aber, J. S., Marzolff, I., & Ries, J. B. (2010). *Gully erosion monitoring* (pp. 193–200). Elsevier BV. Available from 10.1016/b978-0-444-53260-2.10013-4.

Amiri, M., & Pourghasemi, H. R. (2020). *Mapping and preparing a susceptibility map of gully erosion using the mars model. Advances in Science, Technology and Innovation* (pp. 405–413). Iran: SpringerNature. Available from https://www.springer.com/series/15883, 10.1007/978-3-030-23243-6_27.

Arabameri, A., Cerda, A., & Tiefenbacher, J. P. (2019). Spatial pattern analysis and prediction of gully erosion using novel hybrid model of entropy-weight of evidence. *Water, 11*(6), 1129. Available from https://doi.org/10.3390/w11061129.

Arabameri, A., Nalivan, O. A., Pal, S. C., Chakrabortty, R., Saha, A., Lee, S., ... Bui, D. T. (2020). Novel machine learning approaches for modelling the gully erosion susceptibility. *Remote Sensing, 12*(17), 2833. Available from https://doi.org/10.3390/rs12172833.

Arabameri, A., & Pourghasemi, H. R. (2019). *Spatial modeling of gully erosion using linear and quadratic discriminant analyses in GIS and R* (pp. 299–321). Elsevier BV. Available from 10.1016/b978-0-12-815226-3.00013-2.

Arabameri, A., Cerda, A., Pradhan, B., Tiefenbacher, J. P., Lombardo, L., & Bui, D. T. (2020). A methodological comparison of head-cut based gully erosion susceptibility models: combined use of statistical and artificial intelligence. *Geomorphology, 359.* Available from https://doi.org/10.1016/j.geomorph.2020.107136, http://www.elsevier.com/inca/publications/store/5/0/3/3/3/4/.

Arabameri, A., Chandra Pal, S., Costache, R., Saha, A., Rezaie, F., Seyed Danesh, A., ... Hoang, N. D. (2021). Perdition of gully erosion susceptibility mapping using novel ensemble machine learning algorithms. *Natural Hazards and Risk, 12*(1), 469–498. Available from https://doi.org/10.1080/19475705.2021.1880977, http://www.tandfonline.com/toc/tgnh20/current.

Arabameri, A., Chen, W., Blaschke, T., Tiefenbacher, J. P., Pradhan, B., & Bui, D. T. (2020). Gully head-cut distribution modeling using machine learning methods—a case study of N.W. Iran. *Water (Switzerland), 12*(1). Available from https://doi.org/10.3390/

w12010016, https://res.mdpi.com/d_attachment/water/water-12-00016/article_deploy/water-12-00016-v4.pdf.

Arabameri, A., Pradhan, B., Pourghasemi, H. R., Rezaei, K., & Kerle, N. (2018). Spatial modelling of Gully erosion using GIS and R programing: a comparison among three data mining algorithms. *Applied Sciences (Switzerland)*, *8*(8). Available from https://doi.org/10.3390/app8081369, http://www.mdpi.com/2076-3417/8/8/1369/pdf.

Arabameri, A., Pradhan, B., Rezaei, K., Yamani, M., Pourghasemi, H. R., & Lombardo, L. (2018). Spatial modelling of gully erosion using evidential belief function, logistic regression, and a new ensemble of evidential belief function−logistic regression algorithm. *Land Degradation and Development*, *29*(11), 4035−4049. Available from https://doi.org/10.1002/ldr.3151, http://onlinelibrary.wiley.com/journal/10.1002/(ISSN)1099-145X.

Austin, A. M., Ramkumar, N., Gladders, B., Barnes, J. A., Eid, M. A., Moore, K. O., ... Goodney, P. P. (2022). Using a cohort study of diabetes and peripheral artery disease to compare logistic regression and machine learning via random forest modeling. *BMC Medical Research Methodology*, *22*(1). Available from https://doi.org/10.1186/s12874-022-01774-8, https://bmcmedresmethodol.biomedcentral.com/.

Avni, Y. (2005). Gully incision as a key factor in desertification in an arid environment, the Negev highlands. *Catena*, *63*(2−3), 185−220. Available from https://doi.org/10.1016/j.catena.2005.06.004.

Azareh, A., Rahmati, O., Rafiei-Sardooi, E., Sankey, J. B., Lee, S., Shahabi, H., & Ahmad, B. B. (2019). Modelling gully-erosion susceptibility in a semi-arid region, Iran: Investigation of applicability of certainty factor and maximum entropy models. *Science of the Total Environment*, *655*, 684−696. Available from https://doi.org/10.1016/j.scitotenv.2018.11.235, http://www.elsevier.com/locate/scitotenv.

Azedou, A., Lahssini, S., Khattabi, A., Meliho, M., & Rifai, N. (2021). A methodological comparison of three models for gully erosion susceptibility mapping in the rural municipality of El Faid (Morocco). *Sustainability*, *13*(2), 682. Available from https://doi.org/10.3390/su13020682.

Baiddah, A., Krimissa, S., Hajji, S., Ismaili, M., Abdelrahman, K., El Bouzekraoui, M., ... Namous, M. (2023). Head-cut gully erosion susceptibility mapping in semi-arid region using machine learning methods: insight from the high atlas, Morocco. *Frontiers in Earth Science*, *11*, 1184038.

Band, S. S., Janizadeh, S., Pal, S. C., Saha, A., Chakrabortty, R., Shokri, M., & Mosavi, A. (2020). Novel ensemble approach of deep learning neural network (Dlnn) model and particle swarm optimization (PSO) algorithm for prediction of gully erosion susceptibility. *Sensors (Switzerland)*, *20*(19), 1−28. Available from https://doi.org/10.3390/s20195609, https://www.mdpi.com/1424-8220/20/19/5609/pdf.

Busch, R., Hardt, J., Nir, N., & Schütt, B. (2021). Modeling gully erosion susceptibility to evaluate human impact on a local landscape system in Tigray, Ethiopia. *Germany Remote Sensing*, *13*(10). Available from https://doi.org/10.3390/rs13102009, https://www.mdpi.com/2072-4292/13/10/2009/pdf.

Chowdhuri, I., Pal, S. C., Arabameri, A., Saha, A., Chakrabortty, R., Blaschke, T., ... Band, S. S. (2020). Implementation of artificial intelligence based ensemble models for gully erosion susceptibility assessment. *Remote Sensing*, *12*(21), 3620. Available from https://doi.org/10.3390/rs12213620.

Conforti, M., Aucelli, P. P. C., Robustelli, G., & Scarciglia, F. (2011). Geomorphology and GIS analysis for mapping gully erosion susceptibility in the Turbolo stream catchment

(Northern Calabria, Italy). *Natural Hazards, 56*(3), 881–898. Available from https://doi.org/10.1007/s11069-010-9598-2, http://www.wkap.nl/journalhome.htm/0921-030X.

Conoscenti, C., Agnesi, V., Angileri, S., Cappadonia, C., Rotigliano, E., & Märker, M. (2013). A GIS-based approach for gully erosion susceptibility modelling: a test in Sicily, Italy. *Environmental Earth Sciences, 70*(3), 1179–1195. Available from https://doi.org/10.1007/s12665-012-2205-y.

Conoscenti, C., Angileri, S., Cappadonia, C., Rotigliano, E., Agnesi, V., & Märker, M. (2014). Gully erosion susceptibility assessment by means of GIS-based logistic regression: a case of Sicily (Italy). *Geomorphology, 204*, 399–411. Available from https://doi.org/10.1016/j.geomorph.2013.08.021.

Conoscenti, C., & Rotigliano, E. (2020). Predicting gully occurrence at watershed scale: comparing topographic indices and multivariate statistical models. *Geomorphology, 359*, 107123. Available from https://doi.org/10.1016/j.geomorph.2020.107123.

Eke, E. I., & Ogba, K. T. U. (2020). *Challenges of addressing natural disasters in Nigeria through public policy implementation: An examination of Isuikwuato erosion and the ecological fund. Economic effects of natural disasters: Theoretical foundations, methods, and tools* (pp. 397–437). Nigeria: Elsevier. Available from https://www.elsevier.com/books/economic-effects-of-natural-disasters/chaiechi/978-0-12-817465-4, https://doi.org/10.1016/B978-0-12-817465-4.00025-X.

Friedman, J. H. (1991). Multivariate adaptive regression splines. *The Annals of Statistics, 19*(1). Available from https://doi.org/10.1214/aos/1176347963.

Gayen, A., Haque, S. M., & Saha, S. (2020). *Modeling of gully erosion based on random forest using GIS and R. Advances in science, technology and innovation* (pp. 35–44). SpringerNature. Available from https://www.springer.com/series/15883, https://doi.org/10.1007/978-3-030-23243-6_3.

Gómez-Gutiérrez, Á., Conoscenti, C., Angileri, S. E., Rotigliano, E., & Schnabel, S. (2015). Using topographical attributes to evaluate gully erosion proneness (susceptibility) in two mediterranean basins: Advantages and limitations. *Natural Hazards, 79*, 291–314. Available from https://doi.org/10.1007/s11069-015-1703-0, http://www.wkap.nl/journalhome.htm/0921-030X.

Gupta, A. K., Sarkar, K. A., Bhattacharya, D., & Dhakre, D. S. (2022). Potato yield modeling based on meteorological factors using discriminant analysis and artificial neural networks. *International Journal of Vegetable Science, 28*(5), 465–476. Available from https://doi.org/10.1080/19315260.2021.2021342, http://www.tandfonline.com/toc/wijv20/current.

Gupta, A. K., Sarkar, K. A., Dhakre, D. S., & Bhattacharya, D. (2023). Weather based crop yield prediction using artificial neural networks: a comparative study with other approaches. *MAUSAM, 74*(3), 825–832. Available from https://doi.org/10.54302/mausam.v74i3.174.

Gutiérrez, A. G., Schnabel, S., & Felicísimo, A. M. (2009). Modelling the occurrence of gullies in rangelands of southwest Spain. *Earth Surface Processes and Landforms, 34*(14), 1894–1902. Available from https://doi.org/10.1002/esp.1881Spain, http://www3.interscience.wiley.com/cgi-bin/fulltext/122617362/PDFSTART.

Hembram, T. K., Paul, G. C., & Saha, S. (2020). Modelling of gully erosion risk using new ensemble of conditional probability and index of entropy in Jainti River basin of Chotanagpur Plateau Fringe Area, India. *Applied Geomatics, 12*(3), 337–360. Available from https://doi.org/10.1007/s12518-020-00301-y, http://www.springerlink.com/content/1866-9298/.

Hughes, A. O., & Prosser, I. P. (2012). Gully erosion prediction across a large region: Murray−Darling Basin, Australia. *Soil Research*, *50*(4), 267−277.

Iqbal, U., Perez, P., & Barthelemy, J. (2021). A process-driven and need-oriented framework for review of technological contributions to disaster management. *Heliyon*, *7*(11). Available from https://doi.org/10.1016/j.heliyon.2021.e08405, http://www.journals.elsevier.com/heliyon/.

Jahantigh, M., & Pessarakli, M. (2011). Causes and effects of gully erosion on agricultural lands and the environment. *Communications in Soil Science and Plant Analysis*, *42* (18), 2250−2255. Available from https://doi.org/10.1080/00103624.2011.602456.

Kheir, R. B., Wilson, J., & Deng, Y. (2007). Use of terrain variables for mapping gully erosion susceptibility in Lebanon. *Earth Surface Processes and Landforms*, *32*(12), 1770−1782. Available from https://doi.org/10.1002/esp.1501.

Lei, X., Chen, W., Avand, M., Janizadeh, S., Kariminejad, N., Shahabi, H., . . . Mosavi, A. (2020). GIS-based machine learning algorithms for gully erosion susceptibility mapping in a semi-arid region of Iran. *Remote Sensing*, *12*((15), 2478. Available from https://doi.org/10.3390/rs12152478.

Malinowski, R., Heckrath, G., Rybicki, M., & Eltner, A. (2023). Mapping rill soil erosion in agricultural fields with UAV-borne remote sensing data. *Earth Surface Processes and Landforms*, *48*(3), 596−612. Available from https://doi.org/10.1002/esp.5505, http://onlinelibrary.wiley.com/journal/10.1002/(ISSN)1096-9837.

Marjanović, M., Kovačević, M., Bajat, B., & Voženílek, V. (2011). Landslide susceptibility assessment using SVM machine learning algorithm. *Engineering Geology*, *123*(3), 225−234. Available from https://doi.org/10.1016/j.enggeo.2011.09.006.

McIvor, I., Youjun, H., Daoping, L., Eyles, G., & Pu, Z. (2017). *Agroforestry: Conservation trees and erosion prevention*. Elsevier BV. Available from 10.1016/b978-0-08-100596-5.22382-2.

Mokarram, M., & Zarei, A. R. (2021). Determining prone areas to gully erosion and the impact of land use change on it by using multiple-criteria decision-making algorithm in arid and semi-arid regions. *Geoderma*, *403*. Available from https://doi.org/10.1016/j.geoderma.2021.115379, http://www.elsevier.com/inca/publications/store/5/0/3/3/2.

Mountrakis, G., Im, J., & Ogole, C. (2011). Support vector machines in remote sensing: A review. *ISPRS Journal of Photogrammetry and Remote Sensing*, *66*(3), 247−259. Available from https://doi.org/10.1016/j.isprsjprs.2010.11.001.

Naghibi, S. A., Moghaddam, D. D., Kalantar, B., Pradhan, B., & Kisi, O. (2017). A comparative assessment of GIS-based data mining models and a novel ensemble model in groundwater well potential mapping. *Journal of Hydrology*, *548*, 471−483. Available from https://doi.org/10.1016/j.jhydrol.2017.03.020, http://www.elsevier.com/inca/publications/store/5/0/3/3/4/3.

Nachtergaele, J., Poesen, J., Sidorchuk, A., & Torri, D. (2002). Prediction of concentrated flow width in ephemeral gully channels. *Hydrological Processes*, *16*(10), 1935−1953.

Ohlmacher, G. C., & Davis, J. C. (2003). Using multiple logistic regression and GIS technology to predict landslide hazard in northeast Kansas, USA. *Engineering Geology*, *69* (3−4), 331−343. Available from https://doi.org/10.1016/S0013-7952(03)00069-3, http://www.elsevier.com/inca/publications/store/5/0/3/3/0/.

Pal, S. C., Arabameri, A., Blaschke, T., Chowdhuri, I., Saha, A., Chakrabortty, R., . . . Band, S. S. (2020). Ensemble of machine-learning methods for predicting gully erosion susceptibility. *Remote Sensing, 12*(22), 3675. Available from https://doi.org/10.3390/rs12223675.

Patton, P. C., & Schumm, S. A. (1975). Gully erosion, Northwestern Colorado: A threshold phenomenon. *Geology, 3*(2), 88−90. Available from https://pubs.geoscienceworld.org/geology, https://doi.org/10.1130/0091-7613(1975)3 < 88:GENCAT > 2.0.CO;2.

Poesen, J., Nachtergaele, J., Verstraeten, G., & Valentin, C. (2003). Gully erosion and environmental change: Importance and research needs. *Catena, 50*(2−4), 91−133. Available from https://doi.org/10.1016/S0341-8162(02)00143-1.

Poesen, J., Vandaele, K., & van Wesemael, B. (1998). *Gully erosion: Importance and model implications* (pp. 285−311). Springer Nature. Available from https://doi.org/10.1007/978-3-642-58913-3_22.

Poesen, J. W. A., Torri, D. B., & Van Walleghem, T. (2011). *Gully erosion: Procedures to adopt when modelling soil erosion in landscapes affected by gullying. Handbook of erosion modelling* (pp. 360−386). John Wiley and Sons. Available from http://onlinelibrary.wiley.com/book/10.1002/9781444328455, https://doi.org/10.1002/9781444328455.ch19.

Pourghasemi, H. R., Sadhasivam, N., Kariminejad, N., & Collins, A. L. (2020). Gully erosion spatial modelling: Role of machine learning algorithms in selection of the best controlling factors and modelling process. *Geoscience Frontiers, 11*(6), 2207−2219. Available from https://doi.org/10.1016/j.gsf.2020.03.005, https://www.sciencedirect.com/journal/geoscience-frontiers.

Pourghasemi, H. R., Yousefi, S., Kornejady, A., & Cerdà, A. (2017). Performance assessment of individual and ensemble data-mining techniques for gully erosion modeling. *Science of the Total Environment, 609*, 764−775. Available from https://doi.org/10.1016/j.scitotenv.2017.07.198, http://www.elsevier.com/locate/scitotenv.

Rahmati, O., Haghizadeh, A., Pourghasemi, H. R., & Noormohamadi, F. (2016). Gully erosion susceptibility mapping: The role of GIS-based bivariate statistical models and their comparison. *Natural Hazards, 82*(2), 1231−1258. Available from https://doi.org/10.1007/s11069-016-2239-7, http://www.wkap.nl/journalhome.htm/0921-030X.

Rahmati, O., Tahmasebipour, N., Haghizadeh, A., Pourghasemi, H. R., & Feizizadeh, B. (2017). Evaluation of different machine learning models for predicting and mapping the susceptibility of gully erosion. *Geomorphology, 298*, 118−137. Available from https://doi.org/10.1016/j.geomorph.2017.09.006, http://www.elsevier.com/inca/publications/store/5/0/3/3/3/4/.

Razavi-Termeh, S. V., Sadeghi-Niaraki, A., & Choi, S. M. (2020). Gully erosion susceptibility mapping using artificial intelligence and statistical models. *Geomatics, Natural Hazards and Risk, 11*(1), 821−845. Available from https://doi.org/10.1080/19475705.2020.1753824, http://www.tandfonline.com/toc/tgnh20/current.

Roy, J., & Saha, S. (2019). GIS-based gully erosion susceptibility evaluation using frequency ratio, cosine amplitude and logistic regression ensembled with fuzzy logic in Hinglo River Basin, India. *Remote Sensing Applications: Society and Environment, 15*, 100247. Available from https://doi.org/10.1016/j.rsase.2019.100247.

Roy, P., Pal, S. C., Arabameri, A., Chakrabortty, R., Pradhan, B., Chowdhuri, I., … Bui, D. T. (2020). Novel ensemble of multivariate adaptive regression spline with spatial logistic regression and boosted regression tree for gully erosion susceptibility. *Remote Sensing, 12*(20), 1−35. Available from https://doi.org/10.3390/rs12203284, https://www.mdpi.com/2072-4292/12/20/3284/pdf.

Saha, S., Roy, J., Arabameri, A., Blaschke, T., & Bui, D. T. (2020). Machine learning-based gully erosion susceptibility mapping: A case study of eastern India. *Sensors*, *20* (5), 1313. Available from https://doi.org/10.3390/s20051313.

Saha, S., Sarkar, R., Thapa, G., & Roy, J. (2021). Modeling gully erosion susceptibility in Phuentsholing, Bhutan using deep learning and basic machine learning algorithms. *India Environmental Earth Sciences*, *80*(8). Available from https://doi.org/10.1007/s12665-021-09599-2, https://link.springer.com/journal/12665.

Sain, S. R., & Vapnik, V. N. (1996). The nature of statistical learning theory. *Technometrics*, *38*(4), 409. Available from https://doi.org/10.2307/1271324.

Samani, A. N., Rad, F. T., Azarakhshi, M., Rahdari, M. R., & Rodrigo-Comino, J. (2018). Assessment of the sustainability of the territories affected by gully head advancements through aerial photography and modeling estimations: A case study on Samal Watershed, Iran. *Sustainability*, *10*(8), 2909. Available from https://doi.org/10.3390/su10082909.

Setargie, T. A., Tsunekawa, A., Haregeweyn, N., Tsubo, M., Rossi, M., Ardizzone, F., ... Meshesha, T. M. (2023). Modeling of gully erosion in Ethiopia as influenced by changes in rainfall and land use management practices. *Land*, *12*(5). Available from https://doi.org/10.3390/land12050947, http://www.mdpi.com/journal/land/.

Shen, S., Chen, J., Zhang, S., Cheng, D., Wang, Z., & Zhang, T. (2021). Deep Fusion of DOM and DSM Features for Benggang discovery. *ISPRS International Journal of Geo-Information*, *10*(8), 556. Available from https://doi.org/10.3390/ijgi10080556.

Shmueli, G., Bruce, P. C., Yahav, I., Patel, K. C., & Lichtendahl. (2018). *Data mining for business analytics—Concepts, techniques, and applications in R*. John Wiley and Sons, Inc.

Thwaites, R. N., Brooks, A. P., Pietsch, T. J., & Spencer, J. R. (2022). What type of gully is that? The need for a classification of gullies. *Earth Surface Processes and Landforms*, *47*(1), 109–128. Available from https://doi.org/10.1002/esp.5291, http://onlinelibrary.wiley.com/journal/10.1002/(ISSN)1096-9837.

Valentin, C., Poesen, J., & Li, Y. (2005). Laos gully erosion: Impacts, factors and control. *Catena*, *63*(2–3), 132–153. Available from https://doi.org/10.1016/j.catena.2005.06.001.

Vandaele, K., Poesen, J., Govers, G., & van Wesemael, B. (1996). Geomorphic threshold conditions for ephemeral gully incision. *Geomorphology*, *16*(2), 161–173. Available from https://doi.org/10.1016/0169-555x(95)00141-q.

Vandekerckhove, L., Poesen, J., Oostwoud Wijdenes, D., Nachtergaele, J., Kosmas, C., Roxo, M. J., & De Figueiredo, T. (2000). Thresholds for gully initiation and sedimentation in Mediterranean Europe. *Earth Surface Processes and Landforms*, *25*(11), 1201–1220.

Vandekerckhove, L., Poesen, J., Wijdenes, D. O., & De Figueiredo, T. (1998). Topographical thresholds for ephemeral gully initiation in intensively cultivated areas of the Mediterranean. *Catena*, *33*(3–4), 271–292.

Vanmaercke, M., Panagos, P., Vanwalleghem, T., Hayas, A., Foerster, S., Borrelli, P., ... Poesen, J. (2021). Measuring, modelling and managing gully erosion at large scales: A state of the art. *Earth-Science Reviews*, *218*. Available from https://doi.org/10.1016/j.earscirev.2021.103637, http://www.sciencedirect.com/science/journal/00128252.

Vapnik, V. N. (1995). *The nature of statistical learning theory*. New-York: Springer-Verlag.

Wang, B., Zhang, Z., Wang, X., Zhao, X., Yi, L., & Hu, S. (2020). Object-based mapping of gullies using optical images: A case study in the black soil region, northeast of China. *Remote Sensing*, *12*(3), 487. Available from https://doi.org/10.3390/rs12030487.

Yang, A., Wang, C., Pang, G., Long, Y., Wang, L., Cruse, R. M., & Yang, Q. (2021). Gully erosion susceptibility mapping in highly complex terrain using machine learning models. *ISPRS International Journal of Geo-Information*, *10*(10), 680. Available from https://doi.org/10.3390/ijgi10100680.

Zabihi, M., Mirchooli, F., Motevalli, A., Khaledi Darvishan, A., Pourghasemi, H. R., Zakeri, M. A., & Sadighi, F. (2018). Spatial modelling of gully erosion in Mazandaran Province, northern Iran. *Catena*, *161*, 1−13. Available from https://doi.org/10.1016/j.catena.2017.10.010, http://www.elsevier.com/inca/publications/store/5/2/4/6/0/9.

Zhang, G., Eddy Patuwo, B., & Hu, M. Y. (1998). Forecasting with artificial neural networks: The state of the art. *International Journal of Forecasting*, *14*(1), 35−62. Available from https://doi.org/10.1016/S0169-2070(97)00044-7, http://www.elsevier.com/locate/ijforecast.

Community preferences in the adaptation of river neighborhood at Skudai River, Malaysia

22

Rohana Mohd Firdaus[1], Mathanraj Seevarethnam[2], Tika Ainnunisa Fitria[3] and Sara Izrar Aziz[4]

[1]*Program of Landscape Architecture, Universiti Teknologi Malaysia, Skudai, Malaysia*
[2]*Department of Geography, Eastern University, Batticaloa, Sri Lanka*
[3]*Architecture Study Program, Universitas 'Aisyiyah Yogyakarta, Yogyakarta, Indonesia*
[4]*Interdisciplinary Program in Landscape Architecture, Seoul National University, Seoul, South Korea*

22.1 Introduction

Resilience can be widely viewed from different perspectives, such as ecosystem (Holling, 1973), disaster risk (Patel & Gleason, 2018), management (Walker et al., 2002), and community (Esteban, 2020). The latter perspective is associated with the adaptation shaping a community's resilience in a river neighborhood. It is essential not to forget the value of humans in the river landscape because the spatial environment is their home (Hayrol Azril et al., 2011). Hence, this study looked at a community that lived by the riverside as the setting. Adaptation in the river neighborhood is influenced by the social interaction in the open spaces near the water body. The interaction results in different kinds of resilience toward human-made issues such as river pollution. The community was more appreciative of its natural resources despite being the nearest to the area with the highest flood risk. The river environment has become a living place for some communities since the industrial era in the 1970s (Mann, 1973). The communities resided by the riverside mainly for its source of water, use of transportation, fertile soil, and so forth. Over time, the river environment evolved into a neighborhood as the population grew and the surrounding context became modern and urbanized. This history is evident in Kuala Lumpur's development, whereby the city had transformed from the Klang River and Gombak River confluence. Unfortunately, the changes in urban areas have become one of the sources of river pollution threatening the life of communities. This issue happened to the Skudai River in Johor, putting its water quality in class III. Those living in the river neighborhood are the most exposed to the river issues (Chiang, 2018). Despite that, some of them

Applications of Geospatial Technology and Modeling for River Basin Management.
DOI: https://doi.org/10.1016/B978-0-443-23890-1.00022-0

have grown resilient toward river pollution. They remained in the neighborhood for reasons, as shown by the Citarum River, Indonesia, and Pasig River, Philippines, communities. This situation is sometimes mistaken as resilience despite being associated with vulnerability, the opposite of resilience. Hence, this study aims to explore the preferences in their adaptation that have shaped community resilience in the neighborhood.

22.1.1 Adaptation within community resilience and river environment

Between resilience and vulnerability is the process of communities adapting to situations, sometimes called adaptive capacity. The difference is that resilience is adapting to a favorable or desired situation, whereas vulnerability is adapting to an undesired situation (Paul, Deka, Gujre, Rangan, & Mitra, 2019). This research focused on adaptation, which contributes to adaptive capacity. Adaptive capacity is still underresearched due to its complexity of being between two major keywords: resilience and vulnerability (Engle, 2011). Because of this, adaptation is emphasized as one of the keywords in the discussion. The adaptation is apparent through identifying social drivers (Wilson, 2012) contributing to the community's residence in the neighborhood. Although these drivers are not discussed in the findings, they are acknowledged for their importance in the adaptation. In this paper, the scope of resilience is situated in a river environment where communities' livelihood is portrayed. Studies of rivers and the integration of resilience have been increasing (Parsons & Thoms, 2018). Hence, this paper attempted to appreciate the integration between resilience and rivers. In general, resilience is associated with the capacity of a system to overcome disturbances without collapsing. This buzzword was first coined regarding the ecosystem (Holling, 1973). Over time, different fields borrowed and adjusted this word into their scope. For example, community resilience associated with rural community development (Samion, 2023). The author discovered how land use changes affected fishermen and farmers' resilience, yet they were more resilient than urban dwellers. Meanwhile, this study regards adaptive capacity as the bridge for a community to be resilient. The setting for that community was at the riverside, which supplemented its resilience by providing open spaces for the community to socialize. It was later found that social interaction was their way of adapting to the neighborhood.

Perception is culturally driven and may be unclear, resulting in uncertainty in river management. This uncertainty causes residents' preference toward their environment to become difficult to predict (Darby & Sear, 2008). Uncertainty is a part of resilience that relies on the adaptation to stay or move away from the environment. While preference is an essential key to ensuring a community's resilience, the preference also comes from the residents, suggesting what matters to them as the users of the river neighborhood (Mohd Firdaus, 2022). It is often found that preferences in landscapes are associated with esthetics (Le Lay, Piégay, & Rivière-Honegger, 2013; Tieskens, Van Zanten, Schulp, & Verburg, 2018). However, in this research, the focus on preferences is integrated with adaptation in addressing the growing gap in community resilience. Communities' experience of exposure to the river and its disturbances guides their preferences for what matters to their lives. The exposure allows them to understand what matters to them because their home is at risk. Appleton's theory on preference explained that it had been an intuitive characteristic of humankind influenced by the social aspect (Appleton, 1996). In this study's case, the intuition on what was best for the residents appeared through the social interaction within the community. Social interaction is explained further in the following section. Thus this preference is postulated to establish a sense of belonging, which fuels the community's adaptation to remain in the neighborhood.

22.1.2 Adaptation toward floods and river pollution

A flood is an unfortunate disturbance that commonly hits a river neighborhood and spills over to inland towns if left untreated. Over time, governments addressed this disturbance and developed various flood prevention measures to ensure the continuation of the livelihood of communities. This effort can be seen in Malaysia's Department of Irrigation and Drainage (DID), where flood mitigation has been in their scope since the 1970s. Nonetheless, following the phenomenon, communities in river neighborhoods were posited to develop their ability to remain in their living environment because that was their home. This situation showcases the resistance of that community to face and overcome the disturbance by remaining in the same living environment. Additionally, river pollution is also prevalent within the river environment. It detriments the ecosystem until the river can no longer provide cultural services that support the communities (Everard & Moggridge, 2012). River pollution seems to settle in the community's consciousness, exacerbating this situation and resulting in a more damaging outcome as it is against the value of the river. Despite that, both situations represent resistance toward the disturbances, commonly mistaken as resilience. If the resistance is favorable, such as the environment is good for community growth, it is called resilience, whereas if the resistance is unfavorable, such as the environment threatening the community's well-being, it is called vulnerability. These two outcomes are not static and constantly dynamic; they are called the adaptive cycle (Fath, Dean, & Katzmair, 2015). For this article, the cycle is interpreted as adaptation. Thus the term adaptation is used more often to reduce confusion of keywords.

22.1.3 Degradation of the value of river

This decade, rivers have been increasingly damaged by industrialization and pressuring cities (Shafaghat, Mir Ghasemi, Keyvanfar, Lamit, & Ferwati, 2017). The damage includes straightening, enlarging, and armoring rivers (Vietz, Rutherfurd, Fletcher, & Walsh, 2016) and illegal encroachment (Zinia & McShane, 2018). According to a report on Intergraded Basin Management, the latter damage is also common on riversides in Malaysia, giving the impression of vulnerability that leads to river pollution ("Integrated River Basin Management Report," 2017). However, some communities strive for resilience in facing these disturbances. Rural communities are more resilient toward disturbances because they rely heavily on the river for life (Naito, 2008; Yodsurang, Hiromi, & Yasufumi, 2015). In this paper, the attention is focused on urban communities that are more disconnected from the river (Wohl, 2014). A sense of connection with the river is important in restoring a better relationship between humans and the river (Fox et al., 2017). Nowadays, government effort has been moving toward appreciating the social aspect where, in previous decades, river management was lacking (Chan, 2012). The attention is bringing the community as the actor in river management before beautifying the river. This action is becoming more crucial because the river has become the community's living place. Thus it becomes the setting for this study. The Ministry of Environment and Water has structured a program called *Program Denai Sungai Kebangsaan* (DSK), which promotes nature-based solutions for better planetary health. This program centralizes the community as an actor who inspires others to improve and protect the river. An example of activities easily committed by the communities is *Kutip Sampah Sambil Riadah*, an action where people clean up the garbage as they were having recreational activity. This effort has highlighted the importance of community preferences regarding their living environment. These preferences show the community's willingness to take care of its home. Thus this paper aims to explore the community preferences in the adaptation that shapes the community resilience of the river neighborhood. With rivers continuously being degraded and polluted, the lack of social aspect in river management becomes more apparent. In order to restore the value of rivers, social understanding should be addressed to avoid poor efficiency when tackling the issue (Che, Li, Shang, Liu, & Yang, 2014). Community understanding should be integrated into river management and become an inspiration for river environment planning. Hence, Sustainable Development Goal 11: Sustainable Cities and Communities is addressed through the community preferences as an aspiration in managing rivers.

22.2 Rationale of the study

The study on resilience has been increasingly acknowledged, especially in many different dimensions, including the river (Everard & Moggridge, 2012; Parsons & Thoms, 2018) and social (Kondolf & Pinto, 2017; Ling & Chiang, 2018; Rufat, Tate, Burton, & Maroof, 2015). Despite the attention, the lack of respect for nature has grown, especially in urban places where water bodies have turned into a drainage. Rivers becoming drainage invites users to dispose of waste into the channel, resulting in river pollution. This issue is a growing concern in communities because it indicates that humans have mistreated one of the Earth's resources. While the flood requires a structured measure, river pollution requires a nonstructured measure. Thus this paper highlights communities' resilience in restoring awareness toward the river. This awareness is gained from the preferences that assist the communities in their adaptation to the river environment. With preferences taken into consideration, a sense of stewardship is nurtured as it is needed to restore the awareness. Therefore this study aimed to explore the preferences in their adaptation which shaped the community resilience in the neighborhood. In short, the rationale of this study is based on the hope of recovering the value of rivers through the community preferences that are believed to be capable as they deal with the root: communities.

22.3 Materials and methods

The materials for this study came from a river neighborhood community. The community was the residents of Kg. Pertanian/Separa, in Johor, Malaysia. They became the unit of analysis that participated in the brainstorming session to elicit their adaptation to the river environment and drawing activity to enhance the depth of the result. The following sections elaborate on the study area, the unit of analysis, methods, and analysis tools.

Study area

The study area was located at Kg. Pertanian/Separa within the district of Kulai, state of Johor, Malaysia (Fig. 22.1). It was a 40-minute drive to Johor Bahru's city center. Within the study area, out of the 47.4 km of Skudai River, 2 km flows through the neighborhood. This river flows from the western part of Johor (Sedenak) and discharges directly into Johor Strait (Danladi Bello, Mohd Haniffah, Hashim, & Anuar, 2018). The neighborhood was a kampung-like setting where the houses were landed.

The study area was decided based on the presence of a river, exposure to river pollution, and the existing community. These criteria helped to understand community resilience based on how they overcame the disturbances. It is worth noting that the disturbances were not limited to river pollution, as floods were also an issue when deciding the river as a setting for this study. Although floods had been resolved and no longer occurred after 2006, it was still inevitable to be separated from the discussion as it had value in their responses. Kg. Pertanian/Separa's worst flood was in 2004; they had to evacuate the neighborhood because the flood water reached their rooftop. This situation provided a more meaningful appreciation toward the river landscape

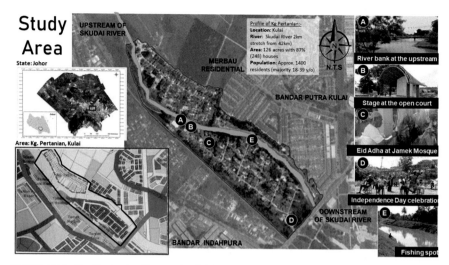

FIGURE 22.1 Kg. Pertanian/Separa is divided into two parts, namely, Pertanian (below) and Separa (upper part).

The formation of the neighborhood started with Pertanian and followed with Separa when the population increased in 1980s.

because the residents stayed by the riverside despite the unfortunate events. In addition, this neighborhood was not excluded from river pollution. According to Malaysian news, the Skudai River was one of the rivers in class III (Abdullah, 2018). According to the Department of Environment Malaysia, this category means that the water requires extensive treatment. This situation suggested that the Skudai River was polluted and threatening the neighborhood communities.

Despite that, the living environment of this neighborhood still indicated potential social bonding among the neighbors. Social bonding is an important factor in the social capital of adaptation (Dressel, Johansson, Ericsson, & Sandström, 2020). This indication was evident through their active participation in communal activities documented in the Jawatankuasa Kemajuan dan Keselamatan Kampung (JKKK) profile book. JKKK was an organization that managed its neighborhood. Each area had its own JKKK. Regarding appreciation, the community could easily access the river without a visual barrier at the riverbank, resulting in maximum exposure for the community to the river. Visual access contributes to feeling connected with the river without physical contact with the water (Durán Vian, Pons Izquierdo, & Serrano Martínez, 2021). This opportunity provided a better connection for the community to appreciate their river environment. Exposure to the reality of the river enhanced the response validity because the responses were based on the residents' experience of living through the disturbance. In addition, the advantage of being near the DID provided this neighborhood with careful supervision from the expert. Thus this study area became one of the references to promote river awareness through campaigns such as Community in River organized by DID. The participation from the community contributed to their cohesion, which helped their adaptation to facing river pollution and floods. Hence, this study area was selected to explore the community preferences toward the river neighborhood.

Unit of analysis

The respondents for this study were the community that resided in the river neighborhood of Kg. Pertanian/Separa, Johor, Malaysia. Around 36 residents gathered in a workshop and participated in the brainstorming session. Despite the population being over 1400, the participants were reliable enough to provide the appropriate response because their participation was based on their willingness to be involved. Their willingness was an early indication of adaptation. A community's motivation for adaptation involves their willingness to engage with disturbances (Wilson, 2012). The willingness carried value because the willingness to participate in the session showed that the river neighborhood had meaning to them. A community is willing to commit to what is meaningful to them (Gottwald & Stedman, 2020). Thus the community became the unit of analysis in this study.

Brainstorming session

This method was inspired from his handbook on community planning, which suggests that an interactive display of a map allows respondents to engage enjoyably, particularly on a post-it board (Wates, 2000). The method of using a map to indicate preferences had also been demonstrated successfully (Tieskens et al., 2018). The authors describe that this method allows for studying social interaction with the residents' landscape. Consistently, this study connected social interaction within the river environment to the community adaptation with the result of resilience. It means that the spaces in the river environment had meaning to the community. This method is also similar to that of Yamashita, who elicited perceptions of a Japanese river environment, where the residents were required to capture pictures that caught their eyes whether they liked it or not (Yamashita, 2002). This method was used to understand their opinion and feelings about the river. On the other hand, this paper elicited the residents' opinions through the meaningful spaces in the river environment, which included the neighborhood. A total of 36 local respondents participated. They were divided into five groups; each consisted of five to six respondents and was assigned a moderator to conduct the session. Each group was given an A1-sized map with sticky notes for the respondents to stick to the chosen river neighborhood spaces. However, their choices were not limited to the neighborhood, as they could pinpoint any spaces outside the boundary that they found

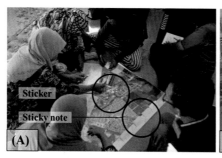

FIGURE 22.2 (A) Respondents indicating their meaningful spaces on the map with justification stuck at each side. (B) Respondents showing the outcome of the pinpointed map.

(A) The respondents were given freedom to indicate the meaningful spaces. (B) Each group was briefed and monitored by a moderator.

meaningful. This expansion of meaningful spaces helped to maximize their expressions, as the meaningful spaces led to the importance of their preferences. Fig. 22.2 demonstrates the process of the activity.

The participants had to use the sticky notes they stuck on the map to write down the reasons behind their choices for a particular space. The notes were numbered to avoid confusion and ensure the accuracy of the spaces and justification during analysis. For example, choice 1 parallels justification 1, and choice 2 parallels justification 2. Ultimately, the respondents produced five pinpointed maps, illustrating the spaces preferred for social interaction. The points were tabulated in Excel and imported into Quantum Geographic Information System (QGIS) to produce an overall of the pinpointed maps. Information on local spaces, such as preferred spaces, is beneficial in gathering landscape value because it involves the participation of the residents. Their involvement is significant and would lead to an aspiration for the river environment, thus enhancing their adaptation according to their preferences in the neighborhood.

Drawing activity

The drawing activity further supported the outcome of the brainstorming session. In response to the participants' pinpointed preferences, an in-depth preference was elicited through this activity. The respondents were asked to draw their preferred elements in the river environment. Their answers were responding to their justifications for the pinpointed spaces. The drawing outcomes were analyzed using Nvivo12 as a supplementary result. However, the drawing skills were varied among the participants. Therefore a text description was also welcomed to describe the specific elements the participants preferred to have in their neighborhood.

Tools of analysis

Two main tools were used to analyze the data: (1) QGIS and (2) Nvivo12. Firstly, the latitude and longitude of the pinpointed locations were identified manually using Google Maps. The locations from the five maps were tabulated in Excel before being exported into QGIS, a free geographic information system software. A map was generated using the software, showing the overall pinpoints by the respondents. Meanwhile, information on the justifications remained in Excel and was categorized based on content analysis. The content analysis was applied to interpret the justifications by categorizing them according to the activities mentioned on the sticky notes. The same approach was used for the drawing data, where the respondents' drawings reflected the elements they preferred per their adaptation.

22.4 Results and discussion

The approach to examining these results to minimize river pollution is holistic from a social perspective. The community's favorable preferences toward the river environment have the potential to foster appreciation within the community for river care. Therefore this research hopes to convey that appreciation plays a significant role in the community's sense of responsibility toward the river's health. A pinpointed map of meaningful spaces according to the community of the river neighborhood was produced (Fig. 22.3). It consisted of the local spatial functions divided into two main findings, namely, categories of preferred spaces and functions of spaces, elaborated in Sections 22.4.1 and 22.4.2, respectively.

24.4.1 Categories of preferred spaces in the river neighborhood

Table 22.1 tabulates the number of pinpoints, number of spaces, and examples of spaces and activities the community provided. Each category of spaces refers to particular areas with similar features. For example, (1) neighborhood facilities are the areas the community utilized for daily activities, and (2) the river environment refers to the river and the riverside. Next, (3) home compound indicates the residents' houses, and (4) greenery is the spaces that showed as green or open spaces as indicated by the map. It was found that the majority of the participants preferred spaces that consisted of facilities, and greenery was the least preferred category. Their justifications were related to social interaction—further elaboration in the following sections.

i Neighborhood facilities

The residents pinpointed spaces with facilities the most: as many as 43% of the total pinpoints. These facilities included the public spaces used by them. That percentage was made up of 13 spaces, namely, (1) the sidewalk, (2) the mosque,

FIGURE 22.3 The pinpointed map generated by QGIS.

The latitude and longitude were detected using Google Map and tabulated into a table before importing them into QGIS.

Table 22.1 Spaces preferred in the river neighborhood.

Category	(1) Neighborhood facilities	(2) River environment	(3) Home compound	(4) Greenery
No. of pinpoints	57 (43%)	47 (36%)	24 (18%)	4 (3%)
No. of spaces	13	2	1	2
Example of spaces	Sidewalk, mosque, sports court, car park, small shop, shop lot, mall, *warung*, residential park, kindergarten, petrol station, community hall, and PAWE hall	River, riverside	Residents' houses	Green or open space
Activities	Independence Day, *Maulidur Rasul*, Eid al-Adha, awareness campaigns, sports practicing, jogging	Fishing	Family gathering, visiting neighbors	Meeting and hanging out

(3) the sports court, (4) the car park, (5) the small shop, (6) the shop lot, (7) the mall, (8) the *warung*, (9) the residential park, (10) the kindergarten, (11) the petrol station, (12) the community hall, and (13) the PAWE hall. Out of these spaces, five were located outside of the neighborhood. They were the car park, the shop

FIGURE 22.4 The neighborhood environment of Kg. Pertanian/Separa.

Basic facilities (e.g., mosque and shop) fulfill the needs of the community in the neighborhood.

lot, the mall, the residential park and the petrol station. Fig. 22.4 illustrates the spaces mentioned around the neighborhood.

The justifications that came with the spaces consisted of the activities that supported their social interactions. The activities mentioned include annual events such as Independence Day, *Maulidur Rasul*, Eid al-Adha, and awareness campaigns. Also included are daily recreational activities such as playing sports and jogging. Fig. 22.5 shows examples of the community's annual events. This finding suggests that their preferences were based on the activities within the spaces they consistently engaged within the river neighborhood. These activities being part of the neighborhood norms indicates that the community had already adapted to them. The kinship developed through the social interactions within the activities allowed the residents to work in unison as a community of the river neighborhood. This finding is similar to that of Nemeth and Olivier, who highlighted that connectedness in a community is one of the keys to adapting or coping with the challenges in a neighborhood (Nemeth & Olivier, 2017). This evidence was shown in their permanency of staying in the neighborhood despite disturbances.

ii River environment

The river environment comprised 36% of the pinpoints and consisted of two spaces: the water body and the riverside. Although the river environment had

Community Sports Event in conjunction with Independence Day Celebration

Preparation for meat distribution during Eid al-Adha

FIGURE 22.5 Examples of community annual events that led to their gathering in the neighborhood.

The gathering of the community symbolized the strength of their togetherness in the neighborhood.

FIGURE 22.6 (A) Residents watching anglers catching fish on a rainy day. (B) A gathering was observed at one of the tributaries. (C) Catfish caught by the anglers.

(A) and (B) The residents have grown immune to the rainy days, even flood. (C) An indicator that the river in the neighborhood is in healthy state where fishes are thriving.

only two spaces, the percentage of pinpoints was the second highest. The sole activity for the river environment was fishing, indicating that recreational activity was preferred in the river neighborhood. The fishing activity in the neighborhood shared a similar outcome to catching fish in rural areas (Abu Samah et al., 2011), except that the residents here fish primarily for recreational purposes instead of generating income that supports the economy. This study found that the community had been performing the activity frequently, regularizing the activity. The regularity of performing that activity suggests that the activity was already within their norm. This finding reveals that the different adaptation to the lifestyle was derived from different needs; one community fished for income and food supply to sustain life (Naito, 2008), while another community fished as part of recreation because of the opportunity brought by the river. Fig. 22.6 shows the community's participation in fishing, even on rainy days.

Bunn et al. stated that fish's existence is one of the indicators of a healthy eco-system; thus it can be said that the Skudai River in the neighborhood was reasonably healthy, which encouraged the residents to fish (Bunn et al., 2010). The encouragement led to the human−nature relationship, which grew through fishing and contributed to the space function. Accordingly, their familiarity with this situation was positioned through their engagement with the river; thus adaptation was manifested. The adaptation was clear, especially when the experience of water disturbances did not discontinue their engagement with the water body for recreational activities. This scene exemplifies the activities that had become part of the residents' norms.

iii Home compound

With 18%, the home compound was the third most-pinpointed category. The residents' houses represented it. Many functions in this category contributed to the socializing function (see Section 22.4.2.1), where their houses became the destination for visiting, gathering, and performing religious acts such as *kenduri*, followed by activities such as eating and buying at the houses that also functioned as shops and dining areas. *Kenduri* is part of the Malay culture performed annually, consisting of a feast and recitation of duas for the deceased. This finding showcases that such activities were meaningful because they were part of the culture of the river neighborhood. Thus these spaces carried meaning according to the activities conducted in them. Furthermore, these activities were the sustenance of their neighborhood's harmonious living. Their way of living overshadowed the disturbances threatening their lives, especially the flood in 2004. Hence, adaptation was always fundamental to their existence in the neighborhood and further strengthened by the social interactions in their gathering activities.

iv Greenery

The last of the categories is greenery, which comprised only 3% of the pinpoints. Greenery refers to any green space without a specific landmark, including trees and open spaces. Unfortunately, this finding was considered insignificant as the justifications provided for the pinpoints were inconsistent. The inconsistency might be due to the inaccuracy in pinpointing the locations. Nevertheless, the following sections identified and discussed functions such as socializing and a sense of familiarity.

Function of spaces in the river neighborhood

The four categories of preferred spaces were further broken down according to functions, such as (1) socializing (Milestad, Westberg, Geber, & Björklund, 2010), (2) sense of familiarity (Verbrugge, Ganzevoort, Fliervoet, Panten, & van den Born, 2017), (3) sense of attachment (Gottwald & Stedman, 2020), (4) response (Ling & Chiang, 2018), and (5) uncertain preferences. Table 22.2 presents the results of the river neighborhood's functions, activities or elements, and spaces.

Table 22.2 Summary of the functions of spaces in the river neighborhood.

Functions	1. Socializing (n = 82, 62%)				2. Sense of familiarity (n = 29, 22%)		3. Sense of attachment (n = 9, 7%)	4. Response (n = 5, 4%)	5. Uncertain preferences (n = 7, 5%)
Activities/elements	a. Meeting and gathering	b. Recreational activities	c. Communal activities	d. Knowledge gaining	a. Surrounding landscape	b. Location and events	Appreciation	Suggestion	Dining at "warung," income, accident, hobby, needs
No. of sticky notes scribed with function	22	30	22	8	10	19	9	5	7
Neighborhood facilities	▣	▣	▣	▣	▣	▣	▣		▣
River environment	▣	▣	▣	▣	▣	▣	▣	▣	▣
Home compound	▣	▣	▣	▣		▣	▣		▣
Greenery		▣	▣			▣			

v Socializing

Socializing is an essential denotation of the space function (Ujang, 2016). Based on the results, it was found that 62% of the functions revolved around socializing in all four spaces. The residents socialized through (1) meetings and gatherings, (2) recreation, (3) communal activities, and (4) knowledge-gaining activities. These activities encouraged the residents to socialize in the spaces with neighborhood facilities, river environment, home compound, and greenery. Recreational activities dominated the river environment, with 30 pinpoints associated with the area (see Figs. 22.7–22.10). The activities included sports, fishing, and leisure activities such as jogging, walking, and relaxing.

The activities mentioned above gathered the residents in one space, allowing them to socialize. This finding shows that it had been part of their norms to

FIGURE 22.7 Examples of spaces where socializing took place in the river environment.
Walkway is where the residents gather for their morning jogging activity.

FIGURE 22.8 Examples of spaces where socializing took place in the river environment.
While residents fish, the shelter was used as part of their resting place.

FIGURE 22.9 Examples of spaces where socializing took place in the river environment.
The residents were able to socialize during fishing in group.

FIGURE 22.10 Examples of spaces where socializing took place in the river environment.
The fishing activity occurred at the decking where the residents were able to meet one another for conversation.

perform recreational activities when they got together. To them, social interaction was typical, as they were comfortable communicating. The social interaction that happened through recreational activities led to the enhancement of their social bonding. This finding is similar to that of Smith et al., who found that socializing strengthens social bonding (Smith, Davenport, Anderson, & Leahy, 2011). From the perspective of adaptation, the prominence of recreational activities superseded the memories of disturbances. It means that adaptation is more associated with socializing rather than disturbances. This finding is in contrast with the findings of McEwen et al., who found that repeated flooding was associated with adaptation (McEwen, Garde-Hansen, Holmes, Jones, & Krause, 2017). For the community in Kg. Pertanian/Separa, the disturbances were not significant, as evident from the results; the socializing function brought more meaning to the community's life than the remembrance of floods. This finding reveals that the social aspect was one of the roles that shaped their adaptation in the river neighborhood. Recreational activities are followed by meetings, gatherings, and communal activities, each with 22 pinpoints and knowledge gaining with eight pinpoints; these also contributed to socializing. Communal activities occurred in all spaces, especially those with facilities (see, Figs. 22.11 and 22.12). For example, the residents had meetings and programs at the community hall. Such meetings and gatherings happened close to the river, home compounds, and spaces with facilities, which included the *warung*. The meetings and gatherings at *warung* were casually termed *ngetih* or *ngeteh* and *membawang*. These terms were defined as conversing while having a drink with friends.

Knowledge gaining happened everywhere, except in green and open spaces. For instance, the residents learned *silat*, a form of Malay martial arts, by the riverside. Another example of knowledge gaining is learning to play the *kompang*, a Malay musical instrument commonly played during weddings. Both *silat* and *kompang* are pertinent to the traditional Malay culture. These examples reflect that in gaining such knowledge, the residents had the opportunity to socialize in

FIGURE 22.11 Examples of communal activities that took place at the futsal court.

This picture was taken during the Independence Day Celebration where the community gather and conduct communal activities that involve children, teenagers, adults, and elders.

FIGURE 22.12 Examples of gathering at the.

warung by the riverside. This *warung* has become one of the attractions for the residents to meet neighbors and *ngetih* (similar to chatting or gossiping).

that situation. Since these activities gathered them for socializing, the residents appreciated their neighborhood. The appreciation was denoted by their pinpoints of the spaces, which shows that they were functioning as a community in the river neighborhood despite disturbances. This situation indicates the process of becoming resilient (Patel & Gleason, 2018). It means that the adaptation consists of not only disturbances but also the life of a community. Due to the norm becoming dominant over the disturbance, this finding discovers that socializing was the root of the residents' persistence in living in the neighborhood. This finding parallels that of Rufat et al. and Parsons et al., who indicated that connectedness within a community contributes to the residents' adaptation to facing disturbances (Parsons et al., 2021; Rufat et al., 2015).

vi Sense of familiarity

From the socializing, the function evolved to the second function: the sense of familiarity, which refers to how used the residents feel to spaces (Southon, Jorgensen, Dunnett, Hoyle, & Evans, 2018). The sense of familiarity comprised 22% of the pinpoints, with (1) "surrounding landscape" contributing 19 pinpoints and (2) "location and events" contributing 10. The surrounding landscape refers to the existing design elements in the river neighborhood, such as ornamental plants and shelter, while the location and events refer to specific spaces, such as the *warung* and community hall, and events, such as open market.

22.4.2.2.1 Location and events

For "location and events," 19 pinpoints were placed at all spaces, and "locations" revolved around the PAWE hall, the *warung* (e.g., Warung Pak Dasit, Warung Imam, and Warung Ijul), and the market. Fig. 22.13 illustrates four locations pinpointed for "location and event."

Concerning socializing, these *warungs* were the location for the meetings and gatherings. The "events" describe what happened in a particular space. For example, the residents tasted food at the *warung*, shopped for goods at the grocery store near the home compound, and held the Saturday market at the shop lot. These examples show that they frequently visited the spaces because they knew the location and the related events. Their visits also allowed social interaction to take place. This finding parallels that of Ujang, who stated that frequent engagement encourages a community to be familiar with their environment (Ujang, 2016). This occurrence means that the sense of familiarity was promoted through socializing, where their neighborliness made them cohesive. The past disturbances did not change their relationship as neighbors. The occurring engagement saturated their adaptation; thus the community could stay and live harmoniously despite disturbances.

22.4.2.2.2 Surrounding landscape

For the "surrounding landscape," 10 pinpoints were marked at the river environment and spaces with facilities (see Fig. 22.14). These spaces were easily

FIGURE 22.13 Example of locations pinpointed for "location and event."

These locations were potential to be nodes of the neighborhood.

FIGURE 22.14 Examples of areas that are referred to as the surrounding landscape.

These landscapes represent the character of the river neighborhood.

accessible to the residents, encouraging engagement within the spaces. It was found that the residents became familiar with the spaces through visits.

Their visits gave meaning to those spaces by them being familiar enough to perform social interactions. In other words, the social interaction consequently led them to enhance their sense of familiarity with the surrounding landscape. This finding parallels that of Verbrugge et al., who discovered that familiarity is associated with community connectedness, particularly in a river landscape (Verbrugge et al., 2017). In this study, the residents were familiar with the surrounding landscape as they frequently engaged with the landscape throughout their stay in the neighborhood. It means that the residents acknowledged the spaces due to the memories of engaging within the spaces in the neighborhood. The familiarity helped them value the surrounding environment that supported their adaptation by appreciating the neighborhood's existing elements. This finding parallels those of Ujang, Kozlowski, and Maulan, who stated that community engagement becomes significant and memorable due to familiarity and engagement in the spaces (Ujang, Kozlowski, & Maulan, 2018). The sense of familiarity leads to the third function: the sense of attachment.

vii Sense of attachment

The sense of attachment comprised 7% of the pinpoints in all spaces except for the greenery. This function is represented by the residents' connectedness and appreciation of engaging with their living environment, which was their home.

1 Connectedness

One of the residents mentioned that her children liked to eat at the riverside *warung*. This exchange shows that the community recognized the *warung* as a place for familial activities. Their gathering at the *warung* and the joy of eating food evoked a sense of attachment to the place and made it a meaningful space. The sense of attachment was evident and strengthened by their revisits. Revisiting was indicated through the continued service of the *warung* in the neighborhood. This finding is similar to that of Ujang et al. (2018), who found that the purpose of residents visiting a space was associated with meaningful time spent with friends or relatives. It means the community favored the spaces because they carried beneficial importance (Appleton, 1996), especially toward the community's social lives. Gottwald and Stedman found that a community was willing to commit if it was meaningful to them (Gottwald & Stedman, 2020). Similarly, the residents of Kg. Pertanian/Separa were willing to revisit *warung* because it had meaning to them. However, this study's finding of using the river environment for social needs differs from the findings of Abu Bakar, Ujang, and Aziz, who suggested that waterfront communities tend to use the spaces for daily functions such as circulation and transition (Abu Bakar, Ujang, & Aziz, 2016). Their way of living also differs from a modern urban society, which is more individualistic (Ujang et al., 2018). The close-knit relationship between the residents of Kg. Pertanian/Separa encouraged them to socially connect while getting together with family or friends. This finding aligns with that of Nemeth and Olivier, who suggested that connectedness in being together is deeply important to flourish resilience (Nemeth & Olivier, 2017). Adaptation to the disturbance was somewhat camouflaged by the sense of attachment, leading to a positive future for the river neighborhood. Hence, kinship and friendship are the roots of neighbors' connection (Chaskin, 1997). In other words, the combination of the residents, spaces, and activities nurtured a sense of attachment.

2 Appreciation

An example of appreciation was expressed as "gaining extra good deeds or *pahala,*" written for the mosque. Most of the residents were Muslims who valued their religious activities by going to the mosque most days to pray congregationally and learn to read the Quran (*mengaji*). Their frequent visits to the mosque show that the religious institution contributed to their sense of attachment. Such activities were already a norm; hence, religiousness was embedded in their lives. This finding aligns with that of Knez et al., who stated that religious value is embodied when humans bond with the physical space (Knez et al., 2018). In other words, spiritual enrichment (La Rosa, Spyra, & Inostroza, 2016) is enhanced when people gather to practice religious activities. Eventually, the spiritual value cultivates into a norm that the residents practice throughout their lives. The sense of attachment that developed made them appreciate the neighborhood despite the floods and river pollution that had occurred. With an average of 21 years of living in the river neighborhood, they adapted to the phenomenon. This norm reflected their identity as united residents of the river neighborhood.

The sense of attachment could also be seen from their appreciation of their home. For instance, "*Rumahku, Syurgaku,*" meaning "My home, My heaven," was written as a justification. This justification suggests that their living environment was the best place to be. They were also fond of the river environment; one of the residents expressed that the space was peaceful and included a slogan stating "Love your river." The river in the neighborhood was still in its natural form as the riverbank was not concreted. The presence of natural elements suggests that the community's good well-being could be affected by those elements. Subsequently, their exposure to those elements greatly promoted their connection to nature as they often used and observed the river. This finding is similar to Frischenbruder and Pellegrino's finding that green and blue features are crucial for connecting humans and nature (Frischenbruder & Pellegrino, 2006). From the connection, both the landscape features and humans could benefit as nature performs its ecological services. The feeling of attachment felt by the community about the river enhanced their adaptation to overlook the risk the water body could bring. Hence, the bond created through the attachment enhanced their ability to appreciate their neighborhood despite the risks of water-related disturbances.

viii *Responses*

Responses refer to the community's suggestions on their preferred designs for the river neighborhood. The responses were only for the river environment, making up 4% of the scribed notes. Their suggestions show that they were still welcoming improvement in the river environment. Table 22.3 summarizes the responses from the community.

Table 22.3 Responses from the community.

Response	Suggestion	Feature	Function(s) involved
Infrastructure for fishing	Build a fishing jetty	Jetty	Recreational activity (socializing)
Connectivity	Build a pedestrian bridge for community recreation	Pedestrian bridge	Sense of familiarity
Recreation	Add shelters for the residents to rest	Shelter	Meeting and gathering, recreational activity (socializing)
	Build a children's playground	Playground	Recreational activity (socializing)
Landscaping	Plant flowers	Softscapes	Environmental landscape (sense of familiarity)
Management	Continuous supervision from responsible authorities	–	Sense of attachment

These suggestions symbolize the meaningfulness of the river to the residents (Gottwald & Stedman, 2020), as they were willing to contribute their preferences, showing their ideas of an ideal river environment. Nonetheless, it was acknowledged that the suggestions were related to the facilities that allowed them to perform recreational activities such as fishing and resting. For example, a resident suggested building a fishing jetty beside the river. Another resident suggested adding shelters that would be a place for meetings, gatherings, or relaxing. Furthermore, the residents suggested a pedestrian bridge, a playground, and planting flowers by the riverside. The latter suggestion indicates that the community appreciated the esthetics of the riverside that was already familiar to them. These elements were already present in the river neighborhood, including supervision, which emphasizes the role of the stakeholders. This finding points out that the community wanted things that would be socially beneficial for them. Their ability to adapt gave them a perspective of an ideal river environment for the neighborhood. Although the stakeholders were the ones who took action, the residents' views were equally essential to be addressed because their living environment received the impact of the actions or disturbances. However, the finding partially contradicts Ling and Chiang, who found that adaptation influences the perceived risk that determines community responses (Ling & Chiang, 2018). The community of Kg. Pertanian/Separa's perceived risk was toward the benefits of well-being, not the disturbances. It means that their responses were rooted in the neighborly life they experienced and not based on disturbances such as floods or river pollution. This discovery shows that community adaptation is manifested when residents cannot change or mitigate such disturbances. Their understanding of the river environment is essential in establishing the ideal environment. To an extent, the community will live and experience the environment.

ix Uncertain preferences

Uncertain preferences refer to activities outside of the residents' neighborhood or more related to individual preferences. Facilities such as the petrol station, car park, and shop lot were outside the river neighborhood, yet they still had meaning to the residents. One resident wrote that they had night markets that sold *satay, nasi lemak, mee rebus*, and more in the car park, while another wrote that the shop lot was a center for generating income. These descriptions suggest that the residents' adapting and sustaining life also involved places outside their living environment. It indicates that the external resources or supplies were necessary for their lives. This finding parallels that of Wilson, who described how a community is resilient only when the capitals (e.g., social, economic, and environmental) are glued together (Wilson, 2012). Nevertheless, the facilities to support their needs also existed in the neighborhood, such as the shop, which eased their grocery shopping because of the proximity. The facilities were not only related to the social aspect but also the economic aspect. Although the latter aspect was not within the scope, it indicates that adaptation does not rely on only one aspect as it is related to other aspects, such as the economy. Although places to generate income were not within the neighborhood, the external resources were still necessary for the community's life. Meanwhile, the events inside the river neighborhood in this function were also found to be more individual. For example, one resident indicated a home compound for sewing activity. His or her hobby and personal preference mattered to him or her. Another example is the memory of an accident at a junction mentioned in another scribed meaning. The experience was memorable and gave meaning to his or her life. The results included the word "bride" written by a resident of the river environment. This result was singular, and hence, it was interpreted as an individual event. Despite that, it was taken into consideration as their memories are believed to be significant individually. Thus it was categorized under this function: uncertain preferences.

22.5 Limitations of the study

There were two limitations in conducting this study. Following the limitations are the recommendations for future study.

1. This study only considered the social aspect with minimal integration with other aspects such as the environment, specifically the river as the setting. The spatial element was not comprehensively measured due to the time constraint as the approach in time horizon is cross-sectional (Saunders, Lewis, & Thornhill, 2019). Thus it is recommended for future studies to include the river as a major focus in addressing adaptation as the subject.

2. The respondents were mainly middle-aged females between 40 and 70 during the data collection. Participation was lacking from the youth due to the time

of the data collection, as most of them were at school at the time. Nonetheless, the data presented was not jeopardized as the respondents fulfilled the requirement of being the key persons with place-based knowledge (Inman, Gosnell, Lach, & Kornhauser, 2018). Future studies can integrate the youth's perspective regarding adaptation within the river environment.

22.6 Conclusion

This research concludes that the four preferred spaces are where social interactions are most dominant, leading to the five functions discussed previously. Based on these findings, it could assist authorities, designers, landscape architects, and planners in making decisions when developing the spatial aspects of a river in a neighborhood. The potential impact of this application is to foster involvement and cooperation between stakeholders at all levels, leading to a better society. Ultimately, it supports SDG 11: Sustainable Cities and Communities. An example of a river community that values its water body could serve as a reference for other communities. Although the preferences of the community do not appear to be a formal composition of design, this research conveys that the river environment can be redefined with deeper meaning when associated with the community. The meaning refers to the community's association with the spaces within its neighborhood. This meaning is established through exposure and interaction within the living environment. With this knowledge, stakeholders such as the DID are able to approach the community with nonstructural measures (e.g., program). In other words, efforts related to the river environment have the potential to foster a strong relationship between the community and the river. This metaphor parallels the efforts by the Ministry of Environment and Water on the National River Trails (DSK), which centralizes the community as the end users. Although the community preferences may seem basic, the importance of their involvement and cooperation in decision-making would impact their adaptation, as they are the "keepers" of the neighborhood. The intangible bond between them and the environment is the hope of recovering the value of the river, as they are the roots that hold the ground. It is rationalized through meaningful spaces where social interactions become the norm for the neighborhood, resulting in a stronger bond between neighbors. It means the more social interaction or usage of the spaces occurs, the more cohesive the community becomes. In other words, social interaction enhances the connectedness within the community. To an extent, this value can supersede the disturbances that affect the community's livelihood. Fig. 22.15 simplifies the sequence from findings to the implications.

The community preferences can be summarized into a sequence of four, starting with disturbances, as they were always exposed to the risk of floods and river pollution. Despite that, neighborly kinship led to social interaction within the neighborhood, making the spaces meaningful as they carried memories.

Embedding memories in the spaces showcases how the community's appreciation of their neighborhood was enough to shape their preferences. This appreciation symbolizes their connectedness with the neighborhood. Meanwhile, their preferences are the representation of an ideal river environment. This ideal river environment consists of the meaningful spaces that resonate the most with them the most as the locals of the neighborhood. Fig. 22.16 summarizes the sequence of community preferences that start with disturbances and end with meaningful spaces, and the ideal river environment is illustrated in Fig. 22.17. It comprises the community's preferences that fit their social interactions. With a careful design that incorporates the community's ideas, the river environment would become part of the meaningful spaces that contribute to the community's adaptation in the neighborhood.

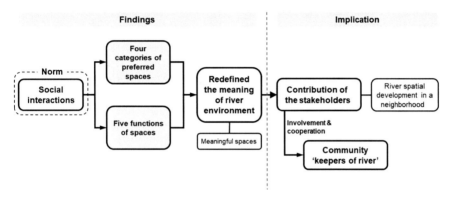

FIGURE 22.15 The sequence from findings to the implications.

The sequence begins with social interaction where it has become a norm in the neighborhood among river community.

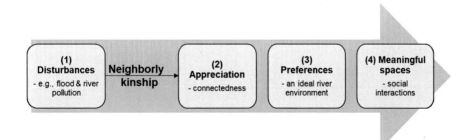

FIGURE 22.16 Sequence of community preferences.

The disturbance initiates the appreciation whereby the kinship among the residents enhances the sense of connectedness leading to preferences and meaningful spaces.

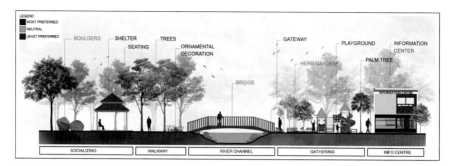

FIGURE 22.17 The idea of preferred landscape elements in a river environment.

The elements were based on the existing and activities highlighted by the residents during the data collection.

The community's adaptation is a subset of community resilience, which can be viewed through different lenses. The lens of this research is the social aspect, with minimal spatial inclusion that can be considered for next exploration with a larger scale of the river environment. This means that considering a larger sample or different settings of communities' livelihoods is necessary to ensure the comprehensiveness of river management. As humans become part of river management, it will lead to an increasing complexity in the decision-making process, but the meaningful decisions made will contribute to a thriving river environment that is favorable to the whole ecosystem.

Acknowledgment

The first author would like to express her gratitude to the community of Kg. Pertanian/Separa, in Johor, Malaysia, for generously giving their insight on the preferences regarding their living environment.

References

Abdullah, R. (2018). *Ini 51 sungai tercemar di Malaysia*. Malaysia. Astro Awani Network Sdn Bhd. https://www.astroawani.com/berita-malaysia/ini-51-sungai-tercemar-di-malaysia-194141.

Abu Bakar, E. A., Ujang, N., & Aziz, F. (2016). Place attachment towards waterfront in Kangar, Perlis, Malaysia. *Alam Pencipta.*, *9*(2), 33–44j.

Abu Samah, B., Sulaiman, M., Shaffril, H., Hassan, M. S., Othman, S., Abu Samah, A., & Ramli, A. (2011). Relationship to the river: The case of the Muar River community. *American Journal of Environmental Sciences*, *7*(4), 362–369. Available from https://doi.org/10.3844/ajessp.2011.362.369Malaysia, http://thescipub.com/pdf/10.3844/ajessp.2011.362.369.

Appleton, J. (1996). *The experience of landscape* (1). Wiley.

Bunn, S. E., Abal, E. G., Smith, M. J., Choy, S. C., Fellows, C. S., Harch, B. D., ... Sheldon, F. (2010). Integration of science and monitoring of river ecosystem health to guide investments in catchment protection and rehabilitation. *Freshwater Biology*, *55*(s1), 223−240. Available from https://doi.org/10.1111/j.1365-2427.2009.02375.x.

Chan, N. W. (2012). Managing urban rivers and water quality in Malaysia for sustainable water resources. *International Journal of Water Resources Development*, *28*(2), 343−354. Available from https://doi.org/10.1080/07900627.2012.668643.

Chaskin, R. J. (1997). Perspectives on neighborhood and community: A review of the literature. *Social Service Review*, *71*(4), 521−547. Available from https://doi.org/10.1086/604277.

Che, Y., Li, W., Shang, Z., Liu, C., & Yang, K. (2014). Residential preferences for river network improvement: An exploration of choice experiments in Zhujiajiao, Shanghai, China. *Environmental Management*, *54*(3), 517−530. Available from https://doi.org/10.1007/s00267-014-0323-x, http://link.springer.de/link/service/journals/00267/index.htm.

Chiang, Y. C. (2018). Exploring community risk perceptions of climate change − A case study of a flood-prone urban area of Taiwan. *Cities (London, England)*, *74*, 42−51. Available from https://doi.org/10.1016/j.cities.2017.11.001, http://www.elsevier.com/inca/publications/store/3/0/3/9/6/.

Danladi Bello, A. A., Mohd Haniffah, M. R., Hashim, N., & Anuar, K. (2018). Estimation of hydrological changes in a tropical watershed using multi-temporal land-use and dynamic modelling. *Jurnal Teknologi*, *80*(3), 123−136. Available from https://doi.org/10.11113/jt.v80.11179, https://jurnalteknologi.utm.my/index.php/jurnalteknologi/article/download/11179/6367.

Darby, S., & Sear, D. (2008). *River restoration: Managing the uncertainty in restoring physical habitat*. Wiley. Available from 10.1002/9780470867082.

Dressel, S., Johansson, M., Ericsson, G., & Sandström, C. (2020). Perceived adaptive capacity within a multi-level governance setting: The role of bonding, bridging, and linking social capital. *Environmental Science and Policy*, *104*, 88−97. Available from https://doi.org/10.1016/j.envsci.2019.11.011, http://www.elsevier.com/wps/find/journaldescription.cws_home/601264/description#description.

Durán Vian, F., Pons Izquierdo, J. J., & Serrano Martínez, M. (2021). River-city recreational interaction: A classification of urban riverfront parks and walks. *Urban Forestry and Urban Greening*, *59*. Available from https://doi.org/10.1016/j.ufug.2021.127042, http://www.elsevier.com/wps/find/journaldescription.cws_home/701803/description#description.

Engle, N. L. (2011). Adaptive capacity and its assessment. *Global Environmental Change*, *21*(2), 647−656. Available from https://doi.org/10.1016/j.gloenvcha.2011.01.019.

Esteban, T. A. O. (2020). Building resilience through collective engagement. *Architecture_MPS*, *17*(1). Available from https://doi.org/10.14324/111.444.amps.2020v17i1.001.

Everard, M., & Moggridge, H. L. (2012). Rediscovering the value of urban rivers. *Urban Ecosystems*, *15*(2), 293−314. Available from https://doi.org/10.1007/s11252-011-0174-7, http://www.kluweronline.com/issn/1083-8155.

Fath, B. D., Dean, C. A., & Katzmair, H. (2015). Navigating the adaptive cycle an approach to managing the resilience of social systems. *Ecology and Society.*, *20*(2).

Fox, C. A., Reo, N. J., Turner, D. A., Cook, J. A., Dituri, F., Fessell, B., ... Wilson, M. (2017). "The river is us; the river is in our veins": Re-defining river restoration in three

Indigenous communities. *Sustainability Science*, *12*(4), 521−533. Available from https://doi.org/10.1007/s11625-016-0421-1, http://www.springer.com/east/home?SGWID = 5-102-70-144940151-0&changeHeader = true&SHORTCUT = http://www.springer.com/journal/11625.

Frischenbruder, M. T. M., & Pellegrino, P. (2006). Using greenways to reclaim nature in Brazilian cities. *Landscape and Urban Planning*, *76*(1−4), 67−78. Available from https://doi.org/10.1016/j.landurbplan.2004.09.043, http://www.elsevier.com/inca/publications/store/5/0/3/3/4/7.

Gottwald, S., & Stedman, R. C. (2020). Preserving ones meaningful place or not? Understanding environmental stewardship behaviour in river landscapes. *Landscape and Urban Planning*, *198*, 103778. Available from https://doi.org/10.1016/j.landurbplan.2020.103778.

Hayrol Azril, M. S., Sulaiman, M. Y., Md. Salleh, H., Mohd Shahwahid, O., Bahaman, A. S., Asnarulkhadi, A. S., & Siti Aisyah, R. (2011). Pahang river community satisfaction towards their quality of life: The case of community in Pekan, Pahang. *Asian Social Science*, *7*(12), 43−55. Available from https://doi.org/10.5539/ass.v7n12p43Malaysia, http://www.ccsenet.org/journal/index.php/ass/article/view/11106/9313.

Holling, C. S. (1973). Resilience and stability of ecological systems. *Annual Review of Ecology and Systematics*, *4*(1), 1−23. Available from https://doi.org/10.1146/annurev.es.04.110173.000245.

Inman, T. B., Gosnell, H., Lach, D. H., & Kornhauser, K. (2018). Social-ecological change, resilience, and adaptive capacity in the McKenzie River Valley, Oregon. *Humboldt Journal of Social Relations*, *1*(40), 68−88. Available from https://doi.org/10.55671/0160-4341.1075.

Integrated River Basin Management Report. (2017). Unpublished content. https://www.water.gov.my/index.php/pages/view/708.

Knez, I., Butler, A., Sang, O., Ångman, E., Sarlöv-Herlin, I., & Åkerskog, A. (2018). Before and after a natural disaster: Disruption in emotion component of place-identity and wellbeing. *Journal of Environmental Psychology*, *55*, 11−17. Available from https://doi.org/10.1016/j.jenvp.2017.11.002, http://www.elsevier.com/inca/publications/store/6/2/2/8/7/2/index.htt.

Kondolf, G. M., & Pinto, P. J. (2017). The social connectivity of urban rivers. *Geomorphology*, *277*, 182−196. Available from https://doi.org/10.1016/j.geomorph.2016.09.028, http://www.elsevier.com/inca/publications/store/5/0/3/3/3/4/.

La Rosa, D., Spyra, M., & Inostroza, L. (2016). Indicators of cultural ecosystem services for urban planning: A review. *Ecological Indicators*, *61*, 74−89. Available from https://doi.org/10.1016/j.ecolind.2015.04.028, http://www.elsevier.com/locate/ecolind.

Le Lay, Y. F., Piégay, H., & Rivière-Honegger, A. (2013). Perception of braided river landscapes: Implications for public participation and sustainable management. *Journal of Environmental Management*, *119*, 1−12. Available from https://doi.org/10.1016/j.jenvman.2013.01.006, https://www.sciencedirect.com/journal/journal-of-environmental-management.

Ling, T. Y., & Chiang, Y. C. (2018). Strengthening the resilience of urban retailers towards flood risks − A case study in the riverbank region of Kaohsiung City. *International Journal of Disaster Risk Reduction*, *27*, 541−555. Available from https://doi.org/10.1016/j.ijdrr.2017.11.020, http://www.journals.elsevier.com/international-journal-of-disaster-risk-reduction/.

Mann, R. (1973). *Rivers in the city*. Praeger.

McEwen, L., Garde-Hansen, J., Holmes, A., Jones, O., & Krause, F. (2017). Sustainable flood memories, lay knowledges and the development of community resilience to future flood risk. *Transactions of the Institute of British Geographers*, *42*(1), 14−28. Available from https://doi.org/10.1111/tran.12149, http://www.blackwellpublishing.com/aims.asp?ref = 0020-2754.

Milestad, R., Westberg, L., Geber, U., & Björklund, J. (2010). Enhancing adaptive capacity in food systems: Learning at farmers' markets in Sweden. *Ecology and Society*, *15*(3). Available from https://doi.org/10.5751/ES-03543-150329, http://www.ecologyandsociety.org/vol15/iss3/art29/ES-2010-3543.pdf.

Mohd Firdaus, R. (2022) *Community adaptive capacity of river neighbourhood* (Doctorate Philosophy). Universiti Teknologi Malaysia, Malaysia.

Naito, D. (2008). *Change of orang Sungai's subsistence activities in the Kinabatangan River Basin in Sabah, Malaysia. Sustainability and biodiversity assessment on forest utilization options* (pp. 357−363). Research Institute for Humanity and Nature.

Nemeth, D. G., & Olivier, T. W. (2017). *Resilience: Defined and explored. Innovative approaches to individual and community resilience* (pp. 1−23). Academic Press.

Parsons, M., Reeve, I., McGregor, J., Hastings, P., Marshall, G. R., McNeill, J., . . . Glavac, S. (2021). Disaster resilience in Australia: A geographic assessment using an index of coping and adaptive capacity. *International Journal of Disaster Risk Reduction*, *62*, 102422. Available from https://doi.org/10.1016/j.ijdrr.2021.102422.

Parsons, M., & Thoms, M. C. (2018). From academic to applied: Operationalising resilience in river systems. *Geomorphology*, *305*, 242−251. Available from https://doi.org/10.1016/j.geomorph.2017.08.040, http://www.elsevier.com/inca/publications/store/5/0/3/3/3/4/.

Patel, R. B., & Gleason, K. M. (2018). The association between social cohesion and community resilience in two urban slums of Port au Prince, Haiti. *International Journal of Disaster Risk Reduction*, *27*, 161−167. Available from https://doi.org/10.1016/j.ijdrr.2017.10.003, http://www.journals.elsevier.com/international-journal-of-disaster-risk-reduction/.

Paul, A., Deka, J., Gujre, N., Rangan, L., & Mitra, S. (2019). Does nature of livelihood regulate the urban community's vulnerability to climate change? Guwahati city, a case study from North East India. *Journal of Environmental Management*, *251*, 109591. Available from https://doi.org/10.1016/j.jenvman.2019.109591.

Rufat, S., Tate, E., Burton, C. G., & Maroof, A. S. (2015). Social vulnerability to floods: Review of case studies and implications for measurement. *International Journal of Disaster Risk Reduction*, *14*, 470−486. Available from https://doi.org/10.1016/j.ijdrr.2015.09.013, http://www.journals.elsevier.com/international-journal-of-disaster-risk-reduction/.

Samion, J. (2023). *Resilience of communities in Pontian district from rapid development of Iskandar Malaysia* (Doctorate Philosophy, Universiti Teknologi Malaysia). *Perpustakaan Raja Zarith Sofiah*.

Saunders, M. N. K., Lewis, P., & Thornhill, A. (2019). *Research methods for business students*. Pearson Education.

Shafaghat, A., Mir Ghasemi, M., Keyvanfar, A., Lamit, H., & Ferwati, M. S. (2017). Sustainable riverscape preservation strategy framework using goal-oriented method: Case of historical heritage cities in Malaysia. *International Journal of Sustainable Built*

Environment, 6(1), 143−159. Available from https://doi.org/10.1016/j.ijsbe.2017.03.003, http://www.journals.elsevier.com/international-journal-of-sustainable-built-environment/.

Smith, J. W., Davenport, M. A., Anderson, D. H., & Leahy, J. E. (2011). Place meanings and desired management outcomes. *Landscape and Urban Planning, 101*(4), 359−370. Available from https://doi.org/10.1016/j.landurbplan.2011.03.002, http://www.elsevier.com/inca/publications/store/5/0/3/3/4/7.

Southon, G. E., Jorgensen, A., Dunnett, N., Hoyle, H., & Evans, K. L. (2018). Perceived species-richness in urban green spaces: Cues, accuracy and well-being impacts. *Landscape and Urban Planning, 172*, 1−10. Available from https://doi.org/10.1016/j.landurbplan.2017.12.002, http://www.elsevier.com/inca/publications/store/5/0/3/3/4/7.

Tieskens, K. F., Van Zanten, B. T., Schulp, C. J. E., & Verburg, P. H. (2018). Aesthetic appreciation of the cultural landscape through social media: An analysis of revealed preference in the Dutch river landscape. *Landscape and Urban Planning, 177*, 128−137. Available from https://doi.org/10.1016/j.landurbplan.2018.05.002, http://www.elsevier.com/inca/publications/store/5/0/3/3/4/7.

Ujang, N., Kozlowski, M., & Maulan, S. (2018). Linking place attachment and social interaction: Towards meaningful public places. *Journal of Place Management and Development, 11*(1), 115−129. Available from https://doi.org/10.1108/JPMD-01-2017-0012, http://emeraldinsight.com/journals.htm?issn = 1753-8335.

Ujang, N. (2016). Defining place attachment in Asian urban places through opportunities for social interactions. *Environment-Behaviour Proceedings Journal, 1*(1), 28−35. Available from https://doi.org/10.21834/e-bpj.v1i1.191.

Verbrugge, L. N. H., Ganzevoort, W., Fliervoet, J. M., Panten, K., & van den Born, R. J. G. (2017). Implementing participatory monitoring in river management: The role of stakeholders' perspectives and incentives. *Journal of Environmental Management, 195*, 62−69. Available from https://doi.org/10.1016/j.jenvman.2016.11.035, https://www.sciencedirect.com/journal/journal-of-environmental-management.

Vietz, G. J., Rutherfurd, I. D., Fletcher, T. D., & Walsh, C. J. (2016). Thinking outside the channel: Challenges and opportunities for protection and restoration of stream morphology in urbanizing catchments. *Landscape and Urban Planning, 145*, 34−44. Available from https://doi.org/10.1016/j.landurbplan.2015.09.004, http://www.elsevier.com/inca/publications/store/5/0/3/3/4/7.

Walker, B., Carpenter, S., Anderies, J., Abel, N., Cumming, G., Janssen, M., . . . Pritchard, R. (2002). Resilience management in social-ecological systems: A working hypothesis for a participatory approach. *Ecology and Society, 6*(1). Available from https://doi.org/10.5751/es-00356-060114, http://www.ecologyandsociety.org.

Wates, N. (2000). *The community planning handbook: How people can shape their cities, towns and villages in any part of the world.* Earthscan.

Wilson, G. A. (2012). *Community resilience and environmental transitions. Community resilience and environmental transitions* (pp. 1−251). Taylor and Francis. Available from http://www.tandfebooks.com/doi/book/10.4324/9780203144916, 10.4324/9780203144916.

Wohl, E. (2014). Time and the rivers flowing: Fluvial geomorphology since 1960. *Geomorphology, 216*, 263−282. Available from https://doi.org/10.1016/j.geomorph.2014.04.012, http://www.elsevier.com/inca/publications/store/5/0/3/3/4/.

Yamashita, S. (2002). Perception and evaluation of water in landscape: Use of photoprojective method to compare child and adult residents' perceptions of a Japanese river environment. *Landscape and Urban Planning*, *62*(1), 3−17. Available from https://doi.org/10.1016/S0169-2046(02)00093-2,   http://www.elsevier.com/inca/publications/store/5/0/3/3/4/7.

Yodsurang, P., Hiromi, M., & Yasufumi, U. (2015). A traditional community in the Chao Phraya River Basin: Classification and characteristics of a waterfront community complex. *Asian Culture and History*, *8*(1), 57. Available from https://doi.org/10.5539/ach.v8n1p57.

Zinia, N. J., & McShane, P. (2018). Ecosystem services management: An evaluation of green adaptations for urban development in Dhaka, Bangladesh. *Landscape and Urban Planning*, *173*, 23−32. Available from https://doi.org/10.1016/j.landurbplan.2018.01.008, http://www.elsevier.com/inca/publications/store/5/0/3/3/4/7.

Estimation of soil erosion risk and vulnerable zone using the revised universal soil loss equation and geographic information system approaches

23

Rahul Kumar[1], Shambhu Nath Mishra[1], Rajiv Pandey[2] and Vijender Pal Panwar[3]

[1]*Forest Ecology and Climate Change Division, ICFRE-Institute of Forest Productivity, Ranchi, Jharkhand, India*
[2]*Indian Council of Forestry Research and Education, Dehradun, Uttarakhand, India*
[3]*ICFRE-Forest Research Institute, Dehradun, Uttarakhand, India*

23.1 Introduction

Erosion is a global environmental and economic issue that is based on the amount of soil that has been removed, and the ability of wind or water power to carry it (Lal, 2017) causes land deterioration by removing fertility and nutrients from the topsoil (Rodrigo Comino et al., 2016). The process of soil erosion is directly related to sediment transport and deposition (Karan, Ghosh, & Samadder, 2019). The erosion vulnerability of a region depends on the rate of weathering, soil particle detachment, and transport, as well as the type and density of vegetation (Borrelli et al., 2020; Islam, Jaafar, Hin, Osman, & Karim, 2020; Vijith, Seling, & Dodge-Wan, 2018). It is estimated that 5−10 million ha of land are affected by soil erosion each year, amounting to 23% of the Earth's land surface (Stavi & Lal, 2015). At the beginning of the 21st century, the global average erosion rate was estimated to be 10.2 t/ha/year, and the global loss of soil due to erosional processes and sediment export to oceans was predicted to be 24 t/ha/year (Dimotta, 2019; Sharda & Ojasvi, 2016) to 75 billion tons of nutrient-rich soil (Rashmi et al., 2022; Rhodes, 2014; Tunçay & Başkan, 2023). Throughout Asia, Africa, and South America, erosion is quite high, averaging 30−40 t/ha/year (Barrow, 1991). Erosion removes around 75 billion tons of soil yearly, primarily from agricultural lands, and about 20 million hectares have already been lost (Pandey, Mathur, Mishra, & Mal, 2009). According to the Soil and Water Conservation Atlas of India, published by the Indian Council of Agricultural

Research, about 147 million hectares (45% of India's total geographical area) are prone to soil erosion. In some areas, soil erosion and desertification have reduced productivity by 50%. From a total area of India (329 Mha), 120.4 Mha of land has been degraded (68% due to water erosion), causing an annual loss of 5.3 Gt of soil (Kar et al., 2022). Water erosion caused by the Himalayan range degrades about one-third of the area in the Eastern Himalayan region of India. An approximation shows that India's soil erosion rate is 16.35 t/ha/year (Vemu & Pinnamaneni, 2011), whereas the permissible erosion rate is 4.5−11.2 t/ha. The process of modeling of soil erosion involves mathematically describing soil particle detachment, transport, and deposition.

23.1.1 Major drivers for soil erosion and its threats

Both past and present management techniques influence the soil's vulnerability to erosion. Historical soil management techniques increase soil erosion risk, including plowing and subsurface loosening, which can cause irreversible soil degradation (Grahmann, Rubio, Perez-Bidegain, & Quincke, 2022). Because runoff from heavy rain and severe precipitation episodes can dislodge and take soil particles away, they can cause erosion. Steeper slopes are more prone to erosion because water runoff can build speed and readily transport dirt away. Humans can disrupt soil erosion prevention mechanisms by altering the natural landscape, such as through deforestation, agriculture, urbanization, and construction. Soil erosion damages the soil structure and affects its water-holding capacity, which decreases the soil's overall fertility (Gu et al., 2018; Zhang et al., 2021). The loss of nutrient-abundant fertile topsoil in the catchment belt, along with sediment deposition and channel silt, minified reservoir capacity (Mullan, 2013). Deterioration of the vegetation accelerates soil degradation and has a significant role in soil deterioration, especially soil erosion. In areas where vegetation cover is lost, soils become more susceptible to wind and water erosion, resulting in the loss of organic material. This decreases soil aggregation and stability, as well as soil fertility (Chalise, Kumar, & Kristiansen, 2019). This can have serious implications for food security. Loss of organic matter reduces soil nutrient content and water-holding capacity, which makes vegetation less resilient (Tran, Chang, & Cho, 2019). Erosive soils remove the nutrient-rich fertile topsoil layers and disrupt the delicate balance of microbial communities, which are vital to nutrient cycling, decomposition, and overall soil health. As a result of erosion, disturbed areas can become more vulnerable to invasion by invasive plants. The invasion of these plants can result in the loss of biodiversity and a change in ecosystem dynamics as native plants are unable to compete with the invaders. Livestock grazing can also result in the obliteration of vegetative cover and soil trampling, which increases the soil's susceptibility to erosion. Major drivers for soil erosion, threats, and consequences are shown in Fig. 23.1.

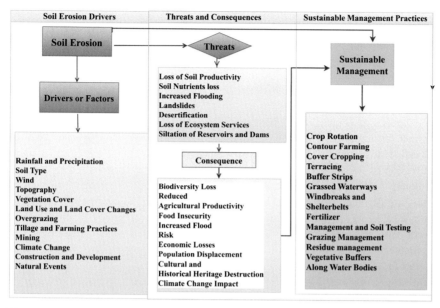

| Soil Erosion Drivers | Threats and Consequences | Sustainable Management Practices |

Soil Erosion → Threats

Drivers or Factors

Loss of Soil Productivity
Soil Nutrients loss
Increased Flooding
Landslides
Desertification
Loss of Ecosystem Services
Siltation of Reservoirs and Dams

Sustainable Management

Consequence

Rainfall and Precipitation
Soil Type
Wind
Topography
Vegetation Cover
Land Use and Land Cover Changes
Overgrazing
Tillage and Farming Practices
Mining
Climate Change
Construction and Development
Natural Events

Biodiversity Loss
Reduced
Agricultural Productivity
Food Insecurity
Increased Flood
Risk
Economic Losses
Population Displacement
Cultural and
Historical Heritage Destruction
Climate Change Impact

Crop Rotation
Contour Farming
Cover Cropping
Terracing
Buffer Strips
Grassed Waterways
Windbreaks and
Shelterbelts
Fertilizer
Management and Soil Testing
Grazing Management
Residue management
Vegetative Buffers
Along Water Bodies

FIGURE 23.1

Factors responsible for soil erosion, its consequences, and management practices.

23.1.2 Soil erosion consequence

The consequences of soil degradation brought on by land usage and land cover changes can be severe, including reduced soil fertility, diminished agricultural productivity, water pollution, increased flooding, and loss of biodiversity. A soil erosion process can be divided into two categories: on-site and off-site effects (Issaka & Ashraf, 2017; Liu, Zhao, & Yu, 2020; Patault et al., 2021) based on whether the outcomes result in reduced yields for agriculture or lower water quality (particle-born pollutants and turbidity). Reservoir sedimentation on-site, disruption of hydrological regimes (e.g., elevated flood risk as a result of sedimentation of riverbeds), and off-site stream plugging (Chomitz & Kumari, 1998; Lal, 2003; Locatelli, Imbach, Vignola, Metzger, & Hidalgo, 2011; Morgan, 2009) are reported in the literature. In addition to evaluating the effects on biophysical soil qualities and geomorphic processes (such as sheet, rill, gully, wind, and landslip erosion), the on-site effects of erosion can also be evaluated regarding the socio-cultural and economic implications. It is crucial to emphasize that unless the lost nutrients are replaced and, in certain situations, additional water is supplied by irrigation, the primary on-site impact of erosion refers to the decline in the organic matter and nutrient content of the soil, which results in reduced water retention capacity and crop production (Ilao & Guzman, 2001). According to Stocking and Clark (1999), the effects of soil erosion off-site are often restricted to reservoir sedimentation, ignoring other significant effects like road damage, harbor siltation, siltation of sewers and basements, disruption of drainage, undermining of pavements and foundations, road gullies, failure of earth dams, the eutrophication of rivers, and channels, loss of wildlife habitat and disturbance of stream ecology, flooding, harm to Public wellness, and increased costs associated with water treatment. It is necessary to consider the advantages and disadvantages of preventative and landscape restoration efforts in addition to the expenses associated with land degradation. The environmental relationships between the impacts of soil erosion on- and off-site made it possible to identify and establish a connection between the socioeconomic context and several static and variable elements that are part of the biophysical context, such as soil (Dimotta, 2019). Fig. 23.2 illustrates that the secondary calculated input parameters are derived directly from the primary measured ones through the negative/positive effect (+/−); From these last ones, the ultimate outcome, including soil erosion, soil resource, and yield, when coupled with biomass—characterizes the "parameters" intended to predict the related long-term trends by taking into account the previously described interactions.

In some cases, soil loss rates may result in silty reservoirs and reduced storage capacity due to reduced storage amplitude. In low-lying locations, soil erosion can cause flash flooding due to increased runoff and the potential for large-scale

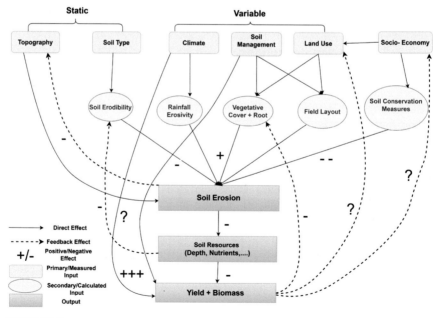

FIGURE 23.2

The primary ways in which the various environmental components interact to forecast long-term trends in soil erosion, soil resources, and soil yields (Vanwalleghem et al., 2017).

sediment mobilization and deposition. Because of their high energy flow and spatial connectedness, flash floods can dissolve and carry large volumes of sediments from channel banks and land. These will have an impact on the productivity of the land in catchments, as well as in the higher and lower portions of hills, which could someday be left as barren land (Bannari, Guédon, & El-Ghmari, 2016; Bracken & Croke, 2007; Trnka et al., 1901; Vanmaercke et al., 2010). In coalfields, extensive surface mining operations (drilling, blasting, coal washing, and material excavation close to watercourses) could be the main contributors to soil erosion. Particularly if these places are close to rivers and reservoirs, erosion in these locations leads to a specific combination of environmental and water resource management issues. Human action on hill incline disrupts current equilibrium circumstances by contouring the incline and removing protecting vegetation from hill slopes for plantation or agricultural purposes, increasing erosion vulnerability (Cerdà et al., 2016; Keesstra et al., 2016; Mekonnen, Keesstra, Stroosnijder, Baartman, & Maroulis, 2015). These actions also reduce the chance of indigenous plant recovery by removing vegetation seeds along the rill and sheet soil loss or eroding (Yu et al., 2017). Soil erosion risk refers to the chance or potential for soil to be eroded or washed away from its original place as a result of a variety of natural or man-made sources. To assess soil erosion risk, several elements such as terrain, rainfall and precipitation, vegetation cover, soil type and texture, land use and land cover (LULC), climate and weather patterns, and processes must be examined (Berberoglu, Cilek, Kirkby, Irvine, & Donmez, 2020) and evaluate the overall risk.

23.1.3 Soil erosion risk and vulnerable zone assessment

There is no doubt that erosion has serious environmental implications, and it is one of the most significant problems of land deterioration and water resources management in river basins and catchments globally (Zhang et al., 2021). In the field of natural resources management and environmental planning, one of the most challenging aspects is the quantification of soil loss (Arekhi, Niazi, & Kalteh, 2012). Different approaches are examined by Ke and Zhang (2022) and Morgan (2009) to predict erosion-affected areas and depict the relationship between land cover, rainfall, slope, and soil and depict the interaction of soil, slope, rainfall, and land cover. This process is complex, so the erosion rate can vary over time and space. Floral diversity and tree density play an essential part in soil erosion control; soil hydrological characteristics also affect susceptibility to erosion (Brevik et al., 2015; Keesstra et al., 2016). A variety of land use and management practices are being used for predicting soil loss using computer simulation models, which have become increasingly popular in recent years. A new development is the use of models in conjunction with remote sensing and geographic information systems (GISs) to estimate watershed erosion (Mishra & Deng, 2009; Pandey, Chowdary, & Mal, 2007; Vijith, Suma, Rekha, Shiju, & Rejith, 2012). Numerous soil erosion prediction models were created like the

"Agricultural Non-Point Source Pollution Model" (Young, Onstad, Bosch, & Anderson, 1989), "the Water Erosion Prediction Project (WEPP)" (Nearing, Foster, Lane, & Finkner, 1989), "the Areal Nonpoint Source Watershed Environment Response Simulation (ANSWERS)" (Beasley & Huggins, 1980), and "Soil and Water Assessment Tool (SWAT)" (Arnold, Srinivasan, Muttiah, & Williams, 1998). One of the primary benefits of these models is their ability to evaluate geographic variation in watershed environments appropriately.

23.1.4 The rationale of the study

Worldwide, land degradation in terms of erosion affects 1094 million hectares (Mha) of land, with 751 Mha severely affected; similarly, wind erosion affects 449 Mha of land, of which 296 Mha is badly impacted (Lal, 2003). The rationale for a study on estimating soil erosion risk and vulnerable zones using various methodologies and tools is multifaceted. An estimated 75 billion tons of rich soil are lost due to global agriculture annually (Eswaran, Lal, & Reich, 2019) and soil costs the globe around 400 billion US dollars each year, or approximately 70 US dollars per person (Blaikie & Brookfield, 2015). An estimated 36 million tons of cereal equivalent are lost to erosion each year in South Asia, with water erosion costing 400 million US dollars and wind erosion costing 1800 million US dollars (Alam, 2014). Historical soil erosion can reduce yields by 2%−40% in Africa, with an average impairment of 8.2% across the continent (Rhodes, 2014). According to Kar et al. (2022), a total of 29% of soil that is eroded is completely lost to the sea, 10% is collected in reservoirs, and 61% is relocated from its parental place. According to another study (Sharda & Ojasvi, 2016), it was found that Indian soil lost 15.59 t ha/year, of which 22.9% was lost to the oceans, 34.1% was contained in reservoirs, and 43.0% was relocated from the source soil.

23.2 Materials and methods

23.2.1 Estimation of soil erosion risk and vulnerable zone

There are many methods for predicting soil erosion worldwide, including physical, empirical, statistical, and process-based models (Dimotta, 2019; Kar et al., 2022; Pandey, Kumar, Zlatic, Nautiyal, & Panwar, 2021). Pan-European Soil Erosion Risk Assessment (PESERA); WEPP; the revised universal soil loss equation (RUSLE); the SWAT, and the erosion potential method are some of the physical, process-based empirical models in order to use estimate soil erosion vulnerability quantitatively. Usually, these can be classified as either empirical models or physical models, depending on how erosion is evaluated Table 23.1 and operational-driven models (Alonso, Requejo, Alvarez, Montes, & M, 1994; De Roo, Hazelhoff, & Burrough, 1989; Favis-Mortlock, Quinton, & Dickinson, 1996; Jetten, Govers, & Hessel, 2003; Mitasova & Mitas, 1998).

Table 23.1 Models employing empirical equations to evaluate erosion.

Models employing empirical equations to evaluate erosion			
Model	**Abbreviations**	**Method**	**References**
"Chemicals Run-off and Erosion from Agricultural Management Systems"	CREAMS	USLE	Knisel (1980)
Erosion/Productivity Impact Calculator	EPIC	RUSLE, MUSLE	Williams, Jones, and Dyke (1984)
"Agricultural Non-Point Source Pollution Model"	AGNPS	RUSLE	Young et al. (1989)
"Matsalu,"	MATSALU	MUSLE	Krysanova, Meiner, Roosaare, and Vasilyev (1989)
"Simulator for Water Resources in Rural Basins-Water Quality"	SWRRBWQ	MUSLE	Williams and Arnold (1996)
"Soil and Water Assessment Tool"	SWAT	USLE	William and Arnold (1993)
"Soil and Water Integrated Model"	SWIM	MUSLE	Krysanova, Wechsung, Arnold, Srinivasan, and Williams (2000)
"Channel-Hill-slope Integrated Landscape Development"	CHILD	USLE	Tucker, Lancaster, Gasparini, and Bras (2001)
"Land Degradation Model"	LDM	USLE	Hootsmans, Bouwman, Leemans, and Kreileman (2001)

MUSLE, *Modified universal soil loss equation;* RUSLE, *revised universal soil loss equation;* USLE, *universal soil loss equation.*

Dabral, Baithuri, and Pandey (2008) utilized the Morgan, Morgan, and Finney (MMF) soil erode model, while Pandey et al. (2009) used the universal soil loss equation (USLE) model for admeasurement of soil erosion in Arunachal Pradesh. The results of both models showed that soil erosion was substantial in Arunachal Pradesh due to the heavy rainfall. Jain, Kumar, and Varghes (2001) also computed soil erosion in a Himalayan watershed using the USLE and Morgan models. It is observed that the estimates derived from the Morgan model yield better results than the USLE estimates, which have resulted in a higher erosion rate. The revised MMF model is applied in a Himalayan catchment in India (Morgan, 2001). In Spain, soil erosion in a catchment of 211 km^2 has been predicted using the Agricultural Nonpoint Source Pollution and RUSLE (Renschler, Mannaerts, & Diekkrüger, 1999). USLE has been used in the Pinto Lake of California by Boyle et al. (2011) to predict sediment yield, while MMF is the only method used to estimate soil erosion in the karst region of Cuba (Febles-González, Vega-Carreño, Tolón-Becerra, & Lastra-Bravo, 2012). Soil erosion estimation has been conducted using two different types of models in some limited studies. Different land characteristics, slopes, and soil types may affect the accuracy of other models, as well as the erosion rate. Many researchers and institutions developed physical models based on the criterion that they estimate erosion more accurately (Table 23.2). Climate models are useful for selecting and developing technologies and strategies that reduce or control processes both under current climate conditions and those forecast to emerge in the future.

Table 23.2 Erosion and water-quality models developed by process simulation.

Abbreviation	Name of model	References
"ANSWERS"	Areal Non-point Source Watershed Environmental Response Simulation	Beasley and Huggins (1980)
"BASINS"	Better Assessment Science Integrating Point and Nonpoint Sources	Kinerson, Kittle, and Duda (2009)
"EROSION 2D"	Erosion modeling or simulation in a two-dimensional (2D) environment	Schmidt (1991)
"EROSION 3D"	Simulating or modeling erosion processes in a three-dimensional (3D)	Werner (1995)
"EUROSEM"	European Soil Erosion Model	Morgan et al. (1998)
"GLEAMS"	Groundwater Loading Effects on Agricultural Management System	Leonard, Knisel, and Still (1987)
"GUEST"	Griffith University Erosion System Template	Rose, Coughlan, and Fentie (1998)
"KINEROS2"	Kinematic Runoff and Erosion Model	Woolhiser (1996)
"LISEM"	Limburg Soil Erosion Model	De Roo, Offermans, and Cremers (1996)
"MEDALUS"	Family of Models: MEDALUS, MEDRUSH; CSEP	Kirkby, Abrahart, McMahon, Shao, and Thornes (1998)
"MMF"	Morgan, Morgan, and Finney Model	Morgan (2001)
"SEMMED"	Soil Erosion Model for Mediterranean Areas	de Jong et al. (1999)
"SHE"	System Hydrologique Europeen	Abbott, Bathurst, Cunge, O'Connell, and Rasmussen (1986)
"SIMWE"	Simulation of Water Erosion	Mitasova, Brown, Johnston, and Mitas (1996)
"TOPMODEL"	Topography-based hydrological model	Beven and Kirkby (1979)
"WEPP"	Water Erosion Prediction Project	Flanagan and Nearing (1995)

Soil erosion models such as "RUSLE"; "SWAT," "ANSWERS," "PESERA," "EUROSEM," "SHE," and "WEPP" allow for quantitative assessment of soil loss (Pandey et al., 2021). Statistical approaches such as "TOPSIS" (Technique for Order of Preference by Similarity to Ideal Solution), "CF" (Compound Factor); VIKOR (Vise Kriterijumska Optimizacija I Kompromisno Resenje), founded on principal component analysis, and the multicriteria decision method provide qualitative examination of the river basin's soil erosion susceptibility. Qualitative and quantitative methods both incorporate multiple parameters such as land terrain, inclination, land pattern use, soil trait, runoff, soil deposit and yield, and so forth, whereas water quality parameters are not included as sign or measure. No approach has been demonstrated to be completely trustworthy in predicting soil loss, and the model's performance is determined by its capacity to achieve the study's purpose (Beven, 2011). Few widely used soil erosion models used to evaluate soil erosion have been discussed briefly.

23.3 Results and discussion

23.3.1 Estimation of soil erosion risk and vulnerable zone using the revised universal soil loss equation model

Many researchers (Balasubramani et al., 2015; Gaubi, Chaabani, Ben Mammou, & Hamza, 2017; Mallick, Alashker, Mohammad, Ahmed, & Hasan, 2014; Ozsoy & Aksoy, 2015; Prasannakumar, Vijith, Abinod, & Geetha, 2012; Renschler, Diekkrüger, & Mannaerts, 1998) have used RUSLE for estimating soil erosion in watersheds. The RUSLE model involves the process of identifying soil erosion using five major factors, which are the rainfall erosivity variables (R^E factor), soil erodibility factor (E^S factor), topographic factor (LS factor), cover management factor (C^m factor), and support practice factor (P^S factor) (Fig. 23.3). Because of its low complexity, the RUSLE model is a frequently used eroding soil model that may be used to a diverse region with many land use categories (Renard, 1997). The RUSLE model can be utilized on a scale that covers the globe (Kumar et al., 2022), on an administrative level (Terranova, Antronico, Coscarelli, & Iaquinta, 2009) and watershed scale (Balasubramani et al., 2015; Toubal, Achite, Ouillon, & Dehni, 2018). The mathematical formula for the RUSLE model in the investigation of soil loss is as follows.

$$A = R^E X \ E^S L^S X \ C^M X \ P^S$$

wherein A is the estimated annual soil loss (t/ha/year), R^E represents the rainfall-runoff erosivity variables (MJ mm/ha/h/year), E^S is the soil's erodibility component (t/ha/MJ/mm), L^S is the sloping length component, C^m is the cover management variable, and P^S is the conservation support practice element for the study area. The L^S, C^m, and P^S factor values are dimensionless values. All factors are combined with GIS tools, and that procedure is used widely.

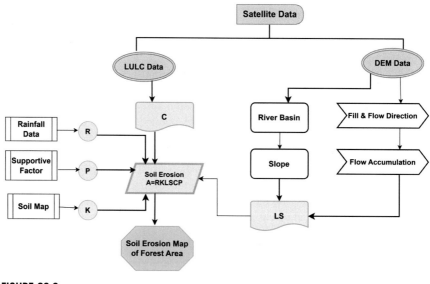

FIGURE 23.3

Methodology of the revised universal soil loss equation (RUSLE) model.

23.3.1.1 Rainfall-runoff erosivity factor (R^E)

The rainfall erosivity factor has been calculated from daily rainfall data on the river basin, and it describes the potential of raindrop impact at a particular location. According to Wischmeier and Smith (1978), the R^E factor measures the consequences of runoff velocity and the quantity of droplets that contribute to rainfall. The yearly erosivity was computed by adding the precipitation erosivity of individual disruptive precipitation throughout the year or season (Wischmeier & Smith, 1978). The parametric connection was established by Choudhury and Nayak (2003), and it was utilized to determine the R^E factor.

$$(R^E) = 79 + 0.363 \times Pa$$

where R^E is the precipitation erosivity parameter in MJ mm/ha/h/year, Pa is the average monthly precipitation in millimeters, and Pa is the yearly average precipitation.

23.3.1.2 Soil erodibility factor (E^S)

In order to determine the soil erodibility element (E^S), the soil properties and the potential of the soil or substrate to withstand erosion are taken into account. The soil erodibility variable measures the quantity of soil deterioration per unit of precipitation erosion score. The term "E^S value" describes how vulnerable certain aspects of soil, including as its texture, depth, structure, drainage, and the resulting organic content, are to soil loss under a typical plot and tillage method

Table 23.3 Soil erodibility factors for different textural classes (Das et al., 2020).

Textural class	Organic matter (%)	
	Below 0.5	2
Loam with a coarse texture	0.27	0.24
Fine loamy	0.35	0.3
Gravelly loam	0.12	0.1
Gravelly sandy loam	0.27	0.24
Loamy	0.38	0.34
Sandy clay loam	0.27	0.25
Sandy loam	0.27	0.24
Landslides	–	–
Rocky Mountain	–	–

Das, B., Bordoloi, R., Thungon, L. T., Paul, A., Pandey, P. K., Mishra, M., & Tripathi, O. P. (2020). An integrated approach of GIS, RUSLE and AHP to model soil erosion in West Kameng watershed, Arunachal Pradesh. Journal of Earth System Science, 129(1). *https://rd.springer.com/journal/12040, https://doi.org/10.1007/s12040-020-1356-6.*

(Wischmeier & Smith, 1968). However, soil texture and organic compounds have the greatest influence on soil erodibility, as reported by Wischmeier and Smith (1978), as shown in Table 23.3, which relate to soil loss in a typical plot and plowing method.

23.3.1.3 Slope length and steepness (LS) factor

There is no doubt that topography plays an important role in soil and water deformation. The topographic factor (*LS*) is a measure of the impacts of slope dimension (*L*) and inclination (*S*) on slope erosion. The length of slope (*L*) and the degree of slope (*S*) are used to calculate the effect of topography on erosion; these two parameters are typically combined to create one factor, that is, *LS* (Perovic et al., 2016). The *LS* factor in RUSLE represents a proportion of soil degradation over a certain slope dimension and steepness to soil removal from a slope (Wu et al., 2018). Slope dimension or length was explained by Wischmeier and Smith (1978) as the distance that runs horizontally between the initial point of the overland flow to the place at which either surface runoff gets concentrated to create a defined channel or the slope gradient declines to the point when deposition starts. In comparison to slope length, soil loss increases more quickly with slope steepness. When all other factors remain the same, the *LS* factor is a percentage of the degradation of soil on a given steep slope's length and inclination to soil erosion from a slope with a length of 22.1 m meters and an inclination of 9% (Renard, 1997), giving important information about the terrain's susceptibility to soil erosion. The *LS* factor is a component of the terrain that indicates how an inclination and its running length respond to soil removal (Moore & Wilson, 1992; Wischmeier & Smith, 1978). The *LS* component was acquired from the digital elevation model built by the

"Topo to Raster" function in the Spatial Analyst extension of ArcGIS, based on contour lines and spot elevations of MRB (in the Survey of India topographic sheets at 1:50,000 level). Additionally, this method was effectively used in hydrologic studies of other watersheds (Jain & Das, 2010; Jain & Kothyari, 2000; Thomas & Prasannakumar, 2015; Wischmeier & Smith, 1978). The *LS* factor was computed after (Renard, 1997); using the following equation:

$$LS = \left(\frac{\lambda}{22.1}\right) m \times \left(65.41 \times \sin^2\theta + 4.65 \times \sin\theta + 0.065\right)$$

Wherein λ is the slope/incline length in meters, θ is the degree of slope, and m is 0.5 if the percentage slope is 5.0 or greater, 0.4 on slopes of 3.6%−4.9%, 0.3 on slopes of 1%−3.5%, and 0.2 on homogeneous gradients of below 1%.

23.3.1.4 Cover management (C^m) factor

The cover and management variable (C^m) quantifies the impact of farming and strategies for runoff and loss of soil (Renard, 1997; Wischmeier & Smith, 1978) and is considered the second major factor (after topography) controlling soil erosion. The proper management of vegetation, plant residue, and tillage can reduce soil erosion (Lee, 2004). There are several methods available for estimating cover and management factors based on the available information. There are many factors that influence crop management, including surface vegetation, land use, roughness, and soil moisture. Satellite photographs can be used to estimate the cover factor successfully and quickly (Vatandaşlar & Yavuz, 2017). It is recommended for the application to obtain the imagery from satellites throughout the rainy period, when soil loss is most aggressive and the canopy of vegetation is at its greatest (Mhangara, Kakembo, & Lim, 2012). The normalized differential vegetation index (NDVI), which is obtained from satellite data and represents the concentration of green vegetation, is used to calculate the C^m factor (Grover & Singh, 2015). Various researchers used different spectral vegetation indices (de Asis & Omasa, 2007; Knijff, Jones, & Montanarella, 2000; Pan & Wen, 2014) in addition to fraction pictures obtained from spectral compound analysis of remotely captured imagery (Lu, Mausel, Brondízio, & Moran, 2004) for computing the C^m factor. NDVI, a measure of plant vitality, was adjusted to estimate the C^m variables (Knijff et al., 2000).

$$C^m = \exp\left[-\alpha \times \frac{NDVI}{\beta - NDVI}\right]$$

The parameters alpha (α) and beta (β) determine the geometry of the NDVI−C^m curve, with α values of 2 and β values of 1 yielding appropriate results (Knijff et al., 2000).

The NDVI was estimated using an integration of red (R) and infrared (NIR) spectra.

$$NDVI = \{(NIR - R)/(NIR + R)\}$$

To determine the quantity of the C^m variable, Toumi, Meddi, Mahé, and Brou (2013) performed regression across two extremes.

$$C^m = 0.9167 - \text{NDVI} \times 1.1667$$

However, many authors (Fayas, Abeysingha, Nirmanee, Samaratunga, & Mallawatantri, 2019; Imamoglu & Dengiz, 2017; Kebede, Endalamaw, Sinshaw, & Atinkut, 2021; Millward & Mersey, 1999; Reusing, Schneider, & Ammer, 2000) used LULC classification for the estimation of C^m assuming that areas with the same amount of land cover have the same C^m values. According to Lillesand and Kiefer (2000), the supervised classification approach (maximum likelihood) was utilized to create the LULC classes in order to calculate the C factor. The C^m factors for the land classes were input as attributes. The LULC outline of the research watershed has been generated, and the reclassification technique in the GIS was used to create a C^m-factor map of the study region.

23.3.1.5 Support practice (P^S) factor

P^S variables link to the agronomical approaches (contour harvesting, strip agriculture, and terrace) that modify the surface runoff, minimizing runoff speed and, therefore, preventing erosion of soils (Renard & Foster, 1983). When conservation practices are implemented, the P^S-factor measures the amount of soil lost compared to regular cultivation up and down slopes (Meshesha & Tripathi, 2016; Wischmeier & Smith, 1978). P^S factor was estimated from the LULC map, and the value varies from zero to one (Ganasri & Ramesh, 2016; Kebede et al., 2021). The score zero implies excellent conservation execution, while a value close to one signifies bad practice in conservation. The P^S value of settlement was lower than that of dense vegetation and barren land, as conservation practices were more numerous, and the P^S value of dense vegetation was higher than that of barren land. It does not seem appropriate to use conventional conservation practices for assessing degradation of soil in the catchment due to poor land management practices. In order to determine the P^S factors, Wischmeier and Smith (1978) suggested classifying the watershed as agricultural as well as other land uses. As shown in Table 23.4, the farming areas were divided into six (6) P^S slope categories, and P-values were allocated to each slope class. However, different land use/ land cover has been allocated a value of one.

The more effectively conservation practices reduce soil loss, the more effective they will be in the factor of soil erosion support practice. The LULC map has been used to develop the support practice factor for watershed areas where soil erosion is more likely to occur.

23.3.2 Revised universal soil loss equation model significance

The RUSLE is a popular model for estimating soil degradation/loss and is known for several advantages compared to other erosion models. RUSLE is relatively simple and user-friendly (Driscoll & Friggens, 2019; Ghosh, 2022; Tamene, Le, & Vlek, 2014),

Table 23.4 25.4 P^S-factor values suggested by various researchers.

Land-use type	Slope (%)	P^S factors	References
Agricultural	0–5	0.1	Wischmeier and Smith (1978), Bewket and Teferi (2009), Ganasri and Ramesh (2016)
	5–10	0.12	
	10–20	0.14	
	20–30	0.19	
	30–50	0.25	
	50–100	0.33	
Another land	All	1	

making it accessible to a wide range of users, including land managers, planners, and researchers. Because it depends on readily available data and simple computations, it can be used efficiently without requiring sophisticated inputs or in-depth technical understanding. On an empirical basis, RUSLE is a more accurate and dependable instrument for predicting soil erosion since it is founded on a substantial quantity of field data and study. The model is suitable for a broad range of scenarios due to its rigorous validation across diverse geographic and climatic settings. Compared to simpler models that might just take a few parameters into account, it provides a more thorough assessment of potential erosion risk by combining various variables. Its adaptability to varied geographic locations and scales makes it suitable for a wide range of applications, from local watersheds to enormous regions (Wang et al., 2016). Its flexibility is increased by users' ability to modify inputs and parameters to fit particular scenarios. Since the information needed for RUSLE inputs such as rainfall data, soil properties, and land cover is frequently easily obtainable from a variety of sources, gathering the information needed for the model application is not too difficult. RUSLE is useful for evaluating the possibility of long-term soil erosion, which is vital for land management and conservation planning (Atoma, Suryabhagavan, & Balakrishnan, 2020; Pan & Wen, 2014; Tang, Xu, Bennett, & Li, 2015). It brings information on how soil loss can accumulate over time under specific conditions, assisting in the formulation of sustainable land use strategies.

23.3.3 Challenges/limitations of the RUSLE model

The "RUSLE" model is valuable tool for assessing soil degradation/loss potential, but it also has its limitations. RUSLE simplifies the complex processes involved in soil erosion, which can lead to inaccuracies in areas with unique conditions. This model makes an inevitable assumption that erosion factors remain constant over time, which may not always hold true (Govers, 2011). The RUSLE erosion model is primarily concerned with sheet and rill soil loss, but does not adequately account for erosion of gullies, which can contribute significantly to soil erosion

(Fagbohun, Anifowose, Odeyemi, Aladejana, & Aladeboyeje, 2016; Tsegaye & Bharti, 2021). The quality and accessibility of input data, such as rainfall data, soil erodibility values, and land cover information, significantly impact RUSLE's accuracy. These data are susceptible to being interpreted incorrectly and may fail to account for these variables' regional and temporal variability. RUSLE is often applied at a relatively coarse spatial resolution, which may not capture localized erosion patterns. Furthermore, short-term fluctuations and extreme events may not be taken into account in this estimation of soil erosion potential (Kinnell, 2010). One of the major drawbacks of the RUSLE model is a lack of sensitivity to land management changes. This model theory assumes that land use and management practices remain constant over time, which may be inaccurate when land use changes or conservation measures are employed (Ganasri & Ramesh, 2016). Urban areas, mining sites, or areas where landscapes are rapidly changing may not be suitable for modeling erosion with this model (Mhaske, Pathak, Dash, & Nayak, 2021). Additionally, it does not consider the impact of climate change on soil erosion, which is becoming an increasingly important factor (Pinson & AuBuchon, 2023). As part of RUSLE, soil loss is estimated from a field or location, but the sediment is not taken into account for delivery to downstream areas, where it can affect the environment. Consequently, the model may be limited in its ability to be used to evaluate broader environmental consequences as a result of erosion. RUSLE may be more suitable for broader-scale applications or where data are limited. However, for detailed, site-specific assessments when high-resolution data are available, MMF is a process-based approach used to estimate soil erosion rates at a fine scale.

23.4 **Morgan–Morgan–Finney**

Several empirical models based on geomorphological parameters have been developed to assess soil erosion and sediment yield (Jose & Das, 1982; Misra, Satyanarayana, & Mukherjee, 1984). As the name MMF suggests, the model is a physical model based on empirical data developed in 1984 by Leonard A. Finney, Arthur L. Morgan, and Donald R. Morgan. This equation has been developed based on Meyer and Wischmeier's concept known as the USLE, which was developed in 1969 (Wischmeier & Smith, 1978). As one of the most widely used hydrological and erosion models, MMF helps estimate soil erosion and runoff in agricultural watersheds on hilly terrain. The model is known for its flexibility and simplicity, as well as its low data requirements, in comparison with more complex erosion models based on processes (Shrestha, 1997). The amount of runoff and the amount of rainfall energy available to separate the soil particles are determined by the water phase. In the sediment phase of the model, soil particle detachment is taken into account as a function of soil erodibility, rainfall energy, and vegetation-affected rainfall interception. According to the MMF model, there are two phases to soil erosion: sediment and water. The annual precipitation is used to calculate the runoff volume and rainfall energy, or kinetic energy, that is available for soil splash detachment in the water phase. The MMF model (Kirkby, Wallings, Yair, & Berkowicz, 1990) is used to predict the annual volume of overland flow. This model implies that runoff occurs when daily rainfall surpasses a critical value that is dependent on the surface soil layer's storage capacity. The following relationship represents the kinetic energy of rainfall (E), which is dependent upon the yearly precipitation amount (R) and the intensity of the rainfall (I) (Wischmeier & Smith, 1978).

$$E = (11.9 + 8.7 \log 10 I)$$

where E represents J/m^2 while I equals mm/h, which is interpreted as 30 mm/h in a strongly seasonal climate. The soil detached velocity can be determined by employing the equation.

$$F = K \times \left(E \times e^{-0.055A} \right) \times 10^{-3}$$

where F is the degree of separation by the impact of precipitation (kg/m^2) and K represents the soil detachability factor referred to as the quantity of soil removed from the mass of soil per unit of precipitation energy, and A represents the rate of precipitation that influences persistent interception and stem flow (%). Splash detachment is modeled in the sediment phase as a role of precipitation energy, soil detachability, along with precipitation interception influence. The overland flow or terrestrial flow pass-on potential is computed by employing the total amount of flow, incline steepness, and the impact of a plant or management of crop cover (Kirkby, 1976). Overland flow has been determined employing the following formulae.

$$Q = R \times \exp\left(-\frac{R_C}{R_0} \right)$$

$$R_C = 1000 \times MS \times BD \times RD \times \left(\frac{E_t}{E_0} \right) 0.5 \text{ and } R_0 = R/R_n$$

Wherein Q is the volume of overland flow (mm), R is yearly precipitation (mm), "R_C" represents soil moisture storage (mm), R_0 represents the average amount of precipitation every precipitation day (mm), R_n is the number of rainfall days within course of a year, E_t/E_0 represents proportion of real (E_t) to potential (E_0) evaporation, MS represents moisture content of the soil at field capacity or 1/3 bar tension (% or w/w), BD represents bulk density of topmost layer (Mg/m^3), while RD represents topsoil layer's rooting depth in meter.

Overland flow's transport capacity can be calculated in the following manner:

$$G = C \times Q^2 \times \sin S \times 10^{-3}$$

where G indicates the overland flow capacity for transport (kg/m^2), C indicates the crop cover management factor, while S represents the terrain's slope's incline symbolized by the slope angle. To assess the spatial pattern of yearly average soil deterioration, MMF parameters, namely, A, C, E_t/E_0, and RD, were calculated. The MMF model evaluates the effects of different land management strategies and conservation measures on soil erosion and sediment yield in agricultural settings. It is important in assessing the efficacy of erosion management tactics. However, for the more comprehensive, reliable, and robust analysis of soil erosion, the analytic hierarchy process (AHP) allows for the decomposition of the problem into smaller, more manageable components, which can help organize and structure the decision-making process in soil erosion studies.

23.5 Analytic hierarchy process

In soil erosion risk modeling, the AHP, often referred to as the Saaty method, is a semiquantitative, statistical, multipurpose technique (Aslam et al., 2021; Mushtaq, Farooq, Tirkey, & Sheikh, 2023; Pradeep, Krishnan, & Vijith, 2015). In order to make an optimal decision for a complex circumstance, AHP utilizes GIS through a hierarchical structure that consists of targets to be reached, decision criteria, as well as possibilities to be considered (Saaty & Vargas, 2001; Saaty, 1990). The AHP process involves "developing the spatial database; establishing evaluation criteria and hierarchical structures for multicriteria questions; calculating the relative importance of each criterion using the AHP method; and finally, estimating soil erosion severity using the weighted sum method." In the ArcGIS environment, databases for numerous theme layers were created. Pairwise comparisons should be done for every level in the hierarchy to determine the items' relative relevance. These comparisons are done using a scale of preference. The Saaty scale, which employs numbers from 1 (equal significance) to 9 (much more significant), is the most widely used scale. After determining the priority, the consistency ratio (CR) is computed to ensure consistency. A low CR shows that the judgments are consistent, whereas a high CR indicates that they are inconsistent. Pairwise comparisons can be improved upon in order to address inconsistency. When all pairwise comparisons have been completed, the relative importance (weight) and alternative for each criterion at each level are determined. The matrices' eigenvectors can be employed for computation purposes. The relative weights of the elements with a "consistency index" (CI) can determine the variable's consistency, as shown below (Saaty, 1990).

Here, λ_{max} represents the greatest or primary eigen component of the examined matrix, while n indicates the sequence of the squared matrices.

The inconsistency index, referred to as the CR, establishes the acceptability of the particular variables and their corresponding classes for inclusion in the analysis provided by

$$CR = \left(\frac{CI}{RI}\right)$$

The random index denotes the CI with a random square matrix having an identical size, as presented by (Saaty, 1980).

Several studies and researches were used to calculate corresponding criteria weights, which could be used to evaluate how effective this method is for controlling soil erosion worldwide. The essential soil erosion-prone locations have been determined through the combination of geo-environmental characteristics (e.g., land use/land cover, geological structure, the density of drainage, drain rate, lineament speed, incline, and relative relief). After evaluating their proportionate impact in conditioning, the topography prone to soil loss is determined using an AHP approach within the raster-based geospatial context. Soil erosion or soil loss intensity maps have been developed by classifying the soil erosion probability map generated via the method known as AHP, which demonstrates regions with varying soil erosion probabilities.

23.6 Conclusion

Soil erosion has recently become an area of investigation and a critical global problem. Remote sensing provides valuable insights into the major drivers of soil erosion by capturing spatial and temporal variations in land cover, land use, topography, and hydrological factors. Integrating remote sensing data with erosion modeling techniques, one can better understand the spatial and temporal dynamics of soil erosion drivers, prioritize conservation efforts, and implement targeted mitigation strategies to reduce erosion risk and promote sustainable land management practices. The study on the estimation of soil erosion risk and vulnerable zones using the RUSLE and GIS approaches provides valuable insights into the identification and management of areas prone to soil erosion. The integration of RUSLE and GIS enables a comprehensive assessment by considering multiple factors such as rainfall, soil erodibility, slope, land cover, and conservation practices. The findings emphasize the importance of adopting a holistic approach for soil erosion risk assessment, allowing for a more accurate identification of vulnerable zones. The study's use of advanced technology, such as GIS, enhances spatial analysis capabilities and facilitates better decision-making for sustainable land management practices. Additionally, the results underscore the need for targeted conservation measures in high-risk areas, promoting effective soil erosion control strategies. Overall, this research contributes significantly to the scientific understanding of soil erosion dynamics and offers practical applications for policy-makers, land managers, and researchers aiming to mitigate the adverse impacts of soil erosion on agricultural productivity and environmental sustainability. The integration of RUSLE and GIS not only enhances the accuracy of soil erosion risk assessment but also provides a robust foundation for developing tailored erosion control strategies, thereby contributing to the overall conservation and sustainable management of soil resources. The current study investigates RUSLE and other GIS-based methods for predicting soil erosion worldwide, including physical, empirical, statistical, and process-based models. In various geographic and climatic settings, the RUSLE model has been rigorously validated for a wide range of scenarios. Although its strengths are considerable, the RUSLE−GIS approach has some limitations and uncertainties, particularly with regard to the accuracy of input data and the assumptions behind it. Small-scale variability in soil erosion and dynamic environmental factors can cause prediction uncertainties. However, there is still much work to be done in order to overcome limitations and improve the accuracy of predictions, ultimately making sustainable land management possible.

References

Abbott, M. B., Bathurst, J. C., Cunge, J. A., O'Connell, P. E., & Rasmussen, J. (1986). An introduction to the European Hydrological System—Systeme Hydrologique Europeen, "SHE," 1: History and philosophy of a physically-based, distributed modelling system. *Journal of Hydrology, 87*(1−2), 45−59. Available from https://doi.org/10.1016/0022-1694(86)90114-9.

Alam, A. (2014). Soil degradation: A challenge to sustainable agriculture. *International Journal of Scientific Research in Agricultural Sciences, 1*, 50−55. Available from https://doi.org/10.12983/ijsras-2014-p0050-0055.

Almorox Alonso, J., De Antonio García, R., Saa Requejo, A., Díaz Alvarez, M. C., & Gascó Montes, J. M. (1994). *Metodos de estimacion de la erosion hidrica*. Editorial Agrícola Española, S.A.

Arekhi, S., Niazi, Y., & Kalteh, A. M. (2012). Soil erosion and sediment yield modeling using RS and GIS techniques: A case study, Iran. *Arabian Journal of Geosciences, 5*(2), 285−296. Available from https://doi.org/10.1007/s12517-010-0220-4, http://www.springer.com/geosciences/journal/12517?cm_mmc = AD-_-enews-_-PSE1892-_-0.

Arnold, J. G., Srinivasan, R., Muttiah, R. S., & Williams, J. R. (1998). Large area hydrologic modeling and assessment Part I: Model development 1. *JAWRA Journal of the American Water Resources Association, 34*(1), 73−89. Available from https://doi.org/10.1111/j.1752-1688.1998.tb05961.x.

Aslam, B., Maqsoom, A., Salah Alaloul, W., Ali Musarat, M., Jabbar, T., & Zafar, A. (2021). Soil erosion susceptibility mapping using a GIS-based multi-criteria decision approach: Case of district Chitral, Pakistan. *Ain Shams Engineering Journal, 12*(2), 1637−1649. Available from https://doi.org/10.1016/j.asej.2020.09.015, http://www.elsevier.com/wps/find/journaldescription.cws_home/724208/description#description.

Atoma, H., Suryabhagavan, K. V., & Balakrishnan, M. (2020). Soil erosion assessment using RUSLE model and GIS in Huluka watershed, Central Ethiopia. *Sustainable Water Resources Management, 6*(1). Available from https://doi.org/10.1007/s40899-020-00365-z.

Balasubramani, A., Larjo, A., Bassein, J. A., Chang, X., Hastie, R. B., Togher, S. M., ... Rao, A. (2015). Cancer-associated ASXL1 mutations may act as gain-of-function mutations of the ASXL1-BAP1 complex. *Nature Communications, 6*. Available from https://doi.org/10.1038/ncomms8307, http://www.nature.com/ncomms/index.html.

Bannari, A., Guédon, A. M., & El-Ghmari, A. (2016). Mapping slight and moderate saline soils in irrigated agricultural land using advanced land imager sensor (EO-1) data and semi-empirical models. *Communications in Soil Science and Plant Analysis*, *47*(16), 1883−1906. Available from https://doi.org/10.1080/00103624.2016.1206919, http://www.tandfonline.com/loi/lcss20.

Barrow, C. J. (1991). *Land degradation: Development and breakdown of terrestrial environments*. Cambridge University Press.

Beasley, D. B., & Huggins, L. F. (1980). *ANSWERS, Areal Nonpoint Source Watershed Environment Response Simulation: User's manual*. US Environmental Protection Agency.

Berberoglu, S., Cilek, A., Kirkby, M., Irvine, B., & Donmez, C. (2020). Spatial and temporal evaluation of soil erosion in Turkey under climate change scenarios using the Pan-European Soil Erosion Risk Assessment (PESERA) model. *Environmental Monitoring and Assessment*, *192*(8). Available from https://doi.org/10.1007/s10661-020-08429-5, https://link.springer.com/journal/10661.

Beven, K. J., & Kirkby, M. J. (1979). A physically based, variable contributing area model of basin hydrology/Un modèle à base physique de zone d'appel variable de l'hydrologie du bassin versant. *Hydrological Sciences Bulletin*, *24*(1), 43−69. Available from https://doi.org/10.1080/02626667909491834.

Beven, K. J. (2011). *Rainfall-runoff modelling: The primer*. John Wiley & Sons.

Bewket, W., & Teferi, E. (2009). Assessment of soil erosion hazard and prioritization for treatment at the watershed level: Case study in the Chemoga watershed, Blue Nile basin, Ethiopia. *Land Degradation & Development*, *20*(6), 609−622. Available from https://doi.org/10.1002/ldr.944.

Blaikie, P., & Brookfield, H. (2015). *Land degradation and society*. Routledge.

Borrelli, P., Robinson, D. A., Panagos, P., Lugato, E., Yang, J. E., Alewell, C., ... Ballabio, C. (2020). Land use and climate change impacts on global soil erosion by water (2015-2070). *Proceedings of the National Academy of Sciences*, *117*(36), 21994−22001. Available from https://doi.org/10.1073/pnas.2001403117.

Boyle, J. F., Plater, A. J., Mayers, C., Turner, S. D., Stroud, R. W., & Weber, J. E. (2011). Land use, soil erosion, and sediment yield at Pinto Lake, California: Comparison of a simplified USLE model with the lake sediment record. *Journal of Paleolimnology*, *45*(2), 199−212. Available from https://doi.org/10.1007/s10933-010-9491-8.

Bracken, L. J., & Croke, J. (2007). The concept of hydrological connectivity and its contribution to understanding runoff-dominated geomorphic systems. *Hydrological Processes*, *21*(13), 1749−1763. Available from https://doi.org/10.1002/hyp.6313.

Brevik, E. C., Cerdà, A., Mataix-Solera, J., Pereg, L., Quinton, J. N., Six, J., & Van Oost, K. (2015). The interdisciplinary nature of SOIL. *SOIL*, *1*(1), 117−129. Available from https://doi.org/10.5194/soil-1-117-2015.

Cerdà, A., González-Pelayo, Ó., Giménez-Morera, A., Jordán, A., Pereira, P., Novara, A., ... Ritsema, C. J. (2016). Use of barley straw residues to avoid high erosion and runoff rates on persimmon plantations in Eastern Spain under low frequency-high magnitude simulated rainfall events. *Soil Research*, *54*(2), 154−165. Available from https://doi.org/10.1071/SR15092, http://www.publish.csiro.au/nid/84/aid/2406.htm.

Chalise, D., Kumar, L., & Kristiansen, P. (2019). Land degradation by soil erosion in Nepal: A review. *Soil Systems*, *3*(1). Available from https://doi.org/10.3390/soilsystems3010012.

Chomitz, K. M., & Kumari, K. (1998). The domestic benefits of tropical forests: A critical review. *The World Bank Research Observer, 13*(1), 13−35. Available from https://doi.org/10.1093/wbro/13.1.13.

Choudhury, M. K., & Nayak, T. (2003). Estimation of soil erosion in Sagar Lake catchment of Central India. *Proceedings of the International Conference on Water and Environment* (pp. 387−392). Bhopal, India.

Dabral, P. P., Baithuri, N., & Pandey, A. (2008). Soil erosion assessment in a hilly catchment of North Eastern India using USLE, GIS and remote sensing. *Water Resources Management, 22*(12), 1783−1798. Available from https://doi.org/10.1007/s11269-008-9253-9.

Das, B., Bordoloi, R., Thungon, L. T., Paul, A., Pandey, P. K., Mishra, M., & Tripathi, O. P. (2020). An integrated approach of GIS, RUSLE and AHP to model soil erosion in West Kameng watershed, Arunachal Pradesh. *Journal of Earth System Science, 129* (1). Available from https://doi.org/10.1007/s12040-020-1356-6, https://rd.springer.com/journal/12040.

de Asis, A. M., & Omasa, K. (2007). Estimation of vegetation parameter for modeling soil erosion using linear Spectral Mixture Analysis of Landsat ETM data. *ISPRS Journal of Photogrammetry and Remote Sensing, 62*(4), 309−324. Available from https://doi.org/10.1016/j.isprsjprs.2007.05.013.

de Jong, S. M., Paracchini, M. L., Bertolo, F., Folving, S., Megier, J., & de Roo, A. P. J. (1999). Regional assessment of soil erosion using the distributed model SEMMED and remotely sensed data. *CATENA, 37*(3-4), 291−308. Available from https://doi.org/10.1016/s0341-8162(99)00038-7.

De Roo, A. P. J., Hazelhoff, L., & Burrough, P. A. (1989). Soil erosion modelling using 'answers' and geographical information systems. *Earth Surface Processes and Landforms, 14*(6), 517−532. Available from https://doi.org/10.1002/esp.3290140608.

De Roo, A. P. J., Offermans, R. J. E., & Cremers, N. H. D. T. (1996). LISEM: A single-event, physically based hydrological and soil erosion model for drainage basins. II: Sensitivity analysis, validation and application. *Hydrological Processes, 10*(8), 1119−1126. Available from https://doi.org/10.1002/(sici)1099-1085(199608)10:8 < 1119::aid-hyp416 > 3.0.co;2-v.

Dimotta, A. (2019). Global soil erosion costs: a critical review of the economic assessment methods. *Soil erosion interdisciplinary overview: modelling approaches, ecosystem services assessment and soil quality restoration. applications and analyses in the Basilicata region. Italy.* Retrieved from https://doi.org/10.13140/RG2.2.13842.30401.

Driscoll, K. P., & Friggens, M. (2019). Assessing risk in a postfire landscape: Are currently available tools good for the local land owner? *Natural Areas Journal, 39*(4), 472−481. Available from https://doi.org/10.3375/043.039.0410, http://www.bioone.org/loi/naar.

Eswaran, H., Lal, R., & Reich, P. F. (2019). *Land degradation: An overview* (pp. 20−35). Informa UK Limited. Available from https://doi.org/10.1201/9780429187957-4.

Fagbohun, B. J., Anifowose, A. Y. B., Odeyemi, C., Aladejana, O. O., & Aladeboyeje, A. I. (2016). GIS-based estimation of soil erosion rates and identification of critical areas in Anambra sub-basin, Nigeria. *Modeling Earth Systems and Environment, 2*(3). Available from https://doi.org/10.1007/s40808-016-0218-3, https://link.springer.com/journal/40808.

Favis-Mortlock, D. T., Quinton, J. N., & Dickinson, W. T. (1996). The GCTE validation of soil erosion models for global change studies. *Journal of Soil and Water Conservation, 51*(5), 397−403.

Fayas, C. M., Abeysingha, N. S., Nirmanee, K. G. S., Samaratunga, D., & Mallawatantri, A. (2019). Soil loss estimation using rusle model to prioritize erosion control in KELANI river basin in Sri Lanka. *International Soil and Water Conservation Research, 7*(2), 130–137. Available from https://doi.org/10.1016/j.iswcr.2019.01.003, http://www.keaipublishing.com/en/journals/international-soil-and-water-conservation-research/.

Febles-González, J. M., Vega-Carreño, M. B., Tolón-Becerra, A., & Lastra-Bravo, X. (2012). Assessment of soil erosion in karst regions of Havana, Cuba. *Land Degradation & Development, 23*(5), 465–474. Available from https://doi.org/10.1002/ldr.1089.

Flanagan, D. C., & Nearing, M. A. (1995). *USDA-water erosion prediction project: Hillslope profile and watershed model documentation* (Nserl Report No. 10).

Ganasri, B. P., & Ramesh, H. (2016). Assessment of soil erosion by RUSLE model using remote sensing and GIS — A case study of Nethravathi Basin. *Geoscience Frontiers, 7* (6), 953–961. Available from https://doi.org/10.1016/j.gsf.2015.10.007.

Gaubi, I., Chaabani, A., Ben Mammou, A., & Hamza, M. H. (2017). A GIS-based soil erosion prediction using the revised universal soil loss equation (RUSLE) (Lebna watershed, Cap Bon, Tunisia). *Natural Hazards, 86*(1), 219–239. Available from https://doi.org/10.1007/s11069-016-2684-3.

Ghosh, S. (2022). *An assessment of RUSLE model and erosion vulnerability in the slopes of Dwarka–Brahmani lateritic interfluve, eastern India* (pp. 475–506). Springer Science and Business Media LLC. Available from https://doi.org/10.1007/978-981-16-6966-8_26.

Govers, G. (2011). *Misapplications and misconceptions of erosion models. Handbook of erosion modelling* (pp. 117–134). Belgium: John Wiley and Sons. Available from http://onlinelibrary.wiley.com/book/10.1002/9781444328455, https://doi.org/10.1002/9781444328455.ch7.

Grahmann, K., Rubio, V., Perez-Bidegain, M., & Quincke, J. A. (2022). Soil use legacy as driving factor for soil erosion under conservation agriculture. *Frontiers in Environmental Science, 10.* Available from https://doi.org/10.3389/fenvs.2022.822967, https://www.frontiersin.org/journals/environmental-science.

Grover, A., & Singh, R. (2015). Analysis of urban heat island (UHI) in relation to normalized difference vegetation index (NDVI): A comparative study of Delhi and Mumbai. *Environments, 2*(4), 125–138. Available from https://doi.org/10.3390/environments2020125.

Gu, Z., Xie, Y., Gao, Y., Ren, X., Cheng, C., & Wang, S. (2018). Quantitative assessment of soil productivity and predicted impacts of water erosion in the black soil region of northeastern China. *Science of the Total Environment, 637-638,* 706–716. Available from https://doi.org/10.1016/j.scitotenv.2018.05.061.

Hootsmans, R. M., Bouwman, A. F., Leemans, R., & Kreileman, G. J. J. (2001). *Modelling land degradation in IMAGE 2, (RIVM / Rapport; No. 481508009). RIVM.* http://www.rivm.nl/bibliotheek/rapporten/481508009.pdf.

Ilao, R. O., & Guzman. (2001). *The on-site cost of soil erosion: The case of Mapawa Catchment, Lantapan, Philippines (No. H029244).* International Water Management Institute.

Imamoglu, A., & Dengiz, O. (2017). Determination of soil erosion risk using RUSLE model and soil organic carbon loss in Alaca catchment (Central Black Sea region, Turkey). *Rendiconti Lincei, 28*(1), 11–23. Available from https://doi.org/10.1007/s12210-016-0556-0.

Islam, M. R., Jaafar, W. Z. W., Hin, L. S., Osman, N., & Karim, M. R. (2020). Development of an erosion model for Langat River Basin, Malaysia, adapting GIS and

RS in RUSLE. *Applied Water Science, 10*(7). Available from https://doi.org/10.1007/s13201-020-01185-4.

Issaka, S., & Ashraf, M. A. (2017). Impact of soil erosion and degradation on water quality: A review. *Geology, Ecology, and Landscapes, 1*(1), 1−11. Available from https://doi.org/10.1080/24749508.2017.1301053.

Jain, M. K., & Das, D. (2010). Estimation of sediment yield and areas of soil erosion and deposition for watershed prioritization using GIS and remote sensing. *Water Resources Management, 24*(10), 2091−2112. Available from https://doi.org/10.1007/s11269-009-9540-0.

Jain, M. K., & Kothyari, U. C. (2000). Estimation of soil erosion and sediment yield using GIS. *Hydrological Sciences Journal, 45*(5), 771−786. Available from https://doi.org/10.1080/02626660009492376.

Jain, S. K., Kumar, S., & Varghese, J. (2001). Estimation of soil erosion for a Himalayan watershed using GIS technique. *Water Resources Management, 15*(1), 41−54. Available from https://doi.org/10.1023/A:1012246029263.

Jetten, V., Govers, G., & Hessel, R. (2003). Erosion models: Quality of spatial predictions. *Hydrological Processes., 17*(5), 887−900. Available from https://doi.org/10.1002/hyp.1168.

Jose, C. S., & Das, D. C. (1982). Geomorphic prediction models for sediment production rate and intensive priorities of watersheds in Mayurakshi catchment. *Proceeding of International Symposium on Hydrological Aspects of Mountainous Watershed* (pp. 15−23).

Kar, S. K., Kumar, S., Sankar, M., Patra, S., Singh, R. M., Shrimali, S. S., & Ojasvi, P. R. (2022). Process-based modelling of soil erosion: Scope and limitation in the Indian context. *Current Science, 122*(5), 533−541. Available from https://doi.org/10.18520/cs/v122/i5/533-541.

Karan, S. K., Ghosh, S., & Samadder, S. R. (2019). Identification of spatially distributed hotspots for soil loss and erosion potential in mining areas of Upper Damodar Basin − India. *CATENA, 182*. Available from https://doi.org/10.1016/j.catena.2019.104144.

Ke, Q., & Zhang, K. (2022). Interaction effects of rainfall and soil factors on runoff, erosion, and their predictions in different geographic regions. *Journal of Hydrology, 605*, 127291. Available from https://doi.org/10.1016/j.jhydrol.2021.127291.

Kebede, Y. S., Endalamaw, N. T., Sinshaw, B. G., & Atinkut, H. B. (2021). Modeling soil erosion using RUSLE and GIS at watershed level in the Upper Beles, Ethiopia. *Environmental Challenges, 2*. Available from https://doi.org/10.1016/j.envc.2020.100009.

Keesstra, S., Pereira, P., Novara, A., Brevik, E. C., Azorin-Molina, C., Parras-Alcántara, L., ... Cerdà, A. (2016). Effects of soil management techniques on soil water erosion in apricot orchards. *Science of the Total Environment, 551−552*, 357−366. Available from https://doi.org/10.1016/j.scitotenv.2016.01.182.

Kinerson, R. S., Kittle, J. L., & Duda, P. B. (2009). *BASINS: Better assessment science integrating point and nonpoint sources. Decision support systems for risk-based management of contaminated sites* (pp. 375−398). United States: Springer US. Available from https://doi.org/10.1007/978-0-387-09722-0_18.

Kinnell, P. I. A. (2010). Event soil loss, runoff and the universal soil loss equation family of models: A review. *Journal of Hydrology, 385*(1−4), 384−397. Available from https://doi.org/10.1016/j.jhydrol.2010.01.024.

Kirkby, M., Wallings, D. E., Yair, A., & Berkowicz, S. (1990). *A simulation model for desert runoff and erosion.* 9780947571375.

Kirkby, M. J., Abrahart, R., McMahon, M. D., Shao, J., & Thornes, J. B. (1998). MEDALUS soil erosion models for global change. *Geomorphology*, *24*(1), 35–49. Available from https://doi.org/10.1016/s0169-555x(97)00099-8.

Kirkby, M. J. (1976). *Hydrological slope models: The influence of climate* (pp. 247–267). Department of Geography, University of Leeds.

Knijff, J. M., Jones, R. J. A., & Montanarella, L. (2000). *Soil erosion risk: Assessment in Europe, (pp. 33). Space Applications Institute, European Soil Bureau. EUR 19044.*

Knisel, W. G. (1980). *CREAMS: A field scale model for chemicals, runoff, and erosion from agricultural management systems (Vol. 26).* Department of Agriculture, Science and Education Administration.

Krysanova, V., Meiner, A., Roosaare, J., & Vasilyev, A. (1989). Simulation modelling of the coastal waters pollution from agricultural watershed. *Ecological Modelling*, *49* (1–2), 7–29. Available from https://doi.org/10.1016/0304-3800(89)90041-0.

Krysanova, V., Wechsung, F., Arnold, J., Srinivasan, R., & Williams, J. (2000). *SWIM (soil and water integrated model)*, Technical Report, PIK-69, pp. 239, Germany.

Kumar, M., Sahu, A. P., Sahoo, N., Dash, S. S., Raul, S. K., & Panigrahi, B. (2022). Global-scale application of the RUSLE model: A comprehensive review. *Hydrological Sciences Journal*, *67*(5), 806–830. Available from https://doi.org/10.1080/02626667.2021.2020277.

Lal, R. (2003). Soil erosion and the global carbon budget. *Environment International*, *29* (4), 437–450. Available from https://doi.org/10.1016/s0160-4120(02)00192-7.

Lal, R. (2017). *Soil erosion by wind and water: Problems and prospects. Soil Erosion Research Methods* (pp. 1–10). United States: CRC Press. Available from https://doi.org/10.1201/9780203739358.

Lee, S. (2004). Soil erosion assessment and its verification using the universal soil loss equation and geographic information system: A case study at Boun, Korea. *Environmental Geology*, *45*(4), 457–465. Available from https://doi.org/10.1007/s00254-003-0897-8.

Leonard, R. A., Knisel, W. G., & Still, D. A. (1987). GLEAMS: Groundwater Loading Effects of Agricultural Management Systems. *Transactions of the ASAE*, *30*(5), 1403–1418. Available from https://doi.org/10.13031/2013.30578.

Lillesand, T. M., & Kiefer, R. W. (2000). *Remote sensing and image interpretation*. John Wiley & Sons.

Liu, Y., Zhao, L., & Yu, X. (2020). A sedimentological connectivity approach for assessing on-site and off-site soil erosion control services. *Ecological Indicators*, *115*. Available from https://doi.org/10.1016/j.ecolind.2020.106434.

Locatelli, B., Imbach, P., Vignola, R., Metzger, M. J., & Hidalgo, E. J. L. (2011). Ecosystem services and hydroelectricity in Central America: Modelling service flows with fuzzy logic and expert knowledge. *Regional Environmental Change*, *11*(2), 393–404. Available from https://doi.org/10.1007/s10113-010-0149-x.

Lu, D., Mausel, P., Brondízio, E., & Moran, E. (2004). Change detection techniques. *International Journal of Remote Sensing*, *25*(12), 2365–2401. Available from https://doi.org/10.1080/0143116031000139863.

Mallick, J., Alashker, Y., Mohammad, S. A. D., Ahmed, M., & Hasan, M. A. (2014). Risk assessment of soil erosion in semi-arid mountainous watershed in Saudi Arabia by RUSLE model coupled with remote sensing and GIS. *Geocarto International*, *29*(8), 915–940. Available from https://doi.org/10.1080/10106049.2013.868044.

Mekonnen, M., Keesstra, S. D., Stroosnijder, L., Baartman, J. E. M., & Maroulis, J. (2015). Soil conservation through sediment trapping: A review. *Land Degradation and Development*, *26*(6), 544−555. Available from https://doi.org/10.1002/ldr.2308.

Meshesha, T. W., & Tripathi, S. K. (2016). Farmer's perception on soil erosion and land degradation problems and management practices in the Beressa Watershed of Ethiopia. *Journal of Water Resources and Ocean Science*, *5*, 64−72.

Mhangara, P., Kakembo, V., & Lim, K. J. (2012). Soil erosion risk assessment of the Keiskamma catchment, South Africa using GIS and remote sensing. *Environmental Earth Sciences*, *65*(7), 2087−2102. Available from https://doi.org/10.1007/s12665-011-1190-x.

Mhaske, S. N., Pathak, K., Dash, S. S., & Nayak, D. B. (2021). Assessment and management of soil erosion in the hilltop mining dominated catchment using GIS integrated RUSLE model. *Journal of Environmental Management*, *294*. Available from https://doi.org/10.1016/j.jenvman.2021.112987.

Millward, A. A., & Mersey, J. E. (1999). Adapting the RUSLE to model soil erosion potential in a mountainous tropical watershed. *CATENA*, *38*(2), 109−129. Available from https://doi.org/10.1016/S0341-8162(99)00067-3.

Mishra, P. K., & Deng, Z. Q. (2009). Sediment TMDL development for the Amite river. *Water Resources Management*, *23*(5), 839−852. Available from https://doi.org/10.1007/s11269-008-9302-4.

Misra, N., Satyanarayana, T., & Mukherjee, R. K. (1984). Effect of top elements on the sediment production rate from Sub-watershed in Upper Damodar Valley. *Journal of Agricultural Engineering*, *21*(3), 65−70.

Mitasova, H., Brown, W. M., Johnston, D., & Mitas, L. (1996). GIS tools for erosion/deposition modelling and multidimensional visualization. Part III: Process based erosion simulation. Geographic Modelling and Systems Laboratory, University of Illinois.

Mitasova, H., & Mitas, L. (1998). *Process modeling and simulations*. NCGIA GISCC Unit.

Moore, I. D., & Wilson, J. P. (1992). Length-slope factors for the revised universal soil loss equation: Simplified method of estimation. *Journal of Soil & Water Conservation*, *47*(5), 423−428.

Morgan, R. P. C., Quinton, J. N., Smith, R. E., Govers, G., Poesen, J. W. A., Auerswald, K., ... Styczen, M. E. (1998). The European Soil Erosion Model (EUROSEM): A dynamic approach for predicting sediment transport from fields and small catchments. *Earth Surface Processes and Landforms*, *23*(6), 527−544. Available from https://doi.org/10.1002/(sici)1096-9837(199806)23:6 < 527::aid-esp868 > 3.0.co;2-5.

Morgan, R. P. C. (2001). A simple approach to soil loss prediction: A revised Morgan−Morgan−Finney model. *CATENA*, *44*(4), 305−322. Available from https://doi.org/10.1016/s0341-8162(00)00171-5.

Morgan, R. P. C. (2009). *Soil erosion and conservation*. John Wiley & Sons.

Mullan, D. (2013). Soil erosion under the impacts of future climate change: Assessing the statistical significance of future changes and the potential on-site and off-site problems. *CATENA*, *109*, 234−246. Available from https://doi.org/10.1016/j.catena.2013.03.007.

Mushtaq, F., Farooq, M., Tirkey, A. S., & Sheikh, B. A. (2023). Analytic hierarchy process (AHP) based soil erosion susceptibility mapping in northwestern Himalayas: A case study of Central Kashmir Province. *Conservation*, *3*(1), 32−52. Available from https://doi.org/10.3390/conservation3010003.

Nearing, M. A., Foster, G. R., Lane, L. J., & Finkner, S. C. (1989). A process-based soil erosion model for USDA-water erosion prediction project technology. *Transactions of the ASAE, 32*(5), 1587−1593. Available from https://doi.org/10.13031/2013.31195.

Ozsoy, G., & Aksoy, E. (2015). Estimation of soil erosion risk within an important agricultural sub-watershed in Bursa, Turkey, in relation to rapid urbanization. *Environmental Monitoring and Assessment, 187*(7). Available from https://doi.org/10.1007/s10661-015-4653-9.

Pan, J., & Wen, Y. (2014). Estimation of soil erosion using RUSLE in Caijiamiao watershed, China. *Natural Hazards, 71*(3), 2187−2205. Available from https://doi.org/10.1007/s11069-013-1006-2.

Pandey, A., Chowdary, V. M., & Mal, B. C. (2007). Identification of critical erosion prone areas in the small agricultural watershed using USLE, GIS and remote sensing. *Water Resources Management, 21*(4), 729−746. Available from https://doi.org/10.1007/s11269-006-9061-z.

Pandey, A., Mathur, A., Mishra, S. K., & Mal, B. C. (2009). Soil erosion modeling of a Himalayan watershed using RS and GIS. *Environmental Earth Sciences, 59*(2), 399−410. Available from https://doi.org/10.1007/s12665-009-0038-0.

Pandey, S., Kumar, P., Zlatic, M., Nautiyal, R., & Panwar, V. P. (2021). Recent advances in assessment of soil erosion vulnerability in a watershed. *International Soil and Water Conservation Research, 9*(3), 305−318. Available from https://doi.org/10.1016/j.iswcr.2021.03.001.

Patault, E., Ledun, J., Landemaine, V., Soulignac, A., Richet, J. B., Fournier, M., ... Laignel, B. (2021). Analysis of off-site economic costs induced by runoff and soil erosion: Example of two areas in the northwestern European loess belt for the last two decades (Normandy, France). *Land Use Policy, 108*. Available from https://doi.org/10.1016/j.landusepol.2021.105541.

Perovic, V., Jaramaz, D., Zivotic, L., Cakmak, D., Mrvic, V., Milanovic, M., & Saljnikov, E. (2016). Design and implementation of WebGIS technologies in evaluation of erosion intensity in the municipality of NIS (Serbia). *Environmental Earth Sciences, 75*(3), 1−12. Available from https://doi.org/10.1007/s12665-015-4857-x.

Pinson, A. O., & AuBuchon, J. S. (2023). A new method for calculating C factor when projecting future soil loss using the revised universal soil loss equation (RUSLE) in semi-arid environments. *CATENA, 226*. Available from https://doi.org/10.1016/j.catena.2023.107067.

Pradeep, G. S., Krishnan, M. V. N., & Vijith, H. (2015). Identification of critical soil erosion prone areas and annual average soil loss in an upland agricultural watershed of Western Ghats, using analytical hierarchy process (AHP) and RUSLE techniques. *Arabian Journal of Geosciences, 8*(6), 3697−3711. Available from https://doi.org/10.1007/s12517-014-1460-5.

Prasannakumar, V., Vijith, H., Abinod, S., & Geetha, N. (2012). Estimation of soil erosion risk within a small mountainous sub-watershed in Kerala, India, using revised universal soil loss equation (RUSLE) and geo-information technology. *Geoscience Frontiers, 3*(2), 209−215. Available from https://doi.org/10.1016/j.gsf.2011.11.003.

Rashmi, I., Karthika, K. S., Roy, T., Shinoji, K. C., Kumawat, A., Kala, S., & Pal, R. (2022). *Soil erosion and sediments: A source of contamination and impact on agriculture productivity. agrochemicals in soil and environment: Impacts and remediation* (pp. 313−345). India: Springer Nature. Available from https://doi.org/10.1007/978-981-16-9310-6_14.

Renard, K. G., & Foster, G. R. (1983). Soil conservation: Principles of erosion by water, Agricultural Research Service, US Department of Agriculture, Tucson, Arizona, USA (pp. 155−176).

Renard, K.G. (1997). *Predicting soil erosion by water: A guide to conservation planning with the revised universal soil loss equation (RUSLE)*. US Department of Agriculture, Agricultural Research Service.

Renschler, C., Diekkrüger, B., & Mannaerts, C. (1998). *Regionalization in surface runoff and soil erosion risk evaluation* (IAHS-AISH Publication 254, pp. 233−241). IAHS Germany.

Renschler, C. S., Mannaerts, C., & Diekkrüger, B. (1999). Evaluating spatial and temporal variability in soil erosion risk—Rainfall erosivity and soil loss ratios in Andalusia, Spain. *CATENA*, *34*(3−4), 209−225. Available from https://doi.org/10.1016/s0341-8162(98)00117-9.

Reusing, M., Schneider, T., & Ammer, U. (2000). Modelling soil loss rates in the Ethiopian Highlands by integration of high resolution MOMS-02/D2-stereo-data in a GIS. *International Journal of Remote Sensing*, *21*(9), 1885−1896. Available from https://doi.org/10.1080/014311600209797.

Rhodes, C. J. (2014). Soil erosion, climate change and global food security: Challenges and strategies. *Science Progress*, *97*(2), 97−153. Available from https://doi.org/10.3184/003685014X13994567941465.

Rodrigo Comino, J., Iserloh, T., Morvan, X., Malam Issa, O., Naisse, C., Keesstra, S., … Ries, J. (2016). Soil erosion processes in European vineyards: A qualitative comparison of rainfall simulation measurements in Germany, Spain and France. *Hydrology*, *3*(1). Available from https://doi.org/10.3390/hydrology3010006.

Rose, C. W., Coughlan, K. J., & Fentie, B. (1998). *Griffith university erosion system template (GUEST)* (pp. 399−412). Springer Science and Business Media LLC. Available from https://doi.org/10.1007/978-3-642-58913-3_30.

Saaty, T. L., & Vargas, L. G. (2001). *How to make a decision* (pp. 1−25). Springer Science and Business Media LLC. Available from https://doi.org/10.1007/978-1-4615-1665-1_1.

Saaty, T. L. (1980). The analytic hierarchy process: Planning, priority setting, resource allocation. *The Journal of the Operational Research Society*, *41*(11), 1073−1076.

Saaty, T. L. (1990). How to make a decision: The analytic hierarchy process. *European Journal of Operational Research*, *48*(1), 9−26. Available from https://doi.org/10.1016/0377-2217(90)90057-I.

Schmidt, J. (1991). A mathematical model to simulate rainfall erosion. *CATENA*, *19*, 101−109, Supplement.

Sharda, V. N., & Ojasvi, P. R. (2016). A revised soil erosion budget for India: Role of reservoir sedimentation and land-use protection measures. *Earth Surface Processes and Landforms*, *41*(14), 2007−2023. Available from https://doi.org/10.1002/esp.3965.

Shrestha, D. P. (1997). *Assessment of soil erosion in the Nepalese Himalaya: A case study in Likhu Khola Valley, Middle Mountain Region* (2). Land Husbandry.

Stavi, I., & Lal, R. (2015). Achieving zero net land degradation: Challenges and opportunities. *Journal of Arid Environments*, *112*, 44−51. Available from https://doi.org/10.1016/j.jaridenv.2014.01.016.

Stocking, M., & Clark, R. (1999). Soil productivity and erosion: Biophysical and farmer-perspective assessment for hillslopes. *Mountain Research and Development*, *19*(3), 191−202.

Tamene, L., Le, Q. B., & Vlek, P. L. G. (2014). A landscape planning and management tool for land and water resources management: An example application in Northern Ethiopia. *Water Resources Management*, *28*(2), 407−424. Available from https://doi.org/10.1007/s11269-013-0490-1.

Tang, Q., Xu, Y., Bennett, S. J., & Li, Y. (2015). Assessment of soil erosion using RUSLE and GIS: A case study of the Yangou watershed in the Loess Plateau, China. *Environmental Earth Sciences*, *73*(4), 1715−1724. Available from https://doi.org/10.1007/s12665-014-3523-z.

Terranova, O., Antronico, L., Coscarelli, R., & Iaquinta, P. (2009). Soil erosion risk scenarios in the Mediterranean environment using RUSLE and GIS: An application model for Calabria (southern Italy). *Geomorphology*, *112*(3−4), 228−245. Available from https://doi.org/10.1016/j.geomorph.2009.06.009.

Thomas, J., & Prasannakumar, V. (2015). Comparison of basin morphometry derived from topographic maps, ASTER and SRTM DEMs: An example from Kerala, India. *Geocarto International*, *30*(3), 346−364. Available from https://doi.org/10.1080/10106049.2014.955063.

Toubal, A. K., Achite, M., Ouillon, S., & Dehni, A. (2018). Soil erodibility mapping using the RUSLE model to prioritize erosion control in the Wadi Sahouat basin, North-West of Algeria. *Environmental Monitoring and Assessment*, *190*(4). Available from https://doi.org/10.1007/s10661-018-6580-z.

Toumi, S., Meddi, M., Mahé, G., & Brou, Y. T. (2013). Cartographie de l'érosion dans le bassin versant de l'Oued Mina en Algérie par télédétection et SIG. *Hydrological Sciences Journal*, *58*(7), 1542−1558. Available from https://doi.org/10.1080/02626667.2013.824088.

Tran, A. T. P., Chang, I., & Cho, G. C. (2019). Soil water retention and vegetation survivability improvement using microbial biopolymers in drylands. *Geomechanics and Engineering*, *17*(5), 475−483. Available from https://doi.org/10.12989/gae.2019.17.5.475.

Trnka, M., Olesen, J. E., Kersebaum, K. C., Rötter, R. P., Brázdil, R., Eitzinger, J., … Balek, J. (1901). Changing regional weather crop yield relationships across Europe between. *Climate Research*, *70*(2−3), 195−214.

Tsegaye, L., & Bharti, R. (2021). Soil erosion and sediment yield assessment using RUSLE and GIS-based approach in Anjeb watershed, Northwest Ethiopia. *SN Applied Sciences*, *3*(5). Available from https://doi.org/10.1007/s42452-021-04564-x, springer.com/snas.

Tucker, G., Lancaster, S., Gasparini, N., & Bras, R. (2001). *The Channel-Hillslope integrated landscape development model (CHILD)* (pp. 349−388). Springer Science and Business Media LLC. Available from https://doi.org/10.1007/978-1-4615-0575-4_12.

Tunçay, T., & Başkan, O. (2023). *Assessment of land degradation factors*. IntechOpen. Available from https://doi.org/10.5772/intechopen.107524.

Vanmaercke, M., Zenebe, A., Poesen, J., Nyssen, J., Verstraeten, G., & Deckers, J. (2010). Sediment dynamics and the role of flash floods in sediment export from medium-sized catchments: A case study from the semi-arid tropical highlands in northern Ethiopia. *Journal of Soils and Sediments*, *10*(4), 611−627. Available from https://doi.org/10.1007/s11368-010-0203-9.

Vanwalleghem, T., Gómez, J. A., Infante Amate, J., González de Molina, M., Vanderlinden, K., Guzmán, G., … Giráldez, J. V. (2017). Impact of historical land use and soil management change on soil erosion and agricultural sustainability during the Anthropocene. *Anthropocene*, *17*, 13−29. Available from https://doi.org/10.1016/j.ancene.2017.01.002.

Vatandaşlar, C., & Yavuz, M. (2017). Modeling cover management factor of RUSLE using very high-resolution satellite imagery in a semiarid watershed. *Environmental Earth Sciences*, *76*(2). Available from https://doi.org/10.1007/s12665-017-6388-0.

Vemu, S., & Pinnamaneni, U. B. (2011). Estimation of spatial patterns of soil erosion using remote sensing and GIS: A case study of Indravati catchment. *Natural Hazards*, *59*(3), 1299−1315. Available from https://doi.org/10.1007/s11069-011-9832-6.

Vijith, H., Seling, L. W., & Dodge-Wan, D. (2018). Estimation of soil loss and identification of erosion risk zones in a forested region in Sarawak, Malaysia, Northern Borneo. *Environment, Development and Sustainability*, *20*(3), 1365−1384. Available from https://doi.org/10.1007/s10668-017-9946-4.

Vijith, H., Suma, M., Rekha, V. B., Shiju, C., & Rejith, P. G. (2012). An assessment of soil erosion probability and erosion rate in a tropical mountainous watershed using remote sensing and GIS. *Arabian Journal of Geosciences*, *5*(4), 797−805. Available from https://doi.org/10.1007/s12517-010-0265-4.

Wang, G., Mang, S., Cai, H., Liu, S., Zhang, Z., Wang, L., & Innes, J. L. (2016). Integrated watershed management: Evolution, development and emerging trends. *Journal of Forestry Research*, *27*(5), 967−994. Available from https://doi.org/10.1007/s11676-016-0293-3.

von Werner, G. (1995). *GIS-orientierte Methoden der digitalen Reliefanalyse zur Modellierung von Bodenerosion in kleinen Einzugsgebieten* (Doctoral dissertation, Verlag nicht ermittelbar).

William, J.R., & Arnold, J.G. (1993). *A system of erosion/sediment yield models*. Reunion Internacional sobre Procesos de Erosion en Tierras de Altas Pendientes: Evaluacion y modelaje, Merida (Venezuela).

Williams, J. R., & Arnold, J. G. (1996). *Water quality models for watershed management* (11). Springer Science and Business Media LLC. Available from https://doi.org/10.1007/978-94-011-0393-0_14.

Williams, J. R., Jones, C. A., & Dyke, P. T. (1984). A modeling approach to determining the relationship between erosion and soil productivity. *Transactions of the ASAE.*, *27*(1), 129−0144.

Wischmeier, W. H., & Smith, D. (1968). Predicting rainfall-erosion losses from cropland east of the Rocky Mountains.Guide for selection of practices for soil and water conservation. No. 282. Agricultural Research Service, US Department of Agriculture.

Wischmeier, W. H., & Smith, D. D. (1978). *Predicting rainfall erosion losses: A guide to conservation planning* (No. 537). Department of Agriculture, Science and Education Administration.

Woolhiser, D. A. (1996). Search for physically based runoff model − A hydrologic El Dorado? *Journal of Hydraulic Engineering*, *122*(3), 122−129. Available from https://doi.org/10.1061/(ASCE)0733-9429(1996)122:3(122).

Wu, X., Wei, Y., Wang, J., Cai, C., Deng, Y., & Xia, J. (2018). RUSLE erodibility of heavy-textured soils as affected by soil type, erosional degradation, and rainfall intensity: A field simulation. *Land Degradation and Development*, *29*(3), 408−421. Available from https://doi.org/10.1002/ldr.2864.

Young, R. A., Onstad, C. A., Bosch, D. D., & Anderson, W. P. (1989). AGNPS: A nonpoint-source pollution model for evaluating agricultural watersheds. *Journal of Soil and Water Conservation*, *44*(2), 168−173.

Yu, W. J., Jiao, J. Y., Chen, Y., Wang, D. L., Wang, N., & Zhao, H. K. (2017). Seed removal due to overland flow on abandoned slopes in the Chinese hilly gullied Loess Plateau region. *Land Degradation and Development*, *28*(1), 274−282. Available from https://doi.org/10.1002/ldr.2519.

Zhang, L., Huang, Y., Rong, L., Duan, X., Zhang, R., Li, Y., & Guan, J. (2021). Effect of soil erosion depth on crop yield based on topsoil removal method: A meta-analysis. *Agronomy for Sustainable Development*, *41*(5). Available from https://doi.org/10.1007/s13593-021-00718-8.

Monitoring land degradation and desertification using the state-of-the-art methods and remote sensing data

24

Debaaditya Mukhopadhyay[1] and Gaurav Mishra[2]

[1]*ICFRE—Rain Forest Research Institute, Jorhat, Assam, India*
[2]*Centre of Excellence on Sustainable Land Management, Indian Council of Forestry Research and Education, Dehradun, Uttarakhand, India*

24.1 Introduction

Desertification happens on all continents except Antarctica, threatening the livelihoods of millions of people worldwide. It is acknowledged as one of the most significant environmental issues, impacting over 250 million people directly and being of worldwide concern (Ajai & Dhinwa, 2017). Numerous writers have studied definitions and the development of the idea of desertification (Ajai, Arya, Dhinwa, Pathan, & Ganesh Raj, 2009; Ajai, 2007; Brabant, 2010; Eswaran, Lal, & Reich, 2019; Puigdefabregas, Barrio, & Hill, 2009; Reynolds et al., 2007, 2011). The most thorough and widely recognized definition of "desertification" is provided by the United Nations Convention to Combat Desertification (UNCCD, 1994), which defines it as degradation of land due to human activity and climate fluctuations in dry sub-humid, semi-arid, and arid regions. There are several definitions of "land degradation" that highlight various facets of the decline in land quality brought about by degradation processes. The process of decreasing or eliminating a land's ability to support cattle, crops, and forests is known as land degradation. It might be a natural occurrence made worse by the effects of human activity, or it can be caused by human activity (Brabant, 2010). People living in rural areas suffer long-term effects from land degradation, which makes them more susceptible over time (Muchena, 2008). "Land degradation is the reduction or loss of agricultural, biological, or economic productivity of rain-fed cropland, irrigated cropland, or range, pasture, forest, and woodlands resulting from land use or from a combination of processes arising from human activities and habitation

Applications of Geospatial Technology and Modeling for River Basin Management.
DOI: https://doi.org/10.1016/B978-0-443-23890-1.00024-4

patterns," according to another definition that is used by both the UNCCD (1994) and the OECD. A productive piece of land may become "degraded land" or "waste land" due to a process known as land degradation (Fig. 24.1).

Numerous economic, political, social, and environmental variables contribute to desertification on a global scale. Land degradation is caused by inadequate infrastructure, unsustainable resource usage, and improper land practices. Desertification is accelerated by human actions such as deforestation, overgrazing, and farming on unstable soil. Poverty, ignorance, and unsustainable "modern" development are the main factors. Climate, periodic droughts, and human intervention intensify the process. Human-caused deforestation affects soil stability and vegetation survival by making soil more susceptible to wind and water erosion (Dregne, 1985; Kishk, 1986). Land degradation poses a major threat to the world's food supply, ecosystems, and human population. Some of the primary causes of land degradation are deforestation, unsustainable agriculture, and climate change. An estimated 35%−55% of arable land is degraded, impacting 20% of the world's agricultural lands. Degradation has affected 29.32% of India, adding 1.87 million hectares in the last 8 years. Desertification, vegetation degradation, and wind and water erosion are major causes (2014).

The evaluation of land use and land cover (LULC) is greatly aided by remote sensing (RS) data, which may also be used to collect data on the kind, degree, and character of land degradation (Dubovyk, 2017; Gibbs & Salmon, 2015). Food

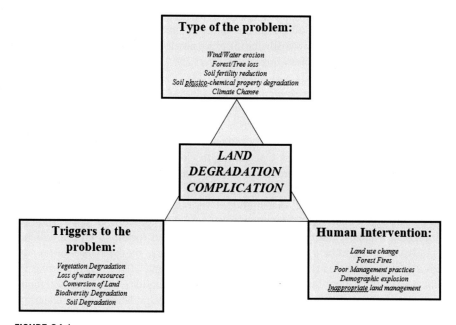

FIGURE 24.1

Land degradation ways and types.

security, organic carbon, nutrients, soil-water storage, and species habitats can all be negatively impacted over an extended period of time by long-term changes in land function brought on by LULC disturbances (Andrew, Wulder, & Nelson, 2014). Therefore, the use of conservation or rehabilitation methods for degraded land can also benefit from the use of geographical data on land use (Eckert, Hüsler, Liniger, & Hodel, 2015; Winslow et al., 2011). As a vital tool for tracking land degradation, RS—especially the study of plant cover dynamics—is frequently employed as a stand-in for in-person observations to evaluate changes in LULC (Easdale et al., 2019). It provides information on the integrated indicators of how the land reacts to both natural and man-made land use activities, such deforestation (Lal, 2001). However, seasonality, location, spatial resolution, and the severity of deteriorated characteristics all have a role in how land degradation is interpreted using RS (Dubovyk, 2017). RS can identify changes in vegetation, which are indicators of various forms of land degradation, such as deforestation, which can harm the environment by increasing surface runoff, causing water erosion, reducing soil water retention, and reducing the amount of organic matter and nutrients in the soil (Lal, 2001). Various degradation patterns occur in various places due to the complexity of interactions between humans and the environment (Kertész & Křeček, 2019).

Even though RS provides insightful information, several problems still exist. It can be challenging to distinguish modest types of deterioration from unaffected areas when using coarse resolution satellite data that integrates spectral signals from many LULC classes (Desprats et al., 2013). Despite being time-consuming, the restrictions need a mix of image-processing algorithms, ground-truth confirmation, and visual interpretation (Ruelland, Tribotte, Puech, & Dieulin, 2011). With applications in India, satellite programs such as Landsat and the Indian Remote Sensing (IRS) are essential for gathering data on land degradation and LULC (Narayan, Rao, & Gautam, 1989). For example, wasteland, often known as "degraded land," makes up around 18% of India's land area (NRSC, 2019a, 2019b). Research in the ecologically significant state of Meghalaya State concentrates on changes in the forest cover, but it also highlights the difficulties and knowledge gaps surrounding long-term land use change in highly degraded areas (Roy & Joshi, 2002). High-resolution picture use in RS approaches is expensive; however, publicly accessible data from sources such as Sentinel satellites offer workable substitutes for LULC mapping (Chen et al., 2014; Khan, Govil, Kumar, & Dave, 2020). In this chapter, the different monitoring frameworks, techniques, and future challenges and limitations shall be discussed.

24.2 Land degradation and desertification assessment frameworks like LADA, UNCCD framework, and others

The evaluation of desertification and land degradation has proven difficult to monitor and comprehend because of its intrinsic complexity. These problems have historically been addressed experimentally, emphasizing observable symptoms. This has resulted in the development of several, nonstandard procedures in recent decades. These approaches can be divided into groups according to their focus on operational or research goals or their geographical scope (global, continental, regional, national, or local). Remarkably, a large number of the monitoring systems in place now track land degradation or evaluate trends; trend analysis is thought to be more useful for spurring action than tracking deterioration as it is at the moment (Millennium Ecosystem Assessment, 2005). Global efforts to monitor and assess desertification have been in place since the United Nations Conference on Desertification (UNCOD) in (1977). Numerous initiatives commissioned by international organizations, such the United Nations Environment Programme and the International Soil Reference and Information Centre, have been crucial in this regard. A few examples include the World Map of Desertification (1977), the Millennium Ecosystem Assessment—Desertification Synthesis (Millennium Ecosystem Assessment, 2005), and the World Atlas of Desertification. These evaluations are helpful in providing broad information about the causes, effects, and scope of desertification; nevertheless, they do not go far enough in providing the specific details required for a successful policy impact assessment (Millennium Ecosystem Assessment, 2005).

The creation of a worldwide map of human-induced soil deterioration at a scale of 1:10,000,000 makes GLASOD (Global Assessment of Human-Induced Soil deterioration) stands out among international programs. By offering a worldwide data source on the state of soil degradation, it has had a substantial impact on environmental policy (Oldeman, Hakkeling, & Sombroek, 1990). However, GLASOD is limited since it only considers soil deterioration, leaving out other important factors like vegetation, climate, and water resources. Additionally, it adds a subjective element to the evaluation by depending on the opinions of experts (Oldeman, 1992; Oldeman et al., 1990; Toulmin, 1997). The usage of extend classes, visual exaggeration, and the lack of consistency testing are among the limitations. Furthermore, it does not take into account all of the factors contributing to land degradation (Sonneveld & Dent, 2009). As follow-up projects carried out under UNEP and FAO, Land Degradation Assessment in Drylands (LADA) and Global Assessment of Land Degradation and Improvement have evolved after GLASOD. The main goal of LADA is to provide instruments and techniques for evaluating and measuring the kind, degree, intensity, and effects of land degradation on ecosystems. In order to direct corrective actions and policies, it functions at the international, national, and subnational levels, detecting degradation status and trends, hotspots, and bright spots (Schwilch et al., 2011). Using the Driving-force, Pressure, State, Impact, and Response architecture, LADA integrates data across scales in a decentralized and interactive manner (Sommer et al., 2011). The participatory approach has drawbacks such as subjectivity and irregular repetition, even if it improves stakeholder acceptability and national understanding (Tengberg & Torheim, 2007).

In order to fight desertification in the circum-Saharan region, Long-Term Ecological Surveillance Observatories Network of the Sahara and Sahel Observatory is another important endeavor. In order to track changes in biophysical characteristics, it uses environmental monitoring techniques and surveys to gather socioeconomic data. However, it is stressed that there is a need for more harmonization of geographic distribution documentation and data gathering methodologies (Sellami, 2009). The Australian federal and state governments work together to manage natural resources in Australia's rangelands through the Australian Collaborative Rangeland Information System (ACRIS), which is not a monitoring system in and of itself. In order to provide a more thorough knowledge of both positive and negative changes connected to human influence, ACRIS gathers and analyses monitoring data at regional to national sizes (Bastin, 2008). Through the identification of data gaps and the provision of a framework for comprehending and analyzing change, this cooperative approach has yielded important insights into Australia's rangelands (Bastin, Smith, Watson, & Fisher, 2009). The frameworks that are being addressed play a key role in helping to comprehend and combat desertification and land degradation. Nonetheless, it is clear that in order to effectively monitor the world and formulate policies, standardized, quantitative, and international collaboration techniques are required. These programs highlight the significance of frequent, thorough, and consistent monitoring to guide sustainable land management practices worldwide. They are significant milestones toward integrated evaluations.

24.3 State-of-the-art methods for monitoring land degradation and desertification

24.3.1 Ground-based monitoring techniques

Classic field survey techniques for charting soil and vegetation and assessing land suitability gave rise to more recent ecological methodologies like the landscape leakiness index, which are used in traditional desertification assessment methods today (Ludwig & Tongway, 1992; Ludwig, Bastin, Chewings, Eager, & Liedloff, 2007; Mouat, Fox, & Rose, 1992). Although these ground-based approaches do not meet the majority of the practical criteria, they can produce highly precise findings in very limited regions if they are grounded in extensive field experience. Land degradation studies have historically placed a great deal of emphasis on field surveys, a practice that has become less important as technology has advanced. Traditional methods have been replaced by emerging techniques including airborne photography and remotely sensed satellite pictures (Lee & Lunetta, 1996). Due to their time-consuming and expensive nature, the conventional methods—which are characterized by labor-intensive procedures including intense fieldwork, supplementary data processing, visual observation, and feature estimation—have become less viable, especially in distant and inaccessible places. The universal soil loss equation (USLE) and its improved version, the revised USLE (RUSLE), are two of the most notable conventional methods. These empirical models rely on data gathered in the field (Wischmeier, 1965, 1978).

Although it solves some of the shortcomings of its predecessor, RUSLE is still restricted to six factors: soil erodibility (K), rainfall erosivity (R), slope length (L), slope steepness (S), soil usage and management (C), and support practices (P). RUSLE is an upgraded version of USLE. Because of this limitation, their conclusions may be incorrect and empirical correlations may change, making them untrustworthy for bigger sizes. They may also fail to notice processes that are important on larger scales but insignificant on smaller ones (Antrop & Van Eetvelde, 2000; Jahun, Ibrahim, Dlamini, & Musa, 2015). Additionally, the first model only provided absolute soil loss estimates due to a lack of spatial distribution information; this constraint was lessened by integrating RS and geographic information systems (GISs) (Fistikoglu & Harmancioglu, 2002).

Survey-based (direct) field observations, or ground-based measurements, encompass methods including expert and land user opinions, field measurements and monitoring, productivity shifts, farm level research, and modeling. These methods are crucial for assessing the national and local processes involved in land degradation (Lynden & Kuhlmann, 2002). However, remotely sensed satellite imaging, radio detection and ranging, and GIS data are used in aboveground measurements (Kapalanga, 2008; Nkonya, Gerber, & Baumgartner, 2011). At the global, regional, national, and local levels, the severity, degree, and extent of land and soil degradation have been assessed by ground-based measures. For instance, data on the worldwide distribution, intensity, and causes of erosional, chemical, and physical deterioration are provided by the expert-based GLASOD (Bridges & Oldeman, 1999; Jones et al., 2003). The impacts of erosion from plowing and harvesting in Kenya have been successfully monitored by direct field observations employing markers of soil erosion, such as degraded clods, flow surfaces, pre-rills, and rills (De Bie, 2005). Such an instance is the Botswana participatory degradation assessment (Reed & Dougill, 2002). This method integrates three approaches: field observations at the farm level, assessment of production changes, and land user opinion. In order to improve the accuracy, scope, and applicability of land degradation assessments in Botswana and Swaziland, a participatory approach that incorporates expert opinions and the experiences of the local land users (through key informant interviews, focus groups, and surveys) was used (Stringer & Reed, 2007).

24.3.2 Remote sensing and geographic information system techniques

The role of RS and GIS in monitoring land degradation and desertification has become increasingly significant, particularly over larger scales ranging from regional to global levels. RS offers a cost-effective and time-efficient means of assessing land degradation over extensive areas, with various techniques employed for identification and mapping. Traditional approaches, such as manual visual methods and the interpretation of aerial photography and satellite imagery, have been utilized in regions like Argentina, sub-Saharan Africa, China, and Spain. These methods, while valuable, have faced criticism for their subjectivity and semiquantitative nature, leading to a shift toward more model-based approaches involving indicators and proxy variables. Notably, the use of spectral indices, such as the Normalized Difference Vegetation Index (NDVI), has gained prominence in assessing vegetation health and land degradation globally (Collado, Chuvieco, & Camarasa, 2002; Gupta, Hock, Xiaojing, & Ping, 2002; Ries & Marzolff, 2003). NDVI, derived from satellite sensors like MODIS, Advanced Very High-Resolution Radiometer (AVHRR), and Landsat, enables reliable spatial and temporal comparisons of photosynthetic activity and canopy structural variations (Running et al., 1994). It has been applied in diverse regions, including Senegal, the Sahel, Zimbabwe, Mozambique, and globally for predicting future land degradation and economic effects (Diouf & Lambin, 2001; Herrmann, Anyamba, & Tucker, 2005; Nicholson, Tucker, & Ba, 1998; Vlek, Le, & Tamene, 2010). An indicator system, a classification system for desertification status mapping (DSM), and the creation of techniques for DSM mapping and tracking using multitemporal imagery from satellites are the fundamental elements of RS-based desertification mapping, monitoring, and assessment (Table 24.1). For DSM and evaluation to be effective, indicators must be defined and identified, and monitoring frequency must be high. Four categories of indicators exist: pressure indicators (climatic and socioeconomic), state indicators (physical, biological, and hydrological), impact indicators (social, migration, and unemployment, among others), and implementation indicators (alterations in land cover as a result of mitigating actions). Indicators are defined as parameters that provide information about the state of a phenomenon. India, aligning its indicator system with Asian TPN-1, applicable to the entire Indian subcontinent, collaborates under the United Nations Conventions to Combat Desertification (UNCCD) to enhance desertification monitoring capabilities. Space Applications Centre (ISRO), Ahmedabad, coordinates these efforts nationally, utilizing satellite images for deciphering and monitoring indicators (Ajai & Dhinwa, 2017).

Researchers may now more effectively and economically examine geographically spread phenomena, such land degradation, thanks to the integration of better sensors and technology tools (Makaya, Mutanga, Kiala, Dube, & Seutloali, 2019; Mansour, Mutanga, Adam, & Abdel-Rahman, 2016; Phinzi & Ngetar, 2017). This development is especially important for mapping isolated areas where intensive, traditional approaches would not be feasible (Seutloali, Beckedahl, Dube, & Sibanda, 2016). Remotely sensed imagery has been used in several researches to identify and map different land degradation processes worldwide. Assessing the geographical distribution of land degradation and mapping changes in land use have both benefited greatly from RS in South Africa (Wessels, Prince, Frost, & Van Zyl, 2004, Wessels et al., 2007). For example, one study evaluated human-induced land degradation in northern South Africa using the NDVI, which is obtained from AVHRR data (Wessels et al., 2007). Nevertheless, it was determined that the 1 km spatial resolution was insufficient to extract specific

Table 24.1 Indicators and limitations to monitor land degradation using Earth observation data.

Applications of earth observation	Indicator of environmental degradation	Data sources	Limitations	Study
AVHRR (GIMMS-3g)	Trends in greenery	Time series linear trend analysis of vegetation greenness	Does not differentiate between crops, trees, grass, and species; heavily reliant on rainfall; unclear driving forces for change; data coarseness blends various processes	Dardel et al. (2014)
AVHRR, VGT, MODIS	Land productivity	Rain use efficiency, NPP	Mixed pixels, unreliable rainfall data, unknown production quality, translation to biomass highly dependent on environmental conditions	Fensholt and Rasmussen (2011)
Landsat, QuickBird, ASTER	Land use/cover change	High-resolution postclassification comparison	Difficulties capturing dynamics and inter/intraannual variability, lack of information on specific species	Mbow, Mertz, Diouf, Rasmussen, and Reenberg (2008)
LTDR (AVHRR)	Water use efficiency, evapotranspiration		Does not distinguish between species; primarily dependent on climate reanalysis (water and energy); unclear driving forces for change; coarse data blends various processes	Marshall, Funk, and Michaelsen (2012)
MODIS, Landsat	Degradation of water resources	Medium resolution time series classification	Cloud cover over water bodies; misclassifications due to similar spectral properties	Moser, Voigt, Schoepfer, and Palmer (2014)
Corona, QuickBird, RapidEye	Wind/water erosion, loss of tree cover	High-resolution visual inspection	Qualitative assessment, challenging to quantify, limited information on specific species	Tappan, Sall, Wood, and Cushing (2004)

Developed from Mbow, Brandt, Ouedraogo, de Leeuw, and Marshall (2015), Mbow et al. (2008), and Toth and Jóźkow (2016).

properties of soil erosion. The application of Landsat images for evaluations of land degradation has also been investigated by researchers. Mambo and Archer (92007) evaluated land degradation in a Zimbabwean watershed using Landsat 5 TM and Landsat 7 ETM, classifying it into five susceptibility categories. Furthermore, Taruvinga (2009) demonstrated the usefulness of Landsat TM in these kinds of applications by mapping gully erosion in KwaZulu-Natal using vegetation indices generated from Landsat TM. Because of its better spatial resolution (15 m panchromatic band), Landsat 8 OLI has been very helpful in tracking the spread of soil erosion (Phinzi & Ngetar, 2017). With the use of Landsat TM, it was able to detect eroded regions and land cover with an overall classification accuracy of 83.94% (Floras & Sgouras, 1999).

Higher spatial resolution imaging sources like WorldView-2, SPOT, QuickBird, and IKONOS have become increasingly well-known because of their improved capacity to identify and map soil erosion features, even if Landsat data is still publicly accessible and provides insightful information (Mayr, Rutzinger, Bremer, & Geitner, 2016). The accuracy of categorization is improved by using object-oriented techniques, which are made easier by these higher resolution pictures. Higher spatial resolution has benefits, but accessibility is still a problem because of expensive acquisition costs. Even with its drawbacks, land cover is still a popular tool since it can be used to map gully erosion and is always becoming better with the introduction of new sensors (Hansen & Loveland, 2012; Turner et al., 2015; Vreiling, 2006). The RS research community has generally approved Sentinel-2 MSI's recent launch. Sentinel-2 provides open availability and comparatively greater geographical and spectral resolutions by acting as a bridge between medium-resolution Landsat imagery and high spatial resolution data sources. Research has shown that Sentinel-2 is more effective than Landsat 8 OLI at mapping a variety of land degradation characteristics (Forkuor, Dimobe, Serme, & Tondoh, 2018; Sibanda, Mutanga, & Rouget, 2016). While Sentinel-2, which was launched in 2015, has opened up new avenues for study in RS, many land cover mapping researchers still choose Landsat to be their first option due to its endurance and continuous advancements.

Nevertheless, there are issues, such as structural mistakes in datasets from RS. Land degradation evaluations must carefully take into account these mistakes, which are caused by different types of sensors, temporal and geographical resolution, and derived data (Le, Nkonya, & Mirzabaev, 2016). A systematic approach to rectify these deficiencies, particularly with variations in net domestic production (NDVI) was carried out (Le Roux & Sumner, 2013; Le et al., 2016). Season, location, spatial resolution, size, and severity of deteriorated features are among the variables that affect how well RS data of land degradation may be interpreted (Dubovyk, 2017). While spectral separability and pattern make it easy to distinguish some types of degraded land, including places impacted by salt or waterlogging, evaluating changes in sparsely vegetated areas is still difficult (Chowdary et al., 2008; Metternicht, Zinck, Blanco, & Valle, 2010; Singh, Bundela, Sethi, Lal, & Kamra, 2010). Limitations in identifying minute alterations such as gullies and rills are another consequence of coarse-resolution satellite data integration (Bai, Dent, Olsson, & Schaepman, 2008; Desprats et al., 2013). Despite these difficulties, ground-truth validation, image processing methods, and visual interpretation continue to be a well-liked and successful combination (Antrop & Van Eetvelde, 2000; Metternicht et al., 2010; Ruelland et al., 2011).

Documentation on LULC and land degradation in India has been gathered mostly by satellite RS, especially with the use of Landsat imagery and the IRS satellite program (Kasturirangan, Aravamudan, Deekshatulu, Joseph, & Chandrasekhar, 1996; Narayan et al., 1989). Wasteland makes up around 18% of the nation's geographical area; RS is useful for locating and tracking these places (Kandekar Avinash, Gaikwad Sunil, Gaikwad Satyajit, Kandekar Smita, & Shelar Avinash, 2021). Furthermore, models and quantification of many biophysical factors associated with drought and desertification have been successfully achieved by RS (Andrew et al., 2014; Kaneko & Kojima, 2010; Roder & Hill, 2005; Vreiling, 2006). Increased spatial resolution, choice of spectral bands, and improved revisit capabilities have enhanced the analysis of land surface variability, erosion, salinization, changes in primary productivity, and groundwater exploitation (Chen et al., 2014; Gibbs & Salmon, 2015; Shruthi et al., 2011). The integration of biophysical and socioeconomic variables within a GIS framework has witnessed advancements, enabling a more comprehensive understanding of the complex interactions driving desertification (Claessens et al., 2009; Rasmy, Feranec, Hazeu, Christensen, & Jaffrain, 2010; Santini, Gibbs, & Salmon, 2010).

Studies in regions like southern Spain, southern Churu in Rajasthan, and the Italian island of Sardinia showcase the application of GIS-based software tools to link overgrazing, vegetation productivity, soil fertility, erosion, and seawater intrusion to produce desertification maps. These tools prioritize simulating the actual states of degradation rather than focusing solely on risk factors, providing valuable insights for land management (Claessens et al., 2009; Ibanez, De, & Devi, 2008; Prokop & Bhattacharyya, 2011; Rasmy et al., 2010; Santini et al., 2010). RS and GIS play a crucial role in monitoring land degradation and desertification by providing a comprehensive and objective assessment of changes over large spatial and temporal scales. Despite challenges, ongoing advancements in technology and methodologies continue to enhance the accuracy and applicability of these tools, contributing to sustainable land management practices globally.

24.4 Remote sensing data sources for land degradation and desertification monitoring

24.4.1 Earth observation satellites

Large datasets in many scientific fields are now much more readily available and plentiful because of technological advancements. Satellite data and photos derived from Earth observation have become essential for both academic study and real-world use.

Through platforms like GEONETCast, a component of the Global Earth Observation System of Systems (GEOSS), agencies including the European Organisation for the Exploitation of Meteorological Satellites (EUMETSAT), NASA, NOAA, and VITO contribute to this abundance of data (Wolf & Williams, 2008). Even with the wealth of remotely sensed data that GEONETCast has made available, many developing nations—especially those in Africa, Asia, and growing economies like Brazil and India—seem to be falling short when it comes to science and technology spending. These nations must have consistent access to accurate Earth observation data in order to pursue sustainable development because of the hazards to the environment and development obstacles they confront (UNEP, 2008).

A worldwide phenomenon, climate change, presents itself in various ways in different places. The Intergovernmental Panel on Climate Change (IPCC) has concluded that the average global surface temperature is rising, which might have an impact on agricultural and human well-being, particularly in poorer nations (Nelson et al., 2009). Due to growing population pressure and restricted livelihood alternatives in drought-prone areas, there is a greater susceptibility to drought and land degradation (Barbosa, Mesquita, & Kumar, 2011; Barbosa & Lakshmi Kumar, 2011). Earth observation satellites are vital in addressing these issues. On a regional and global level, they aid in the observation and evaluation of environmental changes including drought, desertification, and agricultural practices. Satellite-derived data provides information about vegetation dynamics and responses to climatic factors, especially the NDVI (Sellers, 1985). NDVI data is useful for evaluating extensive agricultural droughts and is provided by a number of satellite sensors, including as NOAA AVHRR, MODIS, and SPOT-Vegetation (Barbosa et al., 2011; Lakshmi Kumar, Barbosa, Koteswara Rao, & Prabha Jothi, 2012; Wan, Wang, & Li, 2004).

The Himalayan region, spanning multiple Indian states, is susceptible to various environmental stressors, including soil erosion, deforestation, and changing land use patterns. Climate change further exacerbates these challenges, impacting water resources, biodiversity, and livelihoods. Given the complex terrain and diverse ecosystems, comprehensive monitoring is essential for informed decision-making. Satellite-derived data, particularly the NDVI, is crucial for monitoring vegetation health. Sensors such as MODIS and Landsat provide high-resolution imagery, enabling the assessment of changes in vegetation cover over time (Sarma & Lakshmi Kumar, 2006). Earth observation satellites facilitate the analysis of LULC changes, crucial for understanding the dynamics of urbanization, deforestation, and agricultural expansion. This information is essential for assessing the extent of land degradation (Barbosa et al., 2011). In the hilly terrain of the Indian Himalayas, slope stability is a critical factor. Earth observation satellites equipped with high-resolution imagery and synthetic aperture radar (SAR) can aid in identifying areas prone to erosion and landslides (Lakshmi Kumar et al., 2012). The Himalayan region is home to numerous glaciers, and monitoring changes in snow cover and glacier extent is vital for water resource management.

Satellites like Sentinel-2 contribute to tracking these dynamic processes (Lakshmi Kumar et al., 2012). While Earth observation satellites offer unparalleled advantages, challenges exist, including cloud cover in mountainous regions and the need for high spatial resolution in complex terrains (Table 24.2). Collaborative efforts involving governmental agencies, research institutions, and international organizations are essential for overcoming these challenges and harnessing the full potential of satellite technology.

24.4.2 Remote sensing data providers and platforms

RS relies heavily on the platforms carrying sensors, and their characteristics significantly impact observational efficiency. The use of several sensors on a single platform, such as LiDAR sensors and forward-looking and retrograde cameras, has become commonplace to improve observability (Petrie, 2009). This method differs from conventional single-sensor-based systems and is referred to as multi-sensory systems. As sensor technologies advance and become more affordable, modern RS systems increasingly incorporate multiple sensors, either identical or different (Asner et al., 2012; Nagai, Chen, Shibasaki, Kumagai, & Ahmed, 2009; Paparoditis et al., 2022). One theoretical approach is collaborative sensing, where many platforms collaborate to increase observability. The contemporary multisensory systems are a development from single-sensor-based RS. In order to achieve real-time implementation in cooperative navigation, cooperative sensing makes use of multisensory data collected from many platforms moving in unison (Pages, Nguyen, Priot, Pérennou, & Calmettes, 2015). This concept is exemplified by emerging RS satellite constellations, microsatellite systems, and unmanned aerial systems (UAS) swarms. Satellite-based RS has a history of over 40 years (Table 24.3), with notable systems like Landsat-1, SPOT-1, and Ikonos. As technology advanced, spaceborne sensors transitioned from single-sensor models to cooperative sensing approaches. Satellite constellations, like Landsat, SPOT, GeoEye/WorldView, and RapidEye, offer improved revisit times for comprehensive Earth observation (Murthy et al., 2014). Unique constellations, such as the A-Train and Sentinel family, further contribute to diverse sensor capabilities. UAS, commonly known as drones, have become a vibrant topic in RS. While originally developed for other purposes, UAS now play a crucial role in RS applications. Mobile mapping systems (MMS) employ vehicles equipped with sensors for geospatial data acquisition. The concept has evolved beyond land-based vehicles to include airborne RS systems and indoor mapping. Modern MMS, widely used by major geospatial data providers, utilizes direct georeferencing and fully digital sensor implementations. Static installations of RS systems represent a novel approach, allowing for permanent sensor installations to offer unprecedented temporal resolution and observation capabilities. Examples include high-resolution time-lapse cameras for monitoring environmental changes, such as glacier retreat or growth (Toth & Jóźków, 2016).

Table 24.2 Indicator system for desertification monitoring and assessment.

Category	Subcategory	Specific indicators
Pressure indicators	Climatic	Rainfall patterns, temperature variations, wind intensity, humidity levels, potential evapotranspiration, solar radiation, cloud cover.
	Socioeconomic	Population density, educational attainment, poverty rates, livestock density, forest depletion, fuel and fodder consumption/supply, collection of medicinal plants, shifting cultivation, diminishing water resources, land management practices.
State indicators	Physical and hydrological	Land erosion status, salinity/alkalinity levels, shifting in sand sheets/sand dunes, water logging occurrences, soil moisture levels, soil types and properties, presence of stones, coverage/barren rocky areas, number and spread of water bodies, groundwater status, turbidity of water bodies, soil compaction.
	Biological	Types of vegetation, species composition of vegetation, condition and coverage of vegetation, biomass and productivity of vegetation, crop area and yield.
Impact indicators	Socioeconomic	Income levels, migration patterns, mortality rates, health conditions, unemployment rates, illiteracy levels, food security and malnutrition, prices of food grains, energy consumption, infrastructure development, gender-specific issues, living standards.
	Eco-environmental	Air and water quality, occurrence of dust storms and sandstorms, land pollution levels, degradation of natural resources, provision of ecosystem services.
Implementation Indicators	Action	Economic investments for combating desertification, investment levels, status of the development and implementation of action plans to combat desertification, state of legislation and execution related to combating desertification, people participation, NGO involvement.
	Effect	Proportion of rehabilitated decertified land, socioeconomic standards of the people, improvement in ecosystem services, natural resources, and environmental conditions.

Developed from Ajai & Dhinwa, P. S. (2017). Desertification and land degradation in Indian subcontinent: Issues, present status and future challenges. In Climate variability impacts on land use and livelihoods in drylands *(pp. 181–201). Springer International Publishing. http://www.springer.com/in/book/9783319566801. https://doi.org/10.1007/978-3-319-56681-8_9.*

RS plays a crucial role in the context of India's diverse and dynamic landscape, offering valuable insights for various applications. In the Indian scenario, multiple RS data providers and platforms contribute to effective data acquisition

Table 24.3 Main remote sensing satellite systems.

Satellite name	Launch year	Country	Funding	Operator	Constellation	Sensor	GSD range (m)	Swath width (km)
Ikonos-2	1999	United States	Commercial	Private	Single PAN, 4 MS	0.8, 3.2	11.3	3
QuickBird-2	2001	United States	Commercial	Private	Single PAN, 4 MS	0.7, 2.6	16.8–18, 1–3.5	4
RapidEyea	2008	Germany	Commercial	Private	Five MS	6.5	77	1–5.5
Pleiades 1	2011, 2012	France	Commercial, Govt, Private	Dual PAN, 4 MS	0.5, 2	20	1	
SPOT 6	2012	France	Commercial	Private	Dual PAN, 4 MS	1.5, 6	60	1–5
SPOT 7	2014	France	Commercial	Private	Dual PAN, 4 MS	6	-	-
Landsat-8	2013	United States	Government	Public	Single PAN, 11 MS	15, 30	185	16
SkySat	2013–17	United States	Commercial	Private	PAN Video, 4 MS	0.9–1.1, 2	8	0.12–1
WorldView-3	2014	United States	Commercial, Govt, Private	Private	Single PAN, 8 MS, 8 MS (SWIR), 12 MS	0.3, 1.2, 3.7	13.1, 30	1–4.5
Planet Labs	2014–15	United States	Commercial	Private	Flock of sats., 3 MS	3	Unknown	Unknown
DMC-3	2015	United Kingdom	Commercial	Private	Triple PAN, 4 MS	1, 4	23	1
Sentinel-2	2015–16	EU	Government	Public	Dual, 13 MS	10, 20	290	10–60
Sentinel-3	2015–17	EU	Government	Public	Dual, Triple (planned), 21 MS	300, 500, 1000	1270, 1420	0.25–750
Terra	1999	United States, Japan, and Canada	Government	Public	Single, 14 MS, 36 HSI	15, 30, 90	60	16–250

Aqua	2002	United States	Government	Public	Single, part of A-Train, 36 HSI	250, 500, 1000	2330	1–2
EnMAP	2017	Germany	Government	Public	Single, 232 HSI	30	30	4
ICESat	2003	United States	Government	Public	Single, 2 HSI	70	N/A	8
ICESat-2	2018	United States	Government	Public	Single, 1 HSI, 9-beam	10	N/A	N/A
Envisat	2002	EU	Government	Public	Single, tandem with ERS-2, C-band SAR	28	5	35
RADARSAT-2	2007	Canada	Govt, Private	Public, Private	Single, C-band SAR	3	20	24
TerraSAR-X	2007	Germany	Government	Public	Single, tandem with TanDEM-X, X-band SAR	1, 5, 16	10, 1500	11–100
TanDEM-X	2010	Germany	Govt, Private	Public, private	Single, tandem with TerraSAR-X, X-band SAR	1, 3, 16	10, 30, 100	11–100
Sentinel-1	2014–16	EU	Government	Public	Dual, 5–25 MS	5–40	80–400	12–250

Developed from Toth, C., & Jóźkow G. (2016). Remote sensing platforms and sensors: A survey. Journal of Photogrammetry and Remote Sensing, 115, 22–36. http://www.elsevier.com/inca/publications/store/5/0/3/3/4/0. https://doi.org/10.1016/j.isprsjprs.2015.10.004.

and analysis. Notable among these is the Indian Space Research Organisation (ISRO), which has been at the forefront of advancing satellite-based RS capabilities. ISRO's series of Earth observation satellites, such as the IRS and Resourcesat series, have been instrumental in providing high-resolution imagery and valuable data for applications ranging from agriculture and forestry monitoring to disaster management (Roy & Joshi, 2002). Additionally, collaborations with international partners, like the joint mission of ISRO and NASA, NISAR (NASA-ISRO SAR), further enhance the observational capabilities for a more comprehensive understanding of India's diverse terrains (Rosen & Kumar, 2021). Furthermore, the increasing utilization of UAS for RS applications in India is noteworthy. Organizations like the National Remote Sensing Centre (NRSC) play a pivotal role in coordinating and disseminating RS data to various user communities across the country (Diwakar & Nayak, 2021). The integration of data from diverse platforms, including satellites, UAS, and ground-based sensors, contributes to a holistic and multidimensional approach to RS applications in India. The landscape of RS platforms encompasses satellites, UAS, MMS, and static installations. These platforms leverage advancements in sensor technologies and cooperative sensing to enhance observability and data acquisition for various applications.

24.5 Future trends, limitations and challenges

Monitoring land degradation and desertification is of paramount importance, especially in the context of the Indian subcontinent, which faces significant challenges due to its large dryland areas. The geographic expanse incorporating nations characterized by diverse topographies, including mountains, plains, and coastal regions, is confronted with pervasive manifestations of land degradation. The range of harmful processes in this area includes mining operations, mass wasting, salinization, waterlogging, wind and water erosion, and vegetative degradation (Ajai & Dhinwa, 2017). Deforestation, excessive grazing, inefficient farming methods, agriculture on fringes, industrialization, urbanization, and mining are the main causes of land degradation in these nations, with anthropogenic activity playing a significant role. The escalating human population in the subcontinent, coupled with factors like poverty, illiteracy, and limited resources, intensifies the pressure on land-based natural resources, exacerbating land degradation issues (Murthy et al., 2014). With per capita land availability continuously decreasing and forest cover diminishing, the subcontinent faces a formidable challenge in meeting the growing demands for food, water, and other resources (Singh et al., 2010). Additionally, the looming specter of climate change further compounds the challenges faced by agriculture and allied activities in sustaining the region's population because our food comes from land (Costanza et al., 1997).

To address these challenges and ensure the sustainable use of land resources, there is a critical need to adopt effective land management practices. Sustainable land management involves the development of both water and land resources, with region-specific strategies for soil and water conservation. These strategies must be tailored to the unique socioeconomic and environmental characteristics of each locality. The incorporation of modern technologies, including geospatial technology, is imperative for effective planning and implementation. Monitoring and assessing desertification presents formidable challenges and limitations that impede the development of effective strategies for combatting this complex process of land degradation. One fundamental challenge is establishing a baseline for desertification, serving as the reference frame from which land degradation or improvement is measured. It is challenging to set goals and track the effects of climate change and human activity on vegetation production without a baseline that is both consistently and robustly supported by science (Reed & Dougill, 2002; Reed et al., 2011). Distinguishing the impacts of factors such as overgrazing and drought on vegetation becomes intricate without a clear starting point for evaluation.

Another significant challenge lies in the selection and application of indicator systems for monitoring and assessment. Despite numerous attempts to establish indicator systems, the lack of agreement on their choice and application remains a major handicap. These systems, often region-specific, lack universal applicability, making it challenging to compare desertification status across different areas (Andrew et. al., 2014; Mabbutt, 1986). To map desertification and conduct risk assessments, it is essential to identify pertinent indicators within the physical, biological, and socioeconomic domains (Smith & Reynolds, 2003). However, the difficulty lies in developing indicator systems that are both practical and universally applicable across diverse regions with varying conditions. Temporal and spatial scales add another layer of complexity to desertification assessments. Researchers have conducted studies at different scales—global, continental, regional, national, and local—revealing scale-dependent characteristics that influence desertification trends. Assessments at longer temporal scales may average characteristics, potentially overlooking significant short-term changes. Furthermore, the exact time slice under examination has a substantial influence on identified trends, highlighting the need for careful consideration of temporal and spatial scales in desertification assessments (D'Odorico, Bhattachan, Davis, Ravi, & Runyan, 2013; Xu, Kang, Zhuang, & Pan, 2010). The slow nature of desertification processes occurring at decade to century scales makes long-term observations and monitoring rare, contributing to uncertainties in the assessment process.

Uncertainties also stem from the lack of clarity in the desertification definition widely adopted by scientists. The definition merely categorizes desertification as a form of land degradation without specifying the threshold for degradation or its reversibility properties. This ambiguity results in challenges related to indicator systems in MAD, particularly in distinguishing between fast and slow variables (Vogt et al., 2011). Additionally, the methods employed in MAD contribute to uncertainties. While longer time series datasets have been utilized, they remain too short to examine ecosystem changes effectively. The choice of imaging time introduces uncertainties, with different times yielding variations in desertification identification. Vegetation monitoring presents specific challenges, particularly in areas prone to desertification where vegetation is sparse. RS indicators, such as NDVI, encounter difficulties in accurately revealing changes in sparse vegetation. Accurate categorization is further complicated by the effects of soil foundation on vegetation data as well as the effect of precipitation on vegetation in desertification hazard zones (Edwards, Wellens, & Al-Eisawi, 1999; Leprieur, Kerr, Mastorchio, & Meunier, 2000; Prince, 2002). Comprehensive desertification indexes, including Albedo, LST, and TVDI, face limitations due to a lack of specific biophysical meanings, subjective threshold choices, and uncertainties not strictly associated with real desertification (Liu & Wang, 2007; Zeng, Xiang, Feng, & Xu, 2006).

Addressing spatial and temporal heterogeneity is crucial in understanding vegetation patterns and landscape processes. The significance of vegetation data obtained from RS is closely linked to both regional variability and size. However, spatial and temporal variations, along with the spatial heterogeneity of land systems, make it challenging to identify or diagnose desertification effectively (Schlesinger et al., 1990; Xu et al., 2010). Soil properties, another key indicator of desertification, are limited by challenges in obtaining accurate data at large spatial scales, noise in RS data, and contamination of atmospheric effects (Ben-Dor et al., 2009; Saatchi, Le Vine, & Lang, 1994). In the Indian context, where land degradation is a pressing concern, addressing these challenges requires a collaborative effort involving policymakers, scientists, and local communities. Advancements in RS technologies, including the use of high-resolution satellite imagery and geospatial tools, can enhance the accuracy and efficiency of monitoring land degradation. Furthermore, initiatives for capacity building and awareness among local communities are crucial for the successful implementation of sustainable land management practices. Monitoring land degradation and desertification in the global and Indian context presents both opportunities and challenges. While RS technologies offer valuable tools for spatial inventory and assessment, challenges related to baseline definition, indicator selection, and scale considerations must be addressed. A concerted effort involving international collaborations, technological advancements, and community engagement is essential to effectively combat the multifaceted issues of land degradation and desertification.

24.6 Conclusion

This study underscores the critical importance of employing advanced methods and RS data for precise monitoring of land degradation and desertification. The analysis reveals that desertification is a widespread issue with intricate roots in social, political, economic, and environmental factors, affecting millions globally. The United Nations Convention to Combat Desertification (UNCCD) emphasizes human activities and climatic variations in land degradation. The repercussions of land degradation extend to rural communities, ecosystems, and global food supply. RS, particularly through Earth observation satellites like Landsat and Sentinel, emerges as a powerful, cost-effective tool for large-scale land degradation assessment. Integration with GIS enhances accuracy, utilizing spectral indices like the NDVI. Ongoing technological advancements address challenges such as structural errors and cloud cover. This approach is especially valuable in environmentally stressed regions like the Himalayas. In conclusion, collaborative efforts from international organizations, research institutions, and technological advancements are imperative to advance our understanding and implement sustainable land management practices globally.

References

Ajai. (2007). *Desertification monitoring and assessment using remote sensing and GIS: A pilot project under TPN-1*. Space Applications Centre.

Ajai, R. R., Arya, A. S., Dhinwa, P. S., Pathan, S. K., & Ganesh Raj, K. (2009). Desertification/land degradation status mapping of India. *Current Science*, *97*(10), 1478−1483. Available from http://www.ias.ac.in/currsci/nov252009/1478.pdf.

Ajai., & Dhinwa, P. S. (2017). *Desertification and land degradation in Indian subcontinent: Issues, present status and future challenges. Climate variability impacts on land use and livelihoods in drylands* (pp. 181−201). Springer International Publishing. Available from http://www.springer.com/in/book/9783319566801, https://doi.org/10.1007/978-3-319-56681-8_9.

Andrew, M. E., Wulder, M. A., & Nelson, T. A. (2014). Potential contributions of remote sensing to ecosystem service assessments. *Progress in Physical Geography*, *38*(3), 328−353. Available from https://doi.org/10.1177/0309133314528942, https://journals.sagepub.com/home/PPG.

Antrop, M., & Van Eetvelde, V. (2000). Holistic aspects of suburban landscapes: Visual image interpretation and landscape metrics. *Landscape and Urban Planning*, *50*(1−3), 43−58. Available from https://doi.org/10.1016/S0169-2046(00)00079-7, http://www.elsevier.com/inca/publications/store/5/0/3/3/4/7.

Asner, G. P., Knapp, D. E., Boardman, J., Green, R. O., Kennedy-Bowdoin, T., Eastwood, M., ... Field, C. B. (2012). Carnegie airborne observatory-2: Increasing science data dimensionality via high-fidelity multi-sensor fusion. *Remote Sensing of Environment*, *124*, 454−465. Available from https://doi.org/10.1016/j.rse.2012.06.012.

Bai Z.G., Dent D.L., Olsson L., Schaepman M.E., (2008). Global assessment of land degradation and improvement 1: Identification by remote sensing. ISRIC − World Soil Information. 1.

Barbosa, S., Mesquita, T. V., & Kumar. (2011). What do vegetation indices tell us about the dynamics of the Amazon evergreen forests? *Geophysical Research Abstracts*, *13*.

Bastin, G. (2008). *Rangelands 2008: Taking the pulse*. Published on behalf of the ACRIS Management Committee by the National Land and Water Resources Audit, Canberra.

Bastin, G. N., Smith, D. M. S., Watson, I. W., & Fisher, A. (2009). The Australian Collaborative Rangelands Information System: Preparing for a climate of change. *Rangeland Journal*, *31*(1), 111−125. Available from https://doi.org/10.1071/RJ08072.

Ben-Dor, E., Chabrillat, S., Demattê, J. A. M., Taylor, G. R., Hill, J., Whiting, M. L., & Sommer, S. (2009). Using Imaging Spectroscopy to study soil properties. *Remote Sensing of Environment*, *113*(1), S38−S55. Available from https://doi.org/10.1016/j.rse.2008.09.019.

Brabant, P. (2010). A land degradation assessment and mapping methodology, Standard guideline proposal, Comite' Scientifique Francais De la desertification les dossiers thematiques. *Agropolis International*, *8*.

Bridges, E. M., & Oldeman, L. R. (1999). Global assessment of human-induced soil degradation. *Arid Soil Research and Rehabilitation*, *13*(4), 319−325. Available from https://doi.org/10.1080/089030699263212.

Chen, J., Pan, D., Mao, Z., Chen, N., Zhao, J., & Liu, M. (2014). Land-cover reconstruction and change analysis using multisource remotely sensed imageries in Zhoushan Islands since 1970. *Journal of Coastal Research*, *30*(2), 272−282. Available from https://doi.org/10.2112/JCOASTRES-D-13-00027.1.

Chowdary, V. M., Chandran, R. V., Neeti, N., Bothale, R. V., Srivastava, Y. K., Ingle, P., . . . Singh, R. (2008). Assessment of surface and sub-surface waterlogged areas in irrigation command areas of Bihar state using remote sensing and GIS. *Agricultural Water Management*, *95*(7), 754−766. Available from https://doi.org/10.1016/j.agwat.2008.02.009.

Claessens, L., Chowdary, V. M., Chandran, R. V., Neeti, N., Bothale, R. V., Srivastava, Y. K., & Singh. (2009). Assessment of surface and sub-surface waterlogged areas in irrigation command areas of Bihar state using remote sensing and GIS. *Agricultural Water Management*, *95*, 754−766.

Collado, A. D., Chuvieco, E., & Camarasa, A. (2002). Satellite remote sensing analysis to monitor desertification processes in the crop-rangeland boundary of Argentina. *Journal of Arid Environments*, *52*(1), 121−133. Available from https://doi.org/10.1006/jare.2001.0980.

Costanza, R., D'Arge, R., De Groot, R., Farber, S., Grasso, M., Hannon, B., . . . Van Den Belt, M. (1997). The value of the world's ecosystem services and natural capital. *Nature*, *387*(6630), 253−260. Available from https://doi.org/10.1038/387253a0.

Dardel, C., Kergoat, L., Hiernaux, P., Mougin, E., Grippa, M., & Tucker, C. J. (2014). Re-greening Sahel: 30 years of remote sensing data and field observations (Mali, Niger). *Remote Sensing of Environment*, *140*, 350−364. Available from https://doi.org/10.1016/j.rse.2013.09.011.

De Bie, C. A. J. M. (2005). Assessment of soil erosion indicators for maize-based agro-ecosystems in Kenya. *Catena*, *59*(3), 231−251. Available from https://doi.org/10.1016/j.catena.2004.09.007.

Desprats, J. F., Raclot, D., Rousseau, M., Cerdan, O., Garcin, M., Le Bissonnais, Y., . . . Monfort-Climent, D. (2013). Mapping linear erosion features using high and very high resolution satellite imagery. *Land Degradation and Development*, *24*(1), 22−32. Available from https://doi.org/10.1002/ldr.1094.

Diouf, A., & Lambin, E. F. (2001). Monitoring land-cover changes in semi-arid regions: Remote sensing data and field observations in the Ferlo, Senegal. *Academic Press, Senegal Journal of Arid Environments*, *48*(2), 129−148. Available from https://doi.org/10.1006/jare.2000.0744, http://www.journals.elsevier.com/journal-of-arid-environments/.

Diwakar, P.G., & Nayak, S. (2021). *Satellite remote sensing data policy: Benefits of free & open data*. NIAS Policy Brief.

D'Odorico, P., Bhattachan, A., Davis, K. F., Ravi, S., & Runyan, C. W. (2013). Global desertification: Drivers and feedbacks. *Advances in Water Resources*, *51*, 326−344. Available from https://doi.org/10.1016/j.advwatres.2012.01.013.

Dregne, H. E. (1985). Aridity and land degradation. *Environment*, *27*, 16−20.

Dubovyk, O. (2017). The role of remote sensing in land degradation assessments: Opportunities and challenges. *European Journal of Remote Sensing*, *50*(1), 601−613. Available from https://doi.org/10.1080/22797254.2017.1378926, http://tandfonline.com/toc/tejr20/48/1?nav = tocList.

Easdale, M. H., Fariña, C., Hara, S., Pérez León, N., Umaña, F., Tittonell, P., & Bruzzone, O. (2019). Trend-cycles of vegetation dynamics as a tool for land degradation assessment and monitoring. *Ecological Indicators*, *107*, 105545. Available from https://doi.org/10.1016/j.ecolind.2019.105545.

Eckert, S., Hüsler, F., Liniger, H., & Hodel, E. (2015). Trend analysis of MODIS NDVI time series for detecting land degradation and regeneration in Mongolia. *Journal of Arid Environments*, *113*, 16−28. Available from https://doi.org/10.1016/j.jaridenv.2014.09.001, http://www.journals.elsevier.com/journal-of-arid-environments/.

Edwards, M. C., Wellens, J., & Al-Eisawi, D. (1999). Monitoring the grazing resources of the Badia region, Jordan, using remote sensing. *Applied Geography*, *19*(4), 385−398. Available from https://doi.org/10.1016/S0143-6228(99)00007-7.

Eswaran, H., Lal, R., & Reich, P. F. (2019). *Land degradation: An overview* (pp. 20−35). Informa UK Limited. Available from 10.1201/9780429187957-4.

FAO. (2014). *Land degradation assessment in dryland*. Food and Agriculture Organization Unpublished Content. http://www.fao.org/nr/lada/and http://www.fao.org/nr/land/degradation/en/.

Fensholt, R., & Rasmussen, K. (2011). Analysis of trends in the Sahelian 'rain-use efficiency' using GIMMS NDVI, RFE and GPCP rainfall data. *Remote Sensing of Environment*, *115*(2), 438−451. Available from https://doi.org/10.1016/j.rse.2010.09.014.

Fistikoglu, O., & Harmancioglu, N. B. (2002). Integration of GIS with USLE in assessment of soil erosion. *Water Resources Management*, *16*(6), 447−467. Available from https://doi.org/10.1023/A:1022282125760.

Floras, S. A., & Sgouras, I. D. (1999). Use of geoinformation techniques in identifying and mapping areas of erosion in a hilly landscape of central Greece. *International Journal of Applied Earth Observation and Geoinformation*, *1999*(1), 68−77.

Forkuor, G., Dimobe, K., Serme, I., & Tondoh, J. E. (2018). Landsat-8 vs. Sentinel-2: Examining the added value of sentinel-2's red-edge bands to land-use and land-cover mapping in Burkina Faso. *GIScience & Remote Sensing*, *55*(3), 331−354. Available from https://doi.org/10.1080/15481603.2017.1370169.

Gibbs, H. K., & Salmon, J. M. (2015). Mapping the world's degraded lands. *Applied Geography*, *57*, 12−21. Available from https://doi.org/10.1016/j.apgeog.2014.11.024, http://www.elsevier.com/inca/publications/store/3/0/3/9/0/index.htt.

Gupta, A., Hock, L., Xiaojing, H., & Ping, C. (2002). Evaluation of part of the Mekong river using satellite imagery. *Geomorphology*, *44*(3−4), 221−239. Available from https://doi.org/10.1016/S0169-555X(01)00176-3.

Hansen, M. C., & Loveland, T. R. (2012). A review of large area monitoring of land cover change using Landsat data. *Remote Sensing of Environment*, *122*, 66−74. Available from https://doi.org/10.1016/j.rse.2011.08.024.

Herrmann, S. M., Anyamba, A., & Tucker, C. J. (2005). Recent trends in vegetation dynamics in the African Sahel and their relationship to climate. *Global Environmental Change*, *15*(4), 394−404. Available from https://doi.org/10.1016/j.gloenvcha. 2005.08.004.

Ibanez, J., De, U. K., & Devi, A. (2008). Valuing recreational and conservational benefits of a natural tourist site: The case of Cherrapunjee. *Journal of Quantitative Economics.*, *9*, 154−172.

Jahun, B., Ibrahim, R., Dlamini, N., & Musa, S. (2015). Review of soil erosion assessment using RUSLE model and GIS. *Journal of Biology, Agriculture and Healthcare*, *5*(9), 36−47.

Jones, B.Y. L., Diaz, J.G., Duwel, O., Prasuhn, P.B. V., Yordanov, Y., Strauss, R., Uveges, L., & Vandekerckhove, L. (2003). EU soil thematic strategy: Technical working group on erosion. In *Work Package 2: Nature and extent of soil erosion in Europe*. European Commission.

Kandekar Avinash, M., Gaikwad Sunil, W., Gaikwad Satyajit, K., Kandekar Smita, A., & Shelar Avinash, N. (2021). Mapping of wastelands and significance of morphometric analysis in wasteland management—A remote sensing and GIS approach. *Modern Cartography Series*, *10*. Available from https://doi.org/10.1016/B978-0-12-823895-0.00024-5, http://www.elsevier. com/wps/find/bookdescription.cws_home/BS_MCS/description#description.

Kaneko, B., & Kojima, S. (2010). Tree diversity in homegarden land use of Mawsmai Village karst landscape. *International Journal of Environmental Ecology and Family Urban Studies*, *7*(3), 33−42.

Kapalanga, T.S. (2008). A review of land degradation assessment methods. In Land Restoration Training Programme. Final report, 112. Keldnaholt, Reykjavík, Iceland.

Kasturirangan, K., Aravamudan, R., Deekshatulu, B. L., Joseph, G., & Chandrasekhar, M. G. (1996). Indian Remote Sensing satellite IRS-1C—The beginning of a new era. *Current Science*, *70*, 495−500.

Kertész., & Křeček, J. (2019). Landscape degradation in the world and in Hungary. *Hungarian Geographical Bulletin*, *68*(3), 201−221. Available from https://doi.org/ 10.15201/hungeobull.68.3.1, http://ojs3.mtak.hu/index.php/hungeobull/article/download/ 1576/1316.

Khan, A., Govil, H., Kumar, G., & Dave, R. (2020). Synergistic use of Sentinel-1 and Sentinel-2 for improved LULC mapping with special reference to bad land class: A case study for Yamuna River floodplain, India. *Spatial Information Research*, 1−13. Available from https://doi.org/10.1007/s41324-020-00325-x.

Kishk, M. A. (1986). Land degradation in the Nile Valley (Egypt). *Ambio*, *15*(4), 226−230.

Barbosa, H.A., & Lakshmi Kumar, T.V., (2011). *Strengthening regional capacities for providing remote sensing decision support in drylands in the context of climate variability & change*, pp. 979−9.

Lakshmi Kumar, T. V., Barbosa, H., Koteswara Rao, K., & Prabha Jothi, E. (2012). Some studies on the frequency of extreme weather events over India. *Journal of Agricultural Science and Technology*, *14*(6), 1343−1356. Available from http://jast.journals.modares.ac.ir/? _action = showPDF&article = 601&_ob = 691eee10b72419824597aa2b772450b4&fileName = full_text.pdf.

Lal, R. (2001). Soil degradation by erosion. *Land Degradation and Development*, *12*(6), 519−539. Available from https://doi.org/10.1002/ldr.472.

Le, N. B., Nkonya., & Mirzabaev, A. (2016). *Economics of land degradation and improvement—A global assessment for sustainable development* (pp. 55−84). Springer.

Le Roux, J. J., & Sumner, P. D. (2013). Water erosion risk assessment in South Africa: A proposed methodological framework. *Geografiska Annaler, Series A: Physical Geography*, *95*(4), 323−336. Available from https://doi.org/10.1111/geoa.12018, http://www.blackwellpublishing.com/aims.asp?ref = 0435-3676.

Lee., & Lunetta, R. (1996). *Wetland and environmental application of GIS*. Lewis Publishers.

Leprieur, C., Kerr, Y. H., Mastorchio, S., & Meunier, J. C. (2000). Monitoring vegetation cover across semi-arid regions: Comparison of remote observations from various scales. *International Journal of Remote Sensing*, *21*(2), 281−300. Available from https://doi.org/10.1080/014311600210830.

Liu, S., & Wang, T. (2007). Aeolian desertification from the mid-1970s to 2005 in Otindag Sandy Land, Northern China. *Environmental Geology*, *51*(6), 1057−1064. Available from https://doi.org/10.1007/s00254-006-0375-1.

Ludwig, J. A., Bastin, G. N., Chewings, V. H., Eager, R. W., & Liedloff, A. C. (2007). Leakiness: A new index for monitoring the health of arid and semiarid landscapes using remotely sensed vegetation cover and elevation data. *Ecological Indicators*, *7*(2), 442−454. Available from https://doi.org/10.1016/j.ecolind.2006.05.001.

Ludwig, J. A., & Tongway, D. J. (1992). *Monitoring the condition of Australian arid lands: Linked plant-soil indicators* (pp. 765−772). Springer Science and Business Media LLC. Available from 10.1007/978-1-4615-4659-7_43.

Lynden, & Kuhlmann, T. (2002). *Review of degradation assessment methods*. World Soil Information.

Millennium Ecosystem Assessment. (2005). *Ecosystems and human well-being: Desertification synthesis*. World Resources Institute, Washington, DC. http://www.millenniumassessment.org/documents/document.355.aspx.pdf.

Mabbutt, J. A. (1986). Desertification indicators. *Climatic Change*, *9*(1−2), 113−122. Available from https://doi.org/10.1007/BF00140530.

Makaya, N. P., Mutanga, O., Kiala, Z., Dube, T., & Seutloali, K. E. (2019). Assessing the potential of Sentinel-2 MSI sensor in detecting and mapping the spatial distribution of gullies in a communal grazing landscape. *Physics and Chemistry of the Earth*, *112*, 66−74. Available from https://doi.org/10.1016/j.pce.2019.02.001, http://www.journals.elsevier.com/physics-and-chemistry-of-the-earth/.

Mambo, J., & Archer, E. (2007). An assessment of land degradation in the Save catchment of Zimbabwe. *Area*, *39*(3), 380−391. Available from https://doi.org/10.1111/j.1475-4762.2007.00728.x.

Mansour, K., Mutanga, O., Adam, E., & Abdel-Rahman, E. M. (2016). Multispectral remote sensing for mapping grassland degradation using the key indicators of grass species and edaphic factors. *Geocarto International*, *31*(5), 477−491. Available from https://doi.org/10.1080/10106049.2015.1059898, http://www.tandfonline.com/toc/tgei20/current.

Marshall, M., Funk, C., & Michaelsen, J. (2012). Examining evapotranspiration trends in Africa. *Climate Dynamics*, *38*(9−10), 1849−1865. Available from https://doi.org/10.1007/s00382-012-1299-y.

Mayr, A., Rutzinger, M., Bremer, M., & Geitner, C. (2016). Mapping eroded areas on mountain grassland with terrestrial photogrammetry and object-based image analysis.

ISPRS Annals of the Photogrammetry, Remote Sensing and Spatial Information Sciences, 3, 137−144. Available from https://doi.org/10.5194/isprs-annals-III-5-137-2016, http://www.isprs.org/publications/annals.aspx.

Mbow, C., Brandt, M., Ouedraogo, I., de Leeuw, J., & Marshall, M. (2015). What four decades of earth observation tell us about land degradation in the Sahel? *Remote Sensing, 7*(4), 4048−4067. Available from https://doi.org/10.3390/rs70404048, http://www.mdpi.com/2072-4292/7/4/4048/pdf.

Mbow, C., Mertz, O., Diouf, A., Rasmussen, K., & Reenberg, A. (2008). The history of environmental change and adaptation in eastern Saloum-Senegal-Driving forces and perceptions. *Global and Planetary Change, 64*(3−4), 210−221. Available from https://doi.org/10.1016/j.gloplacha.2008.09.008.

Metternicht, G., Zinck, J. A., Blanco, P. D., & Valle, H. F. D. (2010). Remote sensing of land degradation: Experiences from Latin America and the Caribbean. *Panama Journal of Environmental Quality, 39*(1), 42−61. Available from https://doi.org/10.2134/jeq2009.0127, http://jeq.scijournals.org/cgi/reprint/39/1/42.

Moser, L., Voigt, S., Schoepfer, E., & Palmer, S. (2014). Multitemporal wetland monitoring in sub-Saharan West-Africa using medium resolution optical satellite data. *IEEE Journal of Selected Topics in Applied Earth Observations and Remote Sensing, 7*(8), 3402−3415. Available from https://doi.org/10.1109/JSTARS.2014.2336875, http://ieeexplore.ieee.org/xpl/RecentIssue.jsp?punumber = 4609443.

Mouat, D. A., Fox, C. A., & Rose, M. R. (1992). *Ecological indicator strategy for monitoring arid ecosystems* (pp. 717−737). Springer Science and Business Media LLC. Available from 10.1007/978-1-4615-4659-7_41.

Muchena, F.N. (2008). Indicators for sustainable land management in Kenya's context. In *GEF land degradation focal area indicators*.

Murthy, K., Shearn, M., Smiley, B. D., Chau, A. H., Levine, J., & Robinson, M. D. (2014). Skysat-1: Very high-resolution imagery from a small satellite. *Proceedings of SPIE − The International Society for Optical Engineering, 9241*. Available from https://doi.org/10.1117/12.2074163, http://spie.org/x1848.xml.

Nagai, M., Chen, T., Shibasaki, R., Kumagai, H., & Ahmed, A. (2009). UAV-borne 3-D mapping system by multisensor integration. *IEEE Transactions on Geoscience and Remote Sensing, 47*(3), 701−708. Available from https://doi.org/10.1109/TGRS.2008.2010314.

Narayan, L. R. A., Rao, D. P., & Gautam, N. C. (1989). Wasteland identification in india using satellite remote sensing. *International Journal of Remote Sensing, 10*(1), 93−106. Available from https://doi.org/10.1080/01431168908903850.

Nelson, G., Rosegrant, M., Koo, J., Robertson, R., Sulser, T., Zhu, T., ... Lee D. (2009). *Climate change: Impact on agriculture and cost of adaptation*. IFPRI.

Nicholson, S. E., Tucker, C. J., & Ba, M. B. (1998). Desertification, drought, and surface vegetation: An example from the West African Sahel. *Bulletin of the American Meteorological Society, 79*(5), 815−829. Available from http://ams.allenpress.com, 10.1175/1520-0477(1998)079 < 0815:DDASVA > 2.0.CO;2.

Nkonya, E., Gerber, N., & Baumgartner, P. (2011). *The economics of land degradation: Toward an integrated global assessment. Pinto Graw Kato Kloos Walter 66, Development economics and policy series*. Peter Lang GmbH.

NRSC. (2019a). *Natural resource census land use/land cover analysis − Third cycle. Land use/land cover database for dissemination through bhuvan*. National Remote Sensing Centre, Hyderabad.

NRSC. (2019b). *Wastelands Atlas of India*. National Remote Sensing Centre, Department of Land Resources Ministry of Rural Development, New Delhi, Hyderabad.

Oldeman, L.R. (1992). *Global extent of soil degradation Bi—Annual report*, pp. 19−36.

Oldeman, L.R., Hakkeling, R.T. A., & Sombroek, W.G. (1990). *World map of the status of human-induced soil degradation: An explanatory note* (2nd ed.). International Soil Reference and Information Centre.

Pages, G., Nguyen, A.D., Priot, B., Pérennou, T., & Calmettes, V. (2015). *Proceedings of the Institute of Navigation (ION GNSS +) Tightly coupled INS/DGPS system for collaborative navigation in mobile ad hoc networks*.

Paparoditis, N., Papelard, J.-P., Cannelle, B., Devaux, A., Soheilian, B., David, N., & Houzay, E. (2022). Stereopolis II: A multi-purpose and multi-sensor 3D mobile mapping system for street visualisation and 3D metrology. *Revue Française de Photogrammétrie et de Télédétection*, 200(200), 69−79. Available from https://doi.org/10.52638/rfpt.2012.63.

Petrie, G. (2009). Systematic oblique ae using multiple digitarial photography i frame cameras. *Photogrammetric Engineering and Remote Sensing*, 75(2), 102−107.

Phinzi, K., & Ngetar, N. S. (2017). Mapping soil erosion in a quaternary catchment in Eastern Cape using geographic information system and remote sensing. *South African Journal of Geomatics*, 6(1), 11. Available from https://doi.org/10.4314/sajg.v6i1.2.

Prince. (2002). Spatial and temporal scales for detection of desertification. In *Global desertification: Do humans create deserts*. Dahlem University Press.

Prokop, P., & Bhattacharyya, A. (2011). Reconnaissance of quaternary sediments from Khasi Hills, Meghalaya. *Journal of the Geological Society of India*, 78(3), 258−262. Available from https://doi.org/10.1007/s12594-011-0084-6.

Puigdefabregas, J., Barrio, G., & Hill, J. (2009). Ecosystemic approaches to land degradation. In *Advances in studies on desertification. Contributions to the International Conference in memory of Prof. Johm B. Thornes* (pp. 77−87). Universidad de Murcia, Murcia.

Rasmy, M., Feranec, J., Hazeu, G., Christensen, S., & Jaffrain, G. (2010). Corine land cover change detection in Europe (case studies of The Netherlands and Slovakia). *Land Use Policy*, 24, 234−247.

Reed, M. S., Buenemann, M., Atlhopheng, J., Akhtar-Schuster, M., Bachmann, F., Bastin, G., ... Verzandvoort, S. (2011). Cross-scale monitoring and assessment of land degradation and sustainable land management: A methodological framework for knowledge management. *Land Degradation and Development*, 22(2), 261−271. Available from https://doi.org/10.1002/ldr.1087.

Reed, M. S., & Dougill, A. J. (2002). Participatory selection process for indicators of rangeland condition in the Kalahari. *Geographical Journal*, 168(3), 224−234. Available from https://doi.org/10.1111/1475-4959.00050.

Reynolds, J. F., Grainger, A., Stafford Smith, D. M., Bastin, G., Garcia-Barrios, L., Fernández, R. J., ... Zdruli, P. (2011). Scientific concepts for an integrated analysis of desertification. *Land Degradation and Development*, 22(2), 166−183. Available from https://doi.org/10.1002/ldr.1104.

Reynolds, J. F., Stafford Smith, D. M., Lambin, E. F., Turner, B. L., Mortimore, M., Batterbury, S. P. J., ... Walker, B. (2007). Ecology: Global desertification: Building a science for dryland development. *Science (New York, N.Y.)*, 316(5826), 847−851. Available from https://doi.org/10.1126/science.1131634.

Ries, J. B., & Marzolff, I. (2003). Monitoring of gully erosion in the Central Ebro Basin by large-scale aerial photography taken from a remotely controlled blimp. *Catena*, *50* (2−4), 309−328. Available from https://doi.org/10.1016/S0341-8162(02)00133-9.

Roder, W., & Hill, J. (2005). *Land degradation: Development and breakdown of terrestrial environments*. Cambridge University Press.

Rosen, P.A., & Kumar, R. (2021). NASA-ISRO SAR (NISAR) mission status. In *5 7 IEEE national radar conference − Proceedings*. Institute of Electrical and Electronics Engineers Inc. Available from https://doi.org/10.1109/RadarConf2147009.2021.9455211, http://ieeexplore.ieee.org/xpl/conhome.jsp?punumber = 1001138. 9781728176093.

Roy, P. S., & Joshi, P. K. (2002). Forest cover assessment in north-east India − The potential of temporal wide swath satellite sensor data (IRS-1C WiFS). *International Journal of Remote Sensing*, *23*(22), 4881−4896. Available from https://doi.org/10.1080/01431160110114475, https://www.tandfonline.com/loi/tres20.

Ruelland, D., Tribotte, A., Puech, C., & Dieulin, C. (2011). Comparison of methods for LUCC monitoring over 50 years from aerial photographs and satellite images in a Sahelian catchment. *France International Journal of Remote Sensing*, *32*(6), 1747−1777. Available from https://doi.org/10.1080/01431161003623433, https://www.tandfonline.com/loi/tres20.

Running, S. W., Justice, C. O., Salomonson, V., Hall, D., Barker, J., Kaufmann, Y. J., ... Carneggie, D. (1994). Terrestrial remote sensing science and algorithms planned for EOS/MODIS. *International Journal of Remote Sensing*, *15*(17), 3587−3620. Available from https://doi.org/10.1080/01431169408954346.

Saatchi, S. S., Le Vine, D. M., & Lang, R. H. (1994). Microwave backscattering and emission model for grass canopies. *IEEE Transactions on Geoscience and Remote Sensing*, *32*(1), 177−186. Available from https://doi.org/10.1109/36.285200.

Santini, M., Gibbs, H. K., & Salmon, J. M. (2010). Mapping the world's degraded lands. *Applied Geography*, *57*, 12−21.

Sarma, A. A. L. N., & Lakshmi Kumar, T. V. (2006). Studies on crop growing period and NDVI in relation to water balance components. *Indian Journal of Radio and Space Physics*, *35*(6), 424−434. Available from http://nopr.niscair.res.in/bitstream/123456789/3931/1/IJRSP%2035(6)%20424-434.pdf.

Schlesinger, W. H., Reynolds, J. F., Cunningham, G. L., Huenneke, L. F., Jarrell, W. M., Virginia, R. A., & Whitford, W. G. (1990). Biological feedbacks in global desertification. *Science (New York, N.Y.)*, *247*(4946), 1043−1048. Available from https://doi.org/10.1126/science.247.4946.1043.

Schwilch, G., Bestelmeyer, B., Bunning, S., Critchley, W., Herrick, J., Kellner, K., ... Winslow, M. (2011). Experiences in monitoring and assessment of sustainable land management. *Land Degradation and Development*, *22*(2), 214−225. Available from https://doi.org/10.1002/ldr.1040.

Sellami. (2009). *Thematic and statistical representativeness of local environmental monitoring observatories: Application to the ROSELTOSS network*.

Sellers, P. J. (1985). Canopy reflectance, photosynthesis and transpiration. *International Journal of Remote Sensing*, *6*(8), 1335−1372. Available from https://doi.org/10.1080/01431168508948283.

Seutloali, K. E., Beckedahl, H. R., Dube, T., & Sibanda, M. (2016). An assessment of gully erosion along major armoured roads in south-eastern region of South Africa: A remote sensing and GIS approach. *Geocarto International*, *31*(2), 225−239. Available from https://doi.org/10.1080/10106049.2015.1047412, http://www.tandfonline.com/toc/tgei20/current.

Shruthi, D., Desprats, J. F., Raclot, D., Rousseau, M., Cerdan, O., Garcin, M., & Monfort-Climent, D. (2011). Mapping linear erosion features using high and very high resolution satellite imagery. *Land Degradation & Development, 24*, 22−32.

Sibanda, M., Mutanga, O., & Rouget, M. (2016). Comparing the spectral settings of the new generation broad and narrow band sensors in estimating biomass of native grasses grown under different management practices. *GIScience & Remote Sensing, 53*(5), 614−633. Available from https://doi.org/10.1080/15481603.2016.1221576.

Singh, G., Bundela, D. S., Sethi, M., Lal, K., & Kamra, S. K. (2010). Remote sensing and geographic information system for appraisal of salt-affected soils in India. *Journal of Environmental Quality, 39*(1), 5−15. Available from https://doi.org/10.2134/jeq2009.0032India, http://jeq.scijournals.org/cgi/reprint/39/1/5.

Smith, M. S., & Reynolds, J. F. (2003). The interactive role of human and environmental dimensions in the desertification debate. *Annals of Arid Zone, 42*(3−4), 255−270. Available from http://epubs.icar.org.in/ejournal/index.php/AAZ/issue/archive.

Sommer, S., Zucca, C., Grainger, A., Cherlet, M., Zougmore, R., Sokona, Y., . . . Wang, G. (2011). Application of indicator systems for monitoring and assessment of desertification from national to global scales. *Land Degradation and Development, 22*(2), 184−197. Available from https://doi.org/10.1002/ldr.1084.

Sonneveld, B. G. J. S., & Dent, D. L. (2009). How good is GLASOD? *Journal of Environmental Management, 90*(1), 274−283. Available from https://doi.org/10.1016/j.jenvman.2007.09.008, https://www.sciencedirect.com/journal/journal-of-environmental-management.

Stringer, L. C., & Reed, M. S. (2007). Land degradation assessment in southern Africa: Integrating local and scientific knowledge bases. *Land Degradation and Development, 18*(1), 99−116. Available from https://doi.org/10.1002/ldr.760.

Tappan, G. G., Sall, M., Wood, E. C., & Cushing, M. (2004). Ecoregions and land cover trends in Senegal. *Journal of Arid Environments, 59*(3), 427−462. Available from https://doi.org/10.1016/j.jaridenv.2004.03.018.

Taruvinga, K. (2009). *Gully mapping using remote sensing: Case study in KwaZulu-Natal.*

Tengberg A., Torheim S.I.B., (2007). The role of land degradation in the agriculture and environment nexus. Environmental Science and Engineering (Subseries: Environmental Science). (9783540724377), 267−283, http://www.springer.com/series/7487, https://doi.org/10.1007/978-3-540-72438-4_14.

Toth, C., & Jóźków, G. (2016). Remote sensing platforms and sensors: A survey. *Journal of Photogrammetry and Remote Sensing, 115*, 22−36. Available from https://doi.org/10.1016/j.isprsjprs.2015.10.004, http://www.elsevier.com/inca/publications/store/5/0/3/3/4/0.

Toulmin, C. (1997). Pastures lost. *Geographical Magazine, 69.*

Turner, W., Rondinini, C., Pettorelli, N., Mora, B., Leidner, A. K., Szantoi, Z., . . . Woodcock, C. (2015). Free and open-access satellite data are key to biodiversity conservation. *Biological Conservation, 182*, 173−176. Available from https://doi.org/10.1016/j.biocon.2014.11.048, http://www.elsevier.com/inca/publications/store/4/0/5/8/5/3.

UN. (1977). Status of desertification in the hot arid regions. In *Climatic Aridity Index Map and experimental world scheme of aridity and drought probability at a scale of 1:25,000,000. Explanatory note conference on desertification, A/CONF.74/31.* United Nations: New York, New York.

UNCCD. (1994). *Convention to combat desertification in those coun-tries experiencing serious drought and/or desertification, particularly in Africa*. UNCCD; UNEP. http://www.unccd.int/convention/text/pdf/conv-eng.pdf.

UNEP. (2008). *Climate change strategy*. United Nations Environment Programme.

Vlek, P. L. G., Le, Q. B., & Tamene, L. (2010). *Assessment of land degradation, its possible causes and threat to food security in sub-SaharanAfrica. Food security and soil quality* (pp. 57−86). CRC Press. Available from http://www.tandfebooks.com/doi/book/10.1201/EBK1439800577, 10.1201/EBK1439800577.

Vogt, J. V., Safriel, U., Von Maltitz, G., Sokona, Y., Zougmore, R., Bastin, G., & Hill, J. (2011). Monitoring and assessment of land degradation and desertification: Towards new conceptual and integrated approaches. *Land Degradation and Development, 22*(2), 150−165. Available from https://doi.org/10.1002/ldr.1075.

Vreiling, K. (2006). Holistic aspects of suburban landscapes: Visual image interpretation and landscape metrics. *Landscape and Urban Planning, 50*, 90−99. Available from https://doi.org/10.1016/S0169-2046(00)00099-2.

Wan, Z., Wang, P., & Li, X. (2004). Using MODIS Land Surface Temperature and Normalized Difference Vegetation Index products for monitoring drought in the southern Great Plains, USA. *International Journal of Remote Sensing, 25*(1), 61−72. Available from https://doi.org/10.1080/0143116031000115328, https://www.tandfonline.com/loi/tres20.

Wessels, K. J., Prince, S. D., Frost, P. E., & Van Zyl, D. (2004). Assessing the effects of human-induced land degradation in the former homelands of northern South Africa with a 1 km AVHRR NDVI time-series. *Remote Sensing of Environment, 91*(1), 47−67. Available from https://doi.org/10.1016/j.rse.2004.02.005.

Wessels, K. J., Prince, S. D., Malherbe, J., Small, J., Frost, P. E., & VanZyl, D. (2007). Can human-induced land degradation be distinguished from the effects of rainfall variability? A case study in South Africa. *Journal of Arid Environments, 68*(2), 271−297. Available from https://doi.org/10.1016/j.jaridenv.2006.05.015.

Winslow, M. D., Vogt, J. V., Thomas, R. J., Sommer, S., Martius, C., & Akhtar-Schuster, M. (2011). Science for improving the monitoring and assessment of dryland degradation. *Land Degradation and Development, 22*(2), 145−149. Available from https://doi.org/10.1002/ldr.1044.

Wischmeier, W. H. (1965). Predicting rainfall erosion losses from cropland east of the Rocky Mountain. *Agriculture Handbook, 282*.

Wischmeier, W.H. (1978). Predicting rainfall erosion losses: A guide to conversation planning. In *USDA Agr. handbook*.

Wolf, L., & Williams, M. (2008). GEONETCast − Delivering environmental data to users worldwide (September 2007). *IEEE Systems Journal, 2*(3), 401−405. Available from https://doi.org/10.1109/JSYST.2008.925978.

Xu, D. Y., Kang, X. W., Zhuang, D. F., & Pan, J. J. (2010). Multi-scale quantitative assessment of the relative roles of climate change and human activities in desertification − A case study of the Ordos Plateau, China. *Journal of Arid Environments, 74*(4), 498−507. Available from https://doi.org/10.1016/j.jaridenv.2009.09.030.

Zeng, Y. N., Xiang, N. P., Feng, Z. D., & Xu, H. (2006). Albedo-NDVI space and remote sensing synthesis index models for desertification monitoring. *Scientia Geographica Sinica, 1*, 75−77.

Impact of rapid urbanization on the groundwater environment: a geospatial analysis with a special reference to Bolpur-Sriniketan C.D. Block of Birbhum District in the lower Ajay river basin in the last decade

25

Jayanta Gour

Department of Geography, Sambhu Nath College, Labpur, Birbhum, West Bengal, India

25.1 Introduction

From a global perspective, in almost every continent (excluding Antarctica and uninhabited parts of the world), human beings exploit the water from various sources of their respective river basins, as river basins have been the base of agriculture and industrial activities of all civilizations since prehistoric times. From the catchment to the confluence of the mouths, people utilized the surface of the groundwater resources through different methods for centuries. In due course of time, people learned to store water in various ways to harness groundwater for irrigation and other household purposes during the dry period in their respective basins to sustain their livelihoods. The parts of the world that are devoid of river basins solely focused on groundwater like what used to be done in the Hellenic Civilization (Angelakis et al., 2016) where Minaon settlements mainly used to utilize the groundwater (Angelakis, Voudouris, & Mariolakos, 2016). Ponce et al. (1997) in their article on "Groundwater Utilization And Sustainability" stated that man has a very low percentage of global water resources, that is, only 3% fresh water out of the total water in the earth, out of which 68.7% is permanently stored in icecaps and glaciers, 30.1% is groundwater, 0.3% is surface water, and 0.9% is found in other storage forms (Ponce, Lohani, & Huston, 1997). Mankind has

Applications of Geospatial Technology and Modeling for River Basin Management.
DOI: https://doi.org/10.1016/B978-0-443-23890-1.00025-6

always used and exploited this limited storage of fresh water for its development but has least concentrated on the storage for sustainable development since ancient and medieval times. Ponce also urged in his writings, Hydrological impact on human activities in the Journal of Hydrologic Engineering in 1997, that there is an urgent need for a detailed hydrological study from baseline, and according to them, it should be time-independent on local fresh waterbodies, ecosystems, and geomorphology. It is noteworthy to mention that B.C. Vitruvius from the first century was probably the first one who highlighted the selection of cities by avoiding the marshes. Pliny (Plinius NH, XXXI, xxi–xxiii) also focused on the quality of water man consumes or utilizes (IWA Publishing, 2008). So, the source and the quality of water either on the surface of the river basins or in the groundwater beneath them have always maintained an intricate relationship in natural basin management, and the quality of water also determines the nature and prospects of the livelihoods of any inhabited region in this world. Thus, since the ancient to medieval and medieval to modern times, man has been searching for and managing this limited resource for survival. From regional perspectives, if we look back to water management in ancient India, the "Traditional Water Management Systems of India" by Murthy, Srikonda, and Kasinath (2022) depicts the fact that water management in different parts of ancient India also maintained social and cultural harmony. Thus, it proves that the right over the surface and groundwater belongs to all categories of people and derelict systems must be restored to tackle the water security issues (Murthy et al., 2022). Coming to the medieval times in Bengal, the work "Water Management in Medieval Bengal" by Hasan Sahidul reveals that water management for irrigation and domestic uses has also been changing according to the local and central rulers from time to time (Hasan, 2016).

In this modern age and technical advancements, rural parts are being transformed into urban areas and not only is the human population increasing but also the demands for water in limited areas are increasing rapidly. As we have limited storage of water resources on earth, so every single administrative unit must take responsibility for their microlevel water management system to bring global well-being. Every single block of any nation should be aware of the concept and significance of water resources, and if not nurtured, their river basin may face serious issues shortly. From local perspectives, this chapter has mainly focused on the water management-related issues in such a rapidly developing community development block in West Bengal where the water crisis may be faced shortly due to rapid urbanization. The study area Bolpur-Sriniketan C.D. Block in Birbhum district is well noted not only in West Bengal but in India and abroad for its cultural, social, and academic prosperities. This block became the most important educational and cultural hub after the foundation of Visva-Bharati University founded by Rabindranath Tagore in a lush green ambiance of both rural and urban environments. This block is not only noted for transportation and communication point of view but also noted for tourist attractions. Many scholars, employees, businessmen, traders, and tourists prefer this block for staycation and tourism. Due to the continuous development of roadways and railways networks in the last one and half decades and facilities in rural and urban lush green envirnonments, many people from different parts of India desire to settle down here, and also because this block has seen a rapid change in population growth and density since last few decades. Nowadays, with this surplus population and increasing diversified economic activities, the pressure on drinking water is also increasing rapidly. The number of households, multistory complexes, shopping centers, and hotels, and so forth have been increasing, which is indirectly increasing the number of tap-water connections and borewells pressurizing the groundwater table below this block. Recent samples taken from different sites in and around the Bolpur Municipality area also indicate different parameters, which are also alarming. Unlike some other community development blocks in West Bengal, the Bolpur-Sriniketan has been lagging in preparing a proper groundwater management program for the upcoming years. Side by side, tourism and other sectors are developing at a rapid rate and the population growth rate is supposed to be increased in the next decade. Hence, to keep the pace of urban growth in this block, proper knowledge of the groundwater resource capacity should be increased as global warming may bring more challenges, particularly during dry summers in this block. To find out the groundwater-related issues and sort out upcoming challenges in this developing block located within the Kopai and Ajay River Basins, this work has been carried out with sustainable developmental approaches through local to regional and global approaches.

25.1.1 Objectives

Every research work aims to have some objectives to frame the study to solve the research problems. Likewise, this work aimed at the following objectives:

1. To understand the fluvio-geomorphological environment of the Bolpur-Sriniketan block through Basin Morphology analysis.
2. To know the present status of urbanization and assess the water resources with special attention to groundwater resources.
3. To analyze and assess the water resource potentiality with special reference to the impact of urbanization on groundwater resources and recommend for sustainable development of this block as a model one.

25.1.2 Methodology

This research work has been done through stages following some methodologies, which are as follows:

Prefield stage: Before starting this research work, a thorough detailed knowledge of the administrative and physical setup was obtained from different published secondary sources like books, journals, e-books, e-journals, and websites (both commercial and government portals available). Printed and e-reports, district census handbooks, gazetteers, public notices, and census reports of 2001,

2011, and periodicals were also consulted from time to time. Topographical sheets (NO. 73M9, 73M10,73M13, and 73M14) of the Survey of India were also taken into consideration for delineating the regions of interest. Satellite images (LISS-III, Sentinel-2 download [World], and USGS satellite images) were downloaded from the Bhuban-NRSC Portal and the United States Geological Survey Portal as well for the same region.

Field stage: Due to a lack of raw images, shapefiles, spreadsheet data, and groundwater data of Bolpur Municipality, a field survey was done in Panchayats and sample villages (covering the quadrants of this block) through questionnaires for obtaining the data related to different types of water resources (tubewell/borehole water, tap water, well water, hand pump, river/canal water etc.) and their respective purposes and amount of uses. Telephonic interviews were also carried out to obtain data. The PHE Department of Bolpur Municipality was also covered for obtaining the water resource-related data of this block. Detailed analytical and comparative studies were carried out on-site to compare the groundwater resource status of this block with another block in the Birbhum District of West Bengal.

Postfield stage: Several hydrological maps were prepared with the help of the data obtained in the prefield and field stage of this research work. The Arc GIS 10.8 and Q 3.16 were used for those maps with projection GCS WGS 1984 and datum WGS 1984 UTM 45N, respectively. For a better understanding of land use land cover, vegetation, and water bodies of this block Sentinel-2, world data has also been used. Windows Excel Office 2021 has been used for statistical analysis and chart preparation. The digital elevation model obtained from Bhuban NRSC and USGS has been used for profile and landform analysis (Fig. 25.1).

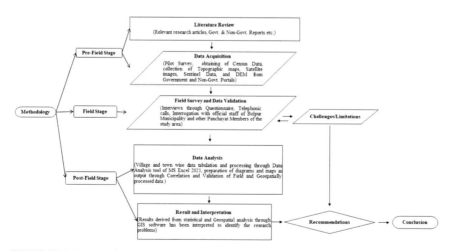

FIGURE 25.1 Schematic representation of the methodology.

The workflow of the study has been schematically represented to understand the prefield, field, and postfield works done during the entire research work.

25.2 Physical background

25.2.1 Geological setup

From the lithostratigraphic point of view, the study area has the geological character of Holocene, Pleistocene (middle to upper), and Pleistocene (lower). Physiographically, the Bolpur-Sriniketan C.D. Block lies in the Rarh region of West Bengal and is an interfluvial part of R. Mayurakshi (to its north) and R. Ajay (its southern boundary). According to the geological map by NATMO, this block is under the quaternary geological environment. The geological formations of Birbhum district are Archaean gneiss, the Gondwana system, laterite, and Gangetic alluvium. The laterite occurs in the form of gravel and rock. The post-trappean Late Cretaceous successions mark the initiation of passive-margin phase in this basin with the deposition of marine Bolpur and Ghatal formations (Prasad & Pundir, 2020). Laterite of Cenozoic age is found in this block as well (2019). The deposition of the subaerial fluvial clastic of the Bolpur Formation and its facies variant of the shell limestone and shale−sandstone of the Ghatal Formation in the shelf area. The Vitrinite reflectance values of the source material within the Gondwana sequences range between 0.47% and 3.29%. Cretaceous and Paleocene Cretaceous sediments show good maturation in Bolpur, Ghatal, and Dhananjaypur Formation Paleocene sediments. Sandstone reservoirs occur in abundance within Bolpur in Birbhum district of West Bengal (Geology Bengal_Basin, 2013). Bolpur-Sriniketan C.D. Block in the southeastern part of Birbhum district is an alluvial plain composed of dark clay, clay, and sandy soil (O'malley, 1910).

25.2.2 Geographical environment

Administratively, Bolpur-Sriniketan C.D. Block lies in the Bolpur Subdivision in Birbhum District of West Bengal, India, with a total population of 202,553 (among which 6% live in urban areas and 94% live in rural areas) and 47,961 households as of the 2011 (Fig. 25.2). The total literacy rate is 70.67%. There are a total of 170 villages in this block with 1 municipality, that is, Bolpur Municipality covering an area of 33.448 km^2 including 22 wards, as of 2023 (Govt. of West Bengal, 2023).

The Sriniketan Santiniketan Development Authority (SSDA) comprised a total of 4 Gram Panchayats (GP) (viz. Kankalitala, Ruppur-Supur, Supur, and Sian-Muluk) in this block (Govt. Of West Bengal, 1997). According to the report of the year 1997 by SSDA, this Sriniketan Santiniketan Development area had a total of 20,705 households, out of which 10,173 households were in urban and 10,532 households were in rural areas. Geospatially, the Bolpur-Sriniketan C.D. Block in Birbhum District of West Bengal, India, lies within 23.5°N−23.7°N latitudes and 87.5°E−87.8°E longitudes covering an area of 334.59 $km^{2degrees}$. Bolpur is the only municipal area and also a Class-II town in West Bengal with a total population of 65,693 (as of the 2001 census data). The undulating plainland

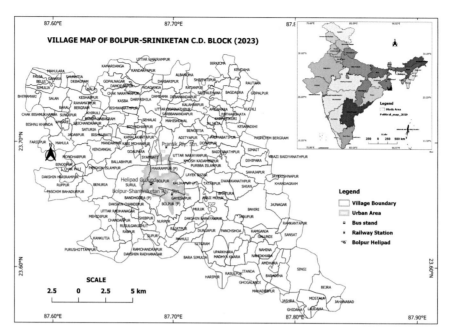

FIGURE 25.2 Location map of Bolpur-Sriniketan C.D. Block (2023).

The location map of any area helps to understand the geospatial locational significance of the study area.

has the highest elevation of 19 m above mean sea level and the lowest value along the R. Kopai flowing from southwest to northwest direction dividing the block into two unequal parts. The western part is relatively higher and covers less area than the eastern part which covers about 3/5th of the area of this block. The eastern part of R. Kopai (flowing for 26.79 km within the block territory from the degrees southwest SW-NE direction) slopes down toward the southeast (toward villages like Bejra, Singi, Jahanabad, and others). The entire southern boundary of this block is drained by R. Ajay (flowing for 26.64 km within the block territory from W−E direction), which inundates the southern lower parts (comprising villages like Purushottampur, Dakshin Radhanagar, Raipur, Rusulganj, Supur, Gitgram, Ghidaha, and others) during heavy rain accompanied by occasional severe cyclonic storms during monsoon season. The central part of this block stretching from southwest to northwest has more elevation than the western and eastern parts and acts as a watershed between the Kopai River Basin and Ajay River Basin. The central part has a maximum slope value (Fig. 25.3) of about $2°$ whereas, the mean slope value of this block is $0.327681°$ with an SD (σ) value of $0.234605°$.

In the floodplain between Suri and Bolpur, the soil varies from red sandy, red loamy, and older alluvium in the south-western to brown and recent alluvium in

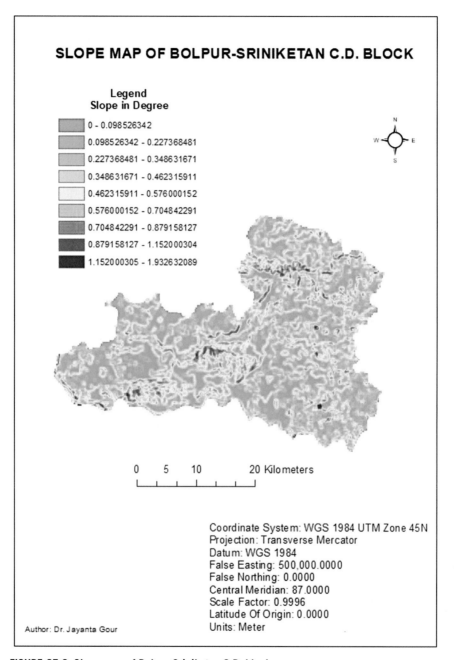

FIGURE 25.3 Slope map of Bolpur-Sriniketan C.D. block.

Generally, the block is a flat land as major parts have an average slope of 0.327681 degree with an SD value of 0.234605 degrees.

the central and south-eastern part of the region (2019) The Bolpur-Sriniketan C. D. Block has one of the five forest ranges in Birbhum district.

25.2.3 Basin morphology and groundwater environment

The Bolpur-Sriniketan C.D. Block in the Birbhum district of West Bengal is actually an interfluvial floodplain of east-flowing R. Mayurakshi and R. Ajay. The left bank of R. Ajay borders this block along the south. The R. Kopai drains this block along its north-central part from southwest to northeast. The block also is served by two major canals, that is, Kopai South Main Canal (for 18.9 km) from west to west and Bakreshwar-Kopai Branch Canal (for 13.9 km). The Kana Ajay Nala (from Kandar Nala near Shibpur, Nurpur to Jahanabad) joins R. Ajay and presently drains into the southeastern part of this block for about 23.5 km. From the hydrological data analysis, it has been found that the highest stream order is 4 (after Strahler) around the north-eastern and south-eastern parts. Another kandar flows parallel to the left bank of R. Ajay along the southern boundary of this block for 8.15 km (from Haripur, Rasulpur to Jashra). Lots of tanks and ponds are also found except in the central part comprising the Bolpur Municipal Area. Lower stream orders are found in the northwestern and central parts of this block. From the focal statistical analysis, it has been found that the value ranges from 1 to 136 (Fig. 25.4). Taking the threshold value at the 5000 level, about 60 tertiary basins (including the canals, kandars, i.e., narrow natural drainage outlet from elevated areas to lowland areas, khals [both natural and artificial], and temporary drainage lines according to the slope value and streamline density) have been noticed here and two major subbasins have been delineated on threshold value 100,000. The streamline density analysis reveals the fact that the Kopai and the southeastern part stretching from northwest to southeast have the maximum line density. The drainage density is moderately higher in this block than in the surrounding blocks of the Birbhum district. Villages like Mahishdhal, Adityapur, Jaljalia, Sarbanandapur, Patharghata, Khudra Amdahara, Laldaha, Sarpalehana, Bisheghata, Sangri, Mahula, Rautara, Keudaha along R. Kopai and Khandagram, Jay Krishnapur, Uparkhara, Madhyakhara Namokhara, and Panchshoa have higher streamline density. Due to moderate to heavy rainfall during monsoon if accompanied by cyclonic storms and water released from Tilpara Mihirlal Barrage at Suri (headquarter of Birbhum district), waterlogging situations are observed in the low-lying villages in the north-eastern, eastern, and south-eastern parts of this block. It takes 2−3 days to come back to normal conditions. In the Bolpur municipal area, although located at a relatively higher elevation than the surrounding villages, several wards suffer from waterlogging conditions, which linger due to very poor drainage management during the monsoon season.

The urban area faces lithological changes abruptly from the rest of the Birbhum district, and an important granular zone occurs between 250 and 450 m below the land surface having a cumulative thickness of around 100−110 m (Chakrabarti & Patra, 2016). The groundwater table is slightly more in elevated

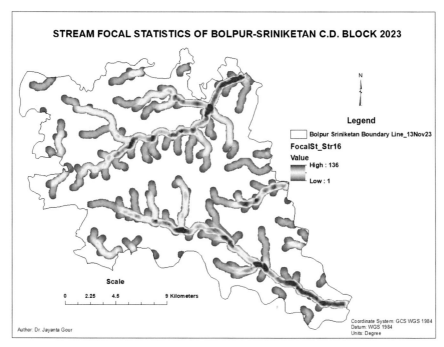

FIGURE 25.4 Stream focal statistics of Bolpur-Sriniketan C.D. block.

The stream focal statistic map helps to understand the magnitude of the different orders of the streams flowing over any region. It also helps validate the correlations between the dependency on different sources of groundwater resources.

areas than in the low-lying areas in this block. According to the residents who possess boreholes at their residences in the Bolpur municipal area, the groundwater table used to reach below 160–170 feet during the last decade has reached up to a depth of 240 feet nowadays. Several boreholes in different wards of Bolpur municipality and shallow pumps in villages along R. Kopai and R. Ajay have been drafting groundwater during dry seasons.

25.3 Urbanization and its impact on groundwater

Bolpur-Sriniketan C.D. Block in Birbhum district of West Bengal comprises rural, semiurban, and urban characteristics to some extent. The major urban periphery includes Bolpur town and presently 22 wards of Bolpur Municipal Authority. Villages nearby this ward bear semiurban characteristics and rural areas dominate the rest part of this block. There are a total of 170 villages including the Bolpur (P). There are several primary schools, government-aided and private higher secondary schools, colleges, one central university, one state university, one private medical college, government and private banks, a sub-divisional court, sub-divisional correctional home, sub-divisional hospital, sub-divisional post office, and almost all institutions which are noticed in any developed community development block of any district in West Bengal. All these institutions require a huge volume of water resources particularly for drinking purposes, whereas people coming from outside for their official and unofficial works also need to be served with water resources not only for drinking purposes, but for other uses also. So, there is a continuous increase in water demand in the urban area of this block. However, visitors and commuters also supplement their water demand through packaged bottles, the selling of which has tremendously increased nowadays. The water demand reaches its peak point during annual fairs and festivals from December to March every year (at the rate of 260 L/persons at the residential complex level).

There were only 15,000 persons in the Bolpur Municipal urban area (Fig. 25.5) in 1951, which rose to 80,210 persons in 2011. This indicates a rapid growth in urban population (434.73%) in Bolpur-Sriniketan C.D. Block over the last 6 census decades (from 1951 to 2011) and maximum decadal growth (22.11% from 2001 to 2011) in population density during the last census decade

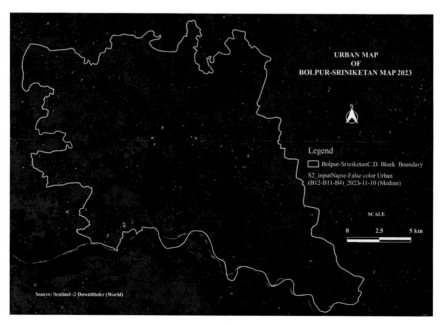

FIGURE 25.5 Urban and rural areas in Bolpur-Sriniketan C.D. block of Birbhum District.

The urban map of any block shows the concentration of households (through False Color Composite) and helps in interpreting the usage of groundwater resources.

(Bolpur Municipality, 2023). When the urban growth rate in West Bengal during 1981−91 was 32.26%, the urban area of Bolpur Municipality, which was a rural village in the 20th century, already reached 37.54%. So, urbanization has already a historical background here. Presently, the number of proposed wards is 22 which started with 11 wards in 1950. The urban area has extended to the outskirt villages after the 6th delimitation of wards (Bolpur Municipality, 2023) under Bolpur Municipality was held in 2021 from the existing ward No 20 to 22 wards with added areas and a total projected population as per household would be approximately 148,520 (total area 35.40 sq. km, number of household 37,230 and projected population 148,520). The municipal area has been increased from 13.13 to 33.448 km^2 (154.68% growth) since 1951−2023. The Urban Index of Bolpur-Sriniketan C.D. Block in Birbhum district as of 2019 was 36.883, which is high in response to other blocks. The HRD Index value is 41.244, which is the second highest (after Suri HRDI-46.471) in the Birbhum district of West Bengal (RSP Green Development & Laboratories Pvt. Ltd, 2019).

25.4 **Analysis**

With regard to the ground water resource of Birbhum District of West Bengal, India, in the year 2020, it was found that 12,620.29 ham of annual replenishable groundwater resource is available in Bolpur-Sriniketan C.D. Block, whereas the annual groundwater draft is 1902.354 ham. The maximum part (1603.6 ham) of the drafted groundwater is used for irrigation purposes and about 298.754 ham is used for domestic and industrial purposes. The net groundwater resource here is 11,358.262 ham, which indicates a safe category (Rank-5 with 18.06%) compared to other blocks in the Birbhum district of West Bengal. Due to the increasing population and households, this block drafts more groundwater (298.754 ham) than adjacent blocks like Illambazar (177.905 ham), Nanoor (216.727 ham), and Labpur C.D. Block (237.449 ham), which draft lesser amount of groundwater annually. As Bolpur-Sriniketan C.D. Block is mainly a growing urban area, the number of households is also higher than the surrounding blocks of Bolpur Subdivision of Birbhum District. Due to the minimum practice of agricultural activities, this block drafts a lesser amount of groundwater (1603.6 ham) for irrigation purposes than the adjacent blocks like Nanoor (4491 ham), Illambazar (2949.1 ham), and Labpur (4350 ham). However, Labpur (196,482 persons as of 2011) and Nanoor C.D. Block (213,387 persons as of 2011), even after having more population than Bolpur-Sriniketan C.D. Block (190,393 persons as of 2011) in Birbhum district, draft less amount (263.201 ham) of groundwater for domestic and industrial purposes. This clearly indicates the fact that usage of groundwater does not solely depend upon the total population and density of the population of any block. Instead, geology, the structure of the groundwater table, the type of economic activity, and urban agglomeration determine the drafting amount of groundwater from place to place. This is because, the same blocks have different amounts of net groundwater availability, that is, Labpur—7339.685 ham, Illambazar—9036.465 ham, Sainthia—10,586.595 ham, and Nanoor—10,450.082 ham of net availability of groundwater resource.

The groundwater resources, which although look safe as per the 2020 report, are being threatened by the increasing number of boreholes in residential areas (particularly during the dry season) and in different housing complexes located in different wards of Bolpur Municipality despite the regular supply of tap water by Indo-German Water Treatment Plant, which is indicating an alarm to review on the overuse of groundwater. Taking the groundwater quality into consideration based on different parameters taken from sample sites (Muluk, Prantik, Supur and Surul) like pH, EC, CO_3, HCO_3, Cl, SO_4, NO_3, pO_4, TH, Ca, Mg, Na, K, F, SiO_2, TDS, and U (ppb), it has been found that Muluk (23.65°N, 87.72°E) has higher electric conductivity (8.28), HCO_3 (372), Na (125), TDS (534), and U 1.13 ppb than Prantik, Supur, and Surul. Prantik (23.68°N, 87.69°E) has higher Cl (167), TH (280), Ca (74), and Mg (23) than Muluk, Supur, and Surul. Supur (23.67°N, 87.70°E) has higher pH (8.28), NO_3 (29), and K (66.7) values than Muluk, Prantik, and Surul. Surul (23.67°N, 87.51°E) has higher SO_4 (15) and low-medium K (10.4) values than Muluk, Prantik, and Supur sample sites (Fig. 25.6).

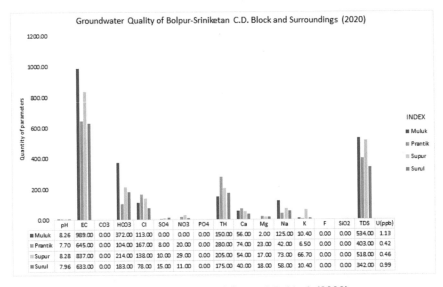

	pH	EC	CO3	HCO3	Cl	SO4	NO3	PO4	TH	Ca	Mg	Na	K	F	SiO2	TDS	U(ppb)
Muluk	8.26	989.00	0.00	372.00	113.00	0.00	0.00	0.00	150.00	56.00	2.00	125.00	10.40	0.00	0.00	534.00	1.13
Prantik	7.70	645.00	0.00	104.00	167.00	8.00	20.00	0.00	280.00	74.00	23.00	42.00	6.50	0.00	0.00	403.00	0.42
Supur	8.28	837.00	0.00	214.00	138.00	10.00	29.00	0.00	205.00	54.00	17.00	73.00	66.70	0.00	0.00	518.00	0.46
Surul	7.96	633.00	0.00	183.00	78.00	15.00	11.00	0.00	175.00	40.00	18.00	58.00	10.40	0.00	0.00	342.00	0.99

FIGURE 25.6 Groundwater quality of Bolpur-Sriniketan C.D. block (2020).

The different levels of parameters of groundwater in different parts of any block help one to understand the comparative knowledge of groundwater quality for consumption.

25.5 **Result and interpretation**

From the data analysis and geographic information system, the groundwater drafting maps show that due to lower agricultural activities, the Bolpur-Sriniketan C. D. Block ranks 10th in the drafting of irrigation water from groundwater among the 19 C.D. Blocks in Birbhum District of West Bengal. However, side by side, it ranks first in drafting groundwater for domestic and industrial purposes. Although there is no large-scale industry in this block, with the increasing number of housing complexes, multistoried shopping complexes, an increasing number of hotels, resorts, guesthouses, and restaurants, and small and cottage industries surrounding the Visva-Bharati University Campus, Sonajhuri Jungle between Bolpur Municipal area and Sriniketan during last one and half decade, the pressure on groundwater drafting is increasing day by day. Hence, as of 2020, from the total draft of groundwater point of view, Bolpur-Sriniketan C.D. Block now ranks ninth among the rest of the blocks in Birbhum district. From the water groundwater quality point of view, it shows that Muluk has exceeded the desired limit (500 mg/L), whereas the total hardness is quite below the desired level (300 mg/L). The nitrate level has also exceeded the permissible limit (100 mg/L) there. Both Muluk and Supur are lying on the marginal level of the desired pH level. The uranium level (U) is 1.13 ppb in groundwater, which is higher than any other parts of this block, which alerts (Tanwer et al., 2023) the future health conditions of the residents of this mouza. According to the Bureau of Indian Standards (BIS), the maximum permissible limit of uranium is 0.03 mg/L (as per WHO provisional guidelines) in all drinking water standards after following due process (Ministry of Jal Shakti, 2020a), whereas most sample sites taken into consideration in Bolpur-Sriniketan C.D. Block have already exceeded the permissible limit. For example, Prantik has a uranium level of 0.41, Supur has 0.46 ppb, and Surul has 0.99 ppb, which is more than the BIS level (Fig. 25.6).

This block stands in the second position in Annual Replenishable Ground Water Resources amounting to 12,620.292 ham after Dubrajpur C.D. Block (13,533.72 ham) in Birbhum District of West Bengal, which is actually due to its basin morphology and geological setup. The replenishing of basin groundwater by the R. Ajay along the south and R. Kopai in its north has kept it in the safe category, but increasing practices of making deep-tubewell and boreholes in urban areas will lower the groundwater table further below. During the pre-monsoon period, that is, during March−May (Chakrabarti & Patra, 2016), several wards in Bolpur Municipality and surrounding villages like Paschim Bahadurpur, Ruppur, Benuria, the eastern part of Paschim Islampur, Surul, Ballabhpur, Syambati, Uttar Natayanpur, Makrampur, Bolpur, Kalikapur, Layek Bazar, Dwarkanathpur, Tatarpur, Araji Muluk, Mahulara, Bheramari, Simulia, Belui, Ganara, Shunutia, and so forth suffer from drinking water also. Bolpur suffers from its elevated location as well as from excess use of groundwater through wells and tubewells. In spite of being nonurban areas, the north-western part suffers due to less streamline density and due to their location in elevated land (Fig. 25.7). The groundwater level which usually was reachable to 170 feet below a decade ago, has now gone down to 240 feet in general. At present, a total (including rural and urban areas) of 5465 households use tubewell/borehole water (Fig. 25.8), 10,66 households use uncovered well water, 520 households use covered well water, 995 households use untreated tap water, 13,057 households use treated tap water, 26,652 households use hand pump water, and 133 households use other sources in Bolpur-Sriniketan C.D. Block (Fig. 25.9). About half of the total villages of this block draft water from rivers and canals for irrigation purposes rather than tube-wells and boreholes (Fig. 25.10). The focal statistics of the stream order (after Strahler) and streamline density of this block (Fig. 25.4) clearly validate the facts that the urban areas, particularly the Bolpur, Surul, and Prantik town areas, use more groundwater through wells, tubewells, hand-pumps, and so forth than other types of resources (Figs. 25.8 and 25.10).

More than 50 flats or apartments are already there in the urban area, that is, within and surrounding the Bolpur municipality, and more are awaited to accommodate to fulfill the demands for flats in and around Bolpur. Side by side, the increasing practices of converting the households into homestay, cafeterias, and

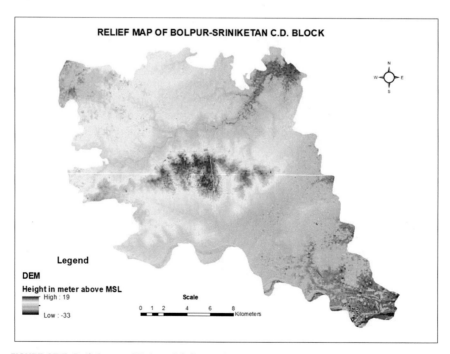

FIGURE 25.7 Relief map of Bolpur-Sriniketan C.D. block.

It shows the change in height (above MSL) of different parts of this block. The central part is higher than the surrounding parts. The north-western part is also a little bit elevated.

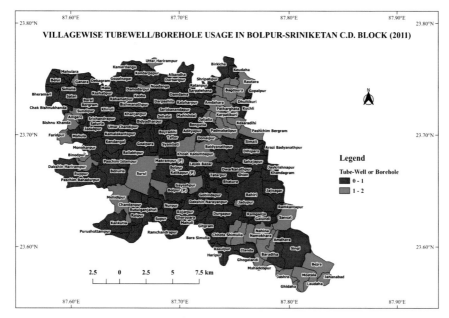

FIGURE 25.8 Tubewell/borehole usage status in Bolpur-Sriniketan C.D. block of Birbhum District of West Bengal.

The tubewell/borehole depicts which parts of the block use more groundwater and future precautions to be taken to avoid future overuse.

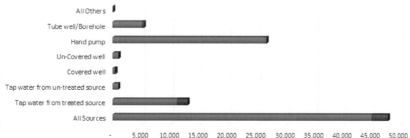

FIGURE 25.9 Different areas of Bolpur-Sriniketan C.D. block with different sources and locations of drinking water.

Rural areas generally depend upon all sources of drinking water rather than concentrating on any particular source.

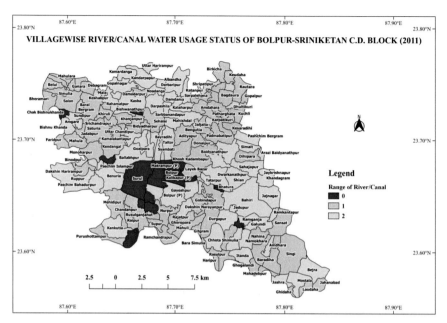

FIGURE 25.10 Villagewise river/canal water usage in Bolpur-Sriniketan C.D. block.

More dependency of the villagers on river water or canal water for irrigation lesser pressure on direct groundwater drafting.

so forth can also be observed in and around Santiniketan premises. An increasing number of students coming from different states for educational purposes and prosperity in tourism have been creating more demands for the construction of hotels and motels as well. All these require daily 260 L/day/person of water to satisfy the demands. As no such other water resources are available in urban areas, people will obviously move to draft groundwater and face health issues in the near future (Fig. 25.11).

Although, the HRD Index has shown a very rapid economic and urban development of the urban area of Bolpur-Sriniketan C.D. Block, but this continuous expansion in the urban area has been increasing the demand for more and more water resources, as a result of which more pressure on groundwater drafting will be challenging in the near future.

25.6 Challenges/limitations of the study

As stated in the methodology (Fig. 25.1), some major challenges were faced during the data collection. There was no recent raw data on the groundwater available in the Bolpur since 2011 as no census was carried out later. So, it was

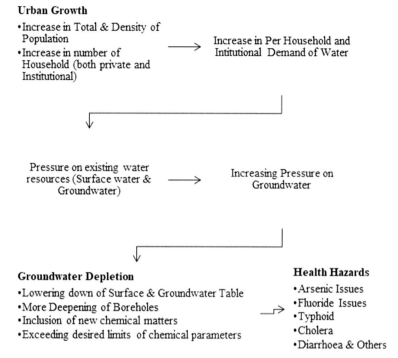

FIGURE 25.11 Impact of urbanization on groundwater resources in Bolpur-Sriniketan C.D. block of Birbhum District.

The flowchart shows how rapid urbanization has an negative impact on groundwater resources in a community development block.

difficult to collect ward-wise official data of the boreholes and actual figures related to the groundwater resources. Besides this, the absence of any groundwater mapping or secondary data published during the last decade also limited this research work to some extent. A proper underground mapping would have helped a lot to prepare the groundwater resource status of this block.

25.7 Recommendations

From the analysis and derived results, some recommendations have been put forth as follows:

1. Although the Bolpur-Sriniketan C.D. Block has shown satisfactory progress in bringing lots of beneficiaries under the Jal Jeevan Mission and Sajal Dhara Scheme of the Government of India and West Bengal, a regular monitoring of misuse of tap water and quality supply of water must be taken

into consideration to restore the groundwater resources. Taps should properly be turned off after use and storing of the flowing water should be drained to avoid unwanted waterlogging.

2. Jin Shao et al. in their case also stated that rooting depth alters the amount of precipitation that is required to produce disconnected recharge and its ratio to total groundwater recharge (Shao, Si, & Jin, 2019). A. Ferro et al. also found that a stand of deep-rooted trees can act as a biological pump, removing a substantial amount of water from the saturated zone (Ferro et al., 2003). So, deep-rooted plants are scarcely found in this block. More deep-rooted trees like Banyan (*Ficus benghalensis* L.), Tamarind (*Tamarindus indica* L.), Peepal (*Ficus religiosa*), Neem (*Azadirachta indica*), and so forth should be planted to raise the groundwater table and to accelerate the evapotranspiration, which will lead to more groundwater recharge through precipitation. Thus, land use land cover plays an important role in deciding water availability and rising or deepening of groundwater table to some extent.

3. All the existing kandars, natural ponds, Kopai, and Ajay rivers should be protected from decaying and the water carrying capacity should be increased through dredging if applicable. This will increase the recharge of groundwater resources and may create less pressure on shallow pumps. Every village which is away from the river water resources must construct sufficiently deep ponds in proper locations not only to store the rainwater that can be used for cultivation in dry seasons but also for natural groundwater recharge.

4. There must be proper guidelines and execution of those guidelines for drawing water from groundwater in urban areas, that is, in Bolpur town and surrounding semitowns. A detailed official record of the total number of tubewells and household boreholes that have been set up already and that are supposed to be set up must be followed.

5. Ideal locations for groundwater recharge should be identified through detailed geospatial and ground-level analysis of the Kopai and Ajay River Basins.

6. The vehicle cleaning or washing centers in urban areas usage of water by the visitors in hotels and resorts must be within limitations to reduce less stress of groundwater.

7. Wastewater recycling process schemes should be implemented in urban areas of Bolpur-Sriniketan C.D. Block. Rooftop Rainwater Harvesting should be an essential condition in the construction plans to be passed officially for Housing Complexes.

8. Ward-wise in municipal areas and panchayat-wise, mass awareness camps or announcements should be carried out during pre-monsoon on how to prevent waste of water. Text messages or leaflets can also be used in this regard.

9. A detailed hydromorphological analysis of aquifers (homogenous and heterogenous), aquicludes, piezometer structure, and perched water tables of

this community development block must be carried out right from now as minimal records are available to calculate the future capacity of groundwater underneath this block.

10. Surveying and designing for water supply should not be restricted to the urban parts of this block. Rather, the increasing population and households in suburban areas should be given more emphasis through a scientific study of groundwater geology.

11. Allowing haphazard construction of boreholes of shallow pump water supply along with free Indo-German Water supply via municipality tap water in various wards of Bolpur Municipality without proper monitoring of overusage can welcome a groundwater table collapse, particularly in densely populated parts of the urban areas.

12. As per the Ministry of Housing and Urban Affairs, 135 L/capita/day is the benchmark for urban water supply. For rural areas, the minimum service delivery of 55 L/capita/day has been fixed under the Jal Jeevan Mission, which may be enhanced to a higher level by states (Ministry of Jal Shakti, 2020b). However, the increasing number of flats or apartments in the urban area of the Bolpur-Sriniketan C.D. Block presently require a per capita demand of more than 200 L of water per day per person (approx.), which can not be fulfilled merely through river/canal water resources or by other sources except the groundwater resources. So, groundwater recharge planning should be implemented soon before it becomes a hazard in dry seasons in the upcoming years.

25.8 Conclusion

In recent days, the Bolpur municipality has already started opening a tender for "Surveying, Design, Execution, and Commissioning for Water Supply Scheme of Bolpur Municipality Including Construction of 57 No. Pump House for 5 Zones 57 No. Dtw Works, Construction of 5 No. Ohr & 5 No. Cwr Including Required Length of Boundary Wall" to mitigate the drinking water-related issues of the urban water supply management (Tender Detail, 2023). Several shallow tube wells are also being executed in different villages to sort out the water crisis. However, a few things should be kept in mind for future groundwater-related issues which are supposed to be inevitable if the global rising of temperature sustains in future decades. The villages encircling the Bolpur municipal area like Goalpara, Surul, Ballabhpur, Shibpur, Nurpur, Muluk, Kalikapur, Tatarpur, Layek Bazar, Khosh Kadambapur, Taltor, Uttar Narayanpur, and Adityapur and villages like Sansat and Singi of this block are rapidly transforming their rural character to urban character in the southeastern part of this block, which indicates more extension of the present urban area of this block in next few decades (Fig. 25.8). Also, the sample site water quality results of this block do not show satisfactory standards, as overdrafting of groundwater does not assure high quality of potable water. Overconsumption due to the increase in total population and population density in the Bolpur municipal area, increased number of vehicles which are washed by drafting groundwater, and increasing number of tourists in different hotels and resorts in and around Bolpur-Santiniketan area as well as in surrounding villages in Bolpur-Sriniketan C.D. Block will certainly lower the groundwater level in the next decade. So, there is an urgent need to prepare a water resource plan to restrict the drafting of groundwater in different wards of Bolpur Municipality and search out an ideal water supply and water treatment plant and policy to fulfill the increasing demand of upcoming households. A proper surveillance and inspection system must be generated by government officials to check whether groundwater is used only for drinking purposes in the right amount. From time to time water quality testing should be done in every part of this block where groundwater is used for livelihood also. Urbanization over the land surface may be a good indicator of economic prosperity and political development of any block in the backward district of India, but it may become a hazard to groundwater just lying invisibly below that urban area. So, developmental approaches must be thought from top to bottom of the relief over which our existence depends upon. Last but not least, more village-wise, tourist site-wise, and ward-wise announcements or arrangments of club-wise awareness camps on proper utilization of groundwater should be carried out, particularly during dry and festival seasons in this block.

References

Angelakis, A., Voudouris, K. S., & Mariolakos, I. (2016). Groundwater utilization through the centuries focusing on the Hellenic civilizations. *Hydrogeology Journal*, *24*(5), 1311−1324. Available from https://doi.org/10.1007/s10040-016-1392-0.

Bolpur Municipality (2023). 11 19. https://bolpurmunicipality.org/about-us/.

Chakrabarti, S., & Patra, P. K. (2016). Chemical analysis of ground water of Bolpur Block, Birbhum, West Bengal, India. *Rasayan Journal of Chemistry*, *9*(4), 627−633. Available from. Available from http://www.rasayanjournal.com.

Geology Bengal_Basin (2013). Directorate general of hydrocarbons. https://www.dghindia.gov.in/assets/downloads/56cfe9cd1a0bfBengal_Basin.pdf.

Ferro, A., Gefell, M., Kjelgren, R., Lipson, D. S., Zollinger, N., & Jackson, S. (2003). Maintaining hydraulic control using deep rooted tree systems. *Advances in Biochemical Engineering/Biotechnology*. Available from https://doi.org/10.1007/3-540-45991-x_5. Available from, https://pubmed.ncbi.nlm.nih.gov/12674401/.

Hasan, S. (2016). Water management in Medieval Bengal. *Bangladesh Historical Studies: Journal of Bangladesh History Association*, *XXIV*, 21−40. Available from https://www.researchgate.net/publication/325689361_Water_Management_in_ Medieval_Bengal.

IWA Publishing. (2008). *A brief history of water and health from ancient civilizations to modern times*. 3 6 2024. IWA Publishing. https://www.iwapublishing.com/news/brief-history-water-and-health-ancient-civilizations-modern-times.

Govt. of West Bengal. (1997). *Land use planning cell & town planning stream SSDA.* Govt.Of West Bengal. Unpublished content land use and development control plan: For Sriniketan Santiniketan Planning Area: Under West Bengal Town & Country (Plng & Dev) Act 1979.

Ministry of Jal Shakti. (2020a). *Uranium contamination in ground water.* https://pib.gov.in/PressReleasePage.aspx?PRID = 1606572.

Murthy, R. N. S., Srikonda, R., & Kasinath, I. V. (2022). Traditional water management systems of India. *ISVS e-Journal*, 9(2), 61−77. Available from. Available from https://isvshome.com/pdf/ISVS_9-2/ISVS9.2.5Satyanarayana.pdf.

O'malley, L. S. S. (1910). *Gazetteer of the Birbhum District. The Bengal Secretrariate Book Depot.* Available from https://archive.org/details/in.ernet.dli.2015.32685/page/n5/mode/2up.

Ministry of Jal Shakti. (2020b). *Per capita availability of water.* 2023 11 19. https://pib.gov.in/PressReleasePage.aspx?PRID = 1604871#: ~ :text = As%20per%20Ministry%20of%20Housing,benchmark%20for%20urban%20water%20supply.

Ponce, V. M., Lohani, A. K., & Huston, P. T. (1997). Surface albedo and water resources: Hydrological impact of human activities. *Journal of Hydrologic Engineering, ASCE, 2*(4), 197−203. Available from http://ponce.sdsu.edu/albedo197.html.

Prasad, B., & Pundir, B. S. (2020). Gondwana biostratigraphy and geology of West Bengal Basin, and its correlation with adjoining Gondwana basins of India and western Bangladesh. *Journal of Earth System Science, 129*(1), 22. Available from https://doi.org/10.1007/s12040-019-1287-2. Available from:, https://ui.adsabs.harvard.edu/abs/2020JESS.129.22P/abstract, https://ui.adsabs.harvard.edu/link_gateway/2020JESS.129.22P/.

RSP Green Development & Laboratories Pvt. Ltd. (2019). *District survey report of Birbhum District, West Bengal.* West Bengal Mineral Development & Trading Corporation Ltd (A Govt. of West Bengal Undertaking).

Shao, J., Si, B., & Jin, J. (2019). Rooting depth and extreme precipitation regulate groundwater recharge in the thick unsaturated zone: A case study. *Water, 11*(6), 1232. Available from https://www.mdpi.com/2073-4441/11/6/1232.

Tanwer, N., Anand, P., Batra, N., Kant, K., Gautam, Y. P., & Sahoo, S. K. (2023). Uranium in groundwater: Distribution and plausible chemo-radiological health risks owing to the long-term consumption of groundwater of Panchkula, Haryana, India. *Pollution, 9*(2), 821−838. Available from https://doi.org/10.22059/POLL.2023.352677.1726. Available from:, https://jpoll.ut.ac.ir/issue_11565_11714.html.

Tender Detail. (2023). https://www.tenderdetail.com/Indian-Tenders/TenderNotice/40874543/8f93c7d64912408b1144d06067776348.

Govt. of West Bengal. (2023). *Urban Development & Municipal Affairs Department.* Govt. of West Bengal. https://www.wburbanservices.gov.in/.

Water management sustainability evaluation at the river basin level: concept, methodology, and application

26

Md Nazirul Islam Sarker[1] and Md Naimur Rahman[2,3,4]

[1]*Miyan Research Institute, International University of Business Agriculture and Technology, Dhaka, Bangladesh*

[2]*Department of Geography, Hong Kong Baptist University, Kowloon, Hong Kong, P.R. China*

[3]*David C Lam Institute for East-West Studies, Hong Kong Baptist University, Kowloon, Hong Kong, P.R. China*

[4]*Department of Development Studies, Daffodil International University, Dhaka, Bangladesh*

26.1 Introduction

Water management sustainability focuses on an all-encompassing assessment of all interlinked variables (social, ecological, and engineering). Water is essential to our existence and forms the cornerstone of virtually all economic activity; therefore, its availability is central to sustainable development efforts (Wang, Davies, & Liu, 2019). The IPCC lists environmental concerns (such as extreme weather events, natural disasters, water crises, and failure of climate change mitigation policics) among its top global risks based on human impact (IPCC, 2014). According to the UN (2019), climate change will exacerbate water stress across more regions and increase water shortages already occurring in these areas. Changes to both the quantity and quality of water cycles, as well as an increase in extreme weather events, are among its primary effects on freshwater resources. Economic inequities, pollution, and geographic locations greatly limit access to essential water services like drinking water for humans (Meshesha & Abdi, 2019). As such, implementation of an integrated water resource management (IWRM) approach to address water shortages remains low across the region. An integrated approach seeks to integrate physical environment management within the larger political and socioeconomic framework (Gooch & Stålnacke, 2010). The river basin model seeks to implement IWRM principles with an emphasis on improving coordination among organizations that operate and manage water in a river basin, with particular attention paid to allocating

Applications of Geospatial Technology and Modeling for River Basin Management.
DOI: https://doi.org/10.1016/B978-0-443-23890-1.00026-8

and providing reliable water-dependent services more equitably (Radif, 1997). Cooperation among water-related participants, as well as identification of root problems and solutions, is necessary for achieving IWRM effectively and must lead to agreement among participants regarding common goals, which is crucial for IWRM implementation success (Komatina, 2011). Water management at a river basin scale enables an analysis of interconnections among land use, water availability, and ecosystem health. River basins facilitate coordinated water management practices across categories and administration boundaries to promote sustainability and efficiency in water usage, leading to greater sustainability and efficiency of usage (Ahmadi, Moridi, & Sarang, 2017; Kolokytha, 2022). River basins play an essential role in dealing with contamination, water shortage, climate change impacts, and more (Loi et al., 2022). Water basin-level assessments offer an invaluable opportunity to integrate ecological, economic, and social concerns holistically into water management decisions (Shafiei et al., 2022). River basins are natural hydrological units characterized by complex interactions between water, land, and human activities. At the basin level, water management practices can more closely align with sustainability concepts and ensure that their use does not negatively affect either ecosystem health or future generations' ability to meet their requirements. As several river basins span multiple jurisdictions and stakeholders, their transboundary nature demonstrates the need for cooperation among various parties involved and stakeholders in order to address issues of sustainable water management on this scale (Richter, Mathews, Harrison, & Wigington, 2016).

River basin water management is essential for many reasons, including holistic resource management by considering all interdependencies between land use, water availability, ecosystem health, and transboundary issues relating to water management as well as achieving the sustainable development goals (SDGs), such as SDG 6, which seeks to ensure equitable and sustainable administration of sanitation and water for all (Evaristo et al., 2023). This method plays a significant role in meeting these goals as it fosters holistic administration of resources like water by taking an ecosystem-centric approach (Everard, Reed, & Kenter, 2016). River basin management is considered to be an efficient solution to water management issues. A river basin refers to an area defined by streams' waters that contains all groundwater resources and surface waters within it (Houdret, Dombrowsky, & Horlemann, 2014). This strategy emphasizes the hydrological, ecological, and administrative significance of water management within the river's natural boundary basins (Wang et al., 2019). It facilitates water resource administration by considering interdependencies among land usage, water availability, and ecosystem health. The river basin approach helps coordinate management methods across different sectors as well as administrative borders to promote the effectiveness and sustainability of water management, which is particularly relevant in transboundary river basins where cooperation among various parties and authorities is critical to effective water management (Aimar & Erabti, 2015; Wang et al., 2021). Evaluating water-related practices for sustainable development presents unique challenges due to all of their interlinked ecological, social, and economic components that need to be considered. Complexity arises due to the necessity of reconciling economic development goals with conservation efforts for natural resources like landscapes and habitats that comprise river basins. Urbanization, climate change, and agricultural activities further complicate this equation, as their combined effects increase both water demand and pollution (Pan et al., 2023). As the assessment of innovation often centers on economic considerations alone, its assessment can overlook social and ecological aspects of sustainability, emphasizing the need for more comprehensive methods of evaluating water management methods (Deng, Guo, Yin, Zeng, & Chen, 2022).

Recent research highlights the complexity of evaluating water management sustainability at the basin level (Aivazidou et al., 2021; Alotaibi, Baig, Najim, Shah, & Alamri, 2023; Bezerra et al., 2022; Bouramdane, 2023), with different evaluation methods, such as indicator-based frameworks (De Stefano, Petersen-Perlman, Sproles, Eynard, & Wolf, 2017; Shafiei et al., 2022), participative approaches (Carr, 2015; Garau, Torralba, & Pueyo-Ros, 2021), and modeling methods (Bradford et al., 2020; Johnston & Smakhtin, 2014; Jumani et al., 2019)—each having their own set of advantages and disadvantages—being employed. Particularly, indicator-based assessments offer a quantitative measure of sustainability but do not accurately capture all aspects of ecological integrity and social equity (Shafiei et al., 2022). Participatory approaches involve people in the evaluation process to increase their value and acceptability; however, these can be time-consuming and difficult to implement on a large scale. The water crisis in our world is one of the greatest concerns of today, affecting millions worldwide. Water scarcity, pollution, and climate change's effect on freshwater sources all play an integral part in this issue. On Earth, approximately 70% of freshwater resources are utilized for agricultural purposes, yet this resource is threatened by overuse, pollution, and climate change (Cosgrove & Loucks, 2015). Climate change exacerbates this situation by altering precipitation patterns, decreasing snowpack and glacier sizes, and intensifying and increasing floods and droughts (Aimar & Erabti, 2015). Wang ct al. (2019) provided evidence of a global increase in river basins facing water shortage due to future pollution, prompting urgent action on water management strategies for the future. Furthermore, the Southern Hemisphere experienced a 20% reduction in available water over 20 years, highlighting global-scale water scarcity (Blöschl & Chaffe, 2023). Though evaluating water sustainability is of great significance to management, assessment methods often do not take into account economic, ecological, and social considerations when making their assessments. Most existing methods focus on one aspect of sustainability—such as water quantity or quality—without taking into account how water systems depend upon each other as well as wider socioeconomic and ecological considerations (Sultana & Mostafa, 2023). No clear consensus has emerged as to the most efficient and flexible approaches for sustainability assessment in various river basin settings. Though a variety of frameworks and tools, such as indicators-based assessments, participatory approaches, and modeling methods, have been proposed in order to address river basin management in different environments (Rodrigues et al., 2023), their effectiveness remains underexamined (Almulla & Nerini, 2021). Furthermore, literature shows a dearth of extensive studies employing these techniques across various geographical and socioeconomic contexts, which hinders our ability to fully comprehend real-world challenges associated with adopting sustainable water management practices or the transferability of effective strategies across basins (Alotaibi et al., 2023).

This study seeks to address several research issues by exploring these research topics in detail.

1. What are the most effective methodologies and indicators for evaluating the sustainability of water management practices at the river basin level?
2. How do policy frameworks and governance structures impact the sustainability of water management in river basins, and what strategies have proven successful in enhancing sustainability outcomes?

This study aims to explore current methods for assessing sustainability in river basin water management, evaluate their integration with sustainability factors, and assess their use across various contexts. This investigation seeks to identify areas of need and opportunities related to improving sustainable water management. This study provides an exhaustive synopsis of assessment methods that can be employed for sustainable water management at the river basin level. This chapter presents an in-depth examination of how sustainability considerations have been integrated into these approaches and features case studies that showcase their application across various scenarios. It stands out from similar works with its holistic approach of synthesizing methodologies and their applications to fill the gap between the theory and practice of sustainability-based water management. By identifying gaps in methodology and suggesting potential avenues for future study or application, the study advances sustainable water management practices at the river basin level.

26.2 Research methodology

26.2.1 Narrative review approach

Narrative review methodology was utilized in order to investigate various aspects of sustainability assessment for water management at the river basin level. A research protocol has been developed for the narrative review approach to guide the study (Table 26.1).

This method provided an in-depth examination of the current state of the research field while simultaneously identifying areas of need and providing suggestions for further study. It proved particularly effective when applied to

Table 26.1 Research protocol for narrative review.

Items	Description	Methodology/ approach	Expected outcome
Formulate the research question	Define the central question guiding the narrative review.	How can water management sustainability at the river basin level be effectively evaluated to incorporate ecological integrity, economic efficiency, and social equity?	A clear research question that directs the scope and focus of the narrative review.
Define inclusion and exclusion criteria	Establish criteria for selecting relevant literature and case studies.	Inclusion: Peer-reviewed articles, books, reports, and case studies published in the last decade, in English, focusing on sustainable water management at the river basin level. Exclusion: Non-peer-reviewed articles, outdated sources, and studies not directly related to the river basin level.	A set of criteria that ensures the relevance, credibility, and contribution of the selected sources to the research question.
Literature search	Conduct a comprehensive search for relevant literature across multiple databases and platforms.	Use of scholarly databases (e.g., Web of Science, Scopus, Google Scholar), search engines, and manual searches with keywords related to sustainable water management, river basin management, and evaluation methodologies.	A collection of literature that provides a broad understanding of the current state of research on the topic.
Literature selection	Apply the inclusion and exclusion criteria to select literature.	Screening of titles, abstracts, and full texts based on the defined criteria to determine the relevance of each source.	A curated list of sources that are pertinent to the research question and meet the established criteria.

(Continued)

Table 26.1 Research protocol for narrative review. *Continued*

Items	Description	Methodology/approach	Expected outcome
Data extraction	Extract key information and data from the selected literature.	Identification and recording of information related to water management sustainability evaluation methodologies, applications, challenges, and outcomes at the river basin level.	A comprehensive dataset that synthesizes the findings from the selected literature.
Data synthesis	Analyze and integrate findings from the collected data.	Thematic analysis to identify patterns, themes, and gaps in the literature. Comparative analysis to examine differences and similarities across studies.	A synthesized overview of the current methodologies, applications, challenges, and gaps in the field of water management sustainability evaluation.
Analysis and interpretation	Interpret the synthesized data to address the research question.	Discussion of the findings in the context of the research question, with consideration of the ecological, social, and economic dimensions of sustainability.	Insights and conclusions that contribute to the understanding of how water management sustainability can be effectively evaluated at the river basin level.
Recommendations for future research	Identify gaps and suggest areas for further investigation.	Based on the analysis and interpretation of the data, outline recommendations for future research directions.	A set of recommendations that address identified gaps and contribute to advancing the field.

sustainable water management—an essential element of environmental science that has to deal with issues of integration across ecological, economic, social, and ecological dimensions. The steps of the narrative review have been described in Table 26.2.

Table 26.2 Major steps of narrative review and description.

Steps of narrative review	Description
1. Identification of research question	How can water management sustainability at the river basin level be effectively evaluated to incorporate ecological integrity, economic efficiency, and social equity?
2. Literature search and selection	Conducting a thorough search of scholarly databases and other credible sources to identify relevant literature on sustainable water management practices, evaluation methodologies, and applications at the river basin level.
3. Data extraction and synthesis	Extracting pertinent information from the selected sources and synthesizing it to provide a coherent overview of current methodologies, their applications, and the challenges faced in sustainable water management at the river basin level.
4. Analysis and interpretation	Analyzing and interpreting the data in relation to the research question, focusing on key findings, significant trends, and the implications of current practices and methodologies for sustainable water management.

26.2.2 Selection criteria

26.2.2.1 Criteria for selecting literature and case studies

Specific criteria were used in selecting the documents to ensure the relevancy, credibility, and value of sources chosen for understanding sustainable water management evaluation (Table 26.3).

26.2.2.2 Specification of types of sources

Research sources included peer-reviewed books, peer-reviewed articles, reports, and case studies selected based on their extensive perspectives, methodologies, and results regarding environmentally sustainable management of water at the river basin level. Peer-reviewed papers were selected because of their rigorous review process, ensuring the validity of research findings and presentations made. Books were selected because of their insightful discussions of water sustainability. Reports provide in-depth analyses and research, offering substantial insight into sustainable practices for managing water. Case studies examine specific cases or practices that promote sustainable use at the river basin level and offer an in-depth understanding of their context.

26.2.3 Data collection

The process of data collection included gathering pertinent case studies and literature related to methods for sustainable water management as well as evaluation methods at the river basin level. The strategy used was multipronged and included

Table 26.3 Criteria for selecting literature and case studies.

Criteria	Description
Relevance	Selection of materials based on their direct connection to the evaluation of water management sustainability at the river basin level.
Credibility	Preference for sources published in peer-reviewed journals, by reputable publishers, or produced by recognized institutions and organizations.
Contribution	Selection of sources that offer valuable insights, facts, or debates enriching the understanding of sustainable water management practices and methodologies.

the use of scholarly databases, search engines, and manual searches in order to gather a vast quantity of pertinent information. Keywords and search terms employed include sustainable water management, river basin management, ecological sustainability, socioeconomic dimensions, and extensive strategies. Filters were employed to refine search results, with special attention paid to sources published within the last decade, peer-reviewed publications from respected publishers, research reports from credible institutions, and publications written exclusively in English for accessibility reasons.

26.2.4 Data analysis

Once all pertinent sources were assembled, data analysis used both thematic analyses to isolate themes prevalent throughout and comparative studies to explore differences and similarities among methods and their applications in river basin environments. As a result of the analyses performed, in-depth knowledge was acquired regarding this subject matter.

26.2.5 Approach to assessing impact, effectiveness, and challenges

Sustainability evaluations for water management methods were examined through an assessment of their application in different cases, challenges encountered when applying them, and results that were achieved. This study sought to identify areas for improvement and offer sustainable practices for water management at the basin level. To assess the efficacy of methods to manage water sustainability at the basin level, we have established four main criteria. They are comprehensiveness, application scalability, applicability, and ease of use. Comprehensiveness refers to the extent to which a method covers economic, environmental, and social dimensions of sustainability. Scope of application examines scenarios where this approach can be effectively utilized, considering various geographic areas and challenges of water management. Scalability assesses its ability to adjust itself without significant modifications for various river basin sizes. Furthermore, implementation ease takes into account practical concerns, including expertise requirements and data availability, resource requirements, and any required training programs or workshops. These assessments combine quantitative scoring methods with qualitative evaluation as part of a systematic research process and expert consultations in order to provide an objective and balanced evaluation of each methodology's strengths and weaknesses.

26.3 Results

26.3.1 Effective methodologies and indicators for evaluating sustainability

26.3.1.1 Overview of identified methodologies

This study has presented several methods uncovered through a literature review to evaluate the effectiveness of water management strategies at a river basin level. Methodologies were judged on factors like comprehensiveness, applicability, and ease of use before outlining their strengths and weaknesses within river basin management contexts (Table 26.4).

26.3.1.2 Key indicators for sustainability evaluation

This research explores the efficacy of various indicators used to assess the sustainability of practices for managing water at the river basin level. Discussion of these indicators includes their ability to capture various aspects of sustainability such as ecological integrity, economic efficiency, and social equity; additionally, it highlights commonly used indicators by providing insights into their prevalence and utility based on literature review and case study analyses (Table 26.5).

Sustainability indicators for water management are integral in evaluating and improving practices at the basin level. Water Quality Index, water use efficiency, and other similar indicators provide powerful measures that help assess and guide the improvement of basin practices, like protecting ecological integrity while increasing economic efficiency (Richter et al., 2016). They offer quantitative measures that are easy to interpret, making them great tools for both policymakers and water managers alike. Indices are essential in developing an in-depth knowledge of sustainability; however, their application poses unique challenges relating to data collection, interpretation, and use. Their efficacy often depends on local environmental conditions as well as accessing accurate ecological and social information; regardless of these limitations, they provide invaluable insight into social and ecological aspects of sustainability, emphasizing the necessity of extensive evaluation frameworks encompassing an array of indicators.

Table 26.4 Comparison of methodologies for evaluating water management sustainability.

Methodology	Comprehensiveness	Applicability	Ease of implementation	Strengths	Limitations	Sources
Quantitative analysis	High	Wide	Moderate	Accurate, objective measurements; facilitates benchmarking and trend analysis	Requires extensive data; may not capture qualitative aspects	Bezerra et al. (2022), Shafiei et al. (2022), and Wang et al. (2014)
Qualitative analysis	Moderate	Specific cases	High	Provides in-depth insights; captures social and cultural dimensions	Subjective; difficult to generalize findings	De Stefano et al. (2017), Kolokytha (2022), and Sultana and Mostafa (2023)
Mixed methods	High	Wide	Moderate	Combines strengths of quantitative and qualitative; holistic understanding	Complex to design and implement; time-consuming	Lebel, Nikitina, Pahl-Wostl, and Knieper (2013) and Segbefia, Honlah, Appiah, and Yildiz (2019)
Life cycle assessment	High	Specific products/processes	Low	Comprehensive environmental impact assessment	Resource-intensive; requires specific expertise	Sharma, Grant, Grant, Pamminger, and Opray (2009)
Participatory approaches	Moderate	Local	High	Engages stakeholders; enhances social acceptability	It may not be comprehensive, influenced by stakeholder biases.	Furber, Medema, Adamowski, Clamen, and Vijay (2016), Garau et al. (2021), and Meshesha and Abdi (2019)

Table 26.5 Effectiveness of key sustainability indicators.

Indicator	Sustainability pillar	Effectiveness	Common usage	Rationale	Sources
Water Quality Index	Ecological integrity	High	Very common	Provides a comprehensive measure of water quality by integrating multiple parameters; widely recognized and easy to communicate.	Islam, Rahman, Ritu, Rahman, and Sarker (2024) and Loi et al. (2022)
Biodiversity-related index	Ecological integrity	Medium	Common	Reflects the health of aquatic ecosystems; however, it requires extensive data collection and expertise.	Jumani et al. (2022) and Sarkar, Gupta, and Lakra (2010)
Water use efficiency	Economic efficiency	High	Very common	Directly relates to the sustainable use of water resources; easily quantifiable and applicable across sectors.	Greve et al. (2018), Li, Wang, Li, and Liu (2022), and Nkhonjera (2017)
Access to clean water	Social equity	High	Very common	A direct indicator of social well-being and equity; straightforward to measure and universally applicable.	Bouramdane (2023) and Evaristo et al. (2023)
Water-related conflict incidence	Social equity	Low	Less common	Important for understanding social tensions; however, it can be not easy to quantify and interpret.	Furber et al. (2016) and Ncube (2022)

26.3.2 Impact of policy frameworks and governance structures

26.3.2.1 Policy frameworks and their impact

Long-term water management sustainability depends heavily on multiple frameworks of policy at both national and local levels (Nikitina, Ostrovskaya, & Fomenko, 2010). These frameworks provide regulatory and legal structures for managing water resources and address questions concerning their quality, allocation to ecosystem conservation efforts, rights of stakeholders, and more. This section highlights the most pertinent policy frameworks identified by research that impact sustainable water management practices within river basins. This section investigates the effects of different policies on sustainable methods of water management within river basins. This analysis includes both positive and negative outcomes, as well as any challenges, by looking at instances from various river basins as examples illustrating actual consequences (Table 26.6).

Analysis has demonstrated that the policy framework is of primary importance in determining sustainable methods of water management across different river basins. International agreements such as those of the UN Watercourse Convention have demonstrated their capacity to encourage cross-border collaboration, such as that found within the Mekong River Basin (Zeitoun, Goulden, & Tickner, 2013). However, they often face difficulty when trying to balance all the needs of their respective nations. Similar to that, the European Water Framework Directive has led to significant improvements in both water quality and ecosystem health in basins such as the Danube. Unfortunately, implementation levels vary among EU member states, which has prevented its full potential from being realized.

26.3.2.2 Governance structures and sustainability outcomes

Decentralized and centralized models of sustainable water management for rivers vary considerably, from community-based management systems and public–private partnerships (PPPs) to community-based models and PPPs (Kauffman, 2015). Each type of strategy plays an essential role in determining efficiency and sustainability. This section presents various research models, which influence outcomes of sustainability in river basins (Table 26.7).

Governance structures governing water resource management within basins have an immense effect on their sustainability outcomes. Centralized models, like those found in the Jordan River Basin, offer numerous advantages when it comes to decision-making processes and implementation policies (Evaristo et al., 2023). However, they do not adequately take into account the unique circumstances and requirements of all stakeholders in each region. Decentralized governance like that seen in Brazil's Piracicaba River Basin encourages local participation while offering more flexibility. However, in order to be truly effective throughout its entirety, an effective mechanism for coordination and support must exist that ensures its overall efficacy. Management practices rooted in community knowledge and sustainable living such as those employed by Indonesia's Brantas River Basin, clearly prove its worth (Evaristo et al., 2023). These models rely on community empowerment and involvement yet can face obstacles due to a lack of

Table 26.6 Analysis of policy frameworks' impact on water management sustainability.

Policy framework	Positive outcomes	Challenges	Sources
UN watercourse convention	Facilitated international cooperation and joint management efforts.	Difficulty in resolving conflicts due to varying national interests.	Zeitoun et al. (2013)
European Water Framework Directive	Improved water quality and ecosystem health through integrated management.	Implementation disparities among member states affect overall effectiveness.	Rahaman et al. (2004) and van der Meulen, van de Ven, van Oel, Rijnaarts, and Sutton (2023)
Water-related law	Enhanced water allocation efficiency and environmental water recovery.	Balancing agricultural demands with environmental needs remains challenging.	Nikitina et al. (2010)
Local water conservation ordinances	Reduced water consumption and increased conservation awareness at the local level.	Insufficient to address broader basin-wide sustainability issues alone.	Nikitina et al. (2010)
Ramsar convention on wetlands	Promoted the conservation and wise use of wetlands through international cooperation.	Challenges in integrating wetland conservation with economic development needs.	Kolokytha (2022)
Convention on Biological Diversity	Supported biodiversity conservation and sustainable use of biological resources.	Difficulty in enforcing measures and coordinating between countries.	Orchard-Webb, Kenter, Bryce, and Church (2016)
SDGs	Encouraged holistic water management practices aligned with global sustainability goals.	Complex socioeconomic dynamics complicate the achievement of SDG targets.	Evaristo et al. (2023) and Shafiei et al. (2022)
Integrated water resources management policies	Fostered a comprehensive approach to water resource management across sectors.	Varied levels of political commitment and resource availability among riparian states.	Savenije and Van der Zaag (2008) and Wang et al. (2019)

SDGs, *Sustainable development goals.*

funds or technical knowledge. PPPs bring the efficiency and creativity of private investment into water management projects (Evaristo et al., 2023). While PPPs can provide infrastructure and capacity, care should be taken to ensure economic interests do not interfere with long-term objectives and public welfare.

Table 26.7 Overview of governance structures and their impact on water management sustainability.

Governance structure	Description	Impact on sustainability	Source
Centralized	A top-down approach where decision-making is concentrated at the national or regional government level.	It can ensure uniform policies and standards but may lack local specificity and stakeholder engagement.	Kauffman (2015) and Megdal, Eden, and Shamir (2017)
Decentralized	Decision-making is distributed among local governments or basin organizations, allowing for more tailored and responsive management.	Promotes local engagement and adaptability but may face challenges in coordination and resource allocation.	Kauffman (2015)
Community-based management	Management by local communities, often through cooperatives or associations, focusing on participatory approaches.	Enhances local stewardship and knowledge but may lack resources and technical expertise.	Bezerra et al. (2022) and Mankad and Tapsuwan (2011)
Public–private partnerships	Collaboration between government entities and the private sector for financing, building, and operating water management projects.	Can leverage private sector efficiency and innovation but risks prioritizing profit over public interest.	Evaristo et al. (2023) and Nikitina et al. (2010)

26.3.2.3 Governance strategies and their impact on sustainability

This research explores governance strategies that have proven successful at increasing sustainability outcomes for water management across different river basins. Of particular interest are strategies involving stakeholders and cross-sector coordination, as well as adaptive management practices that foster sustainable practices (Table 26.8).

26.4 Discussion

26.4.1 Interpretation of findings

26.4.1.1 Contextualizing results

Our findings align with broader discussions surrounding IWRM and sustainability concepts in water management, such as those expressed by Guo's perspective on river basin management (Guo, 2023). Our study's emphasis on cross-sectoral

Table 26.8 Strategies for enhancing sustainability in water management.

Strategy	Description	Impact on sustainability	Source
Stakeholder engagement	Involves the active participation of all stakeholders, including local communities, industries, and governments, in water management decisions.	Enhances the relevance and acceptance of water management policies by ensuring they meet the needs and concerns of all affected parties.	Lim et al. (2022) and Megdal et al. (2017)
Intersectoral coordination	Coordination among different sectors (e.g., agriculture, industry, and urban development) that impact and are impacted by water management.	Prevents conflicting uses and promotes integrated approaches to water resource management, balancing economic development with conservation.	Masthura, Wignyosukarto, Fahriana, and Ardhyan (2023)
Adaptive management	A flexible approach to water management that allows for adjustments based on monitoring outcomes and changing conditions.	Improves the resilience of water management systems to uncertainties and changes, such as climate variability.	Barrios, Rodríguez-Pineda, and De La Maza Benignos (2011) and Pahl-Wostl et al. (2007)
Integrated water resources management	A process that promotes the coordinated development and management of water, land, and related resources.	Encourages sustainable water use, efficient allocation, and protection of water resources across all uses and scales.	Abubakirova et al. (2017) and Wang et al. (2019)
Public–private partnerships	Collaborations between government entities and the private sector for financing, building, and operating water management projects.	Leverages private sector resources and expertise for innovative and efficient water management solutions.	Evaristo et al. (2023) and Nikitina et al. (2010)

coordination and stakeholder participation is in line with the Dublin Principles from the 1992 International Conference on Water and the Environment; they were further expanded through Global Water Partnership to encourage real involvement of stakeholders at every level, along with decision-making at more basic levels (Rahaman, Varis, Kajander, & Water, 2004). This proves the value of our research methodology and findings within these theoretical frameworks, further supporting that our approach to measuring the sustainability of water management is both valid and trustworthy.

26.4.1.2 Innovative practices and lessons learned

This study revealed innovative techniques for adapting adaptive strategies to changes in climate that align with Davis' call for greater participation and local control over water management (Marks & Davis, 2012). These strategies emphasize the necessity of being flexible and responsive to changes in our environment. Meshesha and Abdi (2019) discussed the difficulties and opportunities in implementing IWRM in Ethiopia's Omo-Gibe Basin. Our findings about the efficiency of governance frameworks align perfectly with their findings about implementation difficulties. These insights may prove extremely helpful when improving sustainability evaluation criteria, in particular, including adaptability, stakeholder inclusion, and cross-sector collaboration capabilities, as future criteria should include them explicitly.

26.4.1.3 Implications for sustainability criteria

Analyzing our findings within the existing studies, it becomes evident that water management could become significantly more sustainable through the implementation of strategies for governance involving stakeholder involvement as well as intersectoral coordination and flexible management. Kazakhstan was remarkable in its success at installing IWRMs despite numerous transboundary water issues (Abubakirova, Tanybaeva, Pavlichenko, & Rysmagambetova, 2017), yet 2018 provided further proof of this success story. These cases illustrated how governance structures influence long-term results while changing management concerns presented a distinct set of evaluation challenges, showing why flexible standards that encourage collaboration between governance levels and stakeholder levels for evaluation are vital to sustainable evaluation processes.

26.4.1.4 Integration with global sustainability goals

Our research on sustainable basin-level water supply management will have an enormous effect in meeting several SDGs, such as SDG 6 (Clear Sanitation and Water), 13 (Climate Action), and 14 (Life Below Water) (Biswas, Dandapat, Alam, & Satpati, 2022). It shows just how vital sustainable practices for water management are worldwide, while more effective management practices and innovative water management methods could significantly facilitate their realization.

Our research findings will contribute to meeting SDG 6, which seeks to ensure access to sustainable sanitation and water supply systems for all people worldwide. Biswas et al. (2022) examined India's progress toward SDG 6, which highlighted its critical need to sustainably manage water to ensure access to clean and safe drinking water as well as sanitation. The research also highlighted efforts from both state and national government entities toward meeting this target. Additionally, our research identified challenges and strategies that address water resource sustainability are integral parts of addressing climate change (SDG 13) while safeguarding aquatic life (SDG 14). Evaristo et al. (2023) noted that the challenges associated with meeting SDG 6 may extend far beyond infrastructure: governance, financial policy making, and protecting ecosystems against climate change are vital elements to achieve its fulfillment. Our research indicates that an ecologically sustainable water management system with respect to governance structure and stakeholder engagement can significantly mitigate climate change impacts, while contributing to aquatic ecosystem preservation, furthering global community environmental goals, and contributing to ecological sustainability goals. Our results, informed by global sustainability targets, demonstrate the need to take holistic and integrated approaches to water management. By aligning practices with SDG goals 6–13-14, policymakers, as well as experts, can ensure their efforts enhance river basin sustainability while aiding nations in building more resilient societies.

26.4.2 Theoretical contribution

Water management sustainability analysis helps identify gaps, trends, and emerging research themes that contribute significantly to our understanding of sustainability in water management (Abubakirova et al., 2017). Our findings contribute to the theory of sustainable water management by emphasizing the significance of systems-based methods and multidisciplinary approaches. Elinor Ostrom's framework of social-ecological systems in our study emphasizes the significance of water management within its larger environment of ecological and social systems (Ostrom, 2009). This trend reflects an increased understanding of the necessity for knowledge accumulation and theoretical research that crosses traditional disciplinary boundaries. Through highlighting innovative methods and strategies that enhance sustainability in governance, our study contributes to shaping the ever-evolving theory surrounding water management. It recommends the inclusion of environmental considerations into management research as a means of increasing sustainable management of water supplies, thus contributing to shaping its future direction.

26.4.3 Policy and practice implications

26.4.3.1 Recommendations for policymakers

To improve the sustainability of water management at the basin level, policymakers are advised to adopt IWRM strategies that are inclusive, participative, and adaptive, taking into account both ecological and economic considerations. Strengthening transboundary water cooperation through regional and international agreements that facilitate conflict management jointly, as well as for conflict resolution, is of utmost importance. Enhancing transboundary cooperation on water through international and regional agreements that assist with managing conflicts jointly, as well as conflict resolution, is of utmost importance. Engaging stakeholders such as local communities, industries, and NGOs in decision-making processes will significantly enhance both the efficiency and value of water-management strategies. It is good to promote effective water usage and conservation across industries through incentives and regulations and invest in ecofriendly infrastructure like wastewater treatment or irrigation systems as essential measures to ensure the long-term sustainability of water resources.

26.4.3.2 Strategies for practitioners

Professionals in water management can promote sustainability with strategies of flexible management that allow for informed decisions when monitoring water resources continuously. Community-based management provides local stakeholders with a means of taking on sustainable stewardship locally as well as regionally. Utilizing cutting-edge technologies like remote sensors and GIS will assist in monitoring water and supporting data-driven decision-making processes. Collaboration among sectors can break down silos between agriculture, urban development, and environmental protection to create more integrated solutions to water management. The promotion of policy support enables experts to use their knowledge of sustainable management of water resources to assist with sustainable management strategies for water sources, which helps preserve their long-term health and availability.

26.5 Conclusion

Consideration of the basin area for planning water resources and management is integral in IWRM. IWRM leads to sustainable development, improved allocation, monitoring the use of resources relative to environmental, economic and social objectives, and monitoring performance over time. Isolated management approaches of individual sectors—industrial, domestic, or environmental—often create inconsistency, which ultimately undermines sustainable systems within the basin region. This article reviews various elements of basin-level sustainability assessments of water management by integrating concepts, techniques, and applications through an overview of the literature. Our research has illustrated the significance of adopting IWRM concepts that involve all stakeholders across sectors and adaptive methods in improving the efficiency of water management. Various governance and policy models were evaluated in detail in order to identify strategies that contribute toward more sustainable practices in water management that meet global sustainability goals such as SDGs. This article focused on evaluating the sustainability of water management on a basin level, providing an in-depth examination of current practices, their applications, and governance frameworks used in creating sustainable methods. Through narrative analysis of this topic, we highlighted the delicate balance needed to achieve sustainable water management, keeping ecological integrity intact while simultaneously increasing economic efficiency and social equity. Furthermore, our findings point toward holistic strategies that go beyond traditional management methods by being mindful of how humans and nature coexist within river basins. Frameworks for policies and governance have demonstrated their influence over water management practices. By providing examples of effective techniques, this article offers valuable insights into how policies and governance structures impact sustainability efforts—or even undermine them altogether. Furthermore, their integration into global sustainability goals, such as those outlined by SDGs, illustrates just how essential water management is in achieving wider environmental and societal objectives. Although this study's findings are insightful, its research also acknowledges its limitations, such as narrative review methods' inherent biases or difficulties generalizing findings across socio-political or geographic contexts. To address such limitations, future research should focus on several key issues, such as evaluating the efficacy of identified techniques or governance strategies across settings or researching new issues related to water management, such as urbanization impacts and climate change effects or technological advancements that ensure long-term water resource resilience in all areas that warrant further exploration (Lim, Wong, Elfithri, & Teo, 2022; Wang, Luo, Zhang, & Xia, 2014).

References

Abubakirova, K., Tanybaeva, A., Pavlichenko, L., & Rysmagambetova, A. (2017). Integrated water resources management in the Republic of Kazakhstan: Problems and prospects. *Journal of Geography and Environmental Management*, *47*(4), 23−31. Available from https://doi.org/10.26577/jgem.2018.2.433.

Ahmadi, A., Moridi, A., & Sarang, A. (2017). Integrated planning of water resources based on sustainability indices, a case study: Hamoon-Jazmorian basin. *Environmental Energy and Economic Research*, *1*. Available from https://doi.org/10.22097/eeer.2017.46457.

Aimar, A., & Erabti, H. (2015). Water scarcity and climate change: The search for an optimal and sustainable management (pp. 1−19). Available from https://doi.org/10.12816/0020942.

Aivazidou, E., Banias, G., Lampridi, M., Vasileiadis, G., Anagnostis, A., Papageorgiou, E., & Bochtis, D. (2021). Smart technologies for sustainable water management: An urban analysis. *Sustainability*, *13*(24), 13940. Available from https://doi.org/10.3390/su132413940.

Almulla, Y., & Nerini, F. F. (2021). The role of shared water management to achieve the sustainable development goals. *Sweden Research Square*. Available from https://doi.org/10.21203/rs.3.rs-1000184/v1. Available from, https://www.researchsquare.com/browse.

Alotaibi, B. A., Baig, M. B., Najim, M. M. M., Shah, A. A., & Alamri, Y. A. (2023). Water scarcity management to ensure food scarcity through sustainable water resources management in Saudi Arabia. *Sustainability (Switzerland)*, *15*(13). Available from https://doi.org/10.3390/su151310648. Available from, http://www.mdpi.com/journal/sustainability/.

Barrios, J. E., Rodríguez-Pineda, J. A., & De La Maza Benignos, M. (2011). Integrated river basin management in the Conchos River basin, Mexico: A case study of freshwater climate change adaptation. *Climate and Development*, *1*(3), 249−260. Available from https://doi.org/10.3763/cdev.2009.0024.

Bezerra, M. O., Vollmer, D., Acero, N., Marques, M. C., Restrepo, D., Mendoza, E., ... Serrano, L. (2022). Operationalizing integrated water resource management in Latin America: Insights from application of the freshwater health index. *Environmental Management*, *69*(4), 815−834. Available from https://doi.org/10.1007/s00267-021-01446-1. Available from, http://link.springer.de/link/service/journals/00267/index.htm.

Biswas, S., Dandapat, B., Alam, A., & Satpati, L. (2022). India's achievement towards sustainable Development Goal 6 (Ensure availability and sustainable management of water and sanitation for all) in the 2030 Agenda. *BMC Public Health*, *22*(1). Available from https://doi.org/10.1186/s12889-022-14316-0. Available from, https://bmcpublichealth.biomedcentral.com/.

Blöschl, G., & Chaffe, P. L. B. (2023). Water scarcity is exacerbated in the south. *American Association for the Advancement of Science, Austria Science*, *382*(6670), 512−513. Available from https://doi.org/10.1126/science.adk8164. Available from, https://www.science.org/doi/epdf/10.1126/science.adk8164.

Bouramdane, A. A. (2023). Optimal water management strategies: Paving the way for sustainability in smart cities. *Smart Cities*, *6*(5), 2849−2882. Available from https://doi.org/10.3390/smartcities6050128. Available from, https://www.mdpi.com/journal/smartcities.

Bradford, L., Thapa, A., Duffy, A., Hassanzadeh, E., Strickert, G., Noble, B., & Lindenschmidt, K. E. (2020). Incorporating social dimensions in hydrological and water quality modeling to evaluate the effectiveness of agricultural beneficial management practices in a Prairie River Basin. *Environmental Science and Pollution Research*, *27*(13), 14271−14287. Available from https://doi.org/10.1007/s11356-019-06325-1. Available from, https://link.springer.com/journal/11356.

Carr, G. (2015). Stakeholder and public participation in river basin management—An introduction. *Wiley Interdisciplinary Reviews: Water*, *2*(4), 393−405. Available from https://doi.org/10.1002/WAT2.1086. Available from, http://wires.wiley.com/WileyCDA/WiresJournal/wisId-WAT2.html.

Cosgrove, W. J., & Loucks, D. P. (2015). Water management: Current and future challenges and research directions. *Water Resources Research*, *51*(6), 4823−4839. Available from https://doi.org/10.1002/2014WR016869. Available from, http://onlinelibrary.wiley.com/journal/10.1002/(ISSN)1944-7973.

De Stefano, L., Petersen-Perlman, J. D., Sproles, E. A., Eynard, J., & Wolf, A. T. (2017). Assessment of transboundary river basins for potential hydro-political tensions. *Global Environmental Change*, *45*, 35−46. Available from https://doi.org/10.1016/j.gloenvcha.2017.04.008. Available from, http://www.elsevier.com/inca/publications/store/3/0/4/2/5.

Deng, L., Guo, S., Yin, J., Zeng, Y., & Chen, K. (2022). Multi-objective optimization of water resources allocation in Han River basin (China) integrating efficiency, equity and sustainability. *Nature Research, China Scientific Reports*, *12*(1). Available from https://doi.org/10.1038/s41598-021-04734-2. Available from, http://www.nature.com/srep/index.html.

Evaristo, J., Jameel, Y., Tortajada, C., Wang, R. Y., Horne, J., Neukrug, H., . . . Biswas, A. (2023). Water woes: the institutional challenges in achieving SDG 6. *Sustainable Earth Reviews*, *6*(1). Available from https://doi.org/10.1186/s42055-023-00067-2.

Everard, M., Reed, M. S., & Kenter, J. O. (2016). The ripple effect: Institutionalising pro-environmental values to shift societal norms and behaviours. *Ecosystem Services*, *21*, 230−240. Available from http://www.journals.elsevier.com/ecosystem-services/, 10.1016/j.ecoser.2016.08.001.

Furber, A., Medema, W., Adamowski, J., Clamen, M., & Vijay, M. (2016). Conflict management in participatory approaches to water management: A case study of Lake Ontario and the St. Lawrence River Regulation. *Water*, *8*(7), 280. Available from https://doi.org/10.3390/w8070280.

Garau, E., Torralba, M., & Pueyo-Ros, J. (2021). What is a river basin? Assessing and understanding the sociocultural mental constructs of landscapes from different stakeholders across a river basin. *Landscape and Urban Planning*, *214*, 104192. Available from https://doi.org/10.1016/j.landurbplan.2021.104192.

Gooch, G. D., & Stålnacke, P. (2010). *Science, policy and stakeholders in water management: An integrated approach to river basin management* (pp. 1−179). Taylor and Francis. Available from https://www.routledge.com/Science-Policy-and-Stakeholders-in-Water-Management-An-Integrated-Approach/Gooch-Stalnacke/p/book/9780415853415, 10.4324/9781849775151.

Greve, P., Kahil, T., Mochizuki, J., Schinko, T., Satoh, Y., Burek, P., . . . Wada, Y. (2018). Global assessment of water challenges under uncertainty in water scarcity projections. *Nature Sustainability*, *1*(9), 486−494. Available from https://doi.org/10.1038/s41893-018-0134-9. Available from, http://www.nature.com/natsustain/.

Guo, Q. (2023). Strategies for a resilient, sustainable, and equitable Mississippi River basin. *River*, *2*(3), 336−349. Available from https://doi.org/10.1002/rvr2.60. Available from, https://onlinelibrary.wiley.com/journal/27504867.

Houdret, A., Dombrowsky, I., & Horlemann, L. (2014). The institutionalization of River Basin Management as politics of scale − Insights from Mongolia. *Journal of Hydrology*, *519*, 2392−2404. Available from https://doi.org/10.1016/j.jhydrol.2013.11.037. Available from, http://www.elsevier.com/inca/publications/store/5/0/3/3/4/3.

IPCC. (2014). Full report part A: Global and sectoral aspects. (Contribution of Working Group II to the Fifth Assessment Report of the Intergovernmental Panel on Climate Change). In *Climate change 2014: Impacts, adaptation, and vulnerability*. Available from: https://www.ipcc.ch/pdf/assessment-report/ar5/wg2/WGIIAR5-FrontMatterA_FINAL.pdf.

Islam, M. S., Rahman, M. N., Ritu, N. S., Rahman, M. S., & Sarker, M. N. I. (2024). Impact of COVID-19 on urban environment in developing countries: Case study and environmental sustainability strategy in Bangladesh. *Green Technologies and Sustainability*, 100074.

Johnston, R., & Smakhtin, V. (2014). Hydrological modeling of large river basins: How much is enough? *Water Resources Management*, *28*(10), 2695−2730. Available from https://doi.org/10.1007/s11269-014-0637-8. Available from, http://www.wkap.nl/journalhome.htm/0920-4741.

Jumani, S., Deitch, M. J., Kaplan, D., Anderson, E. P., Krishnaswamy, J., Lecours, V., & Whiles, M. R. (2019). River fragmentation and flow alteration metrics: A review of methods and directions for future research. *Environmental Research Letters*, *15*(12). Available from https://doi.org/10.1088/1748-9326/abcb37. Available from, https://iopscience.iop.org/article/10.1088/1748-9326/abcb37.

Jumani, S., Deitch, M. J., Valle, D., Machado, S., Lecours, V., Kaplan, D., . . . Howard, J. (2022). A new index to quantify longitudinal river fragmentation: Conservation and management implications. *Ecological Indicators*, *136*, 108680. Available from https://doi.org/10.1016/j.ecolind.2022.108680.

Kauffman, G. J. (2015). Governance, policy, and economics of intergovernmental river basin management. *Water Resources Management*, *29*(15), 5689–5712. Available from https://doi.org/10.1007/s11269-015-1141-5. Available from, http://www.wkap.nl/journalhome.htm/0920-4741.

Kolokytha, E. (2022). Adaptation: A vital priority for sustainable water resources management. *Water*, *14*(4), 531. Available from https://doi.org/10.3390/w14040531.

Komatina, D. (2011). *Integrated water resources management as a basis for sustainable development − The case of the Sava River Basin*. InTech. Available from 10.5772/29528.

Lebel, L., Nikitina, E., Pahl-Wostl, C., & Knieper, C. (2013). Institutional fit and river basin governance: A new approach using multiple composite measures. *Ecology and Society*, *18*(1). Available from https://doi.org/10.5751/ES-05097-180101. Available from, http://www.ecologyandsociety.org/vol18/iss1/art1/ES-2012-5097.pdf.

Li, P., Wang, D., Li, W., & Liu, L. (2022). Sustainable water resources development and management in large river basins: an introduction. *Environmental Earth Sciences*, *81*(6). Available from https://doi.org/10.1007/s12665-022-10298-9. Available from, https://link.springer.com/journal/12665.

Lim, C. H., Wong, H. L., Elfithri, R., & Teo, F. Y. (2022). A review of stakeholder engagement in integrated river basin management. *Water (Switzerland)*, *14*(19). Available from https://doi.org/10.3390/w14192973. Available from, http://www.mdpi.com/journal/water.

Loi, J. X., Chua, A. S. M., Rabuni, M. F., Tan, C. K., Lai, S. H., Takemura, Y., & Syutsubo, K. (2022). Water quality assessment and pollution threat to safe water supply for three river basins in Malaysia. *Science of the Total Environment*, *832*. Available from https://doi.org/10.1016/j.scitotenv.2022.155067. Available from, http://www.elsevier.com/locate/scitotenv.

Mankad, A., & Tapsuwan, S. (2011). Review of socio-economic drivers of community acceptance and adoption of decentralised water systems. *Journal of Environmental Management*, *92*(3), 380–391. Available from https://doi.org/10.1016/j.jenvman.2010.10.037. Available from, https://www.sciencedirect.com/journal/journal-of-environmental-management.

Marks, S. J., & Davis, J. (2012). Does user participation lead to sense of ownership for rural water systems? Evidence from Kenya. *World Development*, *40*(8), 1569–1576.

Masthura, L., Wignyosukarto, B. S., Fahriana, N., & Ardhyan, M. Z. (2023). Keterpaduan Lintas Sektoral Dalam Pengembangan Kebijakan Integrated Water Resources Management (IWRM) pada Wilayah Sungai Aceh Meureudu Provinsi Aceh. *Jurnal Daur Lingkungan*, *6*(1), 40. Available from https://doi.org/10.33087/daurling.v6i1.199.

Megdal, S., Eden, S., & Shamir, E. (2017). Water governance, stakeholder engagement, and sustainable water resources management. *Water, 9*(3), 190. Available from https://doi.org/10.3390/w9030190.

Meshesha, Y. B., & Abdi, M. B. (2019). Challenges and opportunities for implementation of integrated water resource management in Omo-Gibe Basin, Ethiopia. *Journal of Ecology and the Natural Environment, 11*(7), 84−97. Available from https://doi.org/10.5897/jene2019.0747.

van der Meulen, E. S., van de Ven, F. H. M., van Oel, P. R., Rijnaarts, H. H. M., & Sutton, N. B. (2023). Improving suitability of urban canals and canalized rivers for transportation, thermal energy extraction and recreation in two European delta cities. *Ambio, 52*(1), 195−209. Available from https://doi.org/10.1007/s13280-022-01759-3. Available from, https://www.springer.com/journal/13280.

Ncube, B. (2022). *Indigenous knowledge perspectives on water management and its challenges in South Africa* (pp. 227−242). Elsevier BV. Available from 10.1016/b978-0-12-824538-5.00012-1.

Nikitina, E., Ostrovskaya, E., & Fomenko, M. (2010). Towards better water governance in river basins: Some lessons learned from the Volga. *Regional Environmental Change, 10*(4), 285−297. Available from https://doi.org/10.1007/s10113-009-0092-x.

Nkhonjera, G. K. (2017). Understanding the impact of climate change on the dwindling water resources of South Africa, focusing mainly on Olifants River basin: A review. *Environmental Science and Policy, 71*, 19−29. Available from https://doi.org/10.1016/j.envsci.2017.02.004. Available from, http://www.elsevier.com/wps/find/journaldescription.cws_home/601264/description#description.

Orchard-Webb, J., Kenter, J. O., Bryce, R., & Church, A. (2016). Deliberative democratic monetary valuation to implement the ecosystem approach. *Ecosystem Services, 21*, 308−318. Available from https://doi.org/10.1016/j.ecoser.2016.09.005. Available from, http://www.journals.elsevier.com/ecosystem-services/.

Ostrom, E. (2009). A general framework for analyzing sustainability of social-ecological systems. *Science (New York, N.Y.), 325*(5939), 419−422. Available from https://doi.org/10.1126/science.1172133.

Pahl-Wostl, C., Sendzimir, J., Jeffrey, P., Aerts, J., Berkamp, G., & Cross, K. (2007). Managing change toward adaptive water management through social learning. *Ecology and Society, 12*(2). Available from https://doi.org/10.5751/ES-02147-120230. Available from, http://www.ecologyandsociety.org/vol12/iss2/art30/ES-2007-2147.pdf.

Pan, Z., Gao, G., Fu, B., Liu, S., Wang, J., He, J., & Liu, D. (2023). Exploring the historical and future spatial interaction relationship between urbanization and ecosystem services in the Yangtze River Basin, China. *Journal of Cleaner Production, 428*, 139401. Available from https://doi.org/10.1016/j.jclepro.2023.139401.

Radif, A. A. (1997). Integrated water resources management (IWRM): an approach to face the challenges of the next century and to avert future crises. *Desalination., 124*, 145−153.

Rahaman, M. M., Varis, O., & Kajander, T. (2004). EU water framework directive vs. integrated water resources management: The seven mismatches. *International Journal of Water Resources Development, 20*(4), 565−575. Available from https://doi.org/10.1080/07900620412331319199.

Richter, B. D., Mathews, R., Harrison, D. L., & Wigington, R. (2016). Ecologically sustainable water management: Managing river flows for ecological integrity. *Ecological Applications, 13*(1), 206−224.

Rodrigues, D., Fonseca, A., Stolarski, O., Freitas, T. R., Guimarães, N., Santos, J. A., & Fraga, H. (2023). Climate change impacts on the Côa Basin (Portugal) and potential impacts on agricultural irrigation. *Water*, *15*(15), 2739. Available from https://doi.org/10.3390/w15152739.

Sarkar, U. K., Gupta, B. K., & Lakra, W. S. (2010). Biodiversity, ecohydrology, threat status and conservation priority of the freshwater fishes of river Gomti, a tributary of river Ganga (India). *The Environmentalist*, *30*(1), 3–17. Available from https://doi.org/10.1007/s10669-009-9237-1.

Savenije, H. H. G., & Van der Zaag, P. (2008). Integrated water resources management: Concepts and issues. *Physics and Chemistry of the Earth*, *33*(5), 290–297. Available from https://doi.org/10.1016/j.pce.2008.02.003.

Segbefia, A. Y., Honlah, E., Appiah, D. O., & Yildiz, F. (2019). Effects of water hyacinth invasion on sustainability of fishing livelihoods along the River Tano and Abby-Tano Lagoon, Ghana. *Cogent Food & Agriculture*, *5*(1), 1654649. Available from https://doi.org/10.1080/23311932.2019.1654649.

Shafiei, M., Rahmani, M., Gharari, S., Davary, K., Abolhassani, L., Teimouri, M. S., & Gharesifard, M. (2022). Sustainability assessment of water management at river basin level: Concept, methodology and application. *Journal of Environmental Management*, *316*. Available from https://doi.org/10.1016/j.jenvman.2022.115201. Available from, https://www.sciencedirect.com/journal/journal-of-environmental-management.

Sharma, A. K., Grant, A. L., Grant, T., Pamminger, F., & Opray, L. (2009). Environmental and economic assessment of urban water services for a greenfield development. *Environmental Engineering Science*, *26*(5), 921–934. Available from https://doi.org/10.1089/ees.2008.0063.

Sultana, M. S., & Mostafa, M. G. (2023). Impact of climate change on economic development associated with water scarcity: A review. *Journal of Sustainability and Environmental Management*, *2*(3), 158–169. Available from https://doi.org/10.3126/josem.v2i3.59105.

UN. (2019). *World population prospects 2019*. Department of Economic and Social Affairs. Available from: https://population.un.org/wpp/publications/files/wpp2019_highlights.pdf.

Wang, K., Davies, E., & Liu, J. (2019). Integrated water resources management and modeling: A case study of Bow river basin, Canada. *Journal of Cleaner Production*, *240*(2006), 1–7. Available from https://doi.org/10.1016/j.jclepro.2019.118242.

Wang, X., Chen, Y., Li, Z., Fang, G., Wang, F., & Hao, H. (2021). Water resources management and dynamic changes in water politics in the transboundary river basins of Central Asia. *Hydrology and Earth System Sciences*, *25*(6), 3281–3299. Available from https://doi.org/10.5194/hess-25-3281-2021. Available from, http://www.hydrol-earth-syst-sci.net/volumes_and_issues.html.

Wang, Zg, Luo, Yz, Zhang, Mh, & Xia, J. (2014). Quantitative evaluation of sustainable development and eco-environmental carrying capacity in water-deficient regions: A case study in the Haihe River Basin, China. *Journal of Integrative Agriculture*, *13*(1), 195–206. Available from https://doi.org/10.1016/S2095-3119(13)60423-2.

Zeitoun, M., Goulden, M., & Tickner, D. (2013). Current and future challenges facing transboundary river basin management. *Wiley Interdisciplinary Reviews: Climate Change*, *4*(5), 331–349. Available from https://doi.org/10.1002/wcc.228. Available from, http://onlinelibrary.wiley.com/journal/10.1002/(ISSN)1757-7799.

A machine learning methodology to calculate the percentage of areas affected by drought in Brazil in map images

Ana Carolina Borges Monteiro, Rodrigo Rodrigo, Reinaldo Padilha França, Herica Fernanda de Sousa Carvalho and Ferrucio de Franco Rosa

Renato Archer Information Technology Center (CTI), Campinas, São Paulo, Brazil

27.1 Introduction

Greenhouse gas emissions are often cited as a major cause of the fast climate change and intensified global warming. These gases come from the generation of electricity and heat by burning fossil fuels. This happens because a large part of electrical energy generation is still dependent on fossil fuels and coal, which results in the release of carbon dioxide and nitrous oxide. With the dismissal of these substances, heat is retained in the earth, which strongly contributes to the worsening of the greenhouse effect (Lineman, Do, Yoon Kim, Joo, & Fowler, 2015; Soeder & Soeder, 2021). Another major problem is deforestation, which further aggravates climate change. Tropical forests are responsible for regulating the climate in different regions of the world, as they normalize the rain cycle and prevent the emergence of new desert regions. When looking at the equator, we can notice that almost all the countries located in this region have desert areas. However, the Amazon region (in South America) is in this same geographic latitude and does not have any desert region. Amazon rainforest also regulates the rain cycle of various regions of Brazil and other countries. This points to the need to preserve forests for the well-being of populations, as well as for the development of a local economy since plant cultivation and animal husbandry are seriously affected by the rain cycles (Oliveira and Fátima, 2021; Nian et al., 2023). Recent data shows that around 12 million hectares of forests are destroyed per year, with most of these areas used for livestock farming and plantations. Accelerated deforestation together with the unbridled emission of greenhouse contributes to climate deregulation, prolonged droughts, flooding, increased sea level temperatures, extinction of fauna and flora species, both aquatic and terrestrial, greater ease of emergence of pandemics, higher incidence of respiratory diseases,

Applications of Geospatial Technology and Modeling for River Basin Management.
DOI: https://doi.org/10.1016/B978-0-443-23890-1.00027-X

among others (Escolhas, 2023). Considering this entire scenario of problems that drive climate change in the world, drought events are part of this context. In Brazil, there is data on this phenomenon available to the population, from the Drought Monitor, which consists of a process of regular and periodic monitoring of the drought situation, whose consolidated results are published through the Drought Monitor Map. This platform is supported by the National Water Agency (in Portuguese called "ANA"), which is a Brazilian government institution. The Drought Monitor aims to enable the translation of information into tools and products usable by decision-making institutions and individuals (ANA, 2023; Marengo et al., 2022). The Drought Monitor aims to integrate technical and scientific knowledge to achieve a common understanding of drought conditions by openly publishing information about the situation every month, data available up to the previous month, with indicators that reflect the short-term (last 3, 4, and 6 months) and long-term (last 12, 18, and 24 months). These data can be used to relate the severity, spatial and temporal evolution, an indication of the evolution of drought in the region, and its impacts on the different sectors involved to strengthen forecasting, monitoring, and early warning mechanisms. This open-source platform provides and allows the free use of your data (images) for the development of scientific research (ANA, 2023). Although image analysis can be carried out through color interpretation, this does not allow for an accurate estimation of the amount of occurrence of a given phenomenon. Therefore, new technologies based on artificial intelligence (AI) techniques, such as machine learning, emerge to solve this type of gap, maximizing the use of data, as well as the ability to model this data later. For example, machine learning reduces data processing time, allows for rapid processing, and supports the analysis of a larger dataset compared to conventional methods (Manley & Egoh, 2022).

The evolution of machine learning algorithms, the increase in processing power, and data availability allowed us to use deep learning algorithms, which have revolutionized image analysis and proven to be an excellent method for quickly evaluating the information inherent in images. Models based on convolutional neural networks (CNNs), according to Li et al. (2020), are particularly successful because they can extract features for classification and effectively use information from images. This algorithm has been successfully used for image classification in various types of applications, including landform, land use, and land cover classification (Li et al., 2020; Zhang et al., 2023). Deep learning has been accepted as a revolutionary technology in machine learning, data mining, and the field of remote sensing research. Deep learning models were employed as a framework that simultaneously collects features and performs classification following approaches, eliminating the need for manual features and predefined functions. Thus, studies that use image classification through deep learning are essential due to their versatility in feature extraction, computational efficiency, and automation (Khan, Vibhute, Mali, & Patil, 2022).

Since the information about drought present in map images does not satisfy the reader in detail about the quantity and distribution of information, there is a need for studies that show and spatially refine existing mapping approaches to fill in the information and improve the readers' understanding. This type of knowledge is important in decision-making, allowing appropriate measures to be used to reduce its effects. In this context, the objective of the work was to use recently available techniques and data sources to improve the interpretation of information regarding the drought climatic phenomenon, to reduce uncertainty about quantity, and to develop a script in the Python language. Map data from 2022 to August 2023 in digital PNG format was used to develop a Python script using TensorFlow, which is an open-source library for machine learning and neural networks, including the use of CNNs to perform segmentation of the images. Regarding CNNs, the segmentation model is based on a neural network designed specifically for image processing and computer vision tasks. The MobileNetV2 model is a foundation, and custom layers are added for the segmentation task. The main objective of the model is to calculate the percentage of areas affected by drought in images, doing this through image segmentation using a CNN-based segmentation model.

27.2 **Overview of drought in Brazil**

Drought is a phenomenon known as a natural disaster without a defined beginning, which can produce a catastrophe when combined with socioeconomic conditions and vulnerability at a global level. Drought definitions can be grouped into two types: conceptual and operational. The conceptual type provides a general description without specifying the quantitative properties of the drought, while the operation aims to determine and analyze the characteristics of a drought event (duration, intensity, frequency, and extension of the territorial area) that are evaluated based on climatological and hydrological data and meteorological conditions of a given location (Gonçalves, das Chagas Vasconcelos, Sakamoto, da Silva Silveira, & Martins, 2021). Different atmospheric systems intensify rain or a lack thereof. In Brazil, some factors and mechanisms influence rainfall and its lack in certain regions. They are El Niño-Southern Oscillation events, sea surface temperature, trade winds, sea level pressure, intertropical convergence zone over the atlantic ocean, cold fronts, vortices high-level cyclonic, lines of instability, and mesoscale convective complexes. Climatologically, drought is characterized when there is a delay of more than 15 days to start the rainy season, when there is a delay of more than fifteen days to the start of the rainy season, and when the average monthly rainfall during the rainy season remains below 60% of long-term monthly averages for the region considered. Drought occurs due to the prolongation and worsening of the drought, causing a brutal reduction in water reserves in the affected area (Santos & Horta, 2020). The lack or decrease in rainfall within a rainy period is usually the beginning of a drought event. This deficiency can decrease the amount of water in the soil, and this can lead to other types of droughts. Terms are defined for drought depending on the sector most affected, namely, meteorological, hydrological, agricultural, and socioeconomic (Belal, El-Ramady, Mohamed, & Saleh, 2014). The occurrence of drought is difficult to assess, as it involves several factors that influence their interrelations to analyze, identify, and obtain significant insights (Balti et al., 2020).

Meteorological drought is characterized by a reduction in rainfall associated with a significant increase in evapotranspiration when compared to the region's historical average. The occurrence of long periods of this condition defines another type of drought called hydrological drought, which is associated with a reduction in surface and underground water resources (Gonçalves et al., 2021; Pontes Filho, Portela, de Carvalho Studart, & de Assis Souza Filho, 2019). According to the latter authors, when a lack of water affects agriculture from planting to harvesting, the usual term is agricultural drought, characterized by a decrease in soil moisture and high evaporative rates of soil water, impacting crops. The combination of the three types of droughts defines socioeconomic drought, especially regarding water supply when it becomes insufficient to meet the supply and demand of living beings. The impacts associated with socioeconomic drought are multisectoral and range from the macro and micro environmental sectors, negatively affecting living beings in terms of food security, water, crop failure, fires, and the emergence of diseases (Balti et al., 2020). Drought can inhibit the regular growth of forests and, if continued, can lead to the death of vegetation due to lack of water, reducing their productivity and potentially leading to the extinction of native vegetation (Jiang et al., 2022). In this sense, recent research (Ekundayo, Abiodun, & Kalumba, 2022) shows an annual growth of more than 7% in drought studies over the last 150 years; in addition, there was an increase in the number of articles published until 2019 using remote sensing technologies. In Brazil, drought is linked to other climate extremes around the globe, considering that in the Northern Hemisphere, record summer temperatures melt as far as Greenland, an Island located in the Arctic Circle, which in a single day (Tuesday, July 27) lost 8.5 billion tons of ice (Boers & Rypdal, 2021; Hanna, 2021). In Canada and the United States, forest fires of unprecedented dimensions are consuming vast regions, a problem that is also observed in Russia, Greece, Italy, and Turkey (Coogan et al., 2021; Woolford et al., 2021). Apocalyptic storms hit Germany and Belgium and killed more than 200 people in Zhengzhou (PUKmedia, 2021).

The consequences of the drought impacted the agricultural sector in dozens of countries, mainly with loss of productivity and difficulties that had repercussions on world markets. This situation favored forest fires and harmed electricity production, mainly hydro and nuclear sources. Due to heat waves in the northern hemisphere (outside tropical regions), the probability of drought increases, which means that in a depth of 1 m, forests and agricultural soils suffer a lack of water that seriously affects plant roots. The influence of human activities is likely higher, demonstrating how human-caused climate change increases the risks of agricultural and ecological drought in agricultural and densely populated regions of the Northern Hemisphere (Liu et al., 2016; Orimoloye, 2022). The scarce rains, for instance, caused a dramatic situation in the south of Minas Gerais (Brazil) state, which is covered in coffee plantations. These plantations are delicate, and it is natural that in some periods, the lack of rain or irregular cold damages the harvest, precisely because of the combination of mild temperature with moist soil, considering that the state harvests half of the national production, which accounts for 40% of the world's supply. However, in recent years, the exceptional drought, interspersed with frosts of rare intensity, has been devastating the coffee farms in Minas Gerais, both due to the persistent drought and the harsh winter (Rural, 2021). The prolonged drought and the lack of rain are part of the consequences of global warming, in which the deforestation of the Amazon is a contributing factor. The Amazon forest is heading toward a breaking point, under threat of large-scale ecological transformation, considering that in the first half of 2021, more than 3000 sq. km of forest disappeared, the equivalent of two cities in São Paulo, which is a consequence of fires. The destruction of the forest has another direct impact on the regulation of rainfall because when preserved, the Amazon biome exudes moisture at a constant rate, allowing for the formation of flying rivers, which are channels of wetness that travel in the upper atmosphere and, upon reaching the Center-South of Brazil (the most populated area), they form heavy clouds that collapse in the form of abundant rain (Amazônia, 2021). Uncontrolled felling of forests has diminished flying rivers and disrupted the water cycle from the atmosphere to the soil and then to the rivers, groundwater, plants, and back to the atmosphere, creating permanent arity, an extreme weather phenomenon. The roots of native vegetation infiltrate the soil below and transport rainwater to deep levels, supplying the water table, resulting in the central plateau, called the water tank of Brazil, being the birthplace of eight of the twelve main Brazilian river basins (Brouwer, Pinto, Dugstad, & Navrud, 2022). The cracked land, the devastated plantations, and the cattle wasting away due to lack of rain are situations that Brazil is familiar with, considering that this landscape can be seen in the Northeast of the country, a semi-arid region with a high incidence of sunshine and low incidence of rain. However, in recent years, drought has affected the South, Southeast, and Central-West regions, which have an abundance of rain. Due to its duration and intensity, it has been a tough drought since the phenomenon began to be measured in 1910. It can be said that climate extremes are already affecting all regions of Brazil (Libonati, 2021; Staal et al., 2020). Since 1950, the world's drylands have increased by almost 2% per decade, and Brazil is no exception. In 2012, around 650 municipalities in the Southern region of Brazil were in drought emergency, that is, 142 in Paraná, 375 in Rio Grande do Sul, and 133 in Santa Catarina. In 2014, the São Paulo state faced the toughest drought in the last 80 years. Problems like those that São Paulo faced, especially water supply, can be attributed not only to climate change but also to urban agglomeration and insufficient supply infrastructure (Empresa Brasileira De Pesquisa Agropecuária, n.d.).

27.3 Methodology: machine learning

MobileNetV2 is an evolution of the original MobileNet, a CNN architecture designed for applications on mobile devices and other platforms with limited computing resources, introducing significant improvements in terms of performance and efficiency. MobileNetV2 is designed to balance computational efficiency and performance in computer vision tasks. This architecture has been widely adopted in object detection applications as it stands out in scenarios where it is crucial to have lightweight and fast models, such as in mobile devices, embedded systems, and other resource-constrained platforms, that is, image segmentation and visual classification in resource-limited environments (Sinha & El-Sharkawy, 2019). MobileNetV2 is a type of supervised neural network, that is, it is trained using input and output pairs, where the desired output (label) is associated with each training example. During training, the network adjusts its parameters to minimize the difference between predicted and actual outputs. In the specific context of MobileNetV2, it is generally trained for supervised tasks such as image classification, object detection, or image segmentation. During training, the network is fed with labeled images, and the network parameters are adjusted to optimize the prediction of these labels. To use MobileNetV2 or any other deep learning architecture, a labeled dataset is required for training, where each example in the dataset has an input (image) associated with a known label. This is critical for supervised training, where the network learns to map inputs to specific outputs based on the examples provided (Sinha & El-Sharkawy, 2019; Souid, Sakli, & Sakli, 2021). MobileNetV2 uses inverted residual layers to increase nonlinear representation without drastically increasing computational complexity. This approach is effective for deeper networks, allowing information to propagate more effectively through the network derived from using inverted residual layers and helping mitigate the problem of performance degradation as networks get deeper (Dong, Zhou, Ruan, & Li, 2020; Sandler, Howard, Zhu, Zhmoginov, & Chen, 2018).

In general, an inverted residual block is composed of linear convolution that considers a 1×1 convolutional layer to expand the dimensionality of features, deep convolution that considers a 3×3 (deep) convolutional layer to capture complex patterns, linear convolution considering a 1×1 convolutional layer to reduce the dimensionality back, and a residual connection that adds the original input (skip connection), which forms a residual block. The linear bottleneck layer introduces a 1×1 linear layer as an expansion before a 3×3 convolutional layer. This feature helps in capturing linear relationships in the features, that is, it helps in capturing linear patterns in the features, improving the representation capacity of the network. The residual inverted block structure, with its combination of linear convolutions, deep convolutions, and residual connection, contributes to a richer and more complex representation of features (Dong et al., 2020). Activation Layer uses ReLU6 activation (Monteiro & Borges, 2023), which is a modified version of the ReLU function to limit values in the range [0, 6], which is especially useful for inference operations on low-precision hardware, as is often found in mobile devices. Separation of convolutions, that is, depthwise separable convolution, reduces the computational burden, which uses depthwise separable convolutions to reduce the number of parameters and, since it performs convolutions separately for each channel and then combines the results, it makes the lightest and most computationally efficient model since it results in fewer operations compared to standard convolutions (Chollet, 2017). The downsampling strategy uses stride 2 convolutions for downsampling helping to reduce the spatial resolution of features efficiently; this is efficient in deeper layers, allowing the network to learn more abstract representations (Zhao, Wang, & Zhang, 2017). Practical applications of MobileNetV2 are related to image classification, often used for image classification tasks on mobile devices and environments with computational restrictions, and can be adapted for image segmentation tasks, where the goal is to identify and delineate specific areas of interest in images (as seen in the case of this chapter) (Dong et al., 2020). MobileNetV2 can be used in object detection tasks, such as SSD (Single Shot Multibox Detector) or YOLO (You Only Look Once) models, for efficient detection in real time. It is also used as a basis for transfer learning in various applications, where prior knowledge acquired on large datasets can be applied to specific tasks with smaller datasets. Also, it is a popular choice, as it is a resource-efficient architecture for computer vision tasks in contexts where computational resources are limited (Dong et al., 2020).

27.3.1 Setup, execution, and results

We acquired and used in this research map data (from 2022 to August 2023) in the PNG format from the Drought Monitor from the Brazilian government institution ANA. The machine learning model was developed using the MobileNetV2 model in Python language using the TensorFlow library and other libraries such as OpenCV (cv2), NumPy, and Matplotlib to segment areas affected by drought in images. OS libraries were imported to provide functions for interacting with the operating system, in this case, for manipulating file paths; cv2 (OpenCV) is used for image manipulation, NumPy (np) for numeric operations, Matplotlib (plt) for creating graphs, and TensorFlow (tf) for machine learning. After defining the variables for the path to the directory containing the images, we list the image filenames in the file directory, the target size of the images after resizing, and list and store the percentages of areas affected by drought. The segmentation model was loaded using the pretrained MobileNetV2 model from the Keras library for image segmentation. Then, we used it to predict drought-affected areas in images and calculate the percentage of these areas. Custom layers were added to adapt the model to the segmentation task. Regarding the model setup, input_shape defines the expected size of the input images, and include_top = False means the fully connected layer on top of MobileNetV2 will not be included, as it will be replaced by layers customized for the specific task. The addition of custom layers for segmentation comes from a transposed convolution layer with 256 filters, a kernel size of (Oliveira & Fátima, 2021), and a stride of (Soeder & Soeder, 2021); it is used to increase the output spatial size. After a convolution layer is added with 128 filters, a kernel size of (Oliveira & Fátima, 2021), ReLU activation, another convolution layer is added transposed with 64 filters, a kernel size of (Oliveira & Fátima, 2021), a stride of (Soeder & Soeder, 2021). Finally, a last convolution layer is added with one filter, a kernel of size (Lineman et al., 2015), and sigmoid activation, common in binary segmentation tasks, such as this one, where the output represents the probability of belonging to the positive class. The model uses the same input as MobileNetV2. The output is now generated by the custom layers added for the specific task, taking the images as input and generating the outcome after the custom segmentation layers. The final layers are replaced with custom layers that are better suited to the specific task of identifying drought-affected areas in images. To calculate the drought percentage, a function was developed that takes an image as input, resizes this image to the target size, preprocesses the image, normalizes and expands the dimensions as necessary, and uses the segmentation model to predict the areas affected by drought. It also calculates the percentage of areas affected by drought based on the segmentation mask. From a processing loop, the function loads each image from the directory, calls the previously described drought percentage calculation function to obtain the percentage of areas affected by drought, and finally adds the percentage to the list of drought percentages. After using the Matplotlib library, a bar chart was created, adding bars to each image with the height representing the percentage of areas affected by drought, adding labels and a title to the chart, and finally displaying the chart, as shown in Fig. 27.1.

Fig. 27.2 Brazilian states affected by drought—timeframe from January 2022 to August 2023.

27.4 Discussion

The increasing incidence of extreme weather events, such as droughts, represents a global challenge with significant implications for food security, water supplies, and environmental health. Given this scenario, the use of AI models to support the identification of drought levels offers several advantages, such as the ability to process large volumes of data quickly and automatically, as the use of advanced AI techniques emerges as a promising tool for the assessment and accurate monitoring of these events (Dikshit & Pradhan, 2021; Jalalkamali, Moradi, & Moradi, 2015; Kikon & Deka, 2022). However, challenges remain as model accuracy significantly depends on the quality of training data, such as the quality of datasets and parameter optimization need to be addressed, including the need for representative datasets and proper tuning of parameters to ensure accurate results,

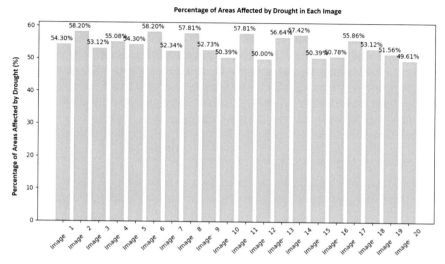

FIGURE 27.1

Percentage of areas affected by drought from the 20 images presented in Fig. 27.2.

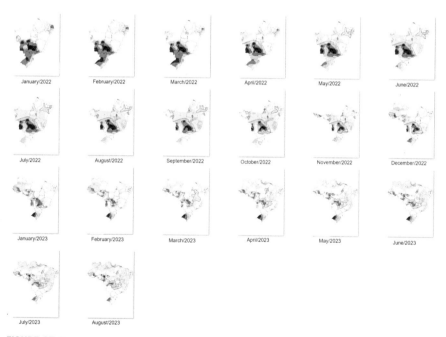

FIGURE 27.2

Brazilian states affected by drought—timeframe from January 2022 to August 2023. It shows one image per month from January 2022 to August 2023.

obtaining representative and well-annotated datasets for droughts can be challenging (França, Borges Monteiro, Arthur, & Iano, 2021; Borges Monteiro, Iano, França, & Arthur, 2021). Furthermore, adapting the model to different geographic and climatic regions can be complex, that is, generalizing to different territories, considering that models trained in a specific area may not generalize well to other sites with different characteristics. As a successful application in one location does not guarantee success in another, careful adaptations are required, transferring knowledge between different climatic and geographic contexts.

Finding the ideal parameters for the model, requiring careful adjustments and validation on diverse datasets, and considering different contexts and geographic scales is a challenge. The ability to interpret model decisions is crucial, and complex models such as neural networks often lack interpretability, especially in contexts where decisions can have significant implications (Monteiro, França, Arthur, Iano, et al., 2021). The model exemplifies the practical application of AI and represents an innovative approach to identifying drought-affected areas, combining a lightweight architecture such as MobileNetV2, custom layers, and image processing techniques. Our proposal provides a scalable and efficient strategy to contribute significantly to environmental monitoring. Interpretation of results provides crucial metrics, and this approach represents a significant step toward automated environmental monitoring and informed decision-making, such as the percentage of areas affected, allowing for a quantitative impact assessment of droughts. It enables continuous monitoring of areas susceptible to droughts. The potential contribution to understanding and mitigating the effects of droughts is substantial, making the application of AI a valuable tool in the sustainable management of natural resources. This also allows for a faster response to extreme weather events. Early identification of drought-affected areas provides opportunities for proactive interventions, that is, natural disaster prevention, such as water-resources reallocation, implementation of drought-resilient agricultural practices, and early warnings for vulnerable communities. Farmers and resource managers can benefit from accurately assessing the impact of droughts in their areas; they could be informed about planting strategies, that is, agricultural planning and resource management, such as water management and agricultural-related decision-making. Automated data collection on drought-affected areas can feed into broader scientific studies on environmental impacts, climate change, and ecological systems modeling. The results provided by AI models through the contribution of scientific research can inform public policies related to natural resource management and climate change adaptation strategies. The use of AI in identifying drought-affected areas represents a powerful tool with broad potential applications. These techniques can contribute to mitigation and adaptation to climate change. To maximize the positive impact, addressing challenges such as data quality or model generalization is imperative, using appropriate data collecting strategies, model optimization, and an ethical approach, promoting sustainable management of natural resources. Critical sectors such as water supply and power generation, enabling the implementation of preventative measures to maintain the stability of these infrastructures, can benefit from early detection of drought-affected areas. Automated analysis of dry areas can inform efficient management of water resources, whereas AI-based systems can be integrated into early warning systems. This includes directing resources to the most affected areas and optimizing water use in agricultural and urban areas. These systems may facilitate preparation and response to water stress situations, providing advance information to vulnerable communities (Rivera & Penalba, 2014). Given the dynamism of climate conditions, implementing mechanisms to incorporate real-time data and considering the need for regular model updates is crucial. Adjusting the model is a challenge to be faced. Access to computational resources for model training and inference may be restricted in areas with limited infrastructure. Practical strategies to deal with these limitations need to be developed. In this scenario, a MobileNetV2 model, like the one presented in this chapter, can be valuable. However, while a common consensus has not been found to solve the challenges related to the deforestation of tropical forests, droughts, and ecological disasters, investment in assertive, low-cost, and easy-to-interpret technologies is required, as predicting natural catastrophes and periods of drought can help disadvantaged populations prepare for periods and shortages and help governments around the world take measures before a dry period sets in. Machine learning can help governments take more effective socioeconomic measures, as dealing with drought requires considerable financial contributions. This happens because drought impacts people's quality of life and the economy. From a humanitarian point of view, drought generates hunger, disease, and malnutrition. Years of prolonged drought can cause severe generational impacts on a population. Drought and other climate disasters force thousands of people to leave their homes. Developing machine learning projects for assertive predictions of environmental catastrophe is critical to protect a generation and favor the socioeconomic development of a region. Malnutrition is a drastic consequence of prolonged drought. This condition leads to slower and impaired child development, which can cause damage to learning and the formation of the physical structure. In more severe and prolonged conditions, malnutrition, lack of access to clean water, and extreme heat can lead to increased child mortality. Another social group harmed by drought, hunger, and high temperatures is the elderly. People who are subjected to these conditions for long periods may suffer from early death. As a result, drought is responsible for reducing the life expectancy of a population.

Countries affected by long periods of climate disasters can lead to a reduction in the Human Development Index. This index considers not only the financial amount that a country has but also the social aspect of the country. In other words, how a country cares for its population is considered. Therefore, a homeless, malnourished population with little prospect of improvements in quality of life impacts the drastic reduction of the Human Development Index. Drought directly impacts agricultural and livestock production. When this happens, the government must act quickly to provide aid and incentive policies for agriculture and livestock farming. Therefore, investing in machine learning is investing in science, technology, economics, and a better-planned future. Drought causes an abrupt increase in the value of food products. This reality drives destruction and hunger. Generally, it is up to the government to intervene with policies to reduce the price of food to ensure that the population can access food. Furthermore, periods of drought require higher investment in technology, as new machinery and techniques can ensure that food production continues. As there is a considerable investment for production to continue, the cost is impacted. Those who are most affected by drought are the population of the country itself. To achieve this objective, regulatory agencies from different countries must collect, identify, and organize climate data objectively and regularly. As the success of any prediction is linked to the quality of data, a broad dataset can benefit different prediction types and provide higher clarification and recognition of climate patterns. Unfortunately, many countries are still not concerned about maintaining good quality climate data for scientific studies, causing a lot of valuable data and information to be lost and hindering the application of more assertive machine learning models. Throughout the centuries and millennia of human history, several people have been affected by periods of prolonged drought and starvation. However, periods of drought and natural disasters tend to intensify due to the impacts of human activities. We hope that machine learning techniques can support us in being more prepared to deal with these issues, preventing us from being surprised by adverse weather conditions, and supporting humanity in discovering more effective and efficient ways to deal with and revert global warming.

27.5 Limitations of the study

The lack of a "gold standard" with labeled maps for this problem makes it difficult to assign comparable accuracy measures and train specific models. We used an already trained model since developing an annotated dataset requires a long-term and costly study. However, our study represents a fundamental step toward the use of AI in drought detection from maps through a model that obtained good accuracy results in segmentation tasks on other types of images (Adikari et al., 2021; Jiang et al., 2022; Sandler et al., 2018). Although promising, the used model (MobileNetV2) presents some limitations that must be considered related to the effectiveness of the model depending on the quality and representativeness of the training dataset. If the dataset is not diverse or does not adequately reflect variations in drought conditions, the model may not generalize well to different contexts. The application of the model can be restricted to specific geographic areas for which it was trained, considering that changes in landscape characteristics, vegetation, and climatology can impact the model's generalization to other regions. Interaction with hydrological conditions, such as the presence of rivers, lakes, and other bodies of water, can present challenges to effective segmentation, potentially confusing areas with water resources with regions affected by drought.

Deep neural networks, such as the one used, often lack interpretability, as understanding how the model makes specific decisions can be challenging and could limit user confidence and applicability in critical scenarios. Model performance can be sensitive to the choice of parameters, including training hyperparameters; therefore, an inadequate selection of these parameters can lead to suboptimal results or even a lack of convergence during training. The quality and availability of satellite images may vary, as in regions where satellite data is limited, the practical application of the model may be compromised. Changes in climatic and environmental conditions over time can affect the relevance of the model. In this sense, keeping the model updated with new data and conditions requires a continuous process, and the lack of this can lead to obsolescence.

The spatial resolution of the model may be limited, especially in detecting drought-affected areas at local or small scales, and this may affect the model's ability to provide detailed insights. Satellite images often face cloud cover or shading, impairing analysis; preprocessing strategies may be needed to deal with these conditions. The use of AI in critical decision-making can introduce bias and ethical challenges, considering that data representativeness and equity in the application of the model are concerns that need to be addressed. In environments with limited computational resources, practical model implementation can be challenging, and real-time inference and scalability can be compromised. Recognizing these limitations is crucial for realistic and ethical model implementation, as strategies to mitigate these limitations include representative data, regular model adjustments, transparency in the interpretation of model decisions, and ethical considerations throughout the process. By proactively addressing these limitations, we can improve the robustness and applicability of the model in identifying areas affected by drought. The model may have difficulty dealing with specific variations in vegetation. As plant types may respond differently to water stress, the generalization to different biomes and vegetation types may be challenging. The model may also not adequately consider the interaction with other climate variables, such as temperature and wind, which also play a role in water stress. The quality of segmentation of drought-affected areas needs to be carefully assessed, as errors in segmentation can lead to inaccurate estimations of the percentage of areas affected. The model may have limitations in dealing with long-term climate change and seasonality, so adapting to weather patterns that vary seasonally or gradually change over the years can be challenging. Atmospheric conditions, such as clouds and snow cover, can impact the correct interpretation of images, as the model may not be fully robust in cloudy environments.

The applicability of the model in urban environments, where there are many artificial structures, may be limited, and the segmentation may not adequately distinguish between urban areas and drought-affected areas. Practical implementation of the model may be restricted by budget constraints and technological limitations, especially in regions with limited resources. In remote or difficult-to-access locations, data collecting for model training and evaluation can be challenging; practical application in these areas may require adaptive strategies. Changes in land usage (e.g., urbanization) or changes in agriculture may affect the model's ability to generalize to evolving conditions, given that these land use dynamics must be considered. Natural disturbances (e.g., forest fires) or anthropogenic activities (e.g., deforestation) can interfere with the model's ability to distinguish areas affected by droughts due to different causes. Soil variability can influence plant response to drought, considering that models that do not assume soil diversity may have limitations in accuracy. These considerations highlight the complexity of applying AI models in drought-related scenarios and the need for a careful and context-aware approach. Validation of the model in the field is essential, and uncertainties associated with remote interpretation and weather conditions can affect confidence in the results. Continuous adaptation, critical assessment, and an in-depth understanding of specific conditions are essential to overcome these limitations and advance the effective use of these models in drought management. Understanding the limitations is crucial to ensuring that practical applications are informed and effective, as they highlight the ongoing need for critical assessment and adjustments as models are implemented in different contexts, considering the complexities of the environment and variabilities in climatic conditions. Continuous development and improvement of these models must be carried out with an iterative approach, considering constant feedback from experts and field validations. Interpretation of drought extent and severity may vary depending on multiple factors, considering the response of plants and the environment to drought can be dynamic and evolve. The model's ability to characterize drought severity may be limited, including soil and vegetation types, as the model can face challenges in efficiently capturing the temporal variability of drought conditions. Changes in climate conditions may alter the distribution and intensity of droughts. The model cannot be able to distinguish between different causes of water stress, such as meteorological or even hydrological, and the model's ability to predict and adapt to future climate change scenarios may be uncertain. We must also consider that human activities, where the landscape includes urban and rural zones, such as intensive irrigation, can also impact the detection of drought-affected areas. Regarding mixed-use areas, segmentation may face difficulties in distinguishing between areas affected by drought and urban infrastructure. Therefore, we highlight no model is free from limitations, as the complexity inherent in detecting drought-affected areas requires a holistic approach, given implementation and interpretation of results must be carried out with due consideration of uncertainties and specific contexts.

27.6 Conclusion

The proposed model uses an architecture based on MobileNetV2 to perform image segmentation and identify areas affected by drought, including preprocessing steps (e.g., resizing and normalization) to ensure that images are at the correct scale for the model. The choice for this type of architecture is strategic, as it is known for its computational efficiency, making it suitable for applications in devices with limited resources, such as automated weather stations and drones. Furthermore, postprocessing involves interpreting the model outputs and converting them into percentages of drought-affected areas. The custom layers added at the end of the model are designed to adapt the network to the specific image segmentation task, considering the use of conventional convolutions and transposed convolutions that reflect the search for a spatially richer representation of the drought-affected areas. The function that calculates the drought percentage interprets the results, and this percentage is a valuable metric to quantify the impact of droughts in each region. It provides crucial information for decision-makers and environmental scientists by calculating the percentage of affected areas based on the segmentation masks generated by the model. The application of AI models to identify areas affected by drought is a multidisciplinary tool with the potential to positively impact different sectors. However, resolving technical, ethical, and operational challenges is essential to ensure effectiveness and equity in its implementation. Collaborative dialog between scientists, policymakers, local communities, and ethicists is critical to building robust and socially responsible solutions. By addressing these challenges holistically, we can fully harness the transformative potential of AI in sustainable drought management and climate change adaptation.

References

Adikari, K. E., Shrestha, S., Ratnayake, D. T., Budhathoki, A., Mohanasundaram, S., & Dailey, M. N. (2021). Evaluation of artificial intelligence models for flood and drought forecasting in arid and tropical regions. *Environmental Modelling & Software, 144*. Available from https://doi.org/10.1016/j.envsoft.2021.105136.

ANA. (2023). *Catálogo de Metadados da, ANA — snirh.*

Amazônia. (2021). Primeiro semestre de 2021 tem maior desmatamento na Amazônia em seis anos.

Balti, H., Abbes, A. B., Mellouli, N., Farah, I. R., Sang, Y., & Lamolle, M. (2020). A review of drought monitoring with big data: Issues, methods, challenges and research directions. *Ecological Informatics, 60*. Available from https://doi.org/10.1016/j.ecoinf.2020.101136.

Belal, A. A., El-Ramady, H. R., Mohamed, E. S., & Saleh, A. M. (2014). Drought risk assessment using remote sensing and GIS techniques. *Arabian Journal of Geosciences, 7*(1), 35−53. Available from https://doi.org/10.1007/s12517-012-0707-2.

Boers, N., & Rypdal, M. (2021). Critical slowing down suggests that the western Greenland Ice Sheet is close to a tipping point. *Proceedings of the National Academy of Sciences, 118*(21). Available from https://doi.org/10.1073/pnas.2024192118.

Brouwer, R., Pinto, R., Dugstad, A., Navrud, S., & E O S. (2022). The economic value of the Brazilian Amazon rainforest ecosystem services: A meta-analysis of the Brazilian literature. *PLoS One, 17*(5). Available from https://doi.org/10.1371/journal.pone.0268425.

Chollet, F. (2017). Xception: Deep learning with depthwise separable convolutions 2017. In *Proceedings – 30th IEEE conference on computer vision and pattern recognition, CVPR 2017* (1800–1807). Available from https://doi.org/10.1109/CVPR.2017.195, 9781538604571. 2017/11/06, Institute of Electrical and Electronics Engineers Inc.

Coogan, S. C. P., Daniels, L. D., Boychuk, D., Burton, P. J., Mike., Flannigan, D., ... Wotton, B. M. (2021). Fifty years of wildland fire science in Canada. *Canadian Journal of Forest Research*, *51*(2), 283–302. Available from https://doi.org/10.1139/cjfr-2020-0314.

Dikshit, A., & Pradhan, B. (2021). Explainable AI in drought forecasting. *Machine Learning with Applications*, *6*. Available from https://doi.org/10.1016/j.mlwa.2021.100192.

Dong, K., Zhou, C., Ruan, Y., & Li, Y. (2020). China MobileNetV2 model for image classification. In *Proceedings – 2020 2nd international conference on Information Technology and Computer Application, ITCA 2020* (476–480). Available from https://doi.org/10.1109/ITCA52113.2020.00106, 9780738111414. 2020/12/01, Institute of Electrical and Electronics Engineers Inc. http://ieeexplore.ieee.org/xpl/mostRecentIssue.jsp?punumber = 9421350.

Ekundayo, O. Y., Abiodun, B. J., & Kalumba, A. M. (2022). Global quantitative and qualitative assessment of drought research from 1861 to 2019. *International Journal of Disaster Risk Reduction*, *70*. Available from https://doi.org/10.1016/j.ijdrr.2021.102770, http://www.journals.elsevier.com/international-journal-of-disaster-risk-reduction/.

Empresa Brasileira De Pesquisa Agropecuária (n.d.).

França, R. P., Borges Monteiro, A. C., Arthur, R., & Iano, Y. (2021). *An overview of deep learning in big data, image, and signal processing in the modern digital age* (pp. 63–87). Elsevier BV. Available from https://doi.org/10.1016/b978-0-12-822226-3.00003-9.

PUKmedia. (2021). *Germany floods: Dozens killed after record rain in Germany and Belgium.*

Gonçalves, S. T. N., das Chagas Vasconcelos, F., Junior, Sakamoto, M. S., da Silva Silveira, C., & Martins, E. S. P. R. (2021). Índices e Metodologias de Monitoramento de Secas: Uma Revisão. *Revista Brasileira de Meteorologia*, *36*(3 suppl), 495–511. Available from https://doi.org/10.1590/0102-77863630007.

Hanna, N. (2021). Greenland is losing ice 7 times faster than in 1990s. In Thanks to climate change indiatimes.

Instituto Escolhas. (2023). Reflorestamento pode gerar R$ 776 bilhões para o Brasil, diz estudo. *CNN Brasil.*

Jalalkamali, A., Moradi, M., & Moradi, N. (2015). Application of several artificial intelligence models and ARIMAX model for forecasting drought using the Standardized Precipitation Index. *International Journal of Environmental Science and Technology*, *12*(4), 1201–1210. Available from https://doi.org/10.1007/s13762-014-0717-6.

Jiang, W., Niu, Z., Wang, L., Yao, R., Gui, X., Xiang, F., & Ji, Y. (2022). Impacts of drought and climatic factors on vegetation dynamics in the Yellow River Basin and Yangtze River Basin, China. *Remote Sensing*, *14*(4). Available from https://doi.org/10.3390/rs14040930.

Khan, A., Vibhute, A. D., Mali, S., & Patil, C. H. (2022). A systematic review on hyperspectral imaging technology with a machine and deep learning methodology for agricultural applications. *Ecological Informatics*, *69*. Available from https://doi.org/10.1016/j.ecoinf.2022.101678.

Kikon, A., & Deka, P. C. (2022). Artificial intelligence application in drought assessment, monitoring and forecasting: A review. *Stochastic Environmental Research and Risk Assessment*, *36*(5), 1197–1214. Available from https://doi.org/10.1007/s00477-021-02129-3.

Li, D., Wang, R., Xie, C., Liu, L., Zhang, J., Li, R., ... Liu, W. (2020). A recognition method for rice plant diseases and pests video detection based on deep convolutional neural network. *Sensors*, *20*(3). Available from https://doi.org/10.3390/s20030578.

Libonati, R. (2021). Twenty-first-century droughts have not increasingly exacerbated fire season severity in the Brazilian Amazon. *Scientific Reports*, *11*, 4400.

Lineman, M., Do, Y., Yoon Kim, J., Joo, G.-J., & Fowler, H. J. (2015). Talking about climate change and global warming. *PLoS One*, *10*(9). Available from https://doi.org/10.1371/journal.pone.0138996.

Liu, X., Zhu, X., Pan, Y., Li, S., Liu, Y., & Ma, Y. (2016). Agricultural drought monitoring: Progress, challenges, and prospects. *Journal of Geographical Sciences*, *26*(6), 750−767. Available from https://doi.org/10.1007/s11442-016-1297-9.

Manley, K., & Egoh, B. N. (2022). Mapping and modeling the impact of climate change on recreational ecosystem services using machine learning and big data. *Environmental Research Letters*, *17*(5). Available from https://doi.org/10.1088/1748-9326/ac65a3.

Marengo, J. A., Galdos, M. V., Challinor, A., Cunha, A. P., Marin, F. R., dos Santos Vianna, M., ... Bender, F. (2022). Drought in Northeast Brazil: A review of agricultural and policy adaptation options for food security. *Climate Resilience and Sustainability*, *1*(1). Available from https://doi.org/10.1002/cli2.17.

Monteiro, A. C. B., França, R. P., Arthur, R., & Iano, Y. (2021). AI approach based on deep learning for classification of white blood cells as a for e-Healthcare solution. *Intelligent Interactive Multimedia Systems for e-Healthcare Applications*, 351−373. Available from https://doi.org/10.1007/978-981-16-6542-4_18, https://doi.org/10.1007/978-981-16-6542-4.

Monteiro, A. C., & Borges, (2023). Proposta de novas metodologias de análise de células sanguíneas por meio dos métodos BSCM (Blood Smear Computational Method) e BSIM (Blood Smear Inteligence Method): informática de baixo custo aplicada à saúde pública.

Borges Monteiro, A. C., Iano, Y., França, R. P., & Arthur, R. (2021). *Deep learning methodology proposal for the classification of erythrocytes and leukocytes* (pp. 129−156). Elsevier BV. Available from https://doi.org/10.1016/b978-0-12-822226-3.00006-4.

Nian, D., Bathiany, S., Ben-Yami, M., Blaschke, L., Hirota, M., Rodrigues, R., & Boers, N. (2023). The combined impact of global warming and AMOC collapse on the Amazon Rainforest. *Research Square, Germany Research Square*. Available from https://doi.org/10.21203/rs.3.rs-2673317/v1, https://www.researchsquare.com/browse.

Oliveira., & Fátima. (2021). Deforestation and climate change are projected to increase heat stress risk in the Brazilian Amazon. *Communications Earth & Environment*, 2.

Orimoloye, I. R. (2022). Agricultural drought and its potential impacts: Enabling decision-support for food security in vulnerable regions. *Frontiers in Sustainable Food Systems*, 6. Available from https://doi.org/10.3389/fsufs.2022.838824.

Pontes Filho, J. D., Portela, M. M., de Carvalho Studart, T. M., & de Assis Souza Filho, F. (2019). A continuous drought probability monitoring system, CDPMS, based on copulas. *Water*, *11*(9). Available from https://doi.org/10.3390/w11091925.

Rivera, J., & Penalba, O. (2014). Trends and spatial patterns of drought affected area in Southern South America. *Climate*, *2*(4), 264−278. Available from https://doi.org/10.3390/cli2040264.

Rural, G. (2021). Colheita de café tem queda em Minas Gerais com seca.

Sandler, M., Howard, A., Zhu, M., Zhmoginov, A., & Chen, L.C. (2018). MobileNetV2: Inverted residuals and linear bottlenecks. In *Proceedings of the IEEE computer society conference on Computer Vision and Pattern Recognition* (pp. 4510−4520). 2018/12/14, IEEE Computer Society. Available from https://doi.org/10.1109/CVPR.2018.00474, 9781538664209.

Santos, B. C., & Horta, I. T. L. G. (2020). *Climatologia dinâmica: conceitos, técnicas e aplicações RiMa Editora Processos e sistemas atmosféricos: contribuições teóricas e aplicadas em climatologia dinâmica.*

Sinha, D., & El-Sharkawy, M. (2019). Thin mobile net: An enhanced mobilenet architecture. In *IEEE 10th annual ubiquitous computing, electronics & mobile communication conference (UEMCON).* IEEE.

Soeder, D. J. (2021). Fracking and the Environment: A scientific assessment of the environmental risks from hydraulic fracturing and fossil fuels (pp. 155−185).

Souid, A., Sakli, N., & Sakli, H. (2021). Classification and predictions of lung diseases from chest X-rays using MobileNet V2. *Applied Sciences*, *11*(6). Available from https://doi.org/10.3390/app11062751.

Staal, A., Flores, B. M., Aguiar, A. P. D., Bosmans, J. H. C., Fetzer, I., & Tuinenburg, O. A. (2020). Feedback between drought and deforestation in the Amazon. *Environmental Research Letters*, *15*(4). Available from https://doi.org/10.1088/1748-9326/ab738e.

Woolford, D. G., Martell, D. L., McFayden, C. B., Evens, J., Stacey, A., Wotton, B. M., & Boychuk, D. (2021). The development and implementation of a human-caused wildland fire occurrence prediction system for the province of Ontario, Canada. *Canadian Journal of Forest Research*, *51*(2), 303−325. Available from https://doi.org/10.1139/cjfr-2020-0313.

Zhang, B., Abu Salem, F. K., Hayes, M. J., Smith, K. H., Tadesse, T., & Wardlow, B. D. (2023). Explainable machine learning for the prediction and assessment of complex drought impacts. *Science of the Total Environment*, *898*. Available from https://doi.org/10.1016/j.scitotenv.2023.165509, http://www.elsevier.com/locate/scitotenv.

Zhao, G., Wang, J., & Zhang, Z. (2017). Random shifting for CNN: A solution to reduce information loss in Down-sampling layers. In *IJCAI International Joint Conferences on Artificial Intelligence China* (3476−3482). Available from https://doi.org/10.24963/ijcai.2017/486, http://www.ijcai.org/, 9780999241103.

Index

Note: Page numbers followed by "*f*" and "*t*" refer to figures and tables, respectively.